Model Categories and Their Localizations

Mathematical
Surveys
and
Monographs

Volume 99

Model Categories and Their Localizations

Philip S. Hirschhorn

American Mathematical Society
Providence, Rhode Island

2000 *Mathematics Subject Classification.* Primary 18G55, 55P60, 55U35; Secondary 18G30.

Library of Congress Cataloging-in-Publication Data

Model categories and their localizations / Philip S. Hirschhorn.
p. cm. (Mathematical surveys and monographs, ISSN 0076-5376; v. 99)
Includes bibliographical references and index.
ISBN 0-8218-3279-4 (alk. paper)
1. Model categories (Mathematics). 2. Homotopy theory. I. Hirschhorn, Philip S. (Philip Steven), 1952–. II. Mathematical surveys and monographs; no. 99.

QA169.M63 2002
512′.55–dc21 2002027794

AMS softcover ISBN: 978-0-8218-4917-0

∞ The paper used in this book is acid-free and falls within the guidelines
established to ensure permanence and durability.
Visit the AMS home page at http://www.ams.org/

12 11 10 9 8 7 6 5 4 3   30 29 28 27 26 25

Contents

Introduction

Model categories and their homotopy categories

A *model category* is Quillen's axiomatization of a place in which you can "do homotopy theory" [**52**]. Homotopy theory often involves treating homotopic maps as though they were the same map, but a homotopy relation on maps is not the starting point for abstract homotopy theory. Instead, homotopy theory comes from choosing a class of maps, called *weak equivalences*, and studying the passage to the *homotopy category*, which is the category obtained by localizing with respect to the weak equivalences, i.e., by making the weak equivalences into isomorphisms (see Definition 8.3.2). A model category is a category together with a class of maps called *weak equivalences* plus two other classes of maps (called *cofibrations* and *fibrations*) satisfying five axioms (see Definition 7.1.3). The cofibrations and fibrations of a model category allow for lifting and extending maps as needed to study the passage to the homotopy category.

The homotopy category of a model category. Homotopy theory originated in the category of topological spaces, which has unusually good technical properties. In this category, the homotopy relation on the set of maps between two objects is always an equivalence relation, and composition of homotopy classes is well defined. In the classical homotopy theory of topological spaces, the passage to the homotopy category was often described as "replacing maps with homotopy classes of maps". Most work was with CW-complexes, though, and whenever a construction led to a space that was not a CW-complex the space was replaced by a weakly equivalent one that was. Thus, weakly equivalent spaces were recognized as somehow "equivalent", even if that equivalence was never made explicit. If instead of starting with a homotopy relation we explicitly cause weak equivalences to become isomorphisms, then homotopic maps do become the same map (see Lemma 8.3.4) and in addition a cell complex weakly equivalent to a space becomes isomorphic to that space, which would not be true if we were simply replacing maps with homotopy classes of maps.

In most model categories, the homotopy relation does not have the good properties that it has in the category of topological spaces unless you restrict yourself to the subcategory of cofibrant-fibrant objects (see Definition 7.1.5). There are actually two different homotopy relations on the set of maps between two objects X and Y: *Left homotopy*, defined using cylinder objects for X, and *right homotopy*, defined using path objects for Y (see Definition 7.3.2). For arbitrary objects X and Y these are different relations, and neither of them is an equivalence relation. However, for cofibrant-fibrant objects, the two homotopy relations are the same, they are equivalence relations, and composition of homotopy classes is well defined (see Theorem 7.4.9 and Theorem 7.5.5). Every object of a model category is weakly equivalent to a cofibrant-fibrant object, and we could thus define a "homotopy category of cofibrant-fibrant objects" by taking the cofibrant-fibrant objects of the model category as our objects and homotopy classes of maps as our morphisms. Since a map between cofibrant-fibrant objects is a weak equivalence if and only if it is a homotopy equivalence (see Theorem 7.5.10 and Theorem 7.8.5), this would send weak equivalences to isomorphisms, and we define the *classical homotopy category* of a model category in exactly this way (see Definition 7.5.8).

The classical homotopy category is inadequate, though, because most work in homotopy theory requires constructions that create objects that may not be cofibrant-fibrant, even if we start out with only cofibrant-fibrant objects. Thus, we need a "homotopy category" containing *all* of the objects of the model category. We define the *Quillen homotopy category* of a model category to be the localization of the category with respect to the class of weak equivalences (see Definition 8.3.2). For the class of weak equivalences of a model category, this always exists (see Remark 8.3.3 and Theorem 8.3.5). Thus, the Quillen homotopy category of a model category contains all of the objects of the model category. The classical homotopy category is a subcategory of the Quillen homotopy category, and the inclusion of the classical homotopy category in the Quillen homotopy category is an equivalence of categories (see Theorem 8.3.9). We refer to the Quillen homotopy category as simply the *homotopy category*.

Homotopy function complexes. Homotopy theory involves the construction of more than just a homotopy category. Dwyer and Kan [**31, 32, 33**] construct the *simplicial localization* of a category with respect to a class of weak equivalences as the derived functor of the functor that constructs the homotopy category. This is a *simplicial category*, i.e., a category enriched over simplicial sets, and so for each pair of objects there is a simplicial set that is the "function complex" of maps between the objects. These function complexes capture the "higher order structure" of the homotopy theory, and taking the set of components of the function complex of maps between two objects yields the set of maps in the homotopy category between those objects.

Dwyer and Kan show that if you start with a *model category*, then simplicial sets weakly equivalent to those function complexes can be constructed using cosimplicial or simplicial resolutions (see Definition 16.1.2) in the model category. We present a self-contained development of these *homotopy function complexes* (see Chapter 17). Constructing homotopy function complexes requires making an arbitrary choice of resolutions, but we show that the category of possible choices has a contractible classifying space (see Theorem 17.5.28), and so there is a distinguished homotopy class of homotopy equivalences between the homotopy function complexes resulting from different choices (see Theorem 17.5.29 and Theorem 17.5.30).

Homotopy theory in model categories. Part 2 of this book studies model categories and techniques of homotopy theory in model categories. Part 2 is intended as a reference, and it logically precedes Part 1. We cover quite a bit of ground, but the topics discussed in Part 2 are only those that are needed for the discussion of localization in Part 1, fleshed out to give a reasonably complete development. We begin Part 2 with the definition of a model category and with the basic results that are by now standard (see, e.g., [**52, 54, 14, 35**]), but we give complete arguments in an attempt to make this accessible to the novice. For a more complete description of the contents of Part 2, see the summary on page 103 and the introductions to the individual chapters. For a description of Part 1, which discusses localizing model category structures, see below, as well as the summary on page 3.

Prerequisites. The category of simplicial sets plays a central role in the homotopy theory of a model category, even for model categories unrelated to simplicial sets. This is because a homotopy function complex between objects in a model category is a simplicial set (see Chapter 17). Thus, we assume that the reader has some familiarity with the homotopy theory of simplicial sets. For readers without the necessary background, we recommend the works by Curtis [**18**], Goerss and Jardine [**39**], and May [**49**].

Localizing model category structures

Localizing a model category with respect to a class of maps does not mean making the maps into isomorphisms; instead, it means making the images of those maps in the homotopy category into isomorphisms (see Definition 3.1.1). Since the image of a map in the homotopy category is an isomorphism if and only if the map is a weak equivalence (see Theorem 8.3.10), localizing a model category with respect to a class of maps means making maps into weak equivalences.

Localized model category structures originated in Bousfield's work on localization with respect to homology ([**8**]). Given a homology theory h_*, Bousfield established a model category structure on the category of simplicial sets in which the weak equivalences were the maps that induced isomorphisms of all homology groups. A space (i.e., a simplicial set) W was defined to be *local* with respect to h_* if it was a Kan complex such that every map $f\colon X \to Y$ that induced isomorphisms $f_*\colon h_*X \approx h_*Y$ of homology groups also induced an isomorphism $f^*\colon \pi(Y,W) \approx \pi(X,W)$ of the sets of homotopy classes of maps to W. In Bousfield's model category structure, a space was fibrant if and only if it was local with respect to h_*.

The problem that led to Bousfield's model category structure was that of constructing a *localization functor* for a homology theory. That is, given a homology theory h_*, the problem was to define for each space X a local space $\mathrm{L}_{h_*}X$ and a natural homology equivalence $X \to \mathrm{L}_{h_*}X$. There had been a number of partial solutions to this problem (perhaps the most complete being that of Bousfield and Kan [**14**]), but each of these was valid only for some special class of spaces, and only for certain homology theories. In [**8**], Bousfield constructed a functorial h_*-localization for an arbitrary homology theory h_* and for every simplicial set. In Bousfield's model category structure, a fibrant approximation to a space (i.e., a weak equivalence from a space to a fibrant space) was exactly a localization of that space with respect to h_*.

Some years later, Bousfield [**9, 10, 11, 12**] and Dror Farjoun [**20, 22, 24**] independently considered the notion of localizing spaces with respect to an arbitrary map, with a definition of "local" slightly different from that used in [**8**]: Given a map of spaces $f\colon A \to B$, a space W was defined to be *f-local* if f induced a weak equivalence of mapping spaces $f^*\colon \mathrm{Map}(B,W) \cong \mathrm{Map}(A,W)$ (rather than just a bijection on components, i.e., an isomorphism of the sets of homotopy classes of maps), and a map $g\colon X \to Y$ was defined to be an *f-local equivalence* if for every f-local space W the induced map of mapping spaces $g^*\colon \mathrm{Map}(Y,W) \to \mathrm{Map}(X,W)$

was a weak equivalence. An *f-localization* of a space X was then an f-local space $\mathrm{L}_f X$ together with an f-local equivalence $X \to \mathrm{L}_f X$. Bousfield and Dror Farjoun constructed f-localization functors for an arbitrary map f of spaces.

Given a map $f\colon A \to B$ of spaces, we construct in Chapters 1 and 2 an f-local model category structure on the category of spaces. That is, we construct a model category structure on the category of spaces in which the weak equivalences are the f-local equivalences, and in which an f-localization functor is a fibrant approximation functor for the f-local model category. In Chapter 4 we extend this to establish S-local model category structures for an arbitrary set S of maps in a left proper (see Definition 13.1.1) *cellular model category* (see page xiii and Chapter 12).

Constructing the localized model category structure. Once we've established the localized model category structure, a localization of an object in the category will be exactly a fibrant approximation to that object in the localized model category, but it turns out that we must first define a natural localization of every object in order to establish the localized model category structure. The reason for this is that we use the localization functor to identify the local equivalences: A map is a local equivalence if and only if its localization is a weak equivalence (see Theorem 3.2.18).

The model categories with which we work are all *cofibrantly generated* model categories (see Definition 11.1.2). That is, there is a set I of cofibrations and a set J of trivial cofibrations such that

- a map is a trivial fibration if and only if it has the right lifting property with respect to every element of I,
- a map is a fibration if and only if it has the right lifting property with respect to every element of J, and
- both of the sets I and J *permit the small object argument* (see Definition 10.5.15).

For example, in the category Top of (unpointed) topological spaces (see Notation 1.1.4), we can take for I the set of inclusions $\mathrm{S}^{n-1} \to \mathrm{D}^n$ for $n \geq 0$ and for J the set of inclusions $|\Lambda[n,k]| \to |\Delta[n]|$ for $n > 0$ and $0 \leq k \leq n$. The left Bousfield localization $\mathrm{L}_f\mathrm{Top}$ of Top with respect to a map f in Top (see Definition 3.3.1) will have the same class of cofibrations as the standard model category structure on Top, and so the set I of generating cofibrations for Top can serve as a set of generating cofibrations for $\mathrm{L}_f\mathrm{Top}$. The difficulty lies in finding a set J_f of generating trivial cofibrations for $\mathrm{L}_f\mathrm{Top}$.

A first thought might be to let J_f be the collection of all cofibrations that are f-local equivalences, since the fibrations of $\mathrm{L}_f\mathrm{Top}$ are defined to be the maps with the right lifting property with respect to all such maps, but then J_f would not be a set. The problem is to find a subcollection J_f of the class of *all* cofibrations that are f-local equivalences such that

- a map has the right lifting property with respect to every element of J_f if and only if it has the right lifting property with respect to every cofibration that is an f-local equivalence, and
- the collection J_f forms a set.

That is the problem that is solved by the Bousfield-Smith cardinality argument.

The Bousfield-Smith cardinality argument. Every map in Top has a cofibrant approximation (see Definition 8.1.22) that is moreover an inclusion of cell complexes (see Definition 10.7.1 and Proposition 11.2.8). Since Top is left proper (see Definition 13.1.1), this implies that for a map to have the right lifting property with respect to all cofibrations that are f-local equivalences, it is sufficient that it have the right lifting property with respect to all inclusions of cell complexes that are f-local equivalences (see Proposition 13.2.1).

If we choose a fixed cardinal γ, then the collection of homeomorphism classes of cell complexes of size no larger than γ forms a set. The cardinality argument shows that there exists a cardinal γ such that a map has the right lifting property with respect to all inclusions of cell complexes that are f-local equivalences if and only if it has the right lifting property with respect to all such inclusions between cell complexes of size no larger than γ. Thus, we can take as our set J_f a set of representatives of the isomorphism classes of such "small enough" inclusions of cell complexes.

Our localization functor L_f is defined by choosing a set of inclusions of cell complexes $\overline{\Lambda\{f\}}$ and then attaching the codomains of the elements of $\overline{\Lambda\{f\}}$ to a space by all possible maps from the domains of the elements of $\overline{\Lambda\{f\}}$, and then repeating this an infinite number of times (see Section 1.3). In order to make the cardinality argument, we need to find a cardinal γ such that

(1) if X is a cell complex, then every subcomplex of its localization $\mathrm{L}_f X$ of size at most γ is contained in the localization of a subcomplex of X of size at most γ, and
(2) if X is a cell complex of size at most γ, then $\mathrm{L}_f X$ is also of size at most γ.

We are able to do this because

(1) every map from a closed cell to a cell complex factors through a finite subcomplex of the cell complex (see Corollary 10.7.7), and
(2) given two cell complexes, there is an upper bound on the cardinal of the set of continuous maps between them, and this upper bound depends only on the size of the cell complexes

(see Section 2.3).

Cellular model categories. Suppose now that $\mathcal{M}$ is a cofibrantly generated model category and that we wish to localize $\mathcal{M}$ with respect to a set S of maps in $\mathcal{M}$ (see Definition 3.3.1). If I is a set of generating cofibrations for $\mathcal{M}$, then

- we define a *relative cell complex* to be a map built by repeatedly attaching codomains of elements of I along maps of their domains (see Definition 10.5.8),
- we define a *cell complex* to be the codomain of a relative cell complex whose domain is the initial object of $\mathcal{M}$, and
- we define an *inclusion of cell complexes* to be a relative cell complex whose domain is a cell complex.

(If $\mathcal{M} = \text{Top}$, the category of topological spaces, our set I of generating cofibrations is the set of inclusions $\mathrm{S}^{n-1} \to \mathrm{D}^n$ for $n \geq 0$, and so a cell complex is a space built by repeatedly attaching disks along maps of their boundary spheres.) In such a model category, every map has a cofibrant approximation (see Definition 8.1.22) that is an inclusion of cell complexes (see Proposition 11.2.8). Thus, if we assume that $\mathcal{M}$ is left proper (see Definition 13.1.1), then for a map to have the right lifting property with respect to all cofibrations that are S-local equivalences, it is sufficient that it have the right lifting property with respect to all inclusions of cell complexes that are S-local equivalences (see Proposition 13.2.1). In order to make the cardinality argument, though, we need to assume that maps between cell complexes in $\mathcal{M}$ are sufficiently well behaved; this leads us to the definition of a *cellular model category* (see Definition 12.1.1).

A cellular model category is a cofibrantly generated model category with additional properties that ensure that

- the intersection of a pair of subcomplexes (see Definition 12.2.5) of a cell complex exists (see Theorem 12.2.6),
- there is a cardinal σ (called the *size of the cells* of $\mathcal{M}$; see Definition 12.3.3) such that if X is a cell complex of size τ, then any map from X to a cell complex factors through a subcomplex of size at most $\sigma\tau$ (see Theorem 12.3.1), and
- if X is a cell complex, then there is a cardinal η such that if Y is a cell complex of size ν ($\nu \geq 2$), then the set $\mathcal{M}(X, Y)$ has cardinal at most ν^η (see Proposition 12.5.1).

Fortunately, these properties follow from a rather minimal set of conditions on the model category $\mathcal{M}$ (see Definition 12.1.1), satisfied by almost all model categories that come up in practice.

Left localization and right localization. There are two types of morphisms of model categories: *Left Quillen functors* and *right Quillen functors* (see Definition 8.5.2). The localizations that we have been discussing are all *left localizations*, because the functor from the original model category to the localized model category is a left Quillen functor that is initial among left Quillen functors whose total left derived functor takes the images of the designated maps into isomorphisms in the homotopy category (see Definition 3.1.1). There is an analogous notion of *right localization.*

Given a CW-complex A, Dror Farjoun [**20, 21, 23, 24**] defines a map of topological spaces $f\colon X \to Y$ to be an *A-cellular equivalence* if the induced map of function spaces $f_*\colon \mathrm{Map}(A, X) \to \mathrm{Map}(A, Y)$ is a weak equivalence. He also defines the class of *A-cellular spaces* to be the smallest class of cofibrant spaces that contains A and is closed under weak equivalences and homotopy colimits. We show in Theorem 5.1.1 and Theorem 5.1.6 that this is an example of a *right localization*, i.e., that there is a model category structure in which the weak equivalences are the A-cellular equivalences and in which the cofibrant objects are the A-cellular spaces. In fact, we do this for an arbitrary right proper cellular model category (see Theorem 5.1.1 and Theorem 5.1.6).

The situation here is not as satisfying as it is for left localizations, though. The left localizations that we construct for left proper cellular model categories are again left proper cellular model categories (see Theorem 4.1.1), but the right localizations that we construct for right proper cellular model categories need not even be cofibrantly generated if not every object of the model category is fibrant. However, if every object is fibrant, then a right localization will again be right proper cellular with every object fibrant; see Theorem 5.1.1.

Acknowledgments

I am grateful to the M.I.T. Mathematics Department for putting up with me over all these years, and making me always feel welcome. Although there has been a gradual change in the group of topologists there since I was a graduate student, what has been constant is the warm, supportive attitude that has allowed me always to think of it as home.

This project grew out of my reading an early draft of Emmanuel Dror Farjoun's book [**24**]; I am grateful to him for his preprints and helpful communications.

Drafts of this book been have been circulating for a number of years, originally under the title *Localization, Cellularization, and Homotopy Colimits* and more recently as *Localization of Model Categories*, and a number of people have provided helpful comments. Paul Goerss helped correct a problem with the proof of the existence of the localization model category structure for cellular model categories. Garth Warner gave several early drafts an amazingly careful review; he found many obscurities and outright errors, and provided a level of analysis that usually takes several years and a group of reviewers. Daniel Biss, Dan Christensen, Dan Dugger, Bill Dwyer, Lars Hesselholt, Mike Hopkins, Mark Hovey, Dan Isaksen, Haynes Miller, George Peschke, Charles Rezk, and Jeff Smith offered corrections and helpful suggestions. Of course, any mistakes or omissions that remain are entirely my responsibility.

Above all, though, I owe a great deal of thanks to Dan Kan. He is responsible for my first becoming interested in this subject, and I am indebted to him for countless helpful conversations and suggestions. His view of mathematics has been a lasting influence upon me, and it is a pleasure to thank him for his support and encouragement over the years.

Part 1

Localization of Model Category Structures

Summary of Part 1

Part 1 contains our discussion of localization of model categories. Throughout Part 1 we freely use the results of Part 2, which is our reference for techniques of homotopy theory in model categories, and which logically precedes Part 1.

In Chapters 1 and 2 we discuss localization in a category of spaces. We work in parallel in four different categories:

- The category of (unpointed) topological spaces.
- The category of pointed topological spaces.
- The category of (unpointed) simplicial sets.
- The category of pointed simplicial sets.

Given a map f, in Chapter 1 we discuss f-local spaces, f-local equivalences, and f-localizations of spaces. We construct an f-localization functor, as well as a continuous version of the f-localization functor. We discuss commuting localizations with the total singular complex and geometric realization functors, and compare localizations in a category of pointed spaces with localizations in a category of unpointed spaces.

In Chapter 2 we establish a model category structure on the category of spaces in which the weak equivalences are the f-local equivalences and the fibrant objects are the f-local spaces. This requires a careful analysis of the cell complexes constructed by the f-localization functor defined in Chapter 1, and the main argument involves studying the cardinality of the set of cells in the localization of a cell complex.

In Chapter 3 we define left and right localizations of a model category $\mathcal{M}$ with respect to a class $\mathcal{C}$ of maps in $\mathcal{M}$. A *left localization* of $\mathcal{M}$ with respect to $\mathcal{C}$ is a left Quillen functor defined on $\mathcal{M}$ that is initial among those that take the images in the homotopy category of the elements of $\mathcal{C}$ into isomorphisms. A *right localization* is the analogous notion for right Quillen functors. We also define *Bousfield localizations*, which are localizations obtained by constructing a new model category structure on the original underlying category. (The localization of Chapter 2 is a left Bousfield localization.) We discuss local objects, local equivalences, and localization functors in this more general context.

Chapter 4 contains our main existence results for left localizations. We show that if $\mathcal{M}$ is a left proper cellular model category, then the left Bousfield localization of $\mathcal{M}$ with respect to an arbitrary set S of maps in $\mathcal{M}$ exists. The proof requires that we first define an S-localization functor for objects of $\mathcal{M}$, and then carefully analyze the cardinality of the set of cells in the localization of a cell complex.

Chapter 5 contains our main existence results for right localizations. If $\mathcal{M}$ is a model category and K is a set of objects of $\mathcal{M}$, then a map $f\colon X \to Y$ in $\mathcal{M}$ is defined to be a *K-colocal equivalence* if for every object A in K the induced map of homotopy function complexes $f_*\colon \operatorname{map}(A,X) \to \operatorname{map}(A,Y)$ is a weak equivalence (see Definition 3.1.8). We show that if $\mathcal{M}$ is a right proper cellular model category and K is an arbitrary set of objects of $\mathcal{M}$, then the right Bousfield localization of $\mathcal{M}$ with respect to the class of K-colocal equivalences exists.

In Chapter 6 we study fiberwise localizations in a category of spaces. If $\mathcal{M}$ is a category of unpointed spaces and S is a set of maps in $\mathcal{M}$, then for every map $p\colon X \to Z$ in $\mathcal{M}$ we construct a "fiberwise S-localization" $\hat{p}\colon \widehat{X} \to Z$ of p, which is a map $X \to \widehat{X}$ over Z such that the induced map from the homotopy fiber of p to that of $\hat{p}$ is an S-localization. We do this by constructing an appropriate localized model category structure on the model category of spaces over Z.

CHAPTER 1

Local Spaces and Localization

We describe our categories of spaces in Section 1.1. In Section 1.2 we define local spaces and local equivalences, and in Sections 1.3 and 1.4 we define a functorial localization and establish some of its properties. In Section 1.5 we show that Postnikov approximations to a space are examples of localizations of the space. In Section 1.6 we investigate localizations in the categories of simplicial sets and of topological spaces, and the relationship between them. In Section 1.7 we construct a continuous variant of our localization functor, and in Section 1.8 we describe the relationship between localizations of pointed spaces and of unpointed spaces.

1.1. Definitions of spaces and mapping spaces

In Chapters 1 and 2 we will be discussing categories of spaces, where by a *space* we mean either a topological space or a simplicial set. We will be working simultaneously in several different categories of spaces (topological spaces or simplicial sets, pointed or unpointed), and a central question will be whether a map of spaces induces a weak equivalence of mapping spaces. In order to discuss all of these categories simultaneously, we will refer uniformly to the *simplicial mapping space* (i.e., the simplicial set of maps) between two spaces no matter what the category of spaces. Section 1.1.1 describes exactly what we will mean by a *topological space*, Section 1.1.3 describes the various categories of topological spaces or of simplicial sets that we will consider, and Definition 1.1.6 describes the simplicial mapping space for each of these categories.

1.1.1. Definition of a topological space. There are several different categories of topological spaces in common use, and any of these is acceptable for our purposes.

NOTATION 1.1.2. We will use Top to denote some category of topological spaces with the following properties:

(1) Top is closed under small colimits and small limits.
(2) Top contains among its objects the geometric realizations of all simplicial sets.
(3) If X and Y are objects of Top and K is a simplicial set, then there is a natural isomorphism of sets

$$\mathrm{Top}(X \times |K|, Y) \approx \mathrm{Top}\big(X, Y^{|K|}\big) \ .$$

Thus, the reader is invited to assume that Top denotes, e.g.,

- the category of compactly generated Hausdorff spaces (see, e.g., [**62**]), or
- the category of compactly generated weak Hausdorff spaces (see, e.g., [**37**, Appendix A1]), or
- some other category of spaces with our three properties (see, e.g., [**63**])

(see also Section 7.10.1).

1.1.3. Our categories of spaces. We will be working with both topological spaces (see Section 1.1.1) and simplicial sets, and for each of these we will consider both the category of pointed spaces and the category of unpointed spaces. In order to keep the terminology concise, the word *space* will be used to mean either a topological space or a simplicial set, and we will use the following notation for our categories of spaces.

NOTATION 1.1.4. We will use the following notation for our categories of spaces:

SS : The category of simplicial sets.
SS_* : The category of pointed simplicial sets.
Top : The category of topological spaces (see Section 1.1.1).
Top_*: The category of pointed topological spaces.

Since much of our discussion will apply to more than one of these categories, we will use the following notation:

$SS_{(*)}$: Either SS or SS_*.
$Top_{(*)}$: Either Top or Top_*.
Spc : A category of unpointed spaces, i.e., either Top or SS.
Spc_* : A category of pointed spaces, i.e., either Top_* or SS_*.
$Spc_{(*)}$: Any of the categories SS, SS_*, Top, or Top_*.

1.1.5. Simplicial mapping spaces. Each of our categories of spaces is a simplicial model category (see Definition 9.1.6), and our localization results will make use of the simplicial mapping space between objects in these categories. We will sometimes refer to the simplicial mapping space between two objects as the *function complex* between those objects.

DEFINITION 1.1.6 (Simplicial mapping spaces).

- If X and Y are objects of SS, then $\mathrm{Map}(X, Y)$ is the simplicial set with n-simplices the simplicial maps $X \times \Delta[n] \to Y$ and face and degeneracy maps induced by the standard maps between the $\Delta[n]$ (see Example 9.1.13).
- If X and Y are objects of SS_*, then $\mathrm{Map}(X, Y)$ is the simplicial set with n-simplices the basepoint preserving simplicial maps $X \wedge \Delta[n]^+ \to Y$ and face and degeneracy maps induced by the standard maps between the $\Delta[n]$ (see Example 9.1.14).
- If X and Y are objects of Top, then $\mathrm{Map}(X, Y)$ is the simplicial set with n-simplices the continuous functions $X \times |\Delta[n]| \to Y$ and face and degeneracy maps induced by the standard maps between the $\Delta[n]$ (see Example 9.1.15).
- If X and Y are objects of Top_*, then $\mathrm{Map}(X, Y)$ is the simplicial set with n-simplices the continuous functions $X \wedge |\Delta[n]|^+ \to Y$ and face and degeneracy maps induced by the standard maps between the $\Delta[n]$ (see Example 9.1.16).

Note that, in all cases, $\mathrm{Map}(X, Y)$ is an *unpointed* simplicial set.

1.1.7. Total singular complex and geometric realization.

DEFINITION 1.1.8. If X and Y are objects of $\mathrm{Spc}_{(*)}$ (see Notation 1.1.4) and K is a simplicial set, then $X \otimes K$ and Y^K will denote the objects of $\mathrm{Spc}_{(*)}$ defined by the simplicial model category structure on $\mathrm{Spc}_{(*)}$ (see Example 9.1.13, Example 9.1.14, Example 9.1.15, and Example 9.1.16), which are characterized by the natural isomorphisms of sets

$$\mathrm{Spc}_{(*)}(X \otimes K, Y) \approx \mathrm{SS}\big(K, \mathrm{Map}(X,Y)\big) \approx \mathrm{Spc}_{(*)}(X, Y^K)$$

(see Definition 9.1.6). Thus,

$$\begin{array}{llll} \text{If } \mathrm{Spc}_{(*)} = \mathrm{SS}, & \text{then } X \otimes K = X \times K & \text{and} & X^K = \mathrm{Map}(K,X). \\ \text{If } \mathrm{Spc}_{(*)} = \mathrm{SS}_*, & \text{then } X \otimes K = X \wedge K^+ & \text{and} & X^K = \mathrm{Map}_*(K^+,X). \\ \text{If } \mathrm{Spc}_{(*)} = \mathrm{Top}, & \text{then } X \otimes K = X \times |K| & \text{and} & X^K = \mathrm{map}\big(|K|,X\big). \\ \text{If } \mathrm{Spc}_{(*)} = \mathrm{Top}_*, & \text{then } X \otimes K = X \wedge |K|^+ & \text{and} & X^K = \mathrm{map}_*\big(|K|^+,X\big). \end{array}$$

(See Definition 18.2.1.)

LEMMA 1.1.9. *Let K be an unpointed simplicial set.*

1. *If X is a topological space (either pointed or unpointed), then there is a natural isomorphism of (pointed or unpointed) simplicial sets $\mathrm{Sing}(X^K) \approx (\mathrm{Sing}\, X)^K$.*
2. *If L is a simplicial set (either pointed or unpointed), then there is a natural homeomorphism of (pointed or unpointed) topological spaces $|L \otimes K| \approx |L| \otimes K$.*

PROOF. If X is a pointed topological space, then there are natural isomorphisms

$$\begin{aligned} \big(\mathrm{Sing}(X^K)\big)_n &= \mathrm{Top}_*\big(|\Delta[n]|^+, X^K\big) \\ &= \mathrm{Top}_*\big(|\Delta[n]|^+, \mathrm{map}_*(|K|^+, X)\big) \\ &\approx \mathrm{Top}_*\big(|\Delta[n]|^+ \wedge |K|^+, X\big) \\ &\approx \mathrm{Top}_*\big(|\Delta[n]^+ \wedge K^+|, X\big) \\ &\approx \mathrm{SS}_*\big(\Delta[n]^+ \wedge K^+, \mathrm{Sing}\, X\big) \\ &\approx \mathrm{SS}_*\big(\Delta[n]^+, (\mathrm{Sing}\, X)^K\big) \\ &\approx \big((\mathrm{Sing}\, X)^K\big)_n \ . \end{aligned}$$

The proof for the unpointed case is similar.

If L is a pointed simplicial set, then there are natural homeomorphisms

$$\begin{aligned} |L \otimes K| &= |L \wedge K^+| \\ &\approx |L| \wedge |K|^+ \\ &= |L| \otimes K \ . \end{aligned}$$

The proof for the unpointed case is similar. □

LEMMA 1.1.10. *If L is a simplicial set and W is a topological space (either both pointed or both unpointed), then the standard adjunction of the geometric realization and total singular complex functors extends to a natural isomorphism of simplicial mapping spaces*

$$\mathrm{Map}\big(|L|, W\big) \approx \mathrm{Map}(L, \mathrm{Sing}\, W) \ .$$

PROOF. This follows from the natural homeomorphism $|L\otimes\Delta[n]| \approx |L|\otimes|\Delta[n]|$ (see Lemma 1.1.9). □

PROPOSITION 1.1.11. *If A and X are objects of $\mathrm{SS}_{(*)}$ and X is fibrant, then there is a natural weak equivalence of simplicial sets*

$$\mathrm{Map}(A, X) \cong \mathrm{Map}(|A|, |X|) .$$

PROOF. Since all simplicial sets are cofibrant, the natural map $X \to \mathrm{Sing}|X|$ induces a weak equivalence $\mathrm{Map}(A, X) \cong \mathrm{Map}(A, \mathrm{Sing}|X|)$ (see Corollary 9.3.3). The proposition now follows from Lemma 1.1.10. □

PROPOSITION 1.1.12. *If A and X are objects of $\mathrm{Top}_{(*)}$ and A is cofibrant, then there is a natural weak equivalence of simplicial sets*

$$\mathrm{Map}(A, X) \cong \mathrm{Map}(\mathrm{Sing}\, A, \mathrm{Sing}\, X) .$$

PROOF. Since all topological spaces are fibrant, the natural map $|\mathrm{Sing}\, A| \to A$ induces a weak equivalence $\mathrm{Map}(A, X) \cong \mathrm{Map}(|\mathrm{Sing}\, A|, X)$ (see Corollary 9.3.3). The proposition now follows from Lemma 1.1.10. □

DEFINITION 1.1.13. Each of our categories of spaces has a functor to SS, and each of these functors has a left adjoint $\mathrm{SS} \to \mathrm{Spc}_{(*)}$, i.e., for an unpointed simplicial set K and an object X of $\mathrm{Spc}_{(*)}$, we have natural isomorphisms

$$\begin{aligned}
\mathrm{SS}(K, X) &\approx \mathrm{SS}(K, X)\\
\mathrm{SS}_*(K^+, X) &\approx \mathrm{SS}(K, X^-)\\
\mathrm{Top}(|K|, X) &\approx \mathrm{SS}(K, \mathrm{Sing}\, X)\\
\mathrm{Top}_*(|K|^+, X) &\approx \mathrm{SS}(K, \mathrm{Sing}\, X^-)
\end{aligned}$$

where "X^-" means "forget the basepoint of X". If K is an (unpointed) simplicial set, then we will use $\mathrm{Spc}_{(*)}(K)$ to denote the image of K in $\mathrm{Spc}_{(*)}$ under this left adjoint. Thus,

If $\mathrm{Spc}_{(*)} = \mathrm{SS}$, then $\mathrm{Spc}_{(*)}(K) = K$.
If $\mathrm{Spc}_{(*)} = \mathrm{SS}_*$, then $\mathrm{Spc}_{(*)}(K) = K^+$.
If $\mathrm{Spc}_{(*)} = \mathrm{Top}$, then $\mathrm{Spc}_{(*)}(K) = |K|$.
If $\mathrm{Spc}_{(*)} = \mathrm{Top}_*$, then $\mathrm{Spc}_{(*)}(K) = |K|^+$.

EXAMPLE 1.1.14. In the standard model category structure on $\mathrm{Spc}_{(*)}$, a map is a fibration if it has the right lifting property (see Definition 7.2.1) with respect to the maps $\mathrm{Spc}_{(*)}(\Lambda[n,k]) \to \mathrm{Spc}_{(*)}(\Delta[n])$ for all $n > 0$ and $0 \le k \le n$ (see Theorem 7.10.10, Theorem 7.10.11, Theorem 7.10.12, and Theorem 7.10.13).

1.2. Local spaces and localization

1.2.1. f-local spaces and f-local equivalences.

DEFINITION 1.2.2. Let $\mathrm{Spc}_{(*)}$ be one of our categories of spaces (see Notation 1.1.4) and let $f\colon A \to B$ be a map between cofibrant spaces in $\mathrm{Spc}_{(*)}$.

(1) A space W is *f-local* if W is fibrant and the induced map of simplicial sets $f^*\colon \mathrm{Map}(B, W) \to \mathrm{Map}(A, W)$ is a weak equivalence. If f is a map $* \to A$, then an f-local space will also be called *A-local* or *A-null*. Bousfield ([**11**]) has used the term *A-periodic* for what we here call A-local.

(2) A map $g\colon X \to Y$ is an *f-local equivalence* if there is a cofibrant approximation $\tilde{g}\colon \widetilde{X} \to \widetilde{Y}$ to g (see Definition 8.1.22) such that for every f-local space W the induced map of simplicial sets $\tilde{g}^*\colon \operatorname{Map}(\widetilde{Y}, W) \to \operatorname{Map}(\widetilde{X}, W)$ is a weak equivalence. (Proposition 9.7.2 implies that if this is true for any one cofibrant approximation to g then it is true for every cofibrant approximation to g.)

If $\mathrm{Spc}_{(*)} = \mathrm{SS}_{(*)}$ then every space is cofibrant, and so a map $g\colon X \to Y$ is an f-local equivalence if and only if for every f-local space W the map $g^*\colon \operatorname{Map}(Y, W) \to \operatorname{Map}(X, W)$ is a weak equivalence. If $\mathrm{Spc}_{(*)} = \mathrm{Top}_{(*)}$ then all CW-complexes are cofibrant, and so a CW-replacement for a space serves as a cofibrant approximation to that space.

A paraphrase of Definition 1.2.2 is that a fibrant space is f-local if it makes f look like a weak equivalence (see Corollary 9.3.3) and a map is an f-local equivalence if all f-local spaces make it look like a weak equivalence. In Theorem 2.1.3 we show that there is a model category structure on $\mathrm{Spc}_{(*)}$ in which the fibrant objects are the local spaces (see Proposition 2.1.4) and the weak equivalences are the f-local equivalences. For a discussion of the relation of our definition of f-local equivalence to earlier definitions, see Remark 1.2.14.

1.2.3. Local spaces.

PROPOSITION 1.2.4. *Let $\mathrm{Spc}_{(*)}$ be one of our categories of spaces (see Notation 1.1.4) and let $f\colon A \to B$ be a map of cofibrant spaces. If X and Y are fibrant spaces and $g\colon X \to Y$ is a weak equivalence, then X is f-local if and only if Y is f-local.*

PROOF. We have a commutative diagram

$$\begin{array}{ccc} \operatorname{Map}(B, X) & \longrightarrow & \operatorname{Map}(A, X) \\ {\scriptstyle\cong}\downarrow & & \downarrow{\scriptstyle\cong} \\ \operatorname{Map}(B, Y) & \longrightarrow & \operatorname{Map}(A, Y) \end{array}$$

in which the vertical maps are weak equivalences (see Corollary 9.3.3). Thus, the top map is a weak equivalence if and only if the bottom map is a weak equivalence. □

PROPOSITION 1.2.5. *Let $\mathrm{Spc}_{(*)}$ be one of our categories of spaces and let $f\colon A \to B$ be a map between cofibrant spaces. If X is an f-local space and Y is a retract of X, then Y is f-local.*

PROOF. Axiom M3 (see Definition 7.1.3) implies that Y is fibrant and the map $f^*\colon \operatorname{Map}(B, Y) \to \operatorname{Map}(A, Y)$ is a retract of the weak equivalence $f^*\colon \operatorname{Map}(B, X) \to \operatorname{Map}(A, X)$ and is thus a weak equivalence. □

LEMMA 1.2.6. *Let $\mathrm{Spc}_{(*)}$ be one of our categories of spaces and let $f\colon A \to B$ be a map between cofibrant spaces. If X is an f-local space, then any space consisting of a nonempty union of path components of X is an f-local space.*

PROOF. A nonempty union of path components of a cofibrant space is a retract of that space, and so the result follows from Proposition 1.2.5. □

1.2.7. Changing the map f.

PROPOSITION 1.2.8. *Let* $\mathrm{Spc}_{(*)}$ *be one of our categories of spaces (see Notation 1.1.4) and let both* f *and* f' *be maps between cofibrant spaces. If the class of* f*-local spaces equals the class of* f'*-local spaces, then the class of* f*-local equivalences equals the class of* f'*-local equivalences.*

PROOF. This follows directly from the definitions. □

EXAMPLE 1.2.9. Let A be a simplicial set if $\mathrm{Spc}_{(*)} = \mathrm{SS}_{(*)}$ or a cell complex if $\mathrm{Spc}_{(*)} = \mathrm{Top}_{(*)}$ (see Notation 1.1.4), and let CA be the cone on A. If $f\colon * \to A$ is the inclusion of a vertex and $f'\colon A \to CA$ is the standard inclusion, then a space is f-local (i.e., A-local; see Definition 1.2.2) if and only if it is f'-local, and so the class of f-local equivalences equals the class of f'-local equivalences.

PROPOSITION 1.2.10. *Let* $\mathrm{Spc}_{(*)}$ *be one of our categories of spaces (see Notation 1.1.4) and let both* $f\colon A \to B$ *and* $f'\colon A' \to B'$ *be maps between cofibrant spaces. If there are weak equivalences* $A \to A'$ *and* $B \to B'$ *such that the square*

$$\begin{array}{ccc} A & \xrightarrow{f} & B \\ {\scriptstyle\cong}\downarrow & & \downarrow{\scriptstyle\cong} \\ A' & \xrightarrow[f']{} & B' \end{array}$$

commutes, then

(1) *the class of* f*-local spaces equals the class of* f'*-local spaces and*
(2) *the class of* f*-local equivalences equals the class of* f'*-local equivalences.*

PROOF. Proposition 1.2.8 implies that part 1 implies part 2, and so it is sufficient to prove part 1.

If W is a fibrant space, then we have the commutative square

$$\begin{array}{ccc} \mathrm{Map}(B', W) & \xrightarrow{(f')^*} & \mathrm{Map}(A', W) \\ {\scriptstyle\cong}\downarrow & & \downarrow{\scriptstyle\cong} \\ \mathrm{Map}(B, W) & \xrightarrow[f^*]{} & \mathrm{Map}(A, W) \end{array}$$

in which the vertical maps are weak equivalences (see Corollary 9.3.3). Thus, f^* is a weak equivalence if and only if $(f')^*$ is a weak equivalence, and so W is f-local if and only if it is f'-local. □

REMARK 1.2.11. Proposition 1.2.10 (see also Proposition 11.2.8) implies that we can always replace our map $f\colon A \to B$ with an inclusion of simplicial sets (if $\mathrm{Spc}_{(*)} = \mathrm{SS}_{(*)}$) or an inclusion of cell complexes (if $\mathrm{Spc}_{(*)} = \mathrm{Top}_{(*)}$) without changing the class of f-local spaces or the class of f-local equivalences. We will often assume that we have done this, and we will summarize this assumption by saying that f is an *inclusion of cell complexes.* (This usage is consistent with the definition of *cell complex* in a cofibrantly generated model category (see Definition 11.1.4).)

1.2.12. f-localization.

DEFINITION 1.2.13. Let $\mathrm{Spc}_{(*)}$ be one of our categories of spaces (see Notation 1.1.4), and let $f\colon A \to B$ be a map between cofibrant spaces.

(1) An *f-localization* of a space X is an f-local space $\widehat{X}$ (see Definition 1.2.2) together with an f-local equivalence $j_X\colon X \to \widehat{X}$. We will sometimes use the phrase *f-localization* to refer to the space $\widehat{X}$ without explicitly mentioning the f-local equivalence j. A *cofibrant f-localization* of X is an f-localization in which the f-local equivalence is also a cofibration.
(2) An *f-localization* of a map $g\colon X \to Y$ is an f-localization $(\widehat{X}, j_X)$ of X, an f-localization $(\widehat{Y}, j_Y)$ of Y, and a map $\hat{g}\colon \widehat{X} \to \widehat{Y}$ such that the square

$$\begin{array}{ccc} X & \xrightarrow{g} & Y \\ {\scriptstyle j_X}\downarrow & & \downarrow{\scriptstyle j_Y} \\ \widehat{X} & \xrightarrow[\hat{g}]{} & \widehat{Y} \end{array}$$

commutes. We will sometimes use the term *f-localization* to refer to the map $\hat{g}$ without explicitly mentioning the f-localizations $(\widehat{X}, j_X)$ of X and $(\widehat{Y}, j_Y)$ of Y.

We will show in Corollary 1.4.13 that all spaces and maps have f-localizations.

The reader should note the similarity between the definitions of f-localization and fibrant approximation (see Definition 8.1.2 and Definition 8.1.22). In Theorem 2.1.3, we prove that there is an f-local model category structure on $\mathrm{Spc}_{(*)}$ in which the fibrant objects are the local spaces and the weak equivalences are the f-local equivalences. In the f-local model category, an f-localization of a space or map is exactly a fibrant approximation to that space or map.

REMARK 1.2.14. In most earlier work on localization [**22, 20, 25, 24, 11, 16**] an f-local equivalence was defined to be a map $g\colon X \to Y$ such that for every f-local space W the map of function spaces $g^*\colon \mathrm{Map}(Y, W) \to \mathrm{Map}(X, W)$ is a weak equivalence. In fact, this earlier work considered only the subcategory of cofibrant spaces. Since a cofibrant space is a cofibrant approximation to itself, this earlier definition coincides with ours.

1.2.15. f-local equivalences.

PROPOSITION 1.2.16. *Let $\mathrm{Spc}_{(*)}$ be one of our categories of spaces (see Notation 1.1.4). If $f\colon A \to B$ is a map between cofibrant spaces, then every weak equivalence is an f-local equivalence.*

PROOF. Since a cofibrant approximation to a weak equivalence must also be a weak equivalence, this follows from Corollary 9.3.3. □

PROPOSITION 1.2.17. *Let $\mathrm{Spc}_{(*)}$ be one of our categories of spaces (see Notation 1.1.4). If $f\colon A \to B$ is a map between cofibrant spaces, then the class of f-local equivalences satisfies the "two out of three" axiom, i.e., if g and h are composable maps and if two of g, h, and hg are f-local equivalences, then so is the third.*

PROOF. Given maps $g\colon X \to Y$ and $h\colon Y \to Z$, we can apply a functorial cofibrant approximation (see Proposition 8.1.17) to g and h to obtain the diagram

$$\begin{array}{ccccc} \widetilde{X} & \xrightarrow{\tilde{g}} & \widetilde{Y} & \xrightarrow{\tilde{h}} & \widetilde{Z} \\ \downarrow & & \downarrow & & \downarrow \\ X & \xrightarrow[g]{} & Y & \xrightarrow[h]{} & Z \end{array}$$

in which $\tilde{g}$, $\tilde{h}$, and $\widetilde{hg}$ are cofibrant approximations to g, h, and hg, respectively. If W is a fibrant space, then two of the maps $\tilde{g}^*\colon \operatorname{Map}(\widetilde{Y}, W) \to \operatorname{Map}(\widetilde{X}, W)$, $\tilde{h}^*\colon \operatorname{Map}(\widetilde{Z}, W) \to \operatorname{Map}(\widetilde{Y}, W)$, and $(\widetilde{hg})^*\colon \operatorname{Map}(\widetilde{Z}, W) \to \operatorname{Map}(\widetilde{X}, W)$ are weak equivalences, and so the third is as well. □

PROPOSITION 1.2.18. *Let $\mathrm{Spc}_{(*)}$ be one of our categories of spaces (see Notation 1.1.4). If $f\colon A \to B$ is a map between cofibrant spaces, then a retract (see Definition 7.1.1) of an f-local equivalence is an f-local equivalence.*

PROOF. If $g\colon X \to Y$ is an f-local equivalence and $h\colon V \to W$ is a retract of g, then we apply a functorial cofibrant approximation (see Proposition 8.1.17) to obtain cofibrant approximations $\tilde{g}\colon \widetilde{X} \to \widetilde{Y}$ to g and $\tilde{h}\colon \widetilde{V} \to \widetilde{W}$ such that $\tilde{h}$ is a retract of $\tilde{g}$. If Z is an f-local space, then $\tilde{h}^*\colon \operatorname{Map}(\widetilde{W}, Z) \to \operatorname{Map}(\widetilde{V}, Z)$ is then a retract of the weak equivalence $\tilde{g}^*\colon \operatorname{Map}(\widetilde{Y}, Z) \to \operatorname{Map}(\widetilde{X}, Z)$, and so $\tilde{h}^*$ is a weak equivalence. □

PROPOSITION 1.2.19. *Let $\mathrm{Spc}_{(*)}$ be one of our categories of spaces (see Notation 1.1.4) and let $f\colon A \to B$ be a map of cofibrant spaces. If $g\colon X \to Y$ is a cofibration of cofibrant spaces, then g is an f-local equivalence if and only if it has the left lifting property (see Definition 7.2.1) with respect to the map $W^{\Delta[n]} \to W^{\partial\Delta[n]}$ for all $n \geq 0$ and all f-local spaces W.*

PROOF. This follows from Proposition 9.4.5 and Lemma 9.4.7. □

PROPOSITION 1.2.20. *Let $\mathrm{Spc}_{(*)}$ be one of our categories of spaces (see Notation 1.1.4) and let $f\colon A \to B$ be a map of cofibrant spaces. If T is a totally ordered set and $\boldsymbol{X}\colon T \to \mathrm{Spc}_{(*)}$ is a functor such that if $s, t \in T$ and $s \leq t$ then $\boldsymbol{X}_s \to \boldsymbol{X}_t$ is a cofibration of cofibrant spaces that is an f-local equivalence, then for every $s \in T$ the map $\boldsymbol{X}_s \to \operatorname{colim}_{t \geq s} \boldsymbol{X}_t$ is an f-local equivalence.*

PROOF. This follows from Proposition 1.2.19, Lemma 10.3.5, and Proposition 10.3.6. □

PROPOSITION 1.2.21. *Let $\mathrm{Spc}_{(*)}$ be one of our categories of spaces (see Notation 1.1.4) and let $f\colon A \to B$ be a map of cofibrant spaces. If $g\colon C \to D$ is a cofibration between cofibrant spaces that is also an f-local equivalence and if the square*

$$\begin{array}{ccc} C & \longrightarrow & X \\ {\scriptstyle g}\downarrow & & \downarrow{\scriptstyle h} \\ D & \longrightarrow & Y \end{array}$$

is a pushout, then h is an f-local equivalence.

PROOF. Factor the map $C \to X$ as $C \xrightarrow{u} P \xrightarrow{v} X$, where u is a cofibration and v is a trivial fibration. If we let Q be the pushout $D \amalg_C P$, then we have the commutative diagram

$$\begin{array}{ccccc} C & \xrightarrow{u} & P & \xrightarrow{v} & X \\ {\scriptstyle g}\downarrow & & \downarrow{\scriptstyle k} & & \downarrow{\scriptstyle h} \\ D & \xrightarrow[s]{} & Q & \xrightarrow[t]{} & Y \end{array}$$

in which u and s are cofibrations, and so P and Q are cofibrant. Since k is a cofibration, we are in a proper model category (see Theorem 13.1.11 and Theorem 13.1.13), and Proposition 7.2.14 implies that Y is the pushout $Q \amalg_P X$, the map t is a weak equivalence. Thus, k is a cofibrant approximation to h (see Definition 8.1.22), and so it is sufficient to show that k induces a weak equivalence of mapping spaces to every f-local space. Since g is a cofibration and an f-local equivalence and k is a cofibration, this follows from Proposition 1.2.19 and Lemma 7.2.11. □

PROPOSITION 1.2.22. *Let $\mathrm{Spc}_{(*)}$ be one of our categories of spaces (see Notation 1.1.4) and let $f\colon A \to B$ be an inclusion of cell complexes (see Remark 1.2.11). If I is a set of f-local equivalences that are cofibrations between cofibrant spaces, then a transfinite composition of pushouts of elements of I is also an f-local equivalence.*

PROOF. Proposition 1.2.21 implies that every pushout of an element of I is an f-local equivalence. If λ is an ordinal and

$$X_0 \to X_1 \to X_2 \to \cdots \to X_\beta \to \cdots \qquad (\beta < \lambda)$$

is a λ-sequence of pushouts of elements of I, then Proposition 17.9.4 implies that we can find a λ-sequence of cofibrations together with a map of λ-sequences

$$\begin{array}{ccccccccc} \widetilde{X}_0 & \longrightarrow & \widetilde{X}_1 & \longrightarrow & \widetilde{X}_2 & \longrightarrow \cdots \longrightarrow & \widetilde{X}_\beta & \longrightarrow & \cdots \\ \downarrow & & \downarrow & & \downarrow & & \downarrow & & \\ X_0 & \longrightarrow & X_1 & \longrightarrow & X_2 & \longrightarrow \cdots \longrightarrow & X_\beta & \longrightarrow & \cdots \end{array}$$

such that

(1) each vertical map $\widetilde{X}_\beta \to X_\beta$ is a cofibrant approximation to X_β,
(2) each map $\widetilde{X}_\beta \to \widetilde{X}_{\beta+1}$ is a cofibration, and
(3) the map $\mathrm{colim}_{\beta<\lambda} \widetilde{X}_\beta \to \mathrm{colim}_{\beta<\lambda} X_\beta$ is a cofibrant approximation to $\mathrm{colim}_{\beta<\lambda} X_\beta$.

If W is an f-local space then the simplicial set $\mathrm{Map}(\mathrm{colim}_{\beta<\lambda} \widetilde{X}_\beta, W)$ is isomorphic to $\lim_{\beta<\lambda} \mathrm{Map}(\widetilde{X}_\beta, W)$. Since each map $X_\beta \to X_{\beta+1}$ is an f-local equivalence and each map $\widetilde{X}_\beta \to \widetilde{X}_{\beta+1}$ is a cofibration, each map $\mathrm{Map}(\widetilde{X}_{\beta+1}, W) \to \mathrm{Map}(\widetilde{X}_\beta, W)$ is a trivial fibration. Thus,

$$\mathrm{Map}(\widetilde{X}_0, W) \leftarrow \mathrm{Map}(\widetilde{X}_1, W) \leftarrow \mathrm{Map}(\widetilde{X}_2, W) \leftarrow \cdots \leftarrow \mathrm{Map}(\widetilde{X}_\beta, W) \leftarrow \cdots$$

is a tower of trivial fibrations of simplicial sets, and so the composition

$$\mathrm{Map}(\operatorname*{colim}_{\beta<\lambda} \widetilde{X}_\beta, W) \to \lim_{\beta<\lambda} \mathrm{Map}(\widetilde{X}_\beta, W) \to \mathrm{Map}(\widetilde{X}_0, W)$$

is a weak equivalence, and so the composition $X_0 \to \operatorname{colim}_{\beta<\lambda} X_\beta$ is an f-local equivalence. □

1.2.23. f-local Whitehead theorems.

LEMMA 1.2.24. *Let $\mathrm{Spc}_{(*)}$ be one of our categories of spaces (see Notation 1.1.4) and let $f: A \to B$ be a map between cofibrant spaces. If W is an f-local space and $g: X \to Y$ is an f-local equivalence of cofibrant spaces, then g induces an isomorphism of the sets of simplicial homotopy classes of maps $g^*: [Y, W] \approx [X, W]$.*

PROOF. This follows from Proposition 9.5.10. □

THEOREM 1.2.25 (Strong f-local Whitehead theorem). *Let $\mathrm{Spc}_{(*)}$ be one of our categories of spaces (see Notation 1.1.4) and let $f: A \to B$ be a map between cofibrant spaces. If X and Y are cofibrant f-local spaces and $g: X \to Y$ is an f-local equivalence, then g is a simplicial homotopy equivalence.*

PROOF. This follows from Lemma 1.2.24 and Proposition 9.6.9. □

THEOREM 1.2.26 (Weak f-local Whitehead theorem). *Let $\mathrm{Spc}_{(*)}$ be one of our categories of spaces (see Notation 1.1.4) and let $f: A \to B$ be a map between cofibrant spaces. If X and Y are f-local spaces and $g: X \to Y$ is an f-local equivalence, then g is a weak equivalence.*

PROOF. Choose a cofibrant approximation $\tilde{g}: \widetilde{X} \to \widetilde{Y}$ to g such that $j_X: \widetilde{X} \to X$ and $j_Y: \widetilde{Y} \to Y$ are trivial fibrations (see Proposition 8.1.23). Proposition 1.2.4 implies that $\widetilde{X}$ and $\widetilde{Y}$ are f-local spaces, and Proposition 1.2.16 and Proposition 1.2.17 imply that $\tilde{g}$ is an f-local equivalence. Theorem 1.2.25 and Theorem 7.8.5 now imply that $\tilde{g}$ is a weak equivalence, which implies that g is a weak equivalence. □

1.2.27. Characterizing f-local spaces and f-local equivalences.

THEOREM 1.2.28. *Let $\mathrm{Spc}_{(*)}$ be one of our categories of spaces (see Notation 1.1.4) and let $f: A \to B$ be a map between cofibrant spaces. If X is a fibrant space and $j: X \to \widehat{X}$ is an f-localization of X (see Definition 1.2.13), then j is a weak equivalence if and only if X is f-local.*

PROOF. If X is f-local, then Theorem 1.2.26 implies that j is a weak equivalence. Conversely, if j is a weak equivalence, then Proposition 1.2.4 implies that X is f-local. □

THEOREM 1.2.29. *Let $\mathrm{Spc}_{(*)}$ be one of our categories of spaces (see Notation 1.1.4) and let $f: A \to B$ be a map between cofibrant spaces. If $\hat{g}: \widehat{X} \to \widehat{Y}$ is an f-localization of $g: X \to Y$ (see Definition 1.2.13), then g is an f-local equivalence if and only if $\hat{g}$ is a weak equivalence.*

PROOF. Proposition 1.2.17 implies that g is an f-local equivalence if and only if $\hat{g}$ is an f-local equivalence, and Theorem 1.2.26 and Proposition 1.2.16 imply that $\hat{g}$ is an f-local equivalence if and only if it is a weak equivalence. □

If $\mathrm{Spc}_{(*)}$ is one of our categories of spaces (see Notation 1.1.4) and $f: A \to B$ is a map between cofibrant spaces, then in Definition 1.4.11 we define a functorial

f-localization (L_f, j). Theorem 1.2.28 then implies that a fibrant space X is f-local if and only if the localization map $j(X)\colon X \to \mathrm{L}_f X$ is a weak equivalence (see Theorem 1.4.14), and Theorem 1.2.29 implies that a map $g\colon X \to Y$ is an f-local equivalence if and only if $\mathrm{L}_f(g)\colon \mathrm{L}_f X \to \mathrm{L}_f Y$ is a weak equivalence (see Theorem 1.4.15).

1.2.30. Topological spaces and simplicial sets.

PROPOSITION 1.2.31. *Let $f\colon A \to B$ be a map between cofibrant spaces in $\mathrm{Top}_{(*)}$ (see Notation 1.1.4).*

1. *A space is f-local if and only if it is $|\mathrm{Sing}\, f|$-local.*
2. *A map $g\colon X \to Y$ is an f-local equivalence if and only if it is a $|\mathrm{Sing}\, f|$-local equivalence.*

PROOF. This follows from Proposition 1.2.10 and Proposition 1.2.8. □

PROPOSITION 1.2.32. *Let $f\colon A \to B$ be a map in $\mathrm{SS}_{(*)}$ (see Notation 1.1.4).*

1. *A space is f-local if and only if it is $(\mathrm{Sing}|f|)$-local.*
2. *A map $g\colon X \to Y$ is an f-local equivalence if and only if it is a $(\mathrm{Sing}|f|)$-local equivalence.*

PROOF. Since every simplicial set is cofibrant, the result follows from Proposition 1.2.10 and Proposition 1.2.8. □

PROPOSITION 1.2.33. *If $f\colon A \to B$ is a map in $\mathrm{SS}_{(*)}$ (see Notation 1.1.4), then a topological space W in $\mathrm{Top}_{(*)}$ is $|f|$-local if and only if $\mathrm{Sing}\, W$ is f-local.*

PROOF. Lemma 1.1.10 gives us the commutative square

$$\begin{array}{ccc} \mathrm{Map}(|B|, W) & \longrightarrow & \mathrm{Map}(|A|, W) \\ \approx\downarrow & & \downarrow\approx \\ \mathrm{Map}(B, \mathrm{Sing}\, W) & \longrightarrow & \mathrm{Map}(A, \mathrm{Sing}\, W) \end{array}$$

in which the vertical maps are isomorphisms, from which the proposition follows. □

PROPOSITION 1.2.34. *If $f\colon A \to B$ is a map in $\mathrm{SS}_{(*)}$ (see Notation 1.1.4) and K is a fibrant simplicial set in $\mathrm{SS}_{(*)}$, then K is f-local if and only if $|K|$ is $|f|$-local.*

PROOF. Since K is fibrant the natural map $K \to \mathrm{Sing}|K|$ is a weak equivalence of fibrant spaces, and so we have the commutative square

$$\begin{array}{ccc} \mathrm{Map}(B, K) & \longrightarrow & \mathrm{Map}(A, K) \\ \approx\downarrow & & \downarrow\approx \\ \mathrm{Map}(B, \mathrm{Sing}|K|) & \longrightarrow & \mathrm{Map}(A, \mathrm{Sing}|K|) \end{array}$$

in which the vertical maps are weak equivalences (see Corollary 9.3.3). Thus, K is f-local if and only if $\mathrm{Sing}|K|$ is f-local, and so the proposition follows from Proposition 1.2.33. □

PROPOSITION 1.2.35. *If $f\colon A \to B$ is a map in $\mathrm{SS}_{(*)}$ (see Notation 1.1.4), then the map $g\colon C \to D$ in $\mathrm{SS}_{(*)}$ is an f-local equivalence if and only if the map $|g|\colon |C| \to |D|$ in $\mathrm{Top}_{(*)}$ is a $|f|$-local equivalence.*

PROOF. Since every simplicial set is cofibrant, g is an f-local equivalence if and only if for every f-local simplicial set K the map of simplicial sets $g^*\colon \mathrm{Map}(D, K) \to \mathrm{Map}(C, K)$ is a weak equivalence. If K is an f-local simplicial set then K is fibrant, and so Corollary 9.3.3 implies that g is an f-local equivalence if and only if, for every f-local simplicial set K, the map of simplicial sets $g^*\colon \mathrm{Map}(D, \mathrm{Sing}|K|) \to \mathrm{Map}(C, \mathrm{Sing}|K|)$ is a weak equivalence. Lemma 1.1.10 implies that this is true if and only if $\mathrm{Map}(|D|, |K|) \to \mathrm{Map}(|C|, |K|)$ is a weak equivalence. Proposition 1.2.33 and Proposition 1.2.34 imply that this is true if and only if for every $|f|$-local topological space W the map $\mathrm{Map}(|D|, W) \to \mathrm{Map}(|C|, W)$ is a weak equivalence. Since $|C|$ and $|D|$ are cofibrant, this is true if and only if $|g|\colon |C| \to |D|$ is a $|f|$-local equivalence. □

PROPOSITION 1.2.36. *If $f\colon A \to B$ is a map in $\mathrm{SS}_{(*)}$ (see Notation 1.1.4), then the map $g\colon X \to Y$ in $\mathrm{Top}_{(*)}$ is a $|f|$-local equivalence if and only if the map $(\mathrm{Sing}\, g)\colon \mathrm{Sing}\, X \to \mathrm{Sing}\, Y$ in $\mathrm{SS}_{(*)}$ is an f-local equivalence.*

PROOF. The map $|\mathrm{Sing}\, g|\colon |\mathrm{Sing}\, X| \to |\mathrm{Sing}\, Y|$ is a cofibrant approximation to g (see Definition 8.1.22), and so g is a $|f|$-local equivalence if and only if, for every $|f|$-local topological space W, the map of simplicial sets $\mathrm{Map}(|\mathrm{Sing}\, Y|, W) \to \mathrm{Map}(|\mathrm{Sing}\, X|, W)$ is a weak equivalence. Lemma 1.1.10 implies that this is true if and only if, for every $|f|$-local topological space W, the map $\mathrm{Map}(\mathrm{Sing}\, Y, \mathrm{Sing}\, W) \to \mathrm{Map}(\mathrm{Sing}\, X, \mathrm{Sing}\, W)$ is a weak equivalence. If K is an f-local simplicial set, then K is fibrant, and so the natural map $K \to \mathrm{Sing}|K|$ is a weak equivalence of fibrant objects. Thus, Corollary 9.3.3 and Proposition 1.2.34 imply that g is a $|f|$-local equivalence if and only if, for every f-local simplicial set K, the map $\mathrm{Map}(\mathrm{Sing}\, Y, \mathrm{Sing}|K|) \to \mathrm{Map}(\mathrm{Sing}\, X, \mathrm{Sing}|K|)$ is a weak equivalence. Since every simplicial set is cofibrant, this completes the proof. □

1.3. Constructing an f-localization functor

If $\mathrm{Spc}_{(*)}$ is one of our categories of spaces (see Notation 1.1.4) and $f\colon A \to B$ is a map between cofibrant spaces, we describe in this section how to construct a functorial f-localization on $\mathrm{Spc}_{(*)}$ (see Definition 1.2.13). The construction that we present is essentially the one used by Bousfield in [**9**].

1.3.1. Horns on f. If $\mathrm{Spc}_{(*)}$ is one of our categories of spaces (see Notation 1.1.4) and $f\colon A \to B$ is a map between cofibrant spaces, we want to construct a functorial f-localization (see Definition 1.2.13) on $\mathrm{Spc}_{(*)}$. That is, for every space X we want to construct a natural f-local space $\widehat{X}$ together with a natural f-local equivalence $X \to \widehat{X}$. Remark 1.2.11 implies that we can assume that f is an inclusion of cell complexes, and we will assume that f is such an inclusion.

If $\widehat{X}$ is to be an f-local space, then it must first of all be fibrant. Thus, the map $\widehat{X} \to *$ must have the right lifting property with respect to the inclusions $\mathrm{Spc}_{(*)}(\Lambda[n,k]) \to \mathrm{Spc}_{(*)}(\Delta[n])$ (see Definition 1.1.13) for all $n > 0$ and $n \geq k \geq 0$.

If $\widehat{X}$ is a fibrant space, then $f^*\colon \operatorname{Map}(B,\widehat{X}) \to \operatorname{Map}(A,\widehat{X})$ is already a fibration of simplicial sets (see Proposition 9.3.1). Thus, if $\widehat{X}$ is fibrant, then the assertion that $\widehat{X}$ is f-local is equivalent to the assertion that f^* is a trivial fibration of simplicial sets. Since a map of simplicial sets is a trivial fibration if and only if it has the right lifting property with respect to the inclusions $\partial\Delta[n] \to \Delta[n]$ for $n \geq 0$, this implies that a fibrant space $\widehat{X}$ is f-local if and only if the dotted arrow exists in every solid arrow diagram of the form

$$\begin{array}{ccc} \partial\Delta[n] & \longrightarrow & \operatorname{Map}(B,\widehat{X}) \\ \downarrow & \nearrow & \downarrow \\ \Delta[n] & \longrightarrow & \operatorname{Map}(A,\widehat{X}) \end{array} ,$$

and the isomorphisms of Definition 1.1.8 imply that this is true if and only if the dotted arrow exists in every solid arrow diagram of the form

$$\begin{array}{ccc} A\otimes\Delta[n] \amalg_{A\otimes\partial\Delta[n]} B\otimes\partial\Delta[n] & \longrightarrow & \widehat{X} \\ \downarrow & \nearrow & \downarrow \\ B\otimes\Delta[n] & \longrightarrow & * \end{array} .$$

Thus, a space $\widehat{X}$ is f-local if and only if the map $\widehat{X} \to *$ has the right lifting property with respect to the maps $\operatorname{Spc}_{(*)}(\Lambda[n,k]) \to \operatorname{Spc}_{(*)}(\Delta[n])$ for all $n > 0$ and $n \geq k \geq 0$ and the maps $A\otimes\Delta[n] \amalg_{A\otimes\partial\Delta[n]} B\otimes\partial\Delta[n] \to B\otimes\Delta[n]$ for all $n \geq 0$. This is the motivation for the following definition of $\overline{\Lambda\{f\}}$, the *augmented set of f-horns*.

DEFINITION 1.3.2. Let $\operatorname{Spc}_{(*)}$ be one of our categories of spaces (see Notation 1.1.4) and let $f\colon A \to B$ be an inclusion of cell complexes (see Remark 1.2.11).

- The set $\Lambda\{f\}$ of *horns on f* is the set of maps
$$\Lambda\{f\} = \{A\otimes\Delta[n] \amalg_{A\otimes\partial\Delta[n]} B\otimes\partial\Delta[n] \to B\otimes\Delta[n] \mid n \geq 0\} .$$
If $\operatorname{Spc}_{(*)} = \operatorname{Spc}_*$ and f is the map $f\colon * \to A$, then $\Lambda\{f\}$ is the set of maps
$$\Lambda\{A\} = \{A\otimes\partial\Delta[n] \to A\otimes\Delta[n] \mid n \geq 0\} ,$$
and it will also be called the set of *horns on A*.
- The *augmented set of f-horns* $\overline{\Lambda\{f\}}$ is the set of maps
$$J_f = \Lambda\{f\} \cup \{\operatorname{Spc}_{(*)}(\Lambda[n,k]) \to \operatorname{Spc}_{(*)}(\Delta[n]) \mid n > 0, n \geq k \geq 0\}$$
(see Definition 1.1.13).

PROPOSITION 1.3.3. *If $\operatorname{Spc}_{(*)}$ is one of our categories of spaces (see Notation 1.1.4) and $f\colon A \to B$ is an inclusion of cell complexes (see Remark 1.2.11), then a space X is f-local if and only if the map $X \to *$ has the right lifting property with respect to every element of the augmented set of f-horns (see Definition 1.3.2).*

PROOF. This follows from the discussion preceding Definition 1.3.2. □

We will construct the map $X \to \widehat{X}$ as a transfinite composition (see Definition 10.2.2) of inclusions of cell complexes $X = \mathrm{E}^0 \to \mathrm{E}^1 \to \mathrm{E}^2 \to \cdots \to \mathrm{E}^\beta \to$

$\cdots \quad (\beta < \lambda)$, where $\widehat{X} = \operatorname{colim}_{\beta<\lambda} \mathrm{E}^\beta$. To ensure that $\widehat{X}$ is f-local, we will construct the E^β so that if the map $C \to D$ is an element of $\overline{\Lambda\{f\}}$ then

(1) for every map $h\colon C \to \widehat{X}$ there is an ordinal $\alpha < \lambda$ such that h factors through the map $\mathrm{E}^\alpha \to \widehat{X}$, and
(2) for every ordinal $\alpha < \lambda$ the dotted arrow exists in every solid arrow diagram of the form

$$\begin{array}{ccccc} C & \longrightarrow & \mathrm{E}^\alpha & \longrightarrow & \mathrm{E}^{\alpha+1} \\ \big\downarrow & & & \nearrow & \\ D & & & & \end{array}$$

Thus, if the map $C \to D$ is an element of $\overline{\Lambda\{f\}}$, then the dotted arrow will exist in every solid arrow diagram of the form

$$\begin{array}{ccc} C & \longrightarrow & \widehat{X} \\ \big\downarrow & \nearrow & \\ D & & \end{array}$$

and so the map $\widehat{X} \to *$ will have the right lifting property with respect to every element of $\overline{\Lambda\{f\}}$ (see Proposition 1.3.3).

1.3.4. Choice of the ordinal λ. If A and B are finite complexes, then we let λ be the first infinite cardinal. Otherwise, we let λ be the first cardinal greater than that of the set of simplices (or cells) of $A \amalg B$ (in which case λ is a successor cardinal). In either case, λ is a *regular cardinal* (see Proposition 10.1.14).

Suppose we now construct a λ-sequence (see Definition 10.2.1) of inclusions of cell complexes

$$X = \mathrm{E}^0 \to \mathrm{E}^1 \to \mathrm{E}^2 \to \cdots \to \mathrm{E}^\beta \to \cdots \qquad (\beta < \lambda)$$

and let $\widehat{X} = \operatorname{colim}_{\beta<\lambda} \mathrm{E}^\beta$. If $A \otimes \Delta[n] \amalg_{A\otimes\partial\Delta[n]} B \otimes \partial\Delta[n] \to \widehat{X}$ is any map, then for each simplex (or cell) of $A \otimes \Delta[n] \amalg_{A\otimes\partial\Delta[n]} B \otimes \partial\Delta[n]$ there is an ordinal $\beta < \lambda$ such that that simplex (or cell) lands in E^β. (If $\mathrm{Spc}_{(*)} = \mathrm{Top}_{(*)}$, then this follows from Corollary 10.7.5.) If we let α be the union of the ordinals β obtained in this way for each simplex (or cell) in $A \otimes \Delta[n] \amalg_{A\otimes\partial\Delta[n]} B \otimes \partial\Delta[n]$, then the regularity of λ ensures that $\alpha < \lambda$. Thus, our map factors through E^α. The same argument applies to maps $\mathrm{Spc}_{(*)}(\Lambda[n,k]) \to \widehat{X}$.

1.3.5. Constructing the sequence. We begin the sequence by letting $\mathrm{E}^0 = X$. If $\beta < \lambda$ and we have constructed the sequence through E^β, we let

$$C_\beta = \coprod_{\substack{(C\to D)\in\overline{\Lambda\{f\}} \\ \mathrm{Spc}_{(*)}(C,\mathrm{E}^\beta)}} C \qquad \text{and} \qquad D_\beta = \coprod_{\substack{(C\to D)\in\overline{\Lambda\{f\}} \\ \mathrm{Spc}_{(*)}(C,\mathrm{E}^\beta)}} D\,.$$

We then have a natural map $C_\beta \to \mathrm{E}^\beta$, and we define $\mathrm{E}^{\beta+1}$ by letting the square

$$\begin{array}{ccc} C_\beta & \longrightarrow & \mathrm{E}^\beta \\ \downarrow & & \vdots \\ D_\beta & \cdots\cdots\to & \mathrm{E}^{\beta+1} \end{array}$$

be a pushout. If γ is a limit ordinal, we let $\mathrm{E}^\gamma = \operatorname{colim}_{\beta<\gamma} \mathrm{E}^\beta$. We let $\widehat{X} = \operatorname{colim}_{\beta<\lambda} \mathrm{E}^\beta$.

It remains only to show that the map $X \to \widehat{X}$ that we have constructed is an f-local equivalence. This will follow from Theorem 1.3.11.

1.3.6. Horns on f and f-local equivalences.

PROPOSITION 1.3.7. *If $\mathrm{Spc}_{(*)}$ is one of our categories of spaces (see Notation 1.1.4) and $f\colon A \to B$ is an inclusion of cell complexes (see Remark 1.2.11), then every horn on f (see Definition 1.3.2) is an f-local equivalence.*

PROOF. Since every horn on f is a cofibration between cofibrant spaces, this follows from Proposition 9.4.5 and Proposition 9.4.8. □

DEFINITION 1.3.8. Let $\mathrm{Spc}_{(*)}$ be one of our categories of spaces (see Notation 1.1.4). If $f\colon A \to B$ is an inclusion of cell complexes (see Remark 1.2.11), then a *relative $\overline{\Lambda\{f\}}$-cell complex* is defined to be a map that can be constructed as a transfinite composition (see Definition 10.2.2) of pushouts (see Definition 7.2.10) of elements of $\overline{\Lambda\{f\}}$ (see Definition 1.3.2). If the map from the initial object to a space X is a relative $\overline{\Lambda\{f\}}$-cell complex, then X will be called a *$\overline{\Lambda\{f\}}$-cell complex.*

THEOREM 1.3.9. *Let $\mathrm{Spc}_{(*)}$ be one of our categories of spaces (see Notation 1.1.4). If $f\colon A \to B$ is an inclusion of cell complexes (see Remark 1.2.11), then every relative $\overline{\Lambda\{f\}}$-cell complex is both a cofibration and an f-local equivalence.*

PROOF. Since every element of $\overline{\Lambda\{f\}}$ is a cofibration and cofibrations are closed under both pushouts and transfinite compositions (see Proposition 10.3.4), every relative $\overline{\Lambda\{f\}}$-cell complex is a cofibration, and Proposition 1.2.22 implies that a relative $\overline{\Lambda\{f\}}$-cell complex is an f-local equivalence. □

PROPOSITION 1.3.10. *Let $\mathrm{Spc}_{(*)}$ be one of our categories of spaces (see Notation 1.1.4). If $f\colon A \to B$ is an inclusion of cell complexes (see Remark 1.2.11), then for every space X the map $X \to \widehat{X}$ constructed in Section 1.3.5 is a relative $\overline{\Lambda\{f\}}$-cell complex.*

PROOF. The map $X \to \widehat{X}$ is constructed as a transfinite composition of pushouts of coproducts of elements of $\overline{\Lambda\{f\}}$, and so the result follows from Proposition 10.2.14. □

THEOREM 1.3.11. *Let $\mathrm{Spc}_{(*)}$ be one of our categories of spaces (see Notation 1.1.4). If $f\colon A \to B$ is an inclusion of cell complexes (see Remark 1.2.11), then for every space X the map $X \to \widehat{X}$ constructed in Section 1.3.5 is a natural f-localization of X.*

PROOF. This follows from Proposition 1.3.10, Theorem 1.3.9, Proposition 1.3.3, and the discussion following Proposition 1.3.3. □

1.4. Concise description of the f-localization

1.4.1. f-cofibrations and f-injectives.

DEFINITION 1.4.2. Let $\mathrm{Spc}_{(*)}$ be one of our categories of spaces (see Notation 1.1.4) and let $f\colon A \to B$ be an inclusion of cell complexes (see Remark 1.2.11).

(1) A $\overline{\Lambda\{f\}}$*-injective* is defined to be a map that has the right lifting property (see Definition 7.2.1) with respect to every element of $\overline{\Lambda\{f\}}$ (see Definition 1.3.2). A space X will be called a $\overline{\Lambda\{f\}}$*-injective* if the map $X \to *$ is a $\overline{\Lambda\{f\}}$-injective. If f is a cofibration $f\colon * \to A$, then a $\overline{\Lambda\{f\}}$-injective will also be called a $\overline{\Lambda\{A\}}$*-injective*.

(2) A $\overline{\Lambda\{f\}}$*-cofibration* is defined to be a map that has the left lifting property with respect to all $\overline{\Lambda\{f\}}$-injectives. If the map from the initial object to a space X is a $\overline{\Lambda\{f\}}$-cofibration, then X will be called $\overline{\Lambda\{f\}}$*-cofibrant.* If f is a cofibration $f\colon * \to A$, then a $\overline{\Lambda\{f\}}$-cofibration will also be called a $\overline{\Lambda\{A\}}$*-cofibration*, and a $\overline{\Lambda\{f\}}$-cofibrant space will also be called a $\overline{\Lambda\{A\}}$*-cofibrant space.*

REMARK 1.4.3. The term $\overline{\Lambda\{f\}}$*-injective* comes from the theory of injective classes ([**36**]). A space X is a $\overline{\Lambda\{f\}}$-injective if and only if it is injective in the sense of [**36**] relative to the elements of $\overline{\Lambda\{f\}}$, and we will show in Proposition 1.4.5 that a map $p\colon X \to Y$ is a $\overline{\Lambda\{f\}}$-injective if and only if, in the category $(\mathrm{Spc}_{(*)} \downarrow Y)$ of spaces over Y (see Definition 11.8.1), the object p is injective relative to the class of maps whose image under the forgetful functor $(\mathrm{Spc}_{(*)} \downarrow Y) \to \mathrm{Spc}_{(*)}$ is a relative $\overline{\Lambda\{f\}}$-cell complex (see Definition 1.3.8).

PROPOSITION 1.4.4. *Let $\mathrm{Spc}_{(*)}$ be one of our categories of spaces (see Notation 1.1.4). If $f\colon A \to B$ is an inclusion of cell complexes (see Remark 1.2.11), then a map $p\colon X \to Y$ is a $\overline{\Lambda\{f\}}$-injective if and only if it is a fibration with the homotopy right lifting property with respect to f.*

PROOF. This follows from Lemma 9.4.7. □

PROPOSITION 1.4.5. *Let $\mathrm{Spc}_{(*)}$ be one of our categories of spaces (see Notation 1.1.4). If $f\colon A \to B$ is an inclusion of cell complexes (see Remark 1.2.11), then every relative $\overline{\Lambda\{f\}}$-cell complex (see Definition 1.3.8) is a $\overline{\Lambda\{f\}}$-cofibration.*

PROOF. This follows from Proposition 1.4.4. □

PROPOSITION 1.4.6. *Let $\mathrm{Spc}_{(*)}$ be one of our categories of spaces (see Notation 1.1.4). If $f\colon A \to B$ is an inclusion of cell complexes (see Remark 1.2.11), then every trivial cofibration is a $\overline{\Lambda\{f\}}$-cofibration.*

PROOF. This follows from Proposition 7.2.3. □

PROPOSITION 1.4.7. *Let $\mathrm{Spc}_{(*)}$ be one of our categories of spaces (see Notation 1.1.4). If $f\colon A \to B$ is an inclusion of cell complexes (see Remark 1.2.11), then a space X is a $\overline{\Lambda\{f\}}$-injective if and only if it is f-local (see Definition 1.2.2).*

PROOF. This follows from Proposition 9.4.5 and Proposition 1.4.4. □

1.4.8. The functorial localization.

PROPOSITION 1.4.9. *Let* $\mathrm{Spc}_{(*)}$ *be one of our categories of spaces (see Notation 1.1.4) and let* $f\colon A \to B$ *be an inclusion of cell complexes (see Remark 1.2.11). If* $j\colon X \to \widehat{X}$ *is a relative* $\overline{\Lambda\{f\}}$*-cell complex and* $\widehat{X}$ *is a* $\overline{\Lambda\{f\}}$*-injective, then the pair* $(\widehat{X}, j)$ *is a cofibrant* f*-localization of* X.

PROOF. This follows from Proposition 1.4.7 and Theorem 1.3.9. □

THEOREM 1.4.10. *Let* $\mathrm{Spc}_{(*)}$ *be one of our categories of spaces (see Notation 1.1.4). If* $f\colon A \to B$ *is an inclusion of cell complexes (see Remark 1.2.11), then there is a natural factorization of every map* $X \to Y$ *as*

$$X \xrightarrow{j} \mathrm{E}_f \xrightarrow{p} Y$$

in which j *is a relative* $\overline{\Lambda\{f\}}$*-cell complex (see Definition 1.3.8) and* p *is a* $\overline{\Lambda\{f\}}$*-injective (see Definition 1.4.2).*

PROOF. This follows from Proposition 10.5.16. □

DEFINITION 1.4.11. Let $\mathrm{Spc}_{(*)}$ be one of our categories of spaces (see Notation 1.1.4) and let $f\colon A \to B$ be an inclusion of cell complexes (see Remark 1.2.11). The *f-localization* of a space X is the space $\mathrm{L}_f X$ obtained by applying the factorization of Theorem 1.4.10 to the map $X \to *$ from X to the terminal object of $\mathrm{Spc}_{(*)}$. This factorization defines a natural transformation $j\colon 1 \to \mathrm{L}_f$ such that $j_X\colon X \to \mathrm{L}_f X$ is a relative $\overline{\Lambda\{f\}}$-cell complex.

THEOREM 1.4.12. *Let* $\mathrm{Spc}_{(*)}$ *be one of our categories of spaces (see Notation 1.1.4). If* $f\colon A \to B$ *is an inclusion of cell complexes (see Remark 1.2.11), then for every space* X *the* f*-localization* $j_X\colon X \to \mathrm{L}_f X$ *(see Definition 1.4.11) is a cofibrant* f*-localization of* X.

PROOF. This follows from Proposition 1.4.9. □

COROLLARY 1.4.13. *Let* $\mathrm{Spc}_{(*)}$ *be one of our categories of spaces (see Notation 1.1.4). If* $f\colon A \to B$ *is an inclusion of cell complexes (see Remark 1.2.11), then every space has an* f*-localization.*

PROOF. This follows from Theorem 1.4.12. □

THEOREM 1.4.14. *Let* $\mathrm{Spc}_{(*)}$ *be one of our categories of spaces (see Notation 1.1.4) and let* $f\colon A \to B$ *be an inclusion of cell complexes (see Remark 1.2.11). If* X *is a fibrant space, then* X *is* f*-local if and only if the* f*-localization map* $j_X\colon X \to \mathrm{L}_f X$ *is a weak equivalence.*

PROOF. This follows from Theorem 1.2.28. □

THEOREM 1.4.15. *Let* $\mathrm{Spc}_{(*)}$ *be one of our categories of spaces (see Notation 1.1.4). If* $f\colon A \to B$ *is an inclusion of cell complexes (see Remark 1.2.11), then a map* $g\colon X \to Y$ *is an* f*-local equivalence if and only if its* f*-localization* $\mathrm{L}_f(g)\colon \mathrm{L}_f X \to \mathrm{L}_f Y$ *is a weak equivalence.*

PROOF. This follows from Theorem 1.2.29. □

PROPOSITION 1.4.16. *Let* $\mathrm{Spc}_{(*)}$ *be one of our categories of spaces (see Notation 1.1.4). If* $f\colon A \to B$ *is an inclusion of cell complexes (see Remark 1.2.11), then every* $\overline{\Lambda\{f\}}$*-cofibration (see Definition 1.4.2) is a retract of a relative* $\overline{\Lambda\{f\}}$*-cell complex.*

PROOF. This follows form Theorem 1.4.10 and the retract argument (see Proposition 7.2.2). □

COROLLARY 1.4.17. *Let* $\mathrm{Spc}_{(*)}$ *be one of our categories of spaces (see Notation 1.1.4). If* $f\colon A \to B$ *is an inclusion of cell complexes (see Remark 1.2.11), then every* $\overline{\Lambda\{f\}}$*-cofibration is an* f*-local equivalence.*

PROOF. This follows from Proposition 1.4.16, Theorem 1.3.9, and Proposition 1.2.18. □

1.4.18. The localization of a cofibration.

LEMMA 1.4.19. *Let* $\mathrm{Spc}_{(*)}$ *be one of our categories of spaces (see Notation 1.1.4) and let* $f\colon A \to B$ *be an inclusion of cell complexes (see Remark 1.2.11). Let* $X \to X'$ *and* $Y \to Y'$ *be cofibrations and let the square*

$$\begin{array}{ccc} X & \longrightarrow & Y \\ \downarrow & & \downarrow \\ X' & \longrightarrow & Y' \end{array}$$

be commutative. If we apply the factorization of Theorem 1.4.10 to each of the horizontal maps to obtain the commutative diagram

$$\begin{array}{ccccc} X & \longrightarrow & \mathrm{E}_f & \longrightarrow & Y \\ \downarrow & & \downarrow & & \downarrow \\ X' & \longrightarrow & \mathrm{E}'_f & \longrightarrow & Y' \end{array},$$

then the map $\mathrm{E}_f \to \mathrm{E}'_f$ *is a cofibration.*

PROOF. Using Lemma 7.2.15, one can check inductively that at each stage in the construction of the factorization we have a cofibration $\mathrm{E}^\beta \to (\mathrm{E}^\beta)'$. □

PROPOSITION 1.4.20. *Let* $\mathrm{Spc}_{(*)}$ *be one of our categories of spaces (see Notation 1.1.4) and let* $f\colon A \to B$ *be an inclusion of cell complexes (see Remark 1.2.11). If* $g\colon X \to Y$ *is a cofibration, then so is* $\mathrm{L}_f(g)\colon \mathrm{L}_f X \to \mathrm{L}_f Y$ *(see Definition 1.4.11).*

PROOF. This follows from Lemma 1.4.19. □

1.5. Postnikov approximations

In this section we show that the Postnikov approximations to a space can be obtained as localizations of that space.

PROPOSITION 1.5.1. *If* $n \geq 0$ *and* $f_n\colon S^{n+1} \to D^{n+2}$ *is the standard inclusion in* Top*, then a space* X *is* f_n*-local if and only if* $\pi_i X \approx 0$ *for* $i > n$ *and every choice of basepoint in* X*.*

PROOF. If $k \geq 0$ then the inclusion $S^{n+1} \otimes \Delta[k] \amalg_{S^{n+1} \otimes \partial\Delta[k]} D^{n+2} \otimes \partial\Delta[k] \to D^{n+2} \otimes \Delta[k]$ is a relative CW-complex that attaches a single cell of dimension $n + k + 2$. Thus, any map $S^{n+1} \otimes \Delta[k] \amalg_{S^{n+1} \otimes \partial\Delta[k]} D^{n+2} \otimes \partial\Delta[k] \to X$ can be extended over $D^{n+2} \otimes \Delta[k]$ if and only if $\pi_{n+k+1}X \approx 0$ for every choice of basepoint in X. The result now follows from Proposition 1.3.3. □

PROPOSITION 1.5.2. *Let $n \geq 0$ and let $f_n \colon S^{n+1} \to D^{n+2}$ be the standard inclusion in* Top. *If a map $g \colon X \to Y$ induces isomorphisms $g_* \colon \pi_i X \approx \pi_i Y$ for $i \leq n$ and every choice of basepoint in X, then it is an f_n-local equivalence.*

PROOF. If $g \colon X \to Y$ induces isomorphisms $g_* \colon \pi_i X \approx \pi_i Y$ for $i \leq n$ and every choice of basepoint in X, then we can choose a cofibrant approximation $\tilde{g} \colon \widetilde{X} \to \widetilde{Y}$ to g such that

(1) $\widetilde{Y}$ is a CW-complex,
(2) $\tilde{g}$ is the inclusion of a subcomplex that contains the n-skeleton of $\widetilde{Y}$, and
(3) every $(n+1)$-cell of $\widetilde{Y} - \widetilde{X}$ is attached via a constant map of S^n.

If $k = 0$ then the map $\widetilde{X} \otimes \Delta[k] \amalg_{\widetilde{X} \otimes \partial\Delta[k]} \widetilde{Y} \otimes \partial\Delta[k] \to \widetilde{Y} \otimes \Delta[k]$ is just the map $\widetilde{X} \to \widetilde{Y}$, and if $k > 0$ it is the inclusion of a subcomplex that contains the $(n + k)$-skeleton. Thus, if Z is an f-local space, then Proposition 1.5.1 implies that every map $\widetilde{X} \otimes \Delta[k] \amalg_{\widetilde{X} \otimes \partial\Delta[k]} \widetilde{Y} \otimes \partial\Delta[k] \to Z$ can be extended over $\widetilde{Y} \otimes \Delta[k]$, and so g is an f_n-local equivalence (see Proposition 9.3.10). □

THEOREM 1.5.3. *If $n > 0$ and $f_n \colon S^{n+1} \to D^{n+2}$ is the standard inclusion in* Top, *then the projection of a space onto its n-th Postnikov approximation is an f_n-localization map.*

PROOF. This follows from Proposition 1.5.1 and Proposition 1.5.2. □

PROPOSITION 1.5.4. *Let $n \geq 0$ and let $f_n \colon S^{n+1} \to D^{n+2}$ be the standard inclusion in* Top. *If $g \colon X \to Y$ is an f_n-local equivalence, then g induces isomorphisms $g_* \colon \pi_i X \approx \pi_i Y$ for $i \leq n$ and every choice of basepoint in X.*

PROOF. Theorem 1.5.3 and Theorem 1.2.29 imply that the induced map of n-th Postnikov approximations $\mathrm{P}_n g \colon \mathrm{P}_n X \to \mathrm{P}_n Y$ is a weak equivalence. Thus, for every $i \leq n$ and every choice of basepoint in X we have a commutative diagram

$$\begin{array}{ccc} \pi_i X & \xrightarrow{g_*} & \pi_i Y \\ \downarrow & & \downarrow \\ \pi_i \mathrm{P}_n X & \longrightarrow & \pi_i \mathrm{P}_n Y \end{array}$$

in which every map except the top one is an isomorphism, and so $g_* \colon \pi_i X \to \pi_i Y$ is also an isomorphism. □

1.5.5. The Postnikov tower.

LEMMA 1.5.6. *Let $f \colon A \to B$ and $\tilde{f} \colon \widetilde{A} \to \widetilde{B}$ be maps between cofibrant spaces. If f is an $\tilde{f}$-local equivalence, then every f-local equivalence is an $\tilde{f}$-local equivalence.*

PROOF. Since f is an $\tilde{f}$-local equivalence, every $\tilde{f}$-local space is f-local. □

LEMMA 1.5.7. *Let $f\colon A \to B$ and $\tilde{f}\colon \widetilde{A} \to \widetilde{B}$ be maps between cofibrant spaces. If f is an $\tilde{f}$-local equivalence, then for every object X the f-localization map $X \to \mathrm{L}_f X$ is an $\tilde{f}$-local equivalence.*

PROOF. This follows from Lemma 1.5.6. □

PROPOSITION 1.5.8. *For all $n \geq 0$ let $f_n\colon S^{n+1} \to D^{n+2}$ be the standard inclusion in* Top. *If $i > j \geq 0$ then for every space X an f_i-localization map is an f_j-local equivalence.*

PROOF. This follows from Lemma 1.5.7 and Theorem 1.5.3. □

PROPOSITION 1.5.9. *If for all $n \geq 0$ we let $\mathrm{P}_n X$ denote the n-th Postnikov approximation of the space X, then for all $i > j \geq 0$ there is a map $\mathrm{P}_i X \to \mathrm{P}_j X$, unique up to simplicial homotopy, that makes the triangle*

$$\begin{array}{ccc} X & \longrightarrow & \mathrm{P}_i X \\ & \searrow & \downarrow \\ & & \mathrm{P}_j X \end{array}$$

commute up to simplicial homotopy.

PROOF. This follows from Proposition 1.5.8 and Lemma 1.2.24. □

1.6. Topological spaces and simplicial sets

The main results of this section (Corollary 1.6.5 and Corollary 1.6.7) imply, roughly speaking, that when using the localization functor of Definition 1.4.11, one can pass freely through the geometric realization and total singular complex functors at the cost of only a natural weak equivalence.

LEMMA 1.6.1. *Let K and C be simplicial sets and let X be is a topological space.*

(1) *A map of topological spaces $|K| \to X$ defines a map of simplicial sets $\mathrm{Map}(C, K) \to \mathrm{Map}(|C|, X)$ that is natural in C and in the map $|K| \to X$.*
(2) *A map of simplicial sets $K \to \mathrm{Sing}\, X$ defines a map of simplicial sets $\mathrm{Map}(C, K) \to \mathrm{Map}(|C|, X)$ that is natural in C and in the map $K \to \mathrm{Sing}\, X$.*

PROOF. The map of part 1 is defined as the composition

$$\mathrm{Map}(C, K) \to \mathrm{Map}(|C|, |K|) \to \mathrm{Map}(|C|, X)$$

and the map of part 2 is defined as the composition

$$\mathrm{Map}(C, K) \to \mathrm{Map}(C, \mathrm{Sing}\, X) \to \mathrm{Map}(|C|, X) \ .$$

□

PROPOSITION 1.6.2. *Let $C \to D$ be a map of simplicial sets. If $K \to L$ is a map of simplicial sets, $X \to Y$ is a map of topological spaces, and*

$$\begin{array}{ccc} |K| & \longrightarrow & X \\ \downarrow & & \downarrow \\ |L| & \longrightarrow & Y \end{array}$$

is a commutative square, then there is a natural map from the geometric realization of the pushout P in the diagram

$$\begin{array}{ccccc} C \times \big(\mathrm{Map}(C,K) \times_{\mathrm{Map}(C,L)} \mathrm{Map}(D,L)\big) & \longrightarrow & K & \dashrightarrow & P \\ \downarrow & & \downarrow & \nearrow & \\ D \times \big(\mathrm{Map}(C,K) \times_{\mathrm{Map}(C,L)} \mathrm{Map}(D,L)\big) & \longrightarrow & L & & \end{array}$$

to the pushout Q in the diagram

$$\begin{array}{ccccc} |C| \times \big|\mathrm{Map}(|C|,X) \times_{\mathrm{Map}(|C|,Y)} \mathrm{Map}(|D|,Y)\big| & \longrightarrow & X & \dashrightarrow & Q \\ \downarrow & & \downarrow & \nearrow & \\ |D| \times \big|\mathrm{Map}(|C|,X) \times_{\mathrm{Map}(|C|,Y)} \mathrm{Map}(|D|,Y)\big| & \longrightarrow & Y & & \end{array}$$

that makes the diagram

$$\begin{array}{ccccc} |K| & \longrightarrow & |P| & \longrightarrow & |L| \\ \downarrow & & \downarrow & & \downarrow \\ X & \longrightarrow & Q & \longrightarrow & Y \end{array}$$

commute.

PROOF. Since the geometric realization functor commutes with pushouts, this follows from Lemma 1.6.1. □

PROPOSITION 1.6.3. *If $K \to L$ is a map of simplicial sets, $X \to Y$ is a map of topological spaces, and*

$$\begin{array}{ccc} K & \longrightarrow & \mathrm{Sing}\, X \\ \downarrow & & \downarrow \\ L & \longrightarrow & \mathrm{Sing}\, Y \end{array}$$

a commutative square, then there is a natural map from the pushout P in the diagram

$$\begin{array}{ccccc} C \times \big(\mathrm{Map}(C,K) \times_{\mathrm{Map}(C,L)} \mathrm{Map}(D,L)\big) & \longrightarrow & K & \dashrightarrow & P \\ \downarrow & & \downarrow & \nearrow & \\ D \times \big(\mathrm{Map}(C,K) \times_{\mathrm{Map}(C,L)} \mathrm{Map}(D,L)\big) & \longrightarrow & L & & \end{array}$$

to the total singular complex of the pushout Q in the diagram

$$\begin{array}{ccc} |C| \times \bigl|\mathrm{Map}(|C|, X) \times_{\mathrm{Map}(|C|,Y)} \mathrm{Map}(|D|, Y)\bigr| & \longrightarrow & X & \dashrightarrow & Q \\ \downarrow & & \downarrow & & \\ |D| \times \bigl|\mathrm{Map}(|C|, X) \times_{\mathrm{Map}(|C|,Y)} \mathrm{Map}(|D|, Y)\bigr| & \longrightarrow & Y & & \end{array}$$

that makes the diagram

$$\begin{array}{ccccc} K & \longrightarrow & P & \longrightarrow & L \\ \downarrow & & \downarrow & & \downarrow \\ \operatorname{Sing} X & \longrightarrow & \operatorname{Sing} Q & \longrightarrow & \operatorname{Sing} Y \end{array}$$

commute.

PROOF. This follows from Lemma 1.6.1, using the natural map from the pushout of the total singular complexes to the total singular complex of the pushout. □

THEOREM 1.6.4. *Let $f\colon A \to B$ be a cofibration of simplicial sets and let $g\colon X \to Y$ be a map of topological spaces. If $\mathrm{E}_f(\operatorname{Sing} g)$ is the simplicial set obtained by applying the factorization of Theorem 1.4.10 to the map $\operatorname{Sing} g\colon \operatorname{Sing} X \to \operatorname{Sing} Y$ and $E_{|f|}g$ is the topological space obtained by applying the factorization of Theorem 1.4.10 (with respect to the map $|f|\colon |A| \to |B|$) to the map g, then there is a natural map $\bigl|\mathrm{E}_f(\operatorname{Sing} g)\bigr| \to \mathrm{E}_{|f|}\, g$ that makes the diagram*

$$\begin{array}{ccccc} |\operatorname{Sing} X| & \longrightarrow & \bigl|\mathrm{E}_f(\operatorname{Sing} g)\bigr| & \longrightarrow & |\operatorname{Sing} Y| \\ \downarrow & & \downarrow & & \downarrow \\ X & \longrightarrow & \mathrm{E}_{|f|}\, g & \longrightarrow & Y \end{array}$$

commute.

PROOF. Using Proposition 1.6.2 we can construct the map inductively at each stage in the construction of the factorization. □

COROLLARY 1.6.5. *If $f\colon A \to B$ is a cofibration of simplicial sets, then for every topological space X there is a natural weak equivalence $\bigl|\mathrm{L}_f \operatorname{Sing} X\bigr| \to \mathrm{L}_{|f|}X$ that makes the square*

$$\begin{array}{ccc} |\operatorname{Sing} X| & \longrightarrow & X \\ \downarrow & & \downarrow \\ |\mathrm{L}_f \operatorname{Sing} X| & \longrightarrow & \mathrm{L}_{|f|}X \end{array}$$

commute.

PROOF. The existence of the natural map follows from Theorem 1.6.4. Proposition 1.2.34 implies that $\bigl|\mathrm{L}_f \operatorname{Sing} X\bigr|$ is $|f|$-local, and so Proposition 1.2.35 implies that our natural map is a $|f|$-localization of the weak equivalence $\bigl|\operatorname{Sing} X\bigr| \to X$ (see Definition 1.2.13). Proposition 1.2.16 and Theorem 1.2.29 now imply that our natural map is a weak equivalence. □

THEOREM 1.6.6. *Let $f\colon A \to B$ be a cofibration of simplicial sets and let $g\colon K \to L$ be a map of simplicial sets. If $\mathrm{E}_f\, g$ is the simplicial set obtained by applying the factorization of Theorem 1.4.10 to the map g and $E_{|f|}|g|$ is the topological space obtained by applying the factorization of Theorem 1.4.10 (with respect to the map $|f|\colon |A| \to |B|$) to the map $|g|\colon |K| \to |L|$, then there is a natural map $\mathrm{E}_f\, g \to \mathrm{E}_{|f|}|g|$ that makes the diagram*

$$\begin{array}{ccccc} K & \longrightarrow & \mathrm{E}_f\, g & \longrightarrow & L \\ \downarrow & & \downarrow & & \downarrow \\ \mathrm{Sing}|K| & \longrightarrow & \mathrm{Sing}\,\mathrm{E}_{|f|}|g| & \longrightarrow & \mathrm{Sing}|Y| \end{array}$$

commute.

PROOF. Using Proposition 1.6.3 we can construct the map inductively at each stage in the construction of the factorization. □

COROLLARY 1.6.7. *If $f\colon A \to B$ is a cofibration of simplicial sets, then for every simplicial set K there is a natural simplicial homotopy equivalence $\mathrm{L}_f K \to \mathrm{Sing}\,\mathrm{L}_{|f|}K$ that makes the square*

$$\begin{array}{ccc} K & \longrightarrow & \mathrm{Sing}|K| \\ \downarrow & & \downarrow \\ \mathrm{L}_f K & \longrightarrow & \mathrm{Sing}\,\mathrm{L}_{|f|}|K| \end{array}$$

commute.

PROOF. The existence of the natural map follows from Theorem 1.6.6. Proposition 1.2.33 implies that $\mathrm{Sing}\,\mathrm{L}_{|f|}|K|$ is f-local, and Proposition 1.2.17, Proposition 1.2.16, and Proposition 1.2.36 imply that our natural map is an f-local equivalence of cofibrant f-local spaces. The result now follows from Theorem 1.2.25. □

PROPOSITION 1.6.8. *If $f\colon A \to B$ is a cofibration in $\mathrm{SS}_{(*)}$ (see Notation 1.1.4), $(\mathrm{M}_f, j\colon 1 \to \mathrm{M}_f)$ is a functorial cofibrant f-localization on $\mathrm{SS}_{(*)}$, and $(\mathrm{N}_{|f|}, k\colon 1 \to \mathrm{N}_{|f|})$ is a functorial cofibrant $|f|$-localization on $\mathrm{Top}_{(*)}$, then for every topological space X there is a map $|\mathrm{M}_f\,\mathrm{Sing}\,X| \to \mathrm{N}_{|f|}X$, unique up to simplicial homotopy, that makes the square*

$$\begin{array}{ccc} |\mathrm{Sing}\,X| & \longrightarrow & X \\ \downarrow & & \downarrow \\ |\mathrm{M}_f\,\mathrm{Sing}\,X| & \longrightarrow & \mathrm{N}_{|f|}X \end{array} \tag{1.6.9}$$

commute, and any such map is a weak equivalence. (Since $|\mathrm{M}_f\,\mathrm{Sing}\,X|$ is cofibrant and $\mathrm{N}_{|f|}X$ is fibrant, all notions of homotopy of maps $|\mathrm{M}_f\,\mathrm{Sing}\,X| \to \mathrm{N}_{|f|}X$ coincide and are equivalence relations (see Proposition 9.5.24).) This map is natural *up to homotopy, i.e., if* $g\colon X \to Y$ *is a map of topological spaces, then the square*

$$\begin{array}{ccc} \left|\mathrm{M}_f \operatorname{Sing} X\right| & \longrightarrow & \mathrm{N}_{|f|}X \\ \big\downarrow & & \big\downarrow \\ \left|\mathrm{M}_f \operatorname{Sing} Y\right| & \longrightarrow & \mathrm{N}_{|f|}Y \end{array}$$

commutes up to homotopy.

PROOF. Since Proposition 1.2.35 implies that the map $\left|\operatorname{Sing} X\right| \to \left|\mathrm{M}_f \operatorname{Sing} X\right|$ is a $\left|f\right|$-local equivalence, the existence and uniqueness of the map follow from Lemma 1.2.24. Since Proposition 1.2.34 implies that $\left|\mathrm{M}_f \operatorname{Sing} X\right|$ is $\left|f\right|$-local, Theorem 1.2.26 implies that the map is a weak equivalence.

For the naturality statement, we note that we have the cube

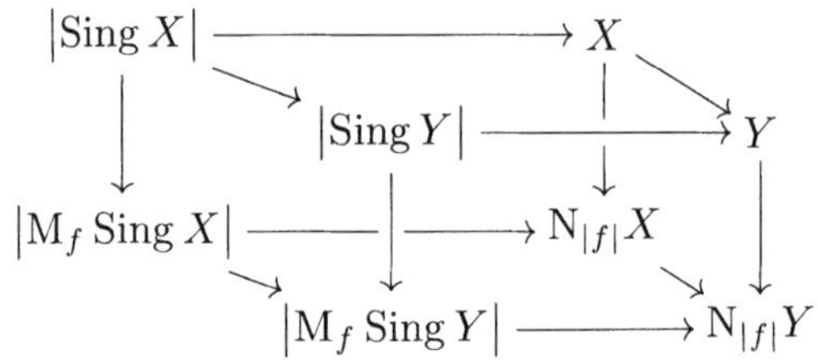

in which the top and side squares commute and the front and back squares commute up to simplicial homotopy. This implies that the composition

$$\left|\operatorname{Sing} X\right| \to \left|\mathrm{M}_f \operatorname{Sing} X\right| \to \left|\mathrm{M}_f \operatorname{Sing} Y\right| \to \mathrm{N}_{|f|}Y$$

is simplicially homotopic to the composition

$$\left|\operatorname{Sing} X\right| \to \left|\mathrm{M}_f \operatorname{Sing} X\right| \to \mathrm{N}_{|f|}X \to \mathrm{N}_{|f|}Y \ ,$$

and so the result follows from Lemma 1.2.24. □

PROPOSITION 1.6.10. *If* $f\colon A \to B$ *is a cofibration in* $\mathrm{SS}_{(*)}$ *(see Notation 1.1.4),* $(\mathrm{M}_f, j\colon 1 \to \mathrm{M}_f)$ *is a functorial cofibrant* f*-localization on* $\mathrm{SS}_{(*)}$*, and* $(\mathrm{N}_{|f|}, k\colon 1 \to \mathrm{N}_{|f|})$ *is a functorial cofibrant* $\left|f\right|$*-localization on* $\mathrm{Top}_{(*)}$*, then for every simplicial set* K *there is a map* $\mathrm{M}_f K \to \operatorname{Sing} \mathrm{N}_{|f|}\left|K\right|$*, unique up to homotopy, that makes the square*

$$\begin{array}{ccc} K & \longrightarrow & \operatorname{Sing}\left|K\right| \\ \big\downarrow & & \big\downarrow \\ \mathrm{M}_f K & \longrightarrow & \operatorname{Sing} \mathrm{N}_{|f|}\left|K\right| \end{array} \tag{1.6.11}$$

commute, and any such map is a homotopy equivalence. (Since every simplicial set is cofibrant and $\operatorname{Sing} \mathrm{N}_{|f|}\left|K\right|$ *is fibrant, all notions of homotopy of maps* $\mathrm{M}_f K \to \operatorname{Sing} \mathrm{N}_{|f|}\left|K\right|$ *coincide and are equivalence relations (see Proposition 9.5.24).) This map is natural up to homotopy, i.e., if* $g\colon K \to L$ *is a map of simplicial sets, then the square*

$$\begin{array}{ccc} \mathrm{M}_f K & \longrightarrow & \operatorname{Sing} \mathrm{N}_{|f|}|K| \\ \big\downarrow & & \big\downarrow \\ \mathrm{M}_f L & \longrightarrow & \operatorname{Sing} \mathrm{N}_{|f|}|L| \end{array}$$

commutes up to homotopy.

PROOF. Proposition 1.2.36 implies that the map $\operatorname{Sing}|K| \to \operatorname{Sing} \mathrm{N}_{|f|}|K|$ is an f-local equivalence and Proposition 1.2.33 implies that $\operatorname{Sing} \mathrm{N}_{|f|}|K|$ is f-local. Since every simplicial set is cofibrant, the existence and uniqueness of the map now follows from Lemma 1.2.24, and Theorem 1.2.25 implies that it is a homotopy equivalence. The naturality statement follows as in the proof of Proposition 1.6.8. □

1.7. A continuous localization functor

Let $\mathrm{Spc}_{(*)}$ be one of our categories of spaces (see Notation 1.1.4), and let $f\colon A \to B$ be an inclusion of cell complexes (see Remark 1.2.11). In this section, we will define a variant $\mathrm{L}_f^{\mathrm{cont}}$ of the f-localization functor L_f (see Definition 1.4.11) that is "continuous". If we were using topological spaces of functions (instead of simplicial sets of functions; see Section 1.1.5) then we would want to define a function

$$\operatorname{Map}(X, Y) \to \operatorname{Map}(\mathrm{L}_f X, \mathrm{L}_f Y) \tag{1.7.1}$$

that is a continuous function of topological spaces. Since we are considering $\mathrm{Spc}_{(*)}$ as a simplicial model category (see Definition 9.1.6), we want to define $\mathrm{L}_f^{\mathrm{cont}}$ to be a *simplicial functor*, i.e., we want a functor $\mathrm{L}_f^{\mathrm{cont}}$ that defines a map of simplicial sets (1.7.1) (see [**52**, Chapter II, Section 1]). Note that not every functor can be extended to a simplicial functor; for a counterexample, see Example 9.8.7.

1.7.2. Construction of the sequence. We follow the procedure described in Section 1.3, using the same ordinal λ, except that we use a new construction to define the space $\mathrm{E}^{\beta+1}$ in terms of the space E^{β} (see Section 1.3.5). We first define a localization functor $\mathrm{L}_f^{\mathrm{cont}}$ that is a variant of the functor L_f defined in Definition 1.4.11 (see Theorem 1.7.4), and then we show that $\mathrm{L}_f^{\mathrm{cont}}$ can be extended to be a simplicial functor (see Theorem 1.7.5).

As in Section 1.3.5, we begin the sequence by letting $\mathrm{E}^0 = X$. If $\beta < \lambda$ and we have constructed the sequence through E^{β}, we let

$$\begin{aligned} C_\beta^{\mathrm{cont}} &= \coprod_{(C\to D)\in\overline{\Lambda\{f\}}} C \otimes \operatorname{Map}(C, \mathrm{E}^{\beta}) \\ D_\beta^{\mathrm{cont}} &= \coprod_{(C\to D)\in\overline{\Lambda\{f\}}} D \otimes \operatorname{Map}(C, \mathrm{E}^{\beta}) \ . \end{aligned}$$

We then have a natural map $C_\beta^{\mathrm{cont}} \to \mathrm{E}^\beta$, and we define $\mathrm{E}^{\beta+1}$ by letting the square

$$\begin{array}{ccc} C_\beta^{\mathrm{cont}} & \longrightarrow & \mathrm{E}^\beta \\ \downarrow & & \vdots \\ D_\beta^{\mathrm{cont}} & \cdots\cdots\rightarrow & \mathrm{E}^{\beta+1} \end{array}$$

be a pushout. If γ is a limit ordinal, we let $\mathrm{E}^\gamma = \mathrm{colim}_{\beta<\gamma} \mathrm{E}^\beta$. We let $\mathrm{L}_f^{\mathrm{cont}} X = \mathrm{colim}_{\beta<\lambda} \mathrm{E}^\beta$.

PROPOSITION 1.7.3. *Let $\mathrm{Spc}_{(*)}$ be one of our categories of spaces (see Notation 1.1.4). If $f\colon A \to B$ is an inclusion of cell complexes and K is a simplicial set, then the maps*

$$\big(A \otimes \Delta[n] \amalg_{A\otimes\partial\Delta[n]} B \otimes \partial\Delta[n]\big) \otimes K \to \big(B \otimes \Delta[n]\big) \otimes K \qquad \text{for } n \geq 0$$

and

$$\mathrm{Spc}_{(*)}(\Lambda[n,k]) \otimes K \to \mathrm{Spc}_{(*)}(\Delta[n]) \otimes K \qquad \text{for } n > 0 \text{ and } 0 \leq k \leq n$$

are all both cofibrations and f-local equivalences.

PROOF. Proposition 9.3.9 implies that each of the maps $\mathrm{Spc}_{(*)}(\Lambda[n,k]) \otimes K \to \mathrm{Spc}_{(*)}(\Delta[n]) \otimes K$ is a trivial cofibration. The result now follows from Proposition 1.2.16, Corollary 10.2.21, and Proposition 1.3.7. □

THEOREM 1.7.4. *Let $\mathrm{Spc}_{(*)}$ be one of our categories of spaces (see Notation 1.1.4) and let $f\colon A \to B$ be an inclusion of cell complexes (see Remark 1.2.11). If X is a space, then the map $X \to \mathrm{L}_f^{\mathrm{cont}} X$ constructed in Section 1.7.2 is a cofibrant f-localization of X.*

PROOF. Proposition 1.7.3 and Proposition 1.2.22 imply that the map $X \to \mathrm{L}_f^{\mathrm{cont}} X$ is both a cofibration and an f-local equivalence, and so it remains only to show that $\mathrm{L}_f^{\mathrm{cont}} X$ is f-local. The 0-skeleton of $\mathrm{Map}(C, \mathrm{E}^\beta)$ is $\mathrm{Spc}_{(*)}(C, \mathrm{E}^\beta)$, and so $C \otimes \mathrm{Map}(C, \mathrm{E}^\beta)$ contains

$$C \otimes \mathrm{Spc}_{(*)}(C, \mathrm{E}^\beta) \approx \coprod_{\mathrm{Spc}_{(*)}(C,\mathrm{E}^\beta)} C$$

as a subcomplex. The discussion in Section 1.3.4 now explains why the space $\mathrm{L}_f^{\mathrm{cont}} X$ is a $\overline{\Lambda\{f\}}$-injective, and so the map $X \to \mathrm{L}_f^{\mathrm{cont}} X$ is a functorial cofibrant f-localization of X. □

THEOREM 1.7.5. *The functor $\mathrm{L}_f^{\mathrm{cont}}$ defined in Section 1.7.2 can be extended to a simplicial functor.*

PROOF. If C and X are spaces and K is a simplicial set, then there is a natural map $\mathrm{Map}(C, X) \otimes K \to \mathrm{Map}(C, X \otimes K)$ that takes the n-simplex $(\alpha\colon C \otimes \Delta[n] \to X, \tau)$ of $\mathrm{Map}(C, X) \otimes K$ to the n-simplex $\sigma(\alpha, \tau)\colon C \otimes \Delta[n] \to X \otimes K$ of $\mathrm{Map}(C, X \otimes K)$ that is the composition

$$C \otimes \Delta[n] \xrightarrow{1_C \otimes D} C \otimes (\Delta[n] \times \Delta[n]) \xrightarrow{\approx} (C \otimes \Delta[n]) \otimes \Delta[n] \xrightarrow{\alpha \otimes i_\tau} X \otimes K$$

where $D\colon \Delta[n] \to \Delta[n] \times \Delta[n]$ is the diagonal map and $i_\tau\colon \Delta[n] \to K$ is the map that takes the nondegenerate n-simplex of $\Delta[n]$ to τ. This natural map σ has the properties required by Theorem 9.8.5, and so we can use it to inductively define σ for all the spaces used in the construction of the localization (see Section 1.7.2). The theorem now follows from Proposition 9.8.9 and Theorem 9.8.5. □

1.8. Pointed and unpointed localization

There is a functor from the category of pointed spaces to the category of unpointed spaces that forgets the basepoint, and so there are two different notions of localization that we can define on a category of pointed spaces. If $f\colon A \to B$ is a cofibration of cofibrant pointed spaces, then

(1) we can consider the notions of *pointed* f-local spaces and *pointed* f-local equivalences in Spc_* (see Notation 1.1.4), or
(2) we can still consider spaces with basepoint (i.e., spaces in Spc_*) but consider the notions of *unpointed* f-local spaces and *unpointed* f-local equivalences in Spc by forgetting the basepoints.

DEFINITION 1.8.1. If $f\colon A \to B$ is a cofibration of cofibrant pointed spaces and X is a pointed space, then we will say that

(1) X is *pointed f-local* if it is an f-local space in Spc_*, and
(2) we will say that X is *unpointed f-local* if X is an f-local space in Spc when we forget the basepoints of all the spaces involved.

Similarly, a map $f\colon X \to Y$ will be called

(1) a *pointed f-local equivalence* if it is an f-local equivalence in Spc_*, and
(2) an *unpointed f-local equivalence* if it is an f-local equivalence in Spc after forgetting all basepoints.

NOTATION 1.8.2. In this section, if X and Y are objects of Spc_* (see Notation 1.1.4), then

(1) $\mathrm{Map}(X, Y)$ will continue to denote the unpointed simplicial set of maps between the pointed spaces X and Y, and
(2) $\mathrm{UMap}(X, Y)$ will denote the unpointed simplicial set of maps between the unpointed spaces obtained from X and Y by forgetting the basepoints.

Thus, Definition 1.8.1 implies that if $f\colon A \to B$ is a cofibration of cofibrant pointed spaces then a fibrant pointed space X is

(1) *pointed f-local* if $f^*\colon \mathrm{Map}(B, X) \to \mathrm{Map}(A, X)$ is a weak equivalence of (unpointed) simplicial sets and it is
(2) *unpointed f-local* if $f^*\colon \mathrm{UMap}(B, X) \to \mathrm{UMap}(A, X)$ is a weak equivalence of (unpointed) simplicial sets.

Similarly, a map $g\colon Y \to Z$ of pointed spaces is

(1) a *pointed f-local equivalence* if there is a cofibrant approximation $\tilde{g}\colon \widetilde{Y} \to \widetilde{Z}$ to g such that for every pointed f-local space W the induced map of simplicial sets $\tilde{g}^*\colon \mathrm{Map}(\widetilde{Z}, W) \to \mathrm{Map}(\widetilde{Y}, W)$ is a weak equivalence and it is
(2) an *unpointed f-local equivalence* if there is a cofibrant approximation $\tilde{g}\colon \widetilde{Y} \to \widetilde{Z}$ to g such that for every unpointed f-local space W the induced map of simplicial sets $\tilde{g}^*\colon \mathrm{UMap}(\widetilde{Z}, W) \to \mathrm{UMap}(\widetilde{Y}, W)$ is a weak equivalence

PROPOSITION 1.8.3. *Let A be a cofibrant object of* Spc_* *and let X be a fibrant object of* Spc_*.

(1) *If* $\mathrm{Spc}_* = \mathrm{SS}_*$, *then there is a natural fibration of unpointed simplicial sets*

$$\mathrm{Map}(A, X) \to \mathrm{UMap}(A, X) \to X \ .$$

(2) *If* $\mathrm{Spc}_* = \mathrm{Top}_*$, *then there is a natural fibration of unpointed simplicial sets*

$$\mathrm{Map}(A, X) \to \mathrm{UMap}(A, X) \to \mathrm{Sing}\, X \ .$$

PROOF. Since $* \to A$ is a cofibration of pointed spaces and X is a fibrant pointed space, $* \to A$ is also a cofibration of unpointed spaces (after forgetting the basepoints) and X is also a fibrant pointed space (after forgetting the basepoint). Thus, Proposition 9.3.1 implies that we have a natural fibration of simplicial sets $\mathrm{UMap}(A, X) \to \mathrm{UMap}(*, X)$. The fiber of this fibration is $\mathrm{Map}(A, X)$. If $\mathrm{Spc}_* = \mathrm{SS}_*$, then $\mathrm{UMap}(*, X)$ is naturally isomorphic to the unpointed simplicial set X. If $\mathrm{Spc}_* = \mathrm{Top}_*$, then $\mathrm{UMap}(*, X)$ is naturally isomorphic to the unpointed simplicial set $\mathrm{Sing}\, X$. □

PROPOSITION 1.8.4. *Let $A \to B$ be a map of cofibrant pointed spaces and let W be a fibrant pointed space.*

(1) *If* $\mathrm{UMap}(B, W) \to \mathrm{UMap}(A, W)$ *(see Notation 1.8.2) is a weak equivalence, then* $\mathrm{Map}(B, W) \to \mathrm{Map}(A, W)$ *is a weak equivalence.*
(2) *If W is path connected and* $\mathrm{Map}(B, W) \to \mathrm{Map}(A, W)$ *is a weak equivalence, then* $\mathrm{UMap}(B, W) \to \mathrm{UMap}(A, W)$ *is a weak equivalence.*

PROOF. This follows from Proposition 1.8.3 and the long exact sequence of homotopy groups of a fibration. □

PROPOSITION 1.8.5. *Let $f\colon A \to B$ be a cofibration of cofibrant pointed spaces and let X be a pointed space.*

(1) *If X is an unpointed f-local space, then it is also a pointed f-local space.*
(2) *If X is a path connected pointed f-local space, then it is also an unpointed f-local space.*

PROOF. This follows from Proposition 1.8.4. □

COROLLARY 1.8.6. *Let $f\colon A \to B$ be a cofibration of cofibrant pointed spaces. If X is a path connected pointed space, then X is pointed f-local if and only if it is unpointed f-local.*

PROOF. This follows from Proposition 1.8.5. □

LEMMA 1.8.7. *If A is a path connected pointed space, X is a pointed space, and X_{b} is the path component of X containing the basepoint, then the natural map* $\mathrm{Map}(A, X_{\mathrm{b}}) \to \mathrm{Map}(A, X)$ *is an isomorphism.*

PROOF. Since the image of a path connected space is path connected, for every $n \geq 0$ the image of a pointed map from $A \otimes \Delta[n]$ to X is contained in X_{b}. □

THEOREM 1.8.8. *If $f\colon A \to B$ is a map of path connected cofibrant pointed spaces and X is a pointed space, then the following are equivalent:*

(1) *X is pointed f-local.*

(2) *Every path component of X is fibrant and the path component of X containing the basepoint is pointed f-local.*
(3) *Every path component of X is fibrant and the path component of X containing the basepoint is unpointed f-local.*

PROOF. This follows from Lemma 1.8.7 and Corollary 1.8.6. □

COROLLARY 1.8.9. *If $f\colon A \to B$ is a map of path connected cofibrant pointed spaces and X is a fibrant pointed space, then X is unpointed f-local if and only if every path component of X is pointed f-local when you choose a basepoint for each path component.*

PROOF. If the path components of X are $\{X_s\}_{s\in S}$, then we have a commutative square

$$\begin{array}{ccc} \mathrm{UMap}(B,X) & \xleftarrow{\approx} & \coprod_{s\in S} \mathrm{UMap}(B,X_s) \\ \downarrow & & \downarrow \\ \mathrm{UMap}(A,X) & \xleftarrow[\approx]{} & \coprod_{s\in S} \mathrm{UMap}(A,X_s) \end{array}$$

(see Notation 1.8.2) in which the horizontal maps are isomorphisms. The result now follows from Corollary 1.8.6. □

COROLLARY 1.8.10. *If $f\colon A \to B$ is a map of path connected cofibrant pointed spaces, X is a pointed space, and X_{b} is the path component of X containing the basepoint, then the natural map*

$$(X - X_{\mathrm{b}}) \amalg \mathrm{L}_f X_{\mathrm{b}} \to \mathrm{L}_f X$$

is a weak equivalence (where L_f denotes pointed f-localization).

PROOF. This follows from Theorem 1.8.8, Lemma 1.8.7, and Theorem 1.2.26. □

PROPOSITION 1.8.11. *Let $f\colon A \to B$ be a map of cofibrant pointed spaces. If $X \to Y$ is an unpointed f-local equivalence of path connected pointed spaces, then it is also a pointed f-local equivalence.*

PROOF. If $\widetilde{X} \to \widetilde{Y}$ is a pointed cofibrant approximation (see Definition 8.1.22) to $X \to Y$, then it is also an unpointed cofibrant approximation. If W is a pointed f-local space, let W_{b} be the path component of W containing the basepoint. Lemma 1.2.6 and Proposition 1.8.5 imply that W_{b} is an unpointed f-local space, and so the map $\mathrm{UMap}(\widetilde{Y}, W_{\mathrm{b}}) \to \mathrm{UMap}(\widetilde{X}, W_{\mathrm{b}})$ is a weak equivalence. Proposition 1.8.4 now implies that the map $\mathrm{Map}(\widetilde{Y}, W_{\mathrm{b}}) \to \mathrm{Map}(\widetilde{X}, W_{\mathrm{b}})$ is a weak equivalence. Lemma 1.8.7 implies that the horizontal maps in the commutative square

$$\begin{array}{ccc} \mathrm{Map}(\widetilde{Y}, W_{\mathrm{b}}) & \longrightarrow & \mathrm{Map}(\widetilde{Y}, W) \\ \downarrow & & \downarrow \\ \mathrm{Map}(\widetilde{X}, W_{\mathrm{b}}) & \longrightarrow & \mathrm{Map}(\widetilde{X}, W) \end{array}$$

are isomorphisms, and so the map $\mathrm{Map}(\widetilde{Y}, W) \to \mathrm{Map}(\widetilde{X}, W)$ is a weak equivalence. □

THEOREM 1.8.12. *If $f\colon A \to B$ is a cofibration of cofibrant pointed spaces and X is a path connected pointed space, then a pointed f-localization of X is also an unpointed f-localization of X.*

PROOF. Let $X \to Y$ be the unpointed f-localization of X. Proposition 1.8.5 implies that Y is pointed f-local and Proposition 1.8.11 implies that the map $X \to Y$ is a pointed f-local equivalence. □

THEOREM 1.8.13. *If $f\colon A \to B$ is a cofibration of path connected cofibrant pointed spaces and X is a pointed space, then the unpointed f-localization of X is weakly equivalent to the space obtained by choosing a basepoint for each path component of X and taking the pointed f-localization of each path component.*

PROOF. Let $\{X_s\}_{s\in S}$ be the set of path components of X. If for every $s \in S$ we choose a basepoint for X_s and let $X_s \to \mathrm{L}_f X_s$ be the pointed f-localization of X_s, then Corollary 1.8.9 implies that $\coprod_{s\in S} \mathrm{L}_f X_s$ is an unpointed f-local space. Theorem 1.8.12 implies that for every $s \in S$ the map $X_s \to \mathrm{L}_f X_s$ is an unpointed f-local equivalence. Let $\widetilde{X}_s \to \widetilde{\mathrm{L}_f X_s}$ be the geometric realization of the total singular complex of $X_s \to \mathrm{L}_f X_s$ if our "spaces" are topological spaces and let it be the map $X_s \to \mathrm{L}_f X_s$ itself if our "spaces" are simplicial sets. In either case, $\coprod_{s\in S} \widetilde{X}_s \to \coprod_{s\in S} \widetilde{\mathrm{L}_f X_s}$ is a cofibrant approximation to $\coprod_{s\in S} X_s \to \coprod_{s\in S} \mathrm{L}_f X_s$, and if W is an unpointed f-local space the map $\mathrm{Map}(\coprod_{s\in S} \widetilde{\mathrm{L}_f X_s}, W) \to \mathrm{Map}(\coprod_{s\in S} \widetilde{X}_s, W)$ is isomorphic to the map $\prod_{s\in S} \mathrm{Map}(\widetilde{\mathrm{L}_f X_s}, W) \to \prod_{s\in S} \mathrm{Map}(\widetilde{X}_s, W)$. This last map is a product of weak equivalences of fibrant simplicial sets and is thus a weak equivalence. Thus, the map $\coprod_{s\in S} X_s \to \coprod_{s\in S} \mathrm{L}_f X_s$ is an unpointed f-local equivalence. □

CHAPTER 2

The Localization Model Category for Spaces

2.1. The Bousfield localization model category structure

In this section, we show that for every map $f\colon A \to B$ in $\mathrm{Spc}_{(*)}$ (see Notation 1.1.4) there is a model category structure on $\mathrm{Spc}_{(*)}$ in which the weak equivalences are the f-local equivalences (see Definition 1.2.2) and the fibrant objects are the f-local spaces (see Theorem 2.1.3 and Proposition 2.1.4). This is a generalization of the h_*-local model category structure for a generalized homology theory h_* on the category of simplicial sets defined by A.K. Bousfield in [**8**]. It is also an example of a *left Bousfield localization* (see Definition 3.1.1). This model category structure has also been obtained by Bousfield in [**12**] for the category of simplicial sets, where he deals as well with localizing certain proper classes of maps of simplicial sets.

2.1.1. Statements of the theorems.

DEFINITION 2.1.2. Let $f\colon A \to B$ be a map between cofibrant spaces in $\mathrm{Spc}_{(*)}$.

(1) An *f-local weak equivalence* is defined to be an f-local equivalence (see Definition 1.2.2).
(2) An *f-local cofibration* is defined to be a cofibration.
(3) An *f-local fibration* is defined to be a map with the right lifting property (see Definition 7.2.1) with respect to all maps that are both f-local cofibrations and f-local weak equivalences. If the map from a space to a point is an f-local fibration, then we will say that the space is *f-local fibrant.*

THEOREM 2.1.3. *If $f\colon A \to B$ is a map between cofibrant spaces in $\mathrm{Spc}_{(*)}$, then there is a simplicial model category structure on $\mathrm{Spc}_{(*)}$ in which the weak equivalences are the f-local weak equivalences, the cofibrations are the f-local cofibrations, the fibrations are the f-local fibrations, and the simplicial structure is the usual simplicial structure on $\mathrm{Spc}_{(*)}$.*

PROPOSITION 2.1.4. *If $f\colon A \to B$ is an inclusion of cell complexes (see Remark 1.2.11), then a space is f-local if and only if it is fibrant in the f-local model category structure of Theorem 2.1.3.*

The main difficulty in the proof of Theorem 2.1.3 lies in finding a set J_f of generating trivial cofibrations for the f-local model category structure. The augmented set of f-horns $\overline{\Lambda\{f\}}$ (see Definition 1.3.2) is a set of cofibrations such that every $\overline{\Lambda\{f\}}$-cofibration is a trivial cofibration in the f-local model category structure (see Corollary 1.4.17), and Proposition 1.4.7 implies that the set $\overline{\Lambda\{f\}}$ *does* suffice to determine the f-local spaces, but it is *not* true that the class of f-local trivial cofibrations must equal the class of $\overline{\Lambda\{f\}}$-cofibrations (see Example 2.1.6). Thus, the proof of Theorem 2.1.3 will use the following proposition, the proof of which we will present in Section 2.3 after some necessary preparatory work in Section 2.2.

PROPOSITION 2.1.5. *If* $f\colon A \to B$ *is a map of cofibrant spaces in* $\mathrm{Spc}_{(*)}$, *then there is a set* J_f *of inclusions of cell complexes (see Remark* 1.2.11*) such that*

(1) *every map in* J_f *is an* f*-local equivalence, and*
(2) *the class of* J_f*-cofibrations (see Definition* 10.5.2*) equals the class of cofibrations that are also* f*-local equivalences.*

We will present the proof of Proposition 2.1.5 in Section 2.3, after some necessary preparatory work in Section 2.2.

We present here an example (due to A. K. Bousfield) of a map f such that, among the cofibrations that are f-local equivalences, there are maps that are not $\overline{\Lambda\{f\}}$-cofibrations.

EXAMPLE 2.1.6. Let $\mathrm{Spc}_{(*)} = \mathrm{Top}_*$, let $n > 0$, and let $f\colon A \to B$ be the inclusion $S^n \to D^{n+1}$. The path space fibration $p\colon \mathrm{P\,K}(\mathbb{Z}, n) \to \mathrm{K}(\mathbb{Z}, n)$ is a $\overline{\Lambda\{f\}}$-injective (see Definition 1.4.2), and so every $\overline{\Lambda\{f\}}$-cofibration has the left lifting property with respect to p. The cofibration $* \to S^n$ does not have the left lifting property with respect to p, and so it is not a $\overline{\Lambda\{f\}}$-cofibration. However, since both the composition $* \to S^n \to D^{n+1}$ and f itself are f-local equivalences (see Proposition 1.2.16), the "two out of three" property of f-local equivalences (see Proposition 1.2.17) implies that the inclusion $* \to S^n$ is an f-local equivalence. Thus, $* \to S^n$ is both a cofibration and an f-local equivalence, but it is not a $\overline{\Lambda\{f\}}$-cofibration.

2.1.7. Proofs.

PROOF OF THEOREM 2.1.3. We begin by using Theorem 11.3.1 to show that there is a cofibrantly generated model category structure on $\mathrm{Spc}_{(*)}$ with weak equivalences, cofibrations, and fibrations as described in the statement of Theorem 2.1.3.

Proposition 1.2.17 implies that the class of f-local equivalences satisfies the "two out of three" axiom, and Proposition 1.2.18 implies that it is closed under retracts.

Let I be the set of maps

$$I = \{\mathrm{Spc}_{(*)}(\partial\Delta[n]) \to \mathrm{Spc}_{(*)}(\Delta[n]) \mid n \geq 0\}$$

(see Definition 1.1.13) and let J_f be the set of maps provided by Proposition 2.1.5. Since every map in either I or J_f is an inclusion of simplicial sets (if $\mathrm{Spc}_{(*)} = \mathrm{SS}_{(*)}$) or an inclusion of cell complexes (if $\mathrm{Spc}_{(*)} = \mathrm{Top}_{(*)}$), Example 10.4.4 and Example 10.4.5 imply that condition 1 of Theorem 11.3.1 is satisfied.

The subcategory of I-cofibrations is the subcategory of cofibrations in the usual model category structure in $\mathrm{Spc}_{(*)}$, and the I-injectives are the usual trivial fibrations. Thus, Proposition 2.1.5 implies that condition 2 of Theorem 11.3.1 is satisfied.

Since the J_f-cofibrations are a subcategory of the I-cofibrations, every I-injective must be a J_f-injective. Proposition 1.2.16 implies that every J_f-injective is an f-local equivalence, and so condition 3 is satisfied.

Proposition 2.1.5 implies that condition 4a of Theorem 11.3.1 is satisfied, and so Theorem 11.3.1 now implies that we have a model category.

To show that our model category is a simplicial model category, we note that, since the simplicial structure is the usual one, axiom M6 of Definition 9.1.6 holds because it does so in the usual simplicial model category structure on $\mathrm{Spc}_{(*)}$. For axiom M7 of Definition 9.1.6, we note that the class of f-local cofibrations equals the usual class of cofibrations and the class of f-local fibrations is contained in the usual class of fibrations. Thus, the first requirement of axiom M7 is clear. In the case that the map p is an f-local equivalence, the rest of axiom M7 follows from the fact that, since the class of f-local cofibrations equals the usual class of cofibrations, the class of f-local trivial fibrations equals the usual class of trivial fibrations (see Proposition 7.2.3).

In the case that the map i is an f-local equivalence, we choose a cofibrant approximation $\tilde{\imath}\colon \widetilde{A} \to \widetilde{B}$ to i such that $\tilde{\imath}$ is a cofibration (see Proposition 8.1.23). Proposition 9.4.5 and Proposition 9.4.8 imply that, for every $n \geq 0$, the map $\widetilde{A} \otimes \Delta[n] \amalg_{\widetilde{A}\otimes\partial\Delta[n]} \widetilde{B} \otimes \partial\Delta[n] \to \widetilde{B} \otimes \Delta[n]$ is also an f-local equivalence, and so it has the left lifting property with respect to the map p. Lemma 9.4.7 now implies that the map $\tilde{\imath}$ has the left lifting property with respect to the map $X^{\Delta[n]} \to Y^{\Delta[n]} \times_{Y^{\partial\Delta[n]}} X^{\partial\Delta[n]}$ for every $n \geq 0$. Since $\mathrm{Spc}_{(*)}$ is a left proper model category (see Theorem 13.1.11 and Theorem 13.1.13), Proposition 13.2.1 implies that the map i has the left lifting property with respect to the map $X^{\Delta[n]} \to Y^{\Delta[n]} \times_{Y^{\partial\Delta[n]}} X^{\partial\Delta[n]}$ for every $n \geq 0$, and so the result follows from Lemma 9.4.7. □

PROOF OF PROPOSITION 2.1.4. If W is fibrant in the f-local model category structure, then the map $W \to *$ has the right lifting property with respect to every cofibration that is an f-local equivalence. Proposition 1.3.7 implies that every horn on f is both a cofibration and an f-local equivalence, and so Proposition 1.3.3 implies that W is f-local.

Conversely, assume that W is f-local. If $i\colon A \to B$ is both a cofibration and an f-local equivalence, then Proposition 8.1.23 implies that there is a cofibrant approximation $\tilde{\imath}\colon \widetilde{A} \to \widetilde{B}$ to i such that $\tilde{\imath}$ is a cofibration, and Proposition 13.2.1 and Proposition 7.2.3 imply that it is sufficient to show that $\tilde{\imath}$ has the left lifting property with respect to the map $W \to *$. Proposition 1.2.16 and Proposition 1.2.17 imply that $\tilde{\imath}$ is an f-local equivalence, and so Proposition 9.4.5 and Proposition 9.4.3 imply that $\tilde{\imath}$ has the left lifting property with respect to the map $W \to *$. □

2.2. Subcomplexes of relative $\overline{\Lambda\{f\}}$-cell complexes

The proof of Proposition 2.1.5 (in Section 2.3) will require a careful analysis of the localization of a space. Since the localization map is a relative $\overline{\Lambda\{f\}}$-cell complex, we need to study subcomplexes of relative $\overline{\Lambda\{f\}}$-cell complexes.

DEFINITION 2.2.1. Let $f\colon A \to B$ be an inclusion of cell complexes (see Remark 1.2.11).

- If $C \to D$ is an element of $\overline{\Lambda\{f\}}$ (see Definition 1.3.2), then D will also be called a *$\overline{\Lambda\{f\}}$-cell*, C will be called the *boundary* of the $\overline{\Lambda\{f\}}$-cell, and $D - C$ will be called the *interior* of the $\overline{\Lambda\{f\}}$-cell. (The interior of a $\overline{\Lambda\{f\}}$-cell is not, in general, a subcomplex.)

- If $C \to D$ is a map in $\overline{\Lambda\{f\}}$ and

$$\begin{array}{ccc} C & \longrightarrow & X \\ \downarrow & & \downarrow \\ D & \longrightarrow & Y \end{array}$$

is a pushout, then we will refer to the image of D in Y as a $\overline{\Lambda\{f\}}$-*cell.*

2.2.2. Presentations of relative $\overline{\Lambda\{f\}}$-cell complexes. A relative $\overline{\Lambda\{f\}}$-cell complex is a map that can be constructed as a transfinite composition of pushouts of elements of $\overline{\Lambda\{f\}}$ (see Definition 1.3.8). To consider subcomplexes of a relative $\overline{\Lambda\{f\}}$-cell complex, we need to choose a particular such construction.

DEFINITION 2.2.3. If $g\colon X \to Y$ is a relative $\overline{\Lambda\{f\}}$-cell complex (see Definition 1.3.8), then a *presentation* of g is a pair consisting of a λ-sequence

$$X = X_0 \to X_1 \to X_2 \to \cdots \to X_\beta \to \cdots \qquad (\beta < \lambda)$$

(for some ordinal λ) and a set of ordered triples

$$\{(T^\beta, e^\beta, h^\beta)\}_{\beta<\lambda}$$

such that

(1) the composition of the λ-sequence is the map $g\colon X \to Y$,
(2) each T^β is a set,
(3) each e^β is a function $e^\beta\colon T^\beta \to \overline{\Lambda\{f\}}$ (see Definition 1.3.2),
(4) for every $\beta < \lambda$, if $i \in T^\beta$ and e^β_i is the $\overline{\Lambda\{f\}}$-cell $C_i \to D_i$, then h^β_i is a map $h^\beta_i\colon C_i \to X_\beta$, and
(5) every $X_{\beta+1}$ is the pushout

$$\begin{array}{ccc} \coprod_{T^\beta} C_i & \longrightarrow & \coprod_{T^\beta} D_i \\ {\scriptstyle \coprod h^\beta_i}\downarrow & & \downarrow \\ X_\beta & \longrightarrow & X_{\beta+1} \end{array} .$$

DEFINITION 2.2.4. Let $g\colon X \to Y$ be a relative $\overline{\Lambda\{f\}}$-cell complex with presentation $\big(X = X_0 \to X_1 \to X_2 \to \cdots \to X_\beta \to \cdots \quad (\beta < \lambda), \{T^\beta, e^\beta, h^\beta\}_{\beta<\lambda}\big)$.

(1) If e is a $\overline{\Lambda\{f\}}$-cell of g (see Definition 1.3.2), the *presentation ordinal* of e is defined to be the first ordinal β such that e is in X_β.
(2) If $\beta < \lambda$, then the β-*skeleton* of g is defined to be X_β. We will sometimes abuse language and refer to the image of X_β in Y as the β-skeleton of g.

2.2.5. Constructing a subcomplex of a relative $\overline{\Lambda\{f\}}$-cell complex.

DEFINITION 2.2.6. If $g\colon X \to Y$ is a relative $\overline{\Lambda\{f\}}$-cell complex with presentation $\big(X = X_0 \to X_1 \to X_2 \to \cdots \to X_\beta \to \cdots \quad (\beta < \lambda), \{T^\beta, e^\beta, h^\beta\}_{\beta<\lambda}\big)$, then a *subcomplex* of g relative to that presentation consists of a family of sets $\{\widetilde{T}^\beta\}_{\beta<\lambda}$ such that

(1) for every $\beta < \lambda$, the set $\widetilde{T}^\beta$ is a subset of T^β,

(2) there is a λ-sequence

$$X = \widetilde{X}_0 \to \widetilde{X}_1 \to \widetilde{X}_2 \to \cdots \to \widetilde{X}_\beta \to \cdots \qquad (\beta < \lambda)$$

(called the λ-sequence *associated* with the subcomplex) and a map of λ-sequences

$$\begin{array}{ccccccccc} X & = & \widetilde{X}_0 & \longrightarrow & \widetilde{X}_1 & \longrightarrow & \widetilde{X}_2 & \longrightarrow & \cdots \\ \| & & \downarrow & & \downarrow & & \downarrow & & \\ X & = & X_0 & \longrightarrow & X_1 & \longrightarrow & X_2 & \longrightarrow & \cdots \end{array}$$

such that, for every $\beta < \lambda$ and every $i \in \widetilde{T}^\beta$, the map $h_i^\beta\colon C_i \to X_\beta$ factors through the map $\widetilde{X}_\beta \to X_\beta$, and

(3) for every $\beta < \lambda$, the square

$$\begin{array}{ccc} \coprod_{\widetilde{T}^\beta} C_i & \longrightarrow & \coprod_{\widetilde{T}^\beta} D_i \\ \downarrow & & \downarrow \\ \widetilde{X}_\beta & \longrightarrow & \widetilde{X}_{\beta+1} \end{array}$$

is a pushout.

REMARK 2.2.7. Although a subcomplex of a relative $\overline{\Lambda\{f\}}$-cell complex can only be defined relative to some particular presentation of that relative $\overline{\Lambda\{f\}}$-cell complex, we will often discuss subcomplexes of a relative $\overline{\Lambda\{f\}}$-cell complex without explicitly mentioning the presentation relative to which the subcomplexes are defined.

REMARK 2.2.8. Although a subcomplex of a relative $\overline{\Lambda\{f\}}$-cell complex with some particular presentation is defined to be a family of sets $\{\widetilde{T}^\beta\}_{\beta<\lambda}$ (see Definition 2.2.6), we will often abuse language and refer to the λ-sequence associated with the subcomplex, or the composition of that λ-sequence, as a "subcomplex".

REMARK 2.2.9. Note that the definition of a subcomplex implies that the maps $\widetilde{X}_\beta \to X_\beta$ are all relative $\overline{\Lambda\{f\}}$-cell complexes. Since a relative $\overline{\Lambda\{f\}}$-cell complex is a monomorphism, the factorization of each h_i^β through $\widetilde{X}_\beta \to X_\beta$ is unique. Thus, a subcomplex of a relative $\overline{\Lambda\{f\}}$-cell complex is itself a relative $\overline{\Lambda\{f\}}$-cell complex with a natural presentation.

PROPOSITION 2.2.10. *Given a relative $\overline{\Lambda\{f\}}$-cell complex $X \to Y$ with presentation $\big(X = X_0 \to X_1 \to X_2 \to \cdots \to X_\beta \to \cdots \quad (\beta < \lambda), \{T^\beta, e^\beta, h^\beta\}_{\beta<\lambda}\big)$, an arbitrary subcomplex can be constructed by the following inductive procedure.*

(1) *Choose an arbitrary subset $\widetilde{T}^0$ of T^0.*
(2) *If $\beta < \lambda$ and we have defined $\{\widetilde{T}^\gamma\}_{\gamma<\beta}$, then we have determined the space $\widetilde{X}_\beta$ and the map $\widetilde{X}_\beta \to X_\beta$ (where $\widetilde{X}_\beta$ is the space that appears in the λ-sequence associated to the subcomplex). Consider the set*

$$\{i \in T^\beta \mid h_i^\beta\colon C_i \to X_\beta \text{ factors through } \widetilde{X}_\beta \to X_\beta\}$$

Choose an arbitrary subset $\widetilde{T}^\beta$ of this set. For every $i \in \widetilde{T}^\beta$, there is a unique map $\tilde{h}_i^\beta : C_i \to \widetilde{X}_\beta$ that makes the diagram

$$\begin{array}{ccc} C_i & & \\ {\scriptstyle \tilde{h}_i^\beta}\downarrow & \searrow^{h_i^\beta} & \\ \widetilde{X}_\beta & \longrightarrow & X_\beta \end{array}$$

commute. We let $\widetilde{X}_{\beta+1}$ be the pushout

$$\begin{array}{ccc} \coprod_{\widetilde{T}^\beta} C_i & \longrightarrow & \coprod_{\widetilde{T}^\beta} D_i \\ {\scriptstyle \coprod \tilde{h}^\beta}\downarrow & & \downarrow \\ \widetilde{X}_\beta & \longrightarrow & \widetilde{X}_{\beta+1} \end{array}$$

PROOF. This follows directly from the definitions. □

PROPOSITION 2.2.11. *Let $g\colon X \to Y$ be a relative $\overline{\Lambda\{f\}}$-cell complex with presentation $\big(X = X_0 \to X_1 \to X_2 \to \cdots \to X_\beta \to \cdots \quad (\beta < \lambda), \{T^\beta, e^\beta, h^\beta\}_{\beta<\lambda}\big)$. If $\big\{\{\widetilde{T}_u^\beta\}_{\beta<\lambda}\big\}_{u\in U}$ is a set of subcomplexes of g, then the intersection $\{\widetilde{T}^\beta\}_{\beta<\lambda}$ of the set of subcomplexes (where $\widetilde{T}^\beta = \bigcap_{u\in U} \widetilde{T}_u^\beta$ for every $\beta < \lambda$) is a subcomplex of g.*

PROOF. It is sufficient to show that, if $\beta < \lambda$ and we have constructed the β-skeleton of the associated λ-sequence $X = \widetilde{X}_0 \to \widetilde{X}_1 \to \widetilde{X}_2 \to \cdots \to \widetilde{X}_\beta$, then, for every $i \in \widetilde{T}^\beta$, the map $h_i^\beta : C_i \to X_\beta$ factors through $\widetilde{X}_\beta \to X_\beta$. If $i \in \widetilde{T}^\beta$, then $i \in \widetilde{T}_u^\beta$ for every $u \in U$, and so h_i^β factors uniquely through $\widetilde{X}_\beta^u \to X_\beta$ for every $u \in U$. Since $\widetilde{X}_\beta$ is the limit of the diagram that contains the map $\widetilde{X}_\beta^u \to X_\beta$ for every $u \in U$, the map h_i^β factors uniquely through $\widetilde{X}_\beta \to X_\beta$. □

COROLLARY 2.2.12. *Let $g\colon X \to Y$ be a relative $\overline{\Lambda\{f\}}$-cell complex with presentation $\big(X = X_0 \to X_1 \to X_2 \to \cdots \to X_\beta \to \cdots \quad (\beta < \lambda), \{T^\beta, e^\beta, h^\beta\}_{\beta<\lambda}\big)$. If e is an f-cell of g, then there is a smallest subcomplex of g that contains e, i.e., a subcomplex of g containing e that is a subcomplex of every subcomplex of g that contains e.*

PROOF. Proposition 2.2.11 implies that we can take the intersection of all subcomplexes of g that contain e. □

DEFINITION 2.2.13. If e is a $\overline{\Lambda\{f\}}$-cell of the relative $\overline{\Lambda\{f\}}$-cell complex $g\colon X \to Y$ with some particular presentation, then the smallest subcomplex of g that contains e (whose existence is guaranteed by Corollary 2.2.12) will be called the subcomplex *generated* by e.

2.2.14. Subcomplexes of the localization. If $f\colon A \to B$ is an inclusion of cell complexes (see Remark 1.2.11), then for every space X, the localization $j_X\colon X \to \mathrm{L}_f X$ has a natural presentation as a relative $\overline{\Lambda\{f\}}$-cell complex. When we discuss subcomplexes of j_X, it will be with respect to that natural presentation.

LEMMA 2.2.15. *Let $f\colon A \to B$ be an inclusion of cell complexes (see Remark 1.2.11), and let X be a simplicial set (or a cell complex). If W is a subcomplex* *of X, then $\mathrm{L}_f W$ is naturally isomorphic (or homeomorphic) to a subcomplex of $\mathrm{L}_f X$ (where by "naturally" we mean that this isomorphism is a functor on the category of subcomplexes of X).*

PROOF. The construction of $\mathrm{L}_f X$ from X defines an obvious presentation of the relative $\overline{\Lambda\{f\}}$-cell complex $j_X : X \to \mathrm{L}_f X$. Since an inclusion of a subcomplex is a monomorphism, the construction of $\mathrm{L}_f W$ from W defines an obvious natural isomorphism of the relative $\overline{\Lambda\{f\}}$-cell complex $W \to \mathrm{L}_f W$ with a subcomplex of $j(X)$. □

PROPOSITION 2.2.16. *Let $f\colon A \to B$ be an inclusion of cell complexes (see Remark 1.2.11). If X is a simplicial set (or a cell complex) and W is a subcomplex of X, then $\mathrm{L}_f W$ is naturally isomorphic (or homeomorphic) to the subcomplex of $\mathrm{L}_f X$ consisting of those $\overline{\Lambda\{f\}}$-cells of $\mathrm{L}_f X$ for which the zero skeleton of the subcomplex of $\mathrm{L}_f X$ generated by that $\overline{\Lambda\{f\}}$-cell (see Definition 2.2.13) is a subcomplex of W.*

PROOF. We identify $\mathrm{L}_f W$ with a subcomplex of $\mathrm{L}_f X$ as in Lemma 2.2.15, and we will show by transfinite induction on the presentation ordinal (see Definition 2.2.4) of the $\overline{\Lambda\{f\}}$-cell that a $\overline{\Lambda\{f\}}$-cell of $\mathrm{L}_f X$ is in $\mathrm{L}_f W$ if and only if the zero skeleton of the subcomplex of $\mathrm{L}_f X$ generated by that $\overline{\Lambda\{f\}}$-cell (see Definition 2.2.13) is a subcomplex of W.

If e is a $\overline{\Lambda\{f\}}$-cell of presentation ordinal 1, then the subcomplex of $\mathrm{L}_f X$ generated by e consists of the union of e and the subcomplexes of X generated by those simplices (or cells) of X whose interiors intersect the image of the attaching map of e. Thus, the zero skeleton of the subcomplex of $\mathrm{L}_f X$ generated by e is a subcomplex of W if and only if the attaching map of e factors through the inclusion $W \to X$, which is true if and only if e is contained in $\mathrm{L}_f W$.

Since there are no $\overline{\Lambda\{f\}}$-cells whose presentation ordinal is a limit ordinal, we assume that $\beta + 1 < \lambda$ and that the assertion is true for all $\overline{\Lambda\{f\}}$-cells of presentation ordinal less than or equal to β. Let e be a $\overline{\Lambda\{f\}}$-cell of presentation ordinal $\beta + 1$. The subcomplex of $\mathrm{L}_f X$ generated by e consists of the union of e and the subcomplexes of $\mathrm{L}_f X$ generated by those $\overline{\Lambda\{f\}}$-cells and simplices (or cells) of X whose interiors intersect the image of the attaching map of e. Each of those $\overline{\Lambda\{f\}}$-cells is of presentation ordinal at most β, and so it is in $\mathrm{L}_f W$ if and only if the zero skeleton of the subcomplex of $\mathrm{L}_f X$ it generates is contained in W, and the inductive hypothesis implies that this is true if and only if that $\overline{\Lambda\{f\}}$-cell is in $\mathrm{L}_f W$. Thus, the subcomplex of $\mathrm{L}_f X$ generated by e is contained in $\mathrm{L}_f W$ if and only if the attaching map for e factors through $W_\beta \to X_\beta$, i.e., if and only if e is in $\mathrm{L}_f W$. □

PROPOSITION 2.2.17. *Let $f\colon A \to B$ be an inclusion of cell complexes (see Remark 1.2.11). If X is a simplicial set (or a cell complex) and $\{W_s\}_{s\in S}$ is a family of subcomplexes of X, then $\mathrm{L}_f(\bigcap_{s\in S} W_s) = \bigcap_{s\in S} \mathrm{L}_f W_s$.*

PROOF. This follows from Proposition 2.2.16. □

PROPOSITION 2.2.18. *Let $f\colon A \to B$ be an inclusion of cell complexes (see Remark 1.2.11). If X is a simplicial set (or a cell complex) and $W_0 \subset W_1 \subset W_2 \subset \cdots \subset W_\beta \subset \cdots \quad (\beta < \lambda)$ is a λ-sequence of subcomplexes of X (where*

λ is the ordinal chosen in Section 1.3.4*), then the natural map* $\operatorname{colim}_{\beta<\lambda} \mathrm{L}_f W_\beta \to \mathrm{L}_f \operatorname{colim}_{\beta<\lambda} W_\beta$ *is an isomorphism (or a homeomorphism).*

PROOF. Proposition 2.2.16 implies that the map is an isomorphism onto a subcomplex; it remains only to show that every $\overline{\Lambda\{f\}}$-cell of $\mathrm{L}_f \operatorname{colim}_{\beta<\lambda} W_\beta$ is contained in some $\mathrm{L}_f W_\beta$. We will do this by a transfinite induction on the presentation ordinal of the $\overline{\Lambda\{f\}}$-cell (see Definition 2.2.4).

If e is a $\overline{\Lambda\{f\}}$-cell of $\mathrm{L}_f \operatorname{colim}_{\beta<\lambda} W_\beta$ of presentation ordinal 1, then its attaching map is a map to $\operatorname{colim}_{\beta<\lambda} W_\beta$, and the discussion in Section 1.3.4 explains why there is an ordinal $\beta < \lambda$ such that the image of the attaching map is contained in W_β. Thus, the $\overline{\Lambda\{f\}}$-cell is in $\mathrm{L}_f W_\beta$.

Since there are no $\overline{\Lambda\{f\}}$-cells of presentation ordinal equal to a limit ordinal, we now let γ be an ordinal such that $\gamma + 1 < \lambda$, and we assume that the assertion is true for all $\overline{\Lambda\{f\}}$-cells of presentation ordinal less than or equal to γ. If e is a $\overline{\Lambda\{f\}}$-cell of presentation ordinal $\gamma+1$, then e has fewer than λ simplices (or cells). Thus, the image of the attaching map of e is contained in the interiors of fewer than λ many $\overline{\Lambda\{f\}}$-cells, each of presentation ordinal less than or equal to γ. (If $\mathrm{Spc}_{(*)} = \mathrm{Top}_{(*)}$, then this follows from Corollary 10.7.5.) The induction hypothesis implies that each of these is contained in some $\mathrm{L}_f W_\beta$. Since λ is a regular cardinal, there must exist $\beta < \lambda$ such that the union of these $\overline{\Lambda\{f\}}$-cells is contained in $\mathrm{L}_f W_\beta$, and so e is also contained in $\mathrm{L}_f W_\beta$. □

2.3. The Bousfield-Smith cardinality argument

The proof of Proposition 2.1.5 is at the end of this section. The cardinality argument that we use here was first used by A. K. Bousfield [**8**] to define a model category structure on the category of simplicial sets in which a weak equivalence was a map that induced a homology isomorphism (for some chosen homology theory). This was extended to more general localizations of cofibrantly generated model categories (see Definition 11.1.2) by J. H. Smith. We are indebted to D. M. Kan for explaining this argument to us.

We will prove Proposition 2.1.5 by showing that there is a set J_f of cofibrations that are f-local equivalences such that every cofibration that is an f-local equivalence has the left lifting property (see Definition 7.2.1) with respect to every J_f-injective. Proposition 2.1.5 will then follow from Corollary 10.5.22.

We will find the set J_f by showing (in Proposition 2.3.8) that there is a cardinal γ such that if a map has the right lifting property with respect to all inclusions of simplicial sets (or of cell complexes) that are f-local equivalences between complexes of size no larger than γ, then it has the right lifting property with respect to all cofibrations that are f-local equivalences. (By the "size" of a simplicial set (or a cell complex) X we mean the cardinal of the set of simplices (or cells) of X; see Definition 2.3.2.) We will then let J_f be a set of representatives of the isomorphism classes of of these "small enough" inclusions of complexes that are f-local equivalences.

We must first deal with an inconvenient aspect of the categories Top and Top_*: Not all spaces are cell complexes. This requires Lemma 2.3.1, which shows that for a fibration to have the right lifting property (see Definition 7.2.1) with respect to all cofibrations that are f-local equivalences, it is sufficient for it to have the right lifting property with respect to all such cofibrations that are inclusions of cell complexes.

LEMMA 2.3.1. *Let $f: A \to B$ be a map of cofibrant spaces in* $\mathrm{Top}_{(*)}$. *If $p: E \to B$ is a fibration with the right lifting property with respect to all inclusions of cell complexes that are f-local equivalences, then it has the right lifting property with respect to all cofibrations that are f-local equivalences.*

PROOF. Let $g: X \to Y$ be a cofibration that is an f-local equivalence. Proposition 11.2.8 implies that there is a cofibrant approximation (see Definition 8.1.22) $\tilde{g}$ to g such that $\tilde{g}$ is an inclusion of cell complexes, and Proposition 1.2.16 and Proposition 1.2.17 imply that $\tilde{g}$ is an f-local equivalence. Since $\mathrm{Top}_{(*)}$ is a left proper model category (see Theorem 13.1.11), the lemma now follows from Proposition 13.2.1. □

We can now restrict our attention to inclusions of simplicial sets (if $\mathrm{Spc}_{(*)} = \mathrm{SS}_{(*)}$) or inclusions of cell complexes (if $\mathrm{Spc}_{(*)} = \mathrm{Top}_{(*)}$). We need to find a cardinal γ with two properties:

(1) The cardinal γ is "large enough" in that for every complex X, every subcomplex of $\mathrm{L}_f X$ of size no greater than γ is contained in the localization of a subcomplex of X of size no greater than γ.
(2) The cardinal γ is "stable" in that if X is a complex of size no greater than γ, then $\mathrm{L}_f X$ will also have size no greater than γ.

Once we have such a cardinal γ, Proposition 2.3.7 (which uses Lemma 2.3.5) will show that any inclusion of complexes that is an f-local equivalence can be built out of ones of size no greater than γ. This will be used in Proposition 2.3.8 to show that if a map has the right lifting property with respect to all "small" inclusions of complexes that are f-local equivalences then it has the right lifting property with respect to all inclusions of complexes that are f-local equivalences. We define our cardinal γ in Definition 2.3.4.

DEFINITION 2.3.2. If the set of simplices (or cells) of the complex X has cardinal κ, then we will say that X is of *size* κ.

LEMMA 2.3.3. *Let the map $f: A \to B$ be an inclusion of cell complexes (see Remark* 1.2.11*) and let λ be the first infinite cardinal greater than that of the simplices (or cells) of $A \amalg B$. For any complex X, we have $\mathrm{L}_f X \approx \operatorname{colim} \mathrm{L}_f X_s$, where X_s varies over the subcomplexes of X of size less than λ.*

PROOF. Proposition 2.2.16 implies that each $\mathrm{L}_f X_s$ is a subcomplex of $\mathrm{L}_f X$, and so we need only show that every $\overline{\Lambda\{f\}}$-cell of $\mathrm{L}_f X$ is contained in $\mathrm{L}_f X_s$ for some small subcomplex X_s of X. We will do this by a transfinite induction on the presentation ordinal of the $\overline{\Lambda\{f\}}$-cell (see Definition 2.2.4). To ease the strain of terminology, for the remainder of this proof, the word "small" will mean "of size less than λ".

The induction is begun by noting that the zero skeleton of $X \to \mathrm{L}_f X$ equals X. Since there are no $\overline{\Lambda\{f\}}$-cells of presentation ordinal equal to a limit ordinal, we need only consider the case of successor ordinals.

Now let $\beta + 1 < \lambda$, and assume that each $\overline{\Lambda\{f\}}$-cell of presentation ordinal less than or equal to β is contained in $\mathrm{L}_f X_s$ for some small subcomplex X_s of X. Any $\overline{\Lambda\{f\}}$-cell of presentation ordinal $\beta + 1$ must be attached by a map of its boundary to the β-skeleton of $\mathrm{L}_f X$ (see Definition 2.2.4). Since the boundary of an $\overline{\Lambda\{f\}}$-cell has size less than λ, the image of the attaching map can intersect the interiors of fewer than λ other simplices (or cells), each of which is either in X or in an $\overline{\Lambda\{f\}}$-cell of sequential dimension less than or equal to β. (If $\mathrm{Spc}_{(*)} = \mathrm{Top}_{(*)}$, then this uses Corollary 10.7.5.) Thus, our $\overline{\Lambda\{f\}}$-cell is attached to the union of X with some $\overline{\Lambda\{f\}}$-cells, each of which is contained in the localization of a small subcomplex of X. If we let Z be the union of those small subcomplexes of X and the subcomplexes of X generated by the (fewer than λ) simplices (or cells) of X in the image of the attaching map of our $\overline{\Lambda\{f\}}$-cell, then Z is the union a collection of size less than λ of subcomplexes of X, each of which is of size less than λ. Since λ is a regular cardinal (see Proposition 10.1.14 and Example 10.1.12), this implies that Z is of size less than λ, and our $\overline{\Lambda\{f\}}$-cell is contained in $\mathrm{L}_f Z$. □

DEFINITION 2.3.4. We let $\mathfrak{c}$ denote the cardinal of the continuum, i.e., $\mathfrak{c}$ is the cardinal of the set of real numbers. We let λ denote the ordinal (which is also a cardinal) selected in Section 1.3.4, i.e., if $f\colon A \to B$, then λ is the first infinite cardinal greater than that of the set of simplices (or cells) of $A \amalg B$. We now define γ as

$$\gamma = \begin{cases} \lambda^{\lambda} & \text{if } \mathrm{Spc}_{(*)} = \mathrm{SS}_{(*)} \\ (\lambda\mathfrak{c})^{\lambda\mathfrak{c}} & \text{if } \mathrm{Spc}_{(*)} = \mathrm{Top}_{(*)} \end{cases} .$$

Thus, if $\mathrm{Spc}_{(*)} = \mathrm{Top}_{(*)}$, then $\gamma = (\lambda\mathfrak{c})^{\lambda\mathfrak{c}} = \max(\lambda^{\lambda}, \mathfrak{c}^{\mathfrak{c}}) = (\lambda^{\lambda})(\mathfrak{c}^{\mathfrak{c}})$ (since the maximum of two infinite cardinals equals their product (see, e.g., [**29**, Chapter 2] or [**17**, page 70])).

LEMMA 2.3.5. *Let the map $f\colon A \to B$ be an inclusion of cell complexes (see Remark* 1.2.11*) and let X be a simplicial set (or a cell complex). If Z is a subcomplex of $\mathrm{L}_f X$ of size less than or equal to γ, then there exists a subcomplex W of X, of size less than or equal to γ, such that $Z \subset \mathrm{L}_f W$.*

PROOF. Lemma 2.3.3 implies that each simplex (or cell) of Z is contained in the localization of some subcomplex of X of size less than λ, and so Proposition 2.2.16 implies that Z is contained in the localization of the union of those subcomplexes. Since $\lambda < \gamma$ (see Definition 2.3.4), $\lambda \times \gamma = \gamma$, and so that union of subcomplexes is of size less than or equal to γ. □

LEMMA 2.3.6. *Let the map $f\colon A \to B$ be an inclusion of cell complexes (see Remark* 1.2.11*). If X is a simplicial set (or a cell complex) of size less than or equal to γ (see Definition* 2.3.4*), then $\mathrm{L}_f X$ has size less than or equal to γ.*

PROOF. Let $X = X_0 \to X_1 \to X_2 \to \cdots \to X_\beta \to \cdots \quad (\beta < \lambda)$ be the λ-sequence that is part of the natural presentation of the relative $\overline{\Lambda\{f\}}$-cell complex $X \to \mathrm{L}_f X$ (see Definition 2.2.3). We will prove by transfinite induction that, for every $\beta < \lambda$, the complex X_β has size less than or equal to γ. Since $\mathrm{L}_f X = \mathrm{colim}_{\beta<\lambda} X_\beta$ and $\mathrm{Succ}(\gamma)$ (see Definition 10.1.10) is a regular cardinal (see Definition 10.1.14), this will imply the lemma.

We begin the induction by noting that $X_0 = X$. If we now assume that X_β has size less than or equal to γ, then (since the boundary of a $\overline{\Lambda\{f\}}$-cell is of size less than λ) there are fewer than $\gamma^{\lambda} = \gamma$ (if $\mathrm{Spc}_{(*)} = \mathrm{SS}_{(*)}$) or $\gamma^{\lambda\mathfrak{c}} = \gamma$ (if $\mathrm{Spc}_{(*)} = \mathrm{Top}_{(*)}$) (see Proposition 10.1.15) many maps from the boundary of a

$\overline{\Lambda\{f\}}$-cell to X_β. Since there are only countably many $\overline{\Lambda\{f\}}$-cells, there are fewer than γ many $\overline{\Lambda\{f\}}$-cells attached to X_β to form $X_{\beta+1}$. Since each $\overline{\Lambda\{f\}}$-cell has fewer than λ many simplices (or cells), $X_{\beta+1}$ has size less than or equal to γ.

If β is a limit ordinal, then X_β is a colimit of complexes, each of which is of size less than or equal to γ. Since $\beta < \lambda < \gamma$, this implies that X_β has size less than or equal to γ. □

The following proposition will be used in Proposition 2.3.8 to extend a map over an arbitrary inclusion of a subcomplex that is an f-local equivalence by extending it over a subcomplex of size no greater than γ.

PROPOSITION 2.3.7. *Let $f\colon A \to B$ be an inclusion of cell complexes (see Remark 1.2.11), and let D be a simplicial set (or a cell complex). If $i\colon C \to D$ is the inclusion of a proper subcomplex and an f-local equivalence, then there is a subcomplex K of D such that*

1. *the subcomplex K is not contained in the subcomplex C,*
2. *the size of K is less than or equal to γ (see Definition 2.3.4), and*
3. *the inclusions $K \cap C \to K$ and $C \to C \cup K$ are both f-local equivalences.*

PROOF. Since $i\colon C \to D$ is the inclusion of a subcomplex and an f-local equivalence, Lemma 2.2.15 and Theorem 1.4.15 imply that $\mathrm{L}_f(i)\colon \mathrm{L}_f C \to \mathrm{L}_f D$ is a trivial cofibration of fibrant spaces, and so it is the inclusion of a strong deformation retract (see Corollary 9.6.5). We choose a strong deformation retraction $R\colon \mathrm{L}_f D \otimes I \to \mathrm{L}_f D$ (where $I = \Delta[1]$), which will remain fixed throughout this proof.

We will show that there exists a subcomplex K of D of size less than or equal to γ such that

1. K is not contained in C,
2. $R|_{\mathrm{L}_f K \otimes I}$ is a deformation retraction of $\mathrm{L}_f K$ to $\mathrm{L}_f(K \cap C)$, and
3. $R|_{\mathrm{L}_f(C\cup K) \otimes I}$ is a deformation retraction of $\mathrm{L}_f(C \cup K)$ to $\mathrm{L}_f C$.

We will do this by constructing a λ-sequence

$$K_0 \subset K_1 \subset K_2 \subset \cdots \subset K_\beta \subset \cdots \qquad (\beta < \lambda)$$

(where λ is as in Definition 2.3.4) of subcomplexes of D such that, for every $\beta < \lambda$,

1. K_β has size less than or equal to γ,
2. $R(\mathrm{L}_f K_\beta \otimes I) \subset \mathrm{L}_f K_{\beta+1}$,

and such that no K_β is contained in C. If we then let $K = \bigcup_{\beta<\lambda} K_\beta$, then Proposition 2.2.18 will imply that K has the properties that we require.

We begin by choosing a simplex (or cell) of D that isn't contained in C, and letting K_0 equal the subcomplex generated by that simplex (or cell).

For successor ordinals, suppose that $\beta + 1 < \gamma$ and that we've constructed K_β. Lemma 2.3.6 implies that $\mathrm{L}_f K_\beta$ has size less than or equal to γ, and so $R(\mathrm{L}_f K_\beta \otimes I)$ is contained in a subcomplex of $\mathrm{L}_f D$ of size less than or equal to γ. (If $\mathrm{Spc}_{(*)} = \mathrm{Top}_{(*)}$, then this uses Corollary 10.7.7.) Lemma 2.3.5 now implies that we can find a subcomplex Z_β of D, of size less than or equal to γ, such that $R(\mathrm{L}_f K_\beta \otimes I) \subset \mathrm{L}_f Z_\beta$. We let $K_{\beta+1} = K_\beta \cup Z_\beta$. It is clear that $K_{\beta+1}$ has the properties required of it, and so the proof is complete. □

PROPOSITION 2.3.8. *Let $f\colon A \to B$ be an inclusion of cell complexes (see Remark 1.2.11). If $p\colon X \to Y$ has the right lifting property with respect to those* *inclusions of subcomplexes $i\colon C \to D$ that are f-local equivalences and such that the size of D is less than or equal to γ (see Definition 2.3.4), then p has the right lifting property with respect to all inclusions of subcomplexes that are f-local equivalences.*

PROOF. Let $i\colon C \to D$ be an inclusion of a subcomplex that is an f-local equivalence, and let the solid arrow diagram

$$\begin{array}{ccc} C & \xrightarrow{h} & X \\ {\scriptstyle i}\downarrow & \nearrow & \downarrow{\scriptstyle p} \\ D & \xrightarrow[k]{} & Y \end{array}$$

be commutative; we must show that there exists a dotted arrow making both triangles commute. To do this, we will consider the subcomplexes of D over which our map can be defined and use Zorn's lemma to show that we can define it over all of D.

Let S be the set of pairs (D_s, g_s) such that

(1) D_s is a subcomplex of D containing C such that the inclusion $i_s\colon C \to D_s$ is an f-local equivalence, and
(2) g_s is a function $D_s \to X$ such that $g_s i_s = h$ and $pg_s = k|_{D_s}$.

We define a preorder on S by defining $(D_s, g_s) < (D_t, g_t)$ if $D_s \subset D_t$ and $g_t|_{D_s} = g_s$.

If $S' \subset S$ is a chain (i.e., a totally ordered subset of S), let $D_u = \operatorname{colim}_{(D_s,g_s)\in S'} D_s$ and define $g_u\colon D_u \to X$ by $g_u = \operatorname{colim}_{(D_s,g_s)\in S'} g_s$. The universal mapping property of the colimit implies that $g_u i_u = h$ and $pg_u = k|_{D_u}$, and Proposition 1.2.20 implies that the map $C \to D_u$ is an f-local equivalence. Thus, (D_u, g_u) is an element of S, and so it is an upper bound for S'. Zorn's lemma now implies that S has a maximal element (D_m, g_m). We will complete the proof by showing that $D_m = D$.

If $D_m \neq D$, then Proposition 2.3.7 implies that there is a subcomplex K of D such that K is not contained in D_m, the size of K is less than or equal to γ, and the inclusions $K \cap D_m \to K$ and $D_m \to D_m \cup K$ are both f-local equivalences. Thus, there is a map $g_K\colon K \to X$ such that $pg_K = k|_K$ and $g_K|_{K\cap D_m} = g_m|_{K\cap D_m}$, and so g_m and g_K combine to define a map $g_{mK}\colon K\cup D_m \to X$ such that $pg_{mK} = k|_{K\cup D_m}$ and $g_{mK} i = h$. Thus, $(K \cup D_m, g_{mK})$ is an element of S strictly greater than (D_m, g_m). This contradicts (D_m, g_m) being a maximal element of S, and so our assumption that $D_m \neq D$ must have been false, and the proof is complete. □

PROOF OF PROPOSITION 2.1.5. Let J_f be a set of representatives of the isomorphism classes of inclusions of subcomplexes that are f-local equivalences of complexes of size less than or equal to γ. Proposition 2.3.8, Corollary 10.5.22 and Lemma 2.3.1 (if $\mathrm{Spc}_{(*)} = \mathrm{Top}_{(*)}$) imply that the J_f-cofibrations are exactly the cofibrations that are f-local equivalences. □

CHAPTER 3

Localization of Model Categories

The purpose of a model category is to serve as a presentation of its homotopy theory, and so a "localization" of a model category should be a construction that adds inverses for maps in the homotopy category, rather than one that adds inverses for maps in the underlying category. If $\mathcal{M}$ is a model category and $\mathcal{C}$ is a class of maps in $\mathcal{M}$, a localization of $\mathcal{M}$ with respect to $\mathcal{C}$ will be a map of model categories $\mathrm{F}\colon \mathcal{M} \to \mathcal{N}$ such that the images in $\mathrm{Ho}\,\mathcal{M}$ of the elements of $\mathcal{C}$ go to isomorphisms in $\mathrm{Ho}\,\mathcal{N}$ and such that F is initial among such maps of model categories. Since there are two different varieties of maps of model categories, *left Quillen functors* and *right Quillen functors* (see Definition 8.5.2), we will define two different varieties of localizations of model categories, *left localizations* and *right localizations* (see Definition 3.1.1).

If $\mathrm{F}\colon \mathcal{M} \to \mathcal{N}$ is a left Quillen functor, $g\colon X \to Y$ is a map in $\mathcal{M}$, and $[g]\colon X \to Y$ is the image of g in $\mathrm{Ho}\,\mathcal{M}$, then the total left derived functor $\mathbf{L}\mathrm{F}\colon \mathrm{Ho}\,\mathcal{M} \to \mathrm{Ho}\,\mathcal{N}$ of F (see Definition 8.4.7) takes $[g]$ to the image in $\mathrm{Ho}\,\mathcal{N}$ of $\mathrm{F}(\tilde{g})$ for some cofibrant approximation $\tilde{g}$ to g. Thus, if $\mathbf{L}\mathrm{F}[g]$ is to be an isomorphism for every element g of $\mathcal{C}$, then Theorem 8.3.10 and Proposition 8.1.24 imply that F must take every cofibrant approximation to an element of $\mathcal{C}$ into a weak equivalence. Thus, if $\mathcal{C}$ is a class of maps in $\mathcal{M}$, then a left localization of $\mathcal{M}$ with respect to $\mathcal{C}$ will be a left Quillen functor that takes cofibrant approximations to elements of $\mathcal{C}$ into weak equivalences and is initial among such left Quillen functors (see Theorem 3.1.6). Similarly, a right localization of $\mathcal{M}$ with respect to $\mathcal{C}$ will be a right Quillen functor that takes fibrant approximations to elements of $\mathcal{C}$ into weak equivalences and is initial among such right Quillen functors.

In Section 3.1 we define left and right localizations of model categories, and explain the connection between left localizations, local objects, and local equivalences (and, dually, the connection between right localizations, colocal objects, and colocal equivalences). In Section 3.2 we establish some properties of (co)local objects and (co)local equivalences, and in Section 3.3 we discuss (left and right) Bousfield localizations of categories, a special case of left and right localizations (see Theorem 3.3.19). (The localizations that we construct in Chapters 4 and 5 are Bousfield localizations.) In Section 3.4 we discuss left Bousfield localizations of left proper model categories (and right Bousfield localizations of right proper model categories), and in Section 3.5 we establish a method for detecting (co)local equivalences.

3.1. Left localization and right localization

DEFINITION 3.1.1. Let $\mathcal{M}$ be a model category and let $\mathcal{C}$ be a class of maps in $\mathcal{M}$.

(1) A *left localization of* $\mathcal{M}$ *with respect to* $\mathcal{C}$ is a model category $L_{\mathcal{C}}\mathcal{M}$ together with a left Quillen functor (see Definition 8.5.2) $j\colon \mathcal{M} \to L_{\mathcal{C}}\mathcal{M}$ such that
   (a) the total left derived functor $\mathbf{L}j\colon \operatorname{Ho}\mathcal{M} \to \operatorname{Ho}L_{\mathcal{C}}\mathcal{M}$ (see Definition 8.4.7) of j takes the images in $\operatorname{Ho}\mathcal{M}$ of the elements of $\mathcal{C}$ into isomorphisms in $\operatorname{Ho}L_{\mathcal{C}}\mathcal{M}$, and
   (b) if $\mathcal{N}$ is a model category and $\varphi\colon \mathcal{M} \to \mathcal{N}$ is a left Quillen functor such that $\mathbf{L}\varphi\colon \operatorname{Ho}\mathcal{M} \to \operatorname{Ho}\mathcal{N}$ takes the images in $\operatorname{Ho}\mathcal{M}$ of the elements of $\mathcal{C}$ into isomorphisms in $\operatorname{Ho}\mathcal{N}$, then there is a unique left Quillen functor $\delta\colon L_{\mathcal{C}}\mathcal{M} \to \mathcal{N}$ such that $\delta j = \varphi$.

(2) A *right localization of* $\mathcal{M}$ *with respect to* $\mathcal{C}$ is a model category $R_{\mathcal{C}}\mathcal{M}$ together with a right Quillen functor $j\colon \mathcal{M} \to R_{\mathcal{C}}\mathcal{M}$ such that
   (a) the total right derived functor $\mathbf{R}j\colon \operatorname{Ho}\mathcal{M} \to \operatorname{Ho}R_{\mathcal{C}}\mathcal{M}$ of j takes the images in $\operatorname{Ho}\mathcal{M}$ of the elements of $\mathcal{C}$ into isomorphisms in $\operatorname{Ho}R_{\mathcal{C}}\mathcal{M}$, and
   (b) if $\mathcal{N}$ is a model category and $\varphi\colon \mathcal{M} \to \mathcal{N}$ is a right Quillen functor such that $\mathbf{R}\varphi\colon \operatorname{Ho}\mathcal{M} \to \operatorname{Ho}\mathcal{N}$ takes the images in $\operatorname{Ho}\mathcal{M}$ of the elements of $\mathcal{C}$ into isomorphisms in $\operatorname{Ho}\mathcal{N}$, then there is a unique right Quillen functor $\delta\colon L_{\mathcal{C}}\mathcal{M} \to \mathcal{N}$ such that $\delta j = \varphi$.

PROPOSITION 3.1.2. *Let* $\mathcal{M}$ *be a model category and let* $\mathcal{C}$ *be a class of maps in* $\mathcal{M}$. *If a (left or right) localization of* $\mathcal{M}$ *with respect to* $\mathcal{C}$ *exists, it is unique up to a unique isomorphism.*

PROOF. The standard argument applies. □

3.1.3. $\mathcal{C}$-local objects and $\mathcal{C}$-local equivalences. Given a left (respectively, right) Quillen functor F, we need to be able to describe when the total left (respectively, right) derived functor of F inverts a map in the homotopy category by examining F itself. We will do this in Theorem 3.1.6, using the notions of $\mathcal{C}$-local object and $\mathcal{C}$-local equivalence (respectively, $\mathcal{C}$-colocal object and $\mathcal{C}$-colocal equivalence) (see Definition 3.1.4).

DEFINITION 3.1.4. Let $\mathcal{M}$ be a model category and let $\mathcal{C}$ be a class of maps in $\mathcal{M}$.

(1) (a) An object W of $\mathcal{M}$ is $\mathcal{C}$-*local* if W is fibrant and for every element $f\colon A \to B$ of $\mathcal{C}$ the induced map of homotopy function complexes $f^*\colon \operatorname{map}(B, W) \to \operatorname{map}(A, W)$ (see Notation 17.4.2) is a weak equivalence. (Theorem 17.5.31 implies that if this is true for any one homotopy function complex then it is true for every homotopy function complex.) If $\mathcal{C}$ consists of the single map $f\colon A \to B$ then a $\mathcal{C}$-local object will also be called f-*local*, and if $\mathcal{C}$ consists of the single map from the initial object of $\mathcal{M}$ to an object A then a $\mathcal{C}$-local object will also be called A-*local* or A-*null*.
   (b) A map $g\colon X \to Y$ in $\mathcal{M}$ is a $\mathcal{C}$-*local equivalence* if for every $\mathcal{C}$-local object W the induced map of homotopy function complexes $g^*\colon \operatorname{map}(Y, W) \to \operatorname{map}(X, W)$ (see Notation 17.4.2) is a weak equivalence. (Theorem 17.5.31 implies that if this is true for any one homotopy function complex then it is true for every homotopy function complex.) If $\mathcal{C}$ consists of the single map $f\colon A \to B$ then a $\mathcal{C}$-local equivalence will also be called an f-*local equivalence*, and if $\mathcal{C}$

consists of the single map from the initial object of $\mathcal{M}$ to an object A then a $\mathcal{C}$-local equivalence will also be called an *A-local equivalence.*

(2) (a) An object W of $\mathcal{M}$ is *$\mathcal{C}$-colocal* if W is cofibrant and for every element $f\colon A \to B$ of $\mathcal{C}$ the induced map of homotopy function complexes $f_*\colon \mathrm{map}(W, A) \to \mathrm{map}(W, B)$ (see Notation 17.4.2) is a weak equivalence. (Theorem 17.5.31 implies that if this is true for any one homotopy function complex then it is true for every homotopy function complex.)

(b) A map $g\colon X \to Y$ in $\mathcal{M}$ is a *$\mathcal{C}$-colocal equivalence* if for every $\mathcal{C}$-colocal object W the induced map of homotopy function complexes $g_*\colon \mathrm{map}(W, X) \to \mathrm{map}(W, Y)$ (see Notation 17.4.2) is a weak equivalence. (Theorem 17.5.31 implies that if this is true for any one homotopy function complex then it is true for every homotopy function complex.)

PROPOSITION 3.1.5. *If $\mathcal{M}$ is a model category and $\mathcal{C}$ is a class of maps in $\mathcal{M}$ then every weak equivalence is both a $\mathcal{C}$-local equivalence and a $\mathcal{C}$-colocal equivalence.*

PROOF. This follows from Theorem 17.6.3. □

THEOREM 3.1.6. *Let $\mathcal{M}$ and $\mathcal{N}$ be model categories and let $\mathrm{F}\colon \mathcal{M} \rightleftarrows \mathcal{N} :\mathrm{U}$ be a Quillen pair.*

(1) *If $\mathcal{C}$ is a class of maps in $\mathcal{M}$, then the following are equivalent:*
   (a) *The total left derived functor $\mathbf{L}\mathrm{F}\colon \mathrm{Ho}\,\mathcal{M} \to \mathrm{Ho}\,\mathcal{N}$ (see Definition 8.4.7) of F takes the images in $\mathrm{Ho}\,\mathcal{M}$ of the elements of $\mathcal{C}$ into isomorphisms in $\mathrm{Ho}\,\mathcal{N}$.*
   (b) *The functor F takes every cofibrant approximation to an element of $\mathcal{C}$ into a weak equivalence in $\mathcal{N}$.*
   (c) *The functor U takes every fibrant object of $\mathcal{N}$ into a $\mathcal{C}$-local object of $\mathcal{M}$.*
   (d) *The functor F takes every $\mathcal{C}$-local equivalence between cofibrant objects into a weak equivalence in $\mathcal{N}$.*

(2) *If $\mathcal{C}$ is a class of maps in $\mathcal{N}$, then the following are equivalent:*
   (a) *The total right derived functor $\mathbf{R}\mathrm{U}\colon \mathrm{Ho}\,\mathcal{N} \to \mathrm{Ho}\,\mathcal{M}$ (see Definition 8.4.7) of U takes the images in $\mathrm{Ho}\,\mathcal{N}$ of the elements of $\mathcal{C}$ into isomorphisms in $\mathrm{Ho}\,\mathcal{M}$.*
   (b) *The functor U takes every fibrant approximation to an element of $\mathcal{C}$ into a weak equivalence in $\mathcal{M}$.*
   (c) *The functor F takes every cofibrant object of $\mathcal{M}$ into a $\mathcal{C}$-colocal object of $\mathcal{N}$.*
   (d) *The functor U takes every $\mathcal{C}$-colocal equivalence between fibrant objects into a weak equivalence in $\mathcal{M}$.*

PROOF. We will prove part 1; the proof of part 2 is dual.

(a) is equivalent to (b): If $g\colon X \to Y$ is a map in $\mathcal{M}$, then the total left derived functor of F takes the image of g in $\mathrm{Ho}\,\mathcal{M}$ to the image in $\mathrm{Ho}\,\mathcal{N}$ of $\mathrm{F}(\tilde{g})$ for some cofibrant approximation $\tilde{g}$ to g (see the proof of Proposition 8.4.4). Since a map in $\mathcal{N}$ is a weak equivalence if and only if its image in $\mathrm{Ho}\,\mathcal{N}$ is an isomorphism (see Theorem 8.3.10), Proposition 8.1.24 implies that (a) is equivalent to (b).

(b) is equivalent to (c): If $f\colon A \to B$ is an element of $\mathcal{C}$, then Proposition 8.1.24 implies that F takes every cofibrant approximation to f into a weak equivalence if and only if F takes at least one cofibrant approximation to f into a weak equivalence. If $\tilde{f}\colon \widetilde{\boldsymbol{A}} \to \widetilde{\boldsymbol{B}}$ is a cosimplicial resolution of f in $\mathcal{M}$, then $\tilde{f}_0\colon \widetilde{\boldsymbol{A}}_0 \to \widetilde{\boldsymbol{B}}_0$ is a cofibrant approximation to f and $\mathrm{F}(\tilde{f})\colon \mathrm{F}(\widetilde{\boldsymbol{A}}) \to \mathrm{F}(\widetilde{\boldsymbol{B}})$ is a cosimplicial resolution in $\mathcal{N}$ of $\mathrm{F}(\tilde{f}_0)\colon \mathrm{F}(\widetilde{\boldsymbol{A}}_0) \to \mathrm{F}(\widetilde{\boldsymbol{B}}_0)$ (see Proposition 16.2.1). Theorem 17.7.7 implies that the map of simplicial sets $\mathcal{N}\bigl(\mathrm{F}(\widetilde{\boldsymbol{B}}), W\bigr) \to \mathcal{N}\bigl(\mathrm{F}(\widetilde{\boldsymbol{A}}), W\bigr)$ is a weak equivalence for every fibrant object W in $\mathcal{N}$ if and only if $\mathrm{F}(\tilde{f}_0)$ is a weak equivalence. The adjointness of F and U now implies that the map of simplicial sets $\mathcal{M}\bigl(\widetilde{\boldsymbol{B}}, \mathrm{U}(W)\bigr) \to \mathcal{M}\bigl(\widetilde{\boldsymbol{A}}, \mathrm{U}(W)\bigr)$ is a weak equivalence for every fibrant object W of $\mathcal{N}$ if and only if $\mathrm{F}(\tilde{f}_0)$ is a weak equivalence, and so (b) is equivalent to (c).

(c) is equivalent to (d): If W is a fibrant object of $\mathcal{N}$ and $\widehat{\boldsymbol{W}}$ is a simplicial resolution of W in $\mathcal{N}$, then $\mathrm{U}(\widehat{\boldsymbol{W}})$ is a simplicial resolution of $\mathrm{U}(W)$ in $\mathcal{M}$ (see Proposition 16.2.1). Thus, $\mathrm{U}(W)$ is $\mathcal{C}$-local if and only if the map of simplicial sets $g^*\colon \mathcal{M}\bigl(B, \mathrm{U}(\widehat{\boldsymbol{W}})\bigr) \to \mathcal{M}\bigl(A, \mathrm{U}(\widehat{\boldsymbol{W}})\bigr)$ is a weak equivalence for every $\mathcal{C}$-local equivalence between cofibrant objects $g\colon A \to B$. Theorem 17.7.7 implies that $\mathrm{F}(g)\colon \mathrm{F}(A) \to \mathrm{F}(B)$ is a weak equivalence if and only if the map of simplicial sets $\mathrm{F}(g)^*\colon \mathcal{N}\bigl(\mathrm{F}(B), \widehat{\boldsymbol{W}}\bigr) \to \mathcal{M}\bigl(\mathrm{F}(A), \widehat{\boldsymbol{W}}\bigr)$ is a weak equivalence for every fibrant object W, and so the result follows from the adjointness of the pair (F, U).

□

3.1.7. Cellularization.

DEFINITION 3.1.8. Let $\mathcal{M}$ be a model category and let K be a set of objects in $\mathcal{M}$.

(1) A map $g\colon X \to Y$ will be called a *K-colocal equivalence* or a *K-cellular equivalence* if for every element A of K the induced map of homotopy function complexes $g_*\colon \mathrm{map}(A, X) \to \mathrm{map}(A, Y)$ (see Notation 17.4.2) is a weak equivalence. (Theorem 17.5.31 implies that if this is true for any one homotopy function complex, then it is true for every homotopy function complex.) If K consists of the single object A, then a K-colocal equivalence will also be called an *A-colocal equivalence* or an *A-cellular equivalence.*

(2) If $\mathcal{C}$ is the class of K-colocal equivalences, then a $\mathcal{C}$-colocal object (see Definition 3.1.4) will also be called *K-colocal.*

REMARK 3.1.9. Earlier work on colocalization was exclusively in a category of pointed spaces ([**20, 21, 23, 24**]) and was called *cellularization*. Given a pointed space A, an A-cellular equivalence of pointed spaces was defined to be a map $g\colon X \to Y$ for which the induced map $g_*\colon \mathrm{Map}(A, X) \to \mathrm{Map}(A, Y)$ (see Definition 1.1.6) is a weak equivalence, and the class of A-cellular spaces was defined to be the smallest class of cofibrant spaces containing A and closed under homotopy colimits and weak equivalences. Since this earlier work considered only the subcategory of fibrant objects (or worked entirely in the category of pointed topological spaces, in which every object is fibrant), this earlier definition of an A-cellular equivalence coincides with our definition of an A-colocal equivalence (see Example 17.1.4 and Example 17.2.4). We will show in Theorem 5.1.5 that this earlier definition of an A-cellular space also coincides with our definition of an A-colocal space.

REMARK 3.1.10. If $\mathcal{M} = \mathrm{Spc}$ (a category of unpointed spaces; see Notation 1.1.4) and A is a non-empty space, then a one point space is a retract of A, and so every space X is a retract of the space of maps X^A. This implies that if K is a set of nonempty spaces, then a K-colocal equivalence of unpointed spaces must actually be a weak equivalence. Thus, to consider the notion of K-colocal equivalence of unpointed spaces would be pointless.[1]

3.1.11. Localization and Quillen functors. Let $\mathcal{M}$ and $\mathcal{N}$ be model categories and let $\mathrm{F}\colon \mathcal{M} \rightleftarrows \mathcal{N} :\mathrm{U}$ be a Quillen pair. If g is a map in $\mathcal{M}$ with respect to which we intend to localize $\mathcal{M}$, then the corresponding localization of $\mathcal{N}$ would *not* be with respect to $\mathrm{F}g$. This is because the image in $\mathrm{Ho}\,\mathcal{N}$ of $\mathrm{F}g$ does not, in general, equal the image in $\mathrm{Ho}\,\mathcal{N}$ under $\mathbf{L}\mathrm{F}$ (see Definition 8.4.7) of the image in $\mathrm{Ho}\,\mathcal{M}$ of g; that is, the square

$$\begin{array}{ccc} \mathcal{M} & \xrightarrow{\mathrm{F}} & \mathcal{N} \\ \downarrow & & \downarrow \\ \mathrm{Ho}\,\mathcal{M} & \xrightarrow[\mathbf{L}\mathrm{F}]{} & \mathrm{Ho}\,\mathcal{N} \end{array}$$

does not, in general, commute. If g is a map in $\mathcal{M}$ and $\tilde{g}$ is a cofibrant approximation to g, then $\mathrm{F}\tilde{g}$ is a map in $\mathcal{N}$ whose image in $\mathrm{Ho}\,\mathcal{N}$ is isomorphic to the image under $\mathbf{L}\mathrm{F}$ of the image of g in $\mathrm{Ho}\,\mathcal{M}$. Thus, if $\mathcal{C}$ is a class of maps in $\mathcal{M}$ with respect to which we will left localize $\mathcal{M}$, then the corresponding class of maps in $\mathcal{N}$ is $\mathbf{L}\mathrm{F}\mathcal{C}$ (see Definition 8.5.11).

PROPOSITION 3.1.12. *Let $\mathcal{M}$ and $\mathcal{N}$ be model categories and let* $\mathrm{F}\colon \mathcal{M} \rightleftarrows \mathcal{N} :\mathrm{U}$ *be a Quillen pair.*

(1) *If $\mathcal{C}$ is a class of maps in $\mathcal{M}$ and W is a fibrant object of $\mathcal{N}$, then W is* $\mathbf{L}\mathrm{F}\mathcal{C}$*-local (see Definition* 8.5.11*) if and only if* $\mathrm{U}W$ *is $\mathcal{C}$-local.*
(2) *If $\mathcal{C}$ is a class of maps in $\mathcal{N}$ and W is a cofibrant object of $\mathcal{M}$, then W is* $\mathbf{R}\mathrm{U}\mathcal{C}$*-colocal if and only if* $\mathrm{F}W$ *is $\mathcal{C}$-colocal.*

PROOF. We will prove part 1; the proof of part 2 is dual.

If $i\colon A \to B$ is a cofibrant approximation to an element of $\mathcal{C}$ and $\tilde{\imath}\colon \widetilde{\boldsymbol{A}} \to \widetilde{\boldsymbol{B}}$ is a cosimplicial resolution of i, then $\mathrm{F}\tilde{\imath}\colon \mathrm{F}\widetilde{\boldsymbol{A}} \to \mathrm{F}\widetilde{\boldsymbol{B}}$ is a cosimplicial resolution of $\mathbf{L}\mathrm{F}i$. The result now follows because the map of simplicial sets $\mathcal{N}(\mathrm{F}\widetilde{\boldsymbol{B}}, W) \to \mathcal{N}(\mathrm{F}\widetilde{\boldsymbol{A}}, W)$ is isomorphic to the map of simplicial sets $\mathcal{M}(\widetilde{\boldsymbol{B}}, \mathrm{U}W) \to \mathcal{M}(\widetilde{\boldsymbol{A}}, \mathrm{U}W)$. □

3.2. $\mathcal{C}$-local objects and $\mathcal{C}$-local equivalences

Theorem 3.1.6 implies that to understand a left localization with respect to $\mathcal{C}$ we must understand $\mathcal{C}$-local objects and $\mathcal{C}$-local equivalences, and to understand a right localization with respect to $\mathcal{C}$ we must understand $\mathcal{C}$-local objects and $\mathcal{C}$-local equivalences.

[1] According to E. Dror Farjoun, this joke is due to W. G. Dwyer.

LEMMA 3.2.1. *Let $\mathcal{M}$ be a model category and let $\mathcal{C}$ be a class of maps in $\mathcal{M}$.*

(1) *If X and Y are fibrant objects of $\mathcal{M}$ and $g\colon X \to Y$ is a weak equivalence, then X is $\mathcal{C}$-local (see Definition 3.1.4) if and only if Y is $\mathcal{C}$-local.*
(2) *If X and Y are cofibrant objects of $\mathcal{M}$ and $g\colon X \to Y$ is a weak equivalence, then X is $\mathcal{C}$-colocal if and only if Y is $\mathcal{C}$-colocal.*

PROOF. We will prove part 1; the proof of part 2 is dual.

If $f\colon A \to B$ is an element of $\mathcal{C}$, then we have the commutative diagram

$$\begin{array}{ccc} \operatorname{map}(B,X) & \longrightarrow & \operatorname{map}(A,X) \\ {\scriptstyle\approx}\big\downarrow & & \big\downarrow{\scriptstyle\approx} \\ \operatorname{map}(B,Y) & \longrightarrow & \operatorname{map}(A,Y) \end{array}$$

in which the vertical maps are weak equivalences (see Theorem 17.6.3). Thus, the top map is a weak equivalence if and only if the bottom map is a weak equivalence. □

PROPOSITION 3.2.2. *Let $\mathcal{M}$ be a model category and let $\mathcal{C}$ be a class of maps in $\mathcal{M}$.*

(1) *If X and Y are fibrant objects that are weakly equivalent (see Definition 7.9.2), then X is $\mathcal{C}$-local if and only if Y is $\mathcal{C}$-local.*
(2) *If X and Y are cofibrant objects that are weakly equivalent, then X is $\mathcal{C}$-colocal if and only if Y is $\mathcal{C}$-colocal.*

PROOF. This follows from Lemma 3.2.1. □

PROPOSITION 3.2.3. *Let $\mathcal{M}$ be a model category and let $\mathcal{C}$ be a class of maps in $\mathcal{M}$.*

(1) *The class of $\mathcal{C}$-local equivalences (see Definition 3.1.4) satisfies the "two out of three" axiom, i.e., if g and h are composable maps and if two of g, h, and hg are $\mathcal{C}$-local equivalences, then so is the third.*
(2) *The class of $\mathcal{C}$-colocal equivalences satisfies the "two out of three" axiom, i.e., if g and h are composable maps and if two of g, h, and hg are $\mathcal{C}$-colocal equivalences, then so is the third.*

PROOF. We will prove part 1; the proof of part 2 is dual.

Given maps $g\colon X \to Y$ and $h\colon Y \to Z$, we can apply a functorial cofibrant approximation (see Proposition 8.1.17) to g and h to obtain the diagram

$$\begin{array}{ccccc} \widetilde{X} & \xrightarrow{\tilde{g}} & \widetilde{Y} & \xrightarrow{\tilde{h}} & \widetilde{Z} \\ \big\downarrow & & \big\downarrow & & \big\downarrow \\ X & \xrightarrow[g]{} & Y & \xrightarrow[h]{} & Z \end{array}$$

in which $\tilde{g}$, $\tilde{h}$, and $\widetilde{hg}$ are cofibrant approximations to g, h, and hg, respectively. If W is a $\mathcal{C}$-local object, $\widehat{\boldsymbol{W}}$ is a simplicial resolution of W, and two of the maps $\tilde{g}^*\colon \mathcal{M}(\widetilde{Y},\widehat{\boldsymbol{W}}) \to \mathcal{M}(\widetilde{X},\widehat{\boldsymbol{W}})$, $\tilde{h}^*\colon \mathcal{M}(\widetilde{Z},\widehat{\boldsymbol{W}}) \to \mathcal{M}(\widetilde{Y},\widehat{\boldsymbol{W}})$, and $(\widetilde{hg})^*\colon \mathcal{M}(\widetilde{Z},\widehat{\boldsymbol{W}}) \to \mathcal{M}(\widetilde{X},\widehat{\boldsymbol{W}})$ are weak equivalences, then the third is as well. □

PROPOSITION 3.2.4. *Let $\mathcal{M}$ be a model category and let $\mathcal{C}$ be a class of maps in $\mathcal{M}$.*

(1) *The class of $\mathcal{C}$-local equivalences is closed under retracts (see Definition 7.1.1).*
(2) *The class of $\mathcal{C}$-colocal equivalences is closed under retracts.*

PROOF. We will prove part 1; the proof of part 2 is dual.

If $g\colon X \to Y$ is a $\mathcal{C}$-local equivalence and $h\colon V \to W$ is a retract of g, then we can apply a functorial factorization to the maps from the initial object to each of X, Y, V, and W to obtain cofibrant approximations $\tilde{g}\colon \widetilde{X} \to \widetilde{Y}$ to g and $\tilde{h}\colon \widetilde{V} \to \widetilde{W}$ to h such that $\tilde{h}$ is a retract of $\tilde{g}$. If Z is a $\mathcal{C}$-local object and $\widehat{\boldsymbol{Z}}$ is a simplicial resolution of Z, then $\tilde{h}^*\colon \mathcal{M}(\widetilde{W}, \widehat{\boldsymbol{Z}}) \to \mathcal{M}(\widetilde{V}, \widehat{\boldsymbol{Z}})$ is a retract of the weak equivalence $\tilde{g}^*\colon \mathcal{M}(\widetilde{Y}, \widehat{\boldsymbol{Z}}) \to \mathcal{M}(\widetilde{X}, \widehat{\boldsymbol{Z}})$, and so $\tilde{h}^*$ is a weak equivalence. $\square$

LEMMA 3.2.5. *Let $\mathcal{M}$ be a model category, let $\mathcal{C}$ be a class of maps in $\mathcal{M}$, let $\mathcal{D}$ be a small category, and let $g\colon \boldsymbol{X} \to \boldsymbol{Y}$ is a map of $\mathcal{D}$-diagrams in $\mathcal{M}$.*

(1) *If the map $g_\alpha\colon \boldsymbol{X}_\alpha \to \boldsymbol{Y}_\alpha$ is a $\mathcal{C}$-local equivalence between cofibrant objects for every object α of $\mathcal{D}$, then the induced map of homotopy colimits* $\operatorname{hocolim} g\colon \operatorname{hocolim} \boldsymbol{X} \to \operatorname{hocolim} \boldsymbol{Y}$ *is a $\mathcal{C}$-local equivalence.*
(2) *If the map $g_\alpha\colon \boldsymbol{X}_\alpha \to \boldsymbol{Y}_\alpha$ is a $\mathcal{C}$-colocal equivalence between fibrant objects for every object α of $\mathcal{D}$, then the induced map of homotopy limits* $\operatorname{holim} g\colon \operatorname{holim} \boldsymbol{X} \to \operatorname{hocolim} \boldsymbol{Y}$ *is a $\mathcal{C}$-colocal equivalence.*

PROOF. We will prove part 1; the proof of part 2 is dual.

Let W be a $\mathcal{C}$-local object. Since $\operatorname{map}(\operatorname{hocolim} \boldsymbol{X}_\alpha, W)$ is naturally weakly equivalent to $\operatorname{holim} \operatorname{map}(\boldsymbol{X}_\alpha, W)$ (see Theorem 19.4.4), our map is naturally weakly equivalent to the map $\operatorname{holim} \operatorname{map}(\boldsymbol{Y}_\alpha, W) \to \operatorname{holim} \operatorname{map}(\boldsymbol{X}_\alpha, W)$. Since for each α the map $\boldsymbol{X}_\alpha \to \boldsymbol{Y}_\alpha$ is a $\mathcal{C}$-local equivalence, the result follows from Theorem 19.4.2. $\square$

LEMMA 3.2.6. *Let $\mathcal{M}$ be a model category and let $\mathcal{C}$ be a class of maps in $\mathcal{M}$.*

(1) *If $i\colon A \to B$ is a cofibration of cofibrant objects, then i is a $\mathcal{C}$-local equivalence if and only if it has the left lifting property (see Definition 7.2.1) with respect to the map $\widehat{\boldsymbol{X}}^{\Delta[n]} \to \widehat{\boldsymbol{X}}^{\partial\Delta[n]}$ for every simplicial resolution $\widehat{\boldsymbol{X}}$ of every $\mathcal{C}$-local object X and every $n \geq 0$.*
(2) *If $p\colon X \to Y$ is a fibration of fibrant objects, then p is a $\mathcal{C}$-colocal equivalence if and only if it has the right lifting property with respect to the map $\widetilde{\boldsymbol{B}} \otimes \partial\Delta[n] \to \widetilde{\boldsymbol{B}} \otimes \Delta[n]$ for every cosimplicial resolution $\widetilde{\boldsymbol{B}}$ of every $\mathcal{C}$-colocal object B and every $n \geq 0$.*

PROOF. This follows from Proposition 17.8.5 and Proposition 17.8.8. $\square$

PROPOSITION 3.2.7. *Let $\mathcal{M}$ be a model category, let $\mathcal{C}$ be a class of maps in $\mathcal{M}$, and let T be a totally ordered set.*

(1) *If $\boldsymbol{W}\colon T \to \mathcal{M}$ is a functor such that, if $s, t \in T$ and $s \leq t$, then $\boldsymbol{W}_s \to \boldsymbol{W}_t$ is both a cofibration of cofibrant objects and a $\mathcal{C}$-local equivalence, then for every $s \in T$ the map $\boldsymbol{W}_s \to \operatorname{colim}_{t \geq s} \boldsymbol{W}_t$ is both a cofibration of cofibrant objects and a $\mathcal{C}$-local equivalence.*
(2) *If $\boldsymbol{W}\colon T^{\mathrm{op}} \to \mathcal{M}$ is a functor such that, if $s, t \in T$ and $s \leq t$, then $\boldsymbol{W}_t \to \boldsymbol{W}_s$ is both a fibration of fibrant objects and a $\mathcal{C}$-colocal equivalence, then*

for every $s \in T$ the map $\lim_{t \geq s} \boldsymbol{W}_t \to \boldsymbol{W}_s$ is both a fibration of fibrant objects and a $\mathcal{C}$-colocal equivalence.

PROOF. We will prove part 1; the proof of part 2 is dual.

Part 1 follows from Lemma 3.2.6, Lemma 10.3.5, and Proposition 10.3.6. □

LEMMA 3.2.8. *Let $\mathcal{M}$ be a simplicial model category and let $\mathcal{C}$ be a class of maps in $\mathcal{M}$. If $f\colon A \to B$ is a cofibration between cofibrant objects in $\mathcal{M}$ that is a $\mathcal{C}$-local equivalence and K is a simplicial set, then the map $f \otimes 1_K\colon A \otimes K \to B \otimes K$ is also a cofibration between cofibrant objects that is a $\mathcal{C}$-local equivalence.*

PROOF. If X is a $\mathcal{C}$-local object then we have a commutative square

$$\begin{array}{ccc} \mathrm{Map}(B \otimes K, X) & \xrightarrow{\approx} & \mathrm{Map}\bigl(K, \mathrm{Map}(B, X)\bigr) \\ \downarrow & & \downarrow \\ \mathrm{Map}(A \otimes K, X) & \xrightarrow[\approx]{} & \mathrm{Map}\bigl(K, \mathrm{Map}(A, X)\bigr) \end{array}$$

in which the horizontal maps are isomorphisms. Since X is $\mathcal{C}$-local the map $\mathrm{Map}(B, X) \to \mathrm{Map}(A, X)$ is a trivial fibration of simplicial sets, and so the map on the right is also a trivial fibration of simplicial sets. □

3.2.9. Left proper model categories.

PROPOSITION 3.2.10. *Let $\mathcal{M}$ be a left proper model category, and let $\mathcal{C}$ be a set of maps in $\mathcal{M}$. If $g\colon C \to D$ is a cofibration that is also a $\mathcal{C}$-local equivalence, then any pushout of g is also a $\mathcal{C}$-local equivalence.*

PROOF. This follows from Proposition 17.8.5 and Proposition 17.8.16. □

PROPOSITION 3.2.11. *If $\mathcal{M}$ is a left proper model category and $\mathcal{C}$ is a class of maps in $\mathcal{M}$, then a transfinite composition of maps, each of which is both a cofibration and a $\mathcal{C}$-local equivalence, is both a cofibration and a $\mathcal{C}$-local equivalence.*

PROOF. Let λ be an ordinal and let

$$X_0 \to X_1 \to X_2 \to \cdots \to X_\beta \to \cdots \qquad (\beta < \lambda)$$

be a λ-sequence of maps that are both cofibrations and $\mathcal{C}$-local equivalences. Proposition 10.3.4 implies that the composition of that λ-sequence is a cofibration, and so it remains only to show that it is a $\mathcal{C}$-local equivalence. Proposition 17.9.4 implies that we can find a λ-sequence of cofibrations together with a map of λ-sequences

$$\begin{array}{ccccccccc} \widetilde{X}_0 & \longrightarrow & \widetilde{X}_1 & \longrightarrow & \widetilde{X}_2 & \longrightarrow \cdots \longrightarrow & \widetilde{X}_\beta & \longrightarrow & \cdots \\ \downarrow & & \downarrow & & \downarrow & & \downarrow & & \\ X_0 & \longrightarrow & X_1 & \longrightarrow & X_2 & \longrightarrow \cdots \longrightarrow & X_\beta & \longrightarrow & \cdots \end{array}$$

such that each vertical map $\widetilde{X}_\beta \to X_\beta$ is a cofibrant approximation to X_β and $\operatorname{colim}_{\beta<\lambda} \widetilde{X}_\beta \to \operatorname{colim}_{\beta<\lambda} X_\beta$ is a cofibrant approximation to $\operatorname{colim}_{\beta<\lambda} X_\beta$. If W is a $\mathcal{C}$-local object and $\widehat{\boldsymbol{W}}$ is a simplicial resolution of W, then, since each $X_\beta \to X_{\beta+1}$ is a $\mathcal{C}$-local equivalence and each $\widetilde{X}_\beta \to \widetilde{X}_{\beta+1}$ is a cofibration, each $\mathcal{M}(\widetilde{X}_{\beta+1}, \widehat{\boldsymbol{W}}) \to$

$\mathcal{M}(\widetilde{X}_\beta, \widehat{\boldsymbol{W}})$ is a trivial fibration of simplicial sets (see Theorem 17.8.4). Thus, we have a tower of trivial fibrations of simplicial sets

$$\mathcal{M}(\widetilde{X}_0, \widehat{\boldsymbol{W}}) \leftarrow \mathcal{M}(\widetilde{X}_1, \widehat{\boldsymbol{W}}) \leftarrow \mathcal{M}(\widetilde{X}_2, \widehat{\boldsymbol{W}}) \leftarrow \cdots \leftarrow \mathcal{M}(\widetilde{X}_\beta, \widehat{\boldsymbol{W}}) \leftarrow \cdots$$

and so the projection $\lim_{\beta<\lambda} \mathcal{M}(\widetilde{X}_\beta, \widehat{\boldsymbol{W}}) \to \mathcal{M}(\widetilde{X}_0, \widehat{\boldsymbol{W}})$ is a weak equivalence. Since $\mathcal{M}(\operatorname{colim}_{\beta<\lambda} \widetilde{X}_\beta, \widehat{\boldsymbol{W}})$ is isomorphic to $\lim_{\beta<\lambda} \mathcal{M}(\widetilde{X}_\beta, \widehat{\boldsymbol{W}})$, this implies that the composition $X_0 \to \operatorname{colim}_{\beta<\lambda} X_\beta$ is a $\mathcal{C}$-local equivalence. □

3.2.12. $\mathcal{C}$-(co)local Whitehead theorems.

THEOREM 3.2.13 (Weak $\mathcal{C}$-(co)local Whitehead theorem). *Let $\mathcal{M}$ is a model category and let $\mathcal{C}$ be a class of maps in $\mathcal{M}$.*

(1) *If X and Y are $\mathcal{C}$-local objects and $g\colon X \to Y$ is a $\mathcal{C}$-local equivalence, then g is a weak equivalence.*
(2) *If X and Y are $\mathcal{C}$-colocal objects and $g\colon X \to Y$ is a $\mathcal{C}$-colocal equivalence, then g is a weak equivalence.*

PROOF. This follows from Proposition 17.7.6. □

THEOREM 3.2.14 (Strong $\mathcal{C}$-(co)local Whitehead theorem). *Let $\mathcal{M}$ be a model category and let $\mathcal{C}$ be a class of maps in $\mathcal{M}$.*

(1) *If X and Y are cofibrant $\mathcal{C}$-local objects and $g\colon X \to Y$ is a $\mathcal{C}$-local equivalence, then g is a homotopy equivalence.*
(2) *If X and Y are fibrant $\mathcal{C}$-colocal objects and $g\colon X \to Y$ is a $\mathcal{C}$-colocal equivalence, then g is a homotopy equivalence.*

PROOF. This follows from Theorem 3.2.13 and Theorem 7.5.10. □

3.2.15. $\mathcal{C}$-localization of objects and maps.

DEFINITION 3.2.16. Let $\mathcal{M}$ be a model category and let $\mathcal{C}$ be a class of maps in $\mathcal{M}$.

(1) (a) A *$\mathcal{C}$-localization* of an object X is a $\mathcal{C}$-local object $\widehat{X}$ (see Definition 3.1.4) together with an $\mathcal{C}$-local equivalence $j\colon X \to \widehat{X}$. We will sometimes use the phrase *$\mathcal{C}$-localization* to refer to the object $\widehat{X}$, without explicitly mentioning the $\mathcal{C}$-local equivalence j. A *cofibrant $\mathcal{C}$-localization* of X is a $\mathcal{C}$-localization in which the $\mathcal{C}$-local equivalence j is also a cofibration.
(b) A *$\mathcal{C}$-localization* of a map $g\colon X \to Y$ is a $\mathcal{C}$-localization $(\widehat{X}, j_X)$ of X, a $\mathcal{C}$-localization $(\widehat{Y}, j_Y)$ of Y, and a map $\hat{g}\colon \widehat{X} \to \widehat{Y}$ such that the square

$$\begin{array}{ccc} X & \xrightarrow{g} & Y \\ {\scriptstyle j_X}\downarrow & & \downarrow{\scriptstyle j_Y} \\ \widehat{X} & \xrightarrow[\hat{g}]{} & \widehat{Y} \end{array}$$

commutes. We will sometimes use the term *$\mathcal{C}$-localization* to refer to the map $\hat{g}$, without explicitly mentioning the $\mathcal{C}$-localizations $(\widehat{X}, j_X)$ of X and $(\widehat{Y}, j_Y)$ of Y.

(2) (a) A $\mathcal{C}$-*colocalization* of an object X is a $\mathcal{C}$-colocal object $\widetilde{X}$ (see Definition 3.1.8) together with a $\mathcal{C}$-colocal equivalence $i\colon \widetilde{X} \to X$. We will sometimes use the phrase $\mathcal{C}$-*colocalization* to refer to the object $\widetilde{X}$, without explicitly mentioning the $\mathcal{C}$-colocal equivalence i. A *fibrant* $\mathcal{C}$-*colocalization* of X is a $\mathcal{C}$-colocalization in which the $\mathcal{C}$-colocal equivalence is also a fibration.

(b) A $\mathcal{C}$-*colocalization* of a map $g\colon X \to Y$ is a $\mathcal{C}$-colocalization $(\widetilde{X}, i_X)$ of X, a $\mathcal{C}$-colocalization $(\widetilde{Y}, i_Y)$ of Y, and a map $\tilde{g}\colon \widetilde{X} \to \widetilde{Y}$ such that the square

$$\begin{array}{ccc} \widetilde{X} & \xrightarrow{\tilde{g}} & \widetilde{Y} \\ {\scriptstyle i_X}\downarrow & & \downarrow{\scriptstyle i_Y} \\ X & \xrightarrow[g]{} & Y \end{array}$$

commutes. We will sometimes use the term $\mathcal{C}$-*colocalization* to refer to the map $\tilde{g}$, without explicitly mentioning the $\mathcal{C}$-colocalizations $(\widetilde{X}, i_X)$ of X and $(\widetilde{Y}, i_Y)$ of Y.

THEOREM 3.2.17. *Let $\mathcal{M}$ be a model category and let $\mathcal{C}$ be a class of maps in $\mathcal{M}$.*

(1) *If X is a fibrant object and $j\colon X \to \widehat{X}$ is a $\mathcal{C}$-localization of X (see Definition 3.2.16), then j is a weak equivalence if and only if X is $\mathcal{C}$-local.*
(2) *If X is a cofibrant object and $i\colon \widetilde{X} \to X$ is a $\mathcal{C}$-colocalization of X (see Definition 3.2.16), then i is a weak equivalence if and only if X is $\mathcal{C}$-colocal.*

PROOF. We will prove part 1; the proof of part 2 is dual.

If X is $\mathcal{C}$-local then Theorem 3.2.13 implies that j is a weak equivalence. Conversely, if j is a weak equivalence then Lemma 3.2.1 implies that X is $\mathcal{C}$-local. □

THEOREM 3.2.18. *Let $\mathcal{M}$ be a model category and let $\mathcal{C}$ be a class of maps in $\mathcal{M}$.*

(1) *If $\hat{g}\colon \widehat{X} \to \widehat{Y}$ is a $\mathcal{C}$-localization of $g\colon X \to Y$, then g is a $\mathcal{C}$-local equivalence if and only if $\hat{g}$ is a weak equivalence.*
(2) *If $\tilde{g}\colon \widetilde{X} \to \widetilde{Y}$ is a $\mathcal{C}$-colocalization of $g\colon X \to Y$, then g is a $\mathcal{C}$-colocal equivalence if and only if $\tilde{g}$ is a weak equivalence.*

PROOF. We will prove part 1; the proof of part 2 is dual.

Proposition 3.1.5 and Proposition 3.2.3 imply that g is a $\mathcal{C}$-local equivalence if and only if $\hat{g}$ is an $\mathcal{C}$-local equivalence. Since $\widehat{X}$ and $\widehat{Y}$ are $\mathcal{C}$-local, Theorem 3.2.13 and Proposition 3.1.5 imply that $\hat{g}$ is a $\mathcal{C}$-local equivalence if and only if it is a weak equivalence. □

If $\mathcal{M}$ is a left proper cellular model category (see Definition 12.1.1) and S is a set of maps in $\mathcal{M}$, then in Definition 4.3.2 we define a functorial S-localization (L_S, j). Theorem 3.2.17 then implies that a fibrant object X is S-local if and only if the S-localization map $j(X)\colon X \to \mathrm{L}_S X$ is a weak equivalence (see Theorem 4.3.5), and Theorem 3.2.18 implies that a map $g\colon X \to Y$ is an S-local equivalence if and only if $\mathrm{L}_S(g)\colon \mathrm{L}_S X \to \mathrm{L}_S Y$ is a weak equivalence (see Theorem 4.3.6).

PROPOSITION 3.2.19. *Let $\mathcal{M}$ be a model category and let $\mathcal{C}$ be a class of maps in $\mathcal{M}$.*

(1) (a) *If both X and $\widehat{X}$ are cofibrant and $j\colon X \to \widehat{X}$ is a $\mathcal{C}$-localization of X, then for every $\mathcal{C}$-local object W the map j induces an isomorphism of the sets of homotopy classes of maps $j^*\colon \pi(\widehat{X}, W) \approx \pi(X, W)$.*
   (b) *If X is cofibrant and $j\colon X \to \widehat{X}$ is a cofibrant $\mathcal{C}$-localization of X, then for every $\mathcal{C}$-local object W and every map $f\colon X \to W$ there is a map $g\colon \widehat{X} \to W$, unique up to homotopy, such that $gj = f$.*

(2) (a) *If both $\widetilde{X}$ and X are fibrant and $i\colon \widetilde{X} \to X$ is a $\mathcal{C}$-colocalization of X, then for every $\mathcal{C}$-colocal object B the map i induces an isomorphism of the sets of homotopy classes of maps $i_*\colon \pi(B, \widetilde{X}) \approx \pi(B, X)$.*
   (b) *If X is fibrant and $i\colon \widetilde{X} \to X$ is a fibrant $\mathcal{C}$-colocalization of X, then for every $\mathcal{C}$-colocal object B and every map $f\colon B \to X$ there is a map $g\colon B \to \widetilde{X}$, unique up to homotopy, such that $ig = f$.*

PROOF. We will prove part 1; the proof of part 2 is dual.

Part 1a follows from Proposition 17.7.4. Part 1b follows from part 1a and Proposition 7.3.13. □

3.3. Bousfield localization

The (left and right) localizations (see Definition 3.1.1) that we will construct will actually be new model category structures on the underlying category of the given model category. Since model category structures such as these were originally constructed in the foundational work of Bousfield [**8, 9, 10, 11, 12**], we call these *(left and right) Bousfield localizations.*

DEFINITION 3.3.1. Let $\mathcal{M}$ be a model category and let $\mathcal{C}$ be a class of maps in $\mathcal{M}$.

(1) The *left Bousfield localization* of $\mathcal{M}$ with respect to $\mathcal{C}$ (if it exists; see Remark 3.3.2) is a model category structure $L_{\mathcal{C}}\mathcal{M}$ on the underlying category of $\mathcal{M}$ such that
   (a) the class of weak equivalences of $L_{\mathcal{C}}\mathcal{M}$ equals the class of $\mathcal{C}$-local equivalences of $\mathcal{M}$,
   (b) the class of cofibrations of $L_{\mathcal{C}}\mathcal{M}$ equals the class of cofibrations of $\mathcal{M}$, and
   (c) the class of fibrations of $L_{\mathcal{C}}\mathcal{M}$ is the class of maps with the right lifting property with respect to those maps that are both cofibrations and $\mathcal{C}$-local equivalences.

(2) The *right Bousfield localization* of $\mathcal{M}$ with respect to $\mathcal{C}$ (if it exists; see Remark 3.3.2) is a model category structure $R_{\mathcal{C}}\mathcal{M}$ on the underlying category of $\mathcal{M}$ such that
   (a) the class of weak equivalences of $R_{\mathcal{C}}\mathcal{M}$ equals the class of $\mathcal{C}$-colocal equivalences of $\mathcal{M}$,
   (b) the class of fibrations of $R_{\mathcal{C}}\mathcal{M}$ equals the class of fibrations of $\mathcal{M}$, and
   (c) the class of cofibrations of $R_{\mathcal{C}}\mathcal{M}$ is the class of maps with the left lifting property with respect to those maps that are both fibrations and $\mathcal{C}$-colocal equivalences.

REMARK 3.3.2. We are not asserting that if $\mathcal{M}$ is a model category and $\mathcal{C}$ is a class of maps in $\mathcal{M}$ then the left (or right) Bousfield localization of $\mathcal{M}$ with respect to $\mathcal{C}$ exists; that is, the three classes of maps described in Definition 3.3.1 part 1 (or part 2) might not constitute a model category structure on $\mathcal{M}$. However, we will show in Theorem 3.3.19 that if a Bousfield localization of $\mathcal{M}$ with respect to $\mathcal{C}$ exists (i.e., if the three classes of maps do constitute a model category structure on $\mathcal{M}$), then it is a localization of $\mathcal{M}$ with respect to $\mathcal{C}$ (see Definition 3.1.1). Our existence theorem for left Bousfield localizations is Theorem 4.1.1 and our existence theorem for right Bousfield localizations is Theorem 5.1.1.

PROPOSITION 3.3.3. *Let $\mathcal{M}$ be a model category, and let $\mathcal{C}$ be a class of maps in $\mathcal{M}$.*

(1) *If $L_{\mathcal{C}}\mathcal{M}$ is the left Bousfield localization of $\mathcal{M}$ with respect to $\mathcal{C}$, then*
 (a) *every weak equivalence of $\mathcal{M}$ is a weak equivalence of $L_{\mathcal{C}}\mathcal{M}$,*
 (b) *the class of trivial fibrations of $L_{\mathcal{C}}\mathcal{M}$ equals the class of trivial fibrations of $\mathcal{M}$,*
 (c) *every fibration of $L_{\mathcal{C}}\mathcal{M}$ is a fibration of $\mathcal{M}$, and*
 (d) *every trivial cofibration of $\mathcal{M}$ is a trivial cofibration of $L_{\mathcal{S}}\mathcal{M}$.*

(2) *If $R_{\mathcal{C}}\mathcal{M}$ is the right Bousfield localization of $\mathcal{M}$ with respect to $\mathcal{C}$, then*
 (a) *every weak equivalence of $\mathcal{M}$ is a weak equivalence of $R_{\mathcal{C}}\mathcal{M}$,*
 (b) *the class of trivial cofibrations of $R_{\mathcal{C}}\mathcal{M}$ equals the class of trivial cofibrations of $\mathcal{M}$,*
 (c) *every cofibration of $R_{\mathcal{C}}\mathcal{M}$ is a cofibration of $\mathcal{M}$, and*
 (d) *every trivial fibration of $\mathcal{M}$ is a trivial fibration of $\mathcal{M}$.*

PROOF. This follows from Proposition 3.1.5 and Proposition 7.2.3. □

PROPOSITION 3.3.4. *Let $\mathcal{M}$ be a model category and let $\mathcal{C}$ be a class of maps in $\mathcal{M}$.*

(1) *If $j\colon \mathcal{M} \to L_{\mathcal{C}}\mathcal{M}$ is the left Bousfield localization of $\mathcal{M}$ with respect to $\mathcal{C}$, then the identity functors $1_{\mathcal{M}}\colon \mathcal{M} \rightleftarrows L_{\mathcal{C}}\mathcal{M} : 1_{\mathcal{M}}$ are a Quillen pair (see Definition 8.5.2).*

(2) *If $j\colon \mathcal{M} \to R_{\mathcal{C}}\mathcal{M}$ is the right Bousfield localization of $\mathcal{M}$ with respect to $\mathcal{C}$, then the identity functors $1_{\mathcal{M}}\colon R_{\mathcal{C}}\mathcal{M} \rightleftarrows \mathcal{M} : 1_{\mathcal{M}}$ are a Quillen pair.*

PROOF. This follows from Proposition 3.3.3. □

PROPOSITION 3.3.5. *Let $\mathcal{M}$ be a model category and let $\mathcal{C}$ be a class of maps in $\mathcal{M}$.*

(1) *If $L_{\mathcal{C}}\mathcal{M}$ is the left Bousfield localization of $\mathcal{M}$ with respect to $\mathcal{C}$, $f\colon X \to Z$ and $g\colon Y \to Z$ are fibrations in $L_{\mathcal{C}}\mathcal{M}$, and $h\colon X \to Y$ is a weak equivalence in $L_{\mathcal{C}}\mathcal{M}$ that makes the triangle*

$$\begin{array}{ccc} X & \xrightarrow{\;h\;} & Y \\ & {\scriptstyle f}\searrow \quad \swarrow{\scriptstyle g} & \\ & Z & \end{array}$$

 commute, then h is a weak equivalence in $\mathcal{M}$.

(2) *If $R_{\mathcal{C}}\mathcal{M}$ is the right Bousfield localization of $\mathcal{M}$ with respect to $\mathcal{C}$, $f\colon A \to B$ and $g\colon A \to C$ are cofibrations in $R_{\mathcal{C}}\mathcal{M}$, and $h\colon B \to C$ is a weak*

equivalence in $\mathrm{R}_{\mathcal{C}}\mathcal{M}$ *that makes the triangle*

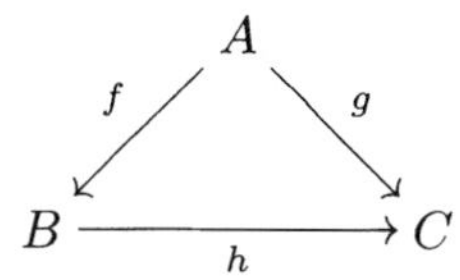

commute, then h *is a weak equivalence in* $\mathcal{M}$.

PROOF. We will prove part 1; the proof of part 2 is dual.

If we let $\tilde{h}\colon \widetilde{X} \to \widetilde{Y}$ be a fibrant cofibrant approximation to h in $\mathcal{M}$ (see Proposition 9.1.9), then we have the diagram

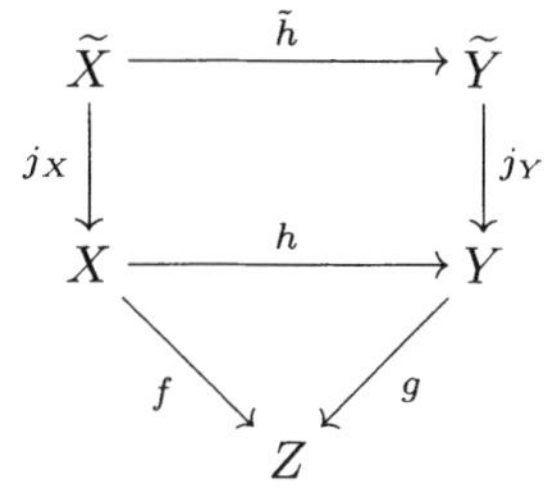

in which $\widetilde{X}$ and $\widetilde{Y}$ are cofibrant in $\mathcal{M}$ and j_X and j_Y are trivial fibrations. Thus, $fj_X\colon \widetilde{X} \to Z$ and $gj_Y\colon \widetilde{Y} \to Z$ are fibrations in $\mathrm{L}_{\mathcal{C}}\mathcal{M}$ (see Proposition 3.3.3), $\tilde{h}$ is a weak equivalence in $\mathrm{L}_{\mathcal{C}}\mathcal{M}$ (see Proposition 3.1.5), and it is sufficient to show that $\tilde{h}$ is a weak equivalence in $\mathcal{M}$.

Since $\tilde{h}$ is a weak equivalence of cofibrant-fibrant objects in $(\mathrm{L}_{\mathcal{C}}\mathcal{M} \downarrow Z)$ (see Theorem 7.6.5), it is a homotopy equivalence in $(\mathrm{L}_{\mathcal{C}}\mathcal{M} \downarrow Z)$ (see Proposition 8.3.26). Thus, there is a map $\tilde{k}\colon \widetilde{Y} \to \widetilde{X}$ in $(\mathrm{L}_{\mathcal{C}}\mathcal{M} \downarrow Z)$ such that $\tilde{h}\tilde{k} \simeq 1_{\widetilde{Y}}$ in $(\mathrm{L}_{\mathcal{C}}\mathcal{M} \downarrow Z)$ and $\tilde{k}\tilde{h} \simeq 1_{\widetilde{X}}$ in $(\mathrm{L}_{\mathcal{C}}\mathcal{M} \downarrow Z)$. Since the trivial fibrations of $\mathrm{L}_{\mathcal{C}}\mathcal{M}$ are the trivial fibrations of $\mathcal{M}$, Proposition 8.3.20 and Proposition 8.4.4 imply that $\tilde{h}\tilde{k} \overset{l}{\simeq} 1_{\widetilde{Y}}$ in $\mathcal{M}$ and $\tilde{k}\tilde{h} \overset{l}{\simeq} 1_{\widetilde{X}}$ in $\mathcal{M}$. If $\gamma\colon \mathcal{M} \to \operatorname{Ho}\mathcal{M}$ is the natural functor from $\mathcal{M}$ to its homotopy category (see Definition 9.6.2), then Lemma 9.6.3 implies that $\gamma(\tilde{k}) \circ \gamma(\tilde{h}) = 1_{\gamma(\widetilde{Y})}$ and $\gamma(\tilde{h}) \circ \gamma(\tilde{k}) = 1_{\gamma(\widetilde{X})}$, and so Theorem 9.6.9 implies that $\tilde{h}$ is a weak equivalence. $\square$

PROPOSITION 3.3.6. *Let* $\mathcal{M}$ *be a model category and let* $\mathcal{C}$ *be a class of maps in* $\mathcal{M}$.

1. *If* X *is a fibrant object of* $\mathcal{M}$, $j\colon X \to \widehat{X}$ *is a cofibrant* $\mathcal{C}$*-localization of* X *(see Definition 3.2.16), and* $\mathrm{L}_{\mathcal{C}}\mathcal{M}$ *is the left Bousfield localization of* $\mathcal{M}$ *with respect to* $\mathcal{C}$*, then the following are equivalent:*
   - (a) *The object* X *is* $\mathcal{C}$*-local.*
   - (b) *The* $\mathcal{C}$*-localization map* $j\colon X \to \widehat{X}$ *is a weak equivalence in* $\mathcal{M}$.
   - (c) *The* $\mathcal{C}$*-localization map* $j\colon X \to \widehat{X}$ *is a homotopy equivalence in* $(X \downarrow \mathcal{M})$.
   - (d) *The* $\mathcal{C}$*-localization map* $j\colon X \to \widehat{X}$ *is the inclusion of a strong deformation retract (see Definition 7.6.10).*
2. *If* X *is a cofibrant object of* $\mathcal{M}$, $i\colon \widetilde{X} \to X$ *is a fibrant* $\mathcal{C}$*-colocalization of* X*, and* $\mathrm{R}_{\mathcal{C}}\mathcal{M}$ *is the right Bousfield localization of* $\mathcal{M}$ *with respect to* $\mathcal{C}$*, then the following are equivalent:*

(a) *The object X is $\mathcal{C}$-colocal.*
(b) *The $\mathcal{C}$-colocalization map $i\colon \widetilde{X} \to X$ is a weak equivalence in $\mathcal{M}$.*
(c) *The $\mathcal{C}$-colocalization map $i\colon \widetilde{X} \to X$ is a homotopy equivalence in $(\mathcal{M} \downarrow X)$.*
(d) *The $\mathcal{C}$-colocalization map $i\colon \widetilde{X} \to X$ has a right inverse $s\colon X \to \widetilde{X}$ such that si is homotopic to $1_{\widetilde{X}}$ in $(\mathcal{M} \downarrow X)$ (that is, it is the dual of a strong deformation retraction).*

PROOF. We will prove part 1; the proof of part 2 is dual.

A cofibrant $\mathcal{C}$-localization of an object is a $\mathcal{C}$-local trivial cofibration to an $\mathcal{C}$-local fibrant object (see Definition 3.3.1 and Proposition 3.4.1). Thus, Proposition 7.6.11 implies that condition 1 implies condition 4. It follows immediately from the definitions that condition 4 implies condition 3 and Theorem 7.8.5 implies that condition 3 implies condition 2. Finally, Lemma 3.2.1 implies that condition 2 implies condition 1. □

3.3.7. $\mathcal{C}$-local objects. For an explanation of the motivation of the definition of a horn, see Section 1.3.

DEFINITION 3.3.8. Let $\mathcal{M}$ be a model category.

(1) If $f\colon A \to B$ is a map in $\mathcal{M}$, then a *horn on f* is a map constructed by
  (a) choosing a cosimplicial resolution $\widetilde{f}\colon \widetilde{\boldsymbol{A}} \to \widetilde{\boldsymbol{B}}$ (see Definition 16.1.20) of f such that $\widetilde{f}$ is a Reedy cofibration,
  (b) choosing an integer $n \geq 0$, and then
  (c) constructing the map $\widetilde{\boldsymbol{A}} \otimes \Delta[n] \amalg_{\widetilde{\boldsymbol{A}} \otimes \partial\Delta[n]} \widetilde{\boldsymbol{B}} \otimes \partial\Delta[n] \to \widetilde{\boldsymbol{B}} \otimes \Delta[n]$.

  If $\mathcal{C}$ is a class of maps in $\mathcal{M}$ then a *horn on $\mathcal{C}$* is a horn on some element of $\mathcal{C}$.
(2) If $f\colon A \to B$ is a map in $\mathcal{M}$, then a *cohorn on f* is a map constructed by
  (a) choosing a simplicial resolution $\widehat{f}\colon \widehat{\boldsymbol{A}} \to \widehat{\boldsymbol{B}}$ (see Definition 16.1.20) of f such that $\widehat{f}$ is a Reedy fibration,
  (b) choosing an integer $n \geq 0$, and then
  (c) constructing the map $\widehat{\boldsymbol{A}}^{\Delta[n]} \to \widehat{\boldsymbol{B}}^{\Delta[n]} \times_{\widehat{\boldsymbol{B}}^{\partial\Delta[n]}} \widehat{\boldsymbol{A}}^{\partial\Delta[n]}$.

  If $\mathcal{C}$ is a class of maps in $\mathcal{M}$ then a *cohorn on $\mathcal{C}$* is a cohorn on some element of $\mathcal{C}$.

PROPOSITION 3.3.9. *If $\mathcal{M}$ is a model category and $\mathcal{C}$ is a class of maps in $\mathcal{M}$, then*

(1) *every horn on $\mathcal{C}$ is a cofibration, and*
(2) *every cohorn on $\mathcal{C}$ is a fibration.*

PROOF. This follows from Proposition 16.3.10. □

PROPOSITION 3.3.10. *Let $\mathcal{M}$ be a model category and let $\mathcal{C}$ be a class of maps in $\mathcal{M}$. If every element of $\mathcal{C}$ is a weak equivalence, then*

(1) *every horn on $\mathcal{C}$ is a trivial cofibration, and*
(2) *every cohorn on $\mathcal{C}$ is a trivial fibration.*

PROOF. This follows from Proposition 16.1.24 and Proposition 16.3.10. □

LEMMA 3.3.11. *Let $\mathcal{M}$ be a model category and let $\mathcal{C}$ be a class of maps in $\mathcal{M}$.*

(1) *An object W of $\mathcal{M}$ is $\mathcal{C}$-local if and only if W is fibrant and the map $W \to *$ (where $*$ is the terminal object of $\mathcal{M}$) has the right lifting property with respect to every horn on $\mathcal{C}$ (see Definition 3.3.8).*
(2) *An object W of $\mathcal{M}$ is $\mathcal{C}$-colocal if and only if W is cofibrant and the map $\emptyset \to W$ (where $\emptyset$ is the initial object of $\mathcal{M}$) has the left lifting property with respect to every cohorn on $\mathcal{C}$ (see Definition 3.3.8).*

PROOF. This follows from Proposition 17.8.5 and Proposition 17.8.8. □

LEMMA 3.3.12. *Let $\mathcal{M}$ and $\mathcal{N}$ be model categories and let $\mathrm{F}\colon \mathcal{M} \rightleftarrows \mathcal{N} :\mathrm{U}$ be a Quillen pair.*

(1) *If $g\colon A \to B$ is a map of cofibrant objects in $\mathcal{M}$ and $h\colon C \to D$ is a horn on g (see Definition 3.3.8), then $\mathrm{F}(h)$ is a horn on $\mathrm{F}(g)$.*
(2) *If $p\colon X \to Y$ is a map of fibrant objects in $\mathcal{N}$ and $q\colon W \to Z$ is a cohorn on p (see Definition 3.3.8), then $\mathrm{U}(q)$ is a cohorn on $\mathrm{U}(p)$.*

PROOF. Since the left adjoint F commutes with colimits and the right adjoint U commutes with limits, this follows from Corollary 16.2.2. □

LEMMA 3.3.13. *Let $\mathcal{M}$ be a model category and let $\mathcal{C}$ be a class of maps in $\mathcal{M}$.*

(1) *If $\mathrm{L}_{\mathcal{C}}\mathcal{M}$ is the left Bousfield localization of $\mathcal{M}$ with respect to $\mathcal{C}$, $f\colon A \to B$ is a map in $\mathcal{M}$, and $\tilde{f}\colon \widetilde{\boldsymbol{A}} \to \widetilde{\boldsymbol{B}}$ is a cosimplicial resolution of f over $\mathcal{M}$ such that $\tilde{f}$ is a Reedy cofibration, then $\tilde{f}$ is also a cosimplicial resolution of f over $\mathrm{L}_{\mathcal{C}}\mathcal{M}$ such that $\tilde{f}$ is a Reedy cofibration.*
(2) *If $\mathrm{R}_{\mathcal{C}}\mathcal{M}$ is the right Bousfield localization of $\mathcal{M}$ with respect to $\mathcal{C}$, $f\colon A \to B$ is a map in $\mathcal{M}$, and $\hat{f}\colon \widehat{\boldsymbol{A}} \to \widehat{\boldsymbol{B}}$ is a simplicial resolution of f over $\mathcal{M}$ such that $\hat{f}$ is a Reedy fibration, then $\hat{f}$ is also a simplicial resolution of f over $\mathrm{R}_{\mathcal{C}}\mathcal{M}$ such that $\hat{f}$ is a Reedy fibration.*

PROOF. Part 1 follows because every cofibration of $\mathcal{M}$ is a cofibration of $\mathrm{L}_{\mathcal{C}}\mathcal{M}$ and every weak equivalence of $\mathcal{M}$ is a weak equivalence of $\mathrm{L}_{\mathcal{C}}\mathcal{M}$. Part 2 follows because every fibration of $\mathcal{M}$ is a fibration of $\mathrm{R}_{\mathcal{C}}\mathcal{M}$ and every weak equivalence of $\mathcal{M}$ is a weak equivalence of $\mathrm{R}_{\mathcal{C}}\mathcal{M}$. □

PROPOSITION 3.3.14. *Let $\mathcal{M}$ be a model category and let $\mathcal{C}$ be a class of maps in $\mathcal{M}$.*

(1) *If $\mathrm{L}_{\mathcal{C}}\mathcal{M}$ is the left Bousfield localization of $\mathcal{M}$ with respect to $\mathcal{C}$, Y is $\mathcal{C}$-local, and there is a map $g\colon X \to Y$ that is a fibration in $\mathrm{L}_{\mathcal{C}}\mathcal{M}$, then X is $\mathcal{C}$-local.*
(2) *If $\mathrm{R}_{\mathcal{C}}\mathcal{M}$ is the right Bousfield localization of $\mathcal{M}$ with respect to $\mathcal{C}$, A is $\mathcal{C}$-colocal, and there is a map $g\colon A \to B$ that is a cofibration in $\mathrm{R}_{\mathcal{C}}\mathcal{M}$, then B is $\mathcal{C}$-colocal.*

PROOF. We will prove part 1; the proof of part 2 is dual.

Since fibrations in $\mathrm{L}_{\mathcal{C}}\mathcal{M}$ are fibrations in $\mathcal{M}$, the composition $X \to Y \to *$ is a fibration in $\mathcal{M}$, and so Lemma 3.3.11 implies that it is sufficient to show that the map $X \to *$ has the right lifting property with respect to every horn on $\mathcal{C}$. If $\alpha\colon A \to B$ is a horn on $\mathcal{C}$ and $s\colon A \to X$ is a map, then we have the solid arrow

diagram

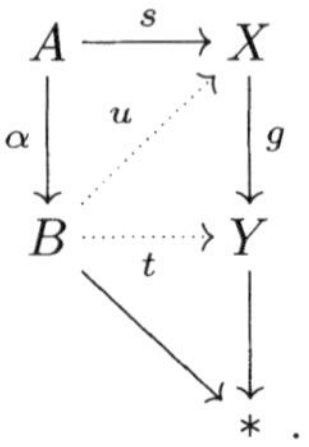

Since Y is $\mathcal{C}$-local, Lemma 3.3.11 implies that there is a map $t\colon B \to Y$ such that $t\alpha = gs$. Proposition 3.3.10 and Lemma 3.3.13 imply that α is a trivial cofibration in $L_{\mathcal{C}}\mathcal{M}$, and so there is a map $u\colon B \to X$ such that $u\alpha = s$ and $gu = t$. □

PROPOSITION 3.3.15. *Let $\mathcal{M}$ be a model category and let $\mathcal{C}$ be a class of maps in $\mathcal{M}$.*

(1) *If $L_{\mathcal{C}}\mathcal{M}$ is the left Bousfield localization of $\mathcal{M}$ with respect to $\mathcal{C}$, $f\colon X \to Z$ is a fibration in $\mathcal{M}$, $g\colon Y \to Z$ is a fibration in $L_{\mathcal{C}}\mathcal{M}$, and $h\colon X \to Y$ is a weak equivalence in $\mathcal{M}$ that makes the triangle*

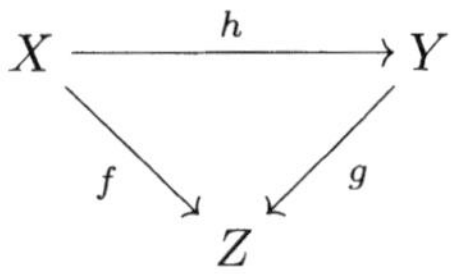

commute, then f is also a fibration in $L_{\mathcal{C}}\mathcal{M}$.

(2) *If $R_{\mathcal{C}}\mathcal{M}$ is the right Bousfield localization of $\mathcal{M}$ with respect to $\mathcal{C}$, $f\colon A \to B$ is a cofibration in $R_{\mathcal{C}}\mathcal{M}$, $g\colon A \to C$ is a cofibration in $\mathcal{M}$, and $h\colon B \to C$ is a weak equivalence in $\mathcal{M}$ that makes the triangle*

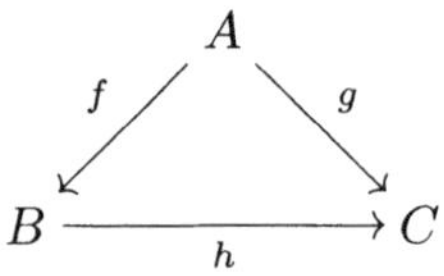

commute, then g is also a cofibration in $R_{\mathcal{C}}\mathcal{M}$.

PROOF. We will prove part 1; the proof of part 2 is dual.

Proposition 7.2.3 implies that it is sufficient to show that if $i\colon A \to B$ is both a cofibration and a $\mathcal{C}$-local equivalence then f has the right lifting property with respect to i. If we have the solid arrow diagram

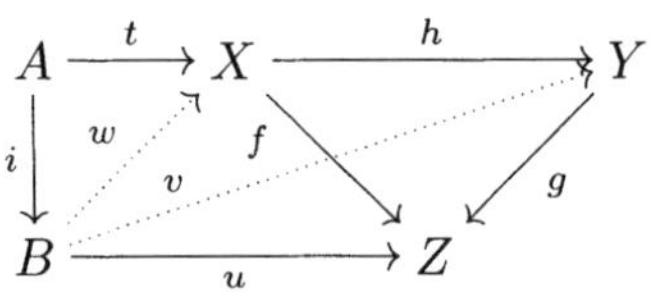

then, since g is a fibration in $L_{\mathcal{C}}\mathcal{M}$, there is a map $v\colon B \to Y$ such that $vi = ht$ and $gv = u$. Thus, in the category $(A \downarrow \mathcal{M} \downarrow Z)$ of objects of $\mathcal{M}$ under A and over Z (see Theorem 7.6.5), there is a map from B to Y. Since B is cofibrant in $(A \downarrow \mathcal{M} \downarrow Z)$ and $h\colon X \to Y$ is a weak equivalence of fibrant objects in $(A \downarrow \mathcal{M} \downarrow Z)$, Corollary 7.7.5 implies that there is also a map $w\colon B \to X$ in $(A \downarrow \mathcal{M} \downarrow Z)$, i.e., a map $w\colon B \to X$ in $\mathcal{M}$ such that $wi = t$ and $fw = u$. Thus, f is a fibration in $L_{\mathcal{C}}\mathcal{M}$. □

PROPOSITION 3.3.16. *Let $\mathcal{M}$ be a model category and let $\mathcal{C}$ be a class of maps in $\mathcal{M}$.*

(1) *If $\mathrm{L}_{\mathcal{C}}\mathcal{M}$ is the left Bousfield localization of $\mathcal{M}$ with respect to $\mathcal{C}$ and X and Y are $\mathcal{C}$-local objects in $\mathcal{M}$ then a map $f\colon X \to Y$ is a fibration in $\mathrm{L}_{\mathcal{C}}\mathcal{M}$ if and only if it is a fibration in $\mathcal{M}$.*
(2) *If $\mathrm{R}_{\mathcal{C}}\mathcal{M}$ is the right Bousfield localization of $\mathcal{M}$ with respect to $\mathcal{C}$ and X and Y are $\mathcal{C}$-colocal objects in $\mathcal{M}$ then a map $f\colon X \to Y$ is a cofibration in $\mathrm{R}_{\mathcal{C}}\mathcal{M}$ if and only if it is a cofibration in $\mathcal{M}$.*

PROOF. We will prove part 1; the proof of part 2 is dual.

Since every fibration in $\mathrm{L}_{\mathcal{C}}\mathcal{M}$ is a fibration in $\mathcal{M}$, we assume that f is a fibration in $\mathcal{M}$ and we will show that it is a fibration in $\mathrm{L}_{\mathcal{C}}\mathcal{M}$. If we factor f as $X \xrightarrow{i} W \xrightarrow{p} Y$ where i is a trivial cofibration in $\mathrm{L}_{\mathcal{C}}\mathcal{M}$ and p is a fibration in $\mathrm{L}_{\mathcal{C}}\mathcal{M}$, then we have the diagram

$$\begin{array}{ccc} X & \xrightarrow{\ i\ } & W \\ & {\scriptstyle f}\searrow \quad \swarrow{\scriptstyle p} & \\ & Y\ . & \end{array}$$

Proposition 3.3.14 implies that W is $\mathcal{C}$-local, and so the weak $\mathcal{C}$-local Whitehead theorem (see Theorem 4.1.10) implies that i is a weak equivalence. The result now follows from Proposition 3.3.15. □

3.3.17. Bousfield localization is a localization.

PROPOSITION 3.3.18. *Let $\mathcal{M}$ be a model category and let $\mathcal{C}$ be a class of maps in $\mathcal{M}$.*

(1) *If $\mathrm{L}_{\mathcal{C}}\mathcal{M}$ is the left Bousfield localization of $\mathcal{M}$ with respect to $\mathcal{C}$, $\mathcal{N}$ is a model category, and $\mathrm{F}\colon \mathcal{M} \to \mathcal{N}$ is a left Quillen functor that takes every cofibrant approximation to an element of $\mathcal{C}$ into a weak equivalence in $\mathcal{N}$, then F is a left Quillen functor when considered as a functor $\mathrm{L}_{\mathcal{C}}\mathcal{M} \to \mathcal{N}$.*
(2) *If $\mathrm{R}_{\mathcal{C}}\mathcal{M}$ is the right Bousfield localization of $\mathcal{M}$ with respect to $\mathcal{C}$, $\mathcal{N}$ is a model category, and $\mathrm{U}\colon \mathcal{M} \to \mathcal{N}$ is a right Quillen functor that takes every fibrant approximation to an element of $\mathcal{C}$ into a weak equivalence in $\mathcal{N}$, then U is a right Quillen functor when considered as a functor $\mathrm{R}_{\mathcal{C}}\mathcal{M} \to \mathcal{N}$.*

PROOF. We will prove part 1; the proof of part 2 is dual.

Since the underlying category of $\mathrm{L}_{\mathcal{C}}\mathcal{M}$ equals that of $\mathcal{M}$, F has a right adjoint whether we consider it to be a functor $\mathrm{F}\colon \mathcal{M} \to \mathcal{N}$ or a functor $\mathrm{F}\colon \mathrm{L}_{\mathcal{C}}\mathcal{M} \to \mathcal{N}$. Let $\mathrm{U}\colon \mathcal{N} \to \mathrm{L}_{\mathcal{C}}\mathcal{M}$ be a right adjoint to F. Proposition 8.5.4 implies that it is sufficient to show that U preserves fibrations between fibrant objects and all trivial fibrations. Since the class of trivial fibrations of $\mathrm{L}_{\mathcal{C}}\mathcal{M}$ equals the class of trivial fibrations of $\mathcal{M}$ and U is a right Quillen functor when viewed as a functor $\mathrm{U}\colon \mathcal{N} \to \mathcal{M}$, U preserves all trivial fibrations when viewed as a functor $\mathrm{U}\colon \mathcal{N} \to \mathrm{L}_{\mathcal{C}}\mathcal{M}$.

If X and Y are fibrant objects of $\mathcal{N}$ and $p\colon X \to Y$ is a fibration in $\mathcal{N}$, then Theorem 3.1.6 implies that $\mathrm{U}X$ and $\mathrm{U}Y$ are $\mathcal{C}$-local objects of $\mathcal{M}$. Since $\mathrm{U}p\colon \mathrm{U}X \to \mathrm{U}Y$ is a fibration in $\mathcal{M}$, Proposition 3.3.16 implies that it is also a fibration in $\mathrm{L}_{\mathcal{C}}\mathcal{M}$. □

THEOREM 3.3.19. *Let $\mathcal{M}$ be a model category and let $\mathcal{C}$ be a class of maps in $\mathcal{M}$.*

(1) *If* $\mathrm{L}_{\mathcal{C}}\mathcal{M}$ *is the left Bousfield localization of* $\mathcal{M}$ *with respect to* $\mathcal{C}$ *then the identity functor* $\mathcal{M} \to \mathrm{L}_{\mathcal{C}}\mathcal{M}$ *is a left localization of* $\mathcal{M}$ *with respect to* $\mathcal{C}$ *(see Definition* 3.1.1*).*
(2) *If* $\mathrm{R}_{\mathcal{C}}\mathcal{M}$ *is the right Bousfield localization of* $\mathcal{M}$ *with respect to* $\mathcal{C}$ *then the identity functor* $\mathcal{M} \to \mathrm{R}_{\mathcal{C}}\mathcal{M}$ *is a right localization of* $\mathcal{M}$ *with respect to* $\mathcal{C}$.

PROOF. We will prove part 1; the proof of part 2 is dual.

Let $\mathrm{L}_{\mathcal{C}}\mathcal{M}$ be the left Bousfield localization of $\mathcal{M}$ with respect to $\mathcal{C}$, let $j\colon \mathcal{M} \to \mathrm{L}_{\mathcal{C}}\mathcal{M}$ be the identity functor, and let $\mathrm{F}\colon \mathcal{M} \rightleftarrows \mathcal{N} :\mathrm{U}$ be a Quillen pair such that the total left derived functor $\mathbf{L}\mathrm{F}\colon \mathrm{Ho}\,\mathcal{M} \to \mathrm{Ho}\,\mathcal{N}$ takes the images in $\mathrm{Ho}\,\mathcal{M}$ of the elements of $\mathcal{C}$ into isomorphisms in $\mathrm{Ho}\,\mathcal{N}$. Since j is the identity functor, the functor $\mathrm{F}\colon \mathrm{L}_{\mathcal{C}}\mathcal{M} \to \mathcal{N}$ is the unique functor such that $\mathrm{F} \circ j = \mathrm{F}$, and Proposition 3.3.18 shows that $\mathrm{F}\colon \mathrm{L}_{\mathcal{C}}\mathcal{M} \to \mathcal{N}$ is a left Quillen functor. □

THEOREM 3.3.20. *Let* $\mathcal{M}$ *and* $\mathcal{N}$ *be model categories and let* $\mathrm{F}\colon \mathcal{M} \rightleftarrows \mathcal{N} :\mathrm{U}$ *be a Quillen pair.*

(1) *If* $\mathcal{C}$ *is a class of maps in* $\mathcal{M}$, $\mathrm{L}_{\mathcal{C}}\mathcal{M}$ *is the left Bousfield localization of* $\mathcal{M}$ *with respect to* $\mathcal{C}$, *and* $\mathrm{L}_{\mathbf{L}\mathrm{F}\mathcal{C}}\mathcal{N}$ *is the left Bousfield localization of* $\mathcal{N}$ *with respect to* $\mathbf{L}\mathrm{F}\mathcal{C}$ *(see Definition* 8.5.11*), then*
   (a) (F,U) *is also a Quillen pair when considered as functors* $\mathrm{F}\colon \mathrm{L}_{\mathcal{C}}\mathcal{M} \rightleftarrows \mathrm{L}_{\mathbf{L}\mathrm{F}\mathcal{C}}\mathcal{N} :\mathrm{U}$ *between the localizations of* $\mathcal{M}$ *and* $\mathcal{N}$, *and*
   (b) *if* (F,U) *is a pair of Quillen equivalences when considered as functors* $\mathrm{F}\colon \mathcal{M} \rightleftarrows \mathcal{N} :\mathrm{U}$, *then* (F,U) *is also a pair of Quillen equivalences when considered as functors* $\mathrm{F}\colon \mathrm{L}_{\mathcal{C}}\mathcal{M} \rightleftarrows \mathrm{L}_{\mathbf{L}\mathrm{F}\mathcal{C}}\mathcal{N} :\mathrm{U}$ *between the localizations of* $\mathcal{M}$ *and* $\mathcal{N}$.
(2) *If* $\mathcal{C}$ *is a class of maps in* $\mathcal{N}$, $\mathrm{R}_{\mathcal{C}}\mathcal{N}$ *is the right Bousfield localization of* $\mathcal{N}$ *with respect to* $\mathcal{C}$, *and* $\mathrm{R}_{\mathbf{R}\mathrm{U}\mathcal{C}}\mathcal{M}$ *is the right Bousfield localization of* $\mathcal{M}$ *with respect to* $\mathbf{R}\mathrm{U}\mathcal{C}$ *(see Definition* 8.5.11*), then*
   (a) (F,U) *is also a Quillen pair when considered as functors* $\mathrm{F} : \mathrm{R}_{\mathbf{R}\mathrm{U}\mathcal{C}}\mathcal{M} \rightleftarrows \mathrm{R}_{\mathcal{C}}\mathcal{N} :\mathrm{U}$ *between the localizations of* $\mathcal{M}$ *and* $\mathcal{N}$, *and*
   (b) *if* (F,U) *is a pair of Quillen equivalences when considered as functors* $\mathrm{F}\colon \mathcal{M} \rightleftarrows \mathcal{N} :\mathrm{U}$, *then* (F,U) *is also a pair of Quillen equivalences when considered as functors* $\mathrm{F}\colon \mathrm{R}_{\mathbf{R}\mathrm{U}\mathcal{C}}\mathcal{M} \rightleftarrows \mathrm{R}_{\mathcal{C}}\mathcal{N} :\mathrm{U}$ *between the localizations of* $\mathcal{M}$ *and* $\mathcal{N}$.

PROOF. We will prove part 1; the proof of part 2 is dual.

Proposition 3.3.18 implies that the composition $\mathcal{M} \xrightarrow{\mathrm{F}} \mathcal{N} \xrightarrow{1_{\mathcal{N}}} \mathrm{L}_{\mathbf{L}\mathrm{F}\mathcal{C}}\mathcal{N}$ is a left Quillen functor when considered as a functor $\mathrm{L}_{\mathcal{C}}\mathcal{M} \to \mathrm{L}_{\mathbf{L}\mathrm{F}\mathcal{C}}\mathcal{N}$, which proves part 1a.

For part 1b, we must show that if X is cofibrant in $\mathrm{L}_{\mathcal{C}}\mathcal{M}$ and Y is fibrant in $\mathrm{L}_{\mathbf{L}\mathrm{F}\mathcal{C}}\mathcal{N}$ then a map $g\colon X \to \mathrm{U}Y$ in $\mathrm{L}_{\mathcal{C}}\mathcal{M}$ is a $\mathcal{C}$-local equivalence if and only if the corresponding map $g^{\sharp}\colon \mathrm{F}X \to Y$ in $\mathrm{L}_{\mathbf{L}\mathrm{F}\mathcal{C}}\mathcal{N}$ is an $\mathbf{L}\mathrm{F}\mathcal{C}$-local equivalence. Given such a map g, we factor it in $\mathcal{M}$ as $X \xrightarrow{h} \widetilde{Y} \xrightarrow{k} \mathrm{U}Y$ where h is a cofibration in $\mathcal{M}$ and k is a trivial fibration in $\mathcal{M}$. Both X and $\widetilde{Y}$ are cofibrant, and since k is a weak equivalence in $\mathcal{M}$, g is a $\mathcal{C}$-local equivalence if and only if h is a $\mathcal{C}$-local equivalence. The corresponding factorization of $g^{\sharp}$ in $\mathcal{N}$ is $\mathrm{F}X \xrightarrow{\mathrm{F}h} \mathrm{F}\widetilde{Y} \xrightarrow{k^{\sharp}} Y$, and since (F,U) is a pair of Quillen equivalences between $\mathcal{M}$ and $\mathcal{N}$, the map $k^{\sharp}$ is a weak equivalence in $\mathcal{N}$. Thus, both $\mathrm{F}X$ and $\mathrm{F}\widetilde{Y}$ are cofibrant, and $g^{\sharp}$ is an $\mathbf{L}\mathrm{F}\mathcal{C}$-local equivalence if and only if $\mathrm{F}h$ is an $\mathbf{L}\mathrm{F}\mathcal{C}$-local equivalence. It remains only to show that h is a $\mathcal{C}$-local equivalence if and only if $\mathrm{F}h$ is an $\mathbf{L}\mathrm{F}\mathcal{C}$-local equivalence.

The map $\mathrm{F}h$ is an $\mathbf{L}\mathrm{F}\mathcal{C}$-local equivalence if and only if for every $\mathbf{L}\mathrm{F}\mathcal{C}$-local object W in $\mathcal{N}$ and every simplicial resolution $\widehat{\boldsymbol{W}}$ of W in $\mathcal{N}$ the map of simplicial sets $\mathcal{N}(\mathrm{F}\widetilde{Y}, \widehat{\boldsymbol{W}}) \to \mathcal{N}(\mathrm{F}X, \widehat{\boldsymbol{W}})$ is a weak equivalence. This map of simplicial sets is isomorphic to the map $\mathcal{M}(\widetilde{Y}, \mathrm{U}\widehat{\boldsymbol{W}}) \to \mathcal{M}(X, \mathrm{U}\widehat{\boldsymbol{W}})$, and so Theorem 17.6.3 implies that it is sufficient to show that every $\mathcal{C}$-local object Z of $\mathcal{M}$ is weakly equivalent in $\mathcal{M}$ to an object of the form $\mathrm{U}W$ for some $\mathbf{L}\mathrm{F}\mathcal{C}$-local object W of $\mathcal{N}$. Thus, Proposition 3.1.12 and Lemma 3.2.1 imply that it is sufficient to show that every $\mathcal{C}$-local object Z of $\mathcal{M}$ is weakly equivalent to an object of the form $\mathrm{U}W$ for some fibrant object W of $\mathcal{N}$. Given such an object Z, we can choose a trivial fibration $\widetilde{Z} \to Z$ in $\mathcal{M}$ with $\widetilde{Z}$ cofibrant in $\mathcal{M}$ and then choose a trivial cofibration $\mathrm{F}\widetilde{X} \to W$ in $\mathcal{N}$ with W fibrant in $\mathcal{N}$. Since $\widetilde{Z}$ is cofibrant in $\mathcal{M}$, W is fibrant in $\mathcal{N}$, and $\mathrm{F}: \mathcal{M} \rightleftarrows \mathcal{N} : \mathrm{U}$ is a pair of Quillen equivalences, the map $\widetilde{Z} \to \mathrm{U}W$ is a weak equivalence in $\mathcal{M}$, and so we have the zig-zag of weak equivalences $Z \leftarrow \widetilde{Z} \to \mathrm{U}W$. □

3.4. Bousfield localization and properness

PROPOSITION 3.4.1. *Let $\mathcal{M}$ be a model category and let $\mathcal{C}$ be a class of maps in $\mathcal{M}$.*

(1) *If $\mathcal{M}$ is left proper and $\mathrm{L}_{\mathcal{C}}\mathcal{M}$ is the left Bousfield localization of $\mathcal{M}$ with respect to $\mathcal{C}$, then an object W of $\mathcal{M}$ is $\mathcal{C}$-local if and only if it is a fibrant object in $\mathrm{L}_{\mathcal{C}}\mathcal{M}$.*
(2) *If $\mathcal{M}$ is right proper and $\mathrm{R}_{\mathcal{C}}\mathcal{M}$ is the right Bousfield localization of $\mathcal{M}$ with respect to $\mathcal{C}$, then an object W of $\mathcal{M}$ is $\mathcal{C}$-colocal if and only if it is a cofibrant object in $\mathrm{R}_{\mathcal{C}}\mathcal{M}$.*

PROOF. We will prove part 1; the proof of part 2 is dual.

If W is a fibrant object of $\mathrm{L}_{\mathcal{C}}\mathcal{M}$ then Theorem 3.1.6 applied to the Quillen pair $1_{\mathcal{M}}: \mathcal{M} \rightleftarrows \mathrm{L}_{\mathcal{C}}\mathcal{M} : 1_{\mathrm{L}_{\mathcal{C}}\mathcal{M}}$ implies that W is $\mathcal{C}$-local.

Conversely, assume that W is $\mathcal{C}$-local. Proposition 7.2.3 implies that it is sufficient to show that if $i: A \to B$ is both a cofibration and an $\mathcal{C}$-local equivalence then the map $W \to *$ has the right lifting property with respect to i. Proposition 16.1.22 implies that we can choose a cosimplicial resolution $\tilde{\imath}: \widetilde{\boldsymbol{A}} \to \widetilde{\boldsymbol{B}}$ of i such that $\tilde{\imath}$ is a Reedy cofibration, and Proposition 17.8.5 and Proposition 17.8.9 imply that the map $W \to *$ has the right lifting property with respect to $\tilde{\imath}^0: \widetilde{\boldsymbol{A}}^0 \to \widetilde{\boldsymbol{B}}^0$. Since $\mathcal{M}$ is left proper, Proposition 13.2.1 and Proposition 16.1.5 now imply that the map $W \to *$ has the right lifting property with respect to i. □

LEMMA 3.4.2. *Let $\mathcal{M}$ be a model category and let $\mathcal{C}$ be a class of maps in $\mathcal{M}$.*

(1) *If $\mathcal{M}$ is left proper and $\mathrm{L}_{\mathcal{C}}\mathcal{M}$ is the left Bousfield localization of $\mathcal{M}$ with respect to $\mathcal{C}$, $g: A \to B$ is a weak equivalence in $\mathrm{L}_{\mathcal{C}}\mathcal{M}$, $h: A \to X$ is a map, at least one of g and h is a cofibration, and the square*

(3.4.3)
$$\begin{array}{ccc} A & \xrightarrow{h} & X \\ {\scriptstyle g}\downarrow & & \downarrow{\scriptstyle k} \\ B & \xrightarrow[j]{} & Y \end{array}$$

is a pushout, then k is a weak equivalence in $\mathrm{L}_{\mathcal{C}}\mathcal{M}$.
(2) *If $\mathcal{M}$ is right proper and $\mathrm{R}_{\mathcal{C}}\mathcal{M}$ is the right Bousfield localization of $\mathcal{M}$ with respect to $\mathcal{C}$, $k: X \to Y$ is a weak equivalence in $\mathrm{R}_{\mathcal{C}}\mathcal{M}$, $j: B \to Y$ is a*

map, at least one of j and k is a fibration, and the square (3.4.3) is a pullback, then g is a weak equivalence in $\mathrm{R}_{\mathcal{C}}\mathcal{M}$.

PROOF. This follows from Proposition 17.8.5 and Proposition 17.8.16. $\square$

PROPOSITION 3.4.4. *Let $\mathcal{M}$ be a model category and let $\mathcal{C}$ be a class of maps in $\mathcal{M}$.*

(1) *If $\mathcal{M}$ is left proper and $\mathrm{L}_{\mathcal{C}}\mathcal{M}$ is the left Bousfield localization of $\mathcal{M}$ with respect to $\mathcal{C}$, then $\mathrm{L}_{\mathcal{C}}\mathcal{M}$ is left proper.*
(2) *If $\mathcal{M}$ is right proper and $\mathrm{R}_{\mathcal{C}}\mathcal{M}$ is the right Bousfield localization of $\mathcal{M}$ with respect to $\mathcal{C}$, then $\mathrm{R}_{\mathcal{C}}\mathcal{M}$ is right proper.*

PROOF. This follows from Lemma 3.4.2. $\square$

3.4.5. Fibrations in $\mathrm{L}_{\mathcal{C}}\mathcal{M}$ and cofibrations in $\mathrm{R}_{\mathcal{C}}\mathcal{M}$.

PROPOSITION 3.4.6. *Let $\mathcal{M}$ be a model category and let $\mathcal{C}$ be a class of maps in $\mathcal{M}$.*

(1) *If $\mathcal{M}$ is left proper, $\mathrm{L}_{\mathcal{C}}\mathcal{M}$ is the left Bousfield localization of $\mathcal{M}$ with respect to $\mathcal{C}$, $f\colon X \to Z$ and $g\colon Y \to Z$ are fibrations in $\mathcal{M}$, and $h\colon X \to Y$ is a weak equivalence in $\mathcal{M}$ that makes the triangle*

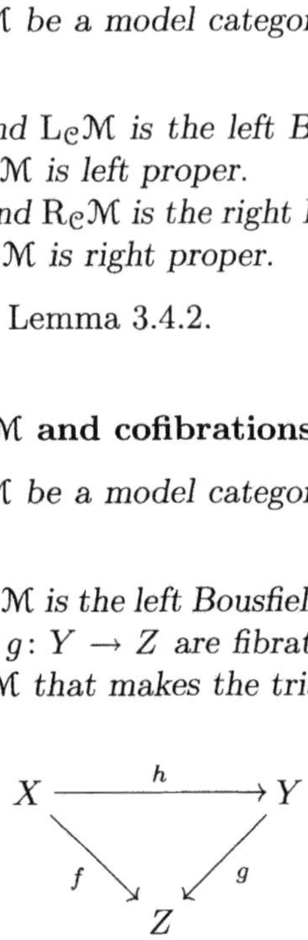

commute, then f is a fibration in $\mathrm{L}_{\mathcal{C}}\mathcal{M}$ if and only if g is a fibration in $\mathrm{L}_{\mathcal{C}}\mathcal{M}$.
(2) *If $\mathcal{M}$ is right proper, $\mathrm{R}_{\mathcal{C}}\mathcal{M}$ is the right Bousfield localization of $\mathcal{M}$ with respect to $\mathcal{C}$, $f\colon A \to B$ and $g\colon A \to C$ are cofibrations in $\mathcal{M}$, and $h\colon B \to C$ is a weak equivalence in $\mathcal{M}$ that makes the triangle*

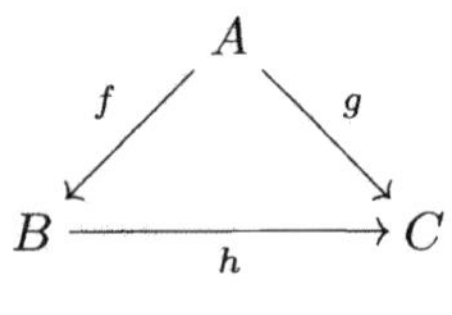

commute, then f is a cofibration in $\mathrm{R}_{\mathcal{C}}\mathcal{M}$ if and only if g is a cofibration in $\mathrm{R}_{\mathcal{C}}\mathcal{M}$.

PROOF. We will prove part 1; the proof of part 2 is dual.

If g is a fibration in $\mathrm{L}_{\mathcal{C}}\mathcal{M}$ then Proposition 3.3.15 implies that f is also a fibration in $\mathrm{L}_{\mathcal{C}}\mathcal{M}$.

Conversely, assume that f is a fibration in $\mathrm{L}_{\mathcal{C}}\mathcal{M}$. Proposition 7.2.3 and Proposition 13.2.1 imply that it is sufficient to show that if $i\colon A \to B$ is a trivial cofibration in $\mathrm{L}_{\mathcal{C}}\mathcal{M}$ and A is cofibrant, then g has the right lifting property with respect to i.

Suppose that we have the solid arrow diagram

$$\begin{array}{ccccc} & & \overset{t}{\frown} & & \\ A & \overset{v}{\dashrightarrow} & X & \xrightarrow{h} & Y \\ {\scriptstyle i}\downarrow & \overset{w}{\nearrow} & & \searrow^{f} \quad {}^{g}\swarrow & \\ B & & \xrightarrow[u]{} & & Z\ . \end{array}$$

Since A is cofibrant in $(\mathcal{M}\downarrow Z)$ and $h\colon X \to Y$ is a weak equivalence of fibrant objects in $(\mathcal{M}\downarrow Z)$, Corollary 8.5.4 implies that there is a map $v\colon A \to X$ in $(\mathcal{M}\downarrow Y)$ such that $hv \simeq t$ in $(\mathcal{M}\downarrow Z)$. Thus, $fv = ui$ in $\mathcal{M}$ and, since f is a fibration in $L_{\mathcal{C}}\mathcal{M}$, there is a map $w\colon B \to X$ in $\mathcal{M}$ such that $wi = v$ and $fw = u$. Since $hwi = hv \simeq t$ in $(\mathcal{M}\downarrow Z)$, $i\colon A \to B$ is a cofibration in $(\mathcal{M}\downarrow Z)$, and Y is fibrant in $(\mathcal{M}\downarrow Z)$, Proposition 8.3.7 implies that we can find a map $s\colon B \to Y$ in $(\mathcal{M}\downarrow Z)$ such that $s \simeq hw$ and $si = t$. Thus, $si = t$ and $gs = u$, and so s is the map we require, and so g is a fibration in $L_{\mathcal{C}}\mathcal{M}$. □

PROPOSITION 3.4.7. *Let $\mathcal{M}$ be a model category and let $\mathcal{C}$ be a class of maps in $\mathcal{M}$.*

(1) *If $\mathcal{M}$ is right proper, $L_{\mathcal{C}}\mathcal{M}$ is the left Bousfield localization of $\mathcal{M}$ with respect to $\mathcal{C}$, $f\colon X \to Y$ is a fibration in $\mathcal{M}$, and there is a homotopy fiber square (see Definition 13.3.12) in $\mathcal{M}$*

$$\begin{array}{ccc} X & \xrightarrow{j_X} & \widehat{X} \\ {\scriptstyle f}\downarrow & & \downarrow{\scriptstyle \hat{f}} \\ Y & \xrightarrow[j_Y]{} & \widehat{Y} \end{array}$$

in which $\widehat{X}$ and $\widehat{Y}$ are $\mathcal{C}$-local and j_X and j_Y are $\mathcal{C}$-local equivalences, then f is a fibration in $L_{\mathcal{C}}\mathcal{M}$.

(2) *If $\mathcal{M}$ is left proper, $R_{\mathcal{C}}\mathcal{M}$ is the right Bousfield localization of $\mathcal{M}$ with respect to $\mathcal{C}$, $f\colon A \to B$ is a cofibration in $\mathcal{M}$, and there is a homotopy cofiber square in $\mathcal{M}$*

$$\begin{array}{ccc} \widetilde{A} & \xrightarrow{i_A} & A \\ {\scriptstyle \tilde{f}}\downarrow & & \downarrow{\scriptstyle f} \\ \widetilde{B} & \xrightarrow[i_B]{} & B \end{array}$$

in which $\widetilde{A}$ and $\widetilde{B}$ are $\mathcal{C}$-colocal and i_A and i_B are $\mathcal{C}$-colocal equivalences, then f is a cofibration in $R_{\mathcal{C}}\mathcal{M}$.

PROOF. We will prove part 1; the proof of part 2 is dual.

If we factor $\hat{f}$ as $\widehat{X} \xrightarrow{i} W \xrightarrow{p} \widehat{Y}$ where i is an trivial cofibration in $L_{\mathcal{C}}\mathcal{M}$ and p is a fibration in $L_{\mathcal{C}}\mathcal{M}$, then p is a fibration in $\mathcal{M}$ and so Proposition 13.3.7 implies that the natural map $X \to Y \times_{\widehat{Y}} W$ is a weak equivalence. The result now follows from Proposition 3.3.15. □

PROPOSITION 3.4.8. *Let $\mathcal{M}$ be a model category and let $\mathcal{C}$ be a class of maps in $\mathcal{M}$.*

(1) *If* $L_{\mathcal{C}}\mathcal{M}$ *is the left Bousfield localization of* $\mathcal{M}$ *with respect to* $\mathcal{C}$*, both* $\mathcal{M}$ *and* $L_{\mathcal{C}}\mathcal{M}$ *are right proper,* $f\colon X \to Y$ *is a fibration in* $\mathcal{M}$*, and* $\hat{f}\colon \widehat{X} \to \widehat{Y}$ *is a* $\mathcal{C}$*-localization of* f *(see Definition 3.2.16), then* f *is a fibration in* $L_{\mathcal{C}}\mathcal{M}$ *if and only if the square*

(3.4.9)
$$\begin{array}{ccc} X & \xrightarrow{j_X} & \widehat{X} \\ {\scriptstyle f}\downarrow & & \downarrow{\scriptstyle \hat{f}} \\ Y & \xrightarrow[j_Y]{} & \widehat{Y} \end{array}$$

is a homotopy fiber square (see Definition 11.2.12) in $\mathcal{M}$.

(2) *If* $R_{\mathcal{C}}\mathcal{M}$ *is the right Bousfield localization of* $\mathcal{M}$ *with respect to* $\mathcal{C}$*, both* $\mathcal{M}$ *and* $R_{\mathcal{C}}\mathcal{M}$ *are left proper,* $f\colon X \to Y$ *is a cofibration in* $\mathcal{M}$*, and* $\tilde{f}\colon \widetilde{X} \to \widetilde{Y}$ *is a* $\mathcal{C}$*-colocalization of* f *(see Definition 3.2.16), then* f *is a cofibration in* $R_{\mathcal{C}}\mathcal{M}$ *if and only if the square*

$$\begin{array}{ccc} \widetilde{X} & \xrightarrow{i_X} & X \\ {\scriptstyle \tilde{f}}\downarrow & & \downarrow{\scriptstyle f} \\ \widetilde{Y} & \xrightarrow[i_Y]{} & Y \end{array}$$

is a homotopy cofiber square in $\mathcal{M}$.

PROOF. We will prove part 1; the proof of part 2 is dual.

If Diagram 3.4.9 is a homotopy fiber square then Proposition 3.4.7 implies that f is a fibration in $L_{\mathcal{C}}\mathcal{M}$.

Conversely, assume that f is a fibration in $L_{\mathcal{C}}\mathcal{M}$. If we factor $\hat{f}$ as $\widehat{X} \xrightarrow{i} W \xrightarrow{p} \widehat{Y}$ where i is a trivial cofibration in $L_{\mathcal{C}}\mathcal{M}$ and p is a fibration in $L_{\mathcal{C}}\mathcal{M}$, then we have the diagram

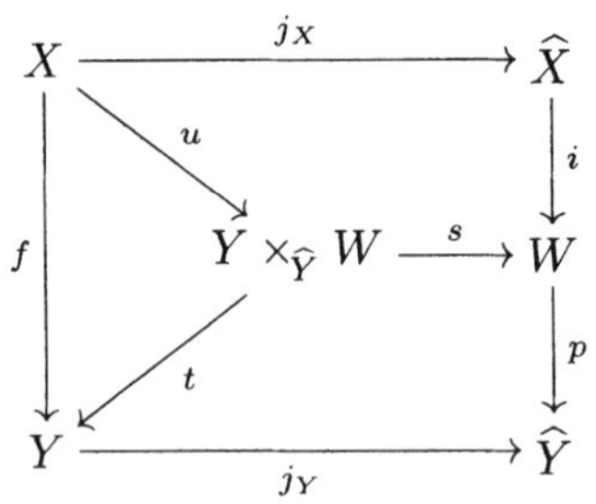

and we must show that the natural map $u\colon X \to Y \times_{\widehat{Y}} W$ is a weak equivalence in $\mathcal{M}$. Since $L_{\mathcal{C}}\mathcal{M}$ is right proper, s is a weak equivalence in $L_{\mathcal{C}}\mathcal{M}$, and so the "two out of three" property of weak equivalences implies that u is a weak equivalence in $L_{\mathcal{C}}\mathcal{M}$. Since t is a pullback of a fibration in $L_{\mathcal{C}}\mathcal{M}$, our result now follows from Proposition 3.3.5. □

3.5. Detecting equivalences

LEMMA 3.5.1. *Let* $\mathcal{M}$ *be a model category and let* $\mathcal{C}$ *be a class of maps in* $\mathcal{M}$.

(1) *If* $L_{\mathcal{C}}\mathcal{M}$ *is the left Bousfield localization of* $\mathcal{M}$ *with respect to* $\mathcal{C}$, X *is a cofibrant object of* $\mathcal{M}$, *and* Y *is a* $\mathcal{C}$*-local object of* $\mathcal{M}$, *then two maps from* X *to* Y *are homotopic in* $\mathcal{M}$ *if and only if they are homotopic in* $L_{\mathcal{C}}\mathcal{M}$.
(2) *If* $R_{\mathcal{C}}\mathcal{M}$ *is the right Bousfield localization of* $\mathcal{M}$ *with respect to* $\mathcal{C}$, A *is a* $\mathcal{C}$*-colocal object of* $\mathcal{M}$, *and* B *is a fibrant object of* $\mathcal{M}$, *then two maps from* A *to* B *are homotopic in* $\mathcal{M}$ *if and only if they are homotopic in* $R_{\mathcal{C}}\mathcal{M}$.

PROOF. This follows from Proposition 8.5.16, Proposition 3.4.1, and Proposition 3.3.4. □

LEMMA 3.5.2. *Let* $\mathcal{M}$ *be a model category and let* $\mathcal{C}$ *be a class of maps in* $\mathcal{M}$.
(1) *If* $L_{\mathcal{C}}\mathcal{M}$ *is the left Bousfield localization of* $\mathcal{M}$ *with respect to* $\mathcal{C}$, X *is a cofibrant object of* $\mathcal{M}$, *and* Y *is a* $\mathcal{C}$*-local object of* $\mathcal{M}$, *then the set* $\pi(X, Y)$ *of homotopy classes of maps from* X *to* Y *is independent of whether we consider the homotopy relation in* $\mathcal{M}$ *or in* $L_{\mathcal{C}}\mathcal{M}$.
(2) *If* $R_{\mathcal{C}}\mathcal{M}$ *is the right Bousfield localization of* $\mathcal{M}$ *with respect to* $\mathcal{C}$, A *is a* $\mathcal{C}$*-colocal object of* $\mathcal{M}$, *and* B *is a fibrant object of* $\mathcal{M}$, *then the set* $\pi(A, B)$ *of homotopy classes of maps from* A *to* B *is independent of whether we consider the homotopy relation in* $\mathcal{M}$ *or in* $R_{\mathcal{C}}\mathcal{M}$.

PROOF. This follows from Lemma 3.5.1. □

PROPOSITION 3.5.3. *Let* $\mathcal{M}$ *be a model category and let* $\mathcal{C}$ *be a class of maps in* $\mathcal{M}$.
(1) *If* $\mathcal{M}$ *is left proper,* $L_{\mathcal{C}}\mathcal{M}$ *is the left Bousfield localization of* $\mathcal{M}$ *with respect to* $\mathcal{C}$, *and* X *and* Y *are cofibrant objects of* $\mathcal{M}$, *then a map* $g\colon X \to Y$ *is a* $\mathcal{C}$*-local equivalence if and only if for every* $\mathcal{C}$*-local object* W *the induced map* $g^*\colon \pi_{\mathcal{M}}(Y, W) \to \pi_{\mathcal{M}}(X, W)$ *of sets of homotopy classes of maps in* $\mathcal{M}$ *is an isomorphism.*
(2) *If* $\mathcal{M}$ *is right proper,* $R_{\mathcal{C}}\mathcal{M}$ *is the right Bousfield localization of* $\mathcal{M}$ *with respect to* $\mathcal{C}$, *and* X *and* Y *are fibrant objects of* $\mathcal{M}$, *then a map* $g\colon X \to Y$ *is a* $\mathcal{C}$*-colocal equivalence if and only if for every* $\mathcal{C}$*-colocal object* W *the induced map* $g_*\colon \pi_{\mathcal{M}}(W, X) \to \pi_{\mathcal{M}}(W, Y)$ *of sets of homotopy classes of maps in* $\mathcal{M}$ *is an isomorphism.*

PROOF. We will prove part 1; the proof of part 2 is dual.

Theorem 7.8.6 and Proposition 3.4.1 imply that g is a $\mathcal{C}$-local equivalence if and only if for every $\mathcal{C}$-local object W the induced map $g^*\colon \pi_{L_{\mathcal{C}}\mathcal{M}}(Y, W) \to \pi_{L_{\mathcal{C}}\mathcal{M}}(X, W)$ of sets of homotopy classes of maps in $L_{\mathcal{C}}\mathcal{M}$ is an isomorphism; the result now follows from Lemma 3.5.2. □

CHAPTER 4

Existence of Left Bousfield Localizations

The main result of this chapter is Theorem 4.1.1, which is our existence theorem for left Bousfield localizations (see Definition 3.3.1). The proof of Theorem 4.1.1 is in Section 4.6.

The main difficulty in establishing the localized model category structure lies in finding a set of generating trivial cofibrations (see Definition 11.1.2). That is, we need to find a set J_S of maps such that a map has the right lifting property with respect to every element of J_S if and only if it has the right lifting property with respect to all cofibrations that are S-local equivalences. Since $\mathcal{M}$ is left proper, it is sufficient to find a set J_S of inclusions of cell complexes (see Definition 11.1.4) such that a map with the right lifting property with respect to every element of J_S will have the right lifting property with respect to every inclusion of cell complexes that is an S-local equivalence (see Lemma 4.5.2). We will do this by showing that there is a cardinal γ such that if a map has the right lifting property with respect to all S-local equivalences that are inclusions of cell complexes of size at most γ, then it has the right lifting property with respect to all S-local equivalences that are inclusions of cell complexes (see Proposition 4.5.6).

In order to make this cardinality argument, we must first define a localization functor for objects in our model category $\mathcal{M}$. Section 4.2 has some technical results (motivated by the discussion of Section 1.3) needed for the construction of a functorial cofibrant localization in Section 4.3 (see Definition 4.3.2 and Theorem 4.3.3). We will then use our localization functor to identify the S-local equivalences (see Theorem 4.3.6). Section 4.4 contains some results about the localization functor and subcomplexes of a cell complex needed for the cardinality argument in Section 4.5, and the proof of Theorem 4.1.1 is in Section 4.6.

Theorem 4.2.9 might lead one to hope that the factorization of Theorem 4.3.1 would serve as the required factorization into an S-local trivial cofibration followed by an S-local fibration (see Definition 7.1.3). Unfortunately, Example 2.1.6 shows that not all S-local trivial cofibrations need be $\widetilde{\Lambda S}$-cofibrations, and so there may be $\widetilde{\Lambda S}$-injectives that are not S-local fibrations. Thus, we must establish Proposition 4.5.1, which shows that there is a *set* J_S of generating trivial cofibrations (see Definition 11.1.2) for the S-local model category structure on $\mathcal{M}$.

4.1. Existence of left Bousfield localizations

Although the axioms for a model category are self dual, the actual model categories in which we work have properties (e.g., cofibrant generation (see Definition 11.1.2)) that are not self dual. Thus, it should not be surprising that our existence theorems for left and right localizations differ.

THEOREM 4.1.1. *Let $\mathcal{M}$ be a left proper cellular model category (see Definition 13.1.1 and Definition 12.1.1) and let S be a set of maps in $\mathcal{M}$.*

(1) *The left Bousfield localization of $\mathcal{M}$ with respect to S exists (see Definition 3.3.1). That is, there is a model category structure $L_S\mathcal{M}$ on the underlying category of $\mathcal{M}$ in which*
   (a) *the class of weak equivalences of $L_{\mathcal{C}}\mathcal{M}$ equals the class of $\mathcal{C}$-local equivalences of $\mathcal{M}$,*
   (b) *the class of cofibrations of $L_{\mathcal{C}}\mathcal{M}$ equals the class of cofibrations of $\mathcal{M}$, and*
   (c) *the class of fibrations of $L_{\mathcal{C}}\mathcal{M}$ is the class of maps with the right lifting property with respect to those maps that are both cofibrations and $\mathcal{C}$-local equivalences.*
(2) *The fibrant objects of $L_S\mathcal{M}$ are the S-local objects of $\mathcal{M}$ (see Definition 3.1.4).*
(3) *$L_S\mathcal{M}$ is a left proper cellular model category.*
(4) *If $\mathcal{M}$ is a simplicial model category, then that simplicial structure gives $L_S\mathcal{M}$ the structure of a simplicial model category.*

The proof of Theorem 4.1.1 is in Section 4.6.

DEFINITION 4.1.2. Let $\mathcal{M}$ be a model category and let S be a set of maps in $\mathcal{M}$.

(1) An *S-local weak equivalence* is defined to be an S-local equivalence (see Definition 3.1.4).
(2) An *S-local cofibration* is defined to be a cofibration.
(3) An *S-local fibration* is defined to be a map with the right lifting property (see Definition 7.2.1) with respect to all maps that are both S-local cofibrations and S-local weak equivalences. If the map $X \to *$ from an object X to the terminal object of $\mathcal{M}$ is an S-local fibration, then we will say that X is *S-local fibrant.*

Thus, Theorem 4.1.1 asserts that if $\mathcal{M}$ is a left proper cellular model category and S is a set of maps in $\mathcal{M}$, then the classes of S-local weak equivalences, S-local cofibrations, and S-local fibrations form a model category structure on $\mathcal{M}$.

4.1.3. Examples of left proper cellular model categories.

PROPOSITION 4.1.4. *The categories* SS, Top, SS_*, *and* Top_* *are left proper cellular model categories.*

PROPOSITION 4.1.5. *If $\mathcal{M}$ is a left proper cellular model category and $\mathcal{C}$ is a small category, then the diagram category $\mathcal{M}^{\mathcal{C}}$ is a left proper cellular model category.*

PROPOSITION 4.1.6. *If $\mathcal{M}$ is a left proper cellular model category and Z is an object of $\mathcal{M}$, then the overcategory $(\mathcal{M}\downarrow Z)$ is a left proper cellular model category.*

PROPOSITION 4.1.7. *If $\mathcal{M}$ is a left proper cellular simplicial model category and $\mathcal{C}$ is a small simplicial category, then the category $\mathcal{M}^{\mathcal{C}}$ of simplicial diagrams is a left proper cellular model category.*

PROPOSITION 4.1.8. *If $\mathcal{M}$ is a pointed left proper cellular model category with an action by pointed simplicial sets, then the category of spectra over $\mathcal{M}$ (as in* [**13**]*) is a left proper cellular model category.*

PROPOSITION 4.1.9. *If $\mathcal{M}$ is a pointed left proper cellular model category with an action by pointed simplicial sets, then J. H. Smith's category of symmetric spectra over $\mathcal{M}$* [**61, 43**] *is a left proper cellular model category.*

4.2. Horns on S and S-local equivalences

DEFINITION 4.2.1. If $\mathcal{M}$ is a model category and $\mathcal{C}$ is a class of maps in $\mathcal{M}$, then a *full class of horns on* $\mathcal{C}$ is a class $\Lambda(\mathcal{C})$ of maps obtained by choosing, for every element $f\colon A \to B$ of $\mathcal{C}$, a cosimplicial resolution $\tilde{f}\colon \widetilde{\boldsymbol{A}} \to \widetilde{\boldsymbol{B}}$ of f (see Definition 16.1.20) such that $\tilde{f}$ is a Reedy cofibration (see Proposition 16.1.22) and letting $\Lambda(\mathcal{C})$ be the class of maps

$$\Lambda(\mathcal{C}) = \{\widetilde{\boldsymbol{A}} \otimes \Delta[n] \amalg_{\widetilde{\boldsymbol{A}} \otimes \partial\Delta[n]} \widetilde{\boldsymbol{B}} \otimes \partial\Delta[n] \to \widetilde{\boldsymbol{B}} \otimes \Delta[n] \mid (A \to B) \in S, n \geq 0\}$$

(see Definition 3.3.8). We will use the symbol $\Lambda(\mathcal{C})$ to denote some full class of horns on $\mathcal{C}$ even though it depends on the choices of cosimplicial resolutions of the elements of $\mathcal{C}$.

DEFINITION 4.2.2. Let $\mathcal{M}$ be a cofibrantly generated model category with generating cofibrations I and generating trivial cofibrations J. If S is a set of maps of $\mathcal{M}$, then an *augmented set of S-horns* is a set $\overline{\Lambda(S)}$ of maps

$$\overline{\Lambda(S)} = \Lambda(S) \cup J$$

for some full set of horns $\Lambda(S)$ on S (see Definition 4.2.1).

PROPOSITION 4.2.3. *If $\mathcal{M}$ is a left proper cellular model category and S is a set of maps in $\mathcal{M}$, then every element of an augmented set of horns on S (see Definition 4.2.2) is an S-local equivalence.*

PROOF. This follows from Proposition 17.8.5 and Proposition 17.8.14. □

PROPOSITION 4.2.4. *Let $\mathcal{M}$ be a left proper cellular model category and let S be a set of maps in $\mathcal{M}$. An object X of $\mathcal{M}$ is S-local if and only if the map $X \to *$ (where $*$ is the terminal object of $\mathcal{M}$) has the right lifting property with respect to every element of an augmented set of S-horns (see Definition 4.2.2).*

PROOF. This follows from Proposition 11.2.1 and Lemma 3.3.11. □

PROPOSITION 4.2.5. *If $\mathcal{M}$ is a left proper cellular model category with generating cofibrations I and S is a set of maps in $\mathcal{M}$, then there is a set $\widetilde{\Lambda S}$ of relative I-cell complexes with cofibrant domains such that*

1. *every element of $\widetilde{\Lambda S}$ is an S-local equivalence, and*
2. *an object X of $\mathcal{M}$ is S-local if and only if the map $X \to *$ (where $*$ is the terminal object of $\mathcal{M}$) is a $\widetilde{\Lambda S}$-injective.*

PROOF. Choose a full set of horns on S (see Definition 4.2.1.) Factor each element $g\colon C \to D$ of $\Lambda(S)$ as $C \xrightarrow{\tilde{g}} \widetilde{D} \xrightarrow{p} D$ where $\tilde{g}$ is a relative I-cell complexes and p is a trivial fibration (see Corollary 11.2.6). The retract argument (see Proposition 7.2.2) implies that g is a retract of $\tilde{g}$. Since p and g are S-local equivalences (see Proposition 3.1.5 and Proposition 4.2.3), Proposition 3.2.3 implies that $\tilde{g}$ is an S-local equivalence.

Proposition 11.2.9 implies that there is a set $\widetilde{J}$ of generating trivial cofibrations for $\mathcal{M}$ such that every element of $\widetilde{J}$ is a relative I-cell complex with cofibrant domain. We let

$$\widetilde{\Lambda S} = \widetilde{J} \cup \{\tilde{g}\}_{g \in \Lambda(S)}.$$

It remains only to show that condition 2 is satisfied. If the map $X \to *$ is a $\widetilde{\Lambda S}$-injective, then Proposition 4.2.4 and Lemma 7.2.8 imply that X is S-local. Conversely, if X is S-local, then X is fibrant and every element of $\widetilde{\Lambda S}$ is a cofibration between cofibrant objects, and so Proposition 17.8.5, Theorem 16.6.9, Proposition 16.6.7, and Proposition 17.8.8 imply that the map $X \to *$ is a $\widetilde{\Lambda S}$-injective. □

DEFINITION 4.2.6. If $\mathcal{M}$ is a left proper cellular model category and S is a set of maps in $\mathcal{M}$, then a *relative $\widetilde{\Lambda S}$-cell complex* is a map that can be constructed as a transfinite composition (see Definition 10.2.2) of pushouts (see Definition 7.2.10) of elements of $\widetilde{\Lambda S}$ (see Proposition 4.2.5).

PROPOSITION 4.2.7. *Let $\mathcal{M}$ be a left proper cellular model category, and let S be a set of maps in $\mathcal{M}$. An object X of $\mathcal{M}$ is S-local if and only if the map $X \to *$ (where $*$ is the terminal object of $\mathcal{M}$) has the right lifting property with respect to all relative $\widetilde{\Lambda S}$-cell complexes.*

PROOF. This follows from Proposition 4.2.5, Lemma 7.2.11, and Lemma 10.3.1. □

4.2.8. Regular $\widetilde{\Lambda S}$-cofibrations and S-local equivalences. The main result of this section is Theorem 4.2.9, which asserts that if $\mathcal{M}$ is a left proper cellular model category and S is a set of maps in $\mathcal{M}$, then every relative $\widetilde{\Lambda S}$-cell complex is an S-local equivalence.

THEOREM 4.2.9. *If $\mathcal{M}$ is a left proper cellular model category and S is a set of maps in $\mathcal{M}$, then every relative $\widetilde{\Lambda S}$-cell complex (see Definition 4.2.6) is both a cofibration and an S-local equivalence.*

PROOF. This follows from Proposition 3.2.10 and Proposition 3.2.11. □

PROPOSITION 4.2.10. *Let $\mathcal{M}$ be a left proper cellular model category, and let S be a set of maps in $\mathcal{M}$. If $j\colon X \to \widehat{X}$ is a relative $\widetilde{\Lambda S}$-cell complex and $\widehat{X}$ is a $\widetilde{\Lambda S}$-injective, then the pair $(\widehat{X}, j)$ is a cofibrant S-localization of X.*

PROOF. This follows from Theorem 4.2.9 and Proposition 4.2.5. □

Theorem 4.2.9, Proposition 3.2.4, and Corollary 10.5.23 imply that every $\widetilde{\Lambda S}$-cofibration is an S-local equivalence. Example 2.1.6 shows that, among the cofibrations that are S-local equivalences, there can be maps that are not $\widetilde{\Lambda S}$-cofibrations.

4.3. A functorial localization

THEOREM 4.3.1. *If $\mathcal{M}$ is a left proper cellular model category and S is a set of maps in $\mathcal{M}$, then there is a natural factorization of every map $X \to Y$ in $\mathcal{M}$ as*

$$X \xrightarrow{j} \mathrm{E}_S \xrightarrow{p} Y$$

in which j is a relative $\widetilde{\Lambda S}$-cell complex (see Definition 4.2.6) and p is a $\widetilde{\Lambda S}$-injective.

PROOF. Proposition 4.2.5 and Theorem 12.4.3 imply that the domains of the elements of $\widetilde{\Lambda S}$ are small relative to the subcategory of relative $\widetilde{\Lambda S}$-cell complexes, and so Lemma 10.4.6 implies that there is a cardinal κ such that the domain of every element of $\widetilde{\Lambda S}$ is κ-small relative to the subcategory of relative $\widetilde{\Lambda S}$-cell complexes. We let $\lambda = \mathrm{Succ}(\kappa)$ (see Definition 10.1.10), so that λ is a regular cardinal (see Proposition 10.1.14). The result now follows from Corollary 10.5.21. □

DEFINITION 4.3.2. Let $\mathcal{M}$ be a left proper cellular model category, and let S be a set of maps in $\mathcal{M}$. The *S-localization* of an object X is the object $\mathrm{L}_S X$ obtained by applying the factorization of Theorem 4.3.1 to the map $X \to *$ (where $*$ is the terminal object of $\mathcal{M}$). This factorization defines a natural transformation $j\colon 1 \to \mathrm{L}_S$ such that $j(X)\colon X \to \mathrm{L}_S X$ is a relative $\widetilde{\Lambda S}$-cell complex for every object X of $\mathcal{M}$.

THEOREM 4.3.3. *If $\mathcal{M}$ is a left proper cellular model category and S is a set of maps in $\mathcal{M}$, then, for every object X, the S-localization $j(X)\colon X \to \mathrm{L}_S X$ (see Definition 4.3.2) is a cofibrant S-localization of X.*

PROOF. This follows from Proposition 4.2.10. □

COROLLARY 4.3.4. *If $\mathcal{M}$ is a left proper cellular model category and S is a set of maps in $\mathcal{M}$, then every object has an S-localization.*

PROOF. This follows from Theorem 4.3.3. □

THEOREM 4.3.5. *Let $\mathcal{M}$ be a left proper cellular model category, and let S be a set of maps in $\mathcal{M}$. If X is a fibrant object, then X is S-local if and only if the S-localization map $j(X)\colon X \to \mathrm{L}_S X$ (see Definition 4.3.2) is a weak equivalence.*

PROOF. This follows from Theorem 3.2.17. □

THEOREM 4.3.6. *Let $\mathcal{M}$ be a left proper cellular model category, and let S be a set of maps in $\mathcal{M}$. The map $g\colon X \to Y$ is an S-local equivalence if and only if its S-localization $\mathrm{L}_S(g)\colon \mathrm{L}_S X \to \mathrm{L}_S Y$ (see Definition 4.3.2) is a weak equivalence.*

PROOF. This follows from Theorem 3.2.18. □

4.3.7. Simplicial localization functors. In this section we show that if $\mathcal{M}$ is a left proper cellular model category that is a *simplicial* model category and if S is a set of maps in $\mathcal{M}$, then we can define a cofibrant S-localization on $\mathcal{M}$ that is a *simplicial functor* (see Section 9.8). To do this, we will modify the construction of our localization functor (see Definition 4.3.2) in a manner analogous to the way in which the functor of Example 9.8.7 was modified to become the simplicial functor of Example 9.8.8. Definition 4.3.2 uses the factorization of Proposition 10.5.16 in the case in which Y is the terminal object of $\mathcal{M}$, which constructs pushouts of diagrams involving a coproduct indexed by the set of maps between objects A_i and E^β (see Diagram 10.5.17). We will construct our simplicial localization functor by replacing the coproduct $\coprod_{\mathcal{M}(A_i,E^\beta)} A_i$ with $A_i \otimes \mathrm{Map}(A_i, E^\beta)$.

THEOREM 4.3.8. *If $\mathcal{M}$ is a left proper simplicial cellular model category and S is a set of maps in $\mathcal{M}$, then there is a cofibrant S-localization functor on $\mathcal{M}$ that is a simplicial functor.*

PROOF. Let $\widetilde{\Lambda S}$ be a set of relative I-cell complexes as in Proposition 4.2.5, and let λ be a regular cardinal such that the domains of the elements of $\widetilde{\Lambda S}$ are λ-small with respect to the subcategory of cofibrations of $\mathcal{M}$ (see Theorem 12.4.3). For each object X of $\mathcal{M}$ we will define a λ-sequence $X = E^0 \to E^1 \to E^2 \to \cdots \to E^\beta \to \cdots$ ($\beta < \lambda$) whose composition will be our simplicial localization functor $X \to \mathrm{L}_S^{\mathrm{cont}} X$.

We begin by letting $E^0 = X$. If $\beta < \lambda$ and we have constructed the sequence through E^β, we let

$$C_\beta^{\mathrm{cont}} = \coprod_{(C \to D) \in \widetilde{\Lambda S}} C \otimes \mathrm{Map}(C, \mathrm{E}^\beta)$$

$$D_\beta^{\mathrm{cont}} = \coprod_{(C \to D) \in \widetilde{\Lambda S}} D \otimes \mathrm{Map}(C, \mathrm{E}^\beta)$$

and we define $E^{\beta+1}$ via the pushout square

$$\begin{array}{ccc} C_\beta^{\mathrm{cont}} & \longrightarrow & \mathrm{E}^\beta \\ \downarrow & & \vdots \\ D_\beta^{\mathrm{cont}} & \cdots\cdots\to & \mathrm{E}^{\beta+1} \end{array}$$

in which the top map on each factor is the natural map that is adjoint to the identity map of $\mathrm{Map}(C, E^\beta)$. If γ is a limit ordinal, we let $E^\gamma = \mathrm{colim}_{\beta<\gamma} E^\beta$. We let $\mathrm{L}_S^{\mathrm{cont}} X = \mathrm{colim}_{\beta<\gamma} E^\beta$.

Lemma 3.2.8, Proposition 3.2.10, and Proposition 3.2.11 imply that the map $X \to \mathrm{L}_S^{\mathrm{cont}} X$ is an S-local equivalence.

For every element $C \to D$ of $\widetilde{\Lambda S}$ and every $\beta < \lambda$, the 0-skeleton of $\mathrm{Map}(C, E^\beta)$ is $\mathcal{M}(C, E^\beta)$, and so for every map $C \to E^\beta$ the composition $C \to E^\beta \to E^{\beta+1}$ can be factored through $C \to D$. Since C is λ-small with respect to the subcategory of cofibrations of $\mathcal{M}$, this implies that $\mathrm{L}_S^{\mathrm{cont}} X$ is a $\widetilde{\Lambda S}$-injective, and so $\mathrm{L}_S^{\mathrm{cont}} X$ is S-local.

The proof that the functor $\mathrm{L}_S^{\mathrm{cont}}$ can be extended to a simplicial functor is as in the proof of Theorem 1.7.5: If C and X are objects in $\mathcal{M}$ and K is a simplicial set, then there is a natural map $\mathrm{Map}(C, X) \otimes K \to \mathrm{Map}(C, X \otimes K)$ that takes the n-simplex $(\alpha\colon C \otimes \Delta[n] \to X, \tau)$ of $\mathrm{Map}(C, X) \otimes K$ to the n-simplex $\sigma(\alpha, \tau)\colon C \otimes \Delta[n] \to X \otimes K$ of $\mathrm{Map}(C, X \otimes K)$ that is the composition

$$C \otimes \Delta[n] \xrightarrow{1_C \otimes D} C \otimes (\Delta[n] \times \Delta[n]) \xrightarrow{\approx} (C \otimes \Delta[n]) \otimes \Delta[n] \xrightarrow{\alpha \otimes i_\tau} X \otimes K$$

(where $D\colon \Delta[n] \to \Delta[n] \times \Delta[n]$ is the diagonal map and $i_\tau\colon \Delta[n] \to K$ is the map that takes the nondegenerate n-simplex of $\Delta[n]$ to τ). This natural map σ has the properties required by Theorem 9.8.5, and so we can use it to inductively define σ for all the objects in the construction of the localization functor. The theorem now follows from Proposition 9.8.9 and Theorem 9.8.5. □

4.4. Localization of subcomplexes

This section contains some technical results on the S-localization functor (see Definition 4.3.2) that are needed for the cardinality argument of Section 4.5.

PROPOSITION 4.4.1. *Let $\mathcal{M}$ be a left proper cellular model category, and let S be a set of maps in $\mathcal{M}$. If $g\colon X \to Y$ is the inclusion of a subcomplex, then so is $\mathrm{L}_S(g)\colon \mathrm{L}_S X \to \mathrm{L}_S Y$ (see Definition 4.3.2).*

PROOF. This follows from Proposition 12.4.7. $\square$

REMARK 4.4.2. If we take S to be the empty set, then $\mathrm{L}_S X$ is a functorial fibrant approximation to X (see Definition 8.1.2). In this case, Proposition 4.4.1 asserts that if W is a subcomplex of X, then this fibrant approximation to W is a subcomplex of this fibrant approximation to X.

PROPOSITION 4.4.3. *Let $\mathcal{M}$ be a left proper cellular model category and let S be a set of maps in $\mathcal{M}$. If $g\colon X \to Y$ is the inclusion of a subcomplex, then it is an S-local equivalence if and only if its localization $\mathrm{L}_S(g)\colon \mathrm{L}_S X \to \mathrm{L}_S Y$ is the inclusion of a strong deformation retract (see Definition 7.6.10).*

PROOF. If $\mathrm{L}_S(g)$ is the inclusion of a strong deformation retract, then it is a weak equivalence, and so Theorem 4.3.6 implies that g is an S-local equivalence.

Conversely, if g is an S-local equivalence, then Theorem 4.3.6 and Proposition 4.4.1 imply that $\mathrm{L}_S(g)$ is a trivial cofibration of fibrant objects, and so Corollary 9.6.5 implies that it is the inclusion of a strong deformation retract. $\square$

PROPOSITION 4.4.4. *Let $\mathcal{M}$ be a left proper cellular model category and let S be a set of maps in $\mathcal{M}$. If X is a cell complex and $K_0 \subset K_1 \subset K_2 \subset \cdots \subset K_\beta \subset \cdots$ $(\beta < \lambda)$ is a λ-sequence of subcomplexes (see Remark 10.6.8) of X (where λ is the ordinal chosen in the proof of Theorem 4.3.1), then the natural map $\operatorname{colim}_{\beta<\lambda} \mathrm{L}_S K_\beta \to \mathrm{L}_S \operatorname{colim}_{\beta<\lambda} K_\beta$ is an isomorphism.*

PROOF. Proposition 4.4.1 implies that the map is an isomorphism onto a subcomplex, and so it remains only to show that every $\widetilde{\Lambda S}$-cell of $\mathrm{L}_S \operatorname{colim}_{\beta<\lambda} K_\beta$ is contained in some $\mathrm{L}_S K_\beta$. We will do this by a transfinite induction on the presentation ordinal of the $\widetilde{\Lambda S}$-cell (see Definition 10.6.4).

Since there are no $\widetilde{\Lambda S}$-cells of presentation ordinal equal to a limit ordinal, we let γ be an ordinal such that $\gamma + 1 < \lambda$ and we assume that the assertion is true for all $\widetilde{\Lambda S}$-cells of presentation ordinal at most γ. This assumption implies that the γ-skeleton of $\mathrm{L}_S \operatorname{colim}_{\beta<\lambda} K_\beta$ is isomorphic to $\operatorname{colim}_{\beta<\lambda}\bigl((\mathrm{L}_S K_\beta)^\gamma\bigr)$. Thus, the γ-skeleta of the $\mathrm{L}_S K_\beta$ form a λ-sequence whose colimit is the γ-skeleton of $\mathrm{L}_S \operatorname{colim}_{\beta<\lambda} K_\beta$. If e is a $\widetilde{\Lambda S}$-cell of $\mathrm{L}_S \operatorname{colim}_{\beta<\lambda} K_\beta$ of presentation ordinal $\gamma + 1$, then the attaching map of e must factor through $(\mathrm{L}_S K_\beta)^\gamma$ for some $\beta < \lambda$, and so e is contained in $\mathrm{L}_S K_\beta$. $\square$

PROPOSITION 4.4.5. *Let $\mathcal{M}$ be a left proper cellular model category and let S be a set of maps in $\mathcal{M}$. If X is a cell complex and A and B are subcomplexes of X, then the natural map $\mathrm{L}_S(A \cap B) \to (\mathrm{L}_S A) \cap (\mathrm{L}_S B)$ (see Proposition 4.4.1 and Theorem 12.2.6) is an isomorphism.*

PROOF. Proposition 4.4.1 implies that the natural map $\mathrm{L}_S(A \cap B) \to (\mathrm{L}_S A) \cap (\mathrm{L}_S B)$ is an isomorphism onto a subcomplex, and so it remains only to show that every $\widetilde{\Lambda S}$-cell of $(\mathrm{L}_S A) \cap (\mathrm{L}_S B)$ is contained in $\mathrm{L}_S(A \cap B)$. We will do this by a transfinite induction on the presentation ordinal of the $\widetilde{\Lambda S}$-cell (see Definition 10.6.4).

Since there are no $\widetilde{\Lambda S}$-cells of presentation ordinal equal to a limit ordinal, we let γ be an ordinal such that $\gamma + 1 < \lambda$ (where λ is the ordinal chosen in the proof of Theorem 4.3.1) and we assume that the assertion is true for all $\widetilde{\Lambda S}$-cells of presentation ordinal at most γ. This assumption implies that the γ-skeleton $\big(\mathrm{L}_S(A \cap B)\big)^{\gamma}$ of $\mathrm{L}_S(A \cap B)$ equals the intersection of γ-skeleta $(\mathrm{L}_S A)^{\gamma} \cap (\mathrm{L}_S B)^{\gamma}$. Thus, if e is a $\widetilde{\Lambda S}$-cell of $(\mathrm{L}_S A) \cap (\mathrm{L}_S B)$ of presentation ordinal $\gamma+1$, then Proposition 12.2.3 implies that the attaching map of e factors through $\big(\mathrm{L}_S(A \cap B)\big)^{\gamma}$, and so e is contained in $\mathrm{L}_S(A \cap B)$. □

4.5. The Bousfield-Smith cardinality argument

The purpose of this section is to prove the following proposition, which will be used in Section 4.6 to prove Theorem 4.1.1.

PROPOSITION 4.5.1. *If $\mathcal{M}$ is a left proper cellular model category and S is a set of maps in $\mathcal{M}$, then there is a set J_S of inclusions of cell complexes such that the class of J_S-cofibrations (see Definition* 10.5.2*) equals the class of cofibrations that are also S-local equivalences.*

The proof of Proposition 4.5.1 is at the end of this section (on page 81). The set J_S will serve as our set of generating trivial cofibrations (see Definition 11.1.2) for the S-local model category structure on $\mathcal{M}$ (see Theorem 4.1.1 and Section 4.6).

We will prove Proposition 4.5.1 by showing that there is a set J_S of cofibrations that are S-local equivalences such that every cofibration that is an S-local equivalence is a J_S-cofibration (see Definition 10.5.2). Proposition 4.5.1 will then follow from Corollary 10.5.23.

We will find the set J_S by showing (in Proposition 4.5.6) that there is a cardinal γ (see Definition 4.5.3) such that if a map has the right lifting property with respect to all inclusions of cell complexes that are S-local equivalences between complexes of size at most γ, then it has the right lifting property with respect to all cofibrations that are S-local equivalences. Since the collection of isomorphism classes of cell complexes of size at most γ is a set, we can then let J_S be a set of representatives of the isomorphism classes of of these "small enough" inclusions of cell complexes that are S-local equivalences.

We begin with the following lemma, which implies that it is sufficient to find a set J_S such that the J_S-injectives have the right lifting property with respect to all inclusions of cell complexes that are S-local equivalences.

LEMMA 4.5.2. *Let $\mathcal{M}$ be a left proper cellular model category and let S be a set of maps in $\mathcal{M}$. If $p\colon E \to B$ is a fibration with the right lifting property with respect to all inclusions of cell complexes that are S-local equivalences, then it has the right lifting property with respect to all cofibrations that are S-local equivalences.*

PROOF. Let $g\colon X \to Y$ be a cofibration that is an S-local equivalence. Proposition 11.2.8 implies that there is a cofibrant approximation (see Definition 8.1.22) $\tilde{g}$ to g such that $\tilde{g}$ is an inclusion of cell complexes. Proposition 3.1.5 and Proposition 3.2.3 imply that $\tilde{g}$ is an S-local equivalence, and so the lemma now follows from Proposition 13.2.1. □

DEFINITION 4.5.3. If $\mathcal{M}$ is a left proper cellular model category and S is a set of maps in $\mathcal{M}$, we let ξ is the smallest cardinal that is at least as large as each of the following cardinals:

(1) the size of the cells of $\mathcal{M}$ (see Definition 12.3.3),

(2) a cardinal η such that the domain of every element of I is η-compact (see Proposition 11.4.6),
(3) the cardinal λ selected in the proof of Theorem 4.3.1,
(4) the cardinal κ described in Proposition 12.5.3 for the set $\widetilde{\Lambda S}$, and
(5) the cardinal κ described in Proposition 12.5.7,

and we let γ denote the cardinal $\gamma = \xi^\xi$.

LEMMA 4.5.4. *Let $\mathcal{M}$ be a left proper cellular model category and let S be a set of maps in $\mathcal{M}$. If X is a cell complex of size at most γ (see Definition 4.5.3), then $\mathrm{L}_S X$ (see Definition 4.3.2) has size at most γ.*

PROOF. This follows from Proposition 12.5.3 and conditions 3 and 4 of Definition 4.5.3. □

The following proposition will be used in Proposition 4.5.6 to show that a fibration that has the right lifting property with respect to all "small enough" inclusions of cell complexes that are S-local equivalences must actually have the right lifting property with respect to *all* inclusions of cell complexes that are S-local equivalences.

PROPOSITION 4.5.5. *Let $\mathcal{M}$ be a left proper cellular model category, let S be a set of maps in $\mathcal{M}$, and let D be a cell complex. If $i\colon C \to D$ is the inclusion of a proper subcomplex and an S-local equivalence, then there is a subcomplex K of D such that*

(1) *the subcomplex K is not contained in the subcomplex C,*
(2) *the size of K is at most γ (see Definition 4.5.3), and*
(3) *the inclusions $K \cap C \to K$ (see Theorem 12.2.6) and $C \to C \cup K$ are both S-local equivalences.*

PROOF. Since $i\colon C \to D$ is an inclusion of a subcomplex and an S-local equivalence, Proposition 4.4.3 implies that $\mathrm{L}_S(i)\colon \mathrm{L}_S C \to \mathrm{L}_S D$ is the inclusion of a deformation retract. Thus, there is a retraction $r\colon \mathrm{L}_S D \to \mathrm{L}_S C$, and Proposition 7.4.7 implies that we can choose a homotopy $R\colon \mathrm{Cyl}^{\mathcal{M}}(\mathrm{L}_S D) \to \mathrm{L}_S D$ (see Definition 12.5.5) from the identity map of $\mathrm{L}_S D$ to $\mathrm{L}_S(i) \circ r$.

We will show that there exists a subcomplex K of D, of size at most γ, such that

(1) K is not contained in C,
(2) the restriction $R|_{\mathrm{Cyl}^{\mathcal{M}}(\mathrm{L}_S K)}$ of R to $\mathrm{L}_S K$ (see Definition 12.5.6) is a deformation retraction of $\mathrm{L}_S K$ onto $\mathrm{L}_S(K \cap C)$, and
(3) the restriction $R|_{\mathrm{Cyl}^{\mathcal{M}}(\mathrm{L}_S(C \cup K))}$ of R to $\mathrm{L}_S(C \cup K)$ is a deformation retraction of $\mathrm{L}_S(C \cup K)$ onto $\mathrm{L}_S C$.

We will do this by constructing a λ-sequence $K_0 \subset K_1 \subset K_2 \subset \cdots \subset K_\beta \subset \cdots$ ($\beta < \lambda$) of subcomplexes of D (where λ is the ordinal selected in the proof of Theorem 4.3.1) such that, for every $\beta < \lambda$,

(1) K_β has size at most γ,
(2) the restriction $R|_{\mathrm{Cyl}^{\mathcal{M}}(\mathrm{L}_S K_\beta)}$ of R to $\mathrm{L}_S K_\beta$ factors through the subcomplex $\mathrm{L}_S K_{\beta+1}$ of $\mathrm{L}_S D$ (see Proposition 4.4.1),

and such that no K_β is contained in C. If we then let $K = \bigcup_{\beta<\lambda} K_\beta$, then Proposition 4.4.4 will imply that $\mathrm{L}_S K \approx \mathrm{colim}_{\beta<\lambda} \mathrm{L}_S K_\beta$. Thus, $R|_{\mathrm{Cyl}^{\mathcal{M}}(\mathrm{L}_S K)}$ will factor through $\mathrm{L}_S K$, $r|_{\mathrm{L}_S K}$ will factor through $(\mathrm{L}_S K) \cap (\mathrm{L}_S C) = \mathrm{L}_S(K \cap C)$ (see Proposition 4.4.5), $R|_{\mathrm{Cyl}^{\mathcal{M}}(\mathrm{L}_S K)}$ will be a deformation retraction of $\mathrm{L}_S K$ onto $\mathrm{L}_S(K \cap C)$, and $R|_{\mathrm{Cyl}^{\mathcal{M}}(C \cup K)}$ will be a deformation retraction of $\mathrm{L}_S(C \cup K)$ onto $\mathrm{L}_S C$.

We begin by choosing a cell of D that isn't contained in C. Since the domains of the elements of I are γ-compact (see condition 2 of Definition 4.5.3), we can choose a subcomplex K_0 of D, of size at most γ, through which the inclusion of that cell factors.

For successor ordinals, suppose that $\beta + 1 < \gamma$ and that we've constructed K_β. Lemma 4.5.4 implies that $\mathrm{L}_S K_\beta$ has size at most γ. Proposition 12.5.7 and condition 5 of Definition 4.5.3 then imply that $\mathrm{Cyl}^{\mathcal{M}}(\mathrm{L}_S K_\beta)$ has size at most γ, and so Definition 12.3.3 implies that $R|_{\mathrm{Cyl}^{\mathcal{M}}(\mathrm{L}_S K_\beta)}$ factors through a subcomplex of $\mathrm{L}_S D$ of size at most $\sigma\gamma = \gamma$ where σ is the size of the cells of $\mathcal{M}$ (see condition 1 of Definition 4.5.3). The zero skeleton of this subcomplex is a subcomplex Z_β of D, of size at most γ, such that $R|_{\mathrm{Cyl}^{\mathcal{M}}(\mathrm{L}_S K_\beta)}$ factors through $\mathrm{L}_S Z_\beta$. We let $K_{\beta+1} = K_\beta \cup Z_\beta$. It is clear that $K_{\beta+1}$ has the properties required of it, and so the proof is complete. □

PROPOSITION 4.5.6. *Let $\mathcal{M}$ be a left proper cellular model category and let S be a set of maps in $\mathcal{M}$. If $p\colon X \to Y$ has the right lifting property with respect to those inclusions of subcomplexes $i\colon C \to D$ that are S-local equivalences and such that the size of D is at most γ (see Definition 4.5.3), then p has the right lifting property with respect to all inclusions of subcomplexes that are S-local equivalences.*

PROOF. Let $i\colon C \to D$ be an inclusion of a subcomplex that is an S-local equivalence, and let the solid arrow diagram

$$\begin{array}{ccc} C & \xrightarrow{h} & X \\ {\scriptstyle i}\downarrow & \nearrow & \downarrow{\scriptstyle p} \\ D & \xrightarrow[k]{} & Y \end{array}$$

be commutative; we must show that there exists a dotted arrow making both triangles commute. To do this, we will consider the subcomplexes of D over which our map can be defined, and use Zorn's lemma to show that it can be defined over all of D.

Let T be the set of pairs (D_t, g_t) such that

(1) D_t is a subcomplex of D containing C such that the inclusion $i_t\colon C \to D_t$ is an S-local equivalence and
(2) g_t is a function $D_t \to X$ such that $g_t i_t = h$ and $pg_t = k|_{D_t}$.

We define a preorder on T by defining $(D_t, g_t) < (D_u, g_u)$ if $D_t \subset D_u$ and $g_u|_{D_t} = g_t$. If $T' \subset T$ is a chain (i.e., a totally ordered subset of T), let $D_u = \mathrm{colim}_{(D_t,g_t)\in T'} D_t$ and define $g_u\colon D_u \to X$ by $g_u = \mathrm{colim}_{(D_t,g_t)\in T'} g_t$. The universal mapping property of the colimit implies that $g_u i_u = h$ and $pg_u = k|_{D_u}$, and Proposition 3.2.7 implies that the map $C \to D_u$ is an S-local equivalence. Thus, (D_u, g_u) is an element of T, and so it is an upper bound for T'. Zorn's lemma now implies that T has a maximal element (D_m, g_m). We will complete the proof by showing that $D_m = D$.

If $D_m \neq D$, then Proposition 4.5.5 implies that there is a subcomplex K of D such that K is not contained in D_m, the size of K is at most γ, and the inclusions $K \cap D_m \to K$ and $D_m \to D_m \cup K$ are both S-local equivalences. Thus, there is a map $g_K\colon K \to X$ such that $pg_K = k|_K$ and $g_K|_{K\cap D_m} = g_m|_{K\cap D_m}$, and so g_m and g_K combine to define a map $g_{mK}\colon K \cup D_m \to X$ such that $pg_{mK} = k|_{K\cup D_m}$ and $g_{mK}i = h$. Thus, $(K \cup D_m, g_{mK})$ is an element of T strictly greater than (D_m, g_m). This contradicts (D_m, g_m) being a maximal element of T, and so our assumption that $D_m \neq D$ must have been false. □

PROOF OF PROPOSITION 4.5.1. Let J_S be a set of representatives of the isomorphism classes of inclusions of subcomplexes that are S-local equivalences of complexes of size at most γ (see Definition 4.5.3). Proposition 4.5.6, Lemma 4.5.2, and Corollary 10.5.22 imply that the J_S-cofibrations are exactly the cofibrations that are S-local equivalences, and so the proof is complete. □

4.6. Proof of the main theorem

This section contains the proof of Theorem 4.1.1.

4.6.1. Proof of part 1. We will use Theorem 11.3.1. Proposition 3.2.3 implies that the class of S-local equivalences satisfies the "two out of three" axiom, and Proposition 3.2.4 implies that it is closed under retracts.

Let J_S be the set of maps provided by Proposition 4.5.1, and let I be the set of generating cofibrations of the original cofibrantly generated model category structure on $\mathcal{M}$. Condition 1 of Theorem 11.3.1 is thus satisfied for I and, since every element of J_S has a cofibrant domain, Theorem 12.4.3 implies that condition 1 of Theorem 11.3.1 is satisfied for J.

The subcategory of I-cofibrations is the subcategory of cofibrations in the given model category structure in $\mathcal{M}$, and the I-injectives are the trivial fibrations in that model category. Thus, Proposition 4.5.1 implies that condition 2 of Theorem 11.3.1 is satisfied.

Since the J_S-cofibrations are a subcategory of the I-cofibrations, every I-injective must be a J_S-injective. Proposition 3.1.5 implies that every J_S-injective is an S-local equivalence, and so condition 3 is satisfied.

Proposition 4.5.1 implies that condition 4a of Theorem 11.3.1 is satisfied, and so Theorem 11.3.1 now implies that we have a model category $\mathrm{L}_S\mathcal{M}$, and the proof of part 1 is complete.

4.6.2. Proof of part 2. This follows from Proposition 3.4.1.

4.6.3. Proof of part 3. Condition 1 of Definition 12.1.1 is satisfied because the class of generating cofibrations of $\mathrm{L}_S\mathcal{M}$ equals that of $\mathcal{M}$. Since the generating trivial cofibrations of $\mathrm{L}_S\mathcal{M}$ are inclusions of cell complexes, condition 2 of Definition 12.1.1 follows from Lemma 12.4.1. Condition 3 is satisfied because the class of cofibrations of $\mathrm{L}_S\mathcal{M}$ equals that of $\mathcal{M}$, and so $\mathrm{L}_S\mathcal{M}$ is cellular. Finally, Proposition 3.4.4 implies that the localization is left proper.

4.6.4. Proof of part 4. Axiom M6 of Definition 9.1.6 holds for $\mathrm{L}_S\mathcal{M}$ because it holds for $\mathcal{M}$.

For axiom M7, if $i\colon A \to B$ is a cofibration in $\mathrm{L}_S\mathcal{M}$ and $p\colon X \to Y$ is a fibration in $\mathrm{L}_S\mathcal{M}$ then i is a cofibration in $\mathcal{M}$ and p is a fibration in $\mathcal{M}$, and so the map $\mathrm{Map}(i,p)\colon \mathrm{Map}(B,X) \to \mathrm{Map}(A,X) \times_{\mathrm{Map}(A,Y)} \mathrm{Map}(B,Y)$ is a fibration of simplicial sets. If p is also a weak equivalence in $\mathrm{L}_S\mathcal{M}$, then p is a trivial fibration in $\mathrm{L}_S\mathcal{M}$, and thus also in $\mathcal{M}$, and so $\mathrm{Map}(i,p)$ is a trivial fibration of simplicial sets. Thus, it remains only to deal with the case in which i is a trivial cofibration in $L_S\mathcal{M}$ and p is a fibration in $L_S\mathcal{M}$.

If i is a trivial cofibration in $L_S\mathcal{M}$ and p is a fibration in $\mathcal{M}$, then Theorem 17.8.10 implies that (i, p) is a homotopy orthogonal pair. Let $\tilde{\imath}\colon \widetilde{A} \to \widetilde{B}$ be a cofibrant approximation to i in $\mathcal{M}$ such that $\tilde{\imath}$ is a cofibration in $\mathcal{M}$. Example 16.6.13 implies that if $\widetilde{\boldsymbol{A}}$ and $\widetilde{\boldsymbol{B}}$ are the cosimplicial objects in $\mathcal{M}$ such that $\widetilde{\boldsymbol{A}}^n = \widetilde{A} \otimes \Delta[n]$ and $\widetilde{\boldsymbol{B}}^n = \widetilde{B} \otimes \Delta[n]$ and $\widetilde{\boldsymbol{A}} \to \widetilde{\boldsymbol{B}}$ is the map induced by $\tilde{\imath}$, then $\widetilde{\boldsymbol{A}} \to \widetilde{\boldsymbol{B}}$ is a cosimplicial resolution of $\tilde{\imath}$ in $\mathcal{M}$ such that $\widetilde{\boldsymbol{A}} \to \widetilde{\boldsymbol{B}}$ is a Reedy cofibration in $\mathcal{M}^{\boldsymbol{\Delta}}$. Corollary 16.2.2 and Proposition 3.3.4 imply that $\widetilde{\boldsymbol{A}} \to \widetilde{\boldsymbol{B}}$ is also a cosimplicial resolution of $\tilde{\imath}$ in $L_S\mathcal{M}$ such that $\widetilde{\boldsymbol{A}} \to \widetilde{\boldsymbol{B}}$ is a Reedy cofibration in $(L_S\mathcal{M})^{\boldsymbol{\Delta}}$, and so Proposition 16.3.10 implies that the map $\widetilde{A} \otimes \Delta[n] \amalg_{\widetilde{A} \otimes \partial\Delta[n]} \widetilde{B} \otimes \partial\Delta[n] \to \widetilde{B} \otimes \Delta[n]$ is a trivial cofibration in $L_S\mathcal{M}$ for every $n \geq 0$. Since $p\colon X \to Y$ is a fibration in $L_S\mathcal{M}$, Lemma 9.4.7 now implies that the map $\tilde{\imath}\colon \widetilde{A} \to \widetilde{B}$ has the homotopy left lifting property with respect to p, and so Corollary 13.2.2 implies that the map $i\colon A \to B$ has the homotopy left lifting property with respect to p.

CHAPTER 5

Existence of Right Bousfield Localizations

This chapter contains the statement and proof of our existence theorem for right Bousfield localization (see Definition 3.3.1). The statement is Theorem 5.1.1, and it is proved in Section 5.4 after some preparatory work in Sections 5.2 and 5.3. Theorem 5.1.5 shows that the class of K-colocal objects (which is the class of cofibrant objects of the localization; see Theorem 5.1.1) equals the class of K-cellular objects of Dror Farjoun ([**20, 21, 23, 24**]).

5.1. Right Bousfield localization: Cellularization

THEOREM 5.1.1. *Let $\mathcal{M}$ be a right proper cellular model category, let K be a set of objects in $\mathcal{M}$, and let $\mathcal{C}$ be the class of K-local equivalences (see Definition 3.1.8).*

1. *The right Bousfield localization of $\mathcal{M}$ with respect to $\mathcal{C}$ exists (see Definition 3.3.1). That is, there is a model category structure $\mathrm{R}_{\mathcal{C}}\mathcal{M}$ on the underlying category of $\mathcal{M}$ in which*
   (a) *the class of weak equivalences of $\mathrm{R}_{\mathcal{C}}\mathcal{M}$ equals the class of $\mathcal{C}$-colocal equivalences of $\mathcal{M}$,*
   (b) *the class of fibrations of $\mathrm{R}_{\mathcal{C}}\mathcal{M}$ equals the class of fibrations of $\mathcal{M}$, and*
   (c) *the class of cofibrations of $\mathrm{R}_{\mathcal{C}}\mathcal{M}$ is the class of maps with the left lifting property with respect to those maps that are both fibrations and $\mathcal{C}$-colocal equivalences.*
2. *The cofibrant objects of $\mathrm{R}_{\mathcal{C}}\mathcal{M}$ are the $\mathcal{C}$-local objects of $\mathcal{M}$ (see Definition 3.1.4).*
3. *$\mathrm{R}_{\mathcal{C}}\mathcal{M}$ is a right proper model category. If every object of $\mathcal{M}$ is fibrant, then $\mathrm{R}_{\mathcal{C}}\mathcal{M}$ is a right proper cellular model category in which every object is fibrant.*
4. *If $\mathcal{M}$ is a simplicial model category, then that simplicial structure gives $\mathrm{R}_{\mathcal{C}}\mathcal{M}$ the structure of a simplicial model category.*

The proof of Theorem 5.1.1 is in Section 5.4. Theorem 5.1.1 for the case in which $\mathcal{M}$ is the category of pointed topological spaces was first obtained by Nofech [**51**].

REMARK 5.1.2. The model category structure $\mathrm{R}_{\mathcal{C}}\mathcal{M}$ of Theorem 5.1.1 exists for a larger class of model categories $\mathcal{M}$ than just the right proper cellular ones. Although the proof of Theorem 5.1.1 does use the right properness of $\mathcal{M}$, the only use made of the assumption that $\mathcal{M}$ is cellular is to deduce that

1. there is a set J of generating trivial cofibrations, and
2. the domains of the elements of $\overline{\Lambda(K)}$ (see Definition 5.2.1) are small relative to $\overline{\Lambda(K)}$, and so $\overline{\Lambda(K)}$ permits the small object argument (see Definition 10.5.15).

Thus, if $\mathcal{M}$ is a right proper model category with a set J of generating trivial cofibrations such that, e.g., every object of $\mathcal{M}$ is small relative to the subcategory of cofibrations, then the model category of Theorem 5.1.1 exists.

DEFINITION 5.1.3. Let $\mathcal{M}$ be a model category, and let K be a set of objects in $\mathcal{M}$.

(1) A *K-colocal weak equivalence* is defined to be a K-colocal equivalence (see Definition 3.1.8).
(2) A *K-colocal fibration* is defined to be a fibration.
(3) A *K-colocal cofibration* is defined to be a map with the left lifting property with respect to all maps that are both fibrations and K-colocal weak equivalences.

Thus, Theorem 5.1.1 asserts that if $\mathcal{M}$ is a right proper cellular model category and K is a set of objects in $\mathcal{M}$, then the classes of K-colocal weak equivalences, K-colocal cofibrations, and K-local fibrations form a model category structure on $\mathcal{M}$.

DEFINITION 5.1.4. Let $\mathcal{M}$ be a model category. If K is a set of cofibrant objects of $\mathcal{M}$, then the class of *K-cellular objects* is defined to be the smallest class of cofibrant objects of $\mathcal{M}$ that contains K and is closed under homotopy colimits and weak equivalences. If K consists of a single object A, then the class of K-cellular objects will also be called the class of *A-cellular objects.*

THEOREM 5.1.5. *Let $\mathcal{M}$ be a model category. If K is a set of cofibrant objects of $\mathcal{M}$, then the class of K-cellular objects (see Definition 5.1.4) equals the class of K-colocal objects (see Definition 3.1.8).*

The proof of Theorem 5.1.5 is in Section 5.5.

THEOREM 5.1.6. *Let $\mathcal{M}$ be a right proper cellular model category and let K be a set of cofibrant objects of $\mathcal{M}$. If $\mathcal{C}$ is the class of K-cellular equivalences, then the class of cofibrant objects of $R_{\mathcal{C}}\mathcal{M}$ (see Theorem 5.1.1) equals the class of K-cellular objects (see Definition 5.1.4).*

PROOF. This follows from Theorem 5.1.5 and part 2 of Theorem 5.1.1. □

5.1.7. Examples of right proper cellular model categories.

PROPOSITION 5.1.8. *The categories SS, Top, SS_*, and Top_* are right proper cellular model categories.*

PROPOSITION 5.1.9. *If $\mathcal{M}$ is a right proper cellular model category and $\mathcal{C}$ is a small category, then the diagram category $\mathcal{M}^{\mathcal{C}}$ is a right proper cellular model category.*

PROPOSITION 5.1.10. *If $\mathcal{M}$ is a right proper cellular model category and Z is an object of $\mathcal{M}$, then the overcategory $(\mathcal{M}\downarrow Z)$ is a right proper cellular model category.*

PROPOSITION 5.1.11. *If $\mathcal{M}$ is a right proper cellular simplicial model category and $\mathcal{C}$ is a small simplicial category, then the category $\mathcal{M}^{\mathcal{C}}$ of simplicial diagrams is a right proper cellular model category.*

5.2. Horns on K and K-colocal equivalences

DEFINITION 5.2.1. Let $\mathcal{M}$ be a right proper cellular model category with generating cofibrations I and generating trivial cofibrations J, and let K be a set of objects of $\mathcal{M}$.

- A *full set of horns on* K is a set $\Lambda(K)$ of maps obtained by choosing a cosimplicial resolution $\widetilde{\boldsymbol{A}}$ of every element A of K and letting
$$\Lambda(K) = \{\widetilde{\boldsymbol{A}} \otimes \partial\Delta[n] \to \widetilde{\boldsymbol{A}} \otimes \Delta[n] \mid A \in S, n \geq 0\}.$$
(This is exactly a full set of horns on the maps from the initial object of $\mathcal{M}$ to the elements of K; see Definition 4.2.1.) If K consists of the single object A, then $\Lambda(K)$ is the set of maps
$$\Lambda\{A\} = \{\widetilde{\boldsymbol{A}} \otimes \partial\Delta[n] \to \widetilde{\boldsymbol{A}} \otimes \Delta[n] \mid n \geq 0\}$$
for some cosimplicial resolution $\widetilde{\boldsymbol{A}}$ of A, and it will also be called a *full set of horns on* A.
- An *augmented set of* K*-horns* is a set $\overline{\Lambda(K)}$ of maps
$$\overline{\Lambda(K)} = \Lambda(K) \cup J$$
for some full set of horns $\Lambda(K)$ on K. If K consists of the single object A, then $\overline{\Lambda(K)}$ will also be denoted $\overline{\Lambda\{A\}}$, and will be called *an augmented set of* A*-horns.*

DEFINITION 5.2.2. Let $\mathcal{M}$ be a right proper cellular model category and let K be a set of objects of $\mathcal{M}$.

- A $\overline{\Lambda(K)}$*-injective* (see Definition 5.2.1) is a map with the right lifting property with respect to every element of $\overline{\Lambda(K)}$.
- A $\overline{\Lambda(K)}$*-cofibration* (see Definition 10.5.2) is a map with the left lifting property with respect to every $\overline{\Lambda(K)}$-injective.
- A *relative* $\overline{\Lambda(K)}$*-cell complex* (see Definition 10.5.8) is a transfinite composition of pushouts of elements of $\overline{\Lambda(K)}$.
- An object of $\mathcal{M}$ is a $\overline{\Lambda(K)}$*-cell complex* if the map to it from the initial object of $\mathcal{M}$ is a relative $\overline{\Lambda(K)}$-cell complex.

PROPOSITION 5.2.3. *Let $\mathcal{M}$ be a right proper cellular model category. If K is a set of objects of $\mathcal{M}$, then there is a functorial factorization of every map $X \to Y$ as $X \xrightarrow{p} W \xrightarrow{q} Y$ where p is a relative $\overline{\Lambda(K)}$-cell complex and q is a $\overline{\Lambda(K)}$-injective.*

PROOF. This follows from Proposition 12.4.6. □

PROPOSITION 5.2.4. *Let $\mathcal{M}$ be a right proper cellular model category and let K be a set of objects of $\mathcal{M}$. If Y is a fibrant object of $\mathcal{M}$, then a map $g\colon X \to Y$ is a $\overline{\Lambda(K)}$-injective if and only if it is both a fibration and a K-colocal equivalence.*

PROOF. Definition 11.1.2 implies that g is a fibration if and only if it is a J-injective. If this is the case, then X is also fibrant, and so Proposition 16.4.5 implies that g is a K-colocal equivalence if and only if it is a $\Lambda(K)$-injective. □

The requirement in Proposition 5.2.4 that Y be fibrant is essential; see Example 5.2.7.

PROPOSITION 5.2.5. *Let $\mathcal{M}$ be a right proper cellular model category. If K is a set of objects of $\mathcal{M}$, then a relative $\overline{\Lambda(K)}$-cell complex is a K-colocal cofibration.*

PROOF. Let $f\colon A \to B$ be a relative $\overline{\Lambda(K)}$-cell complex; we must show that if $g\colon X \to Y$ is both a K-colocal weak equivalence and a K-colocal fibration then f has the left lifting property with respect to g. Proposition 8.1.23 implies that we can choose a fibrant approximation $\hat{g}$ to g such that $\hat{g}$ is a fibration, and Proposition 13.2.1 implies that it is sufficient to show that f has the left lifting property with respect to $\hat{g}$. Since Proposition 5.2.4 implies that $\hat{g}$ is a $\overline{\Lambda(K)}$-injective, the result follows from Proposition 10.5.10. □

PROPOSITION 5.2.6. *Let $\mathcal{M}$ be a right proper cellular model category and let K be a set of objects of $\mathcal{M}$. If every object of $\mathcal{M}$ is fibrant and $\overline{\Lambda(K)}$ is an augmented set of K-horns, then every $\overline{\Lambda(K)}$-injective is both a fibration and a K-colocal equivalence and every relative $\overline{\Lambda(K)}$-cell complex is a K-colocal cofibration.*

PROOF. This follows from Proposition 5.2.4 and Proposition 5.2.5. □

EXAMPLE 5.2.7. We present here an example of an $\overline{\Lambda(K)}$-injective that is not a K-colocal equivalence. Let $\mathcal{M} = \mathrm{SS}_*$ (the category of pointed simplicial sets), and let $K = \{A\}$, where A is the quotient of $\Delta[1]$ obtained by identifying the two vertices of $\Delta[1]$ (so that the geometric realization of A is homeomorphic to a circle). Let Y be $\partial\Delta[2]$, i.e., let Y consist of three 1-simplices with vertices identified so that its geometric realization is homeomorphic to a circle. Let X be the simplicial set built from six 1-simplices by identifying vertices so that the geometric realization of X is homeomorphic to a circle and there is a map $g\colon X \to Y$ whose geometric realization is the double cover of the circle. The map g is a fibration, since it is a fiber bundle with fiber two discrete points (see [**4**, Section IV.2] or [**49**, Lemma 11.9]).

Since no nondegenerate 1-simplex of X has its vertices equal, the only pointed map from A to X is the constant map to the basepoint. One can now show by induction on n that the only pointed map from $A \wedge \Delta[n]^+$ to X is the constant map to the basepoint. Thus, $\mathrm{Map}(A, X)$ has only one simplex in each dimension. Similarly, $\mathrm{Map}(A, Y)$ has only one simplex in each dimension, and so the map $g_*\colon \mathrm{Map}(A, X) \to \mathrm{Map}(A, Y)$ is an isomorphism. Thus, g is a $\overline{\Lambda(K)}$-injective.

To see that g is not an A-colocal equivalence, we note that $\mathrm{Sing}|g|\colon \mathrm{Sing}|X| \to \mathrm{Sing}|Y|$ is a fibrant approximation to g, and the map $\mathrm{Map}(A, \mathrm{Sing}|X|) \to \mathrm{Map}(A, \mathrm{Sing}|Y|)$ is isomorphic to the map $\mathrm{Map}(|A|, |X|) \to \mathrm{Map}(|A|, |Y|)$ (see Lemma 1.1.10). Since the map $|g|\colon |X| \to |Y|$ is homeomorphic to the double covering map of the circle, the induced map $\mathrm{Map}(|A|, |X|) \to \mathrm{Map}(|A|, |Y|)$ is not surjective on the set of components, and so g is not an A-colocal equivalence.

REMARK 5.2.8. Example 5.2.7 shows that, if $\mathcal{M} = \mathrm{SS}_*$, then not every $\overline{\Lambda(K)}$-injective need be a K-colocal weak equivalence. Since the $\overline{\Lambda(K)}$-cofibrations are exactly the maps with the left lifting property with respect to all $\overline{\Lambda(K)}$-injectives, this implies that the K-colocal cofibrations must consist of more than just the $\overline{\Lambda(K)}$-cofibrations (see Proposition 5.2.5). However, if $\mathcal{M}$ is a right proper cellular model category in which every object is fibrant (e.g., if $\mathcal{M} = \mathrm{Top}_*$), then Proposition 5.2.4 implies that the K-colocal cofibrations are exactly the $\overline{\Lambda(K)}$-cofibrations. This is why the K-colocal model category structure on $\mathcal{M}$ is cellular if every object of $\mathcal{M}$ is fibrant (see Theorem 5.1.1 part 3).

5.3. K-colocal cofibrations

The main results of this section are Proposition 5.3.3 and Proposition 5.3.5, which together provide the factorizations needed for the proof of Theorem 5.1.1 in Section 5.4.

LEMMA 5.3.1. *Let $\mathcal{M}$ be a right proper cellular model category. If K is a set of objects of $\mathcal{M}$, then every K-colocal cofibration is a cofibration.*

PROOF. This follows from Proposition 7.2.3 and Proposition 3.1.5. □

LEMMA 5.3.2. *Let $\mathcal{M}$ be a right proper cellular model category. If K is a set of objects of $\mathcal{M}$, then a map $g\colon X \to Y$ is both a K-colocal cofibration and a K-colocal weak equivalence if and only if it is a trivial cofibration.*

PROOF. If g is a trivial cofibration then Proposition 3.1.5 implies that it is a K-colocal weak equivalence and Proposition 7.2.3 implies that it is a K-colocal cofibration.

Conversely, let $g\colon X \to Y$ be both a K-colocal cofibration and a K-colocal weak equivalence. If we factor g as $X \xrightarrow{p} W \xrightarrow{q} Y$ where p is a trivial cofibration and q is a fibration, then Proposition 3.1.5 and the "two out of three" property of K-colocal equivalences (see Proposition 3.2.3) imply that q is a K-colocal equivalence. Thus, p has the left lifting property with respect to q, and so the retract argument (see Proposition 7.2.2) implies that g is a retract of the trivial cofibration p and is thus a trivial cofibration (see axiom M3 of Definition 7.1.3). □

PROPOSITION 5.3.3. *Let $\mathcal{M}$ be a right proper cellular model category. If K is a set of objects of $\mathcal{M}$, then there is a functorial factorization of every map $g\colon X \to Y$ in $\mathcal{M}$ as $X \xrightarrow{p} W \xrightarrow{q} Y$ in which p is both a K-colocal cofibration and a K-colocal weak equivalence and q is a K-colocal fibration.*

PROOF. This follows from Lemma 5.3.2 and the existence of the functorial factorization into a trivial cofibration followed by a fibration. □

LEMMA 5.3.4. *Let $\mathcal{M}$ be a right proper cellular model category and let K be a set of objects of $\mathcal{M}$. If $g\colon A \to B$ is a cofibration, $h\colon B \to C$ is a weak equivalence, and the composition $hg\colon A \to C$ is a K-colocal cofibration, then g is a K-colocal cofibration.*

PROOF. If $f\colon X \to Y$ is both a K-colocal weak equivalence and a K-colocal fibration, then Proposition 8.1.23 implies that we can choose a fibrant approximation $\hat{f}\colon \widehat{X} \to \widehat{Y}$ to f such that $\hat{f}$ is a fibration. Proposition 3.1.5 and Proposition 3.2.3 imply that $\hat{f}$ is a K-colocal weak equivalence, and (since $\mathcal{M}$ is a right proper model category) Proposition 13.2.1 implies that it is sufficient to show that g has the left lifting property with respect to $\hat{f}$.

Suppose that we have the commutative solid arrow diagram

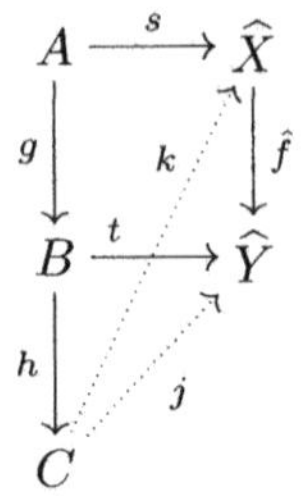

In the category $(A \downarrow \mathcal{M})$ of objects of $\mathcal{M}$ under A, the map h is a weak equivalence of cofibrant objects (see Lemma 5.3.1) and $\widehat{Y}$ is fibrant. Thus, Corollary 7.7.4 implies that there is a map $j\colon C \to \widehat{Y}$ in $(A \downarrow \mathcal{M})$ such that $jh \simeq t$ in $(A \downarrow \mathcal{M})$. Since hg is a K-colocal cofibration and $\hat{f}$ is both a K-colocal weak equivalence and a K-colocal fibration, there exists a map $k\colon C \to \widehat{X}$ such that $khg = s$ and $\hat{f}k = j$.

Since $\hat{f}kh = jh \simeq t$ in $(A \downarrow \mathcal{M})$, if we let $u = kh$, then $u\colon B \to \widehat{X}$, and $\hat{f}u \simeq t$ in $(A \downarrow \mathcal{M})$. Since B is cofibrant in $(A \downarrow \mathcal{M})$ and $\hat{f}$ is a fibration, the homotopy lifting property of fibrations (see Proposition 7.3.11) implies that there is a map $v\colon B \to \widehat{X}$ in $(A \downarrow \mathcal{M})$ such that $v \simeq u$ and $\hat{f}v = t$. The map v satisfies $vg = s$ and $\hat{f}v = t$, and so g has the left lifting property with respect to $\hat{f}$. □

PROPOSITION 5.3.5. *Let $\mathcal{M}$ be a right proper cellular model category. If K is a set of objects of $\mathcal{M}$, then there is a functorial factorization of every map $g\colon X \to Y$ in $\mathcal{M}$ as $X \xrightarrow{p} W \xrightarrow{q} Y$ in which p is a K-colocal cofibration and q is both a K-colocal weak equivalence and a K-colocal fibration.*

PROOF. Choose a functorial cofibrant fibrant approximation $j\colon Y \to \widehat{Y}$ to Y. Proposition 5.2.3 implies that there is a functorial factorization of the composition $jg\colon X \to \widehat{Y}$ as $X \xrightarrow{r} \widehat{W} \xrightarrow{s} \widehat{Y}$ in which r is a relative $\overline{\Lambda(K)}$-cell complex and s is a $\overline{\Lambda(K)}$-injective. If we let Z be the pullback $Y \times_{\widehat{Y}} \widehat{W}$, then we can factor the natural map $X \to Z$ in $\mathcal{M}$ as $X \xrightarrow{p} W \xrightarrow{u} Z$ where p is a cofibration and u is a trivial fibration. If we let $q = vu$, then we have the diagram

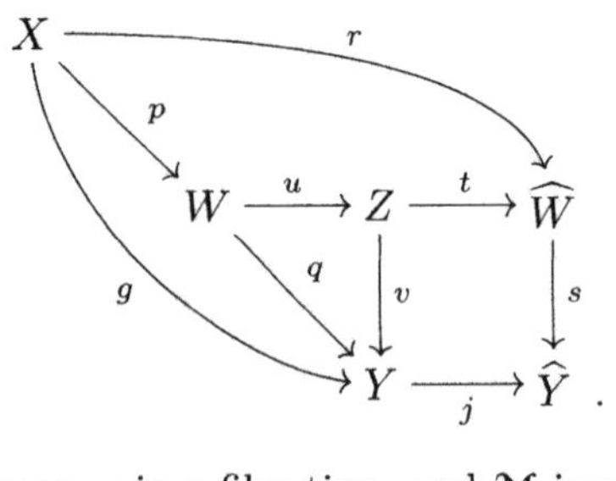

Since j is a weak equivalence, s is a fibration, and $\mathcal{M}$ is a right proper model category, t is a weak equivalence. Thus, the composition tu is a weak equivalence, and so s is a fibrant approximation to q. Since Proposition 5.2.4 implies that s is a K-colocal equivalence, q (which is the composition of two fibrations) is both a K-colocal weak equivalence and a K-colocal fibration. Since r is a K-colocal cofibration (see Proposition 5.2.5), Lemma 5.3.4 implies that p is a K-colocal cofibration. □

PROPOSITION 5.3.6. *Let $\mathcal{M}$ be a right proper cellular model category with generating cofibrations I. If K is a set of objects of $\mathcal{M}$, then a map is a K-colocal cofibration if and only if it is a retract of a relative I-cell complex $X \to Y$ in $\mathcal{M}$ for which there is a weak equivalence $Y \to Z$ in $\mathcal{M}$ such that the composition $X \to Z$ is a relative $\overline{\Lambda(K)}$-cell complex.*

PROOF. This follows from the factorization constructed in the proof of Proposition 5.3.5 and the retract argument (see Proposition 7.2.2). □

COROLLARY 5.3.7. *Let $\mathcal{M}$ be a right proper cellular model category. If K is a set of objects of $\mathcal{M}$, then an object is K-colocal if and only if it is a retract of an object X such that*

(1) *X is a cell complex in $\mathcal{M}$, and*
(2) *there is a $\overline{\Lambda(K)}$-complex Y and a weak equivalence $X \to Y$ in $\mathcal{M}$.*

PROOF. This follows from Proposition 3.4.1 and Proposition 5.3.6. □

5.4. Proof of the main theorem

This section contains the proof of Theorem 5.1.1.

5.4.1. Proof of part 1. We must show that axioms M1 through M5 of Definition 7.1.3 are satisfied.

Axiom M1 is satisfied because it is satisfied in $\mathcal{M}$, axiom M2 follows from Proposition 3.2.3, and axiom M3 follows from Proposition 3.2.4 and Lemma 7.2.8. Axiom M4 part (1) follows from the definition of K-colocal cofibration, and axiom M4 part (2) follows from Lemma 5.3.2. Axiom M5 part (1) follows from Proposition 5.3.5, and axiom M5 part (2) follows from Proposition 5.3.3.

5.4.2. Proof of part 2. This follows from Proposition 3.4.1.

5.4.3. Proof of part 3. Proposition 3.4.4 implies that $R_{\mathcal{C}}\mathcal{M}$ is right proper.

Suppose now that every object of $\mathcal{M}$ is fibrant. Since the classes of fibrations and trivial cofibrations of $R_{\mathcal{C}}\mathcal{M}$ equal those of $\mathcal{M}$, a set J of generating trivial cofibrations of $\mathcal{M}$ serves as a set of generating trivial cofibrations of $R_{\mathcal{C}}\mathcal{M}$. Thus, Proposition 5.2.6 implies that if $\overline{\Lambda(K)}$ is an augmented set of horns on K, then $R_{\mathcal{C}}\mathcal{M}$ is a cofibrantly generated model category with generating cofibrations $\overline{\Lambda(K)}$ and generating trivial cofibrations J (see Definition 11.1.2). Since the class of fibrations of $R_{\mathcal{C}}\mathcal{M}$ equals that of $\mathcal{M}$, every object of $R_{\mathcal{C}}\mathcal{M}$ is fibrant, and so it remains only to show that $R_{\mathcal{C}}\mathcal{M}$ is cellular.

Since the domains and codomains of the elements of $\overline{\Lambda(K)}$ are cofibrant in $\mathcal{M}$ and every cofibration of $R_{\mathcal{C}}\mathcal{M}$ is a cofibration of $\mathcal{M}$, Corollary 12.3.4 implies that condition 1 of Definition 12.1.1 is satisfied. Theorem 12.4.4 implies that condition 2 is satisfied, and condition 3 is satisfied because every cofibration of $R_{\mathcal{C}}\mathcal{M}$ is a cofibration of $\mathcal{M}$.

5.4.4. Proof of part 4. Axiom M6 of Definition 9.1.6 holds in $R_{\mathcal{C}}\mathcal{M}$ because it holds in $\mathcal{M}$.

For axiom M7, if $i\colon A \to B$ is a cofibration in $R_{\mathcal{C}}\mathcal{M}$ and $p\colon X \to Y$ is a fibration in $R_{\mathcal{C}}\mathcal{M}$ then i is a cofibration in $\mathcal{M}$ and p is a fibration in $\mathcal{M}$ and so the map $\mathrm{Map}(i,p)\colon \mathrm{Map}(B,X) \to \mathrm{Map}(A,X) \times_{\mathrm{Map}(A,Y)} \mathrm{Map}(B,Y)$ is a fibration of simplicial sets. If i is also a weak equivalence in $R_{\mathcal{C}}\mathcal{M}$ then i is a trivial cofibration in $R_{\mathcal{C}}\mathcal{M}$ and thus also in $\mathcal{M}$, and so $\mathrm{Map}(i,p)$ is a trivial fibration of simplicial sets. Thus, it remains only to deal with the case in which i is a cofibration in $R_{\mathcal{C}}\mathcal{M}$ and p is a trivial fibration in $R_{\mathcal{C}}\mathcal{M}$.

If i is a cofibration in $R_{\mathcal{C}}\mathcal{M}$ and p is a trivial fibration in $R_{\mathcal{C}}\mathcal{M}$ then Theorem 17.8.10 implies that (i,p) is a homotopy orthogonal pair. Let $\hat{p}\colon \widehat{X} \to \widehat{Y}$ be a fibrant approximation to p in $\mathcal{M}$ such that $\hat{p}$ is a fibration in $\mathcal{M}$. Example 16.6.13 implies that if $\widehat{\boldsymbol{X}}$ and $\widehat{\boldsymbol{Y}}$ are the simplicial objects in $\mathcal{M}$ such that $\widehat{\boldsymbol{X}}_n = \widehat{X}^{\Delta[n]}$ and $\widehat{\boldsymbol{Y}}_n = \widehat{Y}^{\Delta[n]}$ and $\widehat{\boldsymbol{X}} \to \widehat{\boldsymbol{Y}}$ is the map induced by p, then $\widehat{\boldsymbol{X}} \to \widehat{\boldsymbol{Y}}$ is a simplicial resolution of $\hat{p}$ in $\mathcal{M}$ such that $\widehat{\boldsymbol{X}} \to \widehat{\boldsymbol{Y}}$ is a Reedy fibration in $\mathcal{M}^{\boldsymbol{\Delta}^{\mathrm{op}}}$. Corollary 16.2.2 and Proposition 3.3.4 imply that $\widehat{\boldsymbol{X}} \to \widehat{\boldsymbol{Y}}$ is also a simplicial resolution of $\hat{p}$ in $R_{\mathcal{C}}\mathcal{M}$ such that $\widehat{\boldsymbol{X}} \to \widehat{\boldsymbol{Y}}$ is a Reedy fibration in $(R_{\mathcal{C}}\mathcal{M})^{\boldsymbol{\Delta}^{\mathrm{op}}}$, and so Proposition 16.3.10 implies that the map $\widehat{X}^{\Delta[n]} \to \widehat{Y}^{\Delta[n]} \times_{\widehat{Y}^{\partial\Delta[n]}} \widehat{X}^{\partial\Delta[n]}$ is a trivial fibration in $R_{\mathcal{C}}\mathcal{M}$ for every $n \ge 0$. Since $i\colon A \to B$ is a cofibration in $R_{\mathcal{C}}\mathcal{M}$, Lemma 9.4.7 now implies that the map $\hat{p}\colon \widehat{X} \to \widehat{Y}$ has the homotopy right lifting property with respect to i, and so Corollary 13.2.2 implies that the map $p\colon X \to Y$ has the homotopy right lifting property with respect to i.

5.5. K-colocal objects and K-cellular objects

This section contains the proof of Theorem 5.1.5.

PROPOSITION 5.5.1. *Let $\mathcal{M}$ be a right proper cellular model category. If K is a set of objects in $\mathcal{M}$, then the homotopy colimit of a diagram of K-colocal objects is a K-colocal object.*

PROOF. Let $\mathcal{C}$ be a small category and let $\boldsymbol{B}\colon \mathcal{C} \to \mathcal{M}$ be a diagram of K-colocal objects. Theorem 19.4.1 implies that $\operatorname{hocolim} \boldsymbol{B}$ is cofibrant. If $p\colon X \to Y$ is a K-colocal equivalence and $\hat{p}\colon \widehat{X} \to \widehat{Y}$ is a fibrant approximation to p, then Theorem 19.4.4 implies that the map $\mathrm{map}(\operatorname{hocolim} \boldsymbol{B}, \widehat{X}) \to \mathrm{map}(\operatorname{hocolim} \boldsymbol{B}, \widehat{Y})$ is weakly equivalent to the map $\operatorname{holim} \mathrm{map}(\boldsymbol{B}, \widehat{X}) \to \operatorname{holim} \mathrm{map}(\boldsymbol{B}, \widehat{Y})$. Theorem 19.4.2 implies that this map is a weak equivalence, and so Theorem 17.6.3 implies that $\mathrm{map}(\operatorname{hocolim} \boldsymbol{B}, X) \to \mathrm{map}(\operatorname{hocolim} \boldsymbol{B}, Y)$ is a weak equivalence. □

LEMMA 5.5.2. *Let $\mathcal{M}$ be a right proper cellular model category and let K be a set of objects in $\mathcal{M}$. If X is an K-colocal object of $\mathcal{M}$ and L is a simplicial set, then the object $X \otimes L$ is K-colocal.*

PROOF. The result follows from Proposition 5.5.1, Theorem 19.9.1, Proposition 15.10.4, and Lemma 3.2.1. □

PROPOSITION 5.5.3. *Let $\mathcal{M}$ be a right proper cellular model category and let K be a set of cofibrant objects of $\mathcal{M}$. If $\mathcal{C}$ is a class of cofibrant objects of $\mathcal{M}$ that contains K and is closed under homotopy colimits and weak equivalences, then $\mathcal{C}$ contains all $\overline{\Lambda(K)}$-cell complexes (see Definition 5.2.2).*

PROOF. We will prove this by a transfinite induction on the ordinal indexing the λ-sequence whose colimit is the $\overline{\Lambda(K)}$-cell complex. Lemma 5.5.2 implies that $\mathcal{C}$ contains $\widetilde{\boldsymbol{A}} \otimes \partial\Delta[n]$ for every $n \ge 0$ and every cosimplicial resolution $\widetilde{\boldsymbol{A}}$ of every element A of K, and so the inductive step for successor ordinals follows from Proposition 5.5.1 and Proposition 19.9.4. The induction step for limit ordinals follows from Proposition 5.5.1 and Theorem 19.9.1. □

PROOF OF THEOREM 5.1.5. Proposition 5.5.1 and Lemma 3.2.1 imply that the class of K-colocal objects is closed under homotopy colimits and weak equivalences. If $\mathcal{C}$ is a class of cofibrant objects of $\mathcal{M}$ that contains K and is closed under homotopy colimits and weak equivalences, then Proposition 5.5.3 implies that $\mathcal{C}$ contains all $\overline{\Lambda(K)}$-cell complexes, and Proposition 19.9.3 implies that $\mathcal{C}$ contains all retracts of $\overline{\Lambda(K)}$-cell complexes, and so the result follows from Proposition 3.4.1 and Proposition 5.3.6. $\square$

CHAPTER 6

Fiberwise Localization

If $\mathcal{M}$ is one of our categories of spaces (see Section 1.1.3), $\mathcal{C}$ is a class of maps in $\mathcal{M}$, and $p\colon Y \to Z$ is a fibration in $\mathcal{M}$, then a fiberwise $\mathcal{C}$-localization of p should be a map from p to another fibration q over Z

$$\begin{array}{ccc} Y & \xrightarrow{\quad i \quad} & \widehat{Y} \\ & {\scriptstyle p}\searrow \quad \swarrow{\scriptstyle q} & \\ & Z & \end{array}$$

that "localizes the fibers of p", i.e., for every point z in Z the map $p^{-1}(z) \to q^{-1}(z)$ should be a $\mathcal{C}$-localization of $p^{-1}(z)$. The actual definition is a generalization of this that deals with maps p that may not be fibrations (see Definition 6.1.1).

In this chapter, we show that if $\mathcal{M}$ is a category of *unpointed* spaces (see Section 1.1.3) and S is a set of maps in $\mathcal{M}$, then every map $p\colon Y \to Z$ in $\mathcal{M}$ has a natural fiberwise S-localization $Y \to \widehat{Y} \to Z$. We also show that if $p\colon Y \to Z$ is a map in $\mathcal{M}$ and $Y \to \widehat{Y}' \to Z$ is some other fiberwise S-localization of p, then there is a map $\widehat{Y} \to \widehat{Y}'$ under Y and over Z, unique up to simplicial homotopy in $(Y \downarrow \mathcal{M} \downarrow Z)$ (see Definition 11.8.1 and Definition 11.8.3), and any such map is a weak equivalence.

We construct our fiberwise localization for the categories of unpointed spaces Top and SS (see Notation 1.1.4). Since the pointed localization of a connected space is weakly equivalent to its unpointed localization (see Theorem 1.8.12), our construction will also serve as a fiberwise pointed localization for fibrations with connected fibers. This is the strongest possible result; in Proposition 6.1.4, we show that it is not possible to construct a fiberwise pointed localization for fibrations with fibers that are not connected.

6.1. Fiberwise localization

DEFINITION 6.1.1. Let $\mathrm{Spc}_{(*)}$ be one of our categories of spaces (see Notation 1.1.4) and let $\mathcal{C}$ be a class of maps in $\mathrm{Spc}_{(*)}$. If $p\colon Y \to Z$ is a map in $\mathrm{Spc}_{(*)}$, then a *fiberwise* $\mathcal{C}$*-localization* of p is a factorization $Y \xrightarrow{i} \widehat{Y} \xrightarrow{q} Z$ of p such that

(1) q is a fibration, and
(2) for every point z of Z the induced map of homotopy fibers (see Definition 13.4.3) $\mathrm{HFib}_z(p) \to \mathrm{HFib}_z(q)$ is a $\mathcal{C}$-localization of $\mathrm{HFib}_z(p)$ (see Definition 3.2.16).

PROPOSITION 6.1.2. *Let* $\mathrm{Spc}_{(*)}$ *be one of our categories of spaces (see Notation 1.1.4). If* $\mathcal{C}$ *is a class of maps in* $\mathrm{Spc}_{(*)}$*,* $p\colon Y \to Z$ *is a fibration in* $\mathrm{Spc}_{(*)}$*,* *and $Y \xrightarrow{i} \widehat{Y} \xrightarrow{q} Z$ is a factorization of p, then this factorization is a fiberwise $\mathcal{C}$-localization of p if and only if*

(1) *q is a fibration, and*
(2) *for every point z of Z the induced map of fibers $p^{-1}(z) \to q^{-1}(z)$ is a $\mathcal{C}$-localization of $p^{-1}(z)$.*

PROOF. This follows from Proposition 13.4.6. □

The following theorem summarizes the main results of this chapter.

THEOREM 6.1.3. *If Spc is a category of unpointed spaces (see Notation 1.1.4) and S is a set of maps in Spc, then there is a natural factorization of every map $p\colon X \to Z$ as $X \xrightarrow{i} \widetilde{\mathrm{L}}_S X \xrightarrow{q} Z$ such that*

(1) *q is a fibration with S-local fibers,*
(2) *for every point z in Z the induced map of homotopy fibers $\mathrm{HFib}_z(p) \to \mathrm{HFib}_z(q)$ (see Definition 13.4.3) is an S-localization of $\mathrm{HFib}_z(p)$,*
(3) *i is both a cofibration and an S-local equivalence,*
(4) *if we have a solid arrow diagram*

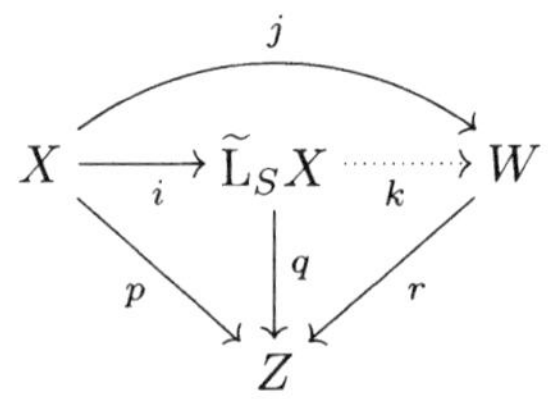

in which r is a fibration with S-local fibers, then there is a map $k\colon \widetilde{\mathrm{L}}_S X \to W$, unique up to simplicial homotopy in $(X \downarrow \mathrm{Spc} \downarrow Z)$, such that $ki = j$, and
(5) *if we have a diagram as in the previous part such that for every point z in Z the map $\mathrm{HFib}_z(p) \to \mathrm{HFib}_z(r)$ of homotopy fibers over z induced by j is an S-local equivalence (i.e., if j is another fiberwise S-localization of p), then the map k is a weak equivalence.*

PROPOSITION 6.1.4. *Let $f\colon A \to B$ be the inclusion $\mathrm{S}^2 \to \mathrm{D}^3$ in Top_*, the category of pointed topological spaces. If $X = \mathrm{S}^2 \times \mathbb{R}$, $Z = \mathrm{S}^1$, and $p\colon X \to Z$ is the composition of the projection $\mathrm{S}^2 \times \mathbb{R} \to \mathbb{R}$ with the universal covering map $\mathbb{R} \to \mathrm{S}^1$, then there is no fiberwise f-localization of p in the category Top_* of pointed spaces.*

PROOF. The fiber F of p is a countable disjoint union of copies of S^2, and so if there were a fiberwise pointed localization of p, its fiber would have countably many path components: one contractible, and the others weakly equivalent to S^2 (see Corollary 1.8.10).

To see that this is not possible, note that $\pi_1 Z$ acts transitively on $\pi_0 F$, and so $\pi_1 Z$ would act transitively on the path components of the fiber of any fiberwise localization of p. Since $\pi_1 Z$ acts on the fiber through (unpointed) weak equivalences, this is impossible, and so there does not exist a fiberwise pointed localization of p. □

6.2. The fiberwise local model category structure

DEFINITION 6.2.1. Let S be a set of maps in Spc. If Z is a space in Spc, then we define $\mathrm{Fib}_Z(S)$ (which we call the *set of elements of S fiberwise over Z*) to be the set of maps in $(\mathrm{Spc}\downarrow Z)$

$$\begin{array}{ccc} A & \xrightarrow{f} & B \\ & \searrow \quad \swarrow & \\ & Z & \end{array}$$

where $f\colon A \to B$ is an element of S and the images of the maps $A \to Z$ and $B \to Z$ are a single point of Z.

PROPOSITION 6.2.2. *If Z is a space in* Spc*, then the category* $(\mathrm{Spc}\downarrow Z)$ *of objects of* Spc *over Z is a left proper cellular model category.*

PROOF. This follows from Proposition 4.1.6. □

THEOREM 6.2.3. *Let Z be a space in* Spc*, and let S be a set of maps in* Spc*. If we define*

(1) a fiberwise over Z S-local equivalence *to be a $\mathrm{Fib}_Z(S)$-local equivalence in* $(\mathrm{Spc}\downarrow Z)$ *(see Definition 3.1.4),*
(2) a fiberwise over Z S-local cofibration *to be a $\mathrm{Fib}_Z(S)$-local cofibration, and*
(3) a fiberwise over Z S-local fibration *to be a $\mathrm{Fib}_Z(S)$-local fibration,*

then there is a simplicial model category structure on $(\mathrm{Spc}\downarrow Z)$ *in which the weak equivalences are the fiberwise over Z S-local equivalences, the cofibrations are the fiberwise over Z S-local cofibrations, and the fibrations are the fiberwise over Z S-local fibrations.*

PROOF. This follows from Theorem 4.1.1 and Proposition 6.2.2. □

PROPOSITION 6.2.4. *If S is a set of maps in* Spc *and Z is a space in* Spc*, then an object of* $(\mathrm{Spc}\downarrow Z)$ *is fibrant in the fiberwise over Z S-local model category structure if and only if it is a fibration and the fiber over every point of Z is an S-local space.*

PROOF. This follows from Proposition 3.4.1. □

6.3. Localizing the fiber

The purpose of this section is to prove the following theorem.

THEOREM 6.3.1. *If S is a set of maps in* Spc*, Z is a space in* Spc*, and*

$$\begin{array}{ccc} X & \longrightarrow & Y \\ & {\scriptstyle p}\searrow \quad \swarrow{\scriptstyle q} & \\ & Z & \end{array}$$

is a $\overline{\Lambda\big(\mathrm{Fib}_Z(S)\big)}$-cofibration (see Definition 4.2.1), then for every point z of Z the induced map of homotopy fibers $\mathrm{HFib}_z(p) \to \mathrm{HFib}_z(q)$ is an S-local equivalence.

The proof of Theorem 6.3.1 is at the end of this section, on page 98.

PROPOSITION 6.3.2. *If $q\colon X \to Z$ is a map of simplicial sets and z is a point in Z, then there is a contractible simplicial set C (which depends naturally on the pair (Z,z)) and a natural (ΔC)-diagram (see Definition 15.1.16) of simplicial sets $\boldsymbol{F}\colon (\Delta C) \to \mathrm{SS}$ such that*

(1) *for every simplex σ of C there is a simplex τ of Z such that $\boldsymbol{F}(\sigma) = \tilde{q}(\tau)$ (see Example 18.9.6), and*
(2) *there is a natural weak equivalence $\operatorname{hocolim} \boldsymbol{F} \cong \mathrm{HFib}_z(q)$ (where $\mathrm{HFib}_z(q)$ is the homotopy fiber of q over z).*

By "natural" we mean that the simplicial set C is a functor of the pair (Z,z) and, for a fixed pair (Z,z), the diagram $\boldsymbol{F}$ is a functor of the object $q\colon X \to Z$ of $(\mathrm{SS}\downarrow Z)$.

PROOF. If $* \to Z$ is the map to the point z in Z, let $* \to C \xrightarrow{p} Z$ be a natural factorization of it into a trivial cofibration followed by a fibration. The homotopy fiber of q over z is then naturally weakly equivalent to the pullback of the diagram $C \xrightarrow{p} Z \xleftarrow{q} X$ (see Proposition 13.3.7). If we let F be that pullback and $r\colon F \to C$ its projection onto C, then the construction of Example 18.9.6 applied to r yields a diagram $\boldsymbol{F}\colon (\Delta C) \to \mathrm{SS}$ that satisfies condition 1. Proposition 18.9.7 implies that $\boldsymbol{F}$ is Reedy cofibrant, and so condition 2 follows from Theorem 19.9.1 and the natural isomorphism $\operatorname{colim} \boldsymbol{F} \approx F$. $\square$

PROPOSITION 6.3.3 (E. Dror Farjoun, [**23**]). *Let S be a set of maps in SS, let Z be a simplicial set, let $p\colon X \to Z$ and $q\colon Y \to Z$ be objects of $(\mathrm{SS}\downarrow Z)$, and let*

$$\begin{array}{ccc} X & \xrightarrow{\quad g \quad} & Y \\ & {\scriptstyle p}\searrow \quad \swarrow{\scriptstyle q} & \\ & Z & \end{array}$$

be a map in $(\mathrm{SS}\downarrow Z)$. If for every simplex σ of Z the induced map $\tilde{p}(\sigma) \to \tilde{q}(\sigma)$ (see Example 18.9.6) is an S-local equivalence, then for every point z in Z the induced map of homotopy fibers $\mathrm{HFib}_z(p) \to \mathrm{HFib}_z(q)$ is an S-local equivalence.

PROOF. This follows from Proposition 6.3.2 and Lemma 3.2.5. $\square$

LEMMA 6.3.4. *If $f\colon A \to B$ is a cofibration in SS, Z is a space in Top,*

$$\begin{array}{ccc} X & \xrightarrow{\quad g \quad} & Y \\ & {\scriptstyle p}\searrow \quad \swarrow{\scriptstyle q} & \\ & Z & \end{array}$$

is a map in $(\mathrm{Top}\downarrow Z)$, and z is a point in Z, then the induced map of homotopy fibers $\mathrm{HFib}_z(p) \to \mathrm{HFib}_z(q)$ is a $|f|$-local equivalence if and only if the induced map of the corresponding homotopy fibers of $(\operatorname{Sing} p)\colon \operatorname{Sing} X \to \operatorname{Sing} Z$ and $(\operatorname{Sing} q)\colon \operatorname{Sing} Y \to \operatorname{Sing} Z$ is an f-local equivalence.

PROOF. Proposition 13.4.10 implies that the "homotopy fiber" and "total singular complex" functors commute up to a natural weak equivalence, and so the result follows from Proposition 1.2.36. $\square$

PROPOSITION 6.3.5. *Let $f\colon A \to B$ be an inclusion of cell complexes in* Spc*, and let Z be a space in* Spc*. If the map*

$$\begin{array}{ccc} X & \xrightarrow{\quad g \quad} & Y \\ & {\scriptstyle p}\searrow \quad \swarrow{\scriptstyle q} & \\ & Z & \end{array}$$

in $(\mathrm{Spc}\downarrow Z)$ *is a pushout of an element of* $\overline{\Lambda(\mathrm{Fib}_Z\{f\})}$ *(see Definition* 4.2.1*), then g is both a cofibration and an f-local equivalence in* Spc*, and for every point z in Z the induced map of homotopy fibers* $\mathrm{HFib}_z(p) \to \mathrm{HFib}_z(q)$ *is an f-local equivalence.*

PROOF. There are two types of maps in the set $\overline{\Lambda(\mathrm{Fib}_Z\{f\})}$. The first type is an element of $\Lambda(\mathrm{Fib}_Z\{f\})$ (see Definition 4.2.1); a map of this type is an S-local equivalence in Spc, and its domain and codomain lie over a single point z of Z. The second type is a generating trivial cofibration of Spc. If Y is obtained from X by pushing out a map of the second type, then the map g is a weak equivalence, and so the induced map of homotopy fibers is a weak equivalence. Thus, we need only consider the case in which Y is obtained from X by pushing out an element of $\Lambda(\mathrm{Fib}_Z\{f\})$.

If $\mathrm{Spc} = \mathrm{SS}$, then for each simplex σ of Z, the map $\tilde{p}(\sigma) \to \tilde{q}(\sigma)$ (see Example 18.9.6) is obtained by pushing out one copy of our element of $\Lambda(\mathrm{Fib}_Z\{f\})$ for each vertex of σ that equals z. Thus, $\tilde{p}(\sigma) \to \tilde{q}(\sigma)$ is an S-local equivalence, and so the lemma follows from Proposition 6.3.3. Thus, we need only consider the case $\mathrm{Spc} = \mathrm{Top}$.

If $\mathrm{Spc} = \mathrm{Top}$, then Proposition 1.2.36 and Proposition 1.2.10 imply that it is sufficient to show that $\mathrm{Sing}(\mathrm{HFib}_z(p)) \to \mathrm{Sing}(\mathrm{HFib}_z(q))$ is a $(\mathrm{Sing}\, f)$-local equivalence, and Proposition 13.4.10 implies that this is equivalent to showing that $\mathrm{HFib}_z(\mathrm{Sing}\, p) \to \mathrm{HFib}_z(\mathrm{Sing}\, q)$ is a $(\mathrm{Sing}\, f)$-local equivalence (where we also use the symbol z to denote the vertex of $\mathrm{Sing}\, Z$ corresponding to the point z of Z).

Let $A \times |\Delta[n]| \amalg_{A\times|\partial\Delta[n]|} B \times |\partial\Delta[n]| \to B \times |\Delta[n]|$ be the element of $\Lambda(\mathrm{Fib}_Z\{f\})$ in the pushout that transforms X into Y. We have a pushout square

$$\begin{array}{ccc} A \times |\Delta[n]| \amalg_{A\times|\partial\Delta[n]|} B \times |\partial\Delta[n]| & \longrightarrow & X \\ \downarrow & & \downarrow \\ B \times |\Delta[n]| & \longrightarrow & Y \end{array}$$

and Proposition 13.5.5 implies that $\mathrm{Sing}\, Y$ is weakly equivalent to the pushout

$$\begin{array}{ccc} \mathrm{Sing}\big(A \times |\Delta[n]| \amalg_{A\times|\partial\Delta[n]|} B \times |\partial\Delta[n]|\big) & \longrightarrow & \mathrm{Sing}\, X \\ \downarrow & & \downarrow \\ \mathrm{Sing}\big(B \times |\Delta[n]|\big) & \longrightarrow & Y' \end{array}$$

If we let $q'\colon Y' \to Z$ be the structure map of Y' in $(\mathrm{SS}\downarrow(\mathrm{Sing}\, Z))$, then for every simplex $\sigma \in \mathrm{Sing}\, Z$ the map $(\widetilde{\mathrm{Sing}\, p})(\sigma) \to \tilde{q}'(\sigma)$ (see Example 18.9.6) is obtained by pushing out one copy of $\mathrm{Sing}\big(A \times |\Delta[n]| \amalg_{A\times|\partial\Delta[n]|} B \times |\partial\Delta[n]|\big) \to \mathrm{Sing}\big(B \times |\Delta[n]|\big)$

for each vertex of σ that equals the image of $\mathrm{Sing}\big(B \times |\Delta[n]|\big)$ in $\mathrm{Sing}\, Z$. Proposition 1.2.36 implies that this is a $(\mathrm{Sing}\, f)$-local equivalence, and so Proposition 6.3.3 implies that $\mathrm{HFib}_z(\mathrm{Sing}\, p) \to \mathrm{HFib}_z(q')$ is a $(\mathrm{Sing}\, f)$-local equivalence. This implies that $\mathrm{HFib}_z(\mathrm{Sing}\, p) \to \mathrm{HFib}_z(\mathrm{Sing}\, q)$ is a $(\mathrm{Sing}\, f)$-local equivalence, and the proof is complete. □

PROOF OF THEOREM 6.3.1. Every $\mathrm{Fib}_Z(S)$-cofibration is a retract of a transfinite composition of pushouts of elements of $\overline{\Lambda\big(\mathrm{Fib}_Z(S)\big)}$ (see Corollary 10.5.22). Since S-local equivalences are closed under retracts, Proposition 13.4.9 implies that a retract of a map in $(\mathrm{Spc} \downarrow Z)$ inducing an S-local equivalence of homotopy fibers over z must also induce an S-local equivalence of homotopy fibers over z. Thus, it is sufficient to show that if

$$\begin{array}{ccccccccc} X_0 & \longrightarrow & X_1 & \longrightarrow & X_2 & \longrightarrow \cdots \longrightarrow & X_\beta & \longrightarrow \cdots & \qquad (\beta < \lambda) \\ & {\scriptstyle p_0} \searrow & & {\scriptstyle p_1} \searrow & \downarrow {\scriptstyle p_2} & \swarrow {\scriptstyle p_\beta} & & & \\ & & & & Z & & & & \end{array}$$

is a transfinite composition of pushouts of elements of $\overline{\Lambda\big(\mathrm{Fib}_Z(S)\big)}$, then the induced map of homotopy fibers $\mathrm{HFib}_z(p_0) \to \mathrm{HFib}_z(\mathrm{colim}_{\beta<\lambda}\, p_\beta)$ is an S-local equivalence.

If $\mathrm{Spc} = \mathrm{SS}$, then we choose a factorization $* \xrightarrow{s} C \xrightarrow{t} Z$ of the map $* \to Z$ whose image is z such that s is a trivial cofibration and t is a fibration, and Proposition 13.4.9 implies that each $\mathrm{HFib}_z(X_\beta)$ is naturally weakly equivalent to $C \times_Z X_\beta$. Each map $C \times_Z X_\beta \to C \times_Z X_{\beta+1}$ is an inclusion (and, thus, a cofibration), and Proposition 6.3.5 implies that it is an S-local equivalence. Thus, it is a trivial cofibration in the S-local model category structure on SS (see Theorem 4.1.1). Proposition 10.3.4 now implies that the transfinite composition $C \times_Z X_0 \to \mathrm{colim}_{\beta<\lambda}(C \times_Z X_\beta) \approx C \times_Z (\mathrm{colim}_{\beta<\lambda} X_\beta)$ is an S-local equivalence, and Proposition 13.4.9 implies that this is weakly equivalent to the map $\mathrm{HFib}_z(p_0) \to \mathrm{HFib}_z(\mathrm{colim}_{\beta<\lambda}\, p_\beta)$.

If $\mathrm{Spc} = \mathrm{Top}$, then Proposition 13.4.10 and Proposition 1.2.36 imply that it is sufficient to show that the induced map of homotopy fibers of total singular complexes $\mathrm{HFib}_z(p_0) \to \mathrm{HFib}_z(\mathrm{colim}_{\beta<\lambda}\, \mathrm{Sing}\, p_\beta) \approx \mathrm{HFib}_z(\mathrm{Sing}\, \mathrm{colim}_{\beta<\lambda}\, p_\beta)$ is a $(\mathrm{Sing}\, S)$-local equivalence (where $(\mathrm{Sing}\, S) = \{\mathrm{Sing}\, f \mid f \in S\}$ and we use the symbol z to also denote the vertex of $\mathrm{Sing}\, Z$ corresponding to z). We choose a factorization $* \xrightarrow{s} C \xrightarrow{t} \mathrm{Sing}\, Z$ in SS of the map $* \to \mathrm{Sing}\, Z$ whose image is z such that s is a trivial cofibration and t is a fibration, and the argument proceeds as in the case $\mathrm{Spc} = \mathrm{SS}$. □

6.4. Uniqueness of the fiberwise localization

PROPOSITION 6.4.1. *If S is a set of maps in $\mathrm{Spc}_{(*)}$, $p\colon X \to Z$ is an object of $(\mathrm{Spc} \downarrow Z)$, $q\colon Y \to Z$ is a fibration with S-local fibers, $g\colon X \to Y$ is a map in $(\mathrm{Spc} \downarrow Z)$ and $X \to \widetilde{\mathrm{L}}_S X$ is the fiberwise S-localization of X over Z, then the dotted* *arrow exists in the diagram*

$$\xymatrix{ X \ar[r] \ar[dr]_{p} \ar@/^2pc/[rr]^{g} & \widetilde{\mathrm{L}}_S X \ar@{.>}[r] \ar[d] & Y \ar[dl]^{q} \\ & Z & }$$

and it is unique up to simplicial homotopy in $(\mathrm{Spc} \downarrow Z)$.

PROOF. Since $q\colon Y \to Z$ is a $(\mathrm{Fib}_Z S)$-injective, this follows from Proposition 9.6.1. □

THEOREM 6.4.2 (Uniqueness of fiberwise localization). *Let S be a set of maps in $\mathrm{Spc}_{(*)}$. If $q\colon Y \to Z$ is a fibration in* Spc *with S-local fibers and*

$$\xymatrix{ X \ar[rr] \ar[dr]_{p} & & Y \ar[dl]^{q} \\ & Z & }$$

is a map in $(\mathrm{Spc} \downarrow Z)$ such that for every point z of Z the induced map of homotopy fibers $\mathrm{HFib}_z(p) \to \mathrm{HFib}_z(q)$ is an S-local equivalence, then the map $\widetilde{\mathrm{L}}_S X \to Y$ of Proposition 6.4.1 *is a weak equivalence.*

PROOF. Since for every point $z \in Z$ the induced map from the homotopy fiber of $\widetilde{\mathrm{L}}_S X \to Z$ over z to the homotopy fiber of q over z is an S-local equivalence between S-local spaces, Theorem 3.2.13 implies that it is a weak equivalence. The theorem now follows from the exact homotopy sequence of a fibration applied over each path component of Z. □

Part 2

Homotopy Theory in Model Categories

Summary of Part 2

In Chapters 7 and 8 we present the basic definitions and ideas of model categories. We begin Chapter 7 with the definition of a model category, and then discuss the lifting and extension properties of maps that follow from the axioms. We define the left and right homotopy relations for maps, and show that for maps between cofibrant-fibrant objects these are the same relation and are equivalence relations. This enables us to define the *classical homotopy category* of a model category as the category whose objects are the cofibrant-fibrant objects and whose maps are homotopy classes of maps. We also show that a map between cofibrant-fibrant objects is a weak equivalence if and only if it is a homotopy equivalence.

The classical homotopy category is often useful, but it is not sufficient for many purposes since it does not contain all of the objects of the model category. In Chapter 8 we discuss cofibrant and fibrant approximations, which we then use to construct the *Quillen homotopy category* of a model category. (The Quillen homotopy category is referred to simply as the *homotopy category.*)

A *cofibrant approximation* to an object is a cofibrant object together with a weak equivalence to the object. Dually, a *fibrant approximation* to an object is a fibrant object together with a weak equivalence from the object. The importance of cofibrant and fibrant approximations to an object is that

- they are isomorphic in the homotopy category to the original objects, and
- maps that are "expected" to exist may exist only when the domain is cofibrant and the codomain is fibrant.

In Chapter 8 we construct the homotopy category of a model category by taking as objects the objects of the model category and as morphisms between objects X and Y the homotopy classes of maps between cofibrant-fibrant objects weakly equivalent to X and Y.

In Chapter 8 we also define *Quillen functors*, which are the interesting functors between model categories. If $\mathcal{M}$ and $\mathcal{N}$ are model categories and $\mathrm{F}\colon \mathcal{M} \rightleftarrows \mathcal{N} :\mathrm{U}$ is an adjoint pair of functors, then the left adjoint F is called a *left Quillen functor* and the right adjoint U is called a *right Quillen functor* if

- the left adjoint F preserves cofibrations and trivial cofibrations and
- the right adjoint U preserves fibrations and trivial fibrations.

We define the *total left derived functor* $\mathbf{L}\mathrm{F}\colon \mathrm{Ho}\,\mathcal{M} \to \mathrm{Ho}\,\mathcal{N}$ of F and the *total right derived functor* $\mathbf{R}\mathrm{U}\colon \mathcal{N} \to \mathcal{M}$ of U, and show that these form an adjoint pair $\mathbf{L}\mathrm{F}\colon \mathrm{Ho}\,\mathcal{M} \rightleftarrows \mathrm{Ho}\,\mathcal{N} :\mathbf{R}\mathrm{U}$. We also define what it means for Quillen functors to be *Quillen equivalences*, and we show that the total derived functors of Quillen equivalences are equivalences of categories between the homotopy categories.

In Chapter 9 we discuss simplicial model categories. A simplicial model category is a model category together with an enrichment of the category over simplicial sets, with a suitable interaction between the model structure and the simplicial structure. Thus, for every pair of objects there is a *simplicial set* of morphisms, the vertices of which are the maps in the underlying category.

In Chapter 10 we discuss several constructions needed for our discussions of cofibrantly generated model categories in Chapter 11 and of cellular model categories in Chapter 12. The main idea is that of a *transfinite composition* of maps; this is the "composition" of an infinitely long sequence of maps indexed by an ordinal. It is used in the *small object argument*, which is a method of factoring a map into factors with specified lifting properties.

If λ is an ordinal, then a *λ-sequence* consists of objects X_α for $\alpha < \lambda$ and maps $X_\alpha \to X_{\alpha+1}$ for all α for which $\alpha + 1 < \lambda$ such that if β is a limit ordinal and $\beta < \lambda$, then $X_\beta = \operatorname{colim}_{\alpha<\beta} X_\alpha$ (see Definition 10.2.1). The natural map $X_0 \to \operatorname{colim}_{\beta<\lambda} X_\beta$ is called the *composition* of the λ-sequence. An object W is said to be *small* with respect to a subcategory $\mathcal{D}$ if for every large enough regular cardinal λ (see Definition 10.1.11) and every λ-sequence $X_0 \to X_1 \to X_2 \to \cdots \to X_\beta \to \cdots$ $(\beta < \lambda)$ in $\mathcal{D}$ (see Definition 10.2.2) the natural map of sets $\operatorname{colim}_{\beta<\lambda} \mathcal{M}(W, X_\beta) \to \mathcal{M}(W, \operatorname{colim}_{\beta<\lambda} X_\beta)$ is an isomorphism (see Definition 10.5.12), that is, if every map from W to the colimit of the λ-sequence factors "essentially uniquely" through some object X_β in the λ-sequence.

If I is a set of maps, then we define a *relative I-cell complex* to be a map that can be constructed by repeatedly attaching codomains of elements of I along maps of their domains, and we define an *I-cell complex* to be an object for which the map from the initial object is a relative I-cell complex (see Definition 10.5.8). For example, in the category Top of topological spaces, if we let I be the set of inclusions $\mathrm{S}^{n-1} \subset \mathrm{D}^n$ for $n \geq 0$, then the I-cell complexes include the CW-complexes, but they also include cell complexes in which the attaching maps of the cells do not factor through a subcomplex of lower dimensional cells. We say that a set I of maps *permits the small object argument* if the domains of the elements of I are small relative to the subcategory of relative I-cell complexes (see Definition 10.5.15), in which case the *small object argument* (see Proposition 10.5.16) constructs a factorization of every map into a relative I-cell complex followed by a map with the right lifting property with respect to every element of I (see Proposition 10.5.16 and its proof).

In Chapter 11 we discuss cofibrantly generated model categories. A *cofibrantly generated model category* (see Definition 11.1.2) is a model category in which

- there is a set I of maps (called a set of *generating cofibrations*) that permits the small object argument and such that a map is a trivial fibration if and only if it has the right lifting property with respect to every element of I and
- there is a set J of maps (called a set of *generating trivial cofibrations*) that permits the small object argument and such that a map is a fibration if and only if it has the right lifting property with respect to every element of J.

In a cofibrantly generated model category, both of the factorizations required by the model category axioms can be constructed using the small object argument (see Proposition 10.5.16). The small object argument and the retract argument (see Proposition 7.2.2) then imply that the cofibrations are the relative I-cell complexes and their retracts and that the trivial cofibrations are the relative J-cell complexes and their retracts. If $\mathcal{M}$ is a cofibrantly generated model category and we have selected a set I of generating cofibrations, then we will refer to relative I-cell complexes simply as *relative cell complexes*, and to I-cell complexes simply as *cell complexes.* We also show in Chapter 11 that there is a cofibrantly generated model category structure on a category of diagrams in a cofibrantly generated model category.

A notion related to smallness is *compactness.* If W is an object, I is a set of maps, and γ is a cardinal, then we will say that W is *γ-compact relative to I* if every map from W to the codomain of a relative I-cell complex factors through a sub-relative I-cell complex of size at most γ, and we will say that it is *compact* relative to I if it is γ-compact for some cardinal γ (see Definition 10.8.1). If I is a set of generating cofibrations, then an object that is compact relative to I will be called simply *compact.* If $\mathcal{M}$ is a cofibrantly generated model category in which cofibrations are monomorphisms, then compact objects are also small relative to I (see Proposition 10.8.7).

In Chapter 12 we discuss cellular model categories. These are cofibrantly generated model categories in which the cell complexes are well enough behaved to allow the localization arguments of Part 1. In particular, we show that in a cellular model category the intersection of a pair of subcomplexes of a cell complex exists, and that there is a cardinal σ such that a cell complex of size τ is $\sigma\tau$-compact.

In Chapter 13 we discuss *properness.* A model category is *left proper* if the pushout of a weak equivalence along a cofibration is a weak equivalence, and it is *right proper* if the pullback of a weak equivalence along a fibration is a weak equivalence. We discuss *homotopy pullbacks* and *homotopy fibers* in right proper model categories and *homotopy pushouts* in left proper model categories.

In Chapter 14 we discuss the classifying space of a small category. Given a small category $\mathcal{C}$ its *classifying space* $\mathrm{B}\mathcal{C}$ is a simplicial set such that

- the vertices of $\mathrm{B}\mathcal{C}$ are the objects of $\mathcal{C}$,
- the 1-simplices of $\mathrm{B}\mathcal{C}$ are the morphisms of $\mathcal{C}$, and
- the n-simplices of $\mathrm{B}\mathcal{C}$ for $n \geq 2$ are the strings of n composable morphisms in $\mathcal{C}$.

If the classifying space of a small category is *contractible*, then any two objects of the category are connected by an essentially unique zig-zag of morphisms in the category (see Definition 14.4.2 and Theorem 14.4.5). There is also an extension of this that applies to categories that may not be small.

Our main use for this will be to prove the essential uniqueness of various constructions. For example, if X is an object of a model category then there is a category whose objects are cofibrant approximations to X and whose morphisms are weak equivalences of cofibrant approximations. We show that this category has a contractible classifying space, which implies that any two cofibrant approximations to X are connected by an essentially unique zig-zag of weak equivalences.

In Chapter 15 we discuss the Reedy model category structure. This is a common generalization of the model categories of simplicial objects in a model category and of cosimplicial objects in a model category. A *Reedy category* is a common generalization of the indexing category for simplicial objects and the indexing category for cosimplicial objects; the objects of a Reedy category have degrees, and the degrees define a filtration of the Reedy category. The Reedy model category structure on a category of diagrams in a model category indexed by a Reedy category is based on defining diagrams inductively over the filtrations of the Reedy category.

In Chapter 16 we discuss cosimplicial and simplicial resolutions. If $\mathcal{M}$ is a simplicial model category and $W \to X$ is a cofibrant approximation to X, then the cosimplicial object $\widetilde{\boldsymbol{X}}$ in which $\widetilde{\boldsymbol{X}}^n = W \otimes \Delta[n]$ is a cosimplicial resolution of X. Dually, if $\mathcal{M}$ is a simplicial model category and $Y \to Z$ is a fibrant approximation to Y, then the simplicial object $\widehat{\boldsymbol{Y}}$ in which $\widehat{\boldsymbol{Y}}_n = Z^{\Delta[n]}$ is a simplicial resolution of Y. In this chapter, we define cosimplicial and simplicial resolutions in an arbitrary model category (see Definition 16.1.2), and establish a number of their technical properties. These will be used in Chapter 17 to define homotopy function complexes and in Chapter 19 to define homotopy colimits and homotopy limits.

In Chapter 17 we define homotopy function complexes. A homotopy function complex between two objects in a model category is a simplicial set whose set of components is the set of maps in the homotopy category between those objects. These serve as replacements in a general model category for the enrichment over simplicial sets that is part of a simplicial model category. There are three types of homotopy function complexes: *left homotopy function complexes*, defined by resolving the first argument (see Definition 17.1.1), *right homotopy function complexes*, defined by resolving the second argument (see Definition 17.2.1), and *two-sided homotopy function complexes*, defined by resolving both arguments (see Definition 17.3.1). Each of these requires making choices, but there is a distinguished transitive homotopy class of homotopy equivalences connecting any two homotopy function complexes (see Theorem 17.5.30).

Chapters 18 and 19 discuss homotopy colimits and homotopy limits. An objectwise weak equivalence between diagrams does not generally induce a weak equivalence of colimits; the homotopy colimit functor repairs this problem, at least for objectwise cofibrant diagrams. Similarly, the homotopy limit functor takes objectwise weak equivalences between objectwise fibrant diagrams into weak equivalences. In Chapter 18 we discuss homotopy colimits and homotopy limits in simplicial model categories, which allows for simpler formulas. In Chapter 19 we generalize this to arbitrary model categories.

CHAPTER 7

Model Categories

We define a model category in Section 7.1 (see Definition 7.1.3). We point out in Proposition 7.1.9 that the axioms for a model category are *self dual*, i.e., if $\mathcal{M}$ is a model category, then there is a model category structure on $\mathcal{M}^{\mathrm{op}}$ in which the weak equivalences are the opposites of the weak equivalences of $\mathcal{M}$, the cofibrations are the opposites of the fibrations of $\mathcal{M}$, and the fibrations are the opposites of the cofibrations of $\mathcal{M}$. Thus, any theorem about model categories implies a "dual theorem" in which cofibrations are replaced by fibrations, fibrations are replaced by cofibrations, colimits are replaced by limits, and limits are replaced by colimits.

In Section 7.2 we discuss lifting and extending maps, including a technique called the *retract argument* (see Proposition 7.2.2). Together with axiom M3 of Definition 7.1.3, this is often used to show that a map is a cofibration, trivial cofibration, fibration, or trivial fibration based on its lifting properties (see Proposition 7.2.3).

In Section 7.3 we discuss the left and right homotopy relations. Left homotopy is defined using a cylinder object (see Definition 7.3.2) for the domain. Cylinder objects exist for any object (see Lemma 7.3.3), but there is no distinguished one. Dually, right homotopy is defined using a path object (see Definition 7.3.2) for the codomain. Path objects exist for any object (see Lemma 7.3.3), but there is no distinguished one. Two maps are called *homotopic* if they are both left homotopic and right homotopic (see Definition 7.3.2). We establish the homotopy extension property of cofibrations for right homotopies when the codomain is fibrant (see Definition 7.1.5 and Proposition 7.3.10) and the homotopy lifting property of fibrations for left homotopies when the domain is cofibrant (see Definition 7.1.5 and Proposition 7.3.11).

If we make no assumptions about our objects being cofibrant or fibrant, then left and right homotopy need not be the same relation, and neither of them need be an equivalence relation. In Section 7.4 we show that:

- If X is cofibrant, then
  - left homotopy is an equivalence relation on the set of maps from X to Y (see Proposition 7.4.5) and
  - if $f, g\colon X \to Y$ are left homotopic then they are also right homotopic, and there exists a right homotopy between them using any path object for Y (see Proposition 7.4.7).
- If Y is fibrant, then
  - right homotopy is an equivalence relation on the set of maps from X to Y (see Proposition 7.4.5) and
  - if $f, g\colon X \to Y$ are right homotopic then they are also left homotopic, and there exists a left homotopy between them using any cylinder object for X (see Proposition 7.4.7).

This implies that if X is cofibrant and Y is fibrant, then the left and right homotopy relations coincide and are equivalence relations on the set of maps from X to Y (see Theorem 7.4.9), and that if $f, g\colon X \to Y$ are homotopic, then there is a left homotopy between them using any cylinder object for X and a right homotopy using any path object for Y (see Proposition 7.4.10).

In Section 7.5 we show that composition of homotopy classes of maps is well defined for maps between cofibrant-fibrant objects (see Theorem 7.5.5). This allows us to define the *classical homotopy category* of a model category $\mathcal{M}$ to be the category with objects the cofibrant-fibrant objects of $\mathcal{M}$ and with morphisms from X to Y the homotopy classes of maps from X to Y (see Definition 7.5.8). (This is *not* the *homotopy category* of $\mathcal{M}$; for that, see Definition 8.3.2). We also prove a Whitehead theorem: If a map between cofibrant-fibrant objects is a weak equivalence, then it is a homotopy equivalence (see Theorem 7.5.10).

In Section 7.6 we discuss the model category of objects *under* a fixed object of a model category (and, dually, the model category of objects *over* a fixed object of a model category). This enables us to prove the uniqueness up to homotopy of the lifts guaranteed by axiom M4 of Definition 7.1.3 (see Proposition 7.6.13).

The main result of Section 7.7 is *Kenny Brown's lemma* (see Lemma 7.7.1). This is a key result that allows us to show that a weak equivalence between cofibrant objects has many of the properties of a trivial cofibration between cofibrant objects (with a dual statement about a weak equivalence between fibrant objects); see, e.g., Corollary 7.7.2, Corollary 7.7.4, Proposition 8.5.7, and Corollary 9.3.3.

In Section 7.8 we show that a homotopy equivalence between cofibrant-fibrant objects is a weak equivalence (see Theorem 7.8.5), and in Section 7.9 we describe the equivalence relation generated by "weak equivalence". Since a weak equivalence need not have an inverse unless its domain and codomain are cofibrant-fibrant, we define a "zig-zag" of weak equivalences, and we use this to say what it means for two functors to a model category to be naturally weakly equivalent (see Definition 7.9.2).

In Section 7.10 we describe the model category structures on the categories of topological spaces and of simplicial sets.

7.1. Model categories

We adopt the definition of a model category used in [**30**]. This is a strengthening of Quillen's axioms for a *closed model category* (see [**54**, page 233]) in that it requires the category to contain all small limits and colimits (rather than just the finite ones), and it requires the factorizations described in the fifth axiom to be functorial.

DEFINITION 7.1.1. If there is a commutative diagram

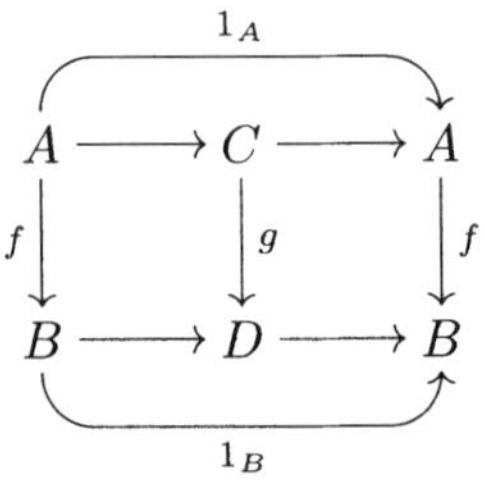

then we will say that the map f is a *retract* of the map g.

DEFINITION 7.1.2. Let $\mathcal{C}$ be a category.

(1) $\mathcal{C}$ is *complete* if it is closed under small limits, i.e., if $\lim_{\mathcal{D}} \mathrm{F}$ exists for every small category $\mathcal{D}$ and every functor $\mathrm{F}\colon \mathcal{D} \to \mathcal{C}$.
(2) $\mathcal{C}$ is *cocomplete* if it is closed under small colimits, i.e., if $\operatorname{colim}_{\mathcal{D}} \mathrm{F}$ exists for every small category $\mathcal{D}$ and every functor $\mathrm{F}\colon \mathcal{D} \to \mathcal{C}$.

DEFINITION 7.1.3. A *model category* is a category $\mathcal{M}$ together with three classes of maps (called the *weak equivalences*, the *cofibrations*, and the *fibrations*), satisfying the following five axioms:

M1: (Limit axiom) The category $\mathcal{M}$ is complete and cocomplete (see Definition 7.1.2).
M2: (Two out of three axiom) If f and g are maps in $\mathcal{M}$ such that gf is defined and two of f, g, and gf are weak equivalences, then so is the third.
M3: (Retract axiom) If f and g are maps in $\mathcal{M}$ such that f is a retract of g (in the category of maps of $\mathcal{M}$; see Definition 7.1.1) and g is a weak equivalence, a fibration, or a cofibration, then so is f.
M4: (Lifting axiom) Given the commutative solid arrow diagram in $\mathcal{M}$

$$\begin{array}{ccc} A & \longrightarrow & X \\ {\scriptstyle i}\big\downarrow & \nearrow & \big\downarrow{\scriptstyle p} \\ B & \longrightarrow & Y \end{array}$$

the dotted arrow exists if either
   (1) i is a cofibration and p is a trivial fibration (i.e., a fibration that is also a weak equivalence) or
   (2) i is a trivial cofibration (i.e., a cofibration that is also a weak equivalence) and p is a fibration.

M5: (Factorization axiom) Every map g in $\mathcal{M}$ has two functorial factorizations:
   (1) $g = qi$, where i is a cofibration and q is a trivial fibration (i.e., a fibration that is also a weak equivalence), and
   (2) $g = pj$, where j is a trivial cofibration (i.e., a cofibration that is also a weak equivalence) and p is a fibration.

REMARK 7.1.4. Once we have defined the homotopy relations (see Definition 7.3.2), the lifting axiom will imply both the homotopy extension property of cofibrations (see Proposition 7.3.10) and the homotopy lifting property of fibrations (see Proposition 7.3.11).

DEFINITION 7.1.5. Let $\mathcal{M}$ be a model category.

(1) A *trivial fibration* is a map that is both a fibration and a weak equivalence.
(2) A *trivial cofibration* is a map that is both a cofibration and a weak equivalence.
(3) An object is *cofibrant* if the map to it from the initial object is a cofibration.
(4) An object is *fibrant* if the map from it to the terminal object is a fibration.
(5) An object is *cofibrant-fibrant* if it is both cofibrant and fibrant.

REMARK 7.1.6. The axioms imply that any two of the three classes of maps cofibrations, fibrations, and weak equivalences determine the third (see Proposition 7.2.7). This was the reason for the use of the name "closed model category" for what we call simply a "model category".

PROPOSITION 7.1.7. *If S is a set and for every element s of S we have a model category $\mathcal{M}_s$, then the category $\prod_{s\in S}\mathcal{M}$ is a model category in which a map is a cofibration, a fibration, or a weak equivalence if each of its components is, respectively, a cofibration, a fibration, or a weak equivalence.*

PROOF. This follows directly from the definitions. □

7.1.8. Duality in model categories. The axioms for a model category are self dual.

PROPOSITION 7.1.9. *If $\mathcal{M}$ is a model category, then its opposite category $\mathcal{M}^{\mathrm{op}}$ is a model category such that*

- *the weak equivalences in $\mathcal{M}^{\mathrm{op}}$ are the opposites of the weak equivalences in $\mathcal{M}$,*
- *the cofibrations in $\mathcal{M}^{\mathrm{op}}$ are the opposites of the fibrations in $\mathcal{M}$, and*
- *the fibrations in $\mathcal{M}^{\mathrm{op}}$ are the opposites of the cofibrations in $\mathcal{M}$.*

PROOF. This follows directly from the definitions. □

REMARK 7.1.10. Proposition 7.1.9 implies that any statement that is proved true for all model categories implies a dual statement in which cofibrations are replaced by fibrations, fibrations are replaced by cofibrations, colimits are replaced by limits, and limits are replaced by colimits.

7.2. Lifting and the retract argument

DEFINITION 7.2.1. If $i\colon A \to B$ and $p\colon X \to Y$ are maps for which the dotted arrow exists in every solid arrow diagram of the form

$$\begin{array}{ccc} A & \longrightarrow & X \\ {\scriptstyle i}\big\downarrow & \nearrow & \big\downarrow{\scriptstyle p} \\ B & \longrightarrow & Y \end{array},$$

then

(1) (i, p) is called a *lifting-extension pair*,
(2) i is said to have the *left lifting property* with respect to p, and
(3) p is said to have the *right lifting property* with respect to i.

Thus, axiom M4 (see Definition 7.1.3) says that cofibrations have the left lifting property with respect to trivial fibrations and that fibrations have the right lifting property with respect to trivial cofibrations. The next proposition is known as the retract argument. Together with axiom M3, it will be used in Proposition 7.2.3 to show that these lifting properties characterize the cofibrations and the fibrations in a model category.

PROPOSITION 7.2.2 (The retract argument). *Let $\mathcal{M}$ be a model category and let $g\colon X \to Y$ be a map in $\mathcal{M}$.*

(1) *If g can be factored as $g = pi$ where p has the right lifting property with respect to g, then g is a retract of i.*
(2) *If g can be factored as $g = pi$ where i has the left lifting property with respect to g, then g is a retract of p.*

PROOF. We will prove part 1; the proof of part 2 is dual.

We have the solid arrow diagram

$$\begin{array}{ccc} X & \xrightarrow{i} & Z \\ {\scriptstyle g}\downarrow & \nearrow{\scriptstyle q} & \downarrow{\scriptstyle p} \\ Y & = & Y \end{array} .$$

Since p has the right lifting property with respect to g, the dotted arrow q exists. This yields the commutative diagram

$$\begin{array}{ccccc} X & = & X & = & X \\ {\scriptstyle g}\downarrow & & \downarrow{\scriptstyle i} & & \downarrow{\scriptstyle g} \\ Y & \xrightarrow[q]{} & Z & \xrightarrow[p]{} & Y \end{array} , \qquad p \circ q = 1_Y$$

and so g is a retract of i. □

PROPOSITION 7.2.3. *Let $\mathcal{M}$ be a model category.*

1. *The map $i\colon A \to B$ is a cofibration if and only if it has the left lifting property with respect to all trivial fibrations.*
2. *The map $i\colon A \to B$ is a trivial cofibration if and only if it has the left lifting property with respect to all fibrations.*
3. *The map $p\colon X \to Y$ is a fibration if and only if it has the right lifting property with respect to all trivial cofibrations.*
4. *The map $p\colon X \to Y$ is a trivial fibration if and only if it has the right lifting property with respect to all cofibrations.*

PROOF. We will prove part 1; the proofs of the other parts are similar.

One direction is part of axiom M4 (see Definition 7.1.3). For the converse, axiom M5 implies that we can factor i as $i = pj$ where p is a trivial fibration and j is a cofibration. Proposition 7.2.2 implies that i is a retract of j, and so the result follows from axiom M3 (see Definition 7.1.3). □

PROPOSITION 7.2.4. *If $\mathcal{M}$ is a model category, then the classes of cofibrations and of fibrations are closed under compositions.*

PROOF. This follows from Proposition 7.2.3. □

PROPOSITION 7.2.5. *Let $\mathcal{M}$ be a model category.*

1. *The class of cofibrations is closed under coproducts.*
2. *The class of trivial cofibrations is closed under coproducts.*
3. *The class of fibrations is closed under products.*
4. *The class of trivial fibrations is closed under products.*

PROOF. This follows from Proposition 7.2.3. □

PROPOSITION 7.2.6. *If $\mathcal{M}$ is a model category, then a map in $\mathcal{M}$ is a weak equivalence if and only if it can be factored as a trivial cofibration followed by a trivial fibration.*

PROOF. Any map that can be factored as a trivial cofibration followed by a trivial fibration is a composition of weak equivalences, and is thus a weak equivalence.

Conversely, if $g\colon X \to Y$ is a weak equivalence, then we can factor it as $X \xrightarrow{h} W \xrightarrow{k} Y$ with h a trivial cofibration and k a fibration. The "two out of three" axiom then implies that k is actually a trivial fibration. □

PROPOSITION 7.2.7. *If $\mathcal{M}$ is a model category, then any two of the classes of cofibrations, fibrations, and weak equivalences determine the third.*

PROOF. Proposition 7.2.3 implies that

- the cofibrations and weak equivalences determine the fibrations, and that
- the fibrations and weak equivalences determine the cofibrations.

Proposition 7.2.4 implies that the trivial cofibrations and trivial fibrations determine the weak equivalences, and Proposition 7.2.3 implies that the cofibrations determine the trivial fibrations and the fibrations determine the trivial cofibrations. Thus,

- the cofibrations and fibrations determine the weak equivalences.

□

LEMMA 7.2.8. *Let $\mathcal{M}$ be a model category, and let $p\colon X \to Y$ is a map in $\mathcal{M}$.*

1. *The class of maps with the left lifting property with respect to p is closed under retracts.*
2. *The class of maps with the right lifting property with respect to p is closed under retracts.*

PROOF. We will prove part 1; the proof of part 2 is dual.

Suppose that $f\colon A \to B$ is a retract of $g\colon C \to D$, and that g has the left lifting property with respect to p; we must show that the dotted arrow ϕ exists in any solid arrow diagram of the form

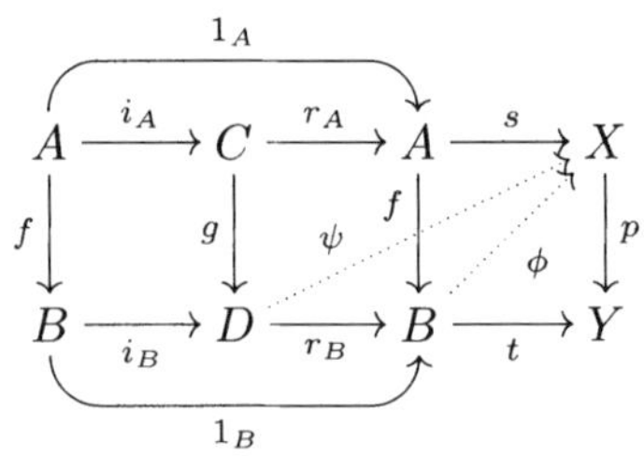

Since g has the left lifting property with respect to p, there exists a map $\psi\colon D \to X$ such that $\psi g = s r_A$ and $p\psi = t r_B$; we define $\phi\colon B \to X$ by letting $\phi = \psi i_B$. We then have $\phi f = \psi i_B f = \psi g i_A = s r_A i_A = s$ and $p\phi = p\psi i_B = t r_B i_B = t$. □

7.2.9. Pushouts and pullbacks.

DEFINITION 7.2.10. If the square

$$\begin{array}{ccc} A & \xrightarrow{h} & C \\ {\scriptstyle f}\downarrow & & \downarrow{\scriptstyle g} \\ B & \xrightarrow[k]{} & D \end{array}$$

is a pushout, then the map g will be called the *pushout* of f along h. If the square is a pullback, then the map f will be called the *pullback* of g along k.

LEMMA 7.2.11. *Let $\mathcal{M}$ be a model category, and let $p\colon X \to Y$ be a map in $\mathcal{M}$.*

(1) *The class of maps with the left lifting property with respect to p is closed under pushouts.*
(2) *The class of maps with the right lifting property with respect to p is closed under pullbacks.*

PROOF. We will prove part 1; the proof of part 2 is dual.

We must show that if $i\colon A \to B$ has the left lifting property with respect to p and we have a solid arrow diagram

$$\begin{array}{ccccc} A & \xrightarrow{s} & C & \xrightarrow{u} & X \\ {\scriptstyle i}\downarrow & \overset{\psi}{\nearrow} & \downarrow{\scriptstyle j} & \overset{\phi}{\nearrow} & \downarrow{\scriptstyle p} \\ B & \xrightarrow[t]{} & D & \xrightarrow[v]{} & Y \end{array}$$

in which the square on the left is a pushout, then the dotted arrow ϕ exists. Since i has the left lifting property with respect to p, there is a map $\psi\colon B \to X$ such that $\psi i = us$ and $p\psi = vt$. Since D is the pushout $B \amalg_A C$, this induces a map $\phi\colon D \to X$ such that $\phi j = u$ and $\phi t = \psi$. We then have $p\phi t = p\psi = vt$ and $p\phi j = pu = vj$, and so the universal mapping property of the pushout implies that $p\phi = v$. □

PROPOSITION 7.2.12. *Let $\mathcal{M}$ be a model category.*

(1) *The class of cofibrations is closed under pushouts.*
(2) *The class of trivial cofibrations is closed under pushouts.*
(3) *The class of fibrations is closed under pullbacks.*
(4) *The class of trivial fibrations is closed under pullbacks.*

PROOF. This follows from Proposition 7.2.3 and Lemma 7.2.11. □

LEMMA 7.2.13. *If $h\colon E \to F$ is a pushout (see Definition 7.2.10) of $g\colon C \to D$ and $k\colon G \to H$ is a pushout of h, then k is a pushout of g.*

PROOF. In the commutative diagram

$$\begin{array}{ccccc} C & \longrightarrow & E & \longrightarrow & G \\ {\scriptstyle g}\downarrow & & \downarrow{\scriptstyle h} & & \downarrow{\scriptstyle k} \\ D & \longrightarrow & F & \longrightarrow & H \end{array}$$

if the two squares are pushouts, then the rectangle is a pushout. □

PROPOSITION 7.2.14. *Consider the commutative diagram*

$$\begin{array}{ccccc} C & \xrightarrow{s} & E & \xrightarrow{t} & G \\ {\scriptstyle f}\downarrow & & \downarrow{\scriptstyle g} & & \downarrow{\scriptstyle h} \\ D & \xrightarrow[u]{} & F & \xrightarrow[v]{} & H \end{array}$$

(1) *If H is the pushout $D \amalg_C G$ and F is the pushout $D \amalg_C E$, then H is the pushout $F \amalg_E G$.*

(2) *If C is the pullback $D \times_H G$ and E is the pullback $F \times_H G$, then C is the pullback $D \times_F E$.*

PROOF. We will prove part 1; the proof of part 2 is dual.

If W is an object and $j\colon F \to W$ and $k\colon G \to W$ are maps such that $jg = kt$, then $kts = jgs = juf$. Since H is the pushout $D \amalg_C G$, there exists a unique map $l\colon H \to W$ such that $lvu = ju$ and $lh = k$. Since F is the pushout $D \amalg_C E$ and the maps j and lv satisfy both $(lv)u = (j)u$ and $(j)g = kt = lht = (lv)g$, we have $j = lv$. Thus, the map l satisfies $lh = k$ and $lv = j$. To see that l is the unique such map, note that if $\tilde{l}$ were another map satisfying $\tilde{l}h = k$ and $\tilde{l}v = j$, then $\tilde{l}vu = ju$, and so the universal property of $D \amalg_C G$ implies that $\tilde{l} = l$. □

LEMMA 7.2.15 (C. L. Reedy, [**57**]). *Let $\mathcal{M}$ be a model category. If we have a commutative diagram in $\mathcal{M}$*

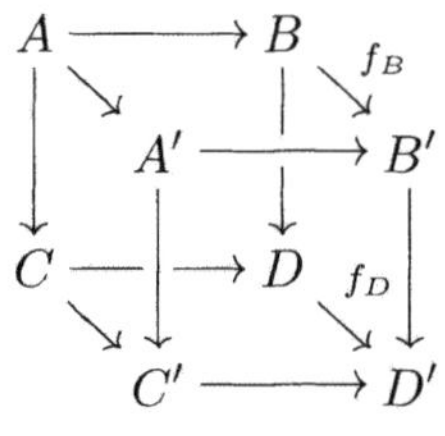

in which the front and back squares are pushouts and both f_B and $C \amalg_A A' \to C'$ are cofibrations, then f_D is a cofibration.

PROOF. It is sufficient to show that f_D has the left lifting property with respect to all trivial fibrations (see Proposition 7.2.3). If we have a commutative diagram

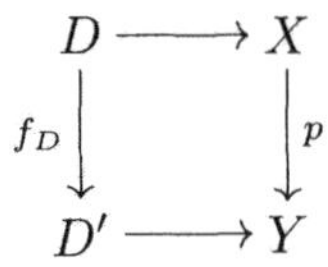

in which p is a trivial fibration, then we also have a similar diagram with f_B in place of f_D. Since f_B is a cofibration, there is a map $h_B\colon B' \to X$ making both triangles commute. Composing h_B with $A' \to B'$ yields a map $h_A\colon A' \to X$ that also makes both triangles commute. This induces a map $C \amalg_A A' \to X$. Since $C \amalg_A A' \to C'$ is a cofibration, there is a map $C' \to X$ making everything commute, and so there is an induced map $D' = C' \amalg_{A'} B' \to X$ making both triangles commute, and the proof is complete. □

7.2.16. Adjointness.

PROPOSITION 7.2.17. *Let $\mathcal{M}$ and $\mathcal{N}$ be categories and let $\mathrm{F}\colon \mathcal{M} \rightleftarrows \mathcal{N} :\mathrm{U}$ be adjoint functors. If $i\colon A \to B$ is a map in $\mathcal{M}$ and $p\colon X \to Y$ is a map in $\mathcal{N}$, then $(\mathrm{F}i, p)$ is a lifting-extension pair (see Definition 7.2.1) if and only if $(i, \mathrm{U}p)$ is a lifting-extension pair.*

PROOF. The adjointness of F and U implies that there is a one to one correspondence between solid arrow diagrams of the forms

$$\begin{array}{ccc} \mathrm{F}A & \longrightarrow & X \\ {\scriptstyle \mathrm{F}i}\downarrow & \overset{h}{\nearrow} & \downarrow{\scriptstyle p} \\ \mathrm{F}B & \longrightarrow & Y \end{array} \qquad \text{and} \qquad \begin{array}{ccc} A & \longrightarrow & \mathrm{U}X \\ {\scriptstyle i}\downarrow & \overset{\tilde{h}}{\nearrow} & \downarrow{\scriptstyle \mathrm{U}p} \\ B & \longrightarrow & \mathrm{U}Y \end{array} .$$

The adjointness also implies that, under this correspondence, the dotted arrow h exists if and only if the dotted arrow $\tilde{h}$ exists. □

PROPOSITION 7.2.18. *Let $\mathcal{M}$ and $\mathcal{N}$ be model categories and let $\mathrm{F}\colon \mathcal{M} \rightleftarrows \mathcal{N} :\mathrm{U}$ be adjoint functors.*

(1) *The left adjoint* F *preserves cofibrations if and only if the right adjoint* U *preserves trivial fibrations.*
(2) *The left adjoint* F *preserves trivial cofibrations if and only if the right adjoint* U *preserves fibrations.*

PROOF. This follows from Proposition 7.2.3 and Proposition 7.2.17. □

7.3. Homotopy

7.3.1. Left homotopy, right homotopy, and homotopy.

DEFINITION 7.3.2. Let $\mathcal{M}$ be a model category and let $f, g\colon X \to Y$ be maps in $\mathcal{M}$.

(1) A *cylinder object* for X is a factorization
$$X \amalg X \xrightarrow{i_0 \amalg i_1} \mathrm{Cyl}(X) \xrightarrow{\;p\;} X$$
of the fold map $1_X \amalg 1_X\colon X \amalg X \to X$ (so that the compositions pi_0 and pi_1 both equal the identity map of X) such that $i_0 \amalg i_1$ is a cofibration and p is a weak equivalence. Note that, although we will often use the notation $\mathrm{Cyl}(X)$ for a cylinder object for X, we do not mean to suggest that this is a functor of X, or that there is any distinguished choice of cylinder object for X.
(2) A *left homotopy* from f to g consists of a cylinder object $X \amalg X \xrightarrow{i_0 \amalg i_1} \mathrm{Cyl}(X) \xrightarrow{p} X$ for X and a map $H\colon \mathrm{Cyl}(X) \to Y$ such that $Hi_0 = f$ and $Hi_1 = g$. If there exists a left homotopy from f to g, then we say that f is *left homotopic* to g (written $f \overset{l}{\simeq} g$).
(3) A *path object* for Y is a factorization
$$Y \xrightarrow{\;s\;} \mathrm{Path}(Y) \xrightarrow{p_0 \times p_1} Y \times Y$$
of the diagonal map (so that the compositions $p_0 s$ and $p_1 s$ both equal the identity map of Y) $1_Y \times 1_Y\colon Y \to Y \times Y$ such that s is a weak equivalence and $p_0 \times p_1$ is a fibration. Note that, although we will often use the notation $\mathrm{Path}(Y)$ for a path object for Y, we do not mean to suggest that this is a functor of Y, or that there is any distinguished choice of path object for Y.

(4) A *right homotopy* from f to g consists of a path object $Y \xrightarrow{s} \mathrm{Path}(Y) \xrightarrow{p_0 \times p_1} Y \times Y$ for Y and a map $H\colon X \to \mathrm{Path}(Y)$ such that $p_0 H = f$ and $p_1 H = g$. If there exists a right homotopy from f to g, then we say that f is right homotopic to g (written $f \overset{r}{\simeq} g$).

(5) If f is both left homotopic and right homotopic to g, then we say that f is *homotopic* to g (written $f \simeq g$).

LEMMA 7.3.3. *Let $\mathcal{M}$ be a model category.*

(1) *Every object X of $\mathcal{M}$ has a cylinder object $X \amalg X \xrightarrow{i_0 \amalg i_1} \mathrm{Cyl}(X) \xrightarrow{p} X$ in which p is a trivial fibration.*

(2) *Every object X of $\mathcal{M}$ has a path object $X \xrightarrow{s} \mathrm{Path}(X) \xrightarrow{p_0 \times p_1} X \times X$ in which s is a trivial cofibration.*

PROOF. For part 1, factor the map $1_X \amalg 1_X \colon X \amalg X \to X$ into a cofibration followed by a trivial fibration. For part 2, factor the map $1_X \times 1_X \colon X \to X \times X$ into a trivial cofibration followed by a fibration. $\square$

PROPOSITION 7.3.4. *Let $\mathcal{M}$ be a model category.*

(1) *If $f, g\colon X \to Y$ are left homotopic and Y is fibrant, then there is a cylinder object $X \amalg X \to \mathrm{Cyl}(X) \xrightarrow{p} X$ in which p is a trivial fibration and a left homotopy $H\colon \mathrm{Cyl}(X) \to Y$ from f to g.*

(2) *If $f, g\colon X \to Y$ are right homotopic and X is cofibrant, then there is a path object $Y \xrightarrow{s} \mathrm{Path}(Y) \to Y \times Y$ in which s is a trivial cofibration and a right homotopy $H\colon X \to \mathrm{Path}(Y)$ from f to g.*

PROOF. We will prove part 1; the proof of part 2 is dual.

If $X \amalg X \to \mathrm{Cyl}(X)' \xrightarrow{p'} X$ is a cylinder object for X such that there is a left homotopy $H'\colon \mathrm{Cyl}(X)' \to Y$ from f to g, then we factor p as $\mathrm{Cyl}(X)' \xrightarrow{j} \mathrm{Cyl}(X) \xrightarrow{p} X$ where j is a cofibration and p is a trivial fibration. The "two out of three" axiom for weak equivalences (see Definition 7.1.3) implies that j is a trivial cofibration, and so the dotted arrow exists in the diagram

$$
\begin{array}{ccc}
\mathrm{Cyl}(X)' & \xrightarrow{H'} & Y \\
{\scriptstyle j}\downarrow & \overset{H}{\nearrow} & \downarrow \\
\mathrm{Cyl}(X) & \longrightarrow & *
\end{array}
$$

which constructs our left homotopy H. $\square$

PROPOSITION 7.3.5. *Let $\mathcal{M}$ be a model category and let $f, g\colon X \to Y$ be maps in $\mathcal{M}$.*

(1) *The maps f and g are left homotopic if and only if there is a factorization $X \amalg X \xrightarrow{i_0 \amalg i_1} C \xrightarrow{p} X$ of the fold map $1_X \amalg 1_X \colon X \amalg X \to X$ such that p is a weak equivalence and a map $H\colon C \to Y$ such that $Hi_0 = f$ and $Hi_1 = g$.*

(2) *The maps f and g are right homotopic if and only if there is a factorization $Y \xrightarrow{s} P \xrightarrow{p_0 \times p_1} Y \times Y$ of the diagonal map $1_Y \times 1_Y \colon Y \to Y \times Y$ such that s is a weak equivalence and a map $H\colon X \to P$ such that $p_0 H = f$ and $p_1 H = g$.*

PROOF. We will prove part 1; the proof of part 2 is dual.

The necessity of the condition follows directly from the definition. Conversely, assume the condition is satisfied. If we factor $i_0 \amalg i_1$ as $X \amalg X \xrightarrow{i'_0 \amalg i'_1} C' \xrightarrow{q} C$ where $i'_0 \amalg i'_1$ is a cofibration and q is a trivial fibration, then $X \amalg X \xrightarrow{i'_0 \amalg i'_1} C' \xrightarrow{pq} X$ is a cylinder object for X and $Hq : C' \to Y$ is a left homotopy from f to g. □

LEMMA 7.3.6. *Let $\mathcal{M}$ be a model category and let X be an object of $\mathcal{M}$.*

(1) *If X is cofibrant, then the injections $i_0, i_1 \colon X \to X \amalg X$ are cofibrations.*
(2) *If X is fibrant, then the projections $p_0, p_1 \colon X \times X \to X$ are fibrations.*

PROOF. We will prove part 1; the proof of part 2 is dual.

Since the diagram

$$\begin{array}{ccc} \emptyset & \longrightarrow & X \\ \downarrow & & \downarrow{\scriptstyle i_1} \\ X & \xrightarrow[i_0]{} & X \amalg X \end{array}$$

(where $\emptyset$ is the initial object of $\mathcal{M}$) is a pushout, the lemma follows from Proposition 7.2.12. □

LEMMA 7.3.7. *Let $\mathcal{M}$ be a model category and let X be an object of $\mathcal{M}$.*

(1) *If $X \amalg X \xrightarrow{i_0 \amalg i_1} \mathrm{Cyl}(X) \xrightarrow{p} X$ is a cylinder object for X, then the injections $i_0, i_1 \colon X \to \mathrm{Cyl}(X)$ are weak equivalences. If X is cofibrant, then they are trivial cofibrations.*
(2) *If $X \xrightarrow{s} \mathrm{Path}(X) \xrightarrow{p_0 \times p_1} X \times X$ is a path object for X, then the projections $p_0, p_1 \colon \mathrm{Path}(X) \to X$ are weak equivalences. If X is fibrant, then they are trivial fibrations.*

PROOF. This follows from the "two out of three" axiom for weak equivalences (see Definition 7.1.3) and Lemma 7.3.6. □

LEMMA 7.3.8. *Let $\mathcal{M}$ and $\mathcal{N}$ be model categories and let $\varphi \colon \mathcal{M} \to \mathcal{N}$ be a functor.*

(1) *If φ takes trivial cofibrations between cofibrant objects in $\mathcal{M}$ to weak equivalences in $\mathcal{N}$, $f, g \colon X \to Y$ are left homotopic maps in $\mathcal{M}$, and X is cofibrant, then $\varphi(f)$ is left homotopic to $\varphi(g)$.*
(2) *If φ takes trivial fibrations between fibrant objects in $\mathcal{M}$ to weak equivalences in $\mathcal{N}$, $f, g \colon X \to Y$ are right homotopic maps in $\mathcal{M}$, and Y is fibrant, then $\varphi(f)$ is right homotopic to $\varphi(g)$.*

PROOF. We will prove part 1; the proof of part 2 is dual.

Since f and g are left homotopic, there is a cylinder object $X \amalg X \xrightarrow{i_0 \amalg i_1} \mathrm{Cyl}(X) \xrightarrow{p} X$ for X and a map $H \colon \mathrm{Cyl}(X) \to Y$ such that $Hi_0 = f$ and $Hi_1 = g$. Since $pi_0 = 1_X$, we have $\varphi(p)\varphi(i_0) = 1_{\varphi(X)}$, and, since i_0 is a trivial cofibration (see Lemma 7.3.7), the "two out of three" property of weak equivalences (see Definition 7.1.3) implies that $\varphi(p)$ is a weak equivalence. The result now follows from Proposition 7.3.5. □

LEMMA 7.3.9. *Let $\mathcal{M}$ be a model category, let $\mathcal{C}$ be a category, and let $\varphi \colon \mathcal{M} \to \mathcal{C}$ be a functor.*

(1) *If φ takes trivial cofibrations between cofibrant objects in $\mathcal{M}$ to isomorphisms in $\mathcal{C}$, $f, g\colon X \to Y$ are left homotopic maps in $\mathcal{M}$, and X is cofibrant, then $\varphi(f) = \varphi(g)$.*

(2) *If φ takes trivial fibrations between fibrant objects in $\mathcal{M}$ to isomorphisms in $\mathcal{C}$, $f, g\colon X \to Y$ are right homotopic maps in $\mathcal{M}$, and Y is fibrant, then $\varphi(f) = \varphi(g)$.*

PROOF. We will prove part 1; the proof of part 2 is dual.

Since f and g are left homotopic, there is a cylinder object $X \amalg X \xrightarrow{i_0 \amalg i_1} \mathrm{Cyl}(X) \xrightarrow{p} X$ for X and a map $H\colon \mathrm{Cyl}(X) \to Y$ such that $Hi_0 = f$ and $Hi_1 = g$. Since $pi_0 = 1_X$, we have $\varphi(p)\varphi(i_0) = 1_{\varphi(X)}$, and, since i_0 is a trivial cofibration (see Lemma 7.3.7), $\varphi(i_0)$ is an isomorphism, and so $\varphi(p)$ is an isomorphism. Since $pi_0 = 1_X = pi_1$, $\varphi(i_0) = \big(\varphi(p)\big)^{-1} = \varphi(i_1)$. Thus, $\varphi(f) = \varphi(H)\varphi(i_0) = \varphi(H)\varphi(i_1) = \varphi(g)$. □

PROPOSITION 7.3.10 (Homotopy extension property of cofibrations). *Let $\mathcal{M}$ be a model category, let X be fibrant, and let $k\colon A \to B$ be a cofibration. If $f\colon A \to X$ is a map, $\tilde{f}\colon B \to X$ is an extension of f, $X \xrightarrow{s} \mathrm{Path}(X) \xrightarrow{p_0 \times p_1} X \amalg X$ is a path object for X, and $H\colon A \to \mathrm{Path}(X)$ is a right homotopy of f (i.e., a map H such that $p_0 H = f$), then there is a map $\widetilde{H}\colon B \to \mathrm{Path}(X)$ such that $p_0\widetilde{H} = \tilde{f}$ and $\widetilde{H}k = H$.*

PROOF. We have the solid arrow diagram

$$\begin{array}{ccc} A & \xrightarrow{H} & \mathrm{Path}(X) \\ {\scriptstyle k}\downarrow & \nearrow_{\widetilde{H}} & \downarrow{\scriptstyle p_0} \\ B & \xrightarrow[\tilde{f}]{} & X \end{array}$$

(the diagonal arrow $\widetilde{H}\colon B \to \mathrm{Path}(X)$ is dotted)

and Lemma 7.3.7 implies that p_0 is a trivial fibration. □

PROPOSITION 7.3.11 (Homotopy lifting property of fibrations). *Let $\mathcal{M}$ be a model category, let A be cofibrant, and let $k\colon X \to Y$ be a fibration. If $f\colon A \to Y$ is a map, $\tilde{f}\colon A \to X$ is a lift of f, $A \amalg A \xrightarrow{i_0 \amalg i_1} \mathrm{Cyl}(A) \xrightarrow{p} A$ is a cylinder object for A, and $H\colon \mathrm{Cyl}(A) \to Y$ is a left homotopy of f (i.e., a map H such that $Hi_0 = f$), then there is a map $\widetilde{H}\colon \mathrm{Cyl}(A) \to X$ such that $\widetilde{H}i_0 = \tilde{f}$ and $k\widetilde{H} = H$.*

PROOF. We have the solid arrow diagram

$$\begin{array}{ccc} A & \xrightarrow{\tilde{f}} & X \\ {\scriptstyle i_0}\downarrow & \nearrow_{\widetilde{H}} & \downarrow{\scriptstyle k} \\ \mathrm{Cyl}(A) & \xrightarrow[H]{} & Y \end{array}$$

(the diagonal arrow $\widetilde{H}\colon \mathrm{Cyl}(A) \to X$ is dotted)

and Lemma 7.3.7 implies that i_0 is a trivial cofibration. □

COROLLARY 7.3.12. *Let $\mathcal{M}$ be a model category.*

(1) *Let X be fibrant and let $k\colon A \to B$ be a cofibration. If $f\colon A \to X$ and $g\colon B \to X$ are maps such that $gk \overset{r}{\simeq} f$, then there is a map $g'\colon B \to X$ such that $g' \overset{r}{\simeq} g$ and $g'k = f$.*

(2) *Let A be cofibrant and let $k\colon X \to Y$ be a fibration. If $f\colon A \to X$ and $g\colon A \to Y$ are maps such that $kf \overset{l}{\simeq} g$, then there is a map $f'\colon A \to X$ such that $f' \overset{l}{\simeq} f$ and $kf' = g$.*

PROOF. Part 1 follows from Proposition 7.3.10 and part 2 follows from Proposition 7.3.11. □

PROPOSITION 7.3.13. *Let $\mathcal{M}$ be a model category.*

(1) *If $i\colon A \to B$ is a cofibration, X is fibrant, and i induces an isomorphism $i^*\colon \pi^r(B,X) \approx \pi^r(A,X)$, then for every map $f\colon A \to X$ there is a map $g\colon B \to X$, unique up to right homotopy, such that $gi = f$.*
(2) *If A is cofibrant, $p\colon X \to Y$ is a fibration, and p induces an isomorphism $p_*\colon \pi^l(A,X) \approx \pi^l(A,Y)$, then for every map $f\colon A \to Y$ there is a map $g\colon A \to X$, unique up to left homotopy, such that $pg = f$.*

PROOF. We will prove part 1; the proof of part 2 is dual.

Since $i^*\colon \pi^r(B,X) \to \pi^r(A,X)$ is surjective there is a map $h\colon B \to X$ such that $hi \overset{r}{\simeq} f$. Corollary 7.3.12 now implies that there exists a map $g\colon B \to X$ such that $gi = f$, and the uniqueness up to right homotopy follows because $i^*\colon \pi^r(B,X) \to \pi^r(A,X)$ is injective. □

7.4. Homotopy as an equivalence relation

The main results of this section are

- Proposition 7.4.5, which asserts that if X is cofibrant, then left homotopy is an equivalence relation on the set of maps from X to Y, and, dually, that if Y is fibrant, then right homotopy is an equivalence relation on the set of maps from X to Y, and
- Theorem 7.4.9, which asserts that if X is cofibrant and Y is fibrant, then the left and right homotopy relations coincide on the set of maps from X to Y.

7.4.1. Left and right homotopy as equivalence relations. We begin with Lemma 7.4.2, which shows that if X is cofibrant, then two cylinder objects for X can be "composed" to produce a cylinder object that we will use in Proposition 7.4.5 to show that left homotopy is an equivalence relation when the domain is cofibrant. Dually, Lemma 7.4.2 also shows that if Y is fibrant, then two path objects for Y can be "composed" to produce a path object that we will use in Proposition 7.4.5 to show that right homotopy is an equivalence relation when the codomain is fibrant.

LEMMA 7.4.2. *Let $\mathcal{M}$ be a model category and let X and Y be objects in $\mathcal{M}$.*

(1) *If X is cofibrant and $X \amalg X \xrightarrow{i_0 \amalg i_1} \mathrm{Cyl}(X) \xrightarrow{p} X$ and $X \amalg X \xrightarrow{i'_0 \amalg i'_1} \mathrm{Cyl}(X)' \xrightarrow{p'} X$ are cylinder objects for X, then there is a third cylinder object $X \amalg X \xrightarrow{i''_0 \amalg i''_1} \mathrm{Cyl}(X)'' \xrightarrow{p''} X$ for X in which*
   (a) *$\mathrm{Cyl}(X)''$ is the pushout of the diagram $\mathrm{Cyl}(X) \xleftarrow{i_1} X \xrightarrow{i'_0} \mathrm{Cyl}(X)'$,*
   (b) *$i''_0\colon X \to \mathrm{Cyl}(X)''$ is the composition $X \xrightarrow{i_0} \mathrm{Cyl}(X) \to \mathrm{Cyl}(X)''$, and*
   (c) *$i''_1\colon X \to \mathrm{Cyl}(X)''$ is the composition $X \xrightarrow{i'_1} \mathrm{Cyl}(X)' \to \mathrm{Cyl}(X)''$.*

(2) *If Y is fibrant and $Y \xrightarrow{s} \mathrm{Path}(Y) \xrightarrow{p_0 \times p_1} Y \times Y$ and $Y \xrightarrow{s'} \mathrm{Path}(Y)' \xrightarrow{p'_0 \times p'_1} Y \times Y$ are path objects for Y, then there is a third path object $Y \xrightarrow{s''} \mathrm{Path}(Y)'' \xrightarrow{p''_0 \times p''_1} Y \times Y$ for Y in which*

   (a) *$\mathrm{Path}(Y)''$ is the pullback of the diagram $\mathrm{Path}(Y) \xrightarrow{p_1} Y \xleftarrow{p'_0} \mathrm{Path}(Y)'$,*

   (b) *$p''_0 \colon \mathrm{Path}(Y)'' \to Y$ is the composition $\mathrm{Path}(Y)'' \to \mathrm{Path}(Y) \xrightarrow{p_0} Y$, and*

   (c) *$p''_1 \colon \mathrm{Path}(Y)'' \to Y$ is the composition $\mathrm{Path}(Y)'' \to \mathrm{Path}(Y)' \xrightarrow{p'_1} Y$.*

PROOF. We will prove part 1; the proof of part 2 is dual.

We have the commutative diagram

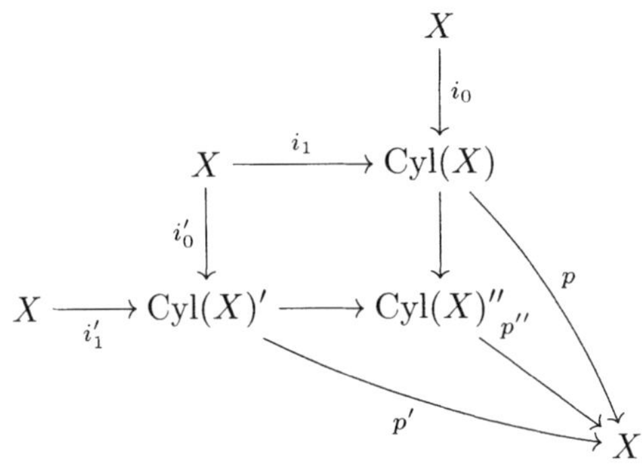

Lemma 7.3.7 and Proposition 7.2.12 imply that i''_0 and i''_1 are trivial cofibrations. Together with the "two out of three" property of weak equivalences (see Definition 7.1.3), this implies that p'' is a weak equivalence.

It remains only to show that the map $X \amalg X \xrightarrow{i''_0 \amalg i''_1} \mathrm{Cyl}(X)''$ is a cofibration. This map equals the composition

$$X \amalg X \xrightarrow{i_0 \amalg 1_X} \mathrm{Cyl}(X) \amalg X \xrightarrow{j_0 \amalg j_1 i'_1} \mathrm{Cyl}(X)''.$$

The first of these is the pushout of $i_0 \colon X \to \mathrm{Cyl}(X)$ along the first inclusion $X \to X \amalg X$, and so Lemma 7.3.7 and Proposition 7.2.12 imply that it is a trivial cofibration. The second is the pushout of $i'_0 \amalg i'_1$ along $i_1 \amalg 1_X \colon X \amalg X \to \mathrm{Cyl}(X) \amalg X$, and so Proposition 7.2.12 implies that it is a cofibration. Proposition 7.2.4 now implies that $i''_0 \amalg i''_1$ is a cofibration. □

DEFINITION 7.4.3. Let $\mathcal{M}$ be a model category and let X and Y be objects in $\mathcal{M}$.

(1) If X is cofibrant, $X \amalg X \xrightarrow{i_0 \amalg i_1} \mathrm{Cyl}(X) \xrightarrow{p} X$ and $X \amalg X \xrightarrow{i'_0 \amalg i'_1} \mathrm{Cyl}(X)' \xrightarrow{p'} X$ are cylinder objects for X, $H \colon \mathrm{Cyl}(X) \to Y$ is a left homotopy from $f \colon X \to Y$ to $g \colon X \to Y$, and $H' \colon \mathrm{Cyl}(X)' \to Y$ is a left homotopy from g to $h \colon X \to Y$, then the *composition* of the left homotopies H and H' is the left homotopy $H \cdot H' \colon \mathrm{Cyl}(X)'' \to Y$ from f to h (where $\mathrm{Cyl}(X)''$ is as in Lemma 7.4.2) defined by H and H'.

(2) If Y is fibrant, $Y \xrightarrow{s} \mathrm{Path}(Y) \xrightarrow{p_0 \times p_1} Y \times Y$ and $Y \xrightarrow{s'} \mathrm{Path}(Y)' \xrightarrow{p'_0 \times p'_1} Y \times Y$ are path objects for Y, $H \colon X \to \mathrm{Path}(Y)$ is a right homotopy from $f \colon X \to Y$ to $g \colon X \to Y$, and $H' \colon X \to \mathrm{Path}(Y)'$ is a right homotopy from g to $h \colon X \to Y$, then the *composition* of the right homotopies H

and H' is the right homotopy $H \cdot H' \colon X \to \mathrm{Path}(Y)''$ from f to h (where $\mathrm{Path}(Y)''$ is as in Lemma 7.4.2) defined by H and H'.

DEFINITION 7.4.4. Let $\mathcal{M}$ be a model category and let X and Y be objects in $\mathcal{M}$.

(1) If $X \amalg X \xrightarrow{i_0 \amalg i_1} \mathrm{Cyl}(X) \xrightarrow{p} X$ is a cylinder object for X and $H \colon \mathrm{Cyl}(X) \to Y$ is a left homotopy from $f \colon X \to Y$ to $g \colon X \to Y$, then the *inverse* of H is the left homotopy $H^{-1} \colon \mathrm{Cyl}(X)^{-1} \to Y$ from g to f where $X \amalg X \xrightarrow{i_0^{-1} \amalg i_1^{-1}} \mathrm{Cyl}(X)^{-1} \xrightarrow{p^{-1}} X$ is the cylinder object for X defined by $\mathrm{Cyl}(X)^{-1} = \mathrm{Cyl}(X)$, $i_0^{-1} = i_1$, $i_1^{-1} = i_0$, and $p^{-1} = p$, and the map H^{-1} equals the map H.
(2) If $Y \xrightarrow{s} \mathrm{Path}(Y) \xrightarrow{p_0 \times p_1} Y \times Y$ is a path object for Y and $H \colon X \to \mathrm{Path}(Y)$ is a right homotopy from $f \colon X \to Y$ to $g \colon X \to Y$, then the *inverse* of H is the right homotopy $H^{-1} \colon X \to \mathrm{Path}(Y)^{-1}$ from g to f where $Y \xrightarrow{s^{-1}} \mathrm{Path}(Y)^{-1} \xrightarrow{p_0^{-1} \times p_1^{-1}} Y \times Y$ is the path object for Y defined by $\mathrm{Path}(Y)^{-1} = \mathrm{Path}(Y)$, $p_0^{-1} = p_1$, $p_1^{-1} = p_0$, and $s^{-1} = s$, and the map H^{-1} equals the map H.

PROPOSITION 7.4.5. *Let $\mathcal{M}$ be a model category, and let X and Y be objects in $\mathcal{M}$.*

(1) *If X is cofibrant, then left homotopy (see Definition 7.3.2) is an equivalence relation on the set of maps from X to Y.*
(2) *If Y is fibrant, then right homotopy (see Definition 7.3.2) is an equivalence relation on the set of maps from X to Y.*
(3) *If X is cofibrant and Y is fibrant, then homotopy (see Definition 7.3.2) is an equivalence relation on the set of maps from X to Y.*

PROOF. We will prove part 1; the proof of part 2 is dual, and part 3 follows from parts 1 and 2.

Since there is a cylinder object for X in which $\mathrm{Cyl}(X) = X$, left homotopy is reflexive. The inverse of a left homotopy (see Definition 7.4.4) implies that left homotopy is symmetric. Finally, the composition of left homotopies (see Definition 7.4.3) implies that left homotopy is transitive. □

7.4.6. Relations between left homotopy and right homotopy.

PROPOSITION 7.4.7. *Let $\mathcal{M}$ be a model category and let $f, g \colon X \to Y$ be maps in $\mathcal{M}$.*

(1) *If X is cofibrant, f is left homotopic to g, and $Y \xrightarrow{s} \mathrm{Path}(Y) \xrightarrow{p_0 \times p_1} Y \times Y$ is a path object for Y, then there is a right homotopy $H \colon X \to \mathrm{Path}(Y)$ from f to g.*
(2) *If Y is fibrant, f is right homotopic to g, and $X \amalg X \xrightarrow{i_0 \amalg i_1} \mathrm{Cyl}(X) \xrightarrow{p} X$ is a cylinder object for X, then there is a left homotopy $H \colon \mathrm{Cyl}(X) \to Y$ from f to g.*

PROOF. We will prove part 1; the proof of part 2 is dual.

Since f is left homotopic to g, there is a cylinder object $X \amalg X \xrightarrow{i_0 \amalg i_1} \mathrm{Cyl}(X) \xrightarrow{p} X$ for X and a left homotopy $G \colon \mathrm{Cyl}(X) \to Y$ from f to g. Thus, we have the solid arrow diagram

$$\begin{array}{ccc} X & \xrightarrow{sf} & \operatorname{Path}(Y) \\ {\scriptstyle i_0}\big\downarrow & \overset{h}{\nearrow} & \big\downarrow{\scriptstyle (p_0,p_1)} \\ \operatorname{Cyl}(X) & \xrightarrow[(fp,G)]{} & Y \times Y \end{array}$$

in which (p_0, p_1) is a fibration. Since X is cofibrant, Lemma 7.3.7 implies that i_0 is a trivial cofibration, and so the dotted arrow h exists. If we let $H = hi_1$, then H is the right homotopy we require. □

PROPOSITION 7.4.8. *Let $\mathcal{M}$ be a model category and let $f, g\colon X \to Y$ be maps in $\mathcal{M}$.*

(1) *If X is cofibrant and $f \overset{l}{\simeq} g$ (see Definition 7.3.2), then $f \overset{r}{\simeq} g$.*
(2) *If Y is fibrant and $f \overset{r}{\simeq} g$, then $f \overset{l}{\simeq} g$.*

PROOF. This follows from Lemma 7.3.3 and Proposition 7.4.7. □

THEOREM 7.4.9. *Let $\mathcal{M}$ be a model category. If X is cofibrant and Y is fibrant, then the left homotopy, right homotopy, and homotopy relations coincide and are equivalence relations on the set of maps from X to Y.*

PROOF. This follows from Proposition 7.4.8 and Proposition 7.4.5. □

PROPOSITION 7.4.10. *Let $\mathcal{M}$ be a model category. If X is cofibrant, Y is fibrant, and $f, g\colon X \to Y$ are homotopic maps,. then*

(1) *if $X \amalg X \to \operatorname{Cyl}(X) \to X$ is a cylinder object for X, then there is a left homotopy $H\colon \operatorname{Cyl}(X) \to Y$ from f to g, and*
(2) *if $Y \to \operatorname{Path}(Y) \to Y \times Y$ is a path object for Y, then there is a right homotopy $H\colon X \to \operatorname{Path}(Y)$ from f to g.*

PROOF. This follows from Proposition 7.4.7. □

7.5. The classical homotopy category

Theorem 7.4.9 implies that for cofibrant-fibrant objects, all notions of homotopy coincide and are equivalence relations. The main result of this section is Theorem 7.5.5, which implies that composition of homotopy classes of maps is well defined for cofibrant-fibrant objects. We also prove a Whitehead theorem, which asserts that a weak equivalence between cofibrant-fibrant objects is a homotopy equivalence (see Theorem 7.5.10).

7.5.1. Composing homotopy classes of maps.

NOTATION 7.5.2. Let $\mathcal{M}$ be a model category and let X and Y be objects of $\mathcal{M}$.

(1) If X is cofibrant, we let $\pi^l(X, Y)$ denote the set of left homotopy classes of maps from X to Y (see Proposition 7.4.5).
(2) If Y is fibrant, we let $\pi^r(X, Y)$ denote the set of right homotopy classes of maps from X to Y (see Proposition 7.4.5).
(3) If X is cofibrant and Y is fibrant, we let $\pi(X, Y)$ denote the set of homotopy classes of maps from X to Y (see Proposition 7.4.5).

PROPOSITION 7.5.3. *Let $\mathcal{M}$ be a model category and let $f, g\colon X \to Y$ be maps in $\mathcal{M}$.*

(1) *If $f \overset{l}{\simeq} g$ (see Definition 7.3.2) and $h\colon Y \to Z$ is a map, then $hf \overset{l}{\simeq} hg$.*
(2) *If $f \overset{r}{\simeq} g$ (see Definition 7.3.2) and $k\colon W \to X$, then $fk \overset{r}{\simeq} gk$.*

PROOF. We will prove part 1; the proof of part 2 is dual.

If $X \amalg X \to \mathrm{Cyl}(X) \to X$ is a cylinder object for X and $F\colon \mathrm{Cyl}(X) \to Y$ is a left homotopy from f to g, then hF is a left homotopy from hf to hg. □

COROLLARY 7.5.4. *Let $\mathcal{M}$ be a model category and let $f, g\colon X \to Y$ be maps in $\mathcal{M}$.*

(1) *If $f \overset{l}{\simeq} g$ (see Definition 7.3.2) and $h\colon Y \to Z$ is a map, then composition with h induces a well defined function $h_*\colon \pi^l(X,Y) \to \pi^l(X,Z)$.*
(2) *If $f \overset{r}{\simeq} g$ (see Definition 7.3.2) and $k\colon W \to X$, then composition with k induces a well defined function $k^*\colon \pi^r(X,Y) \to \pi^r(W,Y)$.*

PROOF. This follows from Proposition 7.5.3. □

THEOREM 7.5.5. *Let $\mathcal{M}$ be a model category, let X, Y, and Z be cofibrant-fibrant objects of $\mathcal{M}$, and let $f, g\colon X \to Y$ and $h, k\colon Y \to Z$ be maps. If $f \simeq g$ and $h \simeq k$, then $hf \simeq kg$, and so composition is well defined on homotopy classes of maps between cofibrant-fibrant objects.*

PROOF. This follows from Corollary 7.5.4 and Theorem 7.4.9. □

7.5.6. The classical homotopy category.

PROPOSITION 7.5.7. *If $\mathcal{M}$ is a model category, then there is a category whose objects are the cofibrant-fibrant objects in $\mathcal{M}$, whose maps are homotopy classes of maps in $\mathcal{M}$, and whose composition of maps is induced by composition of maps in $\mathcal{M}$.*

PROOF. This follows from Theorem 7.5.5. □

DEFINITION 7.5.8. If $\mathcal{M}$ is a model category, then (following D. M. Kan) we define the *classical homotopy category* $\pi\mathcal{M}_{\mathrm{cf}}$ of $\mathcal{M}$ to be the category with objects the cofibrant-fibrant objects of $\mathcal{M}$, and with morphisms from X to Y the homotopy classes of maps from X to Y (see Proposition 7.5.7).

Note that the classical homotopy category of a model category does not contain all of the objects of the model category, and it is *not* what is known as the *homotopy category* of the model category. For the homotopy category of a model category, see Definition 8.3.2.

PROPOSITION 7.5.9. *Let $\mathcal{M}$ be a model category.*

(1) *If A is cofibrant and $p\colon X \to Y$ is a trivial fibration, then p induces an isomorphism of the sets of left homotopy classes of maps $p_*\colon \pi^l(A,X) \to \pi^l(A,Y)$ (see Corollary 7.5.4).*
(2) *If X is fibrant and $i\colon A \to B$ is a trivial cofibration, then i induces an isomorphism of the sets of right homotopy classes of maps $i^*\colon \pi^r(B,X) \to \pi^r(A,X)$ (see Corollary 7.5.4).*

PROOF. We will prove part 1; the proof of part 2 is dual.

If $g\colon A \to Y$ is a map and $\emptyset$ is the initial object of $\mathcal{M}$, then axiom M4 (see Definition 7.1.3) implies that the dotted arrow exists in the diagram

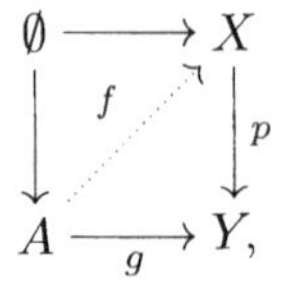

and so p_* is surjective. To see that p_* is injective, let $f, g\colon A \to X$ be maps such that $pf \overset{l}{\simeq} pg$. There is then a cylinder object $A \amalg A \to \mathrm{Cyl}(A) \to A$ for A and a left homotopy $F\colon \mathrm{Cyl}(A) \to Y$ from pf to pg, and so we have the solid arrow diagram

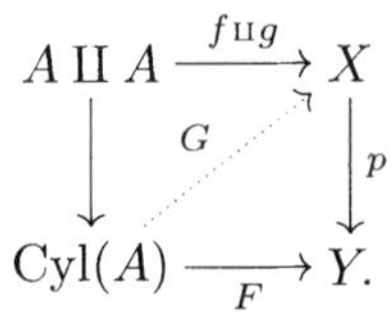

Axiom M4 implies that the dotted arrow G exists, and G is a left homotopy from f to g. □

THEOREM 7.5.10 (Whitehead theorem). *Let $\mathcal{M}$ be a model category. If $f\colon X \to Y$ is a weak equivalence between cofibrant-fibrant objects, then it is a homotopy equivalence.*

PROOF. If we factor f into a cofibration followed by a trivial fibration to obtain $X \xrightarrow{p} W \xrightarrow{q} Y$, then W is also cofibrant-fibrant, and the "two out of three" axiom (see Definition 7.1.3) implies that p is also a weak equivalence. Since a composition of homotopy equivalences between cofibrant-fibrant objects is a homotopy equivalence (see Theorem 7.5.5), it is sufficient to show that a trivial cofibration or trivial fibration between cofibrant-fibrant objects is a homotopy equivalence. We will show this for the trivial cofibration p; the proof for the trivial fibration q is dual.

We have the solid arrow diagram

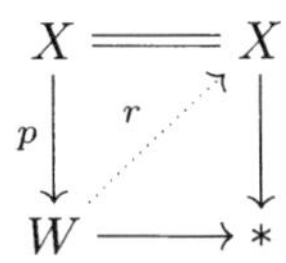

(in which $*$ denotes the terminal object), and so there exists a dotted arrow r such that $rp = 1_X$. Proposition 7.5.9 implies that p induces an isomorphism $p^*\colon \pi^r(W, W) \approx \pi^r(X, W)$, and, since $p^*[pr] = [prp] = [p][rp] = [p][1_X] = [p] = p^*[1_W]$, this implies that $pr \overset{r}{\simeq} 1_W$. Thus, r is a homotopy inverse for p (see Theorem 7.4.9), and so p is a homotopy equivalence. □

PROPOSITION 7.5.11. *Let $\mathcal{M}$ be a model category, let W, X, Y, and Z be cofibrant-fibrant objects, and let $f, g\colon X \to Y$ be a pair of maps.*

(1) *If there is a weak equivalence $h\colon Y \to Z$ such that $hf \simeq hg$, then $f \simeq g$.*
(2) *If there is a weak equivalence $k\colon W \to X$ such that $fk \simeq gk$, then $f \simeq g$.*

PROOF. We will prove part 1; the proof of part 2 is similar.

Theorem 7.5.10 implies that there is a map $\tilde{h}\colon Z \to Y$ such that $\tilde{h}h \simeq 1_Y$. Thus, $f \simeq 1_Y f \simeq \tilde{h}hf \simeq \tilde{h}hg \simeq 1_Y g \simeq g$. □

PROPOSITION 7.5.12. *Let $\mathcal{M}$ be a model category. If X and Y are cofibrant-fibrant objects in $\mathcal{M}$, then a map $g\colon X \to Y$ is a homotopy equivalence if either of the following two conditions is satisfied:*

(1) *The map g induces isomorphisms of the sets of homotopy classes of maps $g_*\colon \pi(X,X) \approx \pi(X,Y)$ and $g_*\colon \pi(Y,X) \approx \pi(Y,Y)$.*
(2) *The map g induces isomorphisms of the sets of homotopy classes of maps $g^*\colon \pi(Y,X) \approx \pi(X,X)$ and $g^*\colon \pi(Y,Y) \approx \pi(X,Y)$.*

PROOF. We will prove this using condition 1; the proof using condition 2 is similar.

The isomorphism $g_*\colon \pi(Y,X) \approx \pi(Y,Y)$ implies that there is a map $h\colon Y \to X$ such that $gh \simeq 1_Y$. Theorem 7.5.5 and the isomorphism $g_*\colon \pi(X,X) \approx \pi(X,Y)$ now imply that h induces an isomorphism $h_*\colon \pi(X,Y) \approx \pi(X,X)$, and so there is a map $k\colon X \to Y$ such that $hk \simeq 1_X$. Thus, h is a homotopy equivalence and g is its inverse, and so g is a homotopy equivalence as well. □

7.6. Relative homotopy and fiberwise homotopy

If $\mathcal{M}$ is a model category and A is an object of $\mathcal{M}$, then the category $(A \downarrow \mathcal{M})$ of *objects of* $\mathcal{M}$ *under* A has objects the maps $A \to X$ and morphisms commutative triangles (see Definition 7.6.1). If $A \to X$ is the inclusion of a subobject, then homotopy of maps from $A \to X$ to $A \to Y$ corresponds to homotopy of maps from X to Y *relative to* A. Dually, the category $(\mathcal{M} \downarrow A)$ of *objects of* $\mathcal{M}$ *over* A has objects the maps $X \to A$ and morphisms commutative triangles (see Definition 7.6.2). If $X \to A$ and $Y \to A$ are fibrations, then homotopy of maps from $X \to A$ to $Y \to A$ corresponds to fiberwise homotopy over A.

DEFINITION 7.6.1. If $\mathcal{M}$ is a category and A is an object of $\mathcal{M}$, then the category $(A \downarrow \mathcal{M})$ of *objects of* $\mathcal{M}$ *under* A is the category in which

- an object is a map $A \to X$ in $\mathcal{M}$,
- a map from $A \to X$ to $A \to Y$ is a map $X \to Y$ in $\mathcal{M}$ such that the triangle

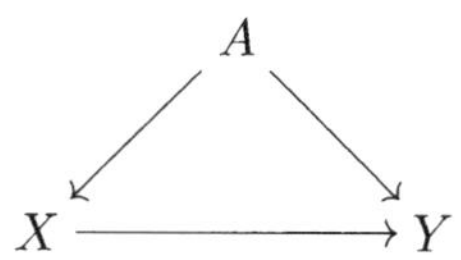

  commutes, and
- composition of maps is defined by composition of maps in $\mathcal{M}$.

DEFINITION 7.6.2. If $\mathcal{M}$ is a category and A is an object of $\mathcal{M}$, then the category $(\mathcal{M} \downarrow A)$ of *objects of* $\mathcal{M}$ *over* A is the category in which

- an object is a map $X \to A$ in $\mathcal{M}$,

- a map from $X \to A$ to $Y \to A$ is a map $X \to Y$ in $\mathcal{M}$ such that the triangle

$$\begin{array}{ccc} X & \longrightarrow & Y \\ & \searrow \quad \swarrow & \\ & A & \end{array}$$

commutes, and
- composition of maps is defined by composition of maps in $\mathcal{M}$.

DEFINITION 7.6.3. If $\mathcal{M}$ is a category and A and B are objects of $\mathcal{M}$, then the category $(A \downarrow \mathcal{M} \downarrow B)$ of *objects of* $\mathcal{M}$ *under* A *and over* B is the category in which

- an object is a diagram $A \to X \to B$ in $\mathcal{M}$,
- a map from $A \to X \to B$ to $A \to Y \to B$ is a map $X \to Y$ in $\mathcal{M}$ such that the diagram

$$\begin{array}{ccc} & A & \\ & \swarrow \quad \searrow & \\ X & \longrightarrow & Y \\ & \searrow \quad \swarrow & \\ & B & \end{array}$$

commutes, and
- composition of maps is defined by composition of maps in $\mathcal{M}$.

Definition 7.6.1 is a special case of Definition 11.8.3, and Definition 7.6.2 is a special case of Definition 11.8.1.

7.6.4. Homotopy in undercategories and overcategories.

THEOREM 7.6.5. *Let* $\mathcal{M}$ *be a model category.*

(1) *If* A *is an object of* $\mathcal{M}$*, then the category* $(A \downarrow \mathcal{M})$ *of objects of* $\mathcal{M}$ *under* A *(see Definition 7.6.1) is a model category in which a map is a weak equivalence, fibration, or cofibration if it is one in* $\mathcal{M}$*.*
(2) *If* X *is an object of* $\mathcal{M}$*, then the category* $(\mathcal{M} \downarrow X)$ *of objects of* $\mathcal{M}$ *over* X *(see Definition 7.6.2) is a model category in which a map is a weak equivalence, fibration, or cofibration if it is one in* $\mathcal{M}$*.*
(3) *If* A *and* B *are objects in* $\mathcal{M}$*, then the category* $(A \downarrow \mathcal{M} \downarrow B)$ *of objects of* $\mathcal{M}$ *under* A *and over* B *(see Definition 7.6.3) is a model category in which a map is a weak equivalence, fibration, or cofibration if it is one in* $\mathcal{M}$*.*

PROOF. This follows directly from the definitions. □

LEMMA 7.6.6. *Let* $\mathcal{C}$ *be a category and let* $g\colon X \to Y$ *be a map in* $\mathcal{C}$*.*

(1) *The functor* $(X \downarrow \mathcal{C}) \to (Y \downarrow \mathcal{C})$ *that takes the object* $X \to Z$ *of* $(X \downarrow \mathcal{C})$ *to its pushout along* g *is left adjoint to the functor* $g^*\colon (Y \downarrow \mathcal{C}) \to (X \downarrow \mathcal{C})$ *that takes the object* $Y \to W$ *of* $(Y \downarrow \mathcal{C})$ *to its composition with* g*.*
(2) *The functor* $(\mathcal{C} \downarrow Y) \to (\mathcal{C} \downarrow X)$ *that takes the object* $W \to Y$ *of* $(\mathcal{C} \downarrow Y)$ *to its pullback along* g *is right adjoint to the functor* $g_*\colon (\mathcal{C} \downarrow X) \to (\mathcal{C} \downarrow Y)$ *that takes the object* $Z \to X$ *of* $(\mathcal{C} \downarrow X)$ *to its composition with* g*.*

PROOF. This follows directly from the universal mapping properties that define the pushout and the pullback. □

DEFINITION 7.6.7. Let $\mathcal{M}$ be a model category, and let A be an object of $\mathcal{M}$.

(1) If $A \to X$ and $A \to Y$ are objects of the category $(A \downarrow \mathcal{M})$ of objects of $\mathcal{M}$ under A, then maps $f, g\colon X \to Y$ in $(A \downarrow \mathcal{M})$ will be called *left homotopic under* A, *right homotopic under* A, or *homotopic under* A if they are, respectively, left homotopic, right homotopic, or homotopic as maps in $(A \downarrow \mathcal{M})$. A map will be called a *homotopy equivalence under* A if it is a homotopy equivalence in the category $(A \downarrow \mathcal{M})$.
(2) If $X \to A$ and $Y \to A$ are objects of the category $(\mathcal{M} \downarrow A)$ of objects of $\mathcal{M}$ over A, then maps $f, g\colon X \to Y$ will be called *left homotopic over* A, *right homotopic over* A, or *homotopic over* A if they are, respectively, left homotopic, right homotopic, or homotopic as maps in $(\mathcal{M} \downarrow A)$. A map will be called a *homotopy equivalence over* A if it is a homotopy equivalence in the category $(\mathcal{M} \downarrow A)$.

PROPOSITION 7.6.8. *Let $\mathcal{M}$ be a model category, and let A be an object of $\mathcal{M}$.*

(1) *If maps are left homotopic, right homotopic, or homotopic under A, then they are, respectively, left homotopic, right homotopic, or homotopic.*
(2) *If maps are left homotopic, right homotopic, or homotopic over A, then they are, respectively, left homotopic, right homotopic, or homotopic.*

PROOF. This follows from Proposition 7.3.5. □

COROLLARY 7.6.9. *Let $\mathcal{M}$ be a model category, and let A be an object of $\mathcal{M}$. If a map is a homotopy equivalence under A or a homotopy equivalence over A, then it is a homotopy equivalence in $\mathcal{M}$.*

PROOF. This follows from Proposition 7.6.8. □

DEFINITION 7.6.10. If $\mathcal{M}$ is a model category, then a map $i\colon A \to B$ will be called *the inclusion of a deformation retract* (and A will be called a *deformation retract* of B) if there is a map $r\colon B \to A$ such that $ri = 1_A$ and $ir \simeq 1_B$. A deformation retract will be called a *strong deformation retract* if $ir \simeq 1_B$ under A.

PROPOSITION 7.6.11. *Let $\mathcal{M}$ be a model category.*

(1) *If $i\colon A \to B$ is a trivial cofibration of fibrant objects, then A is a strong deformation retract of B (see Definition 7.6.10), i.e., there is a map $r\colon B \to A$ such that $ri = 1_A$ and $ir \simeq 1_B$ under A.*
(2) *If $p\colon X \to Y$ is a trivial fibration of cofibrant objects, then there is a map $s\colon Y \to X$ such that $ps = 1_Y$ and $sp \simeq 1_X$ over Y.*

PROOF. We will prove part 1; the proof of part 2 is dual.

We have the solid arrow diagram

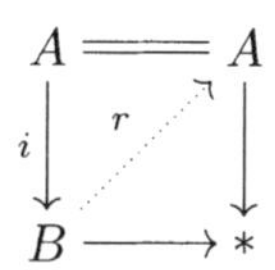

in $(A \downarrow \mathcal{M})$ (see Theorem 7.6.5) in which i is a trivial cofibration and the map on the right is a fibration. Thus, there exists a map $r\colon B \to A$ in $(A \downarrow \mathcal{M})$ such that

$ri = 1_A$. Since $i^*(1_B) = i = iri = i^*(ir)$, Proposition 7.5.9 implies that $1_B \overset{r}{\simeq} ir$ in $(A \downarrow \mathcal{M})$. Since both A and B are both cofibrant-fibrant in $(A \downarrow \mathcal{M})$, Theorem 7.4.9 implies that $1_B \simeq ir$ in $(A \downarrow \mathcal{M})$. □

7.6.12. Homotopy uniqueness of lifts.

PROPOSITION 7.6.13. *Let $\mathcal{M}$ be a model category, and let the solid arrow diagram*

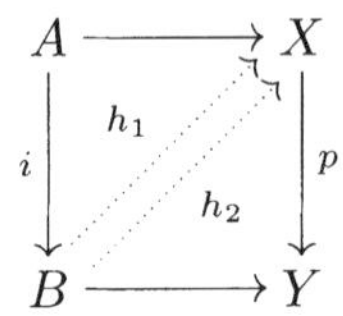

be such that either

(1) *i is a cofibration and p is a trivial fibration, or*
(2) *i is a trivial cofibration and p is a fibration.*

If h_1 and h_2 are maps each of which makes both triangles commute, then $h_1 \simeq h_2$ as maps in $(A \downarrow \mathcal{M} \downarrow Y)$, the category of objects of $\mathcal{M}$ under A and over Y.

PROOF. We will assume that condition 1 holds; the proof in the case that condition 2 holds is similar.

Factor the map $B \amalg_A B \to B$ as $B \amalg_A B \xrightarrow{j} C \xrightarrow{r} B$ where j is a cofibration and r is a trivial fibration. We now have the solid arrow diagram

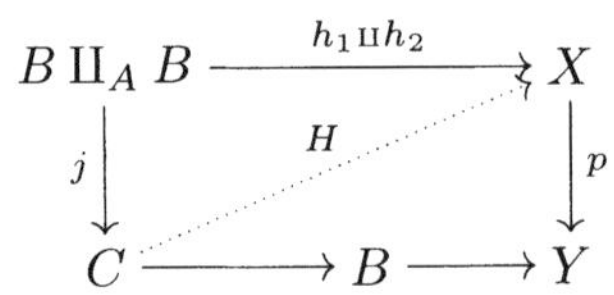

in which j is a cofibration and p is a trivial fibration, and so there exists a dotted arrow H making both triangles commute. In the category $(A \downarrow \mathcal{M} \downarrow Y)$ of objects of $\mathcal{M}$ under A and over Y (see Theorem 7.6.5), $B \amalg_A B \to C \to B$ is a cylinder object for B (see Definition 7.3.2) and H is a left homotopy from h_1 to h_2. Since B is cofibrant and X is fibrant in $(A \downarrow \mathcal{M} \downarrow Y)$, Proposition 7.4.8 implies that h_1 is also right homotopic to h_2, and so h_1 is homotopic to h_2 in $(A \downarrow \mathcal{M} \downarrow Y)$. □

PROPOSITION 7.6.14. *Let $\mathcal{M}$ be a model category. If the solid arrow diagram*

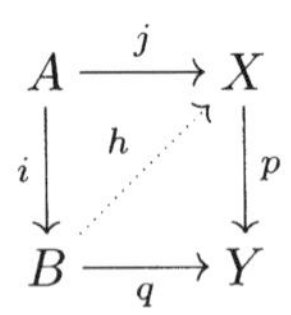

is such that either

(1) *i and j are cofibrations and p and q are trivial fibrations, or*
(2) *i and j are trivial cofibrations and p and q are fibrations,*

then there exists a map $h\colon B \to X$ making both triangles commute, unique up to homotopy in $(A \downarrow \mathcal{M} \downarrow Y)$, and any such map is a homotopy equivalence in $(A \downarrow \mathcal{M} \downarrow Y)$.

PROOF. This follows from Proposition 7.6.13. □

7.7. Weak equivalences

The main result of this section is *Kenny Brown's lemma* (see Lemma 7.7.1). This asserts that a weak equivalence between cofibrant objects can be factored into a trivial cofibration followed by a map that has a trivial cofibration as a one sided inverse (with a dual statement for weak equivalences between fibrant objects). This implies that a functor between model categories that takes trivial cofibrations into weak equivalences must also take weak equivalences between cofibrant objects into weak equivalences (see Corollary 7.7.2), which will be important for our discussion of Quillen functors (see Definition 8.5.2).

LEMMA 7.7.1 (K. S. Brown, [**15**]). *Let $\mathcal{M}$ be a model category.*

(1) *If $g\colon X \to Y$ is a weak equivalence between cofibrant objects in $\mathcal{M}$ then there is a functorial factorization of g as $g = ji$ where i is a trivial cofibration and j is a trivial fibration that has a right inverse that is a trivial cofibration.*
(2) *If $g\colon X \to Y$ is a weak equivalence between fibrant objects in $\mathcal{M}$ then there is a functorial factorization of g as $g = ji$ where i is a trivial cofibration that has a left inverse that is a trivial fibration and j is a trivial fibration.*

PROOF. We will prove part 1; the proof of part 2 is dual.

Since X and Y are cofibrant, both of the injections $X \to X \amalg Y$ and $Y \to X \amalg Y$ are cofibrations. If we factor the map $g \amalg 1_Y \colon X \amalg Y \to Y$ as

$$X \amalg Y \xrightarrow{k} Z \xrightarrow{j} Y$$

where k is a cofibration and j is a trivial fibration, then both compositions $X \to X \amalg Y \to Z$ and $Y \to X \amalg Y \to Z$ are cofibrations. Since g and j are weak equivalences, axiom M2 (see Definition 7.1.3) implies that the cofibration $X \to Z$ is a weak equivalence, and the composition of cofibrations $Y \to X \amalg Y \to Z$ is a right inverse to the trivial fibration j. □

COROLLARY 7.7.2. *Let $\mathcal{M}$ and $\mathcal{N}$ be model categories, and let $\mathrm{F}\colon \mathcal{M} \to \mathcal{N}$ be a functor.*

(1) *If F takes trivial cofibrations between cofibrant objects in $\mathcal{M}$ to weak equivalences in $\mathcal{N}$, then F takes all weak equivalences between cofibrant objects to weak equivalences in $\mathcal{N}$.*
(2) *If F takes trivial fibrations between fibrant objects in $\mathcal{M}$ to weak equivalences in $\mathcal{N}$, then F takes all weak equivalences between fibrant objects to weak equivalences in $\mathcal{N}$.*

PROOF. This follows from Lemma 7.7.1 and the "two out of three" property of weak equivalences. □

COROLLARY 7.7.3. *Let $\mathcal{M}$ be a model category, let $\mathcal{C}$ be a category, and let $\mathrm{F}\colon \mathcal{M} \to \mathcal{C}$ be a functor.*

(1) *If F takes trivial cofibrations between cofibrant objects in $\mathcal{M}$ to isomorphisms in $\mathcal{C}$, then F takes all weak equivalences between cofibrant objects to isomorphisms.*
(2) *If F takes trivial fibrations between fibrant objects in $\mathcal{M}$ to isomorphisms in $\mathcal{C}$, then F takes all weak equivalences between fibrant objects to isomorphisms.*

PROOF. This follows from Lemma 7.7.1. □

COROLLARY 7.7.4. *Let $\mathcal{M}$ be a model category.*

(1) *If $g\colon C \to D$ is a weak equivalence between cofibrant objects in $\mathcal{M}$ and X is a fibrant object of $\mathcal{M}$, then g induces an isomorphism of the sets of homotopy classes of maps $g^*\colon \pi(D,X) \approx \pi(C,X)$.*
(2) *If $g\colon X \to Y$ is a weak equivalence between fibrant objects in $\mathcal{M}$ and C is a cofibrant object of $\mathcal{M}$, then g induces an isomorphism of the sets of homotopy classes of maps $g_*\colon \pi(C,X) \approx \pi(C,Y)$.*

PROOF. This follows from Lemma 7.7.1, Proposition 7.5.9, and Theorem 7.4.9. □

COROLLARY 7.7.5. *Let $\mathcal{M}$ be a model category.*

(1) *If $g\colon C \to D$ is a weak equivalence between cofibrant objects in $\mathcal{M}$ and X is a fibrant object of $\mathcal{M}$, then there is a map $C \to X$ in $\mathcal{M}$ if and only if there is a map $D \to X$ in $\mathcal{M}$.*
(2) *If $g\colon X \to Y$ is a weak equivalence between fibrant objects in $\mathcal{M}$ and C is a cofibrant object of $\mathcal{M}$, then there is a map $C \to X$ in $\mathcal{M}$ if and only if there is a map $C \to Y$ in $\mathcal{M}$.*

PROOF. This follows from Corollary 7.7.4. □

PROPOSITION 7.7.6. *Let $\mathcal{M}$ be a model category, and let $f, g\colon X \to Y$ be maps. If $f \overset{l}{\simeq} g$ or $f \overset{r}{\simeq} g$, then f is a weak equivalence if and only if g is a weak equivalence.*

PROOF. We will consider the case $f \overset{l}{\simeq} g$; the case $f \overset{r}{\simeq} g$ is dual.

Since $f \overset{l}{\simeq} g$, there is a cylinder object $X \amalg X \xrightarrow{i_0 \amalg i_1} \mathrm{Cyl}(X) \xrightarrow{p} X$ for X and a map $H\colon \mathrm{Cyl}(X) \to Y$ such that $hi_0 = f$ and $hi_1 = g$. Lemma 7.3.7 and the "two out of three" property of weak equivalences imply that f is a weak equivalence if and only if H is a weak equivalence, and that this is true if and only if g is a weak equivalence. □

LEMMA 7.7.7. *Let $\mathcal{M}$ and $\mathcal{N}$ be model categories, let $g_0, g_1\colon X \to Y$ be maps in $\mathcal{M}$, and let $\mathrm{F}\colon \mathcal{M} \to \mathcal{N}$ be a functor.*

(1) *If F takes trivial cofibrations between cofibrant objects in $\mathcal{M}$ into weak equivalences in $\mathcal{N}$, the object X is cofibrant, and g_0 is left homotopic to g_1, then $\mathrm{F}(g_0)$ is a weak equivalence if and only if $\mathrm{F}(g_1)$ is a weak equivalence.*
(2) *If F takes trivial fibrations between fibrant objects in $\mathcal{M}$ into weak equivalences in $\mathcal{N}$, the object Y is fibrant, and g_0 is right homotopic to g_1 (see Definition 7.3.2), then $\mathrm{F}(g_0)$ is a weak equivalence if and only if $\mathrm{F}(g_1)$ is a weak equivalence.*

PROOF. This follows from Lemma 7.3.8 and Proposition 7.7.6. □

7.8. Homotopy equivalences

The main result of this section is Theorem 7.8.5, which asserts that a homotopy equivalence between cofibrant-fibrant objects is a weak equivalence. We also prove that a map between cofibrant-fibrant objects that is both a cofibration and a homotopy equivalence is the inclusion of a strong deformation retraction (see Proposition 7.8.2).

LEMMA 7.8.1. *Let $\mathcal{M}$ be a model category and let X and Y be cofibrant-fibrant objects in $\mathcal{M}$.*

(1) *Let $X \amalg X \xrightarrow{i_0 \amalg i_1} \mathrm{Cyl}(X) \xrightarrow{p} X$ be a cylinder object for X and let $H\colon \mathrm{Cyl}(X) \to Y$ be a left homotopy from the map $f\colon X \to Y$ to the map $g\colon X \to Y$. If H'' is the composition (see Definition 7.4.3) of H and H^{-1} (see Definition 7.4.4), then H'' is homotopic in $\big((X \amalg X)\downarrow \mathcal{M}\big)$ to the constant left homotopy (i.e., the composition $\mathrm{Cyl}(X)'' \xrightarrow{p''} X \xrightarrow{f} Y$).*

(2) *Let $Y \xrightarrow{s} \mathrm{Path}(Y) \xrightarrow{p_0 \times p_1} Y \times Y$ be a path object for Y and let $H\colon X \to \mathrm{Path}(Y)$ be a right homotopy from the map $f\colon X \to Y$ to the map $g\colon X \to Y$. If H'' is the composition (see Definition 7.4.3) of H and H^{-1} (see Definition 7.4.4), then H'' is homotopic in $\big(\mathcal{M}\downarrow (Y \amalg Y)\big)$ to the constant right homotopy (i.e., the composition $X \xrightarrow{f} Y \xrightarrow{s''} \mathrm{Cyl}(Y)''$).*

PROOF. We will prove part 1; the proof of part 2 is dual.

Let $Y \xrightarrow{s} \mathrm{Path}(Y) \xrightarrow{p_0 \times p_1} Y \times Y$ be a path object for Y (see Lemma 7.3.3). We have the solid arrow diagram

$$\begin{array}{ccc} X & \xrightarrow{sf} & \mathrm{Path}(Y) \\ {\scriptstyle i_0}\big\downarrow & \overset{K}{\nearrow} & \big\downarrow{\scriptstyle (p_0,p_1)} \\ \mathrm{Cyl}(X) & \xrightarrow[(fp,H)]{} & Y \times Y \end{array}$$

in which i_0 is a trivial cofibration (see Lemma 7.3.7) and (p_0, p_1) is a fibration, and so the dotted arrow K exists. If we let the map $K'\colon \mathrm{Cyl}(X)' \to \mathrm{Path}(Y)$ equal the map K, then K and K' combine to define a map $K''\colon \mathrm{Cyl}(X)'' \to \mathrm{Path}(Y)$ that makes the diagram

$$\begin{array}{ccc} X \amalg X & \xrightarrow{sf \amalg sf} & \mathrm{Path}(Y) \\ {\scriptstyle i_0'' \amalg i_1''}\big\downarrow & \overset{K''}{\nearrow} & \big\downarrow{\scriptstyle (p_0,p_1)} \\ \mathrm{Cyl}(X)'' & \xrightarrow[(fp'',H'')]{} & Y \times Y \end{array}$$

commutes. Thus, K'' is a right homotopy (see Definition 7.3.2) from the map $fp''\colon \mathrm{Cyl}(X)'' \to Y$ to the map $H''\colon \mathrm{Cyl}(X)'' \to Y$ in the category $\big((X \amalg X)\downarrow \mathcal{M}\big)$ of objects of $\mathcal{M}$ under $X \amalg X$. Since $\mathrm{Cyl}(X)''$ is cofibrant in $\big(\mathcal{M}\downarrow (X \amalg X)\big)$ and Y is fibrant in $\big(\mathcal{M}\downarrow (X \amalg X)\big)$, Theorem 7.4.9 implies that fp'' is also left homotopic to H'' in $\big(\mathcal{M}\downarrow (X \amalg X)\big)$, and so fp'' is homotopic to H'' in $\big(\mathcal{M}\downarrow (X \amalg X)\big)$. □

PROPOSITION 7.8.2. *Let $\mathcal{M}$ be a model category and let $f\colon X \to Y$ be a map between cofibrant-fibrant objects.*

(1) *If f is both a cofibration and a homotopy equivalence, then f is the inclusion of a strong deformation retract, i.e., there is a map $g\colon Y \to X$ such that $gf = 1_X$ and $fg \simeq 1_Y$ in $(X \downarrow \mathcal{M})$.*

(2) *If f is both a fibration and a homotopy equivalence, then f is the dual of a strong deformation retract, i.e., there is a map $g\colon Y \to X$ such that $fg = 1_Y$ and $gf \simeq 1_X$ in $(\mathcal{M} \downarrow Y)$.*

PROOF. We will prove part 1; the proof of part 2 is dual.

Since f is a homotopy equivalence, there is a map $h\colon Y \to X$ such that $fh \simeq 1_Y$ and $hf \simeq 1_X$. The homotopy extension property of cofibrations (see Proposition 7.3.10) implies that h is homotopic to a map $g\colon Y \to X$ such that $gf = 1_X$ and $fg \simeq 1_Y$ (see Proposition 7.5.3). Let $Y \xrightarrow{s} \mathrm{Path}(Y) \xrightarrow{p_0 \times p_1} Y \times Y$ be a path object for Y and let $H\colon Y \to \mathrm{Path}(Y)$ be a right homotopy from fg to 1_Y. The composition $Hf\colon X \to \mathrm{Path}(Y)$ is then a right homotopy from $fgf = f$ to $1_Y f = f$. The composite homotopy $(Hfg) \cdot H^{-1}\colon Y \to \mathrm{Path}(Y)''$ (see Definition 7.4.3) composed with f is the composite homotopy $(Hf) \cdot (Hf)^{-1}\colon X \to \mathrm{Path}(Y)''$, and Lemma 7.8.1 implies that $(Hf) \cdot (Hf)^{-1}$ is homotopic in $\big(\mathcal{M} \downarrow (Y \times Y)\big)$ to the constant homotopy $s''f\colon X \to \mathrm{Path}(Y)''$. The homotopy extension property of cofibrations now implies that $(Hfk) \cdot H^{-1}$ is homotopic in $\big(\mathcal{M} \downarrow (Y \times Y)\big)$ to a right homotopy $K\colon Y \to \mathrm{Path}(Y)''$ such that $Kf\colon X \to \mathrm{Path}(Y)''$ equals $s''f$, i.e., K is a homotopy from gf to 1_Y in $(X \downarrow \mathcal{M})$. ☐

PROPOSITION 7.8.3. *Let $\mathcal{M}$ be a model category and let X and Y be cofibrant-fibrant objects in $\mathcal{M}$.*

(1) *If $g\colon X \to Y$ is both a cofibration and a homotopy equivalence, then g is a weak equivalence.*
(2) *If $g\colon X \to Y$ is both a fibration and a homotopy equivalence, then g is a weak equivalence.*

PROOF. We will prove part 1; the proof of part 2 is dual.

If we factor g as $X \xrightarrow{i} W \xrightarrow{p} Y$ where i is a trivial cofibration and p is a fibration, then the retract axiom (see Definition 7.1.3) implies that it is sufficient to show that g is a retract of i. If we can show that the dotted arrow q exists in the diagram

(7.8.4)

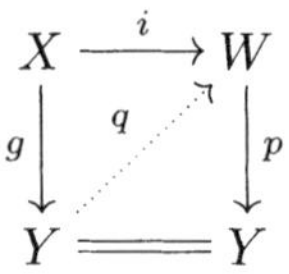

then we would have the diagram

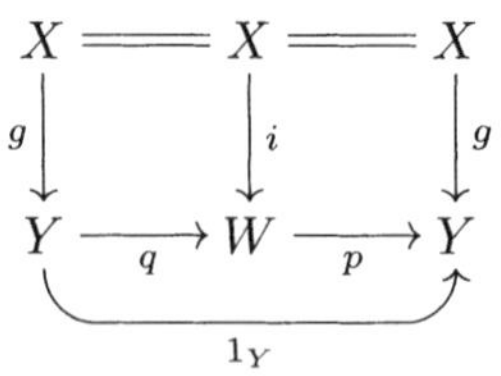

which would show that g is a retract of i. Thus, it is sufficient to find the dotted arrow q in Diagram 7.8.4. Proposition 7.8.2 implies that there is a map $h\colon Y \to X$ such that $hg = 1_X$ and $gh \simeq 1_Y$ in $(X \downarrow \mathcal{M})$. If we let $k\colon Y \to W$ be defined by $k = ih$, then $kg = i$, and $pk = pih = gh \simeq 1_Y$ in $(X \downarrow \mathcal{M})$. The homotopy lifting property (see Proposition 7.3.11) of the fibration p in the category $(X \downarrow \mathcal{M})$ now implies that k is homotopic in $(X \downarrow \mathcal{M})$ to a map $q\colon Y \to W$ such that $pq = 1_Y$. ☐

THEOREM 7.8.5. *Let $\mathcal{M}$ be a model category. If X and Y are cofibrant-fibrant objects in $\mathcal{M}$ and $g\colon X \to Y$ is a homotopy equivalence, then g is a weak equivalence.*

PROOF. If we factor g as $X \xrightarrow{h} W \xrightarrow{k} Y$ where h is a cofibration and k is a trivial fibration, then the "two out of three" property of weak equivalences implies that it is sufficient to show that h is a weak equivalence. Since W is also cofibrant-fibrant, Proposition 7.8.3 implies that it is sufficient to show that h is a homotopy equivalence.

If $g^{-1}\colon Y \to X$ is a homotopy inverse for g, then let $r\colon W \to X$ be defined by $r = g^{-1}k$. Since $rh = g^{-1}kh = g^{-1}g \simeq 1_X$, it is sufficient to show that $hr \simeq 1_W$. Proposition 7.5.9 and Theorem 7.4.9 imply that k induces an isomorphism of sets $k_*\colon \pi(X,W) \approx \pi(X,Y)$. Since $khr = gr = gg^{-1}k \simeq k$, this implies that $hr \simeq 1_W$. □

THEOREM 7.8.6. *Let $\mathcal{M}$ be a model category and let $f\colon X \to Y$ be a map in $\mathcal{M}$.*

(1) *If X and Y are cofibrant, then f is a weak equivalence if and only if for every fibrant object Z of $\mathcal{M}$ the induced map of homotopy classes of maps $f^*\colon \pi(Y,Z) \to \pi(X,Z)$ (see Notation 7.5.2) is an isomorphism.*
(2) *If X and Y are fibrant, then f is a weak equivalence if and only if for every cofibrant object W of $\mathcal{M}$ the induced map of homotopy classes of maps $f_*\colon \pi(W,X) \to \pi(W,Y)$ is an isomorphism.*

PROOF. We will prove part 1; the proof of part 2 is dual.

One direction of part 1 follows from Corollary 7.7.4. For the converse, let $\hat{f}\colon \widehat{X} \to \widehat{Y}$ be a cofibrant fibrant approximation to f (see Definition 8.1.22). Proposition 7.5.9 implies that $\hat{f}\colon \widehat{X} \to \widehat{Y}$ also induces an isomorphism of homotopy classes of maps $\hat{f}^*\colon \pi(\widehat{Y},Z) \to \pi(\widehat{X},Z)$ for every fibrant object Z of $\mathcal{M}$, and the "two out of three" axiom for weak equivalences (see Definition 7.1.3) implies that it is sufficient to show that $\hat{f}$ is a weak equivalence. This follows from Proposition 7.5.12 and Theorem 7.8.5. □

7.9. The equivalence relation generated by "weak equivalence"

The equivalence relation on objects of a model category generated by the relation "there is a weak equivalence from the first object to the second object" is made concrete by the notion of a zig-zag of weak equivalences (see Definition 7.9.1 and Definition 7.9.2). Zig-zags can also used to describe the maps in the localized category (see [**5**, Appendix]).

DEFINITION 7.9.1. Let $\mathcal{K}$ be a category and let $\mathcal{W}$ be a class of maps in $\mathcal{K}$.

(1) If X and Y are objects in $\mathcal{K}$ and $n \geq 0$, then a *zig-zag of elements of $\mathcal{W}$ of length n from X to Y* is a diagram of the form
$$X \xrightarrow{f_1} W_1 \xleftarrow{f_2} W_2 \xrightarrow{f_3} \cdots \xleftarrow{f_{n-1}} W_{n-1} \xrightarrow{f_n} Y$$
where
(a) each f_i is an element of $\mathcal{W}$,
(b) each f_i can point either to the left or to the right, and
(c) consecutive f_i's can point in either the same direction or in opposite directions.

(2) If X, Y, and Z are objects in $\mathcal{K}$ and

$$X \xrightarrow{f_1} W_1 \xleftarrow{f_2} \cdots \xleftarrow{f_{n-1}} W_{n-1} \xrightarrow{f_n} Y \quad \text{and} \quad Y \xrightarrow{g_1} V_1 \xleftarrow{g_2} \cdots \xleftarrow{g_{k-1}} V_{k-1} \xrightarrow{g_k} Z$$

are, respectively, a zig-zag in $\mathcal{W}$ from X to Y and a zig-zag in $\mathcal{W}$ from Y to Z, then the *composition* of those zig-zags is the zig-zag in $\mathcal{W}$ of length $n+k$ from X to Z

$$X \xrightarrow{f_1} W_1 \xleftarrow{f_2} \cdots \xleftarrow{f_{n-1}} W_{n-1} \xrightarrow{f_n} Y \xrightarrow{g_1} V_1 \xleftarrow{g_2} \cdots \xleftarrow{g_{k-1}} V_{k-1} \xrightarrow{g_k} Z$$

DEFINITION 7.9.2. Let $\mathcal{M}$ be a model category.

(1) If X and Y are objects in $\mathcal{M}$, then X and Y are *weakly equivalent* if there is a zig-zag of weak equivalences from X to Y (see Definition 7.9.1).
(2) If $\mathcal{C}$ is a category and F and G are functors from $\mathcal{C}$ to $\mathcal{M}$, then F and G are *naturally weakly equivalent* if there is an integer $n \geq 0$ and functors W_1, W_2, ..., W_n from $\mathcal{C}$ to $\mathcal{M}$ such that for every object A in $\mathcal{C}$ there is a natural zig-zag of weak equivalences

$$\mathrm{F}(A) \xrightarrow{\cong} \mathrm{W}_1(A) \xleftarrow{\cong} \mathrm{W}_2(A) \xrightarrow{\cong} \mathrm{W}_3(A) \xleftarrow{\cong} \cdots \xrightarrow{\cong} \mathrm{W}_n(A) \xleftarrow{\cong} \mathrm{G}(A)$$

from $\mathrm{F}(A)$ to $\mathrm{G}(A)$.

7.10. Topological spaces and simplicial sets

7.10.1. Categories of topological spaces. There are several different categories of topological spaces in common use, and any of these is acceptable for our purposes.

NOTATION 7.10.2. We will use Top to denote some category of topological spaces with the following properties:

(1) Top is closed under small colimits and small limits.
(2) Top contains among its objects the geometric realizations of all simplicial sets.
(3) If X and Y are objects of Top and K is a simplicial set, then there is a natural isomorphism of sets

$$\mathrm{Top}(X \times |K|, Y) \approx \mathrm{Top}(X, Y^{|K|}) .$$

Thus, the reader is invited to assume that Top denotes, e.g.,

- the category of compactly generated Hausdorff spaces (see, e.g., [**62**]), or
- the category of compactly generated weak Hausdorff spaces (see, e.g., [**37**, Appendix A1]), or
- some other category of spaces with our three properties (see, e.g., [**63**]).

The category of *all* topological spaces has Properties 1 and 2 of Notation 7.10.2, but not Property 3. Property 3 is needed only when we want to assume that Top is a *simplicial* model category (see Definition 9.1.6), though, and so if we want to consider Top as only a *model category*, then the reader can also choose to let Top denote the category of all topological spaces. Chapters 1 and 2 assume that Top is a simplicial model category, and so, technically, the category of all topological spaces is not acceptable there. In fact, though, the work in Chapters 1 and 2 only requires that the adjointness isomorphism in Property 3 exists for *finite* simplicial sets K. Since the realization of a finite simplicial set is locally compact, this makes the category of all topological spaces an acceptable definition of Top for Chapters

1 and 2 (see, e.g., [**29**, page 265] or [**50**, page 287]), although many arguments would have to be rephrased so as not to claim that the adjointness isomorphism in Property 3 exists for an arbitrary simplicial set K.

REMARK 7.10.3. Our definition of a *simplicial model category* (see Definition 9.1.6) differs from that of Quillen ([**52**]) in that we require that we have the adjointness isomorphism in Property 3 of Notation 7.10.2 for all simplicial sets K, while Quillen requires it only for finite simplicial sets K. Quillen proves that with his definition the category of *all* topological spaces is a simplicial model category [**52**, Chapter II, Section 3].

7.10.4. The model category structures. We will be working both with topological spaces and with simplicial sets, and for each of these we will consider both the category of pointed spaces and the category of unpointed spaces.

NOTATION 7.10.5. We will use the following notation for our categories of spaces:

SS : The category of (unpointed) simplicial sets.
SS_* : The category of pointed simplicial sets.
Top : The category of (unpointed) topological spaces.
Top_*: The category of pointed topological spaces.

There is a model category structure on each of these categories of spaces:

DEFINITION 7.10.6. If $f\colon X \to Y$ is a map of topological spaces, then

- f is a *weak equivalence* if f induces an isomorphism of path components and an isomorphism of homotopy groups $f_*\colon \pi_n(X, x_0) \approx \pi_n(Y, f(x_0))$ for all $n \geq 1$ and every choice of basepoint x_0 in X,
- f is a *fibration* if it is a Serre fibration, and
- f is a *cofibration* if it has the left lifting property with respect to all maps that are both fibrations and weak equivalences.

DEFINITION 7.10.7. If $f\colon X \to Y$ is a map of pointed topological spaces, then

- f is a *weak equivalence* if it is a weak equivalence of unpointed topological spaces when you forget about the basepoints,
- f is a *fibration* if it is a fibration of unpointed topological spaces when you forget about the basepoints, and
- f is a *cofibration* if it has the left lifting property with respect to all maps that are both fibrations and weak equivalences.

DEFINITION 7.10.8. If $f\colon X \to Y$ is a map of simplicial sets, then

- f is a *weak equivalence* if its geometric realization $|f|\colon |X| \to |Y|$ is a weak equivalence of topological spaces,
- f is a *fibration* if it is a Kan fibration, i.e., if it has the right lifting property with respect to the map $\Lambda[n,k] \to \Delta[n]$ for all $n > 0$ and $0 \leq k \leq n$, and
- f is a *cofibration* if it has the left lifting property with respect to all maps that are both fibrations and weak equivalences.

DEFINITION 7.10.9. If $f\colon X \to Y$ is a map of pointed simplicial sets, then

- f is a *weak equivalence* if it is a weak equivalence of unpointed simplicial sets when you forget about the basepoints,

- f is a *fibration* if it is a fibration of any simplicial set when you forget about the basepoints, and
- f is a *cofibration* if it has the left lifting property with respect to all maps that are both fibrations and weak equivalences.

The following four theorems assert the existence of the standard model category structures for the categories of topological spaces and simplicial sets (both pointed and unpointed). The proofs are surprisingly long, and we will not present them here. The original proofs are due to Quillen [**52**, Chapter II, Section 3], but more detailed and readable versions can be found in the works by Hovey [**42**, Section 2.4 and Chapter 3], Dwyer and Spalinski [**35**, Section 8], and Goerss and Jardine [**39**, Chapter I].

THEOREM 7.10.10. *The category* Top *(see Notation* 7.10.5*) with weak equivalences, fibrations, and cofibrations as in Definition* 7.10.6 *is a model category. In this model category,*

- *a map is a fibration if and only if it has the right lifting property with respect to the maps* $|\Lambda[n,k]| \to |\Delta[n]|$ *for all* $n > 0$ *and* $0 \le k \le n$, *and*
- *a map is a trivial fibration if and only if it has the right lifting property with respect to the maps* $|\partial\Delta[n]| \to |\Delta[n]|$ *for all* $n \ge 0$.

THEOREM 7.10.11. *The category* Top_* *(see Notation* 7.10.5*) with weak equivalences, fibrations, and cofibrations as in Definition* 7.10.7 *is a model category. In this model category,*

- *a map is a fibration if and only if it has the right lifting property with respect to the maps* $|\Lambda[n,k]|^+ \to |\Delta[n]|^+$ *for all* $n > 0$ *and* $0 \le k \le n$, *and*
- *a map is a trivial fibration if and only if it has the right lifting property with respect to the maps* $|\partial\Delta[n]|^+ \to |\Delta[n]|^+$ *for all* $n \ge 0$.

THEOREM 7.10.12. *The category* SS *(see Notation* 7.10.5*) with weak equivalences, fibrations, and cofibrations as in Definition* 7.10.8 *is a model category. In this model category,*

- *a map is a fibration if and only if it has the right lifting property with respect to the maps* $\Lambda[n,k] \to \Delta[n]$ *for all* $n > 0$ *and* $0 \le k \le n$, *and*
- *a map is a trivial fibration if and only if it has the right lifting property with respect to the maps* $\partial\Delta[n] \to \Delta[n]$ *for all* $n \ge 0$.

THEOREM 7.10.13. *The category* SS_* *(see Notation* 7.10.5*) with weak equivalences, fibrations, and cofibrations as in Definition* 7.10.9 *is a model category. In this model category,*

- *a map is a fibration if and only if it has the right lifting property with respect to the maps* $\Lambda[n,k]^+ \to \Delta[n]^+$ *for all* $n > 0$ *and* $0 \le k \le n$, *and*
- *a map is a trivial fibration if and only if it has the right lifting property with respect to the maps* $\partial\Delta[n]^+ \to \Delta[n]^+$ *for all* $n \ge 0$.

CHAPTER 8

Fibrant and Cofibrant Approximations

A *cofibrant approximation* to an object X is a cofibrant object $\widetilde{X}$ weakly equivalent to X; dually, a *fibrant approximation* to an object Y is a fibrant object $\widehat{Y}$ weakly equivalent to Y (see Definition 8.1.2). Cofibrant and fibrant approximations are among the most fundamental tools in homotopy theory because

- maps that are "expected" to exist often exist only when the domain is cofibrant and the codomain is fibrant and,
- since weak equivalences become isomorphisms in the homotopy category, a cofibrant or fibrant approximation to an object is isomorphic to that object in the homotopy category.

For example, (left or right) homotopy is an equivalence relation on the set of maps from X to Y when X is cofibrant and Y is fibrant (see Theorem 7.4.9), and we use this in Section 8.3 to construct the homotopy category $\operatorname{Ho}\mathcal{M}$ of a model category $\mathcal{M}$ by defining $\operatorname{Ho}\mathcal{M}(X,Y)$ to be the set of homotopy classes of maps in $\mathcal{M}$ from X' to Y', where X' and Y' are cofibrant-fibrant objects weakly equivalent to X and Y respectively (see the proof of Theorem 8.3.5).

In the category of topological spaces every object is fibrant, and a CW-approximation to a space X (i.e., a CW-complex weakly equivalent to X) is a cofibrant approximation to X. In the category of simplicial sets every object is cofibrant, and a Kan complex weakly equivalent to X (e.g., the total singular complex of the geometric realization of X) is a fibrant approximation to X. When doing homological algebra, a resolution of an object is a cofibrant or fibrant approximation in a model category of simplicial or cosimplicial objects (see, e.g., [**55**] or [**52**, Chapter II, Section 4]). When constructing function complexes in a model category (see Chapter 17), a resolution of an object is a cofibrant or fibrant approximation in yet a different model category of cosimplicial or simplicial objects (see Definition 16.1.2).

In Section 8.1 we define cofibrant and fibrant approximations and show that they are unique up to a weak equivalence (see Proposition 8.1.9 and Proposition 8.1.19); stronger uniqueness theorems will follow in Chapter 14 (see Proposition 14.6.3 and Theorem 14.6.9). We discuss approximations and homotopy relations in Section 8.2, and in Section 8.3 we construct the homotopy category of a model category.

In Section 8.4 we discuss (left and right) *derived functors*, which are functors induced on the homotopy category of a model category by a functor on the model category. In Section 8.5 we discuss *Quillen functors*, which are the useful functors between model categories. Quillen functors arise in adjoint pairs (see Definition 8.5.2); the left Quillen functor induces a *total left derived functor* (see Definition 8.4.7) between the homotopy categories and the right Quillen functor induces a *total right derived functor* in the opposite direction. We show that the total left derived functor of the left Quillen functor and the total right derived functor of the right Quillen functor are an adjoint pair of functors between the homotopy categories (see Theorem 8.5.18). We also define *Quillen equivalences*, which are Quillen functors satisfying an additional condition (see Definition 8.5.20) that implies that their total derived functors are equivalences of categories between the homotopy categories.

8.1. Fibrant and cofibrant approximations

8.1.1. Approximations to objects.

DEFINITION 8.1.2. Let $\mathcal{M}$ be a model category.

(1) (a) A *cofibrant approximation* to an object X is a pair $(\widetilde{X}, i)$ where $\widetilde{X}$ is a cofibrant object and $i\colon \widetilde{X} \to X$ is a weak equivalence.
   (b) A *fibrant cofibrant approximation* to X is a cofibrant approximation $(\widetilde{X}, i)$ such that the weak equivalence i is a trivial fibration.

   We will sometimes use the term *cofibrant approximation* to refer to the object $\widetilde{X}$ without explicitly mentioning the weak equivalence i.

(2) (a) A *fibrant approximation* to an object X is a pair $(\widehat{X}, j)$ where $\widehat{X}$ is a fibrant object and $j\colon X \to \widehat{X}$ is a weak equivalence.
   (b) A *cofibrant fibrant approximation* to X is a fibrant approximation $(\widehat{X}, j)$ such that the weak equivalence j is a trivial cofibration.

   We will sometimes use the term *fibrant approximation* to refer to the object $\widehat{X}$ without explicitly mentioning the weak equivalence j.

PROPOSITION 8.1.3. *If $\mathcal{M}$ is a model category, then every object X has both a fibrant cofibrant approximation $i\colon \widetilde{X} \to X$ and a cofibrant fibrant approximation $j\colon X \to \widehat{X}$.*

PROOF. Factor the map $\emptyset \to X$ (where $\emptyset$ is the initial object of $\mathcal{M}$) into a cofibration followed by a trivial fibration and factor the map $X \to *$ (where $*$ is the terminal object of $\mathcal{M}$) into a trivial cofibration followed by a fibration. □

DEFINITION 8.1.4. Let $\mathcal{M}$ be a model category and let X be an object of $\mathcal{M}$.

(1) If $(\widetilde{X}, i)$ and $(\widetilde{X}', i')$ are cofibrant approximations to X, a *map of cofibrant approximations* from $(\widetilde{X}, i)$ to $(\widetilde{X}', i')$ is a map $g\colon \widetilde{X} \to \widetilde{X}'$ such that $i'g = i$.
(2) If $(\widehat{X}, j)$ and $(\widehat{X}', j')$ are fibrant approximations to X, a *map of fibrant approximations* from $(\widehat{X}, j)$ to $(\widehat{X}', j')$ is a map $g\colon \widehat{X} \to \widehat{X}'$ such that $gj = j'$.

LEMMA 8.1.5. *Let $\mathcal{M}$ be a model category and let X be an object of $\mathcal{M}$.*

(1) *If $(\widetilde{X}, i)$ and $(\widetilde{X}', i')$ are cofibrant approximations to X and $g\colon \widetilde{X} \to \widetilde{X}'$ is a map of cofibrant approximations, then g is a weak equivalence.*
(2) *If $(\widehat{X}, j)$ and $(\widehat{X}', j')$ are fibrant approximations to X and $g\colon \widehat{X} \to \widehat{X}'$ is a map of fibrant approximations, then g is a weak equivalence.*

PROOF. This follows from the "two out of three" axiom for weak equivalences (see Definition 7.1.3). □

LEMMA 8.1.6. *Let $\mathcal{M}$ be a model category and let X be an object of $\mathcal{M}$.*

(1) *If* $(\widetilde{X}, i)$ *is a fibrant cofibrant approximation to* X *(see Definition 8.1.2) and* $g\colon W \to X$ *is a map from a cofibrant object* W*, then there is a map* $\phi\colon W \to \widetilde{X}$*, unique up to homotopy over* X *(see Definition 7.6.7), such that* $i\phi = g$*.*

(2) *If* $(\widehat{X}, j)$ *is a cofibrant fibrant approximation to* X *and* $g\colon X \to Y$ *is a map to a fibrant object* Y*, then there is a map* $\phi\colon \widehat{X} \to Y$*, unique up to homotopy under* X*, such that* $\phi j = g$*.*

PROOF. This follows from Proposition 7.6.13. □

PROPOSITION 8.1.7. *Let* $\mathcal{M}$ *be a model category and let* X *be an object of* $\mathcal{M}$*.*

(1) *If* $(\widetilde{X}, i)$ *is a cofibrant approximation to* X *and* $(\widetilde{X}', i')$ *is a fibrant cofibrant approximation to* X*, then there is a map of cofibrant approximations* $g\colon \widetilde{X} \to \widetilde{X}'$*, unique up to homotopy over* X *(see Definition 7.6.7), and any such map* g *is a weak equivalence.*

(2) *If* $(\widehat{X}, j)$ *is a cofibrant fibrant approximation to* X *and* $(\widehat{X}', j')$ *is a fibrant approximation to* X*, then there is a map of fibrant approximations* $g\colon \widehat{X} \to \widehat{X}'$*, unique up to homotopy under* X*, and any such map* g *is a weak equivalence.*

PROOF. This follows from Proposition 8.1.6 and Lemma 8.1.5. □

COROLLARY 8.1.8. *Let* $\mathcal{M}$ *be a model category and let* X *be an object of* $\mathcal{M}$*.*

(1) *If* $(\widetilde{X}, i)$ *and* $(\widetilde{X}', i')$ *are fibrant cofibrant approximations to* X*, then there is a map of cofibrant approximations* $g\colon \widetilde{X} \to \widetilde{X}'$*, unique up to homotopy over* X *(see Definition 7.6.7), and any such map* g *is a homotopy equivalence over* X*.*

(2) *If* $(\widehat{X}, j)$ *and* $(\widehat{X}', j')$ *are cofibrant fibrant approximations to* X*, then there is a map of fibrant approximations* $g\colon \widehat{X} \to \widehat{X}'$*, unique up to homotopy under* X*, and any such map* g *is a homotopy equivalence under* X*.*

PROOF. This follows from Proposition 8.1.7. □

PROPOSITION 8.1.9. *Let* $\mathcal{M}$ *be a model category and let* X *be an object of* $\mathcal{M}$*.*

(1) *If* $(\widetilde{X}, i)$ *and* $(\widetilde{X}', i')$ *are cofibrant approximations to* X*, then* $\widetilde{X}$ *and* $\widetilde{X}'$ *are weakly equivalent (see Definition 7.9.2) over* X*.*

(2) *If* $(\widehat{X}, j)$ *and* $(\widehat{X}', j')$ *are fibrant approximations to* X*, then* $\widehat{X}$ *and* $\widehat{X}'$ *are weakly equivalent under* X*.*

PROOF. This follows from Proposition 8.1.3 and Proposition 8.1.7. □

REMARK 8.1.10. We will show in Proposition 14.6.3 that there is an essentially unique zig-zag (see Definition 14.4.2) of weak equivalences between any two cofibrant approximations to the same object (or between any two fibrant approximations to the same object).

8.1.11. Augmented and coaugmented functors.

DEFINITION 8.1.12. Let $\mathcal{M}$ be a model category.

(1) An *augmented functor* on $\mathcal{M}$ is a pair (F, i) where F is a functor $\mathrm{F}\colon \mathcal{M} \to \mathcal{M}$ and i is a natural transformation $i\colon \mathrm{F} \to 1$.

(2) A *coaugmented functor* on $\mathcal{M}$ is a pair (G, j) where G is a functor $\mathrm{G}\colon \mathcal{M} \to \mathcal{M}$ and j is a natural transformation $j\colon 1 \to \mathrm{G}$.

DEFINITION 8.1.13. Let $\mathcal{M}$ be a model category.

(1) An augmented functor (F, i) on $\mathcal{M}$ will be called *homotopy idempotent* if for every object X in $\mathcal{M}$ the natural maps $i_{\mathrm{F}X}, \mathrm{F}(i_X)\colon \mathrm{FF}X \to \mathrm{F}X$ are homotopic over X (see Definition 7.6.7) and are homotopy equivalences over X.
(2) A coaugmented functor (G, j) on $\mathcal{M}$ will be called *homotopy idempotent* if for every object X in $\mathcal{M}$ the natural maps $j_{\mathrm{G}X}, \mathrm{G}(j_X)\colon \mathrm{G}X \to \mathrm{GG}X$ are homotopic under X (see Definition 7.6.7) and are homotopy equivalences under X.

REMARK 8.1.14. Definition 8.1.13 is the lifting to $\mathcal{M}$ of J. F. Adams' notion of an idempotent functor on the homotopy category of $\mathcal{M}$ (see [**2**]).

DEFINITION 8.1.15. Let $\mathcal{M}$ be a model category.

(1) (a) A *functorial cofibrant approximation* on $\mathcal{M}$ is an augmented functor (F, i) on $\mathcal{M}$ such that $i_X\colon \mathrm{F}X \to X$ is a cofibrant approximation to X for every object X of $\mathcal{M}$.
 (b) A *functorial fibrant cofibrant approximation* on $\mathcal{M}$ is a functorial cofibrant approximation such that i_X is a trivial fibration for every object X of $\mathcal{M}$.
 (c) If $\mathcal{K}$ is a subcategory of $\mathcal{M}$, then a *functorial cofibrant approximation* on $\mathcal{K}$ is a pair (F, i) in which $\mathrm{F}\colon \mathcal{K} \to \mathcal{M}$ is a functor and i is a natural transformation such that $i_X\colon \mathrm{F}X \to X$ is a cofibrant approximation to X for every object X of $\mathcal{K}$.
(2) (a) A *functorial fibrant approximation* on $\mathcal{M}$ is a coaugmented functor (G, j) on $\mathcal{M}$ such that $j_X\colon X \to \mathrm{G}X$ is a fibrant approximation to X for every object X of $\mathcal{M}$.
 (b) A *functorial cofibrant fibrant approximation* on $\mathcal{M}$ is a functorial fibrant approximation such that j_X is a trivial cofibration for every object X of $\mathcal{M}$.
 (c) If $\mathcal{K}$ is a subcategory of $\mathcal{M}$, then a *functorial fibrant approximation* on $\mathcal{K}$ is a pair (G, j) in which $\mathrm{G}\colon \mathcal{K} \to \mathcal{M}$ is a functor and j is a natural transformation such that $j_X\colon X \to \mathrm{G}X$ is a fibrant approximation to X for every object X of $\mathcal{K}$.

PROPOSITION 8.1.16. *Let $\mathcal{M}$ be a model category.*

(1) *A functorial fibrant cofibrant approximation (F, i) on $\mathcal{M}$ is homotopy idempotent.*
(2) *A functorial cofibrant fibrant approximation (G, j) on $\mathcal{M}$ is homotopy idempotent.*

PROOF. We will prove part 1; the proof of part 2 is dual.

Since i is a natural transformation, for every object X of $\mathcal{M}$ we have a commutative square

$$\begin{array}{ccc} \mathrm{FF}X & \xrightarrow{i_{\mathrm{F}X}} & \mathrm{F}X \\ {\scriptstyle \mathrm{F}(i_X)}\downarrow & & \downarrow{\scriptstyle i_X} \\ \mathrm{F}X & \xrightarrow[i_X]{} & X \end{array} .$$

Since i_X is a trivial fibration and $\mathrm{FF}X$ is cofibrant, Proposition 7.5.9 implies that $i_{\mathrm{F}X} \overset{l}{\simeq} \mathrm{F}(i_X)$ in $(\mathcal{M}\downarrow X)$. Since both of $\mathrm{F}X$ and $\mathrm{FF}X$ are both cofibrant and fibrant in $(\mathcal{M}\downarrow X)$, Theorem 7.4.9 implies that $i_{\mathrm{F}X} \simeq \mathrm{F}(i_X)$ in $(\mathcal{M}\downarrow X)$, and so Theorem 7.5.10 implies that $i_{\mathrm{F}X}$ and $\mathrm{F}(i_X)$ are homotopy equivalences in $(\mathcal{M}\downarrow X)$. □

PROPOSITION 8.1.17. *If $\mathcal{M}$ is a model category and $\mathcal{K}$ is a subcategory of $\mathcal{M}$, then there is a functorial fibrant cofibrant approximation on $\mathcal{K}$ (see Definition* 8.1.15*) and a functorial cofibrant fibrant approximation on $\mathcal{K}$.*

PROOF. This follows from applying part 1 of the factorization axiom (see Definition 7.1.3) to the map from the initial object and part 2 of the factorization axiom to the map to the terminal object. □

DEFINITION 8.1.18. Let $\mathcal{M}$ be a model category and let $\mathcal{K}$ be a subcategory of $\mathcal{M}$.

(1) If (F, i) and (F', i') are functorial cofibrant approximations on $\mathcal{K}$ (see Definition 8.1.15), a *map of functorial cofibrant approximations* from (F, i) to (F', i') is a natural transformation $\phi\colon \mathrm{F} \to \mathrm{F}'$ such that $i'\phi = i$.
(2) If (G, j) and (G', j') are functorial fibrant approximations on $\mathcal{K}$ (see Definition 8.1.15), a *map of functorial fibrant approximations* from (G, j) to (G', j') is a natural transformation $\phi\colon \mathrm{G} \to \mathrm{G}'$ such that $\phi j = j'$.

PROPOSITION 8.1.19. *Let $\mathcal{M}$ be a model category and let $\mathcal{K}$ be a subcategory of catM.*

(1) *If $i_1(X)\colon \widetilde{\mathrm{C}}_1(X) \to X$ and $i_2(X)\colon \widetilde{\mathrm{C}}_2(X) \to X$ are natural cofibrant approximations defined on $\mathcal{K}$, then $\widetilde{\mathrm{C}}_1(-)$ and $\widetilde{\mathrm{C}}_2(-)$ are naturally weakly equivalent (see Definition* 7.9.2*).*
(2) *If $j_1(X)\colon X \to \widehat{\mathrm{F}}_1(X)$ and $j_2(X)\colon X \to \widehat{\mathrm{F}}_2(X)$ are natural fibrant approximations defined on $\mathcal{K}$, then $\widehat{\mathrm{F}}_1(-)$ and $\widehat{\mathrm{F}}_2(-)$ are naturally weakly equivalent.*

PROOF. We will prove part 1; the proof of part 2 is dual.

If we choose a natural fibrant cofibrant approximation $i(X)\colon \widetilde{\mathrm{C}}(X) \to X$ for every object X in $\mathcal{K}$ (see Proposition 8.1.17), then it is sufficient to show that each of $\widetilde{\mathrm{C}}_1(-)$ and $\widetilde{\mathrm{C}}_2(-)$ is naturally weakly equivalent to $\widetilde{\mathrm{C}}(-)$. We will do this for $\widetilde{\mathrm{C}}_1(-)$; the proof for $\widetilde{\mathrm{C}}_2(-)$ is the same.

For every object X in $\mathcal{K}$, we construct the pullback square

$$\begin{array}{ccc} P_1(X) & \xrightarrow{j(X)} & \widetilde{C}_1(X) \\ {\scriptstyle j_1(X)}\downarrow & & \downarrow{\scriptstyle i_1(X)} \\ \widetilde{C}(X) & \xrightarrow[i(X)]{} & X \end{array}$$

and then we choose a functorial cofibrant approximation $k(X)\colon \widetilde{P}_1(X) \to P_1(X)$ to $P_1(X)$. Since $i(X)$ is a trivial fibration, so is $j(X)$, and so the "two out of three" axiom (see Definition 7.1.3) implies that $j_1(X)$ is also a weak equivalence. Thus, $\widetilde{C}_1(X) \xleftarrow{j(X)k(X)} \widetilde{P}_1(X) \xrightarrow{j_2(X)k(X)} \widetilde{C}(X)$ is a natural zig-zag of weak equivalences of cofibrant approximations to X. □

REMARK 8.1.20. We will show in Theorem 14.6.9 that any two functorial cofibrant approximations are connected by an essentially unique zig-zag (see Definition 14.4.2) of weak equivalences.

8.1.21. Approximations to maps.

DEFINITION 8.1.22. Let $\mathcal{M}$ be a model category.

(1) (a) A *cofibrant approximation* to a map $g\colon X \to Y$ consists of a cofibrant approximation $(\widetilde{X}, i_X)$ tȯ X (see Definition 8.1.2), a cofibrant approximation $(\widetilde{Y}, i_Y)$ to Y, and a map $\tilde{g}\colon \widetilde{X} \to \widetilde{Y}$ such that $i_Y\tilde{g} = gi_X$.
(b) A *fibrant cofibrant approximation* to a map $g\colon X \to Y$ is a cofibrant approximation to g in which the cofibrant approximations $(\widetilde{X}, i_X)$ and $(\widetilde{Y}, i_Y)$ are fibrant cofibrant approximations.

We will sometimes use the term *cofibrant approximation* to refer to the map $\tilde{g}$ without explicitly mentioning the cofibrant approximations $(\widetilde{X}, i_X)$ and $(\widetilde{Y}, i_Y)$.

(2) (a) A *fibrant approximation* to a map $g\colon X \to Y$ consists of a fibrant approximation $(\widehat{X}, j_X)$ to X (see Definition 8.1.2), a fibrant approximation $(\widehat{Y}, j_Y)$ to Y, and a map $\hat{g}\colon \widehat{X} \to \widehat{Y}$ such that $\hat{g}j_X = j_Y g$.
(b) A *cofibrant fibrant approximation* to a map $g\colon X \to Y$ is a fibrant approximation to g in which the fibrant approximations $(\widehat{X}, j_X)$ and $(\widehat{Y}, j_Y)$ are cofibrant fibrant approximations.

We will sometimes use the term *fibrant approximation* to refer to the map $\hat{g}$ without explicitly mentioning the fibrant approximations $(\widehat{X}, j_X)$ and $(\widehat{Y}, j_Y)$.

PROPOSITION 8.1.23. *Let $\mathcal{M}$ be a model category.*

(1) *Every map $g\colon X \to Y$ has a natural fibrant cofibrant approximation $\tilde{g}\colon \widetilde{X} \to \widetilde{Y}$ such that $\tilde{g}$ is a cofibration.*
(2) *Every map $g\colon X \to Y$ has a natural cofibrant fibrant approximation $\hat{g}\colon \widehat{X} \to \widehat{Y}$ such that $\hat{g}$ is a fibration.*

PROOF. We will prove part 1; the proof of part 2 is similar.

Choose a natural fibrant cofibrant approximation $(\widetilde{X}, i_X)$ to X, and then choose a natural factorization of the composition $gi_X\colon \widetilde{X} \to Y$ as $\widetilde{X} \xrightarrow{\tilde{g}} \widetilde{Y} \xrightarrow{i_Y} Y$ where $\tilde{g}$ is a cofibration and i_Y is a trivial fibration. $\square$

PROPOSITION 8.1.24. *Let $\mathcal{M}$ and $\mathcal{N}$ be model categories, let $g\colon X \to Y$ be a map in $\mathcal{M}$, and let $\mathrm{F}\colon \mathcal{M} \to \mathcal{N}$ be a functor.*

1. *If F takes trivial cofibrations between cofibrant objects in $\mathcal{M}$ into weak equivalences in $\mathcal{N}$ and there is a cofibrant approximation $\tilde{g}\colon \widetilde{X} \to \widetilde{Y}$ to g (see Definition 8.1.22) such that $\mathrm{F}(\tilde{g})$ is a weak equivalence, then F takes every cofibrant approximation to g into a weak equivalence.*
2. *If F takes trivial fibrations between fibrant objects in $\mathcal{M}$ into weak equivalences in $\mathcal{N}$ and there is a fibrant approximation $\hat{g}\colon \widehat{X} \to \widehat{Y}$ to g (see Definition 8.1.22) such that $\mathrm{F}(\hat{g})$ is a weak equivalence, then F takes every fibrant approximation to g into a weak equivalence.*

PROOF. We will prove part 1; the proof of part 2 is dual.

Proposition 8.1.23 implies that we can choose a cofibrant approximation $\tilde{g}'\colon \widetilde{X}' \to \widetilde{Y}'$ to g such that the weak equivalences $i'_X\colon \widetilde{X}' \to X$ and $i'_Y\colon \widetilde{Y}' \to Y$ are trivial fibrations. It is sufficient to show that if $\tilde{g}\colon \widetilde{X} \to \widetilde{Y}$ is some other cofibrant approximation to g, then $\mathrm{F}(\tilde{g})$ is a weak equivalence if and only if $\mathrm{F}(\tilde{g}')$ is a weak equivalence.

If $\tilde{g}\colon \widetilde{X} \to \widetilde{Y}$ is some other cofibrant approximation to g, then we have the solid arrow diagram

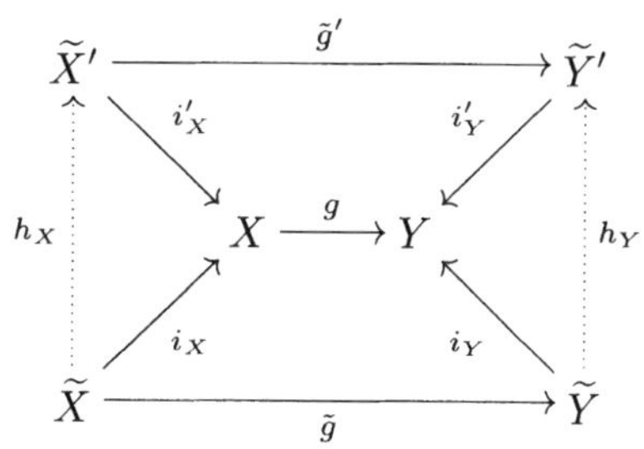

in which i'_X and i'_Y are trivial fibrations and i_X and i_Y are weak equivalences. Proposition 8.1.7 implies that there are weak equivalences $h_X\colon \widetilde{X} \to \widetilde{X}'$ and $h_Y\colon \widetilde{Y} \to \widetilde{Y}'$ such that $i'_X h_X = i_X$ and $i'_Y h_Y = i_Y$. Thus, $i'_Y \tilde{g}' h_X = g i'_X h_X = g i_X = i_Y \tilde{g} = i'_Y h_Y \tilde{g}$. Since i'_Y is a trivial fibration and $\widetilde{X}$ is cofibrant, Proposition 7.5.9 implies that $\tilde{g}' h_X$ is left homotopic to $h_Y \tilde{g}$, and so Lemma 7.7.7 implies that $\mathrm{F}(\tilde{g}' h_X)$ is a weak equivalence if and only if $\mathrm{F}(h_Y \tilde{g})$ is a weak equivalence. Since Corollary 7.7.2 implies that $\mathrm{F}(h_X)$ and $\mathrm{F}(h_Y)$ are weak equivalences, the "two out of three" axiom for weak equivalences (see Definition 7.1.3) implies that $\mathrm{F}(\tilde{g}')$ is a weak equivalence if and only if $\mathrm{F}(\tilde{g})$ is a weak equivalence. $\square$

PROPOSITION 8.1.25. *Let $\mathcal{M}$ be a model category.*

1. *If $g\colon X \to Y$ is a map in $\mathcal{M}$, $\widetilde{X} \to X$ is a cofibrant approximation to X, and $\widetilde{Y} \to Y$ is a fibrant cofibrant approximation to Y, then there exists a cofibrant approximation $\tilde{g}\colon \widetilde{X} \to \widetilde{Y}$ to g, and $\tilde{g}$ is unique up to homotopy over Y.*
2. *If $g\colon X \to Y$ is a map in $\mathcal{M}$, $X \to \widehat{X}$ is a cofibrant fibrant approximation to X, and $Y \to \widehat{Y}$ is a fibrant approximation to Y, then there exists a*

fibrant approximation $\hat{g}\colon \widehat{X} \to \widehat{Y}$ *to* g, *and* $\hat{g}$ *is unique up to homotopy under* X.

PROOF. This follows from Proposition 7.6.13. □

DEFINITION 8.1.26. Let $\mathcal{M}$ be a model category and let $g\colon X \to Y$ be a map in $\mathcal{M}$.

(1) If $\big((\widetilde{X}, i_X), (\widetilde{Y}, i_Y), \tilde{g}\colon \widetilde{X} \to \widetilde{Y}\big)$ and $\big((\widetilde{X}', i'_X), (\widetilde{Y}', i'_Y), \tilde{g}'\colon \widetilde{X}' \to \widetilde{Y}'\big)$ are cofibrant approximations to g, then a *map of cofibrant approximations* from $\big((\widetilde{X}, i_X), (\widetilde{Y}, i_Y), \tilde{g}\colon \widetilde{X} \to \widetilde{Y}\big)$ to $\big((\widetilde{X}', i'_X), (\widetilde{Y}', i'_Y), \tilde{g}'\colon \widetilde{X}' \to \widetilde{Y}'\big)$ consists of maps $h_X\colon \widetilde{X} \to \widetilde{X}'$ and $h_Y\colon \widetilde{Y} \to \widetilde{Y}'$ such that the diagram

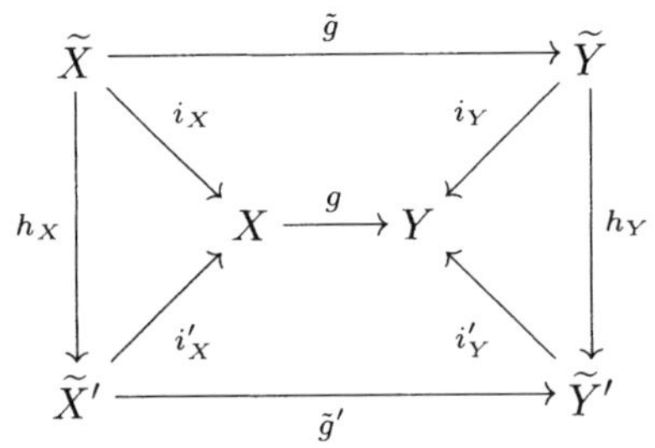

commutes.

(2) If $\big((\widehat{X}, j_X), (\widehat{Y}, j_Y), \hat{g}\colon \widehat{X} \to \widehat{Y}\big)$ and $\big((\widehat{X}', j'_X), (\widehat{Y}', j'_Y), \hat{g}'\colon \widehat{X}' \to \widehat{Y}'\big)$ are fibrant approximations to g, then a *map of fibrant approximations* from $\big((\widehat{X}, j_X), (\widehat{Y}, j_Y), \hat{g}\colon \widehat{X} \to \widehat{Y}\big)$ to $\big((\widehat{X}', j'_X), (\widehat{Y}', j'_Y), \hat{g}'\colon \widehat{X}' \to \widehat{Y}'\big)$ consists of maps $h_X\colon \widehat{X} \to \widehat{X}'$ and $h_Y\colon \widehat{Y} \to \widehat{Y}'$ such that the diagram

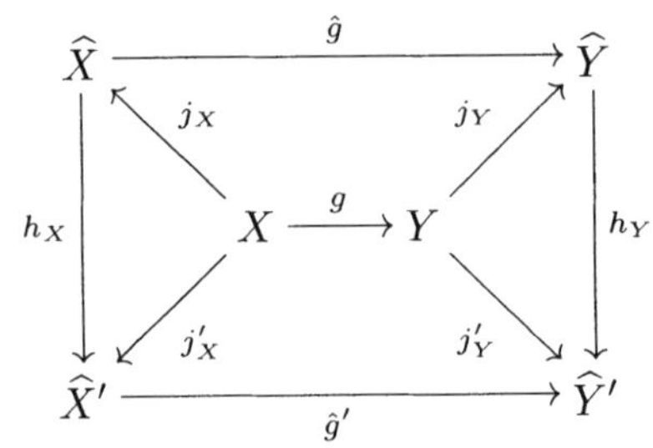

commutes.

REMARK 8.1.27. We will show in Proposition 14.6.6 that any two cofibrant approximations (or fibrant approximations) to a map are connected by an essentially unique zig-zag of weak equivalences.

8.2. Approximations and homotopic maps

LEMMA 8.2.1. *Let* $\mathcal{M}$ *be a model category, let* $X \amalg X \to \mathrm{Cyl}(X) \to X$ *be a cylinder object for* X, *and let* $X \to \mathrm{Path}(X) \to X \times X$ *be a path object for* X.

(1) *If* $i\colon \widetilde{X} \to X$ *is a fibrant cofibrant approximation to* X, *then*

(a) *there is a cylinder object* $\widetilde{X} \amalg \widetilde{X} \to \mathrm{Cyl}(\widetilde{X}) \to \widetilde{X}$ *for* $\widetilde{X}$ *and a diagram*

$$\begin{array}{ccccc}
\widetilde{X} \amalg \widetilde{X} & \longrightarrow & \mathrm{Cyl}(\widetilde{X}) & \longrightarrow & \widetilde{X} \\
\Big\downarrow{\scriptstyle i \amalg i} & & \Big\downarrow{\scriptstyle \mathrm{Cyl}(i)} & & \Big\downarrow{\scriptstyle i} \\
X \amalg X & \longrightarrow & \mathrm{Cyl}(X) & \longrightarrow & X
\end{array}$$

such that $\mathrm{Cyl}(i)\colon \mathrm{Cyl}(\widetilde{X}) \to \mathrm{Cyl}(X)$ *is a fibrant cofibrant approximation to* $\mathrm{Cyl}(X)$, *and*

(b) *there is a path object* $\widetilde{X} \to \mathrm{Path}(\widetilde{X}) \to \widetilde{X} \times \widetilde{X}$ *for* $\widetilde{X}$ *and a diagram*

$$\begin{array}{ccccc}
\widetilde{X} & \longrightarrow & \mathrm{Path}(\widetilde{X}) & \longrightarrow & \widetilde{X} \times \widetilde{X} \\
\Big\downarrow{\scriptstyle i} & & \Big\downarrow{\scriptstyle \mathrm{Path}(i)} & & \Big\downarrow{\scriptstyle i \times i} \\
X & \longrightarrow & \mathrm{Path}(X) & \longrightarrow & X \times X
\end{array} \tag{8.2.2}$$

such that $\mathrm{Path}(i)\colon \mathrm{Path}(\widetilde{X}) \to \mathrm{Path}(X)$ *is a fibrant cofibrant approximation to* $\mathrm{Path}(X)$ *and the right hand square of Diagram* 8.2.2 *is a pullback.*

(2) *If* $j\colon X \to \widehat{X}$ *is a cofibrant fibrant approximation to* X, *then*

(a) *there is a cylinder object* $\widehat{X} \amalg \widehat{X} \to \mathrm{Cyl}(\widehat{X}) \to \widehat{X}$ *for* $\widehat{X}$ *and a diagram*

$$\begin{array}{ccccc}
X \amalg X & \longrightarrow & \mathrm{Cyl}(X) & \longrightarrow & X \\
\Big\downarrow{\scriptstyle j \amalg j} & & \Big\downarrow{\scriptstyle \mathrm{Cyl}(j)} & & \Big\downarrow{\scriptstyle j} \\
\widehat{X} \amalg \widehat{X} & \longrightarrow & \mathrm{Cyl}(\widehat{X}) & \longrightarrow & \widehat{X}
\end{array} \tag{8.2.3}$$

such that $\mathrm{Cyl}(j)\colon \mathrm{Cyl}(X) \to \mathrm{Cyl}(\widehat{X})$ *is a cofibrant fibrant approximation to* $\mathrm{Cyl}(X)$ *and the left hand square of Diagram* 8.2.3 *is a pushout, and*

(b) *there is a path object* $\widehat{X} \to \mathrm{Path}(\widehat{X}) \to \widehat{X} \times \widehat{X}$ *for* $\widehat{X}$ *and a diagram*

$$\begin{array}{ccccc}
X & \longrightarrow & \mathrm{Path}(X) & \longrightarrow & X \times X \\
\Big\downarrow{\scriptstyle j} & & \Big\downarrow{\scriptstyle \mathrm{Path}(j)} & & \Big\downarrow{\scriptstyle j \times j} \\
\widehat{X} & \longrightarrow & \mathrm{Path}(\widehat{X}) & \longrightarrow & \widehat{X} \times \widehat{X}
\end{array}$$

such that $\mathrm{Path}(j)\colon \mathrm{Path}(X) \to \mathrm{Path}(\widehat{X})$ *is a cofibrant fibrant approximation to* $\mathrm{Path}(X)$.

PROOF. We will prove part 1; the proof of part 2 is dual.

Factor the composition $\widetilde{X} \amalg \widetilde{X} \to X \amalg X \to \mathrm{Cyl}(X)$ as $\widetilde{X} \amalg \widetilde{X} \xrightarrow{k} \mathrm{Cyl}(\widetilde{X}) \xrightarrow{\mathrm{Cyl}(i)} \mathrm{Cyl}(X)$ where k is a cofibration and $\mathrm{Cyl}(i)$ is a trivial fibration. Since i is a trivial fibration, the dotted arrow q exists in the solid arrow diagram

$$\begin{array}{ccccc}
\widetilde{X} \amalg \widetilde{X} & & \xrightarrow{\quad 1_{\widetilde{X}} \amalg 1_{\widetilde{X}} \quad} & & \widetilde{X} \\
\Big\downarrow{\scriptstyle k} & & {\scriptstyle q} \nearrow & & \Big\downarrow{\scriptstyle i} \\
\mathrm{Cyl}(\widetilde{X}) & \longrightarrow & \mathrm{Cyl}(X) & \longrightarrow & X
\end{array}$$

and the "two out of three" axiom for weak equivalences (see Definition 7.1.3) implies that q is a weak equivalence.

If we let $\mathrm{Path}(\widetilde{X})$ be the pullback $\mathrm{Path}(X) \times_{(X\times X)} (\widetilde{X} \times \widetilde{X})$, then we have the solid arrow diagram

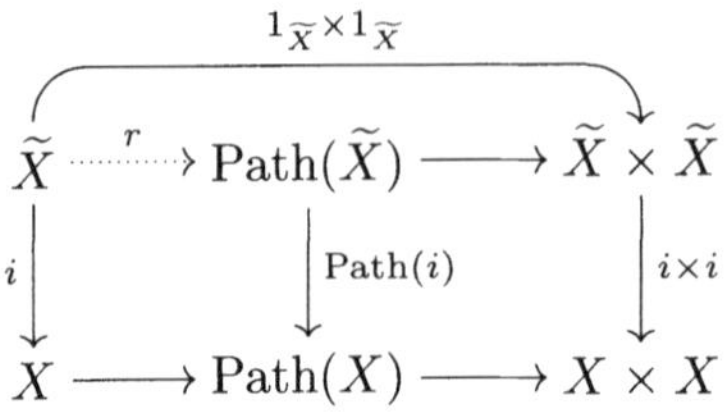

and the universal mapping property of the pullback implies that the dotted arrow r exists. Since i is a trivial fibration, so is $i \times i$, and so $\mathrm{Path}(i)$ (which is a pullback of $i \times i$) is a trivial fibration. The "two out of three" axiom for weak equivalences (see Definition 7.1.3) now implies that r is a weak equivalence. □

PROPOSITION 8.2.4. *Let $\mathcal{M}$ be a model category, and let $f, g\colon X \to Y$ be maps.*

1. *If $\tilde{f}, \tilde{g}\colon \widetilde{X} \to \widetilde{Y}$ are fibrant cofibrant approximations to, respectively, f and g, and if f and g are left homotopic, right homotopic, or homotopic, then $\tilde{f}$ and $\tilde{g}$ are, respectively, left homotopic, right homotopic, or homotopic.*
2. *If $\hat{f}, \hat{g}\colon \widehat{X} \to \widehat{Y}$ are cofibrant fibrant approximations to, respectively, f and g, and if f and g are left homotopic, right homotopic, or homotopic, then $\hat{f}$ and $\hat{g}$ are, respectively, left homotopic, right homotopic, or homotopic.*

PROOF. We will prove part 1; the proof of part 2 is dual.

If f and g are left homotopic, let $X \amalg X \to \mathrm{Cyl}(X) \to X$ be a cylinder object for X such that there is a left homotopy $H\colon \mathrm{Cyl}(X) \to Y$ from f to g. If $\widetilde{X} \amalg \widetilde{X} \to \mathrm{Cyl}(\widetilde{X}) \to \widetilde{X}$ is the cylinder object of Lemma 8.2.1, then we have the solid arrow diagram

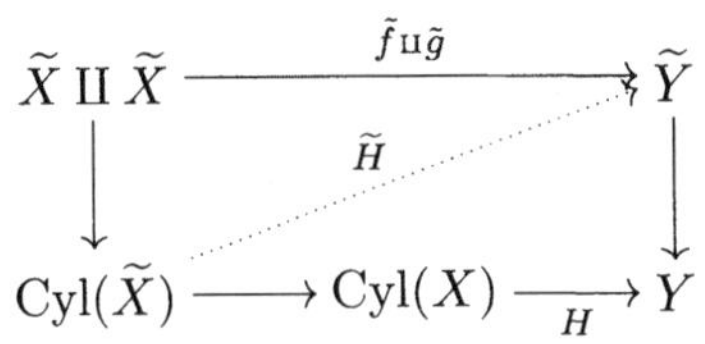

Since $\widetilde{Y} \to Y$ is a trivial fibration, the dotted arrow $\widetilde{H}$ exists, and is a left homotopy from $\tilde{f}$ to $\tilde{g}$.

If f and g are right homotopic, let $Y \to \mathrm{Path}(Y) \to Y \times Y$ be a path object for Y such that there is a right homotopy $K\colon X \to \mathrm{Path}(Y)$ from f to g. If $\widetilde{Y} \to \mathrm{Path}(\widetilde{Y}) \to \widetilde{Y} \times \widetilde{Y}$ is the path object of Lemma 8.2.1, then we have the solid arrow diagram

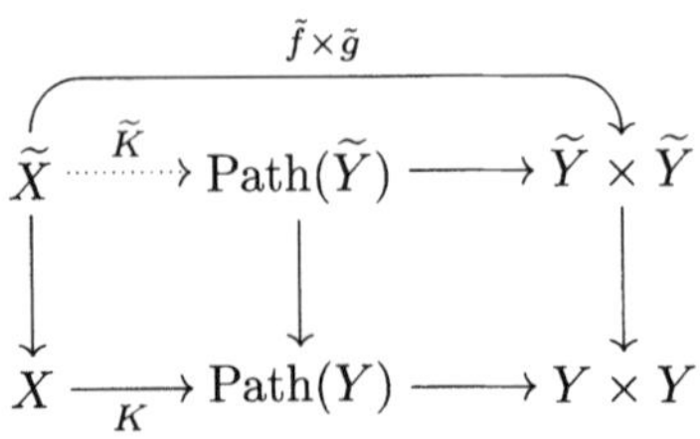

Since the right hand square is a pullback, the dotted arrow $\widetilde{K}$ exists and is a right homotopy from $\tilde{f}$ to $\tilde{g}$. □

8.3. The homotopy category of a model category

DEFINITION 8.3.1. If $\mathcal{M}$ is a category and $\mathcal{W}$ is a class of maps in $\mathcal{M}$, then a *localization* of $\mathcal{M}$ with respect to $\mathcal{W}$ is a category $\mathrm{L}_{\mathcal{W}}\mathcal{M}$ and a functor $\gamma\colon \mathcal{M} \to \mathrm{L}_{\mathcal{W}}\mathcal{M}$ such that

(1) if $w \in \mathcal{W}$, then $\gamma(w)$ is an isomorphism, and
(2) if $\mathcal{N}$ is a category and $\varphi\colon \mathcal{M} \to \mathcal{N}$ is a functor such that $\varphi(w)$ is an isomorphism for every w in $\mathcal{W}$, then there is a unique functor $\delta\colon \mathrm{L}_{\mathcal{W}}\mathcal{M} \to \mathcal{N}$ such that $\delta\gamma = \varphi$.

The usual argument shows that if a localization of $\mathcal{M}$ with respect to $\mathcal{W}$ exists, then it is unique up to a unique isomorphism. Thus, we will speak of *the* localization of $\mathcal{M}$ with respect to $\mathcal{W}$. We will often refer to the category $\mathrm{L}_{\mathcal{W}}\mathcal{M}$ as the localization of $\mathcal{M}$ with respect to $\mathcal{W}$, without explicitly mentioning the functor γ.

DEFINITION 8.3.2. If $\mathcal{M}$ is a model category, then the *Quillen homotopy category* of $\mathcal{M}$ (which we will also call the *homotopy category* of $\mathcal{M}$) is the localization of $\mathcal{M}$ with respect to the class of weak equivalences, which we denote by $\gamma\colon \mathcal{M} \to \mathrm{Ho}\,\mathcal{M}$.

REMARK 8.3.3. If $\mathcal{M}$ is a *small* category, then the localization of $\mathcal{M}$ with respect to any class $\mathcal{W}$ of maps in $\mathcal{M}$ exists. This is because we can construct the maps of the localization using generators and relations to add inverses for the elements of $\mathcal{W}$, and we can be sure that there will only be a set of maps between two objects of $\mathcal{M}$ because there is only a set of maps in all of $\mathcal{M}$ to begin with. If $\mathcal{M}$ is not small, though, then using generators and relations might lead to a proper class of maps between some pair of objects, in which case we would not have a category.

Restating this in terms of universes (see, e.g., [**60**, page 17]): If we start in a fixed universe $\mathcal{U}$, then we can attempt to construct the localization of a $\mathcal{U}$-category $\mathcal{M}$ with respect to a class of maps $\mathcal{W}$ using generators and relations. If $\mathcal{M}$ is not small, though, then we could only be sure of constructing a category in some higher universe $\mathcal{U}'$. The statement that "the localization of $\mathcal{M}$ with respect to $\mathcal{W}$ exists" is the statement that there is a category in our original universe $\mathcal{U}$ that is the localization of $\mathcal{M}$ with respect to $\mathcal{W}$.

We will show that the Quillen homotopy category of a model category $\mathcal{M}$ exists (see Theorem 8.3.5) and that it is equivalent to the classical homotopy category of $\mathcal{M}$ (see Definition 7.5.8 and Theorem 8.3.9). To do this, we will not use the method of generators and relations. Instead, we will construct the set of maps between two objects in the localization by starting with the set of maps between two objects in the original category, and dividing that set by an equivalence relation. Thus, we will be sure of having only a set of maps between any pair of objects.

LEMMA 8.3.4. *Let $\mathcal{M}$ be a model category, let $\mathcal{N}$ be a category, and let $\varphi\colon \mathcal{M} \to \mathcal{N}$ be a functor that takes weak equivalences in $\mathcal{M}$ to isomorphisms in $\mathcal{N}$. If $f, g\colon X \to Y$ are maps in $\mathcal{M}$ such that either $f \overset{l}{\simeq} g$ or $f \overset{r}{\simeq} g$ (see Definition 7.3.2), then $\varphi(f) = \varphi(g)$.*

PROOF. We will consider the case $f \overset{l}{\simeq} g$; the case $f \overset{r}{\simeq} g$ is similar.

If $f \overset{l}{\simeq} g$, then there is a cylinder object (see Definition 7.3.2) $X \amalg X \xrightarrow{i_0 \amalg i_1} \mathrm{Cyl}(X) \xrightarrow{p} X$ for X and a map $H\colon \mathrm{Cyl}(X) \to Y$ such that $Hi_0 = f$ and $Hi_1 = g$. Since p is a weak equivalence, $\varphi(p)$ is an isomorphism. Since $pi_0 = pi_1$, this implies that $\varphi(i_0) = \varphi(i_1)$. Thus, $\varphi(f) = \varphi(H)\varphi(i_0) = \varphi(H)\varphi(i_1) = \varphi(g)$. □

Lemma 8.3.4 implies that a functor $\varphi\colon \mathcal{M} \to \mathcal{N}$ that takes weak equivalences to isomorphisms must identify homotopic maps. Thus, when searching for the Quillen homotopy category of $\mathcal{M}$ (see Definition 8.3.2), a natural object to consider is the classical homotopy category of $\mathcal{M}$ (see Definition 7.5.8). Theorem 7.5.10 implies that if we restrict ourselves to the full subcategory of $\mathcal{M}$ spanned by the cofibrant-fibrant objects, then identifying homotopic maps turns weak equivalences into isomorphisms, and so the classical homotopy category has the required universal property restricted to this subcategory.

To deal with objects that are not cofibrant-fibrant, we note that if $\widetilde{X}$ is weakly equivalent to X and $\widetilde{Y}$ is weakly equivalent to Y, then in any category in which weak equivalences have become isomorphisms the set of maps from X to Y will be isomorphic to the set of maps from $\widetilde{X}$ to $\widetilde{Y}$. This suggests that we should choose $\widetilde{X}$ and $\widetilde{Y}$ to be cofibrant-fibrant objects weakly equivalent to X and Y respectively and define $\mathrm{Ho}\,\mathcal{M}(X,Y)$ to be the set of homotopy classes of maps from $\widetilde{X}$ to $\widetilde{Y}$ in $\mathcal{M}$. This is what we shall do to define $\mathrm{Ho}\,\mathcal{M}$.

THEOREM 8.3.5. *If $\mathcal{M}$ is a model category, then the Quillen homotopy category of $\mathcal{M}$ (see Definition 8.3.2) exists.*

PROOF. Choose a functorial fibrant cofibrant approximation $i_X : \widetilde{\mathrm{C}}X \to X$ and a functorial cofibrant fibrant approximation $j_X : X \to \widehat{\mathrm{F}}X$ for every object X in $\mathcal{M}$ (see Proposition 8.1.17). We define the category $\mathrm{Ho}\,\mathcal{M}$ as follows:

(1) The objects of $\mathrm{Ho}\,\mathcal{M}$ are the objects of $\mathcal{M}$.
(2) If X and Y are objects in $\mathcal{M}$, then $\mathrm{Ho}\,\mathcal{M}(X,Y) = \pi(\widehat{\mathrm{F}}\widetilde{\mathrm{C}}X, \widehat{\mathrm{F}}\widetilde{\mathrm{C}}Y)$ (see Notation 7.5.2).
(3) If X, Y, and Z are objects in $\mathcal{M}$, then the composition

$$\mathrm{Ho}\,\mathcal{M}(Y,Z) \times \mathrm{Ho}\,\mathcal{M}(X,Y) \to \mathrm{Ho}\,\mathcal{M}(X,Z)$$

is the composition of homotopy classes of maps between cofibrant-fibrant objects in $\mathcal{M}$

$$\pi(\widehat{\mathrm{F}}\widetilde{\mathrm{C}}Y, \widehat{\mathrm{F}}\widetilde{\mathrm{C}}Z) \times \pi(\widehat{\mathrm{F}}\widetilde{\mathrm{C}}X, \widehat{\mathrm{F}}\widetilde{\mathrm{C}}Y) \to \pi(\widehat{\mathrm{F}}\widetilde{\mathrm{C}}X, \widehat{\mathrm{F}}\widetilde{\mathrm{C}}Z)$$

(see Theorem 7.5.5).

We define the functor $\gamma\colon \mathcal{M} \to \mathrm{Ho}\,\mathcal{M}$ to be the identity on the class of objects and to take the map $f\colon X \to Y$ to the homotopy class of the map $\widehat{\mathrm{F}}\widetilde{\mathrm{C}}(f)\colon \widehat{\mathrm{F}}\widetilde{\mathrm{C}}X \to \widehat{\mathrm{F}}\widetilde{\mathrm{C}}Y$.

If $f\colon X \to Y$ is a weak equivalence in $\mathcal{M}$, then the "two out of three" property of weak equivalences (see Definition 7.1.3) implies that $\widehat{\mathrm{F}}\widetilde{\mathrm{C}}(f)$ is a weak equivalence, and so Theorem 7.5.10 implies that $\widehat{\mathrm{F}}\widetilde{\mathrm{C}}(f)$ is a homotopy equivalence, i.e., $\gamma(f)$ is an isomorphism.

It remains only to show that if $\varphi\colon \mathcal{M} \to \mathcal{N}$ is a functor that takes weak equivalences in $\mathcal{M}$ to isomorphisms in $\mathcal{N}$, then there is a unique functor $\delta\colon \mathrm{Ho}\,\mathcal{M} \to \mathcal{N}$ such that $\delta\gamma = \varphi$. Let $\varphi\colon \mathcal{M} \to \mathcal{N}$ be such a functor. For every object X of $\mathrm{Ho}\,\mathcal{M}$, we let $\delta(X) = \varphi(X)$. If $g\colon X \to Y$ is a map in $\mathrm{Ho}\,\mathcal{M}$, then g is a homotopy class of maps $\widehat{\mathrm{F}}\widetilde{\mathrm{C}}X \to \widehat{\mathrm{F}}\widetilde{\mathrm{C}}Y$ in $\mathcal{M}$. Lemma 8.3.4 implies that φ takes all elements of that homotopy class to the same map of $\mathcal{N}$, and so we can let

$$\delta(g) = \varphi(i_Y)\big(\varphi(j_{\widetilde{\mathrm{C}}Y})\big)^{-1}\varphi(g)\varphi(j_{\widetilde{\mathrm{C}}X})\big(\varphi(i_X)\big)^{-1}$$

(where by $\varphi(g)$ we mean φ applied to some map in the homotopy class g). To see that δ is a functor, we note that an identity map in $\operatorname{Ho}\mathcal{M}$ is a homotopy class of maps in $\mathcal{M}$ containing an identity map, and composition of maps between cofibrant-fibrant objects of $\mathcal{M}$ is well defined on homotopy classes (see Theorem 7.5.5). Thus, δ is a functor.

To see that $\delta\gamma = \varphi$, we note that γ is the identity on objects, and δ was defined to agree with φ on objects. If $f\colon X \to Y$ is a map in $\mathcal{M}$, then we have the commutative diagram

$$\begin{array}{ccc}
\widehat{\mathrm{F}}\widetilde{\mathrm{C}}X & \xrightarrow{\widehat{\mathrm{F}}\widetilde{\mathrm{C}}(f)} & \widehat{\mathrm{F}}\widetilde{\mathrm{C}}Y \\
{\scriptstyle j_{\widetilde{\mathrm{C}}X}}\big\uparrow & & \big\uparrow{\scriptstyle j_{\widetilde{\mathrm{C}}Y}} \\
\widetilde{\mathrm{C}}X & \xrightarrow{\widetilde{\mathrm{C}}(f)} & \widetilde{\mathrm{C}}Y \\
{\scriptstyle i_X}\big\downarrow & & \big\downarrow{\scriptstyle i_Y} \\
X & \xrightarrow[f]{} & Y.
\end{array}$$

Since φ takes weak equivalences to isomorphisms in $\mathcal{N}$, we have

$$\varphi(f) = \varphi(i_Y)\big(\varphi(j_{\widetilde{\mathrm{C}}Y})\big)^{-1}\varphi\big(\widehat{\mathrm{F}}\widetilde{\mathrm{C}}(f)\big)\varphi(j_{\widetilde{\mathrm{C}}X})\big(\varphi(i_X)\big)^{-1}.$$

Since $\gamma(f)$ is the homotopy class of $\widehat{\mathrm{F}}\widetilde{\mathrm{C}}(f)$, this implies that $\delta\gamma(f) = \varphi(f)$.

Finally, to see that δ is the unique functor satisfying $\delta\gamma = \varphi$, we note that every map of $\operatorname{Ho}\mathcal{M}$ is a composition of maps in the image of γ and inverses of the image under γ of weak equivalences of $\mathcal{M}$. □

THEOREM 8.3.6. *If $\mathcal{M}$ is a model category, then there is a construction of the Quillen homotopy category of $\mathcal{M}$ (see Definition 8.3.2) $\gamma\colon \mathcal{M} \to \operatorname{Ho}\mathcal{M}$ such that if X and Y are cofibrant-fibrant objects in $\mathcal{M}$, then $\operatorname{Ho}\mathcal{M}\big(\gamma(X), \gamma(Y)\big)$ is the set of homotopy classes of maps in $\mathcal{M}$ from X to Y.*

PROOF. For every cofibrant object X, let $\widetilde{\mathrm{C}}X = X$ and let $i_X\colon \widetilde{\mathrm{C}}X \to X$ be the identity map. For every non-cofibrant object X, factor the map from the initial object to X into a cofibration followed by a trivial fibration to obtain a cofibrant object $\widetilde{\mathrm{C}}X$ and a trivial fibration $i_X\colon \widetilde{\mathrm{C}}X \to X$. (In the terminology of Definition 8.1.2, we have chosen a fibrant cofibrant approximation to X.)

For every fibrant object X, let $\widehat{\mathrm{F}}X = X$ and let $j_X\colon X \to \widehat{\mathrm{F}}X$ be the identity map. For every non-fibrant object X, factor the map from X to the terminal object into a trivial cofibration followed by a fibration to obtain a fibrant object $\widehat{\mathrm{F}}X$ and a trivial cofibration $j_X\colon X \to \widehat{\mathrm{F}}X$. (In the terminology of Definition 8.1.2, we have chosen a cofibrant fibrant approximation to X.)

We define the category $\operatorname{Ho}\mathcal{M}$ as follows:

(1) The objects of $\operatorname{Ho}\mathcal{M}$ are the objects of $\mathcal{M}$.
(2) If X and Y are objects in $\mathcal{M}$, then $\operatorname{Ho}\mathcal{M}(X,Y) = \pi(\widehat{\mathrm{F}}\widetilde{\mathrm{C}}X, \widehat{\mathrm{F}}\widetilde{\mathrm{C}}Y)$ (see Notation 7.5.2).

(3) If X, Y, and Z are objects in $\mathcal{M}$, then the composition

$$\operatorname{Ho}\mathcal{M}(Y,Z) \times \operatorname{Ho}\mathcal{M}(X,Y) \to \operatorname{Ho}\mathcal{M}(X,Z)$$

is the composition of homotopy classes of maps between cofibrant-fibrant objects in $\mathcal{M}$

$$\pi(\widehat{\mathrm{F}}\widetilde{\mathrm{C}}Y, \widehat{\mathrm{F}}\widetilde{\mathrm{C}}Z) \times \pi(\widehat{\mathrm{F}}\widetilde{\mathrm{C}}X, \widehat{\mathrm{F}}\widetilde{\mathrm{C}}Y) \to \pi(\widehat{\mathrm{F}}\widetilde{\mathrm{C}}X, \widehat{\mathrm{F}}\widetilde{\mathrm{C}}Z)$$

(see Theorem 7.5.5).

We now define the functor $\gamma\colon \mathcal{M} \to \operatorname{Ho}\mathcal{M}$. We let γ be the identity on the class of objects. For every map $f\colon X \to Y$ in $\mathcal{M}$, we have the solid arrow diagram

$$\begin{array}{ccccc}
\emptyset & & \longrightarrow & & \widetilde{\mathrm{C}}Y \\
\downarrow & & \overset{\widetilde{\mathrm{C}}(f)}{\nearrow} & & \downarrow{\scriptstyle i_Y} \\
\widetilde{\mathrm{C}}X & \xrightarrow[i_X]{} & X & \xrightarrow[f]{} & Y
\end{array}$$

(where $\emptyset$ denotes the initial object of $\mathcal{M}$), and we can choose a dotted arrow $\widetilde{\mathrm{C}}(f)$ that makes the diagram commute. (In the terminology of Definition 8.1.22, $\widetilde{\mathrm{C}}(f)$ is a cofibrant approximation to f.) Proposition 7.5.9 implies that $\widetilde{\mathrm{C}}(f)$ is well defined up to left homotopy, and so Proposition 7.4.8 implies that it is well defined up to right homotopy. We now have the solid arrow diagram

$$\begin{array}{ccccc}
\widetilde{\mathrm{C}}X & \xrightarrow{\widetilde{\mathrm{C}}(f)} & \widetilde{\mathrm{C}}Y & \xrightarrow{j_{\widetilde{\mathrm{C}}Y}} & \widehat{\mathrm{F}}\widetilde{\mathrm{C}}Y \\
{\scriptstyle j_{\widetilde{\mathrm{C}}X}}\downarrow & & \overset{\widehat{\mathrm{F}}\widetilde{\mathrm{C}}(f)}{\nearrow} & & \downarrow \\
\widehat{\mathrm{F}}\widetilde{\mathrm{C}}X & & \longrightarrow & & *
\end{array}$$

(where $*$ denotes the terminal object of $\mathcal{M}$), and we can choose a dotted arrow $\widehat{\mathrm{F}}\widetilde{\mathrm{C}}(f)$ that makes the diagram commute. Proposition 7.5.9 implies that $\widehat{\mathrm{F}}\widetilde{\mathrm{C}}(f)$ is well defined up to homotopy, and we define $\gamma(f)$ to the the element of $\pi(\widehat{\mathrm{F}}\widetilde{\mathrm{C}}X, \widehat{\mathrm{F}}\widetilde{\mathrm{C}}Y)$ represented by $\widehat{\mathrm{F}}\widetilde{\mathrm{C}}(f)$ (see Theorem 7.4.9).

To see that γ is a functor, we note that for every object X in $\mathcal{M}$ Proposition 7.5.9 implies that $\widetilde{\mathrm{C}}(1_X) \overset{l}{\simeq} 1_{\widetilde{\mathrm{C}}X}$, and so $\widetilde{\mathrm{C}}(1_X) \overset{r}{\simeq} 1_{\widetilde{\mathrm{C}}X}$, and so $\widehat{\mathrm{F}}\widetilde{\mathrm{C}}(1_X) \simeq 1_{\widehat{\mathrm{F}}\widetilde{\mathrm{C}}X}$. Similarly, if $f\colon X \to Y$ and $g\colon Y \to Z$ are maps in $\mathcal{M}$, then Proposition 7.5.9 implies that $\widetilde{\mathrm{C}}(g)\widetilde{\mathrm{C}}(f) \overset{l}{\simeq} \widetilde{\mathrm{C}}(gf)$, and so $\widehat{\mathrm{F}}\widetilde{\mathrm{C}}(g)\widehat{\mathrm{F}}\widetilde{\mathrm{C}}(g) \simeq \widehat{\mathrm{F}}\widetilde{\mathrm{C}}(gf)$. Thus, we have defined the category $\operatorname{Ho}\mathcal{M}$ and the functor $\gamma\colon \mathcal{M} \to \operatorname{Ho}\mathcal{M}$. The proof that γ has the required universal property is identical to the proof in the case of Theorem 8.3.5. □

PROPOSITION 8.3.7. *If $\mathcal{M}$ is a model category, then there is a unique isomorphism from the category* $\operatorname{Ho}\mathcal{M}$ *constructed in Theorem* 8.3.5 *to the category* $\operatorname{Ho}\mathcal{M}$ *constructed in Theorem* 8.3.6 *that commutes with the functors from* $\mathcal{M}$.

PROOF. This follows from the universal property of the functors from $\mathcal{M}$. □

THEOREM 8.3.8. *If $\mathcal{M}$ is a model category, then the classical homotopy category of $\mathcal{M}$ (see Definition* 7.5.8*) is naturally isomorphic to the full subcategory of the Quillen homotopy category of $\mathcal{M}$ spanned by the cofibrant-fibrant objects.*

PROOF. This follows from Theorem 8.3.6. □

THEOREM 8.3.9. *If $\mathcal{M}$ is a model category, then the classical homotopy category of $\mathcal{M}$ is equivalent to the Quillen homotopy category of $\mathcal{M}$.*

PROOF. Let ν denote the embedding $\pi\mathcal{M}_{\mathrm{cf}} \to \mathrm{Ho}\,\mathcal{M}$ described in Theorem 8.3.6. To define $\eta\colon \mathrm{Ho}\,\mathcal{M} \to \pi\mathcal{M}_{\mathrm{cf}}$, let $\widetilde{\mathrm{C}}$ and $\widehat{\mathrm{F}}$ be as in the proof of Theorem 8.3.6. If X is an object of $\mathrm{Ho}\,\mathcal{M}$, let $\eta(X) = \widehat{\mathrm{F}}\widetilde{\mathrm{C}}X$. If X and Y are objects of $\mathrm{Ho}\,\mathcal{M}$, then $\mathrm{Ho}\,\mathcal{M}(X,Y) = \pi(\widehat{\mathrm{F}}\widetilde{\mathrm{C}}X, \widehat{\mathrm{F}}\widetilde{\mathrm{C}}Y)$, and we let η be the "identity map" from $\mathrm{Ho}\,\mathcal{M}(X,Y)$ to $\pi\mathcal{M}_{\mathrm{cf}}(\widehat{\mathrm{F}}\widetilde{\mathrm{C}}X, \widehat{\mathrm{F}}\widetilde{\mathrm{C}}Y)$.

Since $\eta\nu$ is the identity functor of $\pi\mathcal{M}_{\mathrm{cf}}$, it remains only to define a natural equivalence θ from the identity functor of $\mathrm{Ho}\,\mathcal{M}$ to $\nu\eta$. If X is an object of $\mathrm{Ho}\,\mathcal{M}$, then $\nu\eta(X) = \widehat{\mathrm{F}}\widetilde{\mathrm{C}}X$, and so $\mathrm{Ho}\,\mathcal{M}\bigl(X, \nu\eta(X)\bigr) = \mathrm{Ho}\,\mathcal{M}(X, \widehat{\mathrm{F}}\widetilde{\mathrm{C}}X) = \pi(\widehat{\mathrm{F}}\widetilde{\mathrm{C}}X, \widehat{\mathrm{F}}\widetilde{\mathrm{C}}\widehat{\mathrm{F}}\widetilde{\mathrm{C}}X) = \pi(\widehat{\mathrm{F}}\widetilde{\mathrm{C}}X, \widehat{\mathrm{F}}\widetilde{\mathrm{C}}X)$; we let $\theta(X)\colon X \to \nu\eta X$ be the homotopy class of the identity map of $\widehat{\mathrm{F}}\widetilde{\mathrm{C}}X$ in $\mathcal{M}$. □

THEOREM 8.3.10. *Let $\mathcal{M}$ be a model category and let $\gamma\colon \mathcal{M} \to \mathrm{Ho}\,\mathcal{M}$ be the canonical functor to it's homotopy category. If $g\colon X \to Y$ is a map in $\mathcal{M}$, then g is a weak equivalence if and only if $\gamma(g)$ is an isomorphism in* $\mathrm{Ho}\,\mathcal{M}$.

PROOF. If g is a weak equivalence, then the definition of $\mathrm{Ho}\,\mathcal{M}$ implies that $\gamma(g)$ is an isomorphism. Conversely, if $\gamma(g)$ is an isomorphism, then $\widehat{\mathrm{F}}\widetilde{\mathrm{C}}(g)$ (see the proof of Theorem 8.3.5) is a homotopy equivalence, and so Theorem 7.8.5 and the "two out of three" property of weak equivalences implies that g is a weak equivalence. □

8.4. Derived functors

DEFINITION 8.4.1. Let $\mathcal{M}$ be a model category, let $\mathcal{C}$ be a category, and let $\varphi\colon \mathcal{M} \to \mathcal{C}$ be a functor.

(1) A *left derived functor* of φ is a functor $\mathrm{L}\varphi\colon \mathrm{Ho}\,\mathcal{M} \to \mathcal{C}$ together with a natural transformation $\varepsilon\colon \mathrm{L}\varphi \circ \gamma \to \varphi$ such that the pair $(\mathrm{L}\varphi, \varepsilon)$ is "closest to φ from the left", i.e., such that if $\mathrm{G}\colon \mathrm{Ho}\,\mathcal{M} \to \mathcal{C}$ is a functor and $\zeta\colon \mathrm{G} \circ \gamma \to \varphi$ is a natural transformation, then there is a unique natural transformation $\theta\colon \mathrm{G} \to \mathrm{L}\varphi$ such that $\zeta = \varepsilon(\theta \circ \gamma)$.
(2) A *right derived functor* of φ is a functor $\mathrm{R}\varphi\colon \mathrm{Ho}\,\mathcal{M} \to \mathcal{C}$ together with a natural transformation $\varepsilon\colon \varphi \to \mathrm{R}\varphi \circ \gamma$ such that the pair $(\mathrm{R}\varphi, \varepsilon)$ is "closest to φ from the right", i.e., such that if $\mathrm{G}\colon \mathrm{Ho}\,\mathcal{M} \to \mathcal{C}$ is a functor and $\zeta\colon \varphi \to \mathrm{G} \circ \gamma$ is a natural transformation, then there is a unique natural transformation $\theta\colon \mathrm{R}\varphi \to \mathrm{G}$ such that $\zeta = (\theta \circ \gamma)\varepsilon$.

REMARK 8.4.2. The usual argument shows that if a left derived functor of φ exists, then it is unique up to a unique natural equivalence. Thus, we will speak of *the* left derived functor of φ. A similar remark applies to right derived functors.

REMARK 8.4.3. The left derived functor of $\varphi\colon \mathcal{M} \to \mathcal{C}$ is also known as the *right Kan extension of φ along* $\gamma\colon \mathcal{M} \to \mathrm{Ho}\,\mathcal{M}$ (see [**47**, page 232–236]). (Note the reversal of left and right.) Similarly, the right derived functor of $\varphi\colon \mathcal{M} \to \mathcal{C}$ is also known as the *left Kan extension of φ along* $\gamma\colon \mathcal{M} \to \mathrm{Ho}\,\mathcal{M}$.

PROPOSITION 8.4.4. *Let $\mathcal{M}$ be a model category, let $\mathcal{C}$ be a category, and let $\varphi\colon \mathcal{M} \to \mathcal{C}$ be a functor.*

(1) *If φ takes trivial cofibrations between cofibrant objects to isomorphisms in $\mathcal{C}$, then the left derived functor of φ exists.*
(2) *If φ takes trivial fibrations between fibrant objects to isomorphisms in $\mathcal{C}$, then the right derived functor of φ exists.*

PROOF. We will prove part 1; the proof of part 2 is dual.

Let $\widetilde{\mathrm{C}}$ be as in the proof of Theorem 8.3.5. We define a functor $\mathrm{D}: \mathcal{M} \to \mathcal{C}$ as follows: If X is an object of $\mathcal{M}$ we let $\mathrm{D}(X) = \varphi(\widetilde{\mathrm{C}}X)$, and if $f: X \to Y$ is a map in $\mathcal{M}$ we let $\mathrm{D}(f) = \varphi\big(\widetilde{\mathrm{C}}(f)\big)$. To see that D is a functor, we note that $\widetilde{\mathrm{C}}(1_X) = 1_{\widetilde{\mathrm{C}}X}$ and so $\mathrm{D}(1_X) = 1_{\mathrm{D}X}$, and if $f: X \to Y$ and $g: Y \to Z$ are maps in $\mathcal{M}$, then $\widetilde{\mathrm{C}}(g)\widetilde{\mathrm{C}}(f) = \widetilde{\mathrm{C}}(gf)$, and so $\mathrm{D}(g)\mathrm{D}(f) = \mathrm{D}(gf)$.

If $f: X \to Y$ is a weak equivalence in $\mathcal{M}$, then $\widetilde{\mathrm{C}}(f)$ is a weak equivalence between cofibrant objects, and so Corollary 7.7.3 implies that $\mathrm{D}(f)$ is an isomorphism. Thus, the universal property of $\mathrm{Ho}\,\mathcal{M}$ (see Definition 8.3.2 and Definition 8.3.1) implies that there is a unique functor $\mathrm{L}\varphi: \mathrm{Ho}\,\mathcal{M} \to \mathcal{C}$ such that $\mathrm{L}\varphi \circ \gamma = \mathrm{D}$. We define a natural transformation $\varepsilon: \mathrm{L}\varphi \circ \gamma \to \varphi$ by letting $\varepsilon(X) = \varphi(i_X): \mathrm{L}\varphi \circ \gamma(X) = \mathrm{D}(X) = \varphi(\widetilde{\mathrm{C}}X) \to \varphi(X)$. We will show that the pair $(\mathrm{L}\varphi, \varepsilon)$ is the left derived functor of φ.

If $\mathrm{G}: \mathrm{Ho}\,\mathcal{M} \to \mathcal{C}$ is a functor and $\zeta: \mathrm{G} \circ \gamma \to \varphi$ is a natural transformation, then we have the solid arrow diagram

$$
\begin{array}{ccc}
\mathrm{G}\circ\gamma(\widetilde{\mathrm{C}}X) & \xrightarrow{\zeta(\widetilde{\mathrm{C}}X)} & \varphi(\widetilde{\mathrm{C}}X) = (\mathrm{L}\varphi\circ\gamma)(X) \\
{\scriptstyle (\mathrm{G}\circ\gamma)(i_X)}\downarrow & \overset{\theta(X)}{\nearrow} & \downarrow{\scriptstyle \varphi(i_X)=\varepsilon(X)} \\
\mathrm{G}\circ\gamma(X) & \xrightarrow[\zeta(X)]{} & \varphi(X)
\end{array}
\tag{8.4.5}
$$

and so we define a natural transformation $\theta: \mathrm{G} \to \mathrm{L}\varphi$ by letting $\theta(X) = \big(\zeta(\widetilde{\mathrm{C}}X)\big) \circ \big((\mathrm{G}\circ\gamma)(i_X)\big)^{-1}$. If X is cofibrant, then $\varphi(i_X)$ is an isomorphism, and so $\theta(X)$ is the only possible map that makes Diagram 8.4.5 commute. Since $\widetilde{\mathrm{C}}X \approx X$ for every object X in $\mathrm{Ho}\,\mathcal{M}$, this implies the uniqueness of θ in general. $\square$

8.4.6. Total derived functors.

DEFINITION 8.4.7. Let $\mathcal{M}$ and $\mathcal{N}$ be model categories and let $\varphi: \mathcal{M} \to \mathcal{N}$ be a functor.

(1) A *total left derived functor of* φ is a left derived functor (see Definition 8.4.1) of the composition $\mathcal{M} \xrightarrow{\varphi} \mathcal{N} \xrightarrow{\nu_{\mathcal{N}}} \mathrm{Ho}\,\mathcal{N}$. Thus, a total left derived functor of φ is a functor $\mathbf{L}\varphi: \mathrm{Ho}\,\mathcal{M} \to \mathrm{Ho}\,\mathcal{N}$ together with a natural transformation $\varepsilon: \mathbf{L}\varphi \circ \nu_{\mathcal{M}} \to \nu_{\mathcal{N}} \circ \varphi$ such that the pair $(\mathbf{L}\varphi, \varepsilon)$ is "closest to $\nu_{\mathcal{N}} \circ \varphi$ from the left" (see Definition 8.4.1). We will often refer to $\mathbf{L}\varphi: \mathrm{Ho}\,\mathcal{M} \to \mathrm{Ho}\,\mathcal{N}$ as the *total left derived functor of* φ, without explicitly mentioning the natural transformation ε.
(2) A *total right derived functor of* φ is a right derived functor of the composition $\mathcal{M} \xrightarrow{\varphi} \mathcal{N} \xrightarrow{\nu_{\mathcal{N}}} \mathrm{Ho}\,\mathcal{N}$. Thus, a total right derived functor of φ is a functor $\mathbf{R}\varphi: \mathrm{Ho}\,\mathcal{M} \to \mathrm{Ho}\,\mathcal{N}$ together with a natural transformation $\varepsilon: \nu_{\mathcal{N}} \circ \varphi \to \mathbf{R}\varphi \circ \nu_{\mathcal{M}}$ such that the pair $(\mathbf{R}\varphi, \varepsilon)$ is "closest to $\nu_{\mathcal{N}} \circ \varphi$ from the right" (see Definition 8.4.1). We will often refer to $\mathbf{R}\varphi: \mathrm{Ho}\,\mathcal{M} \to \mathrm{Ho}\,\mathcal{N}$

as the *total right derived functor of* φ, without explicitly mentioning the natural transformation ε.

PROPOSITION 8.4.8. *Let* $\mathcal{M}$ *and* $\mathcal{N}$ *be model categories and let* $\varphi\colon \mathcal{M} \to \mathcal{N}$ *be a functor.*

(1) *If* φ *takes trivial cofibrations between cofibrant objects in* $\mathcal{M}$ *into weak equivalences in* $\mathcal{N}$*, then the total left derived functor* $\mathbf{L}\varphi\colon \operatorname{Ho}\mathcal{M} \to \operatorname{Ho}\mathcal{N}$ *exists.*
(2) *If* φ *takes trivial fibrations between fibrant objects in* $\mathcal{M}$ *into weak equivalences in* $\mathcal{N}$*, then the total right derived functor* $\mathbf{R}\varphi\colon \operatorname{Ho}\mathcal{M} \to \operatorname{Ho}\mathcal{N}$ *exists.*

PROOF. This follows from Proposition 8.4.4 and Theorem 8.3.10. □

8.5. Quillen functors and total derived functors

8.5.1. Quillen functors.

DEFINITION 8.5.2. Let $\mathcal{M}$ and $\mathcal{N}$ be model categories and let $\mathrm{F}\colon \mathcal{M} \rightleftarrows \mathcal{N} :\mathrm{U}$ be a pair of adjoint functors. We will say that

(1) F *is a left Quillen functor,*
(2) U *is a right Quillen functor,* and
(3) (F, U) *is a Quillen pair,*

if

(1) the left adjoint F preserves both cofibrations and trivial cofibrations, and
(2) the right adjoint U preserves both fibrations and trivial fibrations.

PROPOSITION 8.5.3. *If* $\mathcal{M}$ *and* $\mathcal{N}$ *are model categories and* $\mathrm{F}\colon \mathcal{M} \rightleftarrows \mathcal{N} :\mathrm{U}$ *is a pair of adjoint functors, then the following are equivalent:*

(1) *The pair* (F, U) *is a Quillen pair.*
(2) *The left adjoint* F *preserves both cofibrations and trivial cofibrations.*
(3) *The right adjoint* U *preserves both fibrations and trivial fibrations.*
(4) *The left adjoint* F *preserves cofibrations and the right adjoint* U *preserves fibrations.*
(5) *The left adjoint* F *preserves trivial cofibrations and the right adjoint* U *preserves trivial fibrations.*

PROOF. This follows from Proposition 7.2.18. □

The following strengthening of Proposition 8.5.3, due to D. Dugger [**26**], is useful when dealing with localizations of model category structures (see, e.g., Proposition 3.3.18).

PROPOSITION 8.5.4 (D. Dugger). *If* $\mathcal{M}$ *and* $\mathcal{N}$ *are model categories and* $\mathrm{F} : \mathcal{M} \rightleftarrows \mathcal{N} :\mathrm{U}$ *is a pair of adjoint functors, then the following are equivalent:*

(1) *The pair* (F, U) *is a Quillen pair.*
(2) *The left adjoint* F *preserves cofibrations between cofibrant objects and all trivial cofibrations.*
(3) *The right adjoint* U *preserves fibrations between fibrant objects and all trivial fibrations.*

PROOF. It follows directly from the definition that condition 1 implies both condition 2 and condition 3. We will show that condition 2 implies condition 1; the proof that condition 3 implies condition 1 is dual.

Assume that F preserves cofibrations between cofibrant objects and all trivial cofibrations; Proposition 8.5.3 implies that it is sufficient to show that U preserves all trivial fibrations. Let $p\colon X \to Y$ be a trivial fibration in $\mathcal{N}$; Proposition 7.2.18 implies that $\mathrm{U}p\colon \mathrm{U}X \to \mathrm{U}Y$ is a fibration, and so we need only show that it is a weak equivalence.

Proposition 8.1.23 implies that we can choose a fibrant cofibrant approximation $p'\colon X' \to Y'$ to $\mathrm{U}p$ such that p' is a cofibration, and so we have the diagram

$$\begin{array}{ccc} X' & \xrightarrow{j} & \mathrm{U}X \\ {\scriptstyle p'}\downarrow & & \downarrow{\scriptstyle \mathrm{U}p} \\ Y' & \xrightarrow[k]{} & \mathrm{U}Y \end{array} \tag{8.5.5}$$

in which X' and Y' are cofibrant, j and k are trivial fibrations, and p' is a cofibration. We will complete the proof by showing that the image of p' in $\mathrm{Ho}\,\mathcal{M}$ is an isomorphism. This will imply that p' is a weak equivalence (see Theorem 8.3.10), and the "two out of three" axiom (see Definition 7.1.3) will then imply that $\mathrm{U}p$ is a weak equivalence.

The adjoint of Diagram 8.5.5 is the solid arrow diagram

$$\begin{array}{ccc} \mathrm{F}X' & \xrightarrow{j^\sharp} & X \\ {\scriptstyle \mathrm{F}p'}\downarrow & \overset{g}{\nearrow} & \downarrow{\scriptstyle p} \\ \mathrm{F}Y' & \xrightarrow[k^\sharp]{} & Y \end{array}$$

in which $\mathrm{F}p'$ is a cofibration and p is a trivial fibration. Thus, there is a map $g\colon \mathrm{F}Y' \to X$ such that $g(\mathrm{F}p') = j^\sharp$ and $pg = k^\sharp$. If $g^\flat\colon Y' \to \mathrm{U}X$ is the adjoint of g, then we have the solid arrow diagram

$$\begin{array}{ccc} X' & = & X' \\ {\scriptstyle p'}\downarrow & \overset{s}{\nearrow} & \downarrow{\scriptstyle j} \\ Y' & \xrightarrow[g^\flat]{} & \mathrm{U}X \end{array}$$

in which p' is a cofibration and j is a trivial fibration, and so there is a map $s\colon Y' \to X'$ such that $sp' = 1_{X'}$ and $js = g^\flat$. We also have

$$kp's = (\mathrm{U}p)js = (\mathrm{U}p)g^\flat = k = k1_{Y'} \ .$$

Since k is a trivial fibration and Y' is cofibrant, Proposition 7.5.9 implies that $p's \overset{l}{\simeq} 1_{Y'}$, and so Lemma 8.3.4 implies that the image of $p's$ in $\mathrm{Ho}\,\mathcal{M}$ is the identity map of Y'. Since we also have $sp' = 1_{X'}$, the image of p' in $\mathrm{Ho}\,\mathcal{M}$ is thus an isomorphism, and so p' is a weak equivalence. □

8.5.6. Total derived functors of Quillen functors.

PROPOSITION 8.5.7. *If $\mathcal{M}$ and $\mathcal{N}$ are model categories and $\mathrm{F}\colon \mathcal{M} \rightleftarrows \mathcal{N} :\mathrm{U}$ is a Quillen pair, then*

(1) F *takes weak equivalences between cofibrant objects of* $\mathcal{M}$ *into weak equivalences in* $\mathcal{N}$ *and*
(2) U *takes weak equivalences between fibrant object of* $\mathcal{N}$ *into weak equivalences in* $\mathcal{M}$.

PROOF. This follows from Corollary 7.7.2. □

THEOREM 8.5.8. *If* $\mathcal{M}$ *and* $\mathcal{N}$ *are model categories and* $\mathrm{F}\colon \mathcal{M} \rightleftarrows \mathcal{N} : \mathrm{U}$ *is a Quillen pair, then*

(1) *the total left derived functor (see Definition* 8.4.7*)* $\mathbf{L}\mathrm{F}\colon \operatorname{Ho}\mathcal{M} \to \operatorname{Ho}\mathcal{N}$ *of* F *exists and*
(2) *the total right derived functor (see Definition* 8.4.7*)* $\mathbf{R}\mathrm{F}\colon \operatorname{Ho}\mathcal{M} \to \operatorname{Ho}\mathcal{N}$ *of* U *exists.*

PROOF. This follows from Proposition 8.4.8 and Proposition 8.5.7. □

LEMMA 8.5.9. *Let* $\mathcal{M}$ *and* $\mathcal{N}$ *be model categories.*

(1) *If* $\mathrm{F}\colon \mathcal{M} \to \mathcal{N}$ *is a left Quillen functor and* $g\colon X \to Y$ *is a map in* $\mathcal{M}$, *then the total left derived functor* $\mathbf{L}\mathrm{F}\colon \operatorname{Ho}\mathcal{M} \to \operatorname{Ho}\mathcal{N}$ *of* F *(see Definition* 8.4.7*) takes the image in* $\operatorname{Ho}\mathcal{M}$ *of* g *to the image in* $\operatorname{Ho}\mathcal{N}$ *of* $\mathrm{F}(\tilde{g})$ *for some cofibrant approximation* $\tilde{g}\colon \widetilde{X} \to \widetilde{Y}$ *to* g.
(2) *If* $\mathrm{F}\colon \mathcal{M} \to \mathcal{N}$ *is a right Quillen functor and* $g\colon X \to Y$ *is a map in* $\mathcal{M}$, *then the total right derived functor* $\mathbf{R}\mathrm{F}\colon \operatorname{Ho}\mathcal{M} \to \operatorname{Ho}\mathcal{N}$ *of* F *(see Definition* 8.4.7*) takes the image in* $\operatorname{Ho}\mathcal{M}$ *of* g *to the image in* $\operatorname{Ho}\mathcal{N}$ *of* $\mathrm{F}(\hat{g})$ *for some fibrant approximation* $\hat{g}\colon \widehat{X} \to \widehat{Y}$ *to* g.

PROOF. This follows from the proof of Proposition 8.4.4. □

PROPOSITION 8.5.10. *Let* $\mathcal{M}$ *and* $\mathcal{N}$ *be model categories.*

(1) *If* $\mathrm{F}\colon \mathcal{M} \to \mathcal{N}$ *is a left Quillen functor and* $g\colon X \to Y$ *is a map in* $\mathcal{M}$, *then the following are equivalent:*
    (a) *The total left derived functor* $\mathbf{L}\mathrm{F}\colon \operatorname{Ho}\mathcal{M} \to \operatorname{Ho}\mathcal{N}$ *of* F *(see Theorem* 8.5.18*) of* F *takes the image in* $\operatorname{Ho}\mathcal{M}$ *of* g *to an isomorphism in* $\operatorname{Ho}\mathcal{N}$.
    (b) *The functor* F *takes some cofibrant approximation to* g *to a weak equivalence in* $\mathcal{N}$.
    (c) *The functor* F *takes every cofibrant approximation to* g *to a weak equivalence in* $\mathcal{N}$.
(2) *If* $\mathrm{F}\colon \mathcal{M} \to \mathcal{N}$ *is a right Quillen functor and* $g\colon X \to Y$ *is a map in* $\mathcal{M}$, *then the following are equivalent:*
    (a) *The total right derived functor* $\mathbf{R}\mathrm{F}\colon \operatorname{Ho}\mathcal{M} \to \operatorname{Ho}\mathcal{N}$ *of* F *(see Theorem* 8.5.18*) of* F *takes the image in* $\operatorname{Ho}\mathcal{M}$ *of* g *to an isomorphism in* $\operatorname{Ho}\mathcal{N}$.
    (b) *The functor* F *takes some fibrant approximation to* g *to a weak equivalence in* $\mathcal{N}$.
    (c) *The functor* F *takes every fibrant approximation to* g *to a weak equivalence in* $\mathcal{N}$.

PROOF. This follows from Lemma 8.5.9, Theorem 8.3.10, and Proposition 8.1.24. □

DEFINITION 8.5.11. Let $\mathcal{M}$ and $\mathcal{N}$ be model categories, let $\mathrm{F}\colon \mathcal{M} \to \mathcal{N}$ be a functor, and let $g\colon X \to Y$ be a map in $\mathcal{M}$.

(1) If $\widetilde{\mathrm{C}}$ is the fibrant cofibrant approximation on $\mathcal{M}$ used to construct $\mathrm{Ho}\,\mathcal{M}$ (see the proof of Theorem 8.3.5 and the proof of Theorem 8.3.6), then we will abuse language and let $\mathbf{L}\mathrm{F}(g)$ denote $\mathrm{F}\big(\widetilde{\mathrm{C}}(g)\big)$, and we will call it the *left derived functor of* F *on* g. Note that $\mathbf{L}\mathrm{F}(g)$ is actually a map in $\mathcal{N}$ whose image in $\mathrm{Ho}\,\mathcal{N}$ is isomorphic to the image under the total left derived functor $\mathbf{L}\mathrm{F}\colon \mathrm{Ho}\,\mathcal{M} \to \mathrm{Ho}\,\mathcal{N}$ of F of the image in $\mathrm{Ho}\,\mathcal{M}$ of g, and that $\mathbf{L}\mathrm{F}(g)$ depends on the choice of cofibrant approximation $\widetilde{\mathrm{C}}$.
(2) If $\widehat{\mathrm{F}}$ is the cofibrant fibrant approximation on $\mathcal{M}$ used to construct $\mathrm{Ho}\,\mathcal{M}$, then we will abuse language and let $\mathbf{R}\mathrm{F}(g)$ denote $\mathrm{F}\big(\widehat{\mathrm{F}}(g)\big)$, and we will call it the *right derived functor of* F *on* g. Note that $\mathbf{R}\mathrm{F}(g)$ is actually a map in $\mathcal{N}$ whose image in $\mathrm{Ho}\,\mathcal{N}$ is isomorphic to the image under the total right derived functor $\mathbf{R}\mathrm{F}\colon \mathrm{Ho}\,\mathcal{M} \to \mathrm{Ho}\,\mathcal{N}$ of F of the image in $\mathrm{Ho}\,\mathcal{M}$ of g, and that $\mathbf{R}\mathrm{F}(g)$ depends on the choice of fibrant approximation $\widehat{\mathrm{F}}$.

8.5.12. Quillen functors and homotopy classes of maps.

LEMMA 8.5.13. *Let $\mathcal{M}$ be a model category and let $i_X\colon \widetilde{\mathrm{C}}X \to X$ and $j_X\colon X \to \widehat{\mathrm{F}}X$ be the constructions used in the proof of Theorem 8.3.5.*

(1) *If W is cofibrant and X is fibrant, then i_X induces an isomorphism of the sets of homotopy classes of maps $(i_X)_*\colon \pi(W, \widetilde{\mathrm{C}}X) \to \pi(W, X)$ that is natural in both W and X.*
(2) *If X is cofibrant and Z is fibrant, then j_X induces an isomorphism of the set of homotopy classes of maps $(j_X)^*\colon \pi(\widehat{\mathrm{F}}X, Z) \to \pi(X, Z)$ that is natural in both X and Z.*

PROOF. This follows from Proposition 7.5.9. □

LEMMA 8.5.14. *Let $\mathcal{M}$ and $\mathcal{N}$ be model categories, and let $\mathrm{F}\colon \mathcal{M} \rightleftarrows \mathcal{N} :\mathrm{U}$ be a Quillen pair (see Definition 8.5.2).*

(1) *If B is a cofibrant object of $\mathcal{M}$ and $B \amalg B \to \mathrm{Cyl}(B) \to B$ is a cylinder object for B, then $\mathrm{F}B \amalg \mathrm{F}B \to \mathrm{F}\big(\mathrm{Cyl}(B)\big) \to \mathrm{F}B$ is a cylinder object for $\mathrm{F}B$.*
(2) *If X is a fibrant object of $\mathcal{N}$ and $X \to \mathrm{Path}(X) \to X \times X$ is a path object for X, then $\mathrm{U}X \to \mathrm{U}\big(\mathrm{Path}(X)\big) \to \mathrm{U}X \times \mathrm{U}X$ is a path object for $\mathrm{U}X$.*

PROOF. We will prove part 1; the proof of part 2 is dual.

Since B is cofibrant, Lemma 7.3.7 and the "two out of three" property of weak equivalences (see Definition 7.1.3) imply that the map $\mathrm{F}\big(\mathrm{Cyl}(B)\big) \to \mathrm{F}B$ is a weak equivalence. Since F is a left adjoint, $\mathrm{F}(B \amalg B) \approx \mathrm{F}B \amalg \mathrm{F}B$, and so $\mathrm{F}B \amalg \mathrm{F}B \to \mathrm{F}\big(\mathrm{Cyl}(B)\big) \to \mathrm{F}B$ is a cylinder object for $\mathrm{F}B$. □

LEMMA 8.5.15. *Let $\mathcal{M}$ and $\mathcal{N}$ be model categories and let $\mathrm{F}\colon \mathcal{M} \rightleftarrows \mathcal{N} :\mathrm{U}$ be a Quillen pair (see Definition 8.5.2).*

(1) *If $f, g\colon A \to B$ are left homotopic maps in $\mathcal{M}$ and A is cofibrant, then $\mathrm{F}(f)$ is left homotopic to $\mathrm{F}(g)$.*
(2) *If $f, g\colon X \to Y$ are right homotopic maps in $\mathcal{N}$ and Y is fibrant, then $\mathrm{U}(f)$ is right homotopic to $\mathrm{U}(g)$.*

PROOF. This follows from Lemma 8.5.14. □

PROPOSITION 8.5.16. *Let $\mathcal{M}$ and $\mathcal{N}$ be model categories, and let $F\colon \mathcal{M} \rightleftarrows \mathcal{N} : U$ be a Quillen pair (see Definition 8.5.2). If X is a cofibrant object of $\mathcal{M}$ and Y is a fibrant object of $\mathcal{N}$, then the adjointness isomorphism between F and U induces a natural isomorphism of the sets of homotopy classes of maps $\pi(FX, Y) \approx \pi(X, UY)$.*

PROOF. The adjointness of F and U gives us a natural isomorphism of sets of maps $\mathcal{N}(FX, Y) \approx \mathcal{M}(X, UY)$; we must show that this passes to homotopy classes. Theorem 7.4.9 implies that the left and right homotopy relations coincide for these sets of maps, and Lemma 8.5.14 implies that if two maps $X \to UY$ in $\mathcal{M}$ are left homotopic then the corresponding maps $FX \to Y$ are left homotopic and that if two maps $FX \to Y$ in $\mathcal{N}$ are right homotopic then the corresponding maps $X \to UY$ are right homotopic. □

8.5.17. Adjunction of total derived functors.

THEOREM 8.5.18. *Let $\mathcal{M}$ and $\mathcal{N}$ be model categories. If $F\colon \mathcal{M} \rightleftarrows \mathcal{N} : U$ is a Quillen pair (see Definition 8.5.2), then*

(1) *the total left derived functor $\mathbf{L}F\colon \operatorname{Ho}\mathcal{M} \to \operatorname{Ho}\mathcal{N}$ of F exists,*
(2) *the total right derived functor $\mathbf{R}U\colon \operatorname{Ho}\mathcal{N} \to \operatorname{Ho}\mathcal{M}$ of U exists, and*
(3) *the functors $\mathbf{L}F$ and $\mathbf{R}U$ are an adjoint pair.*

PROOF. The existence of the functors $\mathbf{L}F$ and $\mathbf{R}U$ follows from Theorem 8.5.8. To see that $\mathbf{L}F$ and $\mathbf{R}U$ are adjoint, let X be an object of $\mathcal{M}$, let Y be an object of $\mathcal{N}$, let $\widetilde{C}$ and $\widehat{F}$ be the constructions in $\mathcal{M}$ as in the proof of Theorem 8.3.5, and let $\widetilde{C}'$ and $\widehat{F}'$ be the corresponding constructions in $\mathcal{N}$; then we have natural isomorphisms

$$
\begin{aligned}
\operatorname{Ho}\mathcal{N}(\mathbf{L}FX, Y) &= \operatorname{Ho}\mathcal{N}(F(\widetilde{C}X), Y) \\
&= \pi\big(\widehat{F}'\widetilde{C}'F(\widetilde{C}X), \widehat{F}'\widetilde{C}'Y\big) \\
&\approx \pi\big(F(\widetilde{C}X), \widehat{F}'\widetilde{C}'Y\big) && \text{(see Corollary 7.7.4)} \\
&\approx \pi\big(F(\widetilde{C}X), \widehat{F}'Y\big) && \text{(see Corollary 7.7.4)} \\
&\approx \pi\big(\widetilde{C}X, U(\widehat{F}'Y)\big) && \text{(see Proposition 8.5.16)} \\
&\approx \pi\big(\widehat{F}\widetilde{C}X, U(\widehat{F}'Y)\big) && \text{(see Lemma 8.5.13)} \\
&\approx \pi\big(\widehat{F}\widetilde{C}X, \widehat{F}\widetilde{C}U(\widehat{F}'Y)\big) && \text{(see Corollary 7.7.4)} \\
&= \operatorname{Ho}\mathcal{M}\big(X, U(\widehat{F}'Y)\big) \\
&= \operatorname{Ho}\mathcal{M}(X, \mathbf{R}UY).
\end{aligned}
$$

□

8.5.19. Quillen equivalences.

DEFINITION 8.5.20. Let $\mathcal{M}$ and $\mathcal{N}$ be model categories and let $F\colon \mathcal{M} \rightleftarrows \mathcal{N} : U$ be a Quillen pair (see Definition 8.5.2). We will say that

(1) F *is a left Quillen equivalence,*
(2) U *is a right Quillen equivalence*, and
(3) (F, U) *is a pair of Quillen equivalences*

if for every cofibrant object B in $\mathcal{M}$, every fibrant object X in $\mathcal{N}$, and every map $f\colon B \to \mathrm{U}X$ in $\mathcal{M}$, the map f is a weak equivalence in $\mathcal{M}$ if and only if the corresponding map $f^\sharp\colon \mathrm{F}B \to X$ is a weak equivalence in $\mathcal{N}$.

EXAMPLE 8.5.21. The geometric realization functor from SS to Top (see Notation 7.10.5) and the total singular complex functor from Top to SS are Quillen equivalences.

EXAMPLE 8.5.22. The geometric realization functor from SS_* to Top_* (see Notation 7.10.5) and the total singular complex functor from Top_* to SS_* are Quillen equivalences.

THEOREM 8.5.23. *Let* $\mathcal{M}$ *and* $\mathcal{N}$ *be model categories and let* $\mathrm{F}\colon \mathcal{M} \rightleftarrows \mathcal{N} :\mathrm{U}$ *be a Quillen pair. If* (F,U) *is a pair of Quillen equivalences (see Definition* 8.5.20*), then the total derived functors* $\mathbf{L}\mathrm{F}\colon \mathrm{Ho}\,\mathcal{M} \rightleftarrows \mathrm{Ho}\,\mathcal{N} :\mathbf{R}\mathrm{U}$ *(see Theorem* 8.5.18*) are equivalences of the homotopy categories* $\mathrm{Ho}\,\mathcal{M}$ *and* $\mathrm{Ho}\,\mathcal{N}$.

PROOF. Theorem 8.5.18 implies that we have adjoint functors $\mathbf{L}\mathrm{F}\colon \mathrm{Ho}\,\mathcal{M} \rightleftarrows \mathrm{Ho}\,\mathcal{N} :\mathbf{R}\mathrm{U}$; we must show that

(1) for every object X in $\mathrm{Ho}\,\mathcal{M}$ the natural map $\eta_X\colon X \to \mathbf{R}\mathrm{U}\circ\mathbf{L}\mathrm{F}(X)$ is an isomorphism, and
(2) for every object Y in $\mathrm{Ho}\,\mathcal{N}$, the natural map $\varepsilon_Y\colon \mathbf{L}\mathrm{F}\circ\mathbf{R}\mathrm{U}(Y) \to Y$ is an isomorphism.

We will prove part 1; the proof of part 2 is similar.

If X is an object of $\mathrm{Ho}\,\mathcal{M}$, then the map $\eta_X\colon X \to \mathbf{R}\mathrm{U}\circ\mathbf{L}\mathrm{F}(X)$ corresponds to the identity map $1_{\mathbf{L}\mathrm{F}(X)}\colon \mathbf{L}\mathrm{F}(X) \to \mathbf{L}\mathrm{F}(X)$ under the adjunction of Theorem 8.5.18, and that identity map is the homotopy class of the identity map of $\widehat{\mathrm{F}}'\widetilde{\mathrm{C}}'\mathrm{F}(\widetilde{\mathrm{C}}X)$. The "two out of three" property of weak equivalences implies that the corresponding element of $\pi\big(\mathrm{F}(\widetilde{\mathrm{C}}X), \widehat{\mathrm{F}}'\mathrm{F}(\widetilde{\mathrm{C}}X)\big)$ (see the proof of Theorem 8.5.18) consists of weak equivalences, and so our hypotheses implies that the corresponding element of $\pi\big(\widetilde{\mathrm{C}}X, \mathrm{U}(\widehat{\mathrm{F}}'\mathrm{F}(\widetilde{\mathrm{C}}X))\big)$ also consists of weak equivalences. The "two out of three" property now implies that the corresponding element of $\pi\big(\widehat{\mathrm{F}}\widetilde{\mathrm{C}}X, \widehat{\mathrm{F}}\widetilde{\mathrm{C}}\mathrm{U}(\widehat{\mathrm{F}}'\mathrm{F}(\widetilde{\mathrm{C}}X))\big)$ consists of weak equivalences, and is thus an isomorphism in $\mathrm{Ho}\,\mathcal{M}$. $\square$

CHAPTER 9

Simplicial Model Categories

A *simplicial category* (see Definition 9.1.2) is a category $\mathcal{M}$ that is *enriched over simplicial sets*, i.e., that comes with a simplicial set of maps $\mathrm{Map}(X, Y)$ for every pair of objects X and Y, the vertices of which are the maps from X to Y in $\mathcal{M}$. A *simplicial model category* $\mathcal{M}$ is a simplicial category that is also a model category for which there are natural constructions of objects $X \otimes K$ and X^K in $\mathcal{M}$ for X an object of $\mathcal{M}$ and K a simplicial set, satisfying two axioms (see Definition 9.1.6). The first of these axioms (M6) describes adjointness relations between $X \otimes K$, Y^K, and $\mathrm{Map}(X, Y)$, and the second (M7) is the homotopy lifting extension theorem for the simplicial mapping space $\mathrm{Map}(X, Y)$.

We define simplicial model categories in Section 9.1, and in Section 9.2 we discuss commuting function complexes with colimits and limits. In Section 9.3 we discuss when a map induces a weak equivalences of mapping spaces and, via adjointness, obtain results on the tensor product and exponential constructions. In Section 9.4 we discuss the *homotopy left lifting property* and the *homotopy right lifting property* (see Definition 9.4.2), which are analogous to the left lifting property and the right lifting property in not necessarily simplicial model categories.

In Sections 9.5 and 9.6 we discuss the simplicial homotopy relation for maps in a simplicial model category. Neither the left homotopy relation nor the right homotopy relation is well behaved for maps between objects that are not cofibrant or fibrant. The simplicial homotopy relation, however, is an equivalence relation by definition (see Definition 9.5.2), and it behaves well with respect to composition even when the objects are neither cofibrant nor fibrant (see Corollary 9.5.4). Simplicial homotopy implies both left homotopy and right homotopy (see Proposition 9.5.23), and it agrees with both left homotopy and right homotopy when the domain is cofibrant and the codomain is fibrant (see Proposition 9.5.24).

In Section 9.7 we discuss detecting when a map is a weak equivalence by examining whether it induces weak equivalences of mapping spaces, and in Section 9.8 we discuss when a functor between the underlying categories of two simplicial categories can be extended to a *simplicial functor*.

9.1. Simplicial model categories

9.1.1. Simplicial categories.

DEFINITION 9.1.2. A *simplicial category* (or a *category enriched over simplicial sets*) $\mathcal{M}$ is a category together with

(1) for every two objects X and Y of $\mathcal{M}$ a simplicial set $\mathrm{Map}(X, Y)$ (which we will call the *simplicial set of maps from X to Y* or the *function complex* from X to Y or the *simplicial mapping space* from X to Y),

(2) for every three objects X, Y, and Z of $\mathcal{M}$ a map of simplicial sets

$$c_{X,Y,Z}\colon \operatorname{Map}(Y,Z)\times\operatorname{Map}(X,Y)\to\operatorname{Map}(X,Z)$$

(which we will call the *composition rule*),

(3) for every object X of $\mathcal{M}$ a map of simplicial sets $i_X\colon *\to\operatorname{Map}(X,X)$ (where "$*$" is the simplicial set consisting of a single point), and

(4) for every two objects X and Y of $\mathcal{M}$ an isomorphism $\operatorname{Map}(X,Y)_0\approx\mathcal{M}(X,Y)$ that commutes with the composition rule

such that for all objects W, X, Y, and Z of $\mathcal{M}$ the following three diagrams commute:

(Associativity)

$$\begin{array}{ccc}
\big(\operatorname{Map}(Y,Z)\times\operatorname{Map}(X,Y)\big)\times\operatorname{Map}(W,X) & \xrightarrow{c_{X,Y,Z}\times 1_{\operatorname{Map}(W,X)}} & \operatorname{Map}(X,Z)\times\operatorname{Map}(W,X)\\
\downarrow{\scriptstyle\approx} & & \\
\operatorname{Map}(Y,Z)\times\big(\operatorname{Map}(X,Y)\times\operatorname{Map}(W,X)\big) & & \downarrow{\scriptstyle c_{W,X,Z}}\\
\downarrow{\scriptstyle 1_{\operatorname{Map}(Y,Z)}\times c_{W,X,Y}} & & \\
\operatorname{Map}(Y,Z)\times\operatorname{Map}(W,Y) & \xrightarrow[c_{W,Y,Z}]{} & \operatorname{Map}(W,Z)
\end{array}$$

(Left unit)

$$\begin{array}{ccc}
*\times\operatorname{Map}(X,Y) & \xrightarrow{i_Y\times 1_{\operatorname{Map}(X,Y)}} & \operatorname{Map}(Y,Y)\times\operatorname{Map}(X,Y)\\
& {\scriptstyle\approx}\searrow \quad \swarrow{\scriptstyle c_{X,Y,Y}} & \\
& \operatorname{Map}(X,Y) &
\end{array}$$

(Right unit)

$$\begin{array}{ccc}
\operatorname{Map}(X,Y)\times * & \xrightarrow{1_{\operatorname{Map}(X,Y)}\times i_X} & \operatorname{Map}(X,Y)\times\operatorname{Map}(X,X)\\
& {\scriptstyle\approx}\searrow \quad \swarrow{\scriptstyle c_{X,X,Y}} & \\
& \operatorname{Map}(X,Y) &
\end{array}$$

PROPOSITION 9.1.3. *Let $\mathcal{M}$ be a simplicial category.*

(1) *For each object X of $\mathcal{M}$ the simplicial mapping space defines a functor $\operatorname{Map}(X,-)\colon\mathcal{M}\to\mathrm{SS}$ that takes the object Y of $\mathcal{M}$ to $\operatorname{Map}(X,Y)$ and the map $g\colon Y\to Z$ to the map $g_*\colon\operatorname{Map}(X,Y)\to\operatorname{Map}(X,Z)$ that is the composition*

$$\operatorname{Map}(X,Y)\approx *\times\operatorname{Map}(X,Y)\xrightarrow{i_g\times 1_{\operatorname{Map}(X,Y)}}\operatorname{Map}(Y,Z)\times\operatorname{Map}(X,Y)\xrightarrow{c_{X,Y,Z}}\operatorname{Map}(X,Z)$$

where $i_g\colon *\to\operatorname{Map}(Y,Z)$ takes the vertex of $*$ to g.

(2) *For each object Y of $\mathcal{M}$ the simplicial mapping space defines a functor $\operatorname{Map}(-,Y)\colon\mathcal{M}^{\mathrm{op}}\to\mathrm{SS}$ that takes the object X of $\mathcal{M}$ to $\operatorname{Map}(X,Y)$ and the map $f\colon W\to X$ to the map $f^*\colon\operatorname{Map}(X,Y)\to\operatorname{Map}(W,Y)$ that is*

the composition

$$\operatorname{Map}(X,Y) \approx \operatorname{Map}(X,Y) \times * \xrightarrow{1_{\operatorname{Map}(X,Y)} \times i_f} \operatorname{Map}(X,Y) \times \operatorname{Map}(W,X) \xrightarrow{c_{W,X,Y}} \operatorname{Map}(W,Y)$$

where $i_f \colon * \to \operatorname{Map}(W,X)$ *takes the vertex of* $*$ *to* f.

PROOF. This follows directly from the definitions. □

EXAMPLE 9.1.4. Let SS denote the category of simplicial sets.

- If X and Y are simplicial sets, we let $\operatorname{Map}(X,Y)$ be the simplicial set that in degree n is the set of maps of simplicial sets from $X \times \Delta[n]$ to Y, with face and degeneracy maps induced by the standard maps between the $\Delta[n]$.
- If X is a simplicial set and K is a simplicial set, then we let $X \otimes K$ be $X \times K$ and we let X^K be $\operatorname{Map}(K,X)$.

This gives SS the structure of a simplicial category.

9.1.5. Simplicial model categories.

DEFINITION 9.1.6. A *simplicial model category* is a model category $\mathcal{M}$ that is also a simplicial category (see Definition 9.1.2) such that the following two axioms hold:

M6: For every two objects X and Y of $\mathcal{M}$ and every simplicial set K there are objects $X \otimes K$ and Y^K of $\mathcal{M}$ such that there are isomorphisms of simplicial sets

$$\operatorname{Map}(X \otimes K, Y) \approx \operatorname{Map}\big(K, \operatorname{Map}(X,Y)\big) \approx \operatorname{Map}(X, Y^K)$$

(see Example 9.1.4) that are natural in X, Y, and K.

M7: If $i \colon A \to B$ is a cofibration in $\mathcal{M}$ and $p \colon X \to Y$ is a fibration in $\mathcal{M}$, then the map of simplicial sets

$$\operatorname{Map}(B,X) \xrightarrow{i^* \times p_*} \operatorname{Map}(A,X) \times_{\operatorname{Map}(A,Y)} \operatorname{Map}(B,Y)$$

(see Proposition 9.1.3) is a fibration that is a trivial fibration if either i or p is a weak equivalence.

REMARK 9.1.7. Axiom M7 of Definition 9.1.6 is the *homotopy lifting extension theorem*, which was originally a theorem of D. M. Kan for categories of simplicial objects (see [**45**]).

PROPOSITION 9.1.8. *Let* $\mathcal{M}$ *be a simplicial model category. If* X *and* Y *are objects of* $\mathcal{M}$ *and* K *is a simplicial set, then there are natural isomorphisms of simplicial sets*

$$\operatorname{Map}(X \otimes K, Y) \approx \operatorname{Map}\big(K, \operatorname{Map}(X,Y)\big) \approx \operatorname{Map}(X, Y^K)$$

which, in simplicial degree zero, yield natural isomorphisms of the sets of maps

$$\mathcal{M}(X \otimes K, Y) \approx \operatorname{SS}\big(K, \operatorname{Map}(X,Y)\big) \approx \mathcal{M}(X, Y^K) \ .$$

PROOF. This follows from Definition 9.1.2 and axiom M6 of Definition 9.1.6. □

PROPOSITION 9.1.9. *Let $\mathcal{M}$ be a simplicial model category. If X and Y are objects of $\mathcal{M}$, then for every $n \geq 0$ the set of n-simplices of* $\operatorname{Map}(X, Y)$ *is naturally isomorphic to the set of maps $\mathcal{M}(X \otimes \Delta[n], Y)$.*

PROOF. Since the set of n-simplices of a simplicial set K is naturally isomorphic to the set of maps $\mathrm{SS}(\Delta[n], K)$, axiom M6 of Definition 9.1.6 yields natural isomorphisms

$$\begin{aligned} \operatorname{Map}(X,Y)_n &\approx \mathrm{SS}\big(\Delta[n], \operatorname{Map}(X,Y)\big) \\ &\approx \operatorname{Map}\big(\Delta[n], \operatorname{Map}(X,Y)\big)_0 \\ &\approx \operatorname{Map}(X \otimes \Delta[n], Y)_0 \\ &\approx \mathcal{M}(X \otimes \Delta[n], Y) \ . \end{aligned}$$

□

PROPOSITION 9.1.10. *If $\mathcal{M}$ is a simplicial model category, then for every object X of $\mathcal{M}$ there are natural isomorphisms*

$$X \otimes \Delta[0] \approx X \quad \textit{and} \quad X^{\Delta[0]} \approx X \ .$$

PROOF. There are natural isomorphisms $\mathcal{M}(X \otimes \Delta[0], Y) \approx \operatorname{Map}(X,Y)_0 \approx \mathcal{M}(X,Y)$ for every object Y of $\mathcal{M}$ (see Proposition 9.1.9), and the Yoneda lemma implies that the composition of these is induced by a unique natural isomorphism $X \approx X \otimes \Delta[0]$. The second isomorphism follows in a similar manner. □

PROPOSITION 9.1.11. *If $\mathcal{M}$ is a simplicial model category, then for all objects X of $\mathcal{M}$ and all simplicial sets K and L there are natural isomorphisms*

$$X \otimes (K \times L) \approx (X \otimes K) \otimes L \quad \textit{and} \quad X^{(K \times L)} \approx (X^K)^L \ .$$

PROOF. Proposition 9.1.8 implies that for every object Y of $\mathcal{M}$ we have natural isomorphisms

$$\begin{aligned} \mathcal{M}\big(X \otimes (K \times L), Y\big) &\approx \mathrm{SS}\big(K \times L, \operatorname{Map}(X,Y)\big) \\ &\approx \mathrm{SS}\big(L, \operatorname{Map}\big(K, \operatorname{Map}(X,Y)\big)\big) \\ &\approx \mathrm{SS}\big(L, \operatorname{Map}(X \otimes K, Y)\big) \\ &\approx \mathcal{M}((X \otimes K) \otimes L, Y) \end{aligned}$$

and the Yoneda lemma implies that the composition of these is induced by a unique natural isomorphism $X \otimes (K \times L) \approx (X \otimes K) \otimes L$. The proof for the second isomorphism is similar. □

9.1.12. Examples.

EXAMPLE 9.1.13. Let SS denote the model category of simplicial sets (see Theorem 7.10.12), and give SS the simplicial category structure of Example 9.1.4. With these definitions, SS has the structure of a simplicial model category (see [**42**, Theorem 3.6.5]), or [**39**, Chapter I]), or [**52**, Chapter II, Section 3].

EXAMPLE 9.1.14. Let SS_* denote the model category of pointed simplicial sets (see Theorem 7.10.13).

- If X and Y are pointed simplicial sets, we let $\operatorname{Map}(X, Y)$ be the (unpointed) simplicial set that in degree n is the set of maps of pointed simplicial sets from $X \wedge (\Delta[n]^+)$ to Y, with face and degeneracy maps induced by the standard maps between the $\Delta[n]$.

- If X is a pointed simplicial set and K is a simplicial set, then we let $X \otimes K$ be $X \times K^+$ and we let X^K be $\mathrm{Map}(K^+, X)$.

With these definitions, SS_* has the structure of a simplicial model category (see [**42**, Corollary 3.6.6]), or [**39**, Chapter I]), or [**52**, Chapter II, Section 3].

EXAMPLE 9.1.15. Let Top denote the model category of topological spaces (see Theorem 7.10.10).

- If X and Y are topological spaces, we let $\mathrm{Map}(X, Y)$ be the simplicial set that in degree n is the set of continuous maps from $X \times |\Delta[n]|$ to Y, with face and degeneracy maps induced by the standard maps between the $\Delta[n]$.
- If X is a topological space and K is a simplicial set, then we let $X \otimes K$ be $X \times |K|$ and we let X^K be the space of maps from $|K|$ to X.

With these definitions, Top has the structure of a simplicial model category (see [**42**, Theorem 2.4.19]) or [**52**, Chapter II, Section 3].

EXAMPLE 9.1.16. Let Top_* denote the model category of pointed topological spaces (see Theorem 7.10.11).

- If X and Y are pointed topological spaces, we let $\mathrm{Map}(X, Y)$ be the (unpointed) simplicial set that in degree n is the set of continuous maps from $X \wedge |\Delta[n]^+|$ to Y, with face and degeneracy maps induced by the standard maps between the $\Delta[n]$.
- If X is a pointed topological space and K is a simplicial set, then we let $X \otimes K$ be $X \times |K|^+$ and we let X^K be the space of maps from $|K|^+$ to X.

With these definitions, Top_* has the structure of a simplicial model category (see [**42**, Corollary 2.4.20]) or [**52**, Chapter II, Section 3].

9.2. Colimits and limits

LEMMA 9.2.1. *Let $\mathcal{C}$ be a small category and let $\mathcal{M}$ be a simplicial model category.*

(1) *If $\boldsymbol{X}$ is a $\mathcal{C}$-diagram in $\mathcal{M}$ and K a simplicial set, then there is a natural isomorphism*

$$(\mathrm{colim}\, \boldsymbol{X}) \otimes K \approx \mathrm{colim}(\boldsymbol{X} \otimes K) \ .$$

(2) *If X is an object of $\mathcal{M}$ and $\boldsymbol{K}$ is a $\mathcal{C}$-diagram of simplicial sets, then there is a natural isomorphism*

$$X \otimes (\mathrm{colim}\, \boldsymbol{K}) \approx \mathrm{colim}(X \otimes \boldsymbol{K}) \ .$$

PROOF. We will prove part 1; the proof of part 2 is similar.

If Y is an object of $\mathcal{M}$, then Proposition 9.1.8 implies that we have natural isomorphisms

$$\begin{aligned}\mathcal{M}\big((\operatorname{colim} \boldsymbol{X}) \otimes K, Y\big) &\approx \mathcal{M}(\operatorname{colim} \boldsymbol{X}, Y^K)\\ &\approx \lim \mathcal{M}(\boldsymbol{X}, Y^K)\\ &\approx \lim \mathcal{M}(\boldsymbol{X} \otimes K, Y)\\ &\approx \mathcal{M}\big(\operatorname{colim}(\boldsymbol{X} \otimes K), Y\big) \ ,\end{aligned}$$

and the Yoneda Lemma implies that the composition of these must be induced by a natural isomorphism $(\operatorname{colim} \boldsymbol{X}) \otimes K \approx \operatorname{colim}(\boldsymbol{X} \otimes K)$. □

PROPOSITION 9.2.2. *If $\mathcal{M}$ is a simplicial model category, $\mathcal{C}$ is a small category, $\boldsymbol{X}$ is a $\mathcal{C}$-diagram in $\mathcal{M}$, and Y is an object of $\mathcal{M}$, then there are natural isomorphisms of simplicial sets*

$$\begin{aligned}\operatorname{Map}(\operatorname{colim} \boldsymbol{X}, Y) &\approx \lim \operatorname{Map}(\boldsymbol{X}, Y)\\ \operatorname{Map}(Y, \lim \boldsymbol{X}) &\approx \lim \operatorname{Map}(Y, \boldsymbol{X}) \ .\end{aligned}$$

PROOF. We will prove that the first isomorphism exists; the proof that the second exists is similar.

For every $n \geq 0$, Proposition 9.1.9 and Lemma 9.2.1 yield natural isomorphisms

$$\begin{aligned}\operatorname{Map}(\operatorname{colim} \boldsymbol{X}, Y)_n &\approx \mathcal{M}\big((\operatorname{colim} \boldsymbol{X}) \otimes \Delta[n], Y\big)\\ &\approx \mathcal{M}\big(\operatorname{colim}(\boldsymbol{X} \otimes \Delta[n]), Y\big)\\ &\approx \lim \mathcal{M}(\boldsymbol{X} \otimes \Delta[n], Y)\\ &\approx \lim \operatorname{Map}(\boldsymbol{X}, Y)_n \ .\end{aligned}$$

□

COROLLARY 9.2.3. *Let $\mathcal{M}$ be a simplicial model category and let Y be an object of $\mathcal{M}$. If S is a set and X_s is an object of $\mathcal{M}$ for every $s \in S$, then there is a natural isomorphism of simplicial sets*

$$\operatorname{Map}\Big(\coprod_{s \in S} X_s, Y\Big) \approx \prod_{s \in S} \operatorname{Map}(X_s, Y) \ .$$

PROOF. This follows from Proposition 9.2.2. □

9.3. Weak equivalences of function complexes

PROPOSITION 9.3.1. *Let $\mathcal{M}$ be a simplicial model category.*

(1) *If $i\colon A \to B$ is a cofibration in $\mathcal{M}$ and X is a fibrant object of $\mathcal{M}$, then the map $i^*\colon \operatorname{Map}(B, X) \to \operatorname{Map}(A, X)$ is a fibration of simplicial sets.*
(2) *If A is cofibrant in $\mathcal{M}$ and $p\colon X \to Y$ is a fibration in $\mathcal{M}$, then the map $p_*\colon \operatorname{Map}(A, X) \to \operatorname{Map}(A, Y)$ is a fibration of simplicial sets.*

PROOF. This follows from axiom M7 (see Definition 9.1.6). □

PROPOSITION 9.3.2. *Let $\mathcal{M}$ be a simplicial model category and let X, Y, and Z be objects of $\mathcal{M}$.*

(1) *If X is cofibrant and $g\colon Y \to Z$ is a trivial fibration, then g induces a trivial fibration of simplicial sets $g_*\colon \operatorname{Map}(X, Y) \to \operatorname{Map}(X, Z)$.*

(2) *If Z is fibrant and $h: X \to Y$ is a trivial cofibration, then h induces a trivial fibration of simplicial sets $h^*: \operatorname{Map}(Y, Z) \to \operatorname{Map}(X, Z)$.*

PROOF. This follows from axiom M7 (see Definition 9.1.6). □

COROLLARY 9.3.3. *Let $\mathcal{M}$ be a simplicial model category and let X, Y, and Z be objects of $\mathcal{M}$.*

(1) *If X is cofibrant and $g: Y \to Z$ is a weak equivalence of fibrant objects, then g induces a weak equivalence of simplicial sets $g_*: \operatorname{Map}(X, Y) \to \operatorname{Map}(X, Z)$.*
(2) *If Z is fibrant and $h: X \to Y$ is a weak equivalence of cofibrant objects, then h induces a weak equivalence of simplicial sets $h: \operatorname{Map}(Y, Z) \to \operatorname{Map}(X, Z)$.*

PROOF. This follows from Proposition 9.3.2 and Corollary 7.7.2. □

9.3.4. Consequences of adjointness.

DEFINITION 9.3.5. Let $\mathcal{M}$ be a simplicial model category, let $A \to B$ and $X \to Y$ be maps in $\mathcal{M}$, and let $L \to K$ be a map of simplicial sets.

(1) The map of simplicial sets
$$\operatorname{Map}(B, X) \longrightarrow \operatorname{Map}(A, X) \times_{\operatorname{Map}(A,Y)} \operatorname{Map}(B, Y)$$
will be called the *pullback corner map* of the maps $A \to B$ and $X \to Y$.
(2) The map
$$A \otimes K \amalg_{A \otimes L} B \otimes L \longrightarrow B \otimes K$$
will be called the *pushout corner map* of the maps $A \to B$ and $L \to K$.
(3) The map
$$X^K \longrightarrow X^L \times_{Y^L} Y^K$$
will be called the *pullback corner map* of the maps $X \to Y$ and $L \to K$.

LEMMA 9.3.6. *Let $\mathcal{M}$ be a simplicial model category. If $A \to B$ and $X \to Y$ are maps in $\mathcal{M}$ and $L \to K$ is a map of simplicial sets, then the following are equivalent:*

(1) *The dotted arrow exists in every solid arrow diagram of the form*
$$\begin{array}{ccc} L & \longrightarrow & \operatorname{Map}(B, X) \\ \downarrow & \nearrow & \downarrow \\ K & \longrightarrow & \operatorname{Map}(A, X) \times_{\operatorname{Map}(A,Y)} \operatorname{Map}(B, Y) \end{array}$$
(2) *The dotted arrow exists in every solid arrow diagram of the form*
$$\begin{array}{ccc} A & \longrightarrow & X^K \\ \downarrow & \nearrow & \downarrow \\ B & \longrightarrow & X^L \times_{Y^L} Y^K \end{array}$$
(3) *The dotted arrow exists in every solid arrow diagram of the form*
$$\begin{array}{ccc} A \otimes K \amalg_{A \otimes L} B \otimes L & \longrightarrow & X \\ \downarrow & \nearrow & \downarrow \\ B \otimes K & \longrightarrow & Y \end{array}$$

PROOF. This follows from Definition 9.1.6. □

PROPOSITION 9.3.7. *If $\mathcal{M}$ is a both a model category and a simplicial category and $\mathcal{M}$ satisfies axiom M6 of Definition 9.1.6, then the following are equivalent:*

(1) *$\mathcal{M}$ satisfies axiom M7 of Definition 9.1.6 (i.e., $\mathcal{M}$ is a simplicial model category).*
(2) *If $i\colon A \to B$ is a cofibration in $\mathcal{M}$ and $p\colon X \to Y$ is a fibration in $\mathcal{M}$, then the pullback corner map $\mathrm{Map}(B,X) \to \mathrm{Map}(A,X) \times_{\mathrm{Map}(A,Y)} \mathrm{Map}(B,Y)$ is a fibration of simplicial sets that is a trivial fibration if either i or p is a weak equivalence.*
(3) *If $i\colon A \to B$ is a cofibration in $\mathcal{M}$ and $j\colon L \to K$ is an inclusion of simplicial sets, then the pushout corner map $A\otimes K \amalg_{A\otimes L} B\otimes L \to B\otimes K$ is a cofibration in $\mathcal{M}$ that is a trivial cofibration if either i or j is a weak equivalence.*
(4) *If $j\colon L \to K$ is an inclusion of simplicial sets and $p\colon X \to Y$ is a fibration in $\mathcal{M}$, then the pullback corner map $X^K \to X^L \times_{Y^L} Y^K$ is a fibration in $\mathcal{M}$ that is a trivial fibration if either j or p is a weak equivalence.*

PROOF. Condition 2 is the definition of condition 1. The equivalence of conditions 2, 3, and 4 follows from Proposition 7.2.3 and Lemma 9.3.6. □

PROPOSITION 9.3.8. *Let $\mathcal{M}$ be a simplicial model category.*

(1) *If $i\colon A \to B$ is a cofibration in $\mathcal{M}$ and $j\colon L \to K$ is an inclusion of simplicial sets, then the pushout corner map $A\otimes K \amalg_{A\otimes L} B\otimes L \to B\otimes K$ is a cofibration in $\mathcal{M}$ that is a trivial cofibration if either i or j is a weak equivalence.*
(2) *If $j\colon L \to K$ is an inclusion of simplicial sets and $p\colon X \to Y$ is a fibration in $\mathcal{M}$, then the pullback corner map $X^K \to X^L \times_{Y^L} Y^K$ is a fibration in $\mathcal{M}$ that is a trivial fibration if either j or p is a weak equivalence.*

PROOF. This follows from Proposition 9.3.7. □

PROPOSITION 9.3.9. *Let $\mathcal{M}$ be a simplicial model category.*

(1) (a) *If B is a cofibrant object of $\mathcal{M}$ and $j\colon L \to K$ is an inclusion of simplicial sets, then the map $1_B \otimes j\colon B\otimes L \to B\otimes K$ is a cofibration in $\mathcal{M}$ that is a weak equivalence if j is a weak equivalence.*
    (b) *If $i\colon A \to B$ is a cofibration in $\mathcal{M}$ and K is a simplicial set, then the map $j \otimes 1_K\colon A\otimes K \to B\otimes K$ is a cofibration in $\mathcal{M}$ that is a weak equivalence if i is a weak equivalence.*
(2) (a) *If X is a fibrant object of $\mathcal{M}$ and $j\colon L \to K$ is an inclusion of simplicial sets, then the map $(1_X)^j\colon X^K \to X^L$ is a fibration in $\mathcal{M}$ that is a weak equivalence if j is a weak equivalence.*
    (b) *If $p\colon X \to Y$ is a fibration in $\mathcal{M}$ and K is a simplicial set, then the map $p^{(1_K)}\colon X^K \to Y^K$ is a fibration in $\mathcal{M}$ that is a weak equivalence if p is a weak equivalence.*

PROOF. This follows from Proposition 9.3.8. □

PROPOSITION 9.3.10. *Let $\mathcal{M}$ be a simplicial model category.*

(1) *If $i\colon A \to B$ is a cofibration and X is fibrant, then i induces a weak equivalence of function complexes $i^*\colon \mathrm{Map}(B,X) \approx \mathrm{Map}(A,X)$ if and*

only if every map $A \otimes \Delta[n] \amalg_{A\otimes\partial\Delta[n]} B \otimes \partial\Delta[n] \to X$ *can be extended over* $B \otimes \Delta[n]$ *for every* $n \geq 0$.

(2) *If* B *is cofibrant and* $p\colon X \to Y$ *is a fibration, then* p *induces a weak equivalence of function complexes if and only if every map* $B \to X^{\partial\Delta[n]} \times_{Y^{\partial\Delta[n]}} Y^{\Delta[n]}$ *can be factored through* $X^{\Delta[n]}$ *for every* $n \geq 0$.

PROOF. We will prove part 1; the proof of part 2 is similar.

Proposition 9.3.1 implies that i^* is a fibration, and so it is a weak equivalence if and only if it is a trivial fibration. Since a map of simplicial sets is a trivial fibration if and only if it has the right lifting property with respect to the maps $\partial\Delta[n] \to \Delta[n]$ for all $n \geq 0$, the result follows from Lemma 9.3.6. □

9.4. Homotopy lifting

If $\mathcal{M}$ is a model category, $i\colon A \to B$ is a cofibration in $\mathcal{M}$, $p\colon X \to Y$ is a fibration in $\mathcal{M}$, and at least one of i and p is also a weak equivalence, then axiom M4 (see Definition 7.1.3) implies that (i,p) is a lifting-extension pair (see Definition 7.2.1), i.e., that the map of sets

$$\mathcal{M}(B,X) \longrightarrow \mathcal{M}(A,X) \times_{\mathcal{M}(A,Y)} \mathcal{M}(B,Y)$$

is surjective. If $\mathcal{M}$ is a *simplicial* model category, then a stronger statement is possible: Axiom M7 (see Definition 9.1.6) implies that, under the same hypotheses on i and p, the map of simplicial sets

$$\operatorname{Map}(B,X) \longrightarrow \operatorname{Map}(A,X) \times_{\operatorname{Map}(A,Y)} \operatorname{Map}(B,Y) \tag{9.4.1}$$

is a trivial fibration. This analog for simplicial model categories of being a lifting-extension pair is called being a *homotopy lifting-extension pair* (see Definition 9.4.2).

DEFINITION 9.4.2. Let $\mathcal{M}$ be a simplicial model category. If $i\colon A \to B$ and $p\colon X \to Y$ are maps for which the map of simplicial sets (9.4.1) is a trivial fibration, then

- (i,p) is called a *homotopy lifting extension pair*,
- i is said to have the *homotopy left lifting property* with respect to p, and
- p is said to have the *homotopy right lifting property* with respect to i.

PROPOSITION 9.4.3. *Let* $\mathcal{M}$ *be a simplicial model category. If* $i\colon A \to B$ *and* $p\colon X \to Y$ *are maps in* $\mathcal{M}$ *such that* (i,p) *is a homotopy lifting-extension pair (see Definition 9.4.2), then* (i,p) *is a lifting-extension pair (see Definition 7.2.1).*

PROOF. This follows because a trivial fibration of simplicial sets is surjective on the set of vertices. □

PROPOSITION 9.4.4. *Let* $\mathcal{M}$ *be a simplicial model category.*

(1) *A map is a cofibration if and only if it has the homotopy left lifting property with respect to all trivial fibrations.*
(2) *A map is a trivial cofibration if and only if it has the homotopy left lifting property with respect to all fibrations.*
(3) *A map is a fibration if and only if it has the homotopy right lifting property with respect to all trivial cofibrations.*
(4) *A map is a trivial fibration if and only if it has the homotopy right lifting property with respect to all cofibrations.*

PROOF. This follows from axiom M7 (see Definition 9.1.6), Proposition 9.4.3, and Proposition 7.2.3. □

PROPOSITION 9.4.5. *Let $\mathcal{M}$ be a simplicial model category and let A, B, X, and Y be objects of $\mathcal{M}$.*

(1) *If B is cofibrant and $p\colon X \to Y$ is a fibration, then p has the homotopy right lifting property with respect to the map from the initial object to B if and only if p induces a weak equivalence $p_*\colon \mathrm{Map}(B,X) \cong \mathrm{Map}(B,Y)$.*
(2) *If X is fibrant and $i\colon A \to B$ is a cofibration, then i has the homotopy left lifting property with respect to the map from X to the terminal object if and only if i induces a weak equivalence $i^*\colon \mathrm{Map}(B,X) \cong \mathrm{Map}(A,X)$.*

PROOF. Proposition 9.3.1 implies that both p_* and i^* are fibrations of simplicial sets, and so each of p_* and i^* is a weak equivalence if and only if it is a trivial fibration. □

9.4.6. Closure properties of homotopy lifting properties.

LEMMA 9.4.7. *Let $\mathcal{M}$ be a simplicial model category. If $i\colon A \to B$ and $p\colon X \to Y$ are maps in $\mathcal{M}$, then the following are equivalent:*

(1) *The pair (i,p) is a homotopy lifting-extension pair (see Definition 9.4.2).*
(2) *For every pair of simplicial sets (K,L), the map p has the right lifting property with respect to the pushout corner map*
$$A\otimes K \amalg_{A\otimes L} B\otimes L \longrightarrow B\otimes K\ .$$
(3) *For every $n \geq 0$, the map p has the right lifting property with respect to the pushout corner map*
$$A\otimes \Delta[n] \amalg_{A\otimes\partial\Delta[n]} B\otimes \partial\Delta[n] \longrightarrow B\otimes \Delta[n]\ .$$
(4) *For every pair of simplicial sets (K,L), the map i has the left lifting property with respect to the pullback corner map*
$$X^K \longrightarrow Y^K \times_{Y^L} X^L$$
(5) *For every $n \geq 0$, the map i has the left lifting property with respect to the pullback corner map*
$$X^{\Delta[n]} \longrightarrow Y^{\Delta[n]} \times_{Y^{\partial\Delta[n]}} X^{\partial\Delta[n]}\ .$$

PROOF. Since a map of simplicial sets is a cofibration if and only if it is an inclusion and a trivial fibration if and only if it has the right lifting property with respect to the maps $\partial\Delta[n] \to \Delta[n]$ for $n \geq 0$, this follows from Lemma 9.3.6 and Proposition 7.2.3. □

PROPOSITION 9.4.8. *Let $\mathcal{M}$ be a simplicial model category.*

(1) *If $i\colon A \to B$ has the homotopy left lifting property with respect to $p\colon X \to Y$ and (K,L) is a pair of simplicial sets, then the pushout corner map $A\otimes K \amalg_{A\otimes L} B\otimes L \to B\otimes K$ has the homotopy left lifting property with respect to p.*
(2) *If $p\colon X \to Y$ has the homotopy right lifting property with respect to $i\colon A \to B$ and (K,L) is a pair of simplicial sets, then the pullback corner map $X^K \to Y^K \times_{Y^L} X^L$ has the homotopy right lifting property with respect to i.*

PROOF. We will prove part 2; the proof of part 1 is dual.

Lemma 9.4.7 implies that it is sufficient to show that the map

$$X^K \longrightarrow Y^K \times_{Y^L} X^L$$

has the right lifting property with respect to the map

$$A \otimes \Delta[n] \amalg_{A \otimes \partial\Delta[n]} B \otimes \partial\Delta[n] \longrightarrow B \otimes \Delta[n] \ .$$

Lemma 9.3.6 implies that this is equivalent to showing that the map $p\colon X \to Y$ has the right lifting property with respect to the map

$$(A \otimes \Delta[n] \amalg_{A \otimes \partial\Delta[n]} B \otimes \partial\Delta[n]) \otimes K \amalg_{(A \otimes \Delta[n] \amalg_{A \otimes \partial\Delta[n]} B \otimes \partial\Delta[n]) \otimes L} (B \otimes \Delta[n]) \otimes L \longrightarrow (B \otimes \Delta[n]) \otimes K \ .$$

Lemma 9.2.1 and the isomorphisms of axiom M6 (see Definition 9.1.6) imply that that map is isomorphic to the map

$$B \otimes (\partial\Delta[n] \times K \amalg_{\partial\Delta[n] \times L} \Delta[n] \times L) \amalg_{A \otimes (\partial\Delta[n] \times K \amalg_{\partial\Delta[n] \times L} \Delta[n] \times L)} A \otimes (\Delta[n] \times K) \longrightarrow B \otimes (\Delta[n] \times K) \ ,$$

and Lemma 9.4.7 implies that that map has the left lifting property with respect to p. □

LEMMA 9.4.9. *Let $\mathcal{M}$ be a simplicial model category and let p be a map in $\mathcal{M}$.*

(1) *The class of maps with the homotopy left lifting property with respect to p is closed under pushouts.*
(2) *The class of maps with the homotopy right lifting property with respect to p is closed under pullbacks.*

PROOF. This follows from Lemma 9.4.7 and Lemma 7.2.11. □

LEMMA 9.4.10. *Let $\mathcal{M}$ be a simplicial model category and let p be a map in $\mathcal{M}$.*

(1) *The class of maps with the homotopy left lifting property with respect to p is closed under retracts.*
(2) *The class of maps with the homotopy right lifting property with respect to p is closed under retracts.*

PROOF. This follows from Lemma 9.4.7 and Lemma 7.2.8. □

9.4.11. Homotopy lifting and lifting.

PROPOSITION 9.4.12. *Let $\mathcal{M}$ be a simplicial model category and let $\mathcal{C}$ be a class of maps in $\mathcal{M}$.*

(1) *If every map $g\colon X \to Y$ in $\mathcal{M}$ can be factored as $X \xrightarrow{j} W \xrightarrow{p} Y$ where p is in $\mathcal{C}$ and j has the homotopy left lifting property with respect to every map in $\mathcal{C}$, then a map has the left lifting property with respect to every map in $\mathcal{C}$ if and only if it has the homotopy left lifting property with respect to every map in $\mathcal{C}$.*
(2) *If every map $g\colon X \to Y$ in $\mathcal{M}$ can be factored as $X \xrightarrow{j} W \xrightarrow{p} Y$ where j is in $\mathcal{C}$ and p has the homotopy right lifting property with respect to every map in $\mathcal{C}$, then a map has the right lifting property with respect to every map in $\mathcal{C}$ if and only if it has the homotopy right lifting property with respect to every map in $\mathcal{C}$.*

PROOF. We will prove part 1; the proof of part 2 is dual.

Proposition 9.4.3 implies that if a map has the homotopy left lifting property with respect to every map in $\mathcal{C}$ then it has the left lifting property with respect to every map in $\mathcal{C}$.

Conversely, if the map $g: X \to Y$ has the left lifting property with respect to every map in $\mathcal{C}$, factor g as $X \xrightarrow{j} W \xrightarrow{p} Y$ where p is in $\mathcal{C}$ and j has the homotopy left lifting property with respect to every map in $\mathcal{C}$. The retract argument (see Proposition 7.2.2) implies that g is a retract of j, and so the result follows from Lemma 9.4.10. □

9.5. Simplicial homotopy

9.5.1. Definitions. If X is cofibrant and Y is fibrant then all notions of homotopy for maps from X to Y coincide and are equivalence relations (see Proposition 9.5.24), but this is not true for arbitrary objects X and Y. Thus, it is often useful to consider the *simplicial* homotopy relation (see Definition 9.5.2). The simplicial homotopy relation is an equivalence relation by definition, it is always well behaved under composition (see Corollary 9.5.4), and simplicial homotopy implies both left homotopy and right homotopy (see Proposition 9.5.23). In addition, simplicially homotopic maps of simplicial sets induce simplicially homotopic maps in a simplicial model category (see Lemma 9.5.17).

DEFINITION 9.5.2. Let $\mathcal{M}$ be a simplicial model category, let X and Y be objects of $\mathcal{M}$, and let g and h be maps from X to Y (i.e., vertices of $\mathrm{Map}(X,Y)$; see Definition 9.1.6 and Definition 9.1.2).

(1) g is *strictly simplicially homotopic* to h (denoted $g \overset{ss}{\simeq} h$) if there is a 1-simplex of $\mathrm{Map}(X,Y)$ whose initial vertex is g and whose final vertex is h, i.e., if there is a map $F: X \otimes \Delta[1] \to Y$ such that the composition $X \approx X \otimes \Delta[0] \xrightarrow{1_X \otimes i_0} X \otimes \Delta[1] \xrightarrow{F} Y$ equals g and the composition $X \approx X \otimes \Delta[0] \xrightarrow{1_X \otimes i_1} X \otimes \Delta[1] \xrightarrow{F} Y$ equals h (see Proposition 9.1.9).
(2) g and h are *simplicially homotopic* (denoted $g \overset{s}{\simeq} h$) if they are equivalent under the equivalence relation generated by the relation of strict simplicial homotopy.

PROPOSITION 9.5.3. *Let $\mathcal{M}$ be a simplicial model category and let X and Y be objects of $\mathcal{M}$. If g and h are maps from X to Y, then $g \overset{s}{\simeq} h$ if and only if g and h are in the same component of the simplicial set* $\mathrm{Map}(X,Y)$.

PROOF. This follows directly from the definitions. □

COROLLARY 9.5.4. *Let $\mathcal{M}$ be a simplicial model category and let W, X, Y, and Z be objects of $\mathcal{M}$. If $g, h: X \to Y$ are simplicially homotopic maps and $j: W \to X$ and $k: Y \to Z$ are maps, then $kg \overset{s}{\simeq} kh$ and $gj \overset{s}{\simeq} hj$.*

PROOF. This follows from Proposition 9.5.3. □

DEFINITION 9.5.5. A *generalized interval* is a simplicial set that is a union of finitely many one simplices with vertices identified so that its geometric realization is homeomorphic to a unit interval. If J is a generalized interval, we will let i_0 and i_1 denote the inclusions of $\Delta[0]$ into J at the two end vertices of J.

PROPOSITION 9.5.6. *Let $\mathcal{M}$ be a simplicial model category and let X and Y be objects of $\mathcal{M}$. If g and h are maps from X to Y, then g and h are simplicially homotopic if and only if there is a generalized interval J (see Definition 9.5.5) and a map of simplicial sets $J \to \mathrm{Map}(X,Y)$ taking the ends of J to g and h.*

PROOF. This follows from Proposition 9.5.3. □

DEFINITION 9.5.7. A map $J \to \mathrm{Map}(X,Y)$ as in Proposition 9.5.6 will be called a *simplicial homotopy* from g to h. The maps $X \otimes J \to Y$ and $X \to Y^J$ that correspond under the isomorphisms of Definition 9.1.6 will also be called simplicial homotopies from g to h.

DEFINITION 9.5.8. Let $\mathcal{M}$ be a simplicial model category. The map $g\colon X \to Y$ is a *simplicial homotopy equivalence* if there is a map $h\colon Y \to X$ such that $gh \overset{s}{\simeq} 1_Y$ and $hg \overset{s}{\simeq} 1_X$.

In general, strict simplicial homotopy need not be an equivalence relation, since $\mathrm{Map}(X,Y)$ need not be a fibrant simplicial set. In $\mathrm{Top}_{(*)}$, however, $\mathrm{Map}(X,Y)$ is isomorphic to the total singular complex of the topological space (in our category of spaces) of continuous functions from X to Y, and so it is always a fibrant simplicial set. (Strict simplicial homotopy in $\mathrm{Top}_{(*)}$ is exactly the classical definition of homotopy which is, of course, always an equivalence relation.) In $\mathrm{SS}_{(*)}$ every space is cofibrant, and so $\mathrm{Map}(X,Y)$ will be a fibrant simplicial set if Y is a fibrant space.

9.5.9. Isomorphisms of simplicial homotopy classes of maps.

PROPOSITION 9.5.10. *Let $\mathcal{M}$ be a simplicial model category, let $g\colon X \to Y$ be a map in $\mathcal{M}$, and let W be an object of $\mathcal{M}$.*

(1) *If g induces a weak equivalence of simplicial mapping spaces*
$$g_*\colon \mathrm{Map}(W,X) \cong \mathrm{Map}(W,Y) \ ,$$
then g induces an isomorphism of the sets of simplicial homotopy classes of maps $g_*\colon [W,X] \approx [W,Y]$.

(2) *If g induces a weak equivalence of simplicial mapping spaces*
$$g^*\colon \mathrm{Map}(Y,W) \cong \mathrm{Map}(X,W) \ ,$$
then g induces an isomorphism of the sets of simplicial homotopy classes of maps $g^*\colon [Y,W] \approx [X,W]$.

PROOF. This follows from Proposition 9.5.3. □

COROLLARY 9.5.11. *Let $\mathcal{M}$ be a simplicial model category and let X, Y, and W be objects of $\mathcal{M}$.*

(1) *If W is cofibrant and $g\colon X \to Y$ is a trivial fibration, then g induces an isomorphism of the sets of simplicial homotopy classes of maps $g_*\colon [W,X] \approx [W,Y]$.*
(2) *If W is fibrant and $g\colon X \to Y$ is a trivial cofibration, then g induces an isomorphism of the sets of simplicial homotopy classes of maps $g^*\colon [Y,W] \approx [X,W]$.*

PROOF. This follows from Proposition 9.3.2 and Proposition 9.5.10. □

COROLLARY 9.5.12. *Let $\mathcal{M}$ be a simplicial model category and let X, Y, and W be objects of $\mathcal{M}$.*

(1) *If W is cofibrant and $g\colon X \to Y$ is a weak equivalence of fibrant objects, then g induces an isomorphism of the sets of simplicial homotopy classes of maps $g_*\colon [W, X] \approx [W, Y]$.*
(2) *If W is fibrant and $g\colon X \to Y$ is a weak equivalence of cofibrant objects, then g induces an isomorphism of the sets of simplicial homotopy classes of maps $g^*\colon [Y, W] \approx [X, W]$.*

PROOF. This follows from Corollary 9.3.3 and Proposition 9.5.10. □

9.5.13. Simplicial homotopy, left homotopy, and right homotopy.

LEMMA 9.5.14. *Let $\mathcal{M}$ be a simplicial model category and let J be a generalized interval (see Definition 9.5.5) with endpoint inclusions $i_0\colon \Delta[0] \to J$ and $i_1\colon \Delta[0] \to J$.*

(1) *If X is a cofibrant object of $\mathcal{M}$, then*

$$X \amalg X \approx (X \otimes \Delta[0]) \amalg (X \otimes \Delta[0]) \xrightarrow{(1_X \otimes i_0) \amalg (1_X \otimes i_1)} X \otimes J \to X \otimes \Delta[0] \approx X$$

is a cylinder object for X (see Definition 7.3.2).
(2) *If Y is a fibrant object of $\mathcal{M}$, then*

$$Y \approx Y^{\Delta[0]} \to Y^J \xrightarrow{(1_Y)^{i_0} \times (1_Y)^{i_1}} Y^{\Delta[0]} \times Y^{\Delta[0]} \approx Y \times Y$$

is a path object for Y (see Definition 7.3.2).

PROOF. We will prove part 1; the proof of part 2 is dual.

Since $\Delta[0] \amalg \Delta[0] \xrightarrow{i_0 \amalg i_1} J$ is a cofibration of simplicial sets, Proposition 9.3.9 implies that our map $X \amalg X \to X \otimes J$ is a cofibration. Since the inclusion $i_0\colon \Delta[0] \to J$ is a trivial cofibration of simplicial sets, Proposition 9.3.9 also implies that the map $X \otimes \Delta[0] \to X \otimes J$ is a trivial cofibration, and so the "two out of three" property of weak equivalences (see axiom M2 of Definition 7.1.3) implies that the composition $X \otimes J \to X \otimes \Delta[0] \approx X$ is a weak equivalence. □

LEMMA 9.5.15. *Let $\mathcal{M}$ be a simplicial model category and let X and Y be objects of $\mathcal{M}$. If $f, g\colon X \to Y$ are simplicially homotopic maps, then f and g represent the same map in the homotopy category of $\mathcal{M}$.*

PROOF. If f and g are simplicially homotopic, then there is a generalized interval J (see Definition 9.5.5) and a simplicial homotopy $H\colon X \to Y^J$ such that $\mathrm{ev}_0 H = f$ and $\mathrm{ev}_1 H = g$. If we choose fibrant cofibrant approximations $p_X\colon \widetilde{X} \to X$ and $p_Y\colon \widetilde{Y} \to Y$ (see Definition 8.1.2), then we have the solid arrow diagram

$$\begin{array}{ccccc} & & & & \widetilde{Y}^J \\ & & & \overset{\widetilde{H}}{\nearrow} & \big\downarrow (p_Y)^{(1_J)} \\ \widetilde{X} & \xrightarrow[p_X]{} & X & \xrightarrow[H]{} & Y^J \end{array}$$

in which $(p_Y)^{(1_J)}\colon \widetilde{Y}^J \to Y^J$ is a trivial fibration (see Proposition 9.3.9) and $\widetilde{X}$ is cofibrant. Thus, the dotted arrow $\widetilde{H}$ exists, and we can define $\widetilde{f}, \widetilde{g}\colon \widetilde{X} \to \widetilde{Y}$ by letting $\tilde{f} = \mathrm{ev}_0 \widetilde{H}$ and $\tilde{g} = \mathrm{ev}_1 \widetilde{H}$. We then have $fp_X = \mathrm{ev}_0 Hp_X = \mathrm{ev}_0(p_Y)^{(1_J)}\widetilde{H} = p_Y \mathrm{ev}_0 \widetilde{H} = p_Y\tilde{f}$ and so $fp_X = p_Y\tilde{f}$. Similarly, $gp_X = p_Y\tilde{g}$.

Since $\widetilde{X}$ is cofibrant, $\widetilde{X} \otimes J$ is a cylinder object for $\widetilde{X}$ (see Lemma 9.5.14), and so if we let $\widetilde{H}^{\mathrm{ad}}\colon \widetilde{X} \otimes J \to \widetilde{Y}$ be the map adjoint to $\widetilde{H}\colon \widetilde{X} \to \widetilde{Y}^J$ then $\widetilde{H}^{\mathrm{ad}}$ is a left homotopy from $\tilde{f}$ to $\tilde{g}$. Thus, if we use square brackets to denote the image of a map in $\mathrm{Ho}\,\mathcal{M}$, then Lemma 8.3.4 implies that $[\tilde{f}] = [\tilde{g}]$. Thus, we have $[p_Y]^{-1}[f][p_X] = [p_Y]^{-1}[g][p_X]$ and, since $[p_X]$ and $[p_Y]$ are isomorphisms, $[f] = [g]$. □

PROPOSITION 9.5.16. *If $\mathcal{M}$ is a simplicial model category, then a simplicial homotopy equivalence in $\mathcal{M}$ is a weak equivalence in $\mathcal{M}$.*

PROOF. If $f\colon X \to Y$ is a simplicial homotopy equivalence in $\mathcal{M}$ then there is a map $g\colon Y \to X$ such that $gf \overset{s}{\simeq} 1_X$ and $fg \overset{s}{\simeq} 1_Y$. Lemma 9.5.15 implies that the images of f and g in the homotopy category of $\mathcal{M}$ are isomorphisms, and so Theorem 8.3.10 implies that f and g are weak equivalences. □

LEMMA 9.5.17. *Let $\mathcal{M}$ be a simplicial model category. If K and L are simplicial sets and $f, g\colon K \to L$ are simplicially homotopic maps, then for every object X of $\mathcal{M}$*

1. *the induced maps $1_X \otimes f, 1_X \otimes g\colon X \otimes K \to X \otimes L$ are simplicially homotopic, and*
2. *the induced maps $(1_X)^f, (1_X)^g\colon X^L \to X^K$ are simplicially homotopic.*

PROOF. Let J be a generalized interval (see Definition 9.5.5) such that there is a simplicial homotopy $H\colon K \otimes J \to L$ from f to g (see Definition 9.5.7). The map $1_X \otimes H\colon X \otimes (K \times J) = (X \otimes K) \otimes J \to X \otimes L$ is then a simplicial homotopy from $1_X \otimes f$ to $1_X \otimes g$ and the map $(1_X)^H\colon X^L \to X^{(K\times J)} = (X^K)^J$ is then a simplicial homotopy from $(1_X)^f$ to $(1_X)^g$. □

PROPOSITION 9.5.18. *Let $\mathcal{M}$ be a simplicial model category. If K and L are simplicial sets and $f\colon K \to L$ is a simplicial homotopy equivalence, then for every object X of $\mathcal{M}$*

1. *the induced map $1_X \otimes f\colon X \otimes K \to X \otimes L$ is a simplicial homotopy equivalence, and*
2. *the induced map $(1_X)^f\colon X^L \to X^K$ is a simplicial homotopy equivalence.*

PROOF. This follows from Lemma 9.5.17. □

LEMMA 9.5.19. *If $n \geq 0$, then the inclusion of $\Delta[0]$ into $\Delta[n]$ as either the initial vertex or the final vertex is the inclusion of a simplicial strong deformation retract.*

PROOF. We will prove this for the inclusion as the initial vertex; the proof for the inclusion as the final vertex is similar.

We will define a homotopy $H\colon \Delta[n] \times \Delta[1] \to \Delta[n]$ such that the restriction of H to the initial end of $\Delta[1]$ is the constant map from $\Delta[n]$ to its initial vertex and the restriction of H to the terminal end of $\Delta[1]$ is the identity map of $\Delta[n]$. If $k \geq 0$ then a k-simplex of $\Delta[n]$ is a $(k+1)$-tuple of integers $(i_0, i_1, \ldots, i_k)$ such that $0 \leq i_0 \leq i_1 \leq \cdots \leq i_k \leq n$. Thus, a k-simplex of $\Delta[n] \times \Delta[1]$ is an ordered pair $\bigl((i_0, i_1, \ldots, i_k), (j_0, j_1, \ldots, j_k)\bigr)$ such that $0 \leq i_0 \leq i_1 \leq \cdots \leq i_k \leq n$ and there is an integer ℓ, $0 \leq \ell \leq k+1$, such that $j_m = 0$ for $m < \ell$ and $j_m = 1$ for $m \geq \ell$, and we let $H\big((i_0, i_1, \ldots, i_k), (j_0, j_1, \ldots, j_k)\big) = (0, 0, 0, \ldots, 0, i_\ell, i_{\ell+1}, \ldots, i_k)$. (That is, we replace i_m by 0 for every m such that $j_m = 0$.) It follows directly from the definition that H commutes with face and degeneracy operators and that its restrictions to the two ends of $\Delta[1]$ are as required. □

PROPOSITION 9.5.20. *If $\mathcal{M}$ is a simplicial model category, X is an object of $\mathcal{M}$, and $n \geq 0$, then the maps $X \otimes \Delta[0] \to X \otimes \Delta[n]$ and $X^{\Delta[n]} \to X^{\Delta[0]}$ induced by the inclusion of $\Delta[0]$ as the initial vertex of $\Delta[n]$ are weak equivalences.*

PROOF. Lemma 9.5.19 and Proposition 9.5.18 imply that these maps are simplicial homotopy equivalences, and so the result follows from Proposition 9.5.16. □

LEMMA 9.5.21. *If J is a generalized interval (see Definition 9.5.5) then the inclusion of $\Delta[0]$ into J at any vertex of J is a simplicial homotopy equivalence.*

PROOF. This follows from Lemma 9.5.19 by induction on the number of nondegenerate 1-simplices of J. □

LEMMA 9.5.22. *Let $\mathcal{M}$ be a simplicial model category and let J be a generalized interval (see Definition 9.5.5). If X is an object of $\mathcal{M}$, then*

(1) *a map $X \to X \otimes J$ induced by the inclusion of $\Delta[0]$ as a vertex of J is a weak equivalence, and*
(2) *a map $X^J \to X$ induced by the inclusion of $\Delta[0]$ as a vertex of J is a weak equivalence.*

PROOF. This follows from Lemma 9.5.21, Proposition 9.5.18, and Proposition 9.5.16. □

PROPOSITION 9.5.23. *Let $\mathcal{M}$ be a simplicial model category and let X and Y be objects of $\mathcal{M}$. If $f, g \colon X \to Y$ are maps that are simplicially homotopic, then they are both left homotopic and right homotopic.*

PROOF. If f and g are simplicially homotopic, then there is a generalized interval J and simplicial homotopies $H \colon X \otimes J \to Y$ and $H' \colon X \to Y^J$ from f to g (see Definition 9.5.7). The result now follows from Proposition 7.3.5 and Lemma 9.5.22. □

PROPOSITION 9.5.24. *Let $\mathcal{M}$ be a simplicial model category and let X and Y be objects of $\mathcal{M}$.*

(1) *If $g, h \colon X \to Y$ are simplicially homotopic then they are both left homotopic and right homotopic.*
(2) *If X is cofibrant and Y is fibrant, then the strict simplicial, simplicial, left, and right homotopy relations on the set of maps from X to Y coincide and are equivalence relations.*

PROOF. This follows from Proposition 9.5.23, Lemma 9.5.14, Proposition 7.4.7, and Theorem 7.4.9. □

9.6. Uniqueness of lifts

PROPOSITION 9.6.1. *Let $\mathcal{M}$ be a simplicial model category. If we have the solid arrow diagram*

$$\begin{array}{ccc} A & \longrightarrow & X \\ {\scriptstyle i}\big\downarrow & \overset{h}{\nearrow} & \big\downarrow{\scriptstyle p} \\ B & \longrightarrow & Y \end{array}$$

in $\mathcal{M}$ and if $i\colon A \to B$ has the homotopy left lifting property with respect to $p\colon X \to Y$ (see Definition 9.4.2), then there exists a map $h\colon B \to X$ making both triangles commute, and the map h is unique up to simplicial homotopy.

PROOF. This follows from Definition 9.4.2 and Proposition 9.5.3. □

COROLLARY 9.6.2. *Let $\mathcal{M}$ be a simplicial model category. If we have the solid arrow diagram*

$$\begin{array}{ccc} A & \xrightarrow{\ j\ } & C \\ {\scriptstyle i}\big\downarrow & \overset{h}{\nearrow} & \big\downarrow{\scriptstyle p} \\ B & \xrightarrow[\ q\ ]{} & D \end{array}$$

in $\mathcal{M}$ and if both i and j have the homotopy left lifting property with respect to each of p and q, then there exists a map $h\colon B \to C$, unique up to simplicial homotopy, such that $hi = j$ and $ph = q$, and any such map is a simplicial homotopy equivalence.

PROOF. This follows from Proposition 9.6.1. □

LEMMA 9.6.3. *If $\mathcal{M}$ is a simplicial model category, then an isomorphism in $\mathcal{M}$ has both the homotopy left lifting property and the homotopy right lifting property with respect to every map in $\mathcal{M}$.*

PROOF. This follows from the fact that an isomorphism induces an isomorphism of the simplicial set of maps from (or to) any fixed object. □

PROPOSITION 9.6.4. *Let $\mathcal{M}$ be a simplicial model category and let $g\colon X \to Y$ be a map in $\mathcal{M}$.*

(1) *If g has the homotopy left lifting property with respect to the maps from each of X and Y to the terminal object of the category, then g is the inclusion of a strong deformation retract, i.e., there is a map $r\colon Y \to X$ such that $rg = 1_X$ and $gr \overset{s}{\simeq} 1_Y$, where the simplicial homotopy (see Definition 9.5.7) is constant on X.*
(2) *If g has the homotopy right lifting property with respect to the maps from the initial object of the category to each of X and Y, then there is a map $s\colon Y \to X$ such that $gs = 1_Y$ and $sg \overset{s}{\simeq} 1_X$, where the simplicial homotopy (see Definition 9.5.7) lies over the identity map of Y.*

PROOF. We will prove part 1; the proof of part 2 is similar.

We have the solid arrow diagram

$$\begin{array}{ccc} X & = & X \\ {\scriptstyle g}\downarrow & \overset{r}{\nearrow} & \downarrow \\ Y & \longrightarrow & * \end{array}$$

(in which "$*$" is the terminal object of the category), and so Corollary 9.6.2 and Lemma 9.6.3 imply that there exists a map $r\colon Y \to X$ such that $rg = 1_X$. Thus, we can construct the solid arrow diagram

$$\begin{array}{ccc} X \otimes \Delta[1] \amalg_{X \otimes \partial\Delta[1]} Y \otimes \partial\Delta[1] & \longrightarrow & Y \\ \downarrow & \overset{s}{\nearrow} & \downarrow \\ Y \otimes \Delta[1] & \longrightarrow & * \end{array}$$

in which the top map is the composition $X \otimes \Delta[1] \to X \otimes \Delta[0] \approx X \xrightarrow{g} Y$ on $X \otimes \Delta[1]$ and $gr \amalg 1_Y$ on $Y \otimes \partial\Delta[1]$. Proposition 9.4.8 implies that the vertical map on the left has the homotopy left lifting property with respect to the vertical map on the right, and so Proposition 9.4.3 implies that the dotted arrow $s\colon Y \otimes \Delta[1] \to Y$ exists. □

COROLLARY 9.6.5. *Let $\mathcal{M}$ be a simplicial model category and let $g\colon X \to Y$ be a map in $\mathcal{M}$.*

(1) *If both X and Y are fibrant and g is a trivial cofibration, then g is a simplicial homotopy equivalence. In particular, g is the inclusion of a strong deformation retract.*
(2) *If both X and Y are cofibrant and g is a trivial fibration, then g is a simplicial homotopy equivalence. In particular, g has a right inverse that is a simplicial homotopy inverse.*

PROOF. This follows from Proposition 9.6.4. □

9.6.6. Weak equivalences of simplicial mapping spaces. It should be noted that none of the results of this section make any assumption that any object is cofibrant or fibrant.

PROPOSITION 9.6.7. *Let $\mathcal{M}$ be a simplicial model category and let W, X, and Y be objects of $\mathcal{M}$. If $g, h\colon X \to Y$ are simplicially homotopic maps, then $g_* \overset{s}{\simeq} h_*\colon \mathrm{Map}(W,X) \to \mathrm{Map}(W,Y)$ and $g^* \overset{s}{\simeq} h^*\colon \mathrm{Map}(Y,W) \to \mathrm{Map}(X,W)$.*

PROOF. Let $X \to Y^J$ be a simplicial homotopy from g to h (where J is a generalized interval; see Definition 9.5.7). This induces the map of simplicial sets $\mathrm{Map}(W,X) \to \mathrm{Map}(W,Y^J)$, which corresponds to a map $\mathrm{Map}(W,X) \to \mathrm{Map}(W \otimes J, Y)$, which corresponds to a map $\mathrm{Map}(W,X) \to \mathrm{Map}\big(J, \mathrm{Map}(W,Y)\big)$, which corresponds to a map $\mathrm{Map}(W,X) \otimes J \to \mathrm{Map}(W,Y)$, which is a simplicial homotopy from g_* to h_*.

The second assertion is proved similarly, starting with a simplicial homotopy $X \otimes J \to Y$. □

COROLLARY 9.6.8. *Let $\mathcal{M}$ be a simplicial model category and let X and Y be objects of $\mathcal{M}$. If $g\colon X \to Y$ is a simplicial homotopy equivalence, then for any* *object W of $\mathcal{M}$ the maps $g_*\colon \operatorname{Map}(W,X) \to \operatorname{Map}(W,Y)$ and $g^*\colon \operatorname{Map}(Y,W) \to \operatorname{Map}(X,W)$ are simplicial homotopy equivalences of simplicial sets.*

PROOF. This follows from Proposition 9.6.7. □

PROPOSITION 9.6.9. *Let $\mathcal{M}$ be a simplicial model category. If $g\colon X \to Y$ is a map in $\mathcal{M}$, then g is a simplicial homotopy equivalence if either of the following two conditions is satisfied:*

(1) *The map g induces isomorphisms of the sets of simplicial homotopy classes of maps $g_*\colon [X,X] \approx [X,Y]$ and $g_*\colon [Y,X] \approx [Y,Y]$.*
(2) *The map g induces isomorphisms of the sets of simplicial homotopy classes of maps $g^*\colon [Y,X] \approx [X,X]$ and $g^*\colon [Y,Y] \approx [X,Y]$.*

PROOF. We will prove this using condition 1; the proof using condition 2 is similar.

The isomorphism $g_*\colon [Y,X] \approx [Y,Y]$ implies that there is a map $h\colon Y \to X$ such that $gh \overset{s}{\simeq} 1_Y$. Corollary 9.5.4 and the isomorphism $g_*\colon [X,X] \approx [X,Y]$ now imply that h induces an isomorphism $h_*\colon [X,Y] \approx [X,X]$, and so there is a map $k\colon X \to Y$ such that $hk \overset{s}{\simeq} 1_X$. Thus, $g = g1_X \overset{s}{\simeq} ghk \overset{s}{\simeq} 1_Y k = k$, and so g is a simplicial homotopy equivalence whose inverse is h. □

PROPOSITION 9.6.10. *Let $\mathcal{M}$ be a simplicial model category. If $g\colon X \to Y$ is a map in $\mathcal{M}$, then g is a simplicial homotopy equivalence if either of the following two conditions is satisfied:*

(1) *The map g induces weak equivalences of simplicial sets $g_*\colon \operatorname{Map}(X,X) \cong \operatorname{Map}(X,Y)$ and $g_*\colon \operatorname{Map}(Y,X) \cong \operatorname{Map}(Y,Y)$.*
(2) *The map g induces weak equivalences of simplicial sets $g^*\colon \operatorname{Map}(Y,X) \cong \operatorname{Map}(X,X)$ and $g^*\colon \operatorname{Map}(Y,Y) \cong \operatorname{Map}(X,Y)$.*

PROOF. This follows from Proposition 9.6.9 and Proposition 9.5.10. □

9.7. Detecting weak equivalences

PROPOSITION 9.7.1. *Let $\mathcal{M}$ be a simplicial model category. If $g\colon X \to Y$ is a map in $\mathcal{M}$, then g is a weak equivalence if either of the following two conditions is satisfied:*

(1) *For every fibrant object Z the map of function spaces $g^*\colon \operatorname{Map}(Y,Z) \to \operatorname{Map}(X,Z)$ is a weak equivalence of simplicial sets.*
(2) *For every cofibrant object W the map of function spaces $g_*\colon \operatorname{Map}(W,X) \to \operatorname{Map}(W,Y)$ is a weak equivalence of simplicial sets.*

PROOF. We will prove part 1; the proof of part 2 is dual.

Choose cofibrant fibrant approximations (see Definition 8.1.2) $j_X\colon X \to \widehat{X}$ and $j_Y\colon Y \to \widehat{Y}$ and a fibrant approximation $\hat{g}\colon \widehat{X} \to \widehat{Y}$ to g (see Definition 8.1.22). If Z is a fibrant object, then we have the commutative square

$$\begin{array}{ccc} \operatorname{Map}(\widehat{Y},Z) & \xrightarrow{\hat{g}^*} & \operatorname{Map}(\widehat{X},Z) \\ {\scriptstyle (j_Y)^*}\downarrow & & \downarrow{\scriptstyle (j_X)^*} \\ \operatorname{Map}(Y,Z) & \xrightarrow[g^*]{} & \operatorname{Map}(X,Z) \end{array}$$

in which all the maps except $\hat{g}^*$ are weak equivalences of simplicial sets (see Proposition 9.3.2). This implies that $\hat{g}^*$ is also a weak equivalence, and so Proposition 9.6.10 implies implies that $\hat{g}$ is a simplicial homotopy equivalence. Thus, $\hat{g}$ is a weak equivalence, and so the "two out of three" property implies that g is a weak equivalence. □

PROPOSITION 9.7.2. *Let $\mathcal{M}$ be a simplicial model category, let $g\colon X \to Y$ be a map in $\mathcal{M}$, and let W be an object of $\mathcal{M}$.*

(1) *If W is cofibrant and there is a fibrant approximation $\hat{g}\colon \widehat{X} \to \widehat{Y}$ to g (see Definition 8.1.22) such that the induced map of simplicial sets $\hat{g}_*\colon \mathrm{Map}(W, \widehat{X}) \to \mathrm{Map}(W, \widehat{Y})$ is a weak equivalence, then for any other fibrant approximation $\hat{g}'\colon \widehat{X}' \to \widehat{Y}'$ to g the induced map of simplicial sets $\hat{g}'_*\colon \mathrm{Map}(W, \widehat{X}') \to \mathrm{Map}(W, \widehat{Y}')$ is a weak equivalence.*
(2) *If W is fibrant and there is a cofibrant approximation $\tilde{g}\colon \widetilde{X} \to \widetilde{Y}$ to g (see Definition 8.1.22) such that the induced map of simplicial sets $\tilde{g}^*\colon \mathrm{Map}(\widetilde{Y}, W) \to \mathrm{Map}(\widetilde{X}, W)$ is a weak equivalence, then for any other cofibrant approximation $\tilde{g}'\colon \widetilde{X}' \to \widetilde{Y}'$ to g the induced map of simplicial sets $(\tilde{g}')^*\colon \mathrm{Map}(\widetilde{Y}', W) \to \mathrm{Map}(\widetilde{X}', W)$ is a weak equivalence.*

PROOF. This follows from Proposition 8.1.24 and Proposition 9.3.2. □

PROPOSITION 9.7.3. *Let $\mathcal{M}$ be a simplicial model category, let $f\colon X \to Y$ be a map in $\mathcal{M}$, and let W be an object of $\mathcal{M}$.*

(1) *If X and Y are fibrant and there is a cofibrant approximation $\widetilde{W} \to W$ to W such that the induced map of simplicial sets $f_*\colon \mathrm{Map}(\widetilde{W}, X) \to \mathrm{Map}(\widetilde{W}, Y)$ is a weak equivalence, then for any other cofibrant approximation $\widetilde{W}' \to W$ to W the induced map of simplicial sets $f_*\colon \mathrm{Map}(\widetilde{W}', X) \to \mathrm{Map}(\widetilde{W}', Y)$ is a weak equivalence.*
(2) *If X and Y are cofibrant and there is a fibrant approximation $W \to \widehat{W}$ to W such that the induced map of simplicial sets $f^*\colon \mathrm{Map}(Y, \widehat{W}) \to \mathrm{Map}(X, \widehat{W})$ is a weak equivalence, then for any other fibrant approximation $W \to \widehat{W}'$ to W the induced map of simplicial sets $f^*\colon \mathrm{Map}(Y, \widehat{W}') \to \mathrm{Map}(X, \widehat{W}')$ is a weak equivalence.*

PROOF. We will prove part 1; the proof of part 2 is dual.

Let $\overline{W} \to W$ be a fibrant cofibrant approximation to W (see Proposition 8.1.17). There are maps of cofibrant approximations (see Definition 8.1.4) $\widetilde{W} \to \overline{W}$ and $\widetilde{W}' \to \overline{W}$, both of which are weak equivalences (see Proposition 8.1.7). Thus, we have the diagram

$$\begin{array}{ccccc} \mathrm{Map}(\widetilde{W}', X) & \xleftarrow{\cong} & \mathrm{Map}(\overline{W}, X) & \xrightarrow{\cong} & \mathrm{Map}(\widetilde{W}, X) \\ \downarrow & & \downarrow & & \downarrow \\ \mathrm{Map}(\widetilde{W}', Y) & \xleftarrow[\cong]{} & \mathrm{Map}(\overline{W}, Y) & \xrightarrow[\cong]{} & \mathrm{Map}(\widetilde{W}, Y) \end{array}$$

and Corollary 9.3.3 implies that all the horizontal maps are weak equivalences. □

THEOREM 9.7.4. *If $g\colon X \to Y$ is a map in a simplicial model category, then the following are equivalent:*

1. *The map g is a weak equivalence.*
2. *For some fibrant approximation $\hat{g}\colon \widehat{X} \to \widehat{Y}$ to g (see Definition 8.1.22) and every cofibrant object W the map of simplicial sets $\hat{g}_*\colon \mathrm{Map}(W, \widehat{X}) \to \mathrm{Map}(W, \widehat{Y})$ is a weak equivalence.*
3. *For every fibrant approximation $\hat{g}\colon \widehat{X} \to \widehat{Y}$ to g and every cofibrant object W the map of simplicial sets $\hat{g}_*\colon \mathrm{Map}(W, \widehat{X}) \to \mathrm{Map}(W, \widehat{Y})$ is a weak equivalence.*
4. *For some cofibrant approximation $\tilde{g}\colon \widetilde{X} \to \widetilde{Y}$ to g (see Definition 8.1.22) and every fibrant object Z the map of simplicial sets $\tilde{g}^*\colon \mathrm{Map}(\widetilde{Y}, Z) \to \mathrm{Map}(\widetilde{X}, Z)$ is a weak equivalence.*
5. *For every cofibrant approximation $\tilde{g}\colon \widetilde{X} \to \widetilde{Y}$ to g and every fibrant object Z the map of simplicial sets $\tilde{g}^*\colon \mathrm{Map}(\widetilde{Y}, Z) \to \mathrm{Map}(\widetilde{X}, Z)$ is a weak equivalence.*

PROOF. Proposition 9.7.2 implies that 2 is equivalent to 3 and that 4 is equivalent to 5. Proposition 9.7.1 implies that any of 2, 3, 4, or 5 implies 1, and Corollary 9.3.3 implies that 1 implies both 2 and 4. □

COROLLARY 9.7.5. *Let $\mathcal{M}$ be a simplicial model category and let $g\colon X \to Y$ be a map in $\mathcal{M}$.*

1. *If X and Y are fibrant, then g is a weak equivalence if and only if for every cofibrant object W in $\mathcal{M}$ the map $g_*\colon \mathrm{Map}(W, X) \to \mathrm{Map}(W, Y)$ is a weak equivalence of simplicial sets.*
2. *If X and Y are cofibrant, then g is a weak equivalence if and only if for every fibrant object Z in $\mathcal{M}$ the map $g^*\colon \mathrm{Map}(Y, Z) \to \mathrm{Map}(X, Z)$ is a weak equivalence of simplicial sets.*

PROOF. This follows from Theorem 9.7.4. □

9.8. Simplicial functors

DEFINITION 9.8.1. If $\mathcal{M}$ and $\mathcal{N}$ are simplicial model categories, then a *simplicial functor* F from $\mathcal{M}$ to $\mathcal{N}$ consists of

1. an object $\mathrm{F}(X)$ of $\mathcal{N}$ for every object X of $\mathcal{M}$ and
2. for every two objects X and Y of $\mathcal{M}$ a map of simplicial sets
$$\mathrm{F}_{X,Y}\colon \mathrm{Map}(X, Y) \to \mathrm{Map}(\mathrm{F}X, \mathrm{F}Y)$$
such that for all objects X, Y, and Z of $\mathcal{M}$ the following two diagrams commute:
$$\begin{array}{ccc} \mathrm{Map}(Y,Z)\times \mathrm{Map}(X,Y) & \xrightarrow{c_{X,Y,Z}} & \mathrm{Map}(X,Z) \\ \big\downarrow{\scriptstyle \mathrm{F}_{Y,Z}\times \mathrm{F}_{X,Y}} & & \big\downarrow{\scriptstyle \mathrm{F}_{X,Z}} \\ \mathrm{Map}(\mathrm{F}Y,\mathrm{F}Z)\times \mathrm{Map}(\mathrm{F}X,\mathrm{F}Y) & \xrightarrow[c_{\mathrm{F}X,\mathrm{F}Y,\mathrm{F}Z}]{} & \mathrm{Map}(\mathrm{F}X,\mathrm{F}Z) \end{array}$$

$$\begin{array}{ccc} * & \xrightarrow{i_X} & \mathrm{Map}(X,X) \\ & {\scriptstyle i_{FX}} \searrow \quad \swarrow {\scriptstyle F_{X,X}} & \\ & \mathrm{Map}(FX,FX) & \end{array}$$

If $\mathcal{M}$ and $\mathcal{N}$ are simplicial model categories and $F\colon \mathcal{M} \to \mathcal{N}$ is a functor between the underlying categories of $\mathcal{M}$ and $\mathcal{N}$, then we often want to consider whether F can be extended to a simplicial functor, i.e., whether the definition of F can be extended to define a natural map of simplicial sets

$$\mathrm{Map}(X,Y) \to \mathrm{Map}(FX,FY) \tag{9.8.2}$$

that is compatible with composition and with the isomorphisms $\mathrm{Map}(X,Y)_0 \approx \mathcal{M}(X,Y)$ and $\mathrm{Map}(FX,FY)_0 \approx \mathcal{N}(FX,FY)$.

If F is to be a simplicial functor, then given an n-simplex in $\mathrm{Map}(X,Y)$, i.e., a map $\alpha\colon X \otimes \Delta[n] \to Y$ (see Proposition 9.1.9), we must assign to it an n-simplex of $\mathrm{Map}(FX,FY)$, i.e., a map $FX \otimes \Delta[n] \to FY$. We can attempt to use $F(\alpha)\colon F(X \otimes \Delta[n]) \to FY$, but then we need a map

$$\sigma\colon FX \otimes \Delta[n] \to F(X \otimes \Delta[n])$$

to compose with $F(\alpha)$. If we ensure that σ yields a natural isomorphism $\sigma\colon FX \otimes \Delta[0] \approx F(X \otimes \Delta[0])$ that commutes with the natural isomorphisms $FX \otimes \Delta[0] \approx FX$ and $X \otimes \Delta[0] \approx X$ (see Proposition 9.1.10), then the map (9.8.2) would be an extension of F on $\mathrm{Map}(X,Y)_0 \approx \mathcal{M}(X,Y)$. This would allow us to define the map (9.8.2) for each pair of objects X and Y, but even if we require that σ be natural in both X and $\Delta[n]$, we still could not be sure that the function (9.8.2) commutes with composition of functions, i.e., that the diagram

$$\begin{array}{ccc} \mathrm{Map}(X,Y) \times \mathrm{Map}(Y,Z) & \longrightarrow & \mathrm{Map}(X,Z) \\ \downarrow & & \downarrow \\ \mathrm{Map}(FX,FY) \times \mathrm{Map}(FY,FZ) & \longrightarrow & \mathrm{Map}(FX,FZ) \end{array}$$

commutes. For this, σ must have an additional property.

Given n-simplices $\alpha \in \mathrm{Map}(X,Y)_n$ and $\beta \in \mathrm{Map}(Y,Z)_n$, i.e., functions $\alpha\colon X \otimes \Delta[n] \to Y$ and $\beta\colon Y \otimes \Delta[n] \to Z$, their composition in $\mathrm{Map}(X,Z)_n$ is the composition

$$X \otimes \Delta[n] \xrightarrow{1 \otimes D} X \otimes (\Delta[n] \times \Delta[n]) \approx (X \otimes \Delta[n]) \otimes \Delta[n] \xrightarrow{\alpha \otimes 1} Y \otimes \Delta[n] \xrightarrow{\beta} Z$$

(where $D\colon \Delta[n] \to \Delta[n] \times \Delta[n]$ is the diagonal map). If we apply F and compose with the natural transformation σ, then we get the n-simplex

$$\begin{aligned} FX \otimes \Delta[n] &\xrightarrow{\sigma} F(X \otimes \Delta[n]) \\ &\xrightarrow{F(1 \otimes D)} F\bigl(X \otimes (\Delta[n] \times \Delta[n])\bigr) \approx F\bigl((X \otimes \Delta[n]) \otimes \Delta[n]\bigr) \\ &\xrightarrow{F(\alpha \otimes 1)} F(Y \otimes \Delta[n]) \xrightarrow{F(\beta)} F(Z) \end{aligned}$$

of Map(FX, FZ). Since σ is natural, this can also be written as the composition

$$(9.8.3)\quad \mathrm{F}X \otimes \Delta[n] \xrightarrow{1\otimes D} \mathrm{F}X \otimes (\Delta[n] \times \Delta[n]) \xrightarrow{\sigma} \mathrm{F}\big(X \otimes (\Delta[n] \times \Delta[n])\big) \approx \mathrm{F}\big((X \otimes \Delta[n]) \otimes \Delta[n]\big) \xrightarrow{\mathrm{F}(\alpha\otimes 1)} \mathrm{F}(Y \otimes \Delta[n]) \xrightarrow{\mathrm{F}(\beta)} \mathrm{F}(Z)$$

If we start with the same n-simplices α and β, apply F to each, and compose each with the natural transformation σ, then we get the pair of simplices

$$\mathrm{F}X \otimes \Delta[n] \xrightarrow{\sigma} \mathrm{F}(X \otimes \Delta[n]) \xrightarrow{\mathrm{F}(\alpha)} \mathrm{F}Y$$
$$\mathrm{F}Y \otimes \Delta[n] \xrightarrow{\sigma} \mathrm{F}(Y \otimes \Delta[n]) \xrightarrow{\mathrm{F}(\beta)} \mathrm{F}Z$$

in $\mathrm{Map}(\mathrm{F}X, \mathrm{F}Y)_n \times \mathrm{Map}(\mathrm{F}Y, \mathrm{F}Z)_n$. If we compose these, then we get the element

$$\mathrm{F}X \otimes \Delta[n] \xrightarrow{1\otimes D} \mathrm{F}X \otimes (\Delta[n] \times \Delta[n]) \approx (\mathrm{F}X \otimes \Delta[n]) \otimes \Delta[n] \xrightarrow{\sigma\otimes 1} \mathrm{F}(X \otimes \Delta[n]) \otimes \Delta[n] \xrightarrow{\mathrm{F}(\alpha)\otimes 1} \mathrm{F}Y \otimes \Delta[n] \xrightarrow{\sigma} \mathrm{F}(Y \otimes \Delta[n]) \xrightarrow{\mathrm{F}(\beta)} \mathrm{F}Z$$

of $\mathrm{Map}(\mathrm{F}X, \mathrm{F}Z)_n$. Since σ is natural, this can also be written as the composition

$$(9.8.4)\quad \mathrm{F}X \otimes \Delta[n] \xrightarrow{1\otimes D} \mathrm{F}X \otimes (\Delta[n] \times \Delta[n]) \approx (\mathrm{F}X \otimes \Delta[n]) \otimes \Delta[n] \xrightarrow{\sigma\otimes 1} \mathrm{F}(X \otimes \Delta[n]) \otimes \Delta[n] \xrightarrow{\sigma} \mathrm{F}\big((X \otimes \Delta[n]) \otimes \Delta[n]\big) \xrightarrow{\mathrm{F}(\alpha\otimes 1)} \mathrm{F}(Y \otimes \Delta[n]) \xrightarrow{\mathrm{F}(\beta)} \mathrm{F}Z$$

Since we want the composition (9.8.3) to equal the composition (9.8.4), we must require that the diagram

$$\begin{array}{ccc}
\mathrm{F}X \otimes (\Delta[n] \times \Delta[n]) & \xrightarrow{\approx} & (\mathrm{F}X \otimes \Delta[n]) \otimes \Delta[n] \\
 & & \downarrow{\scriptstyle \sigma\otimes 1} \\
\downarrow{\scriptstyle \sigma} & & \mathrm{F}(X \otimes \Delta[n]) \otimes \Delta[n] \\
 & & \downarrow{\scriptstyle \sigma} \\
\mathrm{F}\big(X \otimes (\Delta[n] \times \Delta[n])\big) & \xrightarrow[\approx]{} & \mathrm{F}\big((X \otimes \Delta[n]) \otimes \Delta[n]\big)
\end{array}$$

commute.

This leads us to the following theorem.

THEOREM 9.8.5. *Let* $\mathcal{M}$ *and* $\mathcal{N}$ *be simplicial model categories. A functor* $\mathrm{F}\colon \mathcal{M} \to \mathcal{N}$ *can be extended to a simplicial functor if and only if for every finite simplicial set* K *and object* X *of* $\mathcal{M}$ *there is a map* $\sigma\colon \mathrm{F}X \otimes K \to \mathrm{F}(X \otimes K)$, *natural in both* X *and* K, *such that*

(1) *for every object X of $\mathcal{M}$, σ defines an isomorphism $\sigma\colon (\mathrm{F}X)\otimes\Delta[0] \approx \mathrm{F}(X\otimes\Delta[0])$ such that the triangle*

$$\begin{array}{ccc} (\mathrm{F}X)\otimes\Delta[0] & \xrightarrow{\approx} & \mathrm{F}(X\otimes\Delta[0]) \\ & {\scriptstyle\approx}\searrow \quad \swarrow{\scriptstyle\approx} & \\ & \mathrm{F}X & \end{array}$$

commutes, and

(2) *for every object X of $\mathcal{M}$ and finite simplicial sets K and L, the diagram*

$$\begin{array}{ccc} \mathrm{F}X\otimes(K\times L) & \xrightarrow{\approx} & (\mathrm{F}X\otimes K)\otimes L \\ & & \downarrow{\scriptstyle\sigma\otimes 1} \\ \downarrow{\scriptstyle\sigma} & & \mathrm{F}(X\otimes K)\otimes L \\ & & \downarrow{\scriptstyle\sigma} \\ \mathrm{F}\bigl(X\otimes(K\times L)\bigr) & \xrightarrow[\approx]{} & \mathrm{F}\bigl((X\otimes K)\otimes L\bigr) \end{array}$$

commutes.

PROOF. We have isomorphisms

$$\mathrm{Map}(X,Y)_n \approx \mathrm{SS}\bigl(\Delta[n],\mathrm{Map}(X,Y)\bigr) \approx \mathcal{M}(X\otimes\Delta[n],Y)$$

that are natural in X, Y, and $\Delta[n]$, and so we can define $\mathrm{F}\colon \mathrm{Map}(X,Y)_n \to \mathrm{Map}(\mathrm{F}X,\mathrm{F}Y)_n$ as the composition

$$\mathcal{M}(X\otimes\Delta[n],Y) \xrightarrow{\mathrm{F}(-,-)} \mathcal{N}\bigl(\mathrm{F}(X\otimes\Delta[n]),\mathrm{F}Y\bigr) \xrightarrow{\sigma^*} \mathcal{N}(\mathrm{F}X\otimes\Delta[n],\mathrm{F}Y)\ .$$

The discussion preceding the statement of the theorem explains why this yields a simplicial functor.

Conversely, if $\mathrm{F}\colon\mathcal{M}\to\mathcal{N}$ is simplicial, then we can define σ as the map corresponding to the composition

$$K \to \mathrm{Map}(X,X\otimes K) \xrightarrow{\mathrm{F}(-,-)} \mathrm{Map}\bigl(\mathrm{F}X,\mathrm{F}(X\otimes K)\bigr)$$

(where the first map above is adjoint to the identity map of $X\otimes K$) under the isomorphism

$$\mathrm{SS}\bigl(K,\mathrm{Map}(\mathrm{F}X,\mathrm{F}(X\otimes K))\bigr) \approx \mathcal{N}\bigl(\mathrm{F}X\otimes K,\mathrm{F}(X\otimes K)\bigr)\ .$$

□

EXAMPLE 9.8.6. Let $\mathcal{M}$ be a simplicial model category. If W is an object of $\mathcal{M}$, then the functor $\mathrm{Map}(W,-)\colon\mathcal{M}\to\mathrm{SS}$ is simplicial. In this case, for $(f,k)\in\bigl(\mathrm{Map}(W,X)\otimes K\bigr)_n$ we have $\sigma(f,k) = f\times\bar{k}$, where $\bar{k}$ is the composition of the projection $W\otimes\Delta[n]\to\Delta[n]$ with the map $\Delta[n]\to K$ that takes the nondegenerate n-simplex of $\Delta[n]$ to k.

EXAMPLE 9.8.7 (Counterexample to continuity). If A is any nonempty space in Top, we define a functor $W_A\colon \mathrm{Top}\to\mathrm{Top}$ by $W_AX = \coprod_{A\to X} A$, that is, we take the disjoint union of one copy of A for each continuous function $g\colon A\to X$. This defines a functor in which the copy of A corresponding to g as above maps under $W_A(f)\colon W_AX\to W_AY$ by the identity map to the copy corresponding to $f\circ g$, but W_A cannot be extended to a simplicial functor. To see this, take $X = A$ and $Y = A \times I$. The simplicial set $\mathrm{Map}(X, Y) = \mathrm{Map}(A, A \times I)$ has vertices (i.e., maps $A \to A \times I$) the inclusions i_0 and i_1 (where $i_0(a) = (a, 0)$ and $i_1(a) = (a, 1)$), and these vertices of $\mathrm{Map}(A, A \times I)$ are connected by a 1-simplex $A \times \Delta[1] \to A \times I$ of $\mathrm{Map}(A, A \times I)$. The functions $W_A(i_0)$ and $W_A(i_1)$, however, take each point of $W_A A$ into different components of $W_A(A \times I)$, and so there can be no 1-simplex of $\mathrm{Map}\big(W_A A, W_A(A \times I)\big)$ connecting these vertices.

EXAMPLE 9.8.8. If we change Example 9.8.7 slightly, we can construct a functor that *is* continuous. Define W_A^c by $W_A^c X = X^A \times A$ (where X^A is the compactly generated topological space of continuous functions $A \to X$). We have a natural transformation $W_A \to W_A^c$ such that $W_A X \to W_A^c X$ is always a continuous bijection, but it is not, in general, a homeomorphism.

PROPOSITION 9.8.9. *Let $\mathcal{M}$ and $\mathcal{N}$ be simplicial model categories, let $\mathcal{C}$ be a small category, and let $\boldsymbol{X}$ be a $\mathcal{C}$-diagram of functors $\mathcal{M} \to \mathcal{N}$ and natural transformations between them. If for each $\alpha \in \mathcal{C}$ there is a map σ_α as in Theorem 9.8.5 that is natural in α and that extends $\boldsymbol{X}_\alpha$ to a simplicial functor, then there is a map σ that extends $\mathrm{colim}_{\alpha \in \mathcal{C}} \boldsymbol{X}_\alpha$ to a simplicial functor.*

PROOF. Let $\sigma = \mathrm{colim}_{\alpha \in \mathcal{C}} \sigma_\alpha$. □

CHAPTER 10

Ordinals, Cardinals, and Transfinite Composition

The main subject of this chapter is a rigorous treatment of the idea of an *infinitely long sequence of maps*, and of the *composition* of such a sequence. These are the ideas used in the *small object argument* (see Proposition 10.5.16), which is a fundamental method of constructing factorizations of maps needed both to establish model category structures (see Theorem 11.3.1) and to construct localization functors (see Section 1.3 and Section 4.3).

If λ is an ordinal, then a λ-*sequence* X in a category $\mathcal{C}$ is a "sequence" of maps in $\mathcal{C}$ indexed by the ordinal λ (see Definition 10.2.1); that is, there are objects X_β of $\mathcal{C}$ for $\beta < \lambda$, for every ordinal β such that $\beta + 1 < \lambda$ there is a map $X_\beta \to X_{\beta+1}$ in $\mathcal{C}$, and we require that if β is a limit ordinal then $X_\beta = \operatorname{colim}_{\alpha<\beta} X_\alpha$. We define the *composition* of the λ-sequence to be the natural map $X_0 \to \operatorname{colim}_{\beta<\lambda} X_\beta$. A composition of a λ-sequence is called a *transfinite composition.*

An object W in a category $\mathcal{C}$ is said to be *small* with respect to a class $\mathcal{D}$ of maps in $\mathcal{C}$ if for every large enough regular cardinal (see Definition 10.1.11) λ and every λ-sequence all of whose maps are elements of $\mathcal{D}$, every map $W \to \operatorname{colim}_{\beta<\lambda} X_\beta$ from W to the colimit of the sequence factors essentially uniquely through some earlier stage X_β in the sequence (see Definition 10.4.1). The *small object argument* is a method of factoring maps into factors that have appropriate lifting properties, and it can be done when the domains of a set of maps are small with respect to the class of pushouts of those maps (see Definition 10.5.15 and Proposition 10.5.16).

We discuss ordinals and cardinals in Section 10.1, transfinite compositions in Section 10.2, and the lifting properties of transfinite compositions in Section 10.3. We discuss small objects in Section 10.4.

Given a set I of maps in a category $\mathcal{C}$, we define a *relative I-cell complex* in Section 10.5 to be a map that can be constructed as a transfinite composition of pushouts of elements of I (see Definition 10.5.8). (If $\mathcal{C} = \mathrm{Top}$, the category of topological spaces, and I is the set of inclusions $\mathrm{S}^{n-1} \to \mathrm{D}^n$ for $n \geq 0$, then a relative I-cell complex is a map that can be constructed as a transfinite composition of maps that attach a single cell along a map of its boundary sphere.) This allows us to describe the *small object argument* in Proposition 10.5.16. We discuss subcomplexes of relative I-cell complexes in Section 10.6. In Section 10.7 we discuss cell complexes in the category of topological spaces. This class of spaces includes the class of CW-complexes, but it also contains spaces built by attaching cells via maps that do not factor through a subcomplex of lower dimensional cells.

In Section 10.8 we discuss *compactness*, which is a variation on the notion of smallness: An object W is *compact* relative to a set I of maps if every map from W to an I-cell complex factors through a subcomplex of bounded size (see Definition 10.8.1). If relative I-cell complexes are monomorphisms, then compactness implies smallness (see Proposition 10.8.7). In Section 10.9 we discuss what it means for a map to be an *effective monomorphism* (see Definition 10.9.1). In the category of sets the effective monomorphisms are the monomorphisms, but the two notions do not agree in general. We will use this to define *cellular model categories* in Chapter 12 (see Definition 12.1.1).

10.1. Ordinals and cardinals

For a thorough discussion of the definitions and basic properties of ordinals and cardinals, see, e.g., [**29**, Chapter II], [**17**, Chapters 4 and 5], or [**40**, Chapter 6].

10.1.1. Ordinals.

DEFINITION 10.1.2.

(1) A *preordered set* is a set with a relation that is reflexive and transitive.
(2) A *partially ordered set* is a preordered set in which the relation is also antisymmetric.
(3) A *totally ordered set* is a partially ordered set in which every pair of elements is comparable.
(4) A *well ordered set* is a totally ordered set in which every nonempty subset has a first element.

We adopt the definition of ordinals that arranges it so that an ordinal is the well ordered set of all lesser ordinals, and every well ordered set is isomorphic to a unique ordinal (see, e.g., [**29**, Chapter II], [**17**, page 47], or [**40**, page 202]). Thus, the union of a set of ordinals is an ordinal, and it is the least upper bound of the set.

REMARK 10.1.3. We will often consider a preordered set to be a small category with objects equal to the elements of the set and a single morphism from the object s to the object t if $s \le t$.

DEFINITION 10.1.4. If S is a totally ordered set and T is a subset of S, then T will be called *right cofinal* (or *terminal*) in S if for every $s \in S$ there exists $t \in T$ such that $s \le t$.

THEOREM 10.1.5. *If $\mathcal{C}$ is a cocomplete category, S is a totally ordered set, T is a right cofinal subset of S, and $X\colon S \to \mathcal{C}$ is a functor, then the natural map $\operatorname{colim}_T X \to \operatorname{colim}_S X$ is an isomorphism.*

PROOF. We will construct a map $\operatorname{colim}_S X \to \operatorname{colim}_T X$ that is an inverse to the natural map $\operatorname{colim}_T X \to \operatorname{colim}_S X$. For every $s \in S$ we choose an element t of T such that $s \le t$ and define a map $X_s \to \operatorname{colim}_T X$ as the composition $X_s \to X_t \to \operatorname{colim}_T X$. If we choose a different element t' of T such that $s \le t'$ then either $t \le t'$ or $t' \le t$, and so our map $X_s \to \operatorname{colim}_T X$ is independent of the choice of the element t. Similarly, if $s' \in S$ is such that $s \le s'$, then for $t \in T$ satisfying $s' \le t$ the composition $X_s \to X_{s'} \to X_t \to \operatorname{colim}_T X$ equals the composition $X_s \to X_t \to \operatorname{colim}_T X$, and so the maps $X_s \to \operatorname{colim}_T X$ combine to define a map $\operatorname{colim}_S X \to \operatorname{colim}_T X$.

If $s \in S$, then the composition $X_s \to \operatorname{colim}_S X \to \operatorname{colim}_T X \to \operatorname{colim}_S X$ equals the map $X_s \to \operatorname{colim}_S X$, and so the composition $\operatorname{colim}_S X \to \operatorname{colim}_T X \to \operatorname{colim}_S X$ is the identity map. Similarly, the composition $\operatorname{colim}_T X \to \operatorname{colim}_S X \to \operatorname{colim}_T X$ is the identity map. □

PROPOSITION 10.1.6. *If S is a totally ordered set, then there is a right cofinal subset T of S that is well ordered.*

PROOF. We will prove the proposition by considering the set of well ordered subsets of S. We will show that this set has a maximal element, and that a maximal element must be right cofinal in S.

Let U be the set of pairs $(\lambda, f\colon \lambda \to S)$ where λ is an ordinal and f is a one to one order preserving function. We define a preorder on U by defining $(\lambda, f) \le (\kappa, g)$ if $\lambda \le \kappa$ and $f = g|_\lambda$. If $U' \subset U$ is a chain (i.e., a totally ordered subset of U), let $\lambda = \bigcup_{(\lambda_u, f_u)\in U'} \lambda_u$ and define $f\colon \lambda \to S$ to be the colimit of the f_u for $(\lambda_u, f_u) \in U'$. The pair (λ, f) is an element of U, and it is an upper bound for the chain. Thus, Zorn's lemma implies that U has a maximal element, and it remains only to show that a maximal element of U must be right cofinal.

If (λ_m, f_m) is a maximal element of U and the image of $f_m\colon \lambda_m \to S$ is not right cofinal, then there is an element s of S such that $f_m(\beta) < s$ for all $\beta \in \lambda_m$. Thus, we can define $g\colon (\lambda_m + 1) \to S$ by extending f_m to include s in its image. This would imply that (λ_m, f_m) was not a maximal element of U, and so the image of $f_m\colon \lambda_m \to S$ must actually be a right cofinal well ordered subset of S. □

10.1.7. Cardinals.

DEFINITION 10.1.8. A *cardinal* is an ordinal that is of greater cardinality than any lesser ordinal.

DEFINITION 10.1.9. If X is a set, then the *cardinal of* X is the unique cardinal whose underlying set has a bijection with X.

DEFINITION 10.1.10. If γ is a cardinal, then by $\mathrm{Succ}(\gamma)$ we will mean the *successor* of γ, i.e., the first cardinal greater then γ.

DEFINITION 10.1.11. A cardinal γ is *regular* if, whenever A is a set whose cardinal is less than γ and for every $a \in A$ there is a set S_a whose cardinal is less than γ, the cardinal of the set $\bigcup_{a\in A} S_a$ is less than γ.

EXAMPLE 10.1.12. The countable cardinal $\aleph_0$ is a regular cardinal. This is just the statement that a finite union of finite sets is finite.

PROPOSITION 10.1.13. *The product of two cardinals, at least one of which is infinite, equals the greater of the two cardinals.*

PROOF. See [**29**, page 53], [**17**, page 69], or [**40**, page 227]. □

PROPOSITION 10.1.14. *If γ is infinite and a successor cardinal, then γ is regular.*

PROOF. Let β be the cardinal such that $\gamma = \mathrm{Succ}(\beta)$; if a set has cardinal less than γ, then its cardinal is less than or equal to β. Let B be a set whose cardinal is β and for every $b \in B$ let S_b be a set whose cardinal is β. Then, if we have sets A and S_a for $a \in A$ all of whose cardinals are less than γ, then $\mathrm{card}(\bigcup_{a\in A} S_a) \le \mathrm{card}(\bigcup_{b\in B} S_b) \le \beta \times \beta = \beta < \gamma$. □

PROPOSITION 10.1.15. *If μ is an infinite cardinal and $\gamma = \mu^\mu$, then $\gamma^\mu = \gamma$.*

PROOF. $\gamma^\mu = (\mu^\mu)^\mu = \mu^{(\mu\times\mu)} = \mu^\mu = \gamma$. □

LEMMA 10.1.16. *Let ν be a cardinal with $\nu \geq 2$ and let S be a set whose cardinal is ν. If μ is a cardinal, then the collection T of subsets of S whose cardinal is at most μ has cardinal at most ν^μ.*

PROOF. Choose a bijection between S and the cardinal ν. This induces a well ordering of every subset of S under which every such subset U is order isomorphic to a unique ordinal $\eta(U) \leq \nu$. This defines a one to one correspondence between T and a subset of the set of functions from μ to S, under which a subset U corresponds to the function that takes $\eta(U)$ isomorphically onto U and takes every element of $\mu - \eta(U)$ to the first element of U. Thus, there is a subset of ν^μ that maps onto T. $\square$

LEMMA 10.1.17. *Let $\mathcal{M}$ be a category and let X, Y, and Z be objects of $\mathcal{M}$. If X is a retract of Y, then*

(1) *the cardinal of $\mathcal{M}(X, Z)$ is less than or equal to the cardinal of $\mathcal{M}(Y, Z)$, and*
(2) *the cardinal of $\mathcal{M}(Z, X)$ is less than or equal to the cardinal of $\mathcal{M}(Z, Y)$.*

PROOF. If $i\colon X \to Y$ and $r\colon Y \to X$ are maps such that $ri = 1_X$, then $(ri)^*\colon \mathcal{M}(X, Z) \to \mathcal{M}(X, Z)$ is the identity map. Thus, $i^*\colon \mathcal{M}(Y, Z) \to \mathcal{M}(X, Z)$ is a surjection. Similarly, $r_*\colon \mathcal{M}(Z, Y) \to \mathcal{M}(Z, X)$ is a surjection. $\square$

10.2. Transfinite composition

DEFINITION 10.2.1. Let $\mathcal{C}$ be a category that is closed under colimits.

(1) If λ is an ordinal, then a *λ-sequence* in $\mathcal{C}$ is a functor $X\colon \lambda \to \mathcal{C}$ (see Remark 10.1.3) (i.e., a diagram

$$X_0 \to X_1 \to X_2 \to \cdots \to X_\beta \to \cdots \qquad (\beta < \lambda)$$

in $\mathcal{C}$) such that for every limit ordinal $\gamma < \lambda$ the induced map

$$\operatorname*{colim}_{\beta<\gamma} X_\beta \to X_\gamma$$

is an isomorphism.
(2) The *composition* of the λ-sequence is the map $X_0 \to \operatorname{colim}_{\beta<\lambda} X_\beta$.

DEFINITION 10.2.2. Let $\mathcal{C}$ be a category that is closed under colimits.

(1) If $\mathcal{D}$ is a class of maps in $\mathcal{C}$ and λ is an ordinal, then a *λ-sequence of maps in $\mathcal{D}$* is a λ-sequence $X_0 \to X_1 \to X_2 \to \cdots \to X_\beta \to \cdots$ $(\beta < \lambda)$ in $\mathcal{C}$ such that the map $X_\beta \to X_{\beta+1}$ is in $\mathcal{D}$ for $\beta + 1 < \lambda$.
(2) If $\mathcal{D}$ is a class of maps in $\mathcal{C}$, then a *transfinite composition* of maps in $\mathcal{D}$ is a map in $\mathcal{C}$ that is the composition of a λ-sequence in $\mathcal{D}$ (for some ordinal λ, possibly finite).
(3) If $\mathcal{D}$ is a subcategory of $\mathcal{C}$, then a *transfinite composition* of maps in $\mathcal{D}$ is a transfinite composition of maps in the class of maps of $\mathcal{D}$.

10.2.3. Reindexing. If λ is an ordinal and $\mathcal{C}$ is a category, then not every functor from λ to $\mathcal{C}$ is a λ-sequence in $\mathcal{C}$, since Definition 10.2.1 requires that at every limit ordinal $\gamma < \lambda$ the functor take the value of the colimit of that functor restricted to γ. However, if λ is a limit ordinal, then any functor from λ to $\mathcal{C}$ can be *reindexed* to define a λ-sequence in $\mathcal{C}$ with the same composition as that of the original functor (see Definition 10.2.5).

LEMMA 10.2.4. *Let $\mathcal{C}$ be a category, let λ be a limit ordinal, and let $X\colon \lambda \to \mathcal{C}$ be a functor. If the functor $Y\colon \lambda \to \mathcal{C}$ is defined by*

$$\begin{aligned} Y_0 &= X_0 \\ Y_{\beta+1} &= X_\beta && \text{if } \beta + 1 < \lambda \\ Y_\beta &= \operatorname*{colim}_{\gamma<\beta} X_\gamma && \text{if } \beta < \lambda \text{ and } \beta \text{ is a limit ordinal} \end{aligned}$$

then Y is a λ-sequence in $\mathcal{C}$ and $\operatorname{colim}_{\beta<\lambda} X_\beta = \operatorname{colim}_{\beta<\lambda} Y_\beta$.

PROOF. This follows from the universal mapping property of the colimit. □

DEFINITION 10.2.5. If $\mathcal{C}$ is a category, λ is a limit ordinal, and $X\colon \lambda \to \mathcal{C}$ is a functor, then the λ-sequence Y obtained from the functor X as in Lemma 10.2.4 will be called the *reindexing* of X.

PROPOSITION 10.2.6. *Let $\mathcal{C}$ be a category that is closed under colimits, let $\mathcal{D}$ be a class of maps in $\mathcal{C}$, and let $f\colon P \to Q$ be a map in $\mathcal{C}$. If there is an ordinal γ and a function $X\colon \gamma \to \mathcal{C}$ such that*

- $X_0 = P$,
- $\operatorname{colim}_{\beta<\gamma} X_\beta = Q$,
- *the natural map to the colimit $P = X_0 \to \operatorname{colim}_{\beta<\gamma} X_\beta = Q$ is the map f, and*
- *for every $\beta + 1 < \gamma$ the map $\operatorname{colim}_{\alpha\le\beta} X_\alpha \to X_{\beta+1}$ is an element of $\mathcal{D}$,*

then the map f is a transfinite composition of elements of $\mathcal{D}$. If γ is infinite, then it is a transfinite composition indexed by an ordinal whose cardinal equals that of γ.

PROOF. If γ is a limit ordinal then let $\delta = \gamma$; otherwise, let δ be the first limit ordinal greater than γ. We can extend X to a functor $X\colon \delta \to \mathcal{C}$ by letting $X_\beta \to X_{\beta+1}$ be an identity map for $\gamma \le \beta + 1 < \delta$ and then reindex X (see Definition 10.2.5) to obtain a δ-sequence Y. If we let λ be the smallest ordinal such that the map $Y_\beta \to Y_{\beta+1}$ is an identity map for $\lambda \le \beta + 1 < \delta$, then the restriction of Y to λ is a λ-sequence whose composition is $f\colon P \to Q$ and is such that $Y_\beta \to Y_{\beta+1}$ is an element of $\mathcal{D}$ for every $\beta + 1 < \lambda$. □

PROPOSITION 10.2.7. *If $\mathcal{C}$ is a category, S is a set, and $g_s\colon C_s \to D_s$ is a map in $\mathcal{C}$ for every $s \in S$, then the coproduct $\amalg g_s\colon \amalg C_s \to \amalg D_s$ is a transfinite composition of pushouts of the g_s. If S is infinite, then it is a transfinite composition indexed by an ordinal whose cardinal equals that of S.*

PROOF. Choose a well ordering of the set S. There is a unique ordinal λ that is isomorphic to the ordered set S (see, e.g., [**29**, Chapter II], [**17**, page 47], or [**40**, page 202]), and we will identify S with λ. We define a functor $X\colon \lambda + 1 \to \mathcal{C}$ by letting

$$X_\beta = \Big(\coprod_{\alpha<\beta} D_\alpha\Big) \amalg \Big(\coprod_{\beta\le\alpha<\lambda+1} C_\alpha\Big)$$

for all $\beta \le \lambda$, with the maps in the sequence being the obvious ones. For each $\beta + 1 < \lambda$ we have a pushout diagram

$$\begin{array}{ccc} C_\beta & \xrightarrow{g_\beta} & D_\beta \\ \downarrow & & \downarrow \\ X_\beta & \longrightarrow & X_{\beta+1} \end{array}$$

and for each limit ordinal $\beta \le \lambda$ we have a pushout diagram

$$\begin{array}{ccc} C_\beta & \xrightarrow{g_\beta} & D_\beta \\ \downarrow & & \downarrow \\ \operatorname{colim}_{\alpha<\beta} X_\alpha & \longrightarrow & X_\beta \end{array} .$$

The result now follows from Proposition 10.2.6. □

PROPOSITION 10.2.8. *Let $\mathcal{C}$ be a category. If the map $X \to Y$ is the composition of the λ-sequence*

$$X = X_0 \to X_1 \to X_2 \to \cdots \to X_\beta \to \cdots \qquad (\beta < \lambda) \tag{10.2.9}$$

(for some ordinal λ) in which each map $X_\beta \to X_{\beta+1}$ is the composition of the γ_β-sequence

$$X_\beta = W_0^\beta \to W_1^\beta \to W_2^\beta \to \cdots \to W_\alpha^\beta \to \cdots \qquad (\alpha < \gamma_\beta) \tag{10.2.10}$$

(for some ordinal γ_β), then the set $P = \{(\beta, \alpha) \mid \beta < \lambda, \alpha < \gamma_\beta\}$ is well ordered by the dictionary order, i.e.,

$$(\beta_1, \alpha_1) < (\beta_2, \alpha_2) \quad \text{if} \quad \beta_1 < \beta_2 \quad \text{or} \quad \beta_1 = \beta_2 \text{ and } \alpha_1 < \alpha_2.$$

We define a quotient $\widetilde{P}$ of P as follows: For each γ_β that is a successor ordinal we let $\bar{\gamma}_\beta$ be the ordinal such that $\bar{\gamma}_\beta + 1 = \gamma_\beta$ (and, thus, $W^\beta_{\bar{\gamma}_\beta} = W_0^{\beta+1}$), and we identify $(\beta, \bar{\gamma}_\beta)$ with $(\beta + 1, 0)$. The well ordering on P induces a well ordering on $\widetilde{P}$, and so there is a unique ordinal κ for which there is an isomorphism of ordered sets $f\colon \kappa \approx \widetilde{P}$, and this isomorphism is also unique. If we define a functor $Y\colon \kappa \to \mathcal{C}$ by $Y(\gamma) = W\big(f(\gamma)\big)$, then Y is a κ-sequence in $\mathcal{C}$.

PROOF. We need only show that if $\gamma < \kappa$ and γ is a limit ordinal, then $Y(\gamma) = \operatorname{colim}_{\alpha<\gamma} Y(\alpha)$. This follows from the universal mapping property of the colimit. □

DEFINITION 10.2.11. The κ-sequence of Proposition 10.2.8 will be said to have been obtained by *interpolating* the sequences (10.2.10) into the sequence (10.2.9).

PROPOSITION 10.2.12. *The λ-sequence of (10.2.9) is right cofinal (see Definition 10.1.4) in the κ-sequence of Proposition 10.2.8.*

PROOF. This follows directly from the definition. □

LEMMA 10.2.13. *Let $\mathcal{C}$ be a category, let $\mathcal{D}$ be a class of maps in $\mathcal{C}$, and let λ be an ordinal. If the map $X \to Y$ is the composition of a λ-sequence $X = X_0 \to X_1 \to X_2 \to \cdots \to X_\beta \to \cdots$ $(\beta < \lambda)$ in which each map $X_\beta \to X_{\beta+1}$ is a transfinite composition of maps in $\mathcal{D}$, then interpolating (see Definition 10.2.11)* *the sequences for each $X_\beta \to X_{\beta+1}$ into the original λ-sequence gives a κ-sequence (for some ordinal κ) of maps in $\mathcal{D}$ whose composition is the map $X \to Y$.*

PROOF. This follows directly from the definitions. □

PROPOSITION 10.2.14. *Let $\mathcal{C}$ be a category, and let $\mathcal{D}$ be a class of maps in $\mathcal{C}$. If the map $g\colon X \to Y$ is a transfinite composition of pushouts of coproducts of elements of $\mathcal{D}$, then g is a transfinite composition of pushouts of elements of $\mathcal{D}$.*

PROOF. This follows from Proposition 10.2.7 and Lemma 10.2.13. □

PROPOSITION 10.2.15. *Let $\mathcal{C}$ be a category, let I be a class of maps in $\mathcal{C}$, and let λ be a regular cardinal (see Definition 10.1.11). If the map $X \to Y$ is the composition of a λ-sequence*

$$X = X_0 \to X_1 \to X_2 \to \cdots \to X_\beta \to \cdots \qquad (\beta < \lambda) \tag{10.2.16}$$

in which each map $X_\beta \to X_{\beta+1}$ is a transfinite composition, indexed by an ordinal whose cardinal is less than λ, of pushouts of coproducts of elements of I, then interpolating the sequences for the $X_\beta \to X_{\beta+1}$ into the sequence (10.2.16) (see Definition 10.2.11) yields a λ-sequence (indexed by the same ordinal λ) of pushouts of coproducts of elements of I.

PROOF. Lemma 10.2.13 implies that there is an ordinal κ such that the map $X \to Y$ is the composition of a κ-sequence of pushouts of coproducts of elements of I, and so it remains only to show that the ordinal κ constructed in the proof of Lemma 10.2.13 equals λ. Since the cardinal of κ equals that of a union, indexed by λ, of sets of cardinal less than λ, the cardinal of κ equals λ. Since any ordinal less than κ is contained within a subunion indexed by an ordinal less than λ of sets of cardinal less than λ, and λ is a regular cardinal, that subunion would have cardinal less than λ, i.e., κ is the first ordinal having its cardinal, and so κ is a cardinal, and so $\kappa = \lambda$. □

10.2.17. Simplicial model categories.

PROPOSITION 10.2.18. *If K is a simplicial set and L is a subcomplex of K, then the inclusion $L \to K$ is a transfinite composition of pushouts of the maps $\{\partial\Delta[n] \to \Delta[n] \mid n \geq 0\}$*

PROOF. For each integer $n \geq 0$ let S_n be the set of nondegenerate n-simplices of K that are not in L, and choose a well ordering of each set S_n. Let $T = \bigcup_{n\geq 0} S_n$ and let T be ordered by the "dictionary order", i.e., if $\sigma, \tau \in T$, then $\sigma < \tau$ if either

- σ is an n-simplex, τ is a k-simplex, and $n < k$, or
- σ and τ are both n-simplices and $\sigma < \tau$ in the well ordering of S_n.

The set T is then a well ordered set and is thus isomorphic to a unique ordinal γ (see, e.g., [**29**, Chapter II], [**17**, page 47], or [**40**, page 202]). We define a functor $X\colon \gamma + 2 \to \mathrm{SS}$ by letting X_β (for $\beta \leq \gamma + 1$) be the union of L with the nondegenerate simplices of $K - L$ of index less than β and their degeneracies. Proposition 10.2.6 now implies that there is an ordinal λ and a λ-sequence $Y_0 \to Y_1 \to Y_2 \to \cdots \to Y_\beta \to \cdots$ $(\beta < \lambda)$ whose composition is the inclusion $L \to K$ and such that each $Y_\beta \to Y_{\beta+1}$ (for $\beta + 1 < \lambda$) attaches a single nondegenerate simplex, and is thus the bottom arrow of a pushout square

$$\begin{array}{ccc} \partial\Delta[n] & \longrightarrow & \Delta[n] \\ \big\downarrow & & \big\downarrow \\ Y_\beta & \longrightarrow & Y_{\beta+1} \end{array} .$$

□

LEMMA 10.2.19. *Let $\mathcal{M}$ be a simplicial category and let $f\colon A \to B$ be a map in $\mathcal{M}$. If $M \to L \to K$ are maps of simplicial sets, then the square*

$$\begin{array}{ccc} A\otimes L \amalg_{A\otimes M} B\otimes M & \longrightarrow & B\otimes L \\ \big\downarrow & & \big\downarrow \\ A\otimes K \amalg_{A\otimes M} B\otimes M & \longrightarrow & A\otimes K \amalg_{A\otimes L} B\otimes L \end{array}$$

is a pushout.

PROOF. For the square to be a pushout, the lower right hand corner must be the colimit of the diagram

$$\begin{array}{ccccc} & \nearrow & B\otimes L & \nwarrow & \\ & & \uparrow & & \\ A\otimes L & \longleftarrow & A\otimes M & \longrightarrow & B\otimes M \\ \big\downarrow & & \| & & \| \\ A\otimes K & \longleftarrow & A\otimes M & \longrightarrow & B\otimes M \end{array} .$$

The colimit of that diagram is isomorphic to the colimit of the diagram

$$\begin{array}{ccccccc} & & & \nearrow & B\otimes L & \nwarrow & \\ & & & & \uparrow & & \\ A\otimes K & \longleftarrow & A\otimes L & \longleftarrow & A\otimes M & \longrightarrow & B\otimes M \end{array} ,$$

which the universal mapping property of the colimit implies is isomorphic to $A\otimes K \amalg_{A\otimes L} B\otimes L$ (see Theorem 14.2.5). □

PROPOSITION 10.2.20. *Let $\mathcal{M}$ be a simplicial model category and let $f\colon A \to B$ be a map in $\mathcal{M}$. If (K, L) is a pair of simplicial sets, then the map*

$$A\otimes K \amalg_{A\otimes L} B\otimes L \to B\otimes K$$

is a transfinite composition of pushouts of the maps $A\otimes\Delta[n] \amalg_{A\otimes\partial\Delta[n]} B\otimes\partial\Delta[n] \to B\otimes\Delta[n]$ for various values of n.

PROOF. The inclusion $L \to K$ can be written as a transfinite composition of pushouts of the inclusions $\partial\Delta[n] \to \Delta[n]$ for various values of n (see Proposition 10.2.18), and so the result follows from Lemma 10.2.19. □

COROLLARY 10.2.21. *Let $\mathcal{M}$ be a simplicial model category. If $f\colon A \to B$ is a map in $\mathcal{M}$, K is a simplicial set, and $n \geq 0$, then the map*

$$\big(A\otimes\Delta[n] \amalg_{A\otimes\partial\Delta[n]} B\otimes\partial\Delta[n]\big)\otimes K \to \big(B\otimes\Delta[n]\big)\otimes K$$

is a transfinite composition of pushouts of the maps $A\otimes\Delta[m]\amalg_{A\otimes\partial\Delta[m]}B\otimes\partial\Delta[m] \to B\otimes\Delta[m]$ *for various values of* m.

PROOF. Lemma 9.2.1 and axiom M6 (see Definition 9.1.6) imply that the map

$$\bigl(A\otimes\Delta[n]\amalg_{A\otimes\partial\Delta[n]} B\otimes\partial\Delta[n]\bigr)\otimes K \to \bigl(B\otimes\Delta[n]\bigr)\otimes K$$

is isomorphic to the map

$$A\otimes\bigl(\Delta[n]\times K\bigr)\amalg_{A\otimes(\partial\Delta[n]\times K)} B\otimes\bigl(\partial\Delta[n]\times K\bigr) \to B\otimes\bigl(\Delta[n]\times K\bigr),$$

and so the result follows from Proposition 10.2.20. □

10.3. Transfinite composition and lifting in model categories

LEMMA 10.3.1. *If* $\mathcal{M}$ *is a category and* $p\colon X\to Y$ *is a map in* $\mathcal{M}$, *then the class of maps with the left lifting property with respect to* p *is closed under transfinite composition (see Definition* 10.2.2*).*

PROOF. Given a λ-sequence of maps with the left lifting property with respect to p and a lifting problem for the composition of the λ-sequence, a lift can be constructed by a transfinite induction. □

PROPOSITION 10.3.2. *If* $\mathcal{M}$ *is a category and* $p\colon X\to Y$ *is a map in* $\mathcal{M}$, *then the class of maps with the left lifting property with respect to* p *is closed under pushouts, transfinite composition, and retracts.*

PROOF. This follows from Lemma 7.2.11, Lemma 10.3.1, and Lemma 7.2.8. □

PROPOSITION 10.3.3. *If* $\mathcal{M}$ *is a simplicial model category and* $\mathcal{C}$ *is a class of maps in* $\mathcal{M}$, *then the class of maps in* $\mathcal{M}$ *that have the homotopy left lifting property with respect to every element of* $\mathcal{C}$ *is closed under pushouts, transfinite compositions, and retracts.*

PROOF. This follows from Lemma 9.4.7 and Proposition 10.3.2. □

PROPOSITION 10.3.4. *If* $\mathcal{M}$ *is a model category, then the classes of cofibrations and of trivial cofibrations are closed under pushouts, transfinite compositions, and retracts.*

PROOF. This follows from Proposition 7.2.3 and Proposition 10.3.2. □

LEMMA 10.3.5. *Let* $\mathcal{M}$ *be a model category and let* $p\colon X\to Y$ *be a map in* $\mathcal{M}$. *If* S *is a totally ordered set and* $\boldsymbol{W}\colon S\to\mathcal{M}$ *is a functor such that if* $s,t\in S$ *and* $s\le t$, *then* $\boldsymbol{W}_s\to\boldsymbol{W}_t$ *has the left lifting property with respect to* p, *then for every* $s\in S$ *the map* $\boldsymbol{W}_s\to\operatorname{colim}_{t\ge s}\boldsymbol{W}_t$ *has the left lifting property with respect to* p.

PROOF. Proposition 10.1.6 implies that we can choose a right cofinal subset T of $\{t\in S \mid t\ge s\}$ such that T is well ordered. There is a unique ordinal λ that is isomorphic to T (see, e.g., [**29**, Chapter II], [**17**, page 47], or [**40**, page 202]), and so we have a right cofinal functor $\lambda\to\mathcal{M}$. If we reindex this functor (see Definition 10.2.5), then we have a λ-sequence of maps with the left lifting property with respect to p. The lemma now follows from Lemma 10.3.1 and Theorem 10.1.5. □

PROPOSITION 10.3.6. *Let $\mathcal{M}$ be a model category, and let S be a totally ordered set. If $\boldsymbol{W}\colon S \to \mathcal{M}$ is a functor such that, if $s, t \in S$ and $s \le t$, then $\boldsymbol{W}_s \to \boldsymbol{W}_t$ is a cofibration, then, for every $s \in S$, the map $\boldsymbol{W}_s \to \operatorname{colim}_{t \ge s} \boldsymbol{W}_t$ is a cofibration.*

PROOF. This follows from Proposition 7.2.3 and Lemma 10.3.5. □

10.4. Small objects

DEFINITION 10.4.1. Let $\mathcal{C}$ be a cocomplete category and let $\mathcal{D}$ be a subcategory of $\mathcal{C}$.

(1) If κ is a cardinal, then an object W in $\mathcal{C}$ is *κ-small relative to* $\mathcal{D}$ if, for every regular cardinal (see Definition 10.1.11) $\lambda \ge \kappa$ and every λ-sequence (see Definition 10.2.1)

$$X_0 \to X_1 \to X_2 \to \cdots \to X_\beta \to \cdots \qquad (\beta < \lambda)$$

in $\mathcal{C}$ such that the map $X_\beta \to X_{\beta+1}$ is in $\mathcal{D}$ for every ordinal β such that $\beta + 1 < \lambda$, the map of sets $\operatorname{colim}_{\beta<\lambda} \mathcal{C}(W, X_\beta) \to \mathcal{C}(W, \operatorname{colim}_{\beta<\lambda} X_\beta)$ is an isomorphism.

(2) An object is *small relative to* $\mathcal{D}$ if it is κ-small relative to $\mathcal{D}$ for some cardinal κ, and it is *small* if it is small relative to $\mathcal{C}$.

EXAMPLE 10.4.2. In the category $\mathrm{SS}_{(*)}$, every simplicial set with finitely many nondegenerate simplices is $\aleph_0$-small relative to the subcategory of inclusions of simplicial sets (where $\aleph_0$ is the first infinite cardinal).

EXAMPLE 10.4.3. Let X be a finite cell complex in $\mathrm{Top}_{(*)}$ (see Definition 10.7.1). Corollary 10.7.7 implies X is $\aleph_0$-small relative to the subcategory of inclusions of cell complexes (where $\aleph_0$ is the first infinite cardinal).

EXAMPLE 10.4.4. Let X be an object of $\mathrm{SS}_{(*)}$. If κ is the first infinite cardinal greater than the cardinal of the set of nondegenerate simplices of X, then X is κ-small relative to the subcategory of inclusions (see Proposition 10.1.14). Thus, every simplicial set is small relative to the subcategory of inclusions.

EXAMPLE 10.4.5. Let X be a cell complex in $\mathrm{Top}_{(*)}$ (see Definition 10.7.1). If κ is the first infinite cardinal greater than the cardinal of the set of cells of X (see Proposition 10.1.14), then Proposition 10.7.4 implies that X is κ-small relative to the subcategory of relative cell complexes. Thus, every cell complex is small relative to the subcategory of relative cell complexes.

LEMMA 10.4.6. *If $\mathcal{C}$ is a cocomplete category, $\mathcal{D}$ is a subcategory of $\mathcal{C}$, and I is a set of objects in $\mathcal{C}$ that are small relative to $\mathcal{D}$, then there is a cardinal κ such that every element of I is κ-small relative to $\mathcal{D}$.*

PROOF. For every object A of I let κ_A be a cardinal such that A is κ_A-small relative to $\mathcal{D}$. If we let κ be the union $\bigcup_{A \in I} \kappa_A$, then every object of I is κ-small relative to $\mathcal{D}$. □

PROPOSITION 10.4.7. *Let $\mathcal{C}$ be a cocomplete category and let $\mathcal{D}$ be a subcategory of $\mathcal{C}$. If κ is a cardinal and X is an object of $\mathcal{C}$ that is κ-small relative to $\mathcal{D}$, then any retract of X is κ-small relative to $\mathcal{D}$.*

PROOF. Let $i\colon W \to X$ and $r\colon X \to W$ be maps such that $ri = 1_W$. If λ is a regular cardinal such that $\lambda \geq \kappa$ and $Z_0 \to Z_1 \to Z_2 \to \cdots \to Z_\beta \to \cdots$ $(\beta < \lambda)$ is a λ-sequence in $\mathcal{D}$, then we have the commutative diagram

$$\begin{array}{ccccc}
& & \xrightarrow{\quad 1_{\operatorname{colim} \mathcal{C}(W, Z_\beta)} \quad} & & \\
\operatorname*{colim}_{\beta<\lambda} \mathcal{C}(W, Z_\beta) & \xrightarrow{\operatorname{colim} r^*} & \operatorname*{colim}_{\beta<\lambda} \mathcal{C}(X, Z_\beta) & \xrightarrow{\operatorname{colim} i^*} & \operatorname*{colim}_{\beta<\lambda} \mathcal{C}(W, Z_\beta) \\
\downarrow & & \downarrow & & \downarrow \\
\mathcal{C}(W, \operatorname*{colim}_{\beta<\lambda} Z_\beta) & \xrightarrow{r^*} & \mathcal{C}(X, \operatorname*{colim}_{\beta<\lambda} Z_\beta) & \xrightarrow{i^*} & \mathcal{C}(W, \operatorname*{colim}_{\beta<\lambda} Z_\beta) \\
& & \xrightarrow[\quad 1_{\mathcal{C}(W, \operatorname{colim} Z_\beta)} \quad]{} & &
\end{array}$$

Thus, the map $\operatorname{colim}_{\beta<\lambda} \mathcal{C}(W, Z_\beta) \to \mathcal{C}(W, \operatorname{colim}_{\beta<\lambda} Z_\beta)$ is a retract of the isomorphism $\operatorname{colim}_{\beta<\lambda} \mathcal{C}(X, Z_\beta) \to \mathcal{C}(X, \operatorname{colim}_{\beta<\lambda} Z_\beta)$, and is thus an isomorphism. $\square$

PROPOSITION 10.4.8. *Let $\mathcal{C}$ be a cocomplete category and let $\mathcal{D}$ be a subcategory of $\mathcal{C}$. If $\mathcal{I}$ is a small category and $\boldsymbol{W}\colon \mathcal{I} \to \mathcal{C}$ is a diagram in $\mathcal{C}$ such that $\boldsymbol{W}_i$ is small relative to $\mathcal{D}$ for every object i in $\mathcal{I}$, then $\operatorname{colim}_{i\in\mathcal{I}} \boldsymbol{W}_i$ is small relative to $\mathcal{D}$.*

PROOF. Let γ be a cardinal such that $\boldsymbol{W}_i$ is γ-small relative to $\mathcal{D}$ for every object i in $\mathcal{I}$ (see Lemma 10.4.6), let δ be the cardinal of the set of morphisms in $\mathcal{I}$, and let κ be the first cardinal greater than both γ and δ; we will show that $\operatorname{colim}_{i\in\mathcal{I}} \boldsymbol{W}_i$ is κ-small relative to $\mathcal{D}$.

Let λ be a regular cardinal such that $\lambda \geq \kappa$, and let

$$X_0 \to X_1 \to X_2 \to \cdots \to X_\beta \to \cdots \qquad (\beta < \lambda)$$

be a λ-sequence in $\mathcal{C}$ such that the map $X_\beta \to X_{\beta+1}$ is in $\mathcal{D}$ for all $\beta < \lambda$. If we have a map $f\colon \operatorname{colim}_{i\in\mathcal{I}} \boldsymbol{W}_i \to \operatorname{colim}_{\beta<\lambda} X_\beta$, then for every object j in $\mathcal{I}$ the composition of f with the natural map $\boldsymbol{W}_j \to \operatorname{colim}_{i\in\mathcal{I}} \boldsymbol{W}_i$ defines a map $f_j\colon \boldsymbol{W}_j \to \operatorname{colim}_{\beta<\lambda} X_\beta$. Since $\boldsymbol{W}_j$ is small relative to $\mathcal{D}$ and λ is a large enough regular cardinal, there exists an ordinal $\beta_j < \lambda$ such that f_j factors through X_{β_j}. If we let $\tilde{\beta} = \bigcup_{j\in\operatorname{Ob}\mathcal{I}} \beta_j$, then (since λ is a regular cardinal) $\tilde{\beta} < \lambda$, and the dotted arrow $\tilde{g}_j$ exists in the diagram

$$\begin{array}{ccc}
\boldsymbol{W}_j & & \\
{\scriptstyle \tilde{g}_j}\,\vdots & \overset{f_j}{\searrow} & \\
X_{\tilde{\beta}} & \longrightarrow & \operatorname*{colim}_{\beta<\lambda} X_\beta
\end{array}$$

for every object j in $\mathcal{I}$.

If $s\colon j \to k$ is a morphism in $\mathcal{I}$, then the composition $\boldsymbol{W}_j \xrightarrow{\boldsymbol{W}_s} \boldsymbol{W}_k \xrightarrow{\tilde{g}_k} X_{\tilde{\beta}}$ need not equal the map $\tilde{g}_j\colon \boldsymbol{W}_j \to X_{\tilde{\beta}}$, but their compositions with the natural map $X_{\tilde{\beta}} \to \operatorname{colim}_{\beta<\lambda} X_\beta$ are equal. Since the natural map of sets $\operatorname{colim}_{\beta<\lambda} \mathcal{C}(\boldsymbol{W}_j, X_\beta) \to \mathcal{C}(\boldsymbol{W}_j, \operatorname{colim}_{\beta<\lambda} X_\beta)$ is an isomorphism, there must exist an ordinal $\hat{\beta}_s < \lambda$ such that their compositions with the map $X_{\tilde{\beta}} \to X_{\hat{\beta}_s}$ are equal. If we let $\hat{\beta} = \bigcup_{(s\colon j\to k)\in\mathcal{I}} \hat{\beta}_s$, then (since λ is a regular cardinal) we have $\hat{\beta} < \lambda$. If, for every object j of $\mathcal{I}$, we let $\hat{g}_j$ equal the composition $\boldsymbol{W}_j \xrightarrow{\tilde{g}_j} X_{\tilde{\beta}} \to X_{\hat{\beta}}$, then for every morphism $s\colon j \to k$ in $\mathcal{J}$ the triangle

$$\begin{array}{ccc} \boldsymbol{W}_j & \xrightarrow{\boldsymbol{W}_s} & \boldsymbol{W}_k \\ & \searrow^{\hat{g}_j} & \downarrow{\scriptstyle \hat{g}_k} \\ & & X_{\hat{\beta}} \end{array}$$

commutes, and so the $\hat{g}_j$ define a map $g\colon \operatorname{colim}_{i\in\mathcal{J}} \boldsymbol{W}_i \to X_{\hat{\beta}}$ whose composition with the natural map $X_{\hat{\beta}} \to \operatorname{colim}_{\beta<\lambda} X_\beta$ equals f. Thus, the map

$$\operatorname*{colim}_{\beta<\lambda} \mathcal{C}(\operatorname*{colim}_{i\in\mathcal{J}} \boldsymbol{W}_i, X_\beta) \to \mathcal{C}(\operatorname*{colim}_{i\in\mathcal{J}} \boldsymbol{W}_i, \operatorname*{colim}_{\beta<\lambda} X_\beta)$$

is surjective.

To show that that map is also injective, let $g'\colon \operatorname{colim}_{i\in\mathcal{J}} \boldsymbol{W}_i \to X_{\bar{\beta}}$ be a map whose composition with the natural map $X_{\bar{\beta}} \to \operatorname{colim}_{\beta<\lambda} X_\beta$ equals f. For every object j in $\mathcal{J}$ the compositions

$$\boldsymbol{W}_j \to \operatorname*{colim}_{i\in\mathcal{J}} \boldsymbol{W}_i \xrightarrow{g} X_{\hat{\beta}} \to \operatorname*{colim}_{\beta<\lambda} X_\beta$$

and

$$\boldsymbol{W}_j \to \operatorname*{colim}_{i\in\mathcal{J}} \boldsymbol{W}_i \xrightarrow{g'} X_{\bar{\beta}} \to \operatorname*{colim}_{\beta<\lambda} X_\beta$$

are equal, and so there exists an ordinal $\alpha_j < \lambda$ such that the compositions

$$\boldsymbol{W}_j \to \operatorname*{colim}_{i\in\mathcal{J}} \boldsymbol{W}_i \xrightarrow{g} X_{\hat{\beta}} \to X_{\alpha_j}$$

and

$$\boldsymbol{W}_j \to \operatorname*{colim}_{i\in\mathcal{J}} \boldsymbol{W}_i \xrightarrow{g'} X_{\bar{\beta}} \to X_{\alpha_j}$$

are equal. If we let $\alpha = \bigcup_{j\in\operatorname{Ob}(\mathcal{J})} \alpha_j$, then $\alpha < \lambda$ and the compositions

$$\operatorname*{colim}_{i\in\mathcal{J}} \boldsymbol{W}_i \to X_{\hat{\beta}} \to X_\alpha \qquad \text{and} \qquad \operatorname*{colim}_{i\in\mathcal{J}} \boldsymbol{W}_i \to X_{\bar{\beta}} \to X_\alpha$$

are equal, and so the map

$$\operatorname*{colim}_{\beta<\lambda} \mathcal{C}(\operatorname*{colim}_{i\in\mathcal{J}} \boldsymbol{W}_i, X_\beta) \to \mathcal{C}(\operatorname*{colim}_{i\in\mathcal{J}} \boldsymbol{W}_i, \operatorname*{colim}_{\beta<\lambda} X_\beta)$$

is an isomorphism. □

COROLLARY 10.4.9. *Let $\mathcal{C}$ be a cocomplete category, let $\mathcal{D}$ be a subcategory of $\mathcal{C}$, and let I be a set of maps in $\mathcal{C}$ whose domains and codomains are small relative to $\mathcal{D}$. If X is small relative to $\mathcal{D}$ and the map $X \to Y$ is a transfinite composition of pushouts of elements of I, then Y is small relative to $\mathcal{D}$.*

PROOF. This follows from Proposition 10.4.8, using a transfinite induction. □

10.5. The small object argument

10.5.1. Injectives, cofibrations, and relative cell complexes.

DEFINITION 10.5.2. Let $\mathcal{C}$ be a category, and let I be a set of maps in $\mathcal{C}$.

(1) The subcategory of *I-injectives* is the subcategory of maps that have the right lifting property (see Definition 7.2.1) with respect to every element of I.

(2) The subcategory of *I-cofibrations* is the subcategory of maps that have the left lifting property (see Definition 7.2.1) with respect to every I-injective. An object is *I-cofibrant* if the map to it from the initial object of $\mathcal{C}$ is an I-cofibration.

REMARK 10.5.3. The term *I-injective* comes from the theory of injective classes ([**36**]). The map $p\colon X \to Y$ is an I-injective if and only if, in the category $(\mathcal{C}\downarrow Y)$ of objects over Y, the object p is injective relative to the class of maps whose image under the forgetful functor $(\mathcal{C}\downarrow Y) \to \mathcal{C}$ is an element of I. The term *I-cofibration* comes from Proposition 11.2.1, which asserts that if I is the set of generating cofibrations of a cofibrantly generated model category (see Definition 11.1.2), then the I-cofibrations are the cofibrations of the model category.

EXAMPLE 10.5.4. If I is the set of inclusions $\partial\Delta[n] \to \Delta[n]$ in SS, then the I-injectives are the trivial fibrations, and the I-cofibrations are the inclusions of simplicial sets (see Proposition 7.2.3).

EXAMPLE 10.5.5. If J is the set of inclusions $\Lambda[n,k] \to \Delta[n]$ in SS, then the J-injectives are the Kan fibrations, and the J-cofibrations are the trivial cofibrations (see Proposition 7.2.3).

PROPOSITION 10.5.6. *Let $\mathcal{C}$ be a category and let J and K be sets of maps in $\mathcal{C}$. If every J-injective is a K-injective, then every K-cofibration is a J-cofibration.*

PROOF. This follows directly from the definitions (see Definition 10.5.2). □

PROPOSITION 10.5.7. *Let $\mathcal{C}$ be a category and let J and K be sets of maps in $\mathcal{C}$. If the subcategory of J-injectives equals the subcategory of K-injectives, then the subcategory of J-cofibrations equals the subcategory of K-cofibrations.*

PROOF. This follows from Proposition 10.5.6. □

DEFINITION 10.5.8. If $\mathcal{C}$ is a category that is closed under small colimits and I is a set of maps in $\mathcal{C}$, then

(1) the subcategory of *relative I-cell complexes* (also known as the subcategory of *regular I-cofibrations*) is the subcategory of maps that can be constructed as a transfinite composition (see Definition 10.2.2) of pushouts (see Definition 7.2.10) of elements of I,
(2) an object is an *I-cell complex* if the map to it from the initial object of $\mathcal{C}$ is a relative I-cell complex, and
(3) a map is an *inclusion of I-cell complexes* if it is a relative I-cell complex whose domain is an I-cell complex.

REMARK 10.5.9. Note that Definition 10.5.8 defines a relative I-cell complex to be a map that can be constructed as a transfinite composition of pushouts of elements of I, but it does not assume that there is any preferred such construction. In Definition 10.6.3 we define a *presented relative I-cell complex* to be a relative I-cell complex together with a choice of such a construction.

PROPOSITION 10.5.10. *If $\mathcal{C}$ is a category and I is a set of maps in $\mathcal{C}$, then every relative I-cell complex is an I-cofibration (see Definition* 10.5.2*).*

PROOF. This follows from Lemma 7.2.11 and Lemma 10.3.1. □

PROPOSITION 10.5.11. *If $\mathcal{M}$ is a category and I is a set of maps in $\mathcal{M}$, then a retract of a relative I-cell complex is an I-cofibration.*

PROOF. This follows from Proposition 10.5.10 and Lemma 7.2.8. □

DEFINITION 10.5.12. Let $\mathcal{M}$ be a cocomplete category and let I be a set of maps in $\mathcal{M}$.

(1) If κ is a cardinal, then an object is *κ-small relative to I* if it is κ-small relative to the subcategory of relative I-cell complexes (see Definition 10.4.1 and Definition 10.5.8).
(2) An object is *small relative to I* if it is κ-small relative to I for some cardinal κ.

PROPOSITION 10.5.13. *Let $\mathcal{C}$ be a cocomplete category and let I be a set of maps in $\mathcal{C}$. If K is a set of relative I-cell complexes, then an object that is small relative to I is also small relative to K.*

PROOF. Since every relative K-cell complex is also a relative I-cell complex, a λ-sequence of relative K-cell complexes is also a λ-sequence of relative I-cell complexes. The result now follows from Proposition 10.2.12 and Theorem 10.1.5. □

10.5.14. The small object argument.

DEFINITION 10.5.15. If $\mathcal{M}$ is a category and I is a set of maps in $\mathcal{M}$, then (following D. M. Kan) we say that I *permits the small object argument* if the domains of the elements of I are small relative to I (see Definition 10.5.12 and Definition 10.5.8).

PROPOSITION 10.5.16 (The small object argument). *If $\mathcal{C}$ is a cocomplete category and I is a set of maps in $\mathcal{C}$ that permits the small object argument (see Definition 10.5.15), then there is a functorial factorization of every map in $\mathcal{C}$ into a relative I-cell complex (see Definition 10.5.8) followed by an I-injective (see Definition 10.5.2).*

PROOF. Lemma 10.4.6 implies that we can choose a regular cardinal λ such that every domain of an element of I is λ-small relative to the subcategory of relative I-cell complexes. If $g\colon X \to Y$ is a map in $\mathcal{C}$, then we will factor g as $X \xrightarrow{j} \mathrm{E}_I \xrightarrow{p} Y$, where j is the transfinite composition of a λ-sequence

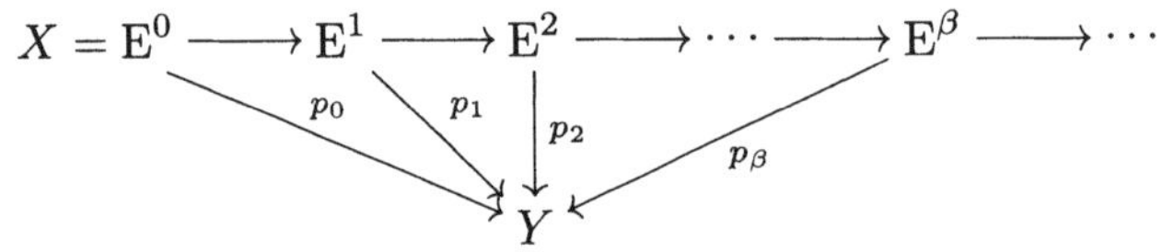

in which each $\mathrm{E}^\beta \to \mathrm{E}^{\beta+1}$ is a pushout of a coproduct of elements of I, each E^β comes with a map $p_\beta\colon \mathrm{E}^\beta \to Y$ such that all the triangles commute, and $p = \operatorname{colim}_{\beta<\lambda} p_\beta$.

We begin by letting $\mathrm{E}^0 = X$ and letting $p_0\colon \mathrm{E}^0 \to Y$ equal g. Given E^β, we have the solid arrow diagram

$$\begin{array}{ccccc} \coprod_{\substack{(A_i \to B_i)\in I \\ \mathcal{M}(A_i,\mathrm{E}^\beta)\times_{\mathcal{M}(A_i,Y)}\mathcal{M}(B_i,Y)}} A_i & \longrightarrow & \mathrm{E}^\beta & \dashrightarrow & \mathrm{E}^{\beta+1} \\ \Big\downarrow & & \Big\downarrow{\scriptstyle p_\beta} & \swarrow{\scriptstyle p_{\beta+1}} & \\ \coprod_{\substack{(A_i \to B_i)\in I \\ \mathcal{M}(A_i,\mathrm{E}^\beta)\times_{\mathcal{M}(A_i,Y)}\mathcal{M}(B_i,Y)}} B_i & \longrightarrow & Y & & \end{array} \tag{10.5.17}$$

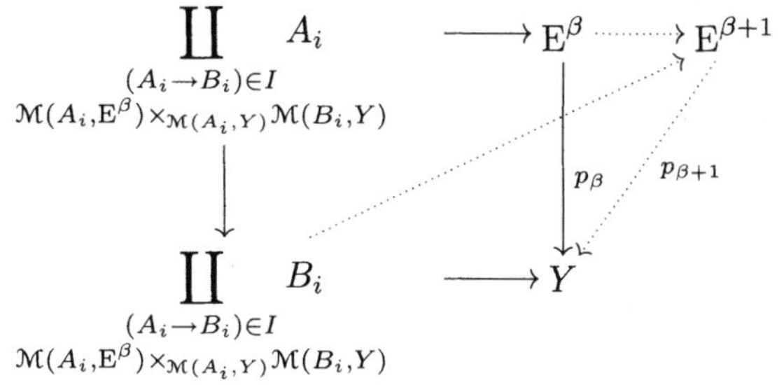

and we let $\mathrm{E}^{\beta+1}$ be the pushout $(\coprod B_i) \amalg_{(\coprod A_i)} \mathrm{E}^\beta$. If γ is a limit ordinal, we let $\mathrm{E}^\gamma = \operatorname{colim}_{\beta<\gamma} \mathrm{E}^\beta$, and we let $\mathrm{E}_I = \operatorname{colim}_{\beta<\lambda} \mathrm{E}^\beta$. The construction of the factorization $X \to \mathrm{E}_I \to Y$ makes it clear that it is functorial. Proposition 10.2.7, Lemma 7.2.13, and Lemma 10.2.13 imply that $X \to \mathrm{E}_I$ is a relative I-cell complex, and so it remains only to show that $\mathrm{E}_I \to Y$ is an I-injective.

Given an element $A \to B$ of I and a solid arrow diagram

$$\begin{array}{ccc} A & \longrightarrow & \mathrm{E}_I \\ \Big\downarrow & \nearrow & \Big\downarrow \\ B & \longrightarrow & Y \end{array} \tag{10.5.18}$$

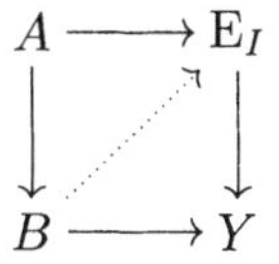

we must show that the dotted arrow exists. Since $\mathrm{E}_I = \operatorname{colim}_{\beta<\lambda} \mathrm{E}^\beta$ and A is λ-small relative to I, the natural map of sets $\operatorname{colim}_{\beta<\lambda} \mathcal{M}(A, \mathrm{E}^\beta) \to \mathcal{M}(A, \mathrm{E}_I)$ is an isomorphism. Thus, the map $A \to \mathrm{E}_I$ factors through $\mathrm{E}^\beta \to \mathrm{E}_I$ for some $\beta < \lambda$, and we have the solid arrow diagram

$$\begin{array}{ccccccc} A & \longrightarrow & \mathrm{E}^\beta & \longrightarrow & \mathrm{E}^{\beta+1} & \longrightarrow & \mathrm{E}_I \\ \Big\downarrow & & \Big\downarrow & \swarrow & & \swarrow & \\ B & \longrightarrow & Y & & & & \end{array}$$

(with a dotted arrow $B \dashrightarrow \mathrm{E}^{\beta+1}$)

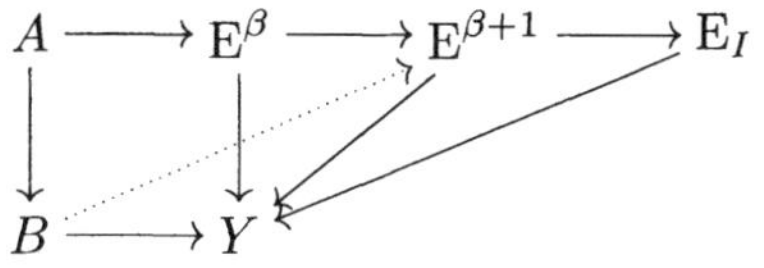

The construction of $\mathrm{E}^{\beta+1}$ implies that the dotted arrow exists, and this dotted arrow defines the dotted arrow in Diagram 10.5.18. □

DEFINITION 10.5.19. Let $\mathcal{C}$ be a cocomplete category, let I be a set of maps in $\mathcal{C}$, and let λ be an ordinal. If we apply the construction in the proof of Proposition 10.5.16 to a map $g\colon X \to Y$ using the set I and the ordinal λ to obtain the factorization $X \to E_I \to Y$, then we will call E_I the *object obtained by applying the small object factorization with the set I and the ordinal λ to the map g.*

PROPOSITION 10.5.20. *Let $\mathcal{C}$ be a cocomplete category, let I be a set of maps in $\mathcal{C}$, and let λ be an ordinal. If the map $g\colon X \to Y$ is a retract of the map $\tilde{g}\colon \widetilde{X} \to \widetilde{Y}$ and we apply the small object factorization to both g and $\tilde{g}$ using the set I and the ordinal λ (see Definition 10.5.19), then the factorization $X \to E_I \to Y$ obtained from g is a retract of the factorization $\widetilde{X} \to \widetilde{E}_I \to \widetilde{Y}$ obtained from $\tilde{g}$.*

PROOF. At each step in the construction of E_I and $\widetilde{E}_I$, the factorization $X \to E^\beta \to Y$ is a retract of the factorization $\widetilde{X} \to \widetilde{E}^\beta \to \widetilde{Y}$. □

COROLLARY 10.5.21. *Let $\mathcal{C}$ be a cocomplete category and let I be a set of maps in $\mathcal{C}$. If κ is a regular cardinal such that the domains of the elements of I are κ-small relative to I, then there is a functorial factorization of every map in $\mathcal{C}$ into the composition of a κ-sequence of pushouts of coproducts of elements of I followed by an I-injective.*

PROOF. This is identical to the proof of Proposition 10.5.16 if we choose the cardinal λ in that proof to equal κ. □

COROLLARY 10.5.22. *If $\mathcal{C}$ is a cocomplete category, I is a set of maps in $\mathcal{C}$ that permits the small object argument, and $g\colon X \to Y$ is an I-cofibration (see Definition 10.5.2), then g is a retract of a relative I-cell complex.*

PROOF. If we apply the factorization of Proposition 10.5.16 to g, we obtain $X \xrightarrow{j} \mathrm{E}_I \xrightarrow{p} Y$ in which j is a relative I-cell complex and p is an I-injective. The result now follows from the retract argument (see Proposition 7.2.2). □

COROLLARY 10.5.23. *Let $\mathcal{C}$ be a cocomplete category. If I is a set of maps in $\mathcal{C}$ that permits the small object argument, then the class of I-cofibrations (see Definition 10.5.2) equals the class of retracts of relative I-cell complexes (see Definition 10.5.8).*

PROOF. This follows from Proposition 10.5.10, Proposition 10.5.11, and Corollary 10.5.22. □

10.5.24. Smallness and cofibrations.

LEMMA 10.5.25. *Let $\mathcal{C}$ be a cocomplete category, let I be a set of maps in $\mathcal{M}$ that permits the small object argument (see Definition 10.5.15), and let κ be a regular cardinal such that the domain of every element of I is κ-small relative to I (see Lemma 10.4.6). If λ is an ordinal and $X_0 \to X_1 \to X_2 \to \cdots \to X_\beta \to \cdots$ $(\beta < \lambda)$ is a λ-sequence of I-cofibrations, then there is a λ-sequence $\widetilde{X}_0 \to \widetilde{X}_1 \to \widetilde{X}_2 \to \cdots \to \widetilde{X}_\beta \to \cdots$ $(\beta < \lambda)$ of relative I-cell complexes and maps of λ-sequences*

$$
\begin{array}{ccccccccc}
X_0 & \xrightarrow{\sigma_0} & X_1 & \xrightarrow{\sigma_1} & X_2 & \xrightarrow{\sigma_2} \cdots \longrightarrow & X_\beta & \xrightarrow{\sigma_\beta} & \cdots \\
\downarrow{\scriptstyle i_0} & & \downarrow{\scriptstyle i_1} & & \downarrow{\scriptstyle i_2} & & \downarrow{\scriptstyle i_\beta} & & \\
\widetilde{X}_0 & \xrightarrow{\tau_0} & \widetilde{X}_1 & \xrightarrow{\tau_1} & \widetilde{X}_2 & \xrightarrow{\tau_2} \cdots \longrightarrow & \widetilde{X}_\beta & \xrightarrow{\tau_\beta} & \cdots \\
\downarrow{\scriptstyle r_0} & & \downarrow{\scriptstyle r_1} & & \downarrow{\scriptstyle r_2} & & \downarrow{\scriptstyle r_\beta} & & \\
X_0 & \xrightarrow{\sigma_0} & X_1 & \xrightarrow{\sigma_1} & X_2 & \xrightarrow{\sigma_2} \cdots \longrightarrow & X_\beta & \xrightarrow{\sigma_\beta} & \cdots
\end{array}
\tag{10.5.26}
$$

such that, for every $\beta < \lambda$,

1. *the composition $r_\beta i_\beta$ is the identity map of X_β, and*
2. *the map $\tau_\beta\colon \widetilde{X}_\beta \to \widetilde{X}_{\beta+1}$ is the composition of a κ-sequence of pushouts of coproducts of elements of I.*

PROOF. We let $\widetilde{X}_0 = X_0$, and we let both i_0 and r_0 be the identity map of X_0. If β is an ordinal such that $\beta + 1 < \lambda$ and we've defined the sequence through $\widetilde{X}_\beta$, then we apply the factorization of Corollary 10.5.21 to the map $\sigma_\beta r_\beta\colon \widetilde{X}_\beta \to$

$X_{\beta+1}$ to obtain $\widetilde{X}_\beta \xrightarrow{\tau_\beta} \widetilde{X}_{\beta+1} \xrightarrow{r_{\beta+1}} X_{\beta+1}$, in which τ_β is the composition of a κ-sequence of pushouts of coproducts of elements of I and $r_{\beta+1}$ is an I-injective. Since $r_{\beta+1}\tau_\beta i_\beta = \sigma_\beta r_\beta i_\beta = \sigma_\beta$, we now have the solid arrow diagram

$$\begin{array}{ccc} X_\beta & \xrightarrow{\tau_\beta i_\beta} & \widetilde{X}_{\beta+1} \\ {\scriptstyle \sigma_\beta}\downarrow & \overset{i_{\beta+1}}{\nearrow} & \downarrow{\scriptstyle r_{\beta+1}} \\ X_{\beta+1} & = & X_{\beta+1} \end{array}$$

in which σ_β is an I-cofibration and $r_{\beta+1}$ is an I-injective, and so there exists a dotted arrow $i_{\beta+1}$ such that $i_{\beta+1}\sigma_\beta = \tau_\beta i_\beta$ and $r_{\beta+1}i_{\beta+1} = 1_{X_{\beta+1}}$.

For every limit ordinal γ such that $\gamma < \lambda$, we let $\widetilde{X}_\gamma = \operatorname{colim}_{\beta<\gamma} \widetilde{X}_\beta$, $i_\gamma = \operatorname{colim}_{\beta<\gamma} i_\beta$, and $r_\gamma = \operatorname{colim}_{\beta<\lambda} r_\beta$. □

THEOREM 10.5.27. *Let $\mathcal{C}$ be a cocomplete category and let I be a set of maps in $\mathcal{C}$ that permits the small object argument (see Definition 10.5.15). If W is an object that is small relative to I, then it is small relative to the subcategory of all I-cofibrations.*

PROOF. Let μ be a cardinal such that W is μ-small relative to I. Lemma 10.4.6 implies that there is a cardinal κ such that the domain of every element of I is κ-small relative to I. If ν is the first cardinal greater than both μ and κ, then we will show that W is ν-small relative to the subcategory of I-cofibrations.

Let λ be a regular cardinal such that $\lambda \geq \nu$ and let $X_0 \to X_1 \to X_2 \to \cdots \to X_\beta \to \cdots$ $(\beta < \lambda)$ be a λ-sequence of I-cofibrations. Lemma 10.5.25 implies that there is a λ-sequence $\widetilde{X}_0 \to \widetilde{X}_1 \to \widetilde{X}_2 \to \cdots \to \widetilde{X}_\beta \to \cdots$ $(\beta < \lambda)$ of relative I-cell complexes and maps of λ-sequences as in Diagram 10.5.26 satisfying the conclusion of Lemma 10.5.25. Proposition 10.2.15 implies that, after interpolations, the λ-sequence $\widetilde{X}_0 \to \widetilde{X}_1 \to \widetilde{X}_2 \to \cdots \to \widetilde{X}_\beta \to \cdots$ $(\beta < \lambda)$ is a λ-sequence of relative I-cell complexes, and so Proposition 10.2.12 and Theorem 10.1.5 imply that the map of sets $\operatorname{colim}_{\beta<\lambda} \mathcal{M}(W, \widetilde{X}_\beta) \to \mathcal{M}(W, \operatorname{colim}_{\beta<\lambda} \widetilde{X}_\beta)$ is an isomorphism. Since the map of sets $\operatorname{colim}_{\beta<\lambda} \mathcal{M}(W, X_\beta) \to \mathcal{M}(W, \operatorname{colim}_{\beta<\lambda} X_\beta)$ is a retract of this isomorphism, it is also an isomorphism. □

10.6. Subcomplexes of relative I-cell complexes

If $\mathcal{C}$ is a cocomplete category and I is a set of maps in $\mathcal{C}$, then a relative I-cell complex is a map that can be constructed as a transfinite composition of pushouts of coproducts of elements of I (see Definition 10.5.8 and Proposition 10.2.14). To consider "subcomplexes" of a relative I-cell complex, we need to choose a "presentation" of it (see Definition 10.6.2), i.e., a particular such construction. In Definition 10.6.3, we define a *presented relative I-cell complex* to be a relative I-cell complex together with a chosen presentation.

10.6.1. Presentations of relative I-cell complexes.

DEFINITION 10.6.2. Let $\mathcal{C}$ be a cocomplete category and let I be a set of maps in $\mathcal{C}$. If $f\colon X \to Y$ is a relative I-cell complex (see Definition 10.5.8), then a *presentation* of f is a pair consisting of a λ-sequence

$$X = X_0 \to X_1 \to X_2 \to \cdots \to X_\beta \to \cdots \qquad (\beta < \lambda)$$

(for some ordinal λ) and a sequence of ordered triples

$$\{(T^\beta, e^\beta, h^\beta)\}_{\beta<\lambda}$$

such that

(1) the composition of the λ-sequence is isomorphic to f, and
(2) for every $\beta < \lambda$
 - T^β is a set,
 - e^β is a function $e^\beta\colon T^\beta \to I$,
 - if $i \in T^\beta$ and e_i^β is the element $C_i \to D_i$ of I, then h_i^β is a map $h_i^\beta\colon C_i \to X_\beta$ such that there is a pushout diagram

$$\begin{array}{ccc} \coprod_{T^\beta} C_i & \longrightarrow & \coprod_{T^\beta} D_i \\ {\scriptstyle \amalg h_i^\beta}\big\downarrow & & \big\downarrow \\ X_\beta & \longrightarrow & X_{\beta+1} \end{array} .$$

If the map $f\colon \emptyset \to Y$ (where $\emptyset$ is the initial object of $\mathcal{C}$) is a relative I-cell complex, then a presentation of f will also be called a *presentation of* Y.

DEFINITION 10.6.3. If $\mathcal{C}$ is a cocomplete category and I is a set of maps in $\mathcal{C}$, then a *presented relative I-cell complex* is a relative I-cell complex $f\colon X \to Y$ together with a particular presentation $\bigl(X = X_0 \to X_1 \to X_2 \to \cdots \to X_\beta \to \cdots \quad (\beta < \lambda), \{T^\beta, e^\beta, h^\beta\}_{\beta<\lambda}\bigr)$ of it (see Definition 10.6.2). A presented relative I-cell complex in which $X = \emptyset$ (the initial object of $\mathcal{C}$) will be called a *presented I-cell complex.*

DEFINITION 10.6.4. Let $\mathcal{C}$ be a cocomplete category and let I be a set of maps in $\mathcal{C}$. If $(f\colon X \to Y, X = X_0 \to X_1 \to X_2 \to \cdots \to X_\beta \to \cdots \quad (\beta < \lambda), \{T^\beta, e^\beta, h^\beta\}_{\beta<\lambda})$ is a presented relative I-cell complex (see Definition 10.6.3), then

(1) the *presentation ordinal* of f is λ,
(2) the *set of cells* of f is $\coprod_{\beta<\lambda} T^\beta$,
(3) the *size* of f is the cardinal of the set of cells of f,
(4) if e is a cell of f, the *presentation ordinal* of e is the ordinal β such that $e \in T^\beta$, and
(5) if $\beta < \lambda$, then the *β-skeleton* of f is X_β.

PROPOSITION 10.6.5. *If $\mathcal{C}$ is a cocomplete category and I is a set of maps in $\mathcal{C}$, then a presented relative I-cell complex is entirely determined by its presentation ordinal λ (see Definition 10.6.4) and its sequence of triples $\{(T, e^\beta, h^\beta)\}_{\beta<\lambda}$.*

PROOF. This follows directly from the definitions. □

10.6.6. Subcomplexes of relative I-cell complexes.

DEFINITION 10.6.7. Let $\mathcal{C}$ be a cocomplete category and let I be a set of maps in $\mathcal{C}$. If $(f\colon X \to Y, X = X_0 \to X_1 \to X_2 \to \cdots \to X_\beta \to \cdots \quad (\beta < \lambda), \{T^\beta, e^\beta, h^\beta\}_{\beta<\lambda})$ is a presented relative I-cell complex, then a *subcomplex* of f consists of a presented relative I-cell complex $(\tilde{f}\colon X \to \tilde{Y}, X = \tilde{X}_0 \to \tilde{X}_1 \to \tilde{X}_2 \to \cdots \to \tilde{X}_\beta \to \cdots \quad (\beta < \lambda), \{\tilde{T}^\beta, \tilde{e}^\beta, \tilde{h}^\beta\}_{\beta<\lambda})$ such that

(1) for every $\beta < \lambda$ the set $\widetilde{T}^\beta$ is a subset of T^β and $\tilde{e}^\beta$ is the restriction of e^β to $\widetilde{T}^\beta$, and
(2) there is a map of λ-sequences

$$\begin{array}{ccccccccc}
X & \xrightarrow{1_X} & \widetilde{X}_0 & \longrightarrow & \widetilde{X}_1 & \longrightarrow & \widetilde{X}_2 & \longrightarrow & \cdots \\
{\scriptstyle 1_X}\downarrow & & \downarrow & & \downarrow & & \downarrow & & \\
X & \xrightarrow[1_X]{} & X_0 & \longrightarrow & X_1 & \longrightarrow & X_2 & \longrightarrow & \cdots
\end{array}$$

such that, for every $\beta < \lambda$ and every $i \in \widetilde{T}^\beta$, the map $\tilde{h}_i^\beta \colon C_i \to \widetilde{X}_\beta$ is a factorization of the map $h_i^\beta \colon C_i \to X_\beta$ through the map $\widetilde{X}_\beta \to X_\beta$.

REMARK 10.6.8. Although a subcomplex of a cell complex is defined to be a presented relative I-cell complex, we will often abuse language and refer to the λ-sequence associated with the subcomplex, or the colimit of that λ-sequence, as the subcomplex.

10.6.9. The case of monomorphisms.

PROPOSITION 10.6.10. *If $\mathcal{C}$ is a cocomplete category and I is a set of maps in $\mathcal{C}$ such that relative I-cell complexes (see Definition* 10.5.8 *and Proposition* 10.2.14*) are monomorphisms, then a subcomplex of a presented relative I-cell complex (see Definition* 10.6.7*) is entirely determined by its set of cells $\{\widetilde{T}^\beta\}_{\beta<\lambda}$ (see Definition* 10.6.4*).*

PROOF. The definition of a subcomplex implies that the maps $\widetilde{X}_\beta \to X_\beta$ are all inclusions of subcomplexes. Since inclusions of subcomplexes are monomorphisms, there is at most one possible factorization $\tilde{h}_i^\beta$ of each h_i^β through $\widetilde{X}_\beta \to X_\beta$. □

PROPOSITION 10.6.11. *Let $\mathcal{C}$ be a cocomplete category and let I be a set of maps in $\mathcal{C}$ such that relative I-cell complexes are monomorphisms. If $(f \colon X \to Y, X = X_0 \to X_1 \to X_2 \to \cdots \to X_\beta \to \cdots \quad (\beta < \lambda), \{T^\beta, e^\beta, h^\beta\}_{\beta<\lambda})$ is a presented relative I-cell complex, then an arbitrary subcomplex of f can be constructed by the following inductive procedure:*

(1) *Choose an arbitrary subset $\widetilde{T}^0$ of T^0.*
(2) *If $\beta < \lambda$ and we have defined $\{\widetilde{T}^\gamma\}_{\gamma<\beta}$, then we have determined the object $\widetilde{X}_\beta$ and the map $\widetilde{X}_\beta \to X_\beta$ (where $\widetilde{X}_\beta$ is the object that appears in the λ-sequence associated with the subcomplex). Consider the set*

$$\{i \in T^\beta \mid h_i^\beta \colon C_i \to X_\beta \text{ factors through } \widetilde{X}_\beta \to X_\beta\}$$

Choose an arbitrary subset $\widetilde{T}^\beta$ of this set. For every $i \in \widetilde{T}^\beta$ there is a unique map $\tilde{h}_i^\beta \colon C_i \to \widetilde{X}_\beta$ that makes the diagram

$$\begin{array}{ccc}
C_i & & \\
{\scriptstyle \tilde{h}_i^\beta}\downarrow & \searrow^{h_i^\beta} & \\
\widetilde{X}_\beta & \longrightarrow & X_\beta
\end{array}$$

commute. Let $\widetilde{X}_{\beta+1}$ be defined by the pushout diagram

$$\begin{array}{ccc} \coprod_{\widetilde{T}^\beta} C_i & \longrightarrow & \coprod_{\widetilde{T}^\beta} D_i \\ {\scriptstyle \amalg \tilde{h}^\beta}\downarrow & & \downarrow \\ \widetilde{X}_\beta & \longrightarrow & \widetilde{X}_{\beta+1} \end{array} .$$

PROOF. This follows directly from the definitions. □

REMARK 10.6.12. If $\mathcal{C}$ is a cocomplete category, I is a set of maps in $\mathcal{C}$ such that relative I-cell complexes are monomorphisms, and $(f\colon X \to Y, X = X_0 \to X_1 \to X_2 \to \cdots \to X_\beta \to \cdots \quad (\beta < \lambda), \{T^\beta, e^\beta, h^\beta\}_{\beta<\lambda})$ is a presented relative I-cell complex, then not every sequence $\{\widetilde{T}^\beta\}_{\beta<\lambda}$ of subsets of $\{T^\beta\}_{\beta<\lambda}$ determines a subcomplex of f. Given such a sequence $\{\widetilde{T}^\beta\}_{\beta<\lambda}$, it determines a subcomplex of f if and only if it satisfies the inductive conditions described in Proposition 10.6.11.

10.7. Cell complexes of topological spaces

A cell complex in $\mathrm{Top}_{(*)}$ is a topological space built by a sequential process of attaching cells. The class of cell complexes includes the class of CW-complexes, but the attaching map of a cell in a cell complex need not be contained in a union of cells of lower dimension. Thus, while a CW-complex can be built by a countable process of attaching coproducts of cells, a general cell complex may require an arbitrarily long transfinite construction.

The main disadvantage of using cell complexes that are not CW-complexes is that the cell structure cannot be used to compute the homology groups of the space. Cell complexes, however, have all the convenient mapping properties of CW-complexes, and the small object factorization (see Definition 10.5.19) produces cell complexes. Cell complexes and their retracts are the cofibrant objects in the standard model category of topological spaces (see Example 11.1.8, Example 11.1.9, and Corollary 11.2.2).

DEFINITION 10.7.1. Let Top denote our category of (unpointed) topological spaces and let Top_* denote our category of pointed topological spaces.

- A *relative cell complex* in Top is a map that is a transfinite composition (see Definition 10.2.2) of pushouts (see Definition 7.2.10) of maps of the form $|\partial\Delta[n]| \to |\Delta[n]|$ for $n \geq 0$. The topological space X in Top is a *cell complex* if the map $\emptyset \to X$ is a relative cell complex, and it is a *finite cell complex* if the map $\emptyset \to X$ is a finite composition of pushouts of maps of the form $|\partial\Delta[n]| \to |\Delta[n]|$ for $n \geq 0$.
- A *relative cell complex* in Top_* is a map that is a transfinite composition of pushouts of maps of the form $|\partial\Delta[n]|^+ \to |\Delta[n]|^+$ for $n \geq 0$. The topological space X in Top_* is a *cell complex* if the map $* \to X$ is a relative cell complex, and it is a *finite cell complex* if the map $* \to X$ is a finite composition of pushouts of maps of the form $|\partial\Delta[n]|^+ \to |\Delta[n]|^+$ for $n \geq 0$.

EXAMPLE 10.7.2. A relative CW-complex in $\mathrm{Top}_{(*)}$ is a relative cell complex, and a CW-complex in $\mathrm{Top}_{(*)}$ is a cell complex.

REMARK 10.7.3. Definition 10.7.1 implies that a relative cell complex in $\mathrm{Top}_{(*)}$ is a map that can be constructed as a transfinite composition of pushouts of inclusions of the boundary of a cell into that cell, but there will generally be many different possible such constructions. When dealing with a topological space that is a cell complex or a map that is a relative cell complex, we will often assume that we have chosen some specific such construction. Furthermore, we may choose a construction of the map as a transfinite composition of pushouts of *coproducts* of cells, i.e., we will consider constructions as transfinite compositions in which more than one cell is attached at a time (see Proposition 10.2.7).

PROPOSITION 10.7.4. *If $X \to Y$ is a relative cell complex in $\mathrm{Top}_{(*)}$, then a compact subset of Y can intersect the interiors of only finitely many cells of $Y - X$.*

PROOF. Let C be a subset of Y; we will show that if C intersects the interiors of infinitely many cells of $Y - X$, then C has an infinite subset that has no accumulation point (which implies that C is not compact).

Suppose now that C intersects the interiors of infinitely many cells of $Y - X$. We construct a subset P of C by choosing one point of C from the interior of each cell whose interior intersects C. We will now show that this infinite subset P of C has no accumulation point in C. We will do this by showing that for every point $c \in C$ there is an open subset U of Y such that $c \in U$ and $U \cap P$ is either empty or contains the one point c.

Let e_c be the unique cell of $Y - X$ that contains c in its interior. Since there is at most one point of P in the interior of any cell of $Y - X$, we can choose an open subset U_c of the interior of e_c that contains no points of P (except for c, if $c \in P$). We will use Zorn's lemma to show that we can enlarge U_c to an open subset of Y that contains no points of P (except for c, if $c \in P$).

Let α be the presentation ordinal (see Definition 10.6.4) of the cell e_c. If the presentation ordinal of the relative cell complex $X \to Y$ is γ, consider the set T of ordered pairs (β, U) where $\alpha \le \beta \le \gamma$ and U is an open subset of Y^{β} such that $U \cap Y^{\alpha} = U_c$ and U contains no points of P except possibly c. We define a preorder on T by defining $(\beta_1, U_1) < (\beta_2, U_2)$ if $\beta_1 < \beta_2$ and $U_2 \cap Y^{\beta_1} = U_1$.

If $\{(\beta_s, U_s)\}_{s\in S}$ is a chain in T, then $(\bigcup_{s\in S} \beta_s, \bigcup_{s\in S} U_s)$ (see Section 10.1.1) is an upper bound in T for the chain, and so Zorn's lemma implies that T has a maximal element (β_m, U_m). We will complete the proof by showing that $\beta_m = \gamma$.

If $\beta_m < \gamma$, then consider the cells of presentation ordinal $\beta_m + 1$. Since Y has the weak topology determined by X and the cells of $Y - X$, we need only enlarge U_m so that its intersection with each cell of presentation ordinal $\beta_m + 1$ is open in that cell, and so that it still contains no points of P except possibly c. If $h\colon S^{n-1} \to Y^{\beta_m}$ is the attaching map for a cell of presentation ordinal $\beta_m + 1$, then $h^{-1}U_m$ is open in S^{n-1}, and so we can "thicken" $h^{-1}U_m$ to an open subset of D^n, avoiding the (at most one) point of P that is in the interior of the cell. If we let U' equal the union of U_m with these thickenings in the interiors of the cells of presentation ordinal $\beta_m + 1$, then the pair $(\beta_m + 1, U')$ is an element of T greater than the maximal element (β_m, U_m) of T. This contradiction implies that $\beta_m = \gamma$, and so the proof is complete. □

COROLLARY 10.7.5. *A compact subset of a cell complex in $\mathrm{Top}_{(*)}$ can intersect the interiors of only finitely many cells.*

PROOF. This follows from Proposition 10.7.4. □

PROPOSITION 10.7.6. *Every cell of a cell complex in* $\mathrm{Top}_{(*)}$ *is contained in a finite subcomplex of the cell complex.*

PROOF. If we choose a presentation of the cell complex X (see Definition 10.6.2), then the proposition follows from Corollary 10.7.5, using a transfinite induction on the presentation ordinal of the cell. The attaching map of any cell intersects the interiors of only finitely many cells, each of which (by the induction hypothesis) is contained in a finite subcomplex of X. □

COROLLARY 10.7.7. *A compact subset of a cell complex in* $\mathrm{Top}_{(*)}$ *is contained in a finite subcomplex of the cell complex.*

PROOF. This follows from Corollary 10.7.5 and Proposition 10.7.6. □

10.8. Compactness

DEFINITION 10.8.1. Let $\mathcal{C}$ be a cocomplete category and let I be a set of maps in $\mathcal{C}$.

(1) If γ is a cardinal, then an object W in $\mathcal{C}$ is *γ-compact relative to I* if, for every presented relative I-cell complex $f\colon X \to Y$ (see Definition 10.6.3), every map from W to Y factors through a subcomplex of f of size (see Definition 10.6.4) at most γ.
(2) An object W in $\mathcal{C}$ is *compact relative to I* if it is γ-compact relative to I for some cardinal γ.

EXAMPLE 10.8.2. If $\mathcal{C} = \mathrm{SS}_{(*)}$ and I is the set of inclusions $\{\partial\Delta[n] \to \Delta[n] \mid n \geq 0\}$, then every finite simplicial set is ω-compact relative to I (where ω is the countable cardinal). If γ is an infinite cardinal and X is a simplicial set of size γ, then X is γ-compact relative to I.

EXAMPLE 10.8.3. If $\mathcal{C} = \mathrm{Top}_{(*)}$ and I is the set of inclusions $\bigl\{|\partial\Delta[n]| \to |\Delta[n]| \bigm| n \geq 0\bigr\}$, then Corollary 10.7.7 implies that every finite cell complex is ω-compact relative to I (where ω is the countable cardinal). If γ is an infinite cardinal and X is a cell complex of size γ, then Corollary 10.7.7 implies that X is γ-compact relative to I.

PROPOSITION 10.8.4. *Let $\mathcal{C}$ be a cocomplete category and let I be a set of maps in $\mathcal{C}$. If γ is a cardinal and an object W is γ-compact relative to I, then any retract of W is γ-compact relative to I.*

PROOF. Let $i\colon V \to W$ and $r\colon W \to V$ be maps such that $ri = 1_V$. If $f\colon X \to Y$ is a relative I-cell complex and $f\colon V \to Y$ is a map, then the map $fr\colon W \to Y$ must factor through some subcomplex Z of Y of size at most γ. Thus, $fri\colon V \to Y$ factors through Z, and $fri = f$. □

PROPOSITION 10.8.5. *Let $\mathcal{C}$ be a cocomplete category and let I be a set of maps in $\mathcal{C}$. If κ and λ are cardinals such that $\kappa < \lambda$, then any object that is κ-compact relative to I is also λ-compact relative to I.*

PROOF. This follows directly from the definitions. □

PROPOSITION 10.8.6. *If $\mathcal{C}$ is a cocomplete category, I is a set of maps in $\mathcal{C}$, and S is a set of objects that are compact relative to I, then there is a cardinal γ such that every element of S is γ-compact relative to I.*

PROOF. For each element X of S, let γ_X be a cardinal such that X is γ_X-compact relative to I. If γ is the cardinal of $\bigcup_{X \in S} \gamma_X$, then Proposition 10.8.5 implies that every element of S is γ-compact relative to I. $\square$

PROPOSITION 10.8.7. *Let $\mathcal{C}$ be a cocomplete category and let I be a set of maps in $\mathcal{C}$ such that relative I-cell complexes are monomorphisms. If γ is a cardinal and W is an object that is γ-compact relative to I (see Definition 10.8.1), then W is $(\gamma + 1)$-small relative to I.*

PROOF. Let λ be a regular cardinal such that $\lambda > \gamma$ and let $X_0 \to X_1 \to X_2 \to \cdots \to X_\beta \to \cdots$ $(\beta < \lambda)$ be a λ-sequence of inclusions of relative I-cell complexes. Since inclusions of relative I-cell complexes are monomorphisms, the map $\operatorname{colim}_{\beta<\lambda} \mathcal{C}(W, X_\beta) \to \mathcal{C}(W, \operatorname{colim}_{\beta<\lambda} X_\beta)$ is injective, and it remains only to show that it is surjective.

If $W \to \operatorname{colim}_{\beta<\lambda} X_\beta$ is a map, then (since W is γ-compact) there is a subcomplex K of $\operatorname{colim} X_\beta$, of size at most γ, such that the map factors through K. For each cell of K there is an ordinal $\beta < \lambda$ such that that cell is contained in X_β. Since λ is a regular cardinal, the union μ of these β is less than λ, and K is contained in X_μ. $\square$

PROPOSITION 10.8.8. *Let $\mathcal{C}$ be a cocomplete category and let I be a set of maps in $\mathcal{C}$ such that relative I-cell complexes are monomorphisms. If $\mathcal{D}$ is a small category and $\boldsymbol{X}\colon \mathcal{D} \to \mathcal{C}$ is a diagram such that $\boldsymbol{X}_\alpha$ is compact relative to I for every object α of $\mathcal{D}$, then $\operatorname{colim}_{\mathcal{D}} \boldsymbol{X}$ is compact relative to I.*

PROOF. Let κ be a cardinal such that $\boldsymbol{X}_\alpha$ is κ-compact relative to I for every object α of $\mathcal{D}$ (see Proposition 10.8.6), let μ be the cardinal of the set of objects of $\mathcal{D}$, and let $\gamma = \kappa\mu$; we will show that $\operatorname{colim}_{\mathcal{D}} \boldsymbol{X}$ is γ-compact relative to I.

If $f\colon X \to Y$ is a relative I-cell complex and $g\colon \operatorname{colim}_{\mathcal{D}} \boldsymbol{X} \to Y$ is a map, then for every object β of $\mathcal{D}$ the composition $\boldsymbol{X}_\beta \to \operatorname{colim}_{\mathcal{D}} \boldsymbol{X} \xrightarrow{g} Y$ factors through some sub relative I-cell complex $X \to W_\beta$ of f of size at most κ. If W is the union of the W_β, then W is of size at most $\kappa\mu = \gamma$. If $s\colon \alpha \to \beta$ is a map in $\mathcal{D}$, then the triangle

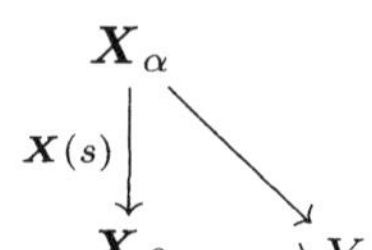

commutes; since the inclusion $W \to Y$ is a monomorphism, the triangle

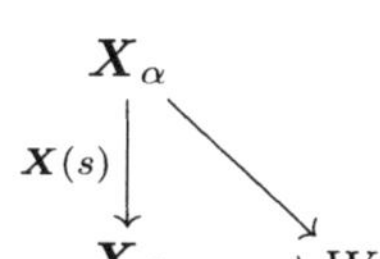

must commute as well, and so the maps $\boldsymbol{X}_\alpha \to W$ define the map $\operatorname{colim}_{\mathcal{D}} \boldsymbol{X} \to W$ that we require. $\square$

10.9. Effective monomorphisms

DEFINITION 10.9.1. Let $\mathcal{C}$ be a category that is closed under pushouts. The map $f\colon A \to B$ in $\mathcal{C}$ is an *effective monomorphism* if f is the equalizer of the pair of natural inclusions $B \rightrightarrows B \amalg_A B$.

REMARK 10.9.2. An effective monomorphism is dual to what Quillen has called an effective epimorphism (see [**52**, Part II, page 4.1]). Effective monomorphisms have also been called *regular monomorphisms* (see [**1**, page 2]), and effective epimorphisms have also been called *regular epimorphisms* (see [**6**, Definition 4.3.1]).

EXAMPLE 10.9.3. If $\mathcal{C}$ is the category of sets, then the class of effective monomorphisms is the class of injective maps.

PROPOSITION 10.9.4. *If $\mathcal{C}$ is a category that is closed under pushouts, then a map is an effective monomorphism if and only if it is the equalizer of some pair of parallel maps.*

PROOF. If $f\colon A \to B$ is an effective monomorphism, then it is defined to be the equalizer of a particular pair of maps. Conversely, if $f\colon A \to B$ is the equalizer of the maps $B \overset{g}{\underset{h}{\rightrightarrows}} W$, then the maps g and h factor as

$$B \overset{i_0}{\underset{i_1}{\rightrightarrows}} B \amalg_A B \xrightarrow{g \amalg h} W \ ,$$

and we must show that f is the equalizer of i_0 and i_1. Since $(g \amalg h)i_0 = g$ and $(g \amalg h)i_1 = h$, this follows directly from the definitions. □

PROPOSITION 10.9.5. *if $\mathcal{C}$ is a category that is closed under pushouts, then an effective monomorphism is a monomorphism.*

PROOF. Let $f\colon A \to B$ be an effective monomorphism and let $g\colon W \to A$ and $h\colon W \to A$ be maps such that $fg = fh$. If i_0 and i_1 are the natural maps from B to $B \amalg_A B$, then $i_0 f = i_1 f$, and so $i_0 fg = i_1 fg$ and $i_0 fh = i_1 fh$. The uniqueness requirement in the definition of equalizer now implies that $g = h$. □

PROPOSITION 10.9.6. *If $\mathcal{C}$ is a category that is closed under pushouts, then the class of effective monomorphisms is closed under retracts.*

PROOF. If $f\colon A \to B$ is a retract of $g\colon C \to D$, then we have the diagram

$$\begin{array}{ccccc}
A & \longrightarrow & C & \longrightarrow & A \\
{\scriptstyle f}\downarrow & & \downarrow{\scriptstyle g} & & \downarrow{\scriptstyle f} \\
B & \longrightarrow & D & \longrightarrow & B \\
{\scriptstyle i_0}\downdownarrows{\scriptstyle i_1} & & {\scriptstyle j_0}\downdownarrows{\scriptstyle j_1} & & {\scriptstyle i_0}\downdownarrows{\scriptstyle i_1} \\
B \amalg_A B & \longrightarrow & D \amalg_C D & \longrightarrow & B \amalg_A B
\end{array}$$

in which all of the horizontal compositions are identity maps. If g is an effective monomorphism then g is the equalizer of j_0 and j_1, and a diagram chase then shows that f is the equalizer of i_0 and i_1. □

CHAPTER 11

Cofibrantly Generated Model Categories

A model category structure on a category consists of three classes of maps (the weak equivalences, the cofibrations, and the fibrations) satisfying five axioms (see Definition 7.1.3). Any two of these classes determine the third (see Proposition 7.2.7), but there are other ways to determine the three classes of maps as well. For example, the fibrations are the maps with the right lifting property (see Definition 7.2.1) with respect to all trivial cofibrations (see Proposition 7.2.3), and so the class of trivial cofibrations determines the class of fibrations. Similarly, the trivial fibrations are the maps with the right lifting property with respect to all cofibrations (see Proposition 7.2.3), and so the class of cofibrations determines the class of trivial fibrations. Since the weak equivalences are the maps that can be written as the composition of a trivial cofibration followed by a trivial fibration (see Proposition 7.2.6), this shows that the classes of cofibrations and of trivial cofibrations entirely determine the model category structure. In some model categories, this leads to a convenient description of the model category structure.

For example, the standard model category structure on the category of simplicial sets can be described as follows:

- A map is a cofibration if it is a retract of a transfinite composition (see Definition 10.2.2) of pushouts of the maps $\partial\Delta[n] \to \Delta[n]$ for all $n \geq 0$.
- A map is a trivial fibration if it has the right lifting property with respect to the maps $\partial\Delta[n] \to \Delta[n]$ for all $n \geq 0$.
- A map is a trivial cofibration if it is a retract of a transfinite composition (see Definition 10.2.2) of pushouts of the maps $\Lambda[n,k] \to \Delta[n]$ for all $n \geq 1$ and $0 \leq k \leq n$.
- A map is a fibration if it has the right lifting property with respect to the maps $\Lambda[n,k] \to \Delta[n]$ for all $n \geq 1$ and $0 \leq k \leq n$.
- A map is a weak equivalence if it is the composition of a trivial cofibration followed by a trivial fibration.

These ideas lead to the notion (due to D. M. Kan) of a *cofibrantly generated model category* (see Definition 11.1.2).

We define cofibrantly generated model categories in Section 11.1; this requires the notions of transfinite composition (see Definition 10.2.2) and smallness (see Definition 10.4.1) discussed in Chapter 10. In Section 11.2 we discuss cofibrations, trivial cofibrations, and smallness in cofibrantly generated model categories.

In Section 11.3 we prove two theorems useful for establishing cofibrantly generated model category structures: The first is a set of sufficient conditions to have a cofibrantly generated model category structure on a category, and the second provides for "lifting" a cofibrantly generated model category structure from one category to another via a pair of adjoint functors. In Section 11.4 we discuss compactness (see Definition 10.8.1) in a cofibrantly generated model category.

In the remainder of the chapter we study categories of diagrams, i.e., functor categories (see Definition 11.5.2). If $\mathcal{M}$ is a cofibrantly generated model category and $\mathcal{C}$ is a small category, then there is a cofibrantly generated model category structure on the category of $\mathcal{C}$-diagrams in $\mathcal{M}$, i.e., the category of functors from $\mathcal{C}$ to $\mathcal{M}$ (see Theorem 11.6.1). In Section 11.5 we describe *free diagrams*, which are diagrams constructed via the left adjoint to the functor that evaluates a $\mathcal{C}$-diagram at a fixed object of $\mathcal{C}$ (see Proposition 11.5.8 and Proposition 11.5.26), and *free cells*, which are maps of $\mathcal{C}$-diagrams in $\mathcal{M}$ constructed by applying the free diagram functor to the generating cofibrations of $\mathcal{M}$ (see Definition 11.5.30). The free cells will be the generating cofibrations for the model category of $\mathcal{C}$-diagrams in $\mathcal{M}$. In Section 11.6 we establish the model category of diagrams in a cofibrantly generated model category, and in Section 11.7 we show that if $\mathcal{M}$ is a cofibrantly generated model category that is also a *simplicial category*, then the model category of $\mathcal{C}$-diagrams in $\mathcal{M}$ is also simplicial.

In Section 11.8 we define overcategories and undercategories, and we use them in Section 11.9 to define extensions of diagrams, which generalize the idea of a free diagram (see Definition 11.9.1).

11.1. Cofibrantly generated model categories

11.1.1. Definitions.

DEFINITION 11.1.2. A *cofibrantly generated model category* is a model category $\mathcal{M}$ such that

(1) there exists a set I of maps (called a set of *generating cofibrations*) that permits the small object argument (see Definition 10.5.15) and such that a map is a trivial fibration if and only if it has the right lifting property with respect to every element of I, and
(2) there exists a set J of maps (called a set of *generating trivial cofibrations*) that permits the small object argument and such that a map is a fibration if and only if it has the right lifting property with respect to every element of J.

REMARK 11.1.3. Although the sets I and J of Definition 11.1.2 are not part of the structure of a cofibrantly generated model category, we will often assume that some particular set I of generating cofibrations has been chosen.

DEFINITION 11.1.4. Let $\mathcal{M}$ be a cofibrantly generated model category with generating cofibrations I (see Remark 11.1.3).

(1) A relative I-cell complex (see Definition 10.5.8) will be called a *relative cell complex*, and an I-cell complex (see Definition 10.5.8) will be called a *cell complex*.
(2) If X is a cell complex and $g\colon X \to Y$ is a relative cell complex, then g will be called an *inclusion of a subcomplex*.
(3) If $\emptyset \to X$ (where $\emptyset$ is the initial object of $\mathcal{M}$) is a finite composition of pushouts of elements of I, then X will be called a *finite cell complex*.

We will show in Proposition 11.2.1 that in a cofibrantly generated model category the class of cofibrations equals the class of retracts of relative cell complexes and the class of trivial cofibrations equals the class of retracts of relative J-cell complexes.

11.1.5. Examples.

EXAMPLE 11.1.6. The model category SS is cofibrantly generated. The generating cofibrations are the inclusions $\partial\Delta[n] \to \Delta[n]$ for $n \geq 0$, and the generating trivial cofibrations are the inclusions $\Lambda[n,k] \to \Delta[n]$ for $n > 0$ and $0 \leq k \leq n$.

EXAMPLE 11.1.7. The model category SS_* is cofibrantly generated. The generating cofibrations are the inclusions $\partial\Delta[n]^+ \to \Delta[n]^+$ for $n \geq 0$, and the generating trivial cofibrations are the inclusions $\Lambda[n,k]^+ \to \Delta[n]^+$ for $n > 0$ and $0 \leq k \leq n$.

EXAMPLE 11.1.8. The model category Top is cofibrantly generated. The generating cofibrations are the inclusions $|\partial\Delta[n]| \to |\Delta[n]|$ for $n \geq 0$, and the generating trivial cofibrations are the inclusions $|\Lambda[n,k]| \to |\Delta[n]|$ for $n > 0$ and $0 \leq k \leq n$.

EXAMPLE 11.1.9. The model category Top_* is cofibrantly generated. The generating cofibrations are the inclusions $|\partial\Delta[n]|^+ \to |\Delta[n]|^+$ for $n \geq 0$, and the generating trivial cofibrations are the inclusions $|\Lambda[n,k]|^+ \to |\Delta[n]|^+$ for $n > 0$ and $0 \leq k \leq n$.

PROPOSITION 11.1.10. *If S is a set and for every element s of S we have a cofibrantly generated model category $\mathcal{M}_s$ with generating cofibrations I_s and generating trivial cofibrations J_s, then the model category structure on $\prod_{s\in S}\mathcal{M}_s$ of Proposition 7.1.7 is cofibrantly generated with generating cofibrations I and generating trivial cofibrations J where*

$$I = \bigcup_{s\in S}(I_s \times \prod_{t\neq s} 1_{\phi_t})$$

$$J = \bigcup_{s\in S}(J_s \times \prod_{t\neq s} 1_{\phi_t})$$

and where 1_{ϕ_t} is the identity map of the initial object of $\mathcal{M}_t$.

PROOF. This follows directly from the definition (see Definition 11.1.2), since a map has the right lifting property with respect to $I_s \times \prod_{t\neq s} 1_{\phi_t}$ if and only if its s-component is a trivial fibration and a map has the right lifting property with respect to $J_s \times \prod_{t\neq s} 1_{\phi_t}$ if and only if its s-component is a fibration. □

11.2. Cofibrations in a cofibrantly generated model category

PROPOSITION 11.2.1. *Let $\mathcal{M}$ be a cofibrantly generated model category (see Definition 11.1.2) with generating cofibrations I and generating trivial cofibrations J.*

1. *The class of cofibrations of $\mathcal{M}$ equals the class of retracts of relative I-cell complexes (see Definition 10.5.8), which equals the class of I-cofibrations (see Definition 10.5.2).*
2. *The class of trivial fibrations of $\mathcal{M}$ equals the class of I-injectives (see Definition 10.5.2).*
3. *The class of trivial cofibrations of $\mathcal{M}$ equals the class of retracts of relative J-cell complexes, which equals the class of J-cofibrations (see Definition 10.5.2).*

(4) *The class of fibrations of* $\mathcal{M}$ *equals the class of* J*-injectives.*

PROOF. This follows from Proposition 7.2.3, Proposition 10.5.11, and Corollary 10.5.22. □

COROLLARY 11.2.2. *If* $\mathcal{M}$ *is a cofibrantly generated model category with generating cofibrations* I*, then every cofibrant object of* $\mathcal{M}$ *is a retract of a cell complex (see Definition* 11.1.4*).*

PROOF. This follows from Proposition 11.2.1. □

PROPOSITION 11.2.3. *Let* $\mathcal{M}$ *be a cofibrantly generated model category with generating cofibrations* I*. If* W *is an object that is small relative to* I*, then it is small relative to the subcategory of all cofibrations.*

PROOF. This follows from Theorem 10.5.27 and Proposition 11.2.1. □

COROLLARY 11.2.4. *Let* $\mathcal{M}$ *be a cofibrantly generated model category with generating cofibrations* I*. If the codomains of the elements of* I *are small relative to* I*, then every cofibrant object of* $\mathcal{M}$ *is small relative to the subcategory of all cofibrations.*

PROOF. This follows from Corollary 10.4.9, Corollary 11.2.2, Proposition 10.4.7, and Proposition 11.2.3. □

PROPOSITION 11.2.5. *If* $\mathcal{M}$ *is a cofibrantly generated model category and* I *is a set of generating cofibrations for* $\mathcal{M}$*, then there is a regular cardinal* κ *such that the domain of every element of* I *is* κ*-small relative to* I*.*

PROOF. This follows from Lemma 10.4.6. □

COROLLARY 11.2.6. *Let* $\mathcal{M}$ *be a cofibrantly generated model category. If* I *is a set of generating cofibrations for* $\mathcal{M}$ *and* κ *is a regular cardinal such that the domain of every element of* I *is* κ*-small relative to* I *(see Proposition* 11.2.5*), then there is a functorial factorization of every map in* $\mathcal{M}$ *into a cofibration that is the composition of a* κ*-sequence of pushouts of coproducts of elements of* I *followed by a trivial fibration.*

PROOF. This follows from Corollary 10.5.21 and Proposition 11.2.1. □

PROPOSITION 11.2.7. *If* $\mathcal{M}$ *is a cofibrantly generated model category with generating cofibrations* I*, then every object* X *has a fibrant cofibrant approximation* $\tilde{\imath}\colon \widetilde{X} \to X$ *such that* $\widetilde{X}$ *is a cell complex.*

PROOF. This follows from Proposition 10.5.16, Proposition 10.5.10, and Proposition 11.2.1. □

PROPOSITION 11.2.8. *If* $\mathcal{M}$ *is a cofibrantly generated model category with generating cofibrations* I*, then every map* $g\colon X \to Y$ *has a cofibrant approximation* $\tilde{g}\colon \widetilde{X} \to \widetilde{Y}$ *such that* $\tilde{g}\colon \widetilde{X} \to \widetilde{Y}$ *is an inclusion of a subcomplex and both* $i_X : \widetilde{X} \to X$ *and* $i_Y : \widetilde{Y} \to Y$ *are trivial fibrations.*

PROOF. Choose a cofibrant approximation $i_X : \widetilde{X} \to X$ such that $\widetilde{X}$ is a cell complex and i_X is a trivial fibration (see Proposition 11.2.7). We can then factor the composition $\widetilde{X} \xrightarrow{i_X} X \xrightarrow{g} Y$ as $\widetilde{X} \xrightarrow{\tilde{g}} \widetilde{Y} \xrightarrow{i_Y} Y$ where $\tilde{g}$ is a relative I-cell complex and i_Y is a trivial fibration (see Proposition 10.5.16). The result now follows from Proposition 10.5.10 and Proposition 11.2.1. □

PROPOSITION 11.2.9. *Let $\mathcal{M}$ be a cofibrantly generated model category, and let I be a set of generating cofibrations for $\mathcal{M}$. If J is a set of generating trivial cofibrations for $\mathcal{M}$, then there is a set $\widetilde{J}$ of generating trivial cofibrations for $\mathcal{M}$ such that*

(1) *there is a bijection between the sets J and $\widetilde{J}$ under which corresponding elements have the same domain, and*
(2) *the elements of $\widetilde{J}$ are relative I-cell complexes.*

PROOF. Factor each element $j\colon C \to D$ of J as $C \xrightarrow{\tilde{j}} \widetilde{D} \xrightarrow{p} D$ where $\tilde{j}$ is a relative I-cell complex and p is a trivial fibration (see Corollary 11.2.6). The retract argument (see Proposition 7.2.2) implies that j is a retract of $\tilde{j}$. Since j and p are weak equivalences, $\tilde{j}$ is also a weak equivalence, and so $\tilde{j}$ is a trivial cofibration. Thus, if we let $\widetilde{J} = \{\tilde{j}\}_{j\in J}$, then $\widetilde{J}$ satisfies conditions 1 and 2, and it remains only to show that $\widetilde{J}$ is a set of generating trivial cofibrations for $\mathcal{M}$.

Proposition 10.5.7 implies that it is sufficient to show that the subcategory of $\widetilde{J}$-injectives equals the subcategory of J-injectives (i.e., of fibrations). Since every $\tilde{j}$ is a trivial cofibration, Proposition 7.2.3 implies that every J-injective is a $\widetilde{J}$-injective, and since every j is a retract of $\tilde{j}$, Lemma 7.2.8 implies that every $\widetilde{J}$-injective is a J-injective. □

PROPOSITION 11.2.10. *Let $\mathcal{M}$ be a cofibrantly generated model category with generating cofibrations I. If relative I-cell complexes are effective monomorphisms (see Definition 10.9.1), then all cofibrations are effective monomorphisms.*

PROOF. This follows from Proposition 11.2.1 and Proposition 10.9.6. □

PROPOSITION 11.2.11. *Let $\mathcal{M}$ be a cofibrantly generated model category with generating cofibrations I. If relative I-cell complexes are monomorphisms, then all cofibrations are monomorphisms.*

PROOF. Since a retract of a monomorphism is a monomorphism, this follows from Proposition 11.2.1. □

11.3. Recognizing cofibrantly generated model categories

In this section we present two theorems of D. M. Kan that are used to establish a cofibrantly generated model category structure on a category. Theorem 11.3.1 is a recognition theorem that gives sufficient conditions to have a cofibrantly generated model category structure on a category. Theorem 11.3.2 is a lifting theorem that gives sufficient conditions for a pair of adjoint functors to "lift" a cofibrantly generated model category structure from one category to another.

THEOREM 11.3.1 (D. M. Kan). *Let $\mathcal{M}$ be a category that is closed under small limits and colimits and let W be a class of maps in $\mathcal{M}$ that is closed under retracts and satisfies the "two out of three" axiom (axiom M2 of Definition 7.1.3). If I and J are sets of maps in $\mathcal{M}$ such that*

(1) *both I and J permit the small object argument (see Definition 10.5.15),*
(2) *every J-cofibration is both an I-cofibration and an element of W,*

(3) *every I-injective is both a J-injective and an element of W, and*
(4) *one of the following two conditions holds:*
  (a) *a map that is both an I-cofibration and an element of W is a J-cofibration, or*
  (b) *a map that is both a J-injective and an element of W is an I-injective,*

then there is a cofibrantly generated model category structure (see Definition 11.1.2) on $\mathcal{M}$ in which W is the class of weak equivalences, I is a set of generating cofibrations, and J is a set of generating trivial cofibrations.

PROOF. We define the weak equivalences to be the elements of W, the cofibrations to be the I-cofibrations, and the fibrations to be the J-injectives. We must show that axioms M1 through M5 are satisfied (see Definition 7.1.3).

Axioms M1 and M2 are part of our assumptions, and axiom M3 follows from the assumptions on W, the definition of I-cofibration (see Definition 10.5.2), and Lemma 7.2.8.

If we apply the small object argument (Proposition 10.5.16) to the set I, then assumption 3 implies that this satisfies axiom M5 part 1, and if we apply the small object argument to the set J, then assumption 2 implies that this satisfies axiom M5 part 2.

It remains only to show that axiom M4 is satisfied. The proof of axiom M4 depends on which part of assumption 4 is satisfied. If assumption 4a is satisfied, then axiom M4 part 2 is clear. For axiom M4 part 1, if $f\colon X \to Y$ is both a fibration and a weak equivalence, we can factor it as $X \xrightarrow{g} Z \xrightarrow{h} Y$ where g is an I-cofibration and h is an I-injective. Axiom M2 and assumption 3 imply that g is also a weak equivalence, and so assumption 4a implies that g is a J-cofibration. Since f is a J-injective, the retract argument (Proposition 7.2.2) implies that f is a retract of h, and is thus an I-injective (see Lemma 7.2.8). This proves axiom M4 part 1, and so the proof in the case that assumption 4a is satisfied is complete. The proof in the case in which assumption 4b is satisfied is similar. □

THEOREM 11.3.2 (D. M. Kan). *Let $\mathcal{M}$ be a cofibrantly generated model category (see Definition 11.1.2) with generating cofibrations I and generating trivial cofibrations J. Let $\mathcal{N}$ be a category that is closed under small limits and colimits, and let $\mathrm{F}\colon \mathcal{M} \rightleftarrows \mathcal{N} :\mathrm{U}$ be a pair of adjoint functors. If we let $\mathrm{F}I = \{\mathrm{F}u \mid u \in I\}$ and $\mathrm{F}J = \{\mathrm{F}v \mid v \in J\}$ and if*

(1) *both of the sets $\mathrm{F}I$ and $\mathrm{F}J$ permit the small object argument (see Definition 10.5.15) and*
(2) *U takes relative $\mathrm{F}J$-cell complexes (see Definition 10.5.8) to weak equivalences,*

then there is a cofibrantly generated model category structure on $\mathcal{N}$ in which $\mathrm{F}I$ is a set of generating cofibrations, $\mathrm{F}J$ is a set of generating trivial cofibrations, and the weak equivalences are the maps that U takes into a weak equivalence in $\mathcal{M}$. Furthermore, with respect to this model category structure, (F,U) is a Quillen pair (see Definition 8.5.2).

PROOF. Let W be the class of maps in $\mathcal{N}$ that U takes to a weak equivalence in $\mathcal{M}$; we will show that W, $\mathrm{F}I$, and $\mathrm{F}J$ satisfy the conditions of Theorem 11.3.1.

Since any functor preserves retract and compositions, W is closed under retracts and satisfies the "two out of three" axiom.

Part 1 of Theorem 11.3.1 is one of our assumptions about the sets FI and FJ.

For part 2 of Theorem 11.3.1, our assumptions imply that relative FJ-cell complexes are elements of W, and so Corollary 10.5.22 implies that every FJ-cofibration is an element of W. Since every I-injective is a J-injective, Proposition 7.2.17 implies that every FI-injective is an FJ-injective, and so Proposition 10.5.6 implies that every FJ-cofibration is an FI-cofibration.

For part 3 of Theorem 11.3.1, we showed in the last paragraph that every FI-injective is an FJ-injective, and Proposition 7.2.17 implies that U takes every FI-injective to a trivial fibration in $\mathcal{M}$.

For part 4 of Theorem 11.3.1, we will show that condition b holds. If $g\colon X \to Y$ is both an FJ-injective and an element of W, then Proposition 7.2.17 implies that Ug is both a J-injective and a weak equivalence in $\mathcal{M}$. Thus, Ug is a trivial fibration in $\mathcal{M}$, and so it is an I-injective. Proposition 7.2.17 now implies that g is an FI-injective.

Finally, since left adjoints preserve all colimits, F takes all relative I-cell complexes to relative FI-cell complexes and all relative J-cell complexes to relative FJ-cell complexes. Since every functor preserves retracts, Proposition 11.2.1 implies that F is a left Quillen functor, and so Proposition 8.5.3 implies that (F, U) is a Quillen pair. □

11.4. Compactness

DEFINITION 11.4.1. Let $\mathcal{M}$ be a cofibrantly generated model category with generating cofibrations I.

(1) If γ is a cardinal, then an object W in $\mathcal{M}$ is *γ-compact* if it is γ-compact relative to I (see Definition 10.8.1).
(2) An object W in $\mathcal{M}$ is *compact* if there is a cardinal γ for which it is γ-compact.

EXAMPLE 11.4.2. If $\mathcal{M} = \mathrm{SS}_{(*)}$, then every finite simplicial set is ω-compact (where ω is the countable cardinal). If γ is an infinite cardinal and X is a simplicial set of size γ, then X is γ-compact.

EXAMPLE 11.4.3. If $\mathcal{M} = \mathrm{Top}_{(*)}$, then Corollary 10.7.7 implies that every finite cell complex is ω-compact (where ω is the countable cardinal). If γ is an infinite cardinal and X is a cell complex of size γ, then Corollary 10.7.7 implies that X is γ-compact.

PROPOSITION 11.4.4. *Let $\mathcal{M}$ be a cofibrantly generated model category with generating cofibrations I. If γ is a cardinal and an object W in $\mathcal{M}$ is γ-compact, then any retract of W is γ-compact.*

PROOF. This follows from Proposition 10.8.4. □

PROPOSITION 11.4.5. *Let $\mathcal{M}$ be a cofibrantly generated model category with generating cofibrations I. If κ and λ are cardinals such that $\kappa < \lambda$, then any object of $\mathcal{M}$ that is κ-compact is also λ-compact.*

PROOF. This follows directly from the definitions. □

PROPOSITION 11.4.6. *If $\mathcal{M}$ is a cofibrantly generated model category with generating cofibrations I and S is a set of objects that are compact, then there is a cardinal γ such that every element of S is γ-compact.*

PROOF. This follows from Proposition 10.8.6. □

PROPOSITION 11.4.7. *Let $\mathcal{M}$ be a cofibrantly generated model category in which cofibrations are monomorphisms and let I be a set of generating cofibrations for $\mathcal{M}$. If the domains and codomains of the elements of I are compact (see Definition 11.4.1), then every cofibrant object is compact.*

PROOF. Since an I-cell complex is the colimit of a λ-sequence of codomains of elements of I, the result follows from Proposition 10.8.8, Corollary 11.2.2, and Proposition 11.4.4. □

PROPOSITION 11.4.8. *Let $\mathcal{M}$ be a cofibrantly generated model category in which cofibrations are monomorphisms. If γ is a cardinal and K is a set of cofibrations whose domains are γ-compact (see Definition 11.4.1), then every K-cell of a relative K-cell complex (see Definition 10.5.8) is contained in a sub relative K-cell complex (see Definition 10.6.7) of size (see Definition 10.6.4) at most γ.*

PROOF. If $f\colon X \to Y$ is a relative K-cell complex, then we can write f as the composition of a λ-sequence $X = X_0 \to X_1 \to X_2 \to \cdots \to X_\beta \to \cdots$ $(\beta < \lambda)$ in which each map $X_\beta \to X_{\beta+1}$ is a pushout of an element of K. We will show by induction on β that the attaching map of each K-cell factors through a sub relative K-cell complex of size at most γ. The induction is begun because the attaching map of the K-cell of presentation ordinal 1 has codomain $X = X_0$.

We now assume that $\alpha < \lambda$ and that every K-cell of $X \to X_\alpha$ is contained in a subcomplex of size at most γ. Let $C \to D$ be the element of K such that $X_{\alpha+1}$ is constructed as a pushout

$$\begin{array}{ccc} C & \longrightarrow & D \\ {\scriptstyle h^\alpha}\downarrow & & \vdots \\ X_\alpha & \cdots\rightarrow & X_{\alpha+1} \end{array} ;$$

we must show that h^α factors through a sub relative K-cell complex of size at most γ. Lemma 10.5.25 implies that we can find a diagram

$$\begin{array}{ccccccc} X_0 & \xrightarrow{\sigma_0} & X_1 & \xrightarrow{\sigma_1} & X_2 & \xrightarrow{\sigma_2} \cdots \longrightarrow & X_\alpha \\ {\scriptstyle i_0}\downarrow & & {\scriptstyle i_1}\downarrow & & {\scriptstyle i_2}\downarrow & & {\scriptstyle i_\alpha}\downarrow \\ \widetilde{X}_0 & \xrightarrow{\tau_0} & \widetilde{X}_1 & \xrightarrow{\tau_1} & \widetilde{X}_2 & \xrightarrow{\tau_2} \cdots \longrightarrow & \widetilde{X}_\alpha \\ {\scriptstyle r_0}\downarrow & & {\scriptstyle r_1}\downarrow & & {\scriptstyle r_2}\downarrow & & {\scriptstyle r_\alpha}\downarrow \\ X_0 & \xrightarrow{\sigma_0} & X_1 & \xrightarrow{\sigma_1} & X_2 & \xrightarrow{\sigma_2} \cdots \longrightarrow & X_\alpha \end{array}$$

such that $r_\beta i_\beta = 1_{X_\beta}$ for $\beta \le \alpha$ and every τ_β is a relative I-cell complex (where I is the set of generating cofibrations for $\mathcal{M}$). Thus, the composition $\widetilde{X}_0 \to \widetilde{X}_\alpha$ is a relative I-cell complex, and so the composition $i_\alpha h^\alpha\colon C \to \widetilde{X}_\alpha$ must factor through some sub relative I-cell complex $\widetilde{X}_0 \to C$ of $\widetilde{X}_0 \to \widetilde{X}_\alpha$ of size at most γ. We will complete the proof by showing that the composition $V \to \widetilde{X}_\alpha \to X_\alpha$ factors through a sub relative K-cell complex of $X_0 \to X_\alpha$ of size at most γ.

For each I-cell of V there is exactly one $\beta < \alpha$ such that that cell is a part of τ_β, and (by the induction hypothesis) the corresponding relative K-cell $\sigma_\beta\colon X_\beta \to X_{\beta+1}$ is contained in a sub relative K-cell complex of $X_0 \to X_\alpha$ of size at most

γ. If we take the union Z of these sub relative K-cell complexes, then the relative K-cell complex $X_0 \to Z$ has size at most $\gamma \times \gamma = \gamma$ (since γ is infinite), and the composition $V \to \widetilde{X}_\alpha \to X_\alpha$ factors through the inclusion $Z \to X_\alpha$. □

PROPOSITION 11.4.9. *Let $\mathcal{M}$ be a cofibrantly generated model category in which cofibrations are monomorphisms and let K be a set of cofibrations with compact domains (see Definition 11.4.1). If an object W of $\mathcal{M}$ is compact, then it is compact relative to K.*

PROOF. Let γ be an infinite cardinal such that W and the domains of the elements of K are γ-compact (see Proposition 11.4.6); we will show that W is γ-compact relative to K.

Let $X = X_0 \to X_1 \to X_2 \to \cdots \to X_\beta \to \cdots$ $(\beta < \lambda)$ be a λ-sequence of pushouts of elements of K, let $X_\lambda = \operatorname{colim}_{\beta<\lambda} X_\beta$ be the colimit of that sequence, and let $g\colon W \to X_\lambda$ be a map; we will show that g factors through a sub relative K-cell complex of $f\colon X_0 \to X_\lambda$ of size at most γ.

Lemma 10.5.25 implies that we can find a diagram

$$\begin{array}{ccccccccccc}
X_0 & \xrightarrow{\sigma_0} & X_1 & \xrightarrow{\sigma_1} & X_2 & \xrightarrow{\sigma_2} & \cdots & \longrightarrow & X_\beta & \longrightarrow & \cdots \\
\downarrow{\scriptstyle i_0} & & \downarrow{\scriptstyle i_1} & & \downarrow{\scriptstyle i_2} & & & & \downarrow{\scriptstyle i_\beta} & & \\
\widetilde{X}_0 & \xrightarrow{\tau_0} & \widetilde{X}_1 & \xrightarrow{\tau_1} & \widetilde{X}_2 & \xrightarrow{\tau_2} & \cdots & \longrightarrow & \widetilde{X}_\beta & \longrightarrow & \cdots \\
\downarrow{\scriptstyle r_0} & & \downarrow{\scriptstyle r_1} & & \downarrow{\scriptstyle r_2} & & & & \downarrow{\scriptstyle r_\beta} & & \\
X_0 & \xrightarrow{\sigma_0} & X_1 & \xrightarrow{\sigma_1} & X_2 & \xrightarrow{\sigma_2} & \cdots & \longrightarrow & X_\beta & \longrightarrow & \cdots
\end{array}$$

such that $r_\beta i_\beta = 1_{X_\beta}$ for $\beta \le \lambda$ and every τ_β is a relative I-cell complex (where I is the set of generating cofibrations for $\mathcal{M}$). Thus, the composition $\widetilde{X}_0 \to \widetilde{X}_\lambda$ is a relative I-cell complex, and so the composition $i_\lambda g\colon W \to \widetilde{X}_\lambda$ must factor through some sub relative I-cell complex $\widetilde{X}_0 \to V$ of $\widetilde{X}_0 \to \widetilde{X}_\lambda$ of size at most γ. We will complete the proof by showing that the composition $V \to \widetilde{X}_\lambda \to X_\lambda$ factors through a sub relative K-cell complex of $X_0 \to X_\lambda$ of size at most γ.

For each I-cell of V there is exactly one $\beta < \lambda$ such that that cell is a part of τ_β, and Proposition 11.4.8 implies that the corresponding relative K-cell $\sigma_\beta\colon X_\beta \to X_{\beta+1}$ is contained in a sub relative K-cell complex of $X_0 \to X_\lambda$ of size at most γ. If we take the union Z of these sub relative K-cell complexes, then the relative K-cell complex $X_0 \to Z$ has size at most $\gamma \times \gamma = \gamma$ (since γ is infinite), and the composition $V \to \widetilde{X}_\lambda \to X_\lambda$ factors through the inclusion $Z \to X_\lambda$. □

11.5. Free cell complexes

11.5.1. Diagram categories.

DEFINITION 11.5.2. Let $\mathcal{C}$ and $\mathcal{M}$ be categories.

(1) A *$\mathcal{C}$-diagram in $\mathcal{M}$* is a functor from $\mathcal{C}$ to $\mathcal{M}$.
(2) a *map of $\mathcal{C}$-diagrams in $\mathcal{M}$* from the diagram $\boldsymbol{X}$ to the diagram $\boldsymbol{Y}$ is a natural transformation of functors from $\boldsymbol{X}$ to $\boldsymbol{Y}$.
(3) If $\mathcal{C}$ is a small category, then the *category of $\mathcal{C}$-diagrams in $\mathcal{M}$* is the category in which the class of objects is the class of functors from $\mathcal{C}$ to

$\mathcal{M}$ and in which the set of morphisms between two functors is the set of natural transformations between those functors.

REMARK 11.5.3. If $\mathcal{C}$ is not small, then there may be a proper class of natural transformations between two functors from $\mathcal{C}$ to $\mathcal{M}$, and so the collection of all functors from $\mathcal{C}$ to $\mathcal{M}$ and all natural transformations between them may not form a category (except possibly in some higher universe; see, e.g., [**60**, page 17]).

DEFINITION 11.5.4. Let $\mathcal{C}$ be a category and let $\mathcal{M}$ be a model category.

(1) If $\boldsymbol{X}$ is a $\mathcal{C}$-diagram in $\mathcal{M}$, then $\boldsymbol{X}$ is
   - *objectwise cofibrant* if $\boldsymbol{X}_\alpha$ is a cofibrant object of $\mathcal{M}$ for every object α of $\mathcal{C}$ and
   - *objectwise fibrant* if $\boldsymbol{X}_\alpha$ is a fibrant object of $\mathcal{M}$ for every object α of $\mathcal{C}$.

(2) If $\boldsymbol{X}$ and $\boldsymbol{Y}$ are $\mathcal{C}$-diagrams in $\mathcal{M}$, then a map of diagrams $f\colon \boldsymbol{X} \to \boldsymbol{Y}$ is
   - an *objectwise cofibration* if $f_\alpha\colon \boldsymbol{X}_\alpha \to \boldsymbol{Y}_\alpha$ is a cofibration for every object α of $\mathcal{C}$,
   - an *objectwise fibration* if $f_\alpha\colon \boldsymbol{X}_\alpha \to \boldsymbol{Y}_\alpha$ is a fibration for every object α of $\mathcal{C}$, and
   - an *objectwise weak equivalence* if $f_\alpha\colon \boldsymbol{X}_\alpha \to \boldsymbol{Y}_\alpha$ is a weak equivalence for every object α of $\mathcal{C}$,

DEFINITION 11.5.5. Let $\mathcal{M}$ be a category, let $\mathcal{C}$ and $\mathcal{D}$ be small categories, and let $\mathrm{F}\colon \mathcal{C} \to \mathcal{D}$ be a functor. If $\boldsymbol{X}$ is a $\mathcal{D}$-diagram in $\mathcal{M}$ (see Definition 11.5.2), then composition with F defines a $\mathcal{C}$-diagram $\mathrm{F}^*\boldsymbol{X} = \boldsymbol{X} \circ \mathrm{F}$ in $\mathcal{M}$, which we will call the $\mathcal{C}$-diagram in $\mathcal{M}$ *induced* by F:

- If α is an object of $\mathcal{C}$ then $(\mathrm{F}^*\boldsymbol{X})_\alpha = \boldsymbol{X}_{\mathrm{F}\alpha}$, and
- if $\sigma\colon \alpha \to \alpha'$ is a map in $\mathcal{C}$ then $(\mathrm{F}^*\boldsymbol{X})_\sigma = \boldsymbol{X}_{\mathrm{F}\sigma}\colon \boldsymbol{X}_{\mathrm{F}\alpha} \to \boldsymbol{X}_{\mathrm{F}\alpha'}$.

In Section 11.6 we will show that if $\mathcal{C}$ is a small category and $\mathcal{M}$ is a cofibrantly generated model category, then there is a model category structure on the category of $\mathcal{C}$-diagrams in $\mathcal{M}$ (and that this model category of diagrams is also cofibrantly generated). The cofibrant objects in this model category will be the free cell complexes (see Definition 11.5.35) and their retracts. Among the examples of free cell complexes are the $\mathcal{C}^{\mathrm{op}}$-diagram of simplicial sets $\mathrm{B}(-\downarrow \mathcal{C})^{\mathrm{op}}$ (see Definition 14.7.2) and the $\mathcal{C}$-diagram of simplicial sets $\mathrm{B}(\mathcal{C}\downarrow -)$ (see Definition 14.7.8), and the fact that these are cofibrant diagrams will imply the homotopy invariance of the homotopy colimit and homotopy limit functors (see Theorem 18.5.3 and Theorem 19.4.2).

11.5.6. Free diagrams of sets. In this section, we define *free diagrams of sets*. This will be used in the next section to define free diagrams in a category of diagrams, which will be used in Section 11.5.29 to define free cell complexes in a category of diagrams in a cofibrantly generated model category.

DEFINITION 11.5.7. Let $\mathcal{C}$ be a small category.

(1) If α is an object of $\mathcal{C}$, then the *free $\mathcal{C}$-diagram of sets generated at position α* is the $\mathcal{C}$-diagram of sets $\mathbf{F}^\alpha_*$ for which
   - $\mathbf{F}^\alpha_*(\beta) = \mathcal{C}(\alpha,\beta)$ for β an object of $\mathcal{C}$ and
   - $\big(\mathbf{F}^\alpha_*(g)\big)(h) = gh$ for $h \in \mathbf{F}^\alpha_*(\beta)$ and $g\colon \beta \to \gamma$ in $\mathcal{C}$.

(2) A *free $\mathcal{C}$-diagram of sets* is a $\mathcal{C}$-diagram of sets that is a coproduct of $\mathcal{C}$-diagrams of the form $\mathbf{F}^\alpha_*$.

PROPOSITION 11.5.8 (The Yoneda lemma). *If $\mathcal{C}$ is a small category and α is an object of $\mathcal{C}$, then for every $\mathcal{C}$-diagram of sets $\boldsymbol{S}$ there is a natural isomorphism*

$$\mathrm{Set}^{\mathcal{C}}(\mathbf{F}_*^\alpha, \boldsymbol{S}) \approx \boldsymbol{S}_\alpha \ .$$

PROOF. This is the Yoneda lemma (see, e.g., [**6**, page 11] or [**47**, page 61]).

If $g \in \mathrm{Set}^{\mathcal{C}}(\mathbf{F}_*^\alpha, \boldsymbol{S})$, then g is a map of diagrams from $\mathbf{F}_*^\alpha$ to $\boldsymbol{S}$, and so g_α is a function from $\mathbf{F}_*^\alpha(\alpha) = \mathcal{C}(\alpha,\alpha)$ to $\boldsymbol{S}_\alpha$; we define a function $\phi\colon \mathrm{Set}^{\mathcal{C}}(\mathbf{F}_*^\alpha, \boldsymbol{S}_\alpha) \to \boldsymbol{S}_\alpha$ by letting $\phi(g) = g_\alpha(1_\alpha)$.

To see that ϕ is injective, let g and h be elements of $\mathrm{Set}^{\mathcal{C}}(\mathbf{F}_*^\alpha, \boldsymbol{S})$ such that $\phi(g) = \phi(h)$. If β is an object of $\mathcal{C}$ and $\sigma \in \mathbf{F}_*^\alpha(\beta) = \mathcal{C}(\alpha,\beta)$, then

$$\begin{aligned}
g_\beta(\sigma) &= g_\beta(\sigma \circ 1_\alpha)\\
&= g_\beta\big((\mathbf{F}_*^\alpha(\sigma))(1_\alpha)\big)\\
&= \big(\boldsymbol{S}(\sigma)\big)\big(g_\alpha(1_\alpha)\big)\\
&= \big(\boldsymbol{S}(\sigma)\big)\big(\phi(g)\big)\\
&= \big(\boldsymbol{S}(\sigma)\big)\big(\phi(h)\big)\\
&= \big(\boldsymbol{S}(\sigma)\big)\big(h_\alpha(1_\alpha)\big)\\
&= h_\beta\big((\mathbf{F}_*^\alpha(\sigma))(1_\alpha)\big)\\
&= h_\beta(\sigma \circ 1_\alpha)\\
&= h_\beta(\sigma) \ .
\end{aligned}$$

Thus, $g = h$.

To see that ϕ is surjective, let $s \in \boldsymbol{S}_\alpha$. If β is an object of $\mathcal{C}$ and $\sigma \in \mathbf{F}_*^\alpha(\beta) = \mathcal{C}(\alpha,\beta)$, then $\boldsymbol{S}(\sigma)$ is a function from $\boldsymbol{S}_\alpha$ to $\boldsymbol{S}_\beta$. Thus, we can define $g_\beta\colon \mathbf{F}_*^\alpha(\beta) \to \boldsymbol{S}_\beta$ by letting $g_\beta(\sigma) = \big(\boldsymbol{S}(\sigma)\big)(s)$. If β and γ are objects of $\mathcal{C}$ and $\tau \in \mathcal{C}(\beta,\gamma)$, then for every $\sigma \in \mathbf{F}_*^\alpha(\beta)$ we have

$$\begin{aligned}
\big(\boldsymbol{S}(\tau)\big)\big(g_\beta(\sigma)\big) &= \big(\boldsymbol{S}(\tau)\big)\big((\boldsymbol{S}(\sigma))(s)\big)\\
&= \boldsymbol{S}(\tau\sigma)(\sigma)\\
&= g_\gamma(\tau\sigma)\\
&= g_\gamma\big((\mathbf{F}_*^\alpha(\tau))(\sigma)\big)
\end{aligned}$$

and so we have a well defined map of diagrams $g\colon \mathbf{F}_*^\alpha \to \boldsymbol{S}$ for which $\phi(g) = g_\alpha(1_\alpha) = \big(\boldsymbol{S}(1_\alpha)\big)(s) = 1_{\boldsymbol{S}_\alpha}(s) = s$. □

EXAMPLE 11.5.9. The diagram of sets $A \to B$ is free if and only if the map $A \to B$ is an inclusion.

EXAMPLE 11.5.10. The diagram of sets $A \to C \leftarrow B$ is free if and only if the maps $A \to C$ and $B \to C$ are inclusions with disjoint images in C.

EXAMPLE 11.5.11. The diagram of sets $A_1 \to A_2 \to A_3 \to \cdots$ is free if and only if all of the maps in the diagram are inclusions.

EXAMPLE 11.5.12. The diagram of sets $A_1 \leftarrow A_2 \leftarrow A_3 \leftarrow \cdots$ is free if and only if all of the maps are inclusions and the inverse limit of the diagram is empty.

EXAMPLE 11.5.13. If a discrete group G is considered to be a category with one object and morphisms equal to the elements of G, then a free G-diagram of sets is what classically is called a free G-set.

EXAMPLE 11.5.14. If $\mathcal{C}$ is a small category and $\boldsymbol{P}\colon \mathcal{C} \to \mathrm{Set}$ is the constant diagram at a point, then $\boldsymbol{P}$ is free if and only if each connected component of $\mathcal{C}$ has an initial object.

EXAMPLE 11.5.15. Let $\boldsymbol{\Delta}$ be the cosimplicial indexing category (i.e., for every nonnegative integer n let $[n]$ denote the ordered set $(0, 1, 2, \ldots, n)$ and let $\boldsymbol{\Delta}$ be the category with objects $\{[n] \mid n \geq 0\}$ and with $\boldsymbol{\Delta}\bigl([n],[k]\bigr)$ the weakly monotone maps from $[n]$ to $[k]$). If $\mathcal{C} = \boldsymbol{\Delta}^{\mathrm{op}}$, then a $\mathcal{C}$-diagram of sets is a simplicial set. The free $\mathcal{C}$-diagram of sets generated at position $[n]$ is the standard n-simplex $\Delta[n]$. Thus, the set of k-simplices of $\Delta[n]$ equals the set $\boldsymbol{\Delta}^{\mathrm{op}}\bigl([n],[k]\bigr) = \boldsymbol{\Delta}\bigl([k],[n]\bigr)$.

EXAMPLE 11.5.16. If $\mathcal{C}$ is the category $\boldsymbol{\Delta}^{\mathrm{op}}$, so that $\boldsymbol{F}^{[n]}_{*}$ is the standard n-simplex $\Delta[n]$ (see Example 11.5.15), then Proposition 11.5.8 is the statement that for every simplicial set X the set of simplicial maps $\mathrm{SS}(\Delta[n], X)$ is naturally isomorphic to the set of n-simplices of X.

DEFINITION 11.5.17. If $\mathcal{C}$ is a small category and S is a set, the *free $\mathcal{C}$-diagram of sets on the set S generated at position α* is the $\mathcal{C}$-diagram of sets $\mathbf{F}^{\alpha}_{S} = \coprod_{S} \mathbf{F}^{\alpha}_{*}$. Thus, for every object β in $\mathcal{C}$,

$$\mathbf{F}^{\alpha}_{S}(\beta) = \coprod_{s\in S} \mathcal{C}(\alpha,\beta).$$

PROPOSITION 11.5.18. *If $\mathcal{C}$ is a small category and α is an object of $\mathcal{C}$, then the functor $\mathbf{F}^{\alpha}_{-}\colon \mathrm{Set} \to \mathrm{Set}^{\mathcal{C}}$ (see Definition 11.5.17) is left adjoint to the functor $\mathrm{Set}^{\mathcal{C}} \to \mathrm{Set}$ that evaluates at α, i.e., for every set S and every $\mathcal{C}$-diagram of sets $\boldsymbol{T}$ there is a natural isomorphism $\mathrm{Set}^{\mathcal{C}}(\mathbf{F}^{\alpha}_{S}, \boldsymbol{T}) \approx \mathrm{Set}(S, \boldsymbol{T}_{\alpha})$.*

PROOF. Since $\mathbf{F}^{\alpha}_{S}$ is a coproduct of diagrams of the form $\mathbf{F}^{\alpha}_{*}$, this follows from Proposition 11.5.8. □

DEFINITION 11.5.19. If $\mathcal{C}$ is a small category, $\mathcal{C}^{\mathrm{disc}}$ is the discrete category with objects equal to the objects of $\mathcal{C}$, and $\boldsymbol{S}$ is an object of $\mathrm{Set}^{(\mathcal{C}^{\mathrm{disc}})}$, then the *free $\mathcal{C}$-diagram of sets generated by $\boldsymbol{S}$* is defined by

$$\mathbf{F}(\boldsymbol{S}) = \coprod_{\alpha\in\mathrm{Ob}(\mathcal{C})} \mathbf{F}^{\alpha}_{\boldsymbol{S}_{\alpha}}$$

(see Definition 11.5.17), so that for every object β of $\mathcal{C}$ we have

$$\bigl(\mathbf{F}(\boldsymbol{S})\bigr)_{\beta} = \coprod_{\alpha\in\mathrm{Ob}(\mathcal{C})} \coprod_{s\in\boldsymbol{S}_{\alpha}} \mathcal{C}(\alpha,\beta) .$$

THEOREM 11.5.20. *If $\mathcal{C}$ is a small category and $\mathcal{C}^{\mathrm{disc}}$ is the discrete category with objects equal to the objects of $\mathcal{C}$, then the functor $\mathbf{F}\colon \mathrm{Set}^{(\mathcal{C}^{\mathrm{disc}})} \to \mathrm{Set}^{\mathcal{C}}$ of Definition 11.5.19 is left adjoint to the forgetful functor $\mathbf{U}\colon \mathrm{Set}^{\mathcal{C}} \to \mathrm{Set}^{(\mathcal{C}^{\mathrm{disc}})}$, i.e., if $\boldsymbol{S}$ is an object of $\mathrm{Set}^{(\mathcal{C}^{\mathrm{disc}})}$ and $\boldsymbol{T}$ is an object of $\mathrm{Set}^{\mathcal{C}}$, there is a natural isomorphism*

$$\mathrm{Set}^{\mathcal{C}}\bigl(\mathbf{F}(\boldsymbol{S}), \boldsymbol{T}\bigr) \approx \mathrm{Set}^{(\mathcal{C}^{\mathrm{disc}})}\bigl(\boldsymbol{S}, \mathbf{U}(\boldsymbol{T})\bigr) .$$

PROOF. Since $\mathbf{F}(\boldsymbol{S})$ is a coproduct of diagrams of the form $\mathbf{F}^{\alpha}_{\boldsymbol{S}_{\alpha}}$, this follows from Proposition 11.5.18. □

11.5.21. Free diagrams. In this section, we define *free diagrams* in a category of diagrams (see Definition 11.5.25). In section Section 11.5.29, we will apply this to the generating cofibrations (see Definition 11.1.2) of a cofibrantly generated model category $\mathcal{M}$ to obtain the *free cells*, which are the generating cofibrations in the category of $\mathcal{C}$-diagrams in $\mathcal{M}$.

DEFINITION 11.5.22. Let $\mathcal{M}$ be a cocomplete category (see Definition 7.1.2). If X is an object of $\mathcal{M}$ and S is a set, then by $X \otimes S$ we will mean the object of $\mathcal{M}$ obtained by taking the coproduct, indexed by S, of copies of X; i.e., $X \otimes S \approx \coprod_S X$.

REMARK 11.5.23. If the cocomplete category $\mathcal{M}$ of Definition 11.5.22 is actually a *simplicial* category (see Definition 9.1.2) and we view the set S as a discrete simplicial set, then the object $X \otimes S$ defined by the simplicial structure is naturally isomorphic to the object defined in Definition 11.5.22.

DEFINITION 11.5.24. If $\mathcal{C}$ is a small category, $\mathcal{M}$ is a cocomplete category, $\boldsymbol{S}$ is a $\mathcal{C}$-diagram of sets, and X is an object of $\mathcal{M}$, then by $X \otimes \boldsymbol{S}$ we will mean the $\mathcal{C}$-diagram in $\mathcal{M}$ such that

$$(X \otimes \boldsymbol{S})_\alpha = X \otimes \boldsymbol{S}_\alpha$$

for every object α in $\mathcal{C}$ (see Definition 11.5.22).

DEFINITION 11.5.25. If $\mathcal{C}$ is a small category, α is an object of $\mathcal{C}$, $\mathcal{M}$ is a cocomplete category, and X is an object of $\mathcal{M}$, then the *free diagram on X generated at α* is the $\mathcal{C}$-diagram in $\mathcal{M}$ defined by $\mathbf{F}^\alpha_X = X \otimes \mathbf{F}^\alpha_*$ (see Definition 11.5.24 and Definition 11.5.7). Thus, if β is an object of $\mathcal{C}$, then $\mathbf{F}^\alpha_X(\beta) = \coprod_{\mathcal{C}(\alpha,\beta)} X$.

We have the following variant of the Yoneda lemma (see Proposition 11.5.8) for free diagrams in $\mathcal{M}$.

PROPOSITION 11.5.26. *If $\mathcal{C}$ is a small category, α is an object of $\mathcal{C}$, and $\mathcal{M}$ is a cocomplete category, then the functor $\mathbf{F}^\alpha_-\colon \mathcal{M} \to \mathcal{M}^\mathcal{C}$ (see Definition 11.5.25) is left adjoint to the functor $\mathcal{M}^\mathcal{C} \to \mathcal{M}$ that evaluates at α, i.e., for every object X in $\mathcal{M}$ and every diagram $\boldsymbol{Y}$ in $\mathcal{M}^\mathcal{C}$ there is a natural isomorphism*

$$\mathcal{M}^\mathcal{C}(\mathbf{F}^\alpha_X, \boldsymbol{Y}) \approx \mathcal{M}(X, \boldsymbol{Y}_\alpha) \ .$$

PROOF. We define $\phi\colon \mathcal{M}^\mathcal{C}(\mathbf{F}^\alpha_X, \boldsymbol{Y}) \to \mathcal{M}(X, \boldsymbol{Y}_\alpha)$ by letting $\phi(g)$ be the composition

$$X \xrightarrow{i_{(1_\alpha)}} \coprod_{\mathcal{C}(\alpha,\alpha)} X = \mathbf{F}^\alpha_X(\alpha) \xrightarrow{g_\alpha} \boldsymbol{Y}_\alpha \ .$$

The remainder of the proof is similar to that of the Yoneda lemma (see Proposition 11.5.18). □

DEFINITION 11.5.27. If $\mathcal{C}$ is a small category, $\mathcal{C}^{\mathrm{disc}}$ is the discrete category with objects equal to the objects of $\mathcal{C}$, $\mathcal{M}$ is a cocomplete category, and $\boldsymbol{X}$ is an object of $\mathcal{M}^{(\mathcal{C}^{\mathrm{disc}})}$, then the *free $\mathcal{C}$-diagram in $\mathcal{M}$ generated by $\boldsymbol{X}$* is defined by

$$\mathbf{F}(\boldsymbol{X}) = \coprod_{\alpha \in \mathrm{Ob}(\mathcal{C})} \mathbf{F}^\alpha_{\boldsymbol{X}_\alpha} = \coprod_{\alpha \in \mathrm{Ob}(\mathcal{C})} \boldsymbol{X}_\alpha \otimes \mathbf{F}^\alpha_*$$

(see Definition 11.5.25), so that for every object β of $\mathcal{C}$ we have

$$\big(\mathbf{F}(\boldsymbol{X})\big)_\beta = \coprod_{\alpha \in \mathrm{Ob}(\mathcal{C})} \coprod_{\mathcal{C}(\alpha,\beta)} \boldsymbol{X}_\alpha \ .$$

THEOREM 11.5.28. *If $\mathcal{C}$ is a small category and $\mathcal{M}$ is a cocomplete category, then the functor* $\mathbf{F}\colon \mathcal{M}^{(\mathcal{C}^{\mathrm{disc}})} \to \mathcal{M}^{\mathcal{C}}$ *of Definition 11.5.27 is left adjoint to the forgetful functor* $\mathbf{U}\colon \mathcal{M}^{\mathcal{C}} \to \mathcal{M}^{(\mathcal{C}^{\mathrm{disc}})}$, *i.e., if* $\boldsymbol{X}$ *is an object of* $\mathcal{M}^{(\mathcal{C}^{\mathrm{disc}})}$ *and* $\boldsymbol{Y}$ *is an object of* $\mathcal{M}^{\mathcal{C}}$, *then there is a natural isomorphism*

$$\mathcal{M}^{\mathcal{C}}\big(\mathbf{F}(\boldsymbol{X}),\boldsymbol{Y}\big) \approx \mathcal{M}^{(\mathcal{C}^{\mathrm{disc}})}(\boldsymbol{X},\mathbf{U}\boldsymbol{Y}) \ .$$

PROOF. Since $\mathbf{F}(\boldsymbol{X})$ is a coproduct of diagrams of the form $\mathbf{F}^{\alpha}_{X}$, this follows from Proposition 11.5.26. □

11.5.29. Free cell complexes. Relative free cell complexes are the analogues for diagrams of topological spaces of relative cell complexes for topological spaces (see Definition 10.7.1). In a cofibrantly generated model category with generating cofibrations I, the relative I-cell complexes play that role, as do the free relative I-cell complexes for a category of diagrams in a cofibrantly generated model category. Relative free cell complexes and their retracts will be the cofibrations in the model category of $\mathcal{C}$-diagrams in a cofibrantly generated model category (see Theorem 11.6.1).

We first describe *free cells*, which will be the generating cofibrations (see Definition 11.1.2) in this model category structure.

DEFINITION 11.5.30. Let $\mathcal{C}$ be a small category and let α be an object of $\mathcal{C}$. If $\mathcal{M}$ is a model category and I is a set of maps in $\mathcal{M}$, then a *free I-cell generated at α* in $\mathcal{M}^{\mathcal{C}}$ is a map of the form

$$\mathbf{F}^{\alpha}_{A} \to \mathbf{F}^{\alpha}_{B}$$

(see Definition 11.5.25) where $A \to B$ is an element of I. At every object β in $\mathcal{C}$, this is the map

$$\coprod_{\mathcal{C}(\alpha,\beta)} A \to \coprod_{\mathcal{C}(\alpha,\beta)} B.$$

EXAMPLE 11.5.31. Let $\mathcal{C}$ be a small category and let α be an object of $\mathcal{C}$.

- A *free cell generated at α* in $\mathrm{Top}^{\mathcal{C}}$ is a map of the form
$$\big|\partial\Delta[n]\big| \otimes \mathbf{F}^{\alpha}_{*} \to \big|\Delta[n]\big| \otimes \mathbf{F}^{\alpha}_{*}$$
(see Definition 11.5.24) for some $n \geq 0$.
- A *free cell generated at α* in $\mathrm{Top}_{*}^{\mathcal{C}}$ is a map of the form
$$\big|\partial\Delta[n]\big|^{+} \otimes \mathbf{F}^{\alpha}_{*} \to \big|\Delta[n]\big|^{+} \otimes \mathbf{F}^{\alpha}_{*}$$
(see Definition 11.5.24) for some $n \geq 0$.
- A *free cell generated at α* in $\mathrm{SS}^{\mathcal{C}}$ is a map of the form
$$\partial\Delta[n] \otimes \mathbf{F}^{\alpha}_{*} \to \Delta[n] \otimes \mathbf{F}^{\alpha}_{*}$$
(see Definition 11.5.24) for some $n \geq 0$.
- A *free cell generated at α* in $\mathrm{SS}_{*}^{\mathcal{C}}$ is a map of the form
$$\partial\Delta[n]^{+} \otimes \mathbf{F}^{\alpha}_{*} \to \Delta[n]^{+} \otimes \mathbf{F}^{\alpha}_{*}$$
(see Definition 11.5.24) for some $n \geq 0$.

LEMMA 11.5.32. *Let $\mathcal{C}$ be a small category and let $\mathcal{M}$ be a model category. If $g\colon A \to B$ is a map in $\mathcal{M}$, α is an object of $\mathcal{C}$, and the square*

$$\begin{array}{ccc} \mathbf{F}^{\alpha}_{A} & \longrightarrow & \mathbf{F}^{\alpha}_{B} \\ \downarrow & & \downarrow \\ \boldsymbol{X} & \longrightarrow & \boldsymbol{Y} \end{array}$$

is a pushout diagram in $\mathcal{M}^{\mathcal{C}}$, then for every object β of $\mathcal{C}$ there is a pushout diagram in $\mathcal{M}$

$$\begin{array}{ccc} C & \longrightarrow & D \\ \downarrow & & \downarrow \\ \boldsymbol{X}_{\beta} & \longrightarrow & \boldsymbol{Y}_{\beta} \end{array}$$

in which the map $C \to D$ is a coproduct of copies of g.

PROOF. This follows because pushouts in $\mathcal{M}^{\mathcal{C}}$ are constructed componentwise and the map $(\mathbf{F}^{\alpha}_{A})_{\beta} \to (\boldsymbol{F}^{\alpha}_{B})_{\beta}$ is the map $\coprod_{\mathcal{C}(\alpha,\beta)} A \to \coprod_{\mathcal{C}(\alpha,\beta)} B$. □

DEFINITION 11.5.33. If $\mathcal{M}$ is a model category, $\mathcal{C}$ is a small category, and K is a set of maps of $\mathcal{M}$, then $\mathbf{F}^{\mathcal{C}}_{K}$ will denote the set of maps of $\mathcal{M}^{\mathcal{C}}$ of the form

$$\mathbf{F}^{\alpha}_{A_k} \to \mathbf{F}^{\alpha}_{B_k}$$

(see Definition 11.5.25) where $A_k \to B_k$ is an element of K and α is an object of $\mathcal{C}$.

PROPOSITION 11.5.34. *If $\mathcal{M}$ is a category, $\mathcal{C}$ is a small category, and K is a set of maps in $\mathcal{M}$, then the map $g\colon \boldsymbol{X} \to \boldsymbol{Y}$ in $\mathcal{M}^{\mathcal{C}}$ is an $\mathbf{F}^{\mathcal{C}}_{K}$-injective (see Definition 10.5.2) if and only if $g_{\alpha}\colon \boldsymbol{X}_{\alpha} \to \boldsymbol{Y}_{\alpha}$ is a K-injective for every object α of $\mathcal{C}$.*

PROOF. This follows from Proposition 11.5.26. □

DEFINITION 11.5.35. If $\mathcal{M}$ is a cofibrantly generated model category and $\mathcal{C}$ is a small category, then

- a *relative free cell complex* in $\mathcal{M}^{\mathcal{C}}$ is a map that is a transfinite composition (see Definition 10.2.2) of pushouts (see Definition 7.2.10) of free cells (see Definition 11.5.30),
- a *free cell complex* in $\mathcal{M}^{\mathcal{C}}$ is a diagram $\boldsymbol{X}$ such that the map from the initial object of $\mathcal{M}^{\mathcal{C}}$ to $\boldsymbol{X}$ is a relative free cell complex, and
- an *inclusion of free cell complexes* is a relative free cell complex whose domain is a free cell complex.

The relative free cell complexes and their retracts will be the cofibrations in the model category of $\mathcal{C}$-diagrams in a cofibrantly generated model category $\mathcal{M}$ (see Theorem 11.6.1).

PROPOSITION 11.5.36. *If $\mathcal{M}$ is a cofibrantly generated model category, $\mathcal{C}$ is a small category, and $f\colon \boldsymbol{X} \to \boldsymbol{Y}$ is a relative free cell complex in $\mathcal{M}^{\mathcal{C}}$, then $f_{\alpha}\colon \boldsymbol{X}_{\alpha} \to \boldsymbol{Y}_{\alpha}$ is a cofibration in $\mathcal{M}$ for every object α of $\mathcal{C}$.*

PROOF. This follows from Lemma 11.5.32, Proposition 7.2.5, and Proposition 7.2.12. □

11.6. Diagrams in a cofibrantly generated model category

THEOREM 11.6.1. *If $\mathcal{C}$ is a small category and $\mathcal{M}$ is a cofibrantly generated model category (see Definition 11.1.2) with generating cofibrations I and generating trivial cofibrations J, then the category $\mathcal{M}^{\mathcal{C}}$ of $\mathcal{C}$-diagrams in $\mathcal{M}$ is a cofibrantly generated model category with generating cofibrations $\mathbf{F}_I^{\mathcal{C}}$ (see Definition 11.5.33) and generating trivial cofibrations $\mathbf{F}_J^{\mathcal{C}}$. In this model category structure, a map $\boldsymbol{X} \to \boldsymbol{Y}$ is*

- *a weak equivalence if $\boldsymbol{X}_\alpha \to \boldsymbol{Y}_\alpha$ is a weak equivalence in $\mathcal{M}$ for every object α of $\mathcal{C}$,*
- *a fibration if $\boldsymbol{X}_\alpha \to \boldsymbol{Y}_\alpha$ is a fibration in $\mathcal{M}$ for every object α of $\mathcal{C}$, and*
- *a cofibration if it is a retract of a transfinite composition of pushouts of elements of $\mathbf{F}_I^{\mathcal{C}}$.*

PROOF. If $\mathcal{C}^{\mathrm{disc}}$ is the discrete category with objects equal to the objects of $\mathcal{C}$, then $\mathcal{M}^{(\mathcal{C}^{\mathrm{disc}})} = \prod_{\mathrm{Ob}(\mathcal{C})} \mathcal{M}$, and so Proposition 11.1.10 gives us a cofibrantly generated model category structure on $\mathcal{M}^{(\mathcal{C}^{\mathrm{disc}})}$. We will show that the adjoint functors of Theorem 11.5.28 satisfy the conditions of Theorem 11.3.2 and thus define a cofibrantly generated model category structure on $\mathcal{M}^{\mathcal{C}}$.

For every object α of $\mathcal{C}$ let I_α be the set of maps in $\mathcal{M}^{(\mathcal{C}^{\mathrm{disc}})}$ given by

$$I_\alpha = I \times \prod_{\substack{\beta \in \mathrm{Ob}(\mathcal{C}) \\ \beta \neq \alpha}} 1_\phi$$

that is, the product of the identity map of the initial object of $\mathcal{M}$ at every object β of $\mathcal{C}$ other than α with an element of I at α. If we let $I_{\mathrm{Ob}(\mathcal{C})} = \bigcup_{\alpha \in \mathrm{Ob}(\mathcal{C})} I_\alpha$ and $J_{\mathrm{Ob}(\mathcal{C})} = \bigcup_{\alpha \in \mathrm{Ob}(\mathcal{C})} J_\alpha$ (where the definition of J_α is analogous to that of I_α), then Proposition 11.1.10 implies that the cofibrantly generated model category structure on $\mathcal{M}^{(^{\mathrm{disc}}\mathcal{C})}$ has $I_{\mathrm{Ob}(\mathcal{C})}$ as a set of generating cofibrations and $J_{\mathrm{Ob}(\mathcal{C})}$ as a set of generating trivial cofibrations. If $g\colon A \to B$ is a map in $\mathcal{M}$, then the functor $\boldsymbol{F}$ of Definition 11.5.27 takes $g \times \prod_{\beta \neq \alpha} 1_\phi$ to the map $\mathbf{F}_A^\alpha \to \mathbf{F}_B^\alpha$, and so $\boldsymbol{F}(I_{\mathrm{Ob}(\mathcal{C})}) = \mathbf{F}_I^{\mathcal{C}}$ and $\boldsymbol{F}(J_{\mathrm{Ob}(\mathcal{C})}) = \mathbf{F}_J^{\mathcal{C}}$.

Proposition 11.5.26, Lemma 11.5.32, and Theorem 10.5.27 imply that $\mathbf{F}_I^{\mathcal{C}}$ permits the small object argument. Similarly, $\mathbf{F}_J^{\mathcal{C}}$ permits the small object argument. Finally, Lemma 11.5.32 implies that a relative $\mathbf{F}_J^{\mathcal{C}}$-cell complex is a relative J-cell complex at every object α of $\mathcal{C}$, and is thus a weak equivalence at every object α of $\mathcal{C}$. □

PROPOSITION 11.6.2. *If $\mathcal{C}$ is a small category and $\mathcal{M}$ is a cofibrantly generated model category, then a free cell complex (see Definition 11.5.35) in $\mathcal{M}^{\mathcal{C}}$ is cofibrant in the model category structure of Theorem 11.6.1.*

PROOF. This follows from Theorem 11.6.1. □

PROPOSITION 11.6.3. *If $\mathcal{C}$ is a small category and is a cofibrantly generated model category, then a cofibration in the model category structure on $\mathcal{M}^{\mathcal{C}}$ of Theorem 11.6.1 is also an objectwise cofibration.*

PROOF. This follows from Theorem 11.6.1 and Proposition 11.5.36. □

LEMMA 11.6.4. *Let $\mathcal{M}$ and $\mathcal{N}$ be categories and let* $\mathrm{F}\colon \mathcal{M} \rightleftarrows \mathcal{N} :\mathrm{U}$ *be a pair of adjoint functors. If $\mathcal{C}$ is a small category, then there is a pair of adjoint functors between diagram categories* $\mathrm{F}^{\mathcal{C}}\colon \mathcal{M}^{\mathcal{C}} \rightleftarrows \mathcal{N}^{\mathcal{C}} :\mathrm{U}^{\mathcal{C}}$ *where* $\mathrm{F}^{\mathcal{C}}(\boldsymbol{X}) = \mathrm{F}\circ \boldsymbol{X}$ *for* $\boldsymbol{X}\colon \mathcal{C} \to \mathcal{M}$ *and* $\mathrm{U}^{\mathcal{C}}(\boldsymbol{Y}) = \mathrm{U}\circ\boldsymbol{Y}$ *for* $\boldsymbol{Y}\colon \mathcal{C}\to\mathcal{N}$.

PROOF. Let $\phi_{X,Y}\colon \mathcal{M}(X,\mathrm{U}Y) \to \mathcal{N}(\mathrm{F}X,Y)$ be an adjunction isomorphism (where X is an object of $\mathcal{M}$ and Y is an object of $\mathcal{N}$). We define an adjunction isomorphism $\phi^{\mathcal{C}}\colon \mathcal{M}^{\mathcal{C}}(\boldsymbol{X},\mathrm{U}^{\mathcal{C}}\boldsymbol{Y}) \to \mathcal{N}^{\mathcal{C}}(\mathrm{F}^{\mathcal{C}}\boldsymbol{X},\boldsymbol{Y})$ (where $\boldsymbol{X}\colon \mathcal{C}\to\mathcal{M}$ and $\boldsymbol{Y}\colon \mathcal{C}\to\mathcal{N}$ are diagrams) by letting $\phi^{\mathcal{C}}(f)$ on an object α of $\mathcal{C}$ be $(\phi^{\mathcal{C}}f)_\alpha = \phi_{\boldsymbol{X}_\alpha,\boldsymbol{Y}_\alpha}(f_\alpha)\colon (\mathrm{F}^{\mathcal{C}}\boldsymbol{X})_\alpha = \mathrm{F}(\boldsymbol{X}_\alpha) \to \boldsymbol{Y}_\alpha$ for $f\in \mathcal{M}^{\mathcal{C}}(\boldsymbol{X},\mathrm{U}^{\mathcal{C}}\boldsymbol{Y})$. □

THEOREM 11.6.5. *Let $\mathcal{M}$ and $\mathcal{N}$ be cofibrantly generated model categories and let* $\mathrm{F}\colon \mathcal{M}\rightleftarrows\mathcal{N} :\mathrm{U}$ *be a Quillen pair (see Definition 8.5.2).*

(1) *If $\mathcal{C}$ is a small category, then the adjoint pair* $\mathrm{F}^{\mathcal{C}}\colon \mathcal{M}^{\mathcal{C}} \rightleftarrows \mathcal{N}^{\mathcal{C}} :\mathrm{U}^{\mathcal{C}}$ *(see Lemma 11.6.4) is a Quillen pair (see Theorem 11.6.1).*
(2) *If $\mathcal{C}$ is a small category and* (F,U) *is a pair of Quillen equivalences (see Definition 8.5.20), then* $(\mathrm{F}^{\mathcal{C}},\mathrm{U}^{\mathcal{C}})$ *is a pair of Quillen equivalences.*

PROOF. Since fibrations and weak equivalences in $\mathcal{M}^{\mathcal{C}}$ and $\mathcal{N}^{\mathcal{C}}$ are defined objectwise (see Theorem 11.6.1), $\mathrm{U}^{\mathcal{C}}$ preserves both fibrations and trivial fibrations, and so part 1 follows from Proposition 8.5.3. Since weak equivalences in $\mathcal{M}^{\mathcal{C}}$ and $\mathcal{N}^{\mathcal{C}}$ are defined objectwise, part 2 follows from Proposition 11.5.36. □

EXAMPLE 11.6.6. If $\mathcal{C}$ is a small category, then the geometric realization and total singular complex functors extend to Quillen equivalences between $\mathrm{SS}^{\mathcal{C}}$ and $\mathrm{Top}^{\mathcal{C}}$ (see Notation 7.10.5).

EXAMPLE 11.6.7. If $\mathcal{C}$ is a small category, then the geometric realization and total singular complex functors extend to Quillen equivalences between $\mathrm{SS}_*^{\mathcal{C}}$ and $\mathrm{Top}_*^{\mathcal{C}}$ (see Notation 7.10.5).

THEOREM 11.6.8. *If $\mathcal{C}$ is a small category and $\mathcal{M}$ is a cofibrantly generated model category, then*

(1) *the colimit functor $\mathcal{M}^{\mathcal{C}} \to \mathcal{M}$ and the constant diagram functor $\mathcal{M}\to\mathcal{M}^{\mathcal{C}}$ are a Quillen pair, and*
(2) *the colimit functor $\mathcal{M}^{\mathcal{C}}\to\mathcal{M}$ takes objectwise weak equivalences between cofibrant $\mathcal{C}$-diagrams in $\mathcal{M}$ into weak equivalences between cofibrant objects in $\mathcal{M}$.*

PROOF. The colimit and constant diagram functors are an adjoint pair for all categories $\mathcal{M}$ and small categories $\mathcal{C}$. Since fibrations and weak equivalences are defined objectwise in $\mathcal{M}^{\mathcal{C}}$, the constant diagram functor preserves both fibrations and trivial fibrations, and so Proposition 8.5.3 implies that this adjoint pair is a Quillen pair. Part 2 follows from part 1 and Proposition 8.5.7. □

11.7. Diagrams in a simplicial model category

DEFINITION 11.7.1. Let $\mathcal{M}$ be a simplicial model category, let $\mathcal{C}$ be a small category, let $\boldsymbol{X}\colon \mathcal{C}\to\mathcal{M}$ be a $\mathcal{C}$-diagram in $\mathcal{M}$, and let K be a simplicial set.

(1) The $\mathcal{C}$-diagram $\boldsymbol{X}\otimes K$ in $\mathcal{M}$ is defined by letting $(\boldsymbol{X}\otimes K)_\alpha = \boldsymbol{X}_\alpha\otimes K$ for every object α of $\mathcal{C}$ and, if $\sigma\colon \alpha\to\alpha'$ is a map in $\mathcal{C}$, by letting $(\boldsymbol{X}\otimes K)_\sigma = \boldsymbol{X}_\sigma\otimes 1_K$.

(2) The $\mathcal{C}$-diagram $\boldsymbol{X}^K$ in $\mathcal{M}$ is defined by letting $(\boldsymbol{X}^K)_\alpha = (\boldsymbol{X}_\alpha)^K$ for every object α of $\mathcal{C}$ and, if $\sigma\colon \alpha \to \alpha'$ is a map in $\mathcal{C}$, by letting $(\boldsymbol{X}^K)_\sigma = (\boldsymbol{X}_\sigma)^{(1_K)}$.

DEFINITION 11.7.2. Let $\mathcal{M}$ be a simplicial model category. If $\mathcal{C}$ is a small category and $\boldsymbol{X}, \boldsymbol{Y}\colon \mathcal{C} \to \mathcal{M}$ are $\mathcal{C}$-diagrams in $\mathcal{M}$, then $\mathrm{Map}(\boldsymbol{X}, \boldsymbol{Y})$ is defined to be the simplicial set whose set of n-simplices is the set of maps of diagrams $\boldsymbol{X} \otimes \Delta[n] \to \boldsymbol{Y}$ (see Definition 11.7.1) and whose face and degeneracy maps are induced by the standard maps between the $\Delta[n]$.

THEOREM 11.7.3. *If $\mathcal{C}$ is a small category and $\mathcal{M}$ is a simplicial cofibrantly generated model category, then the model category structure of Theorem* 11.6.1 *with the simplicial structure of Definition* 11.7.1 *and Definition* 11.7.2 *makes $\mathcal{M}^{\mathcal{C}}$ a simplicial model category.*

PROOF. Definition 11.7.1 and Definition 11.7.2 satisfy axiom M6 (see Definition 9.1.6) because the constructions are all done objectwise and $\mathcal{M}$ is a simplicial model category. For axiom M7, Proposition 9.3.7 implies that it is sufficient to show that if $j\colon K \to L$ is an inclusion of simplicial sets and $p\colon \boldsymbol{X} \to \boldsymbol{Y}$ is a fibration in $\mathcal{M}^{\mathcal{C}}$, then $\boldsymbol{X}^L \to \boldsymbol{X}^K \times_{\boldsymbol{Y}^K} \boldsymbol{Y}^L$ is a fibration that is also a weak equivalence if either j or p is a weak equivalence. Since both fibrations and weak equivalences in $\mathcal{M}^{\mathcal{C}}$ are defined objectwise, this follows from the assumption that $\mathcal{M}$ is a simplicial model category. $\square$

11.8. Overcategories and undercategories

If $\mathcal{C}$ and $\mathcal{D}$ are categories and $\mathrm{F}\colon \mathcal{C} \to \mathcal{D}$ is a functor, then for each object α of $\mathcal{D}$ we define the category $(\alpha\downarrow \mathrm{F})$ of *objects of* $\mathcal{C}$ *under* α and the category $(\mathrm{F}\downarrow \alpha)$ of *objects of* $\mathcal{C}$ *over* α. These reduce to Definition 7.6.2 and Definition 7.6.1 when $\mathcal{C} = \mathcal{D}$ and F is the identity functor.

These more general notions will be used in Section 11.9 to define extensions of diagrams (see Definition 11.9.1), in Chapter 15 to define the Reedy model category structure (see Section 15.2), and in in Chapters 18 and 19 to define homotopy colimit and homotopy limit functors (see Definition 18.1.2, Definition 18.1.8, Definition 19.1.2, and Definition 19.1.5).

DEFINITION 11.8.1. If $\mathcal{C}$ and $\mathcal{D}$ are categories, $\mathrm{F}\colon \mathcal{C} \to \mathcal{D}$ is a functor, and α is an object of $\mathcal{D}$, then the category $(\mathrm{F}\downarrow \alpha)$ of *objects of* $\mathcal{C}$ *over* α is the category in which an object is a pair (β, σ) where β is an object of $\mathcal{C}$ and σ is a map $\mathrm{F}\beta \to \alpha$ in $\mathcal{D}$, and a morphism from the object (β, σ) to the object (β', σ') is a map $\tau\colon \beta \to \beta'$ in $\mathcal{C}$ such that the triangle

$$\begin{array}{ccc} \mathrm{F}\beta & \xrightarrow{\mathrm{F}\tau} & \mathrm{F}\beta' \\ & {\scriptstyle\sigma}\searrow \quad \swarrow{\scriptstyle\sigma'} & \\ & \alpha & \end{array}$$

commutes.

If $\mathcal{C} = \mathcal{D}$ and F is the identity functor, then we use $(\mathcal{C}\downarrow \alpha)$ to denote the category $(\mathrm{F}\downarrow \alpha)$. An object of $(\mathcal{C}\downarrow \alpha)$ is a map $\beta \to \alpha$ in $\mathcal{C}$, and a morphism from

$\beta \to \alpha$ to $\beta' \to \alpha$ is a map $\beta \to \beta'$ in $\mathcal{C}$ such that the triangle

$$\begin{array}{ccc} \beta & \longrightarrow & \beta' \\ & \searrow \quad \swarrow & \\ & \alpha & \end{array}$$

commutes.

EXAMPLE 11.8.2. Let $\mathcal{C}$ and $\mathcal{D}$ be categories and let $F\colon \mathcal{C} \to \mathcal{D}$ be a functor. If α is an object of $\mathcal{C}$, then there is a functor $F_*\colon (\mathcal{C} \downarrow \alpha) \to (F \downarrow F\alpha)$ that takes the object $\sigma\colon \beta \to \alpha$ of $(\mathcal{C} \downarrow \alpha)$ to the object $(\beta, F\sigma\colon F\beta \to F\alpha)$ of $(F \downarrow F\alpha)$ and the morphism $\tau\colon \beta \to \beta'$ from $\sigma\colon \beta \to \alpha$ to $\sigma'\colon \beta' \to \alpha$ to the morphism $\tau\colon \beta \to \beta'$ from $(\beta, F\sigma\colon F\beta \to \alpha)$ to $(\beta', F\sigma'\colon F\beta' \to F\alpha)$.

DEFINITION 11.8.3. If $\mathcal{C}$ and $\mathcal{D}$ are categories, $F\colon \mathcal{C} \to \mathcal{D}$ is a functor, and α is an object of $\mathcal{D}$, then the category $(\alpha \downarrow F)$ of *objects of* $\mathcal{C}$ *under* α is the category in which an object is a pair (β, σ) where β is an object of $\mathcal{C}$ and σ is a map $\alpha \to F\beta$ in $\mathcal{D}$, and a morphism from the object (β, σ) to the object (β', σ') is a map $\tau\colon \beta \to \beta'$ in $\mathcal{C}$ such that the triangle

$$\begin{array}{ccc} & \alpha & \\ & {}^{\sigma}\swarrow \quad \searrow^{\sigma'} & \\ F\beta & \xrightarrow[F\tau]{} & F\beta' \end{array}$$

commutes. The opposite $(\alpha \downarrow F)^{\mathrm{op}}$ is the category in which an object is a pair (β, σ) where β is an object of $\mathcal{C}$ and σ is a map $\alpha \to F\beta$ in $\mathcal{D}$, and a morphism from the object (β, σ) to the object (β', σ') is a map $\tau\colon \beta' \to \beta$ in $\mathcal{C}$ such that the triangle

$$\begin{array}{ccc} & \alpha & \\ & {}^{\sigma}\swarrow \quad \searrow^{\sigma'} & \\ F\beta & \xleftarrow[F\tau]{} & F\beta' \end{array}$$

commutes.

If $\mathcal{C} = \mathcal{D}$ and F is the identity functor, then we use $(\alpha \downarrow \mathcal{C})$ to denote the category $(\alpha \downarrow F)$. An object of $(\alpha \downarrow \mathcal{C})$ is a map $\alpha \to \beta$ in $\mathcal{C}$, and a morphism from $\alpha \to \beta$ to $\alpha \to \beta'$ is a map $\beta \to \beta'$ in $\mathcal{C}$ such that the triangle

$$\begin{array}{ccc} & \alpha & \\ & \swarrow \quad \searrow & \\ \beta & \longrightarrow & \beta' \end{array}$$

commutes. The opposite $(\alpha \downarrow \mathcal{C})^{\mathrm{op}}$ is the category in which an object is a map $\alpha \to \beta$ in $\mathcal{C}$, and a morphism from $\alpha \to \beta$ to $\alpha \to \beta'$ is a map $\beta' \to \beta$ in $\mathcal{C}$ such that the triangle

$$\begin{array}{ccc} & \alpha & \\ & \swarrow \quad \searrow & \\ \beta & \longleftarrow & \beta' \end{array}$$

commutes.

EXAMPLE 11.8.4. Let $\mathcal{C}$ and $\mathcal{D}$ be categories and let $\mathrm{F}\colon \mathcal{C} \to \mathcal{D}$ be a functor. If α is an object of $\mathcal{C}$, then there is a functor $\mathrm{F}_*\colon (\alpha \downarrow \mathcal{C})^{\mathrm{op}} \to (\mathrm{F}\alpha \downarrow \mathrm{F})^{\mathrm{op}}$ that takes the object $\sigma\colon \alpha \to \beta$ of $(\alpha \downarrow \mathcal{C})^{\mathrm{op}}$ to the object $(\beta, \mathrm{F}\sigma\colon \mathrm{F}\alpha \to \mathrm{F}\beta)$ of $(\mathrm{F}\alpha \downarrow \mathrm{F})^{\mathrm{op}}$ and the morphism $\tau\colon \beta' \to \beta$ from $\sigma\colon \alpha \to \beta$ to $\sigma'\colon \alpha \to \beta'$ to the morphism $\tau\colon \beta \to \beta'$ from $(\beta, \mathrm{F}\sigma\colon \mathrm{F}\alpha \to \mathrm{F}\beta)$ to $(\beta', \mathrm{F}\sigma'\colon \mathrm{F}\alpha \to \mathrm{F}\beta')$.

PROPOSITION 11.8.5. *If $\mathcal{C}$ and $\mathcal{D}$ are small categories, $\mathrm{F}\colon \mathcal{C} \to \mathcal{D}$ is a functor, and $\mathrm{F}^{\mathrm{op}}\colon \mathcal{C}^{\mathrm{op}} \to \mathcal{D}^{\mathrm{op}}$ is the opposite of F, then for every object α of $\mathcal{D}$ there is a natural isomorphism of categories*

$$(\alpha \downarrow \mathrm{F})^{\mathrm{op}} \approx (\mathrm{F}^{\mathrm{op}} \downarrow \alpha) \ .$$

PROOF. An object of $(\mathrm{F}^{\mathrm{op}} \downarrow \alpha)$ is a map $\alpha \to \mathrm{F}\beta$ in $\mathcal{D}$ for some object β of $\mathcal{C}$, and a morphism in $(\mathrm{F}^{\mathrm{op}} \downarrow \alpha)$ from $\alpha \to \mathrm{F}\beta$ to $\alpha \to \mathrm{F}\beta'$ is a map $\sigma\colon \beta' \to \beta$ in $\mathcal{C}$ such that the triangle

(11.8.6)
$$\begin{array}{ccc} & \alpha & \\ \swarrow & & \searrow \\ \mathrm{F}\beta & \xleftarrow[\mathrm{F}\sigma]{} & \mathrm{F}\beta' \end{array}$$

commutes. An object of $(\alpha \downarrow \mathrm{F})$ is a map $\alpha \to \mathrm{F}\beta$ in $\mathcal{D}$ for some object β of $\mathcal{C}$, and a morphism in $(\alpha \downarrow \mathrm{F})$ from $\alpha \to \mathrm{F}\beta$ to $\alpha \to \mathrm{F}\beta'$ is a map $\tau\colon \beta \to \beta'$ in $\mathcal{C}$ such that the triangle

$$\begin{array}{ccc} & \alpha & \\ \swarrow & & \searrow \\ \mathrm{F}\beta & \xrightarrow[\mathrm{F}\tau]{} & \mathrm{F}\beta' \end{array}$$

commutes. Thus, an object of $(\alpha \downarrow \mathrm{F})^{\mathrm{op}}$ is a map $\alpha \to \mathrm{F}\beta$ in $\mathcal{D}$ and a morphism in $(\alpha \downarrow \mathrm{F})^{\mathrm{op}}$ from $\alpha \to \mathrm{F}\beta$ to $\alpha \to \mathrm{F}\beta'$ is a map $\sigma\colon \beta' \to \beta$ in $\mathcal{C}$ such that the triangle (11.8.6) commutes. □

COROLLARY 11.8.7. *If $\mathcal{C}$ is a small category and α is an object of $\mathcal{C}$, then there is a natural isomorphism of categories*

$$(\alpha \downarrow \mathcal{C})^{\mathrm{op}} \approx (\mathcal{C}^{\mathrm{op}} \downarrow \alpha)$$

PROOF. This follows from Proposition 11.8.5. □

11.9. Extending diagrams

If $\mathbf{1}$ is the category with one object and with no non-identity maps, then an object X of a category $\mathcal{M}$ can be identified with a functor $i_X\colon \mathbf{1} \to \mathcal{M}$. If $\mathcal{C}$ is a small category, α is an object of $\mathcal{C}$, and $i_\alpha\colon \mathbf{1} \to \mathcal{C}$ is the functor that takes the object of $\mathbf{1}$ to α, then, for a diagram $\boldsymbol{X}\colon \mathcal{C} \to \mathcal{M}$, evaluation of $\boldsymbol{X}$ at α is equivalent to composing $\boldsymbol{X}$ with i_α. In this setting, Proposition 11.5.26 says that there is a natural isomorphism

$$\mathcal{M}^{\mathcal{C}}(\mathbf{F}^{\alpha}_{X}, \boldsymbol{Y}) \approx \mathcal{M}^{\mathbf{1}}(X, \boldsymbol{Y} \circ i_\alpha) \ .$$

In this section, we will obtain a similar result for functors of indexing categories more general than $i_\alpha\colon \mathbf{1} \to \mathcal{C}$.

Let $\mathcal{C}$ be a small category, let $\mathcal{B}$ be a subcategory of $\mathcal{C}$, and let $i\colon \mathcal{B} \to \mathcal{C}$ be the inclusion functor. If $\mathcal{M}$ is a cocomplete category (see Definition 7.1.2) and

$\boldsymbol{X}\colon \mathcal{B} \to \mathcal{M}$ is a diagram, we want to "extend" $\boldsymbol{X}$ to a diagram $\mathbf{L}\boldsymbol{X}\colon \mathcal{C} \to \mathcal{M}$ so that if $\boldsymbol{Y}\colon \mathcal{C} \to \mathcal{M}$ is a diagram and $i^*\boldsymbol{Y} = \boldsymbol{Y} \circ i$ is its "restriction" to $\boldsymbol{B}$, we have a natural isomorphism

$$\mathcal{M}^{\mathcal{C}}(\mathbf{L}\boldsymbol{X}, \boldsymbol{Y}) \approx \mathcal{M}^{\mathcal{B}}(\boldsymbol{X}, i^*\boldsymbol{Y}) \ .$$

If α is an object of $\mathcal{C}$, then we must define $(\mathbf{L}\boldsymbol{X})_\alpha$ so that for every object β of $\mathcal{B}$ and every map $\sigma\colon i(\beta) \to \alpha$ in $\mathcal{C}$ we have a map $(\mathbf{L}\boldsymbol{X})_\sigma\colon \boldsymbol{X}_\beta \to (\mathbf{L}\boldsymbol{X})_\alpha$. If $\tau\colon \beta' \to \beta$ is a map in $\mathcal{B}$, then we must ensure that the triangle

$$\begin{array}{ccc} \boldsymbol{X}_{\beta'} & \xrightarrow{\boldsymbol{X}_\tau} & \boldsymbol{X}_\beta \\ & {\scriptstyle (\mathbf{L}\boldsymbol{X})_{(\sigma\tau)}} \searrow \quad \swarrow {\scriptstyle (\mathbf{L}\boldsymbol{X})_\sigma} & \\ & (\mathbf{L}\boldsymbol{X})_\alpha & \end{array}$$

commutes. This suggests that we define $(\mathbf{L}\boldsymbol{X})_\alpha$ to be a colimit indexed by $(i \downarrow \alpha)$, the category of objects of $\mathcal{B}$ over α (see Definition 11.8.1). In fact, this construction works well for an arbitrary functor between small categories $i\colon \mathcal{B} \to \mathcal{C}$.

DEFINITION 11.9.1. Let $i\colon \mathcal{B} \to \mathcal{C}$ be a functor between small categories, let $\mathcal{M}$ be a cocomplete category (see Definition 7.1.2), and let $\boldsymbol{X}\colon \mathcal{B} \to \mathcal{M}$ be a functor. The *extension* $\mathbf{L}\boldsymbol{X}$ *of* $\boldsymbol{X}$ *to* $\mathcal{C}$ is the functor $\mathbf{L}\boldsymbol{X}\colon \mathcal{C} \to \mathcal{M}$ that on an object α of $\mathcal{C}$ is defined by

$$(\mathbf{L}\boldsymbol{X})_\alpha = \operatorname*{colim}_{(\beta,\sigma)\in \mathrm{Ob}\,(i\downarrow\alpha)} \boldsymbol{X}_\beta$$

(see Definition 11.8.1) and on a map $\tau\colon \alpha \to \alpha'$ in $\mathcal{C}$ is the natural map of colimits induced by $\tau_*\colon (i \downarrow \alpha) \to (i \downarrow \alpha')$ (where τ_* takes the object (β, σ) of $(i \downarrow \alpha)$ to the object $(\beta, \tau\sigma)$ of $(i \downarrow \alpha')$).

REMARK 11.9.2. In the setting of Definition 11.9.1, the functor $\mathbf{L}\boldsymbol{X}$ is known as the *left Kan extension* of $\boldsymbol{X}$ along i (see, e.g., [**6**, Section 3.7] or [**47**, Chapter X]).

THEOREM 11.9.3. *Let* $i\colon \mathcal{B} \to \mathcal{C}$ *be a functor between small categories, let* $\mathcal{M}$ *be a cocomplete category, and let* $\boldsymbol{X}\colon \mathcal{B} \to \mathcal{M}$ *be a functor. If* $\mathbf{L}\boldsymbol{X}$ *is the extension of* $\boldsymbol{X}$ *to* $\mathcal{C}$ *(see Definition* 11.9.1*), then for every functor* $\boldsymbol{Y}\colon \mathcal{C} \to \mathcal{M}$ *there is an isomorphism*

$$\mathcal{M}^{\mathcal{C}}(\mathbf{L}\boldsymbol{X}, \boldsymbol{Y}) \approx \mathcal{M}^{\mathcal{B}}(\boldsymbol{X}, i^*\boldsymbol{Y})$$

(where $i^*\boldsymbol{Y} = \boldsymbol{Y} \circ i$*) that is natural in both* $\boldsymbol{X}$ *and* $\boldsymbol{Y}$*.*

PROOF. We will define natural maps $\phi\colon \mathcal{M}^{\mathcal{C}}(\mathbf{L}\boldsymbol{X}, \boldsymbol{Y}) \to \mathcal{M}^{\mathcal{B}}(\boldsymbol{X}, i^*\boldsymbol{Y})$ and $\psi\colon \mathcal{M}^{\mathcal{B}}(\boldsymbol{X}, i^*\boldsymbol{Y}) \to \mathcal{M}^{\mathcal{C}}(\mathbf{L}\boldsymbol{X}, \boldsymbol{Y})$ that are inverses to each other. If $\mathrm{F}\colon \mathbf{L}\boldsymbol{X} \to \boldsymbol{Y}$ is a natural transformation, we let $(\phi\mathrm{F})\colon \boldsymbol{X} \to i^*\boldsymbol{Y}$ be the natural transformation that on an object γ of $\mathcal{B}$ is the composition

$$\boldsymbol{X}_\gamma \xrightarrow{i_{(\gamma, 1_{i(\gamma)})}} \operatorname*{colim}_{(\beta,\sigma)\in \mathrm{Ob}\,(i\downarrow i(\gamma))} \boldsymbol{X}_\beta = (\mathbf{L}\boldsymbol{X})_{i(\gamma)} \xrightarrow{\mathrm{F}} \boldsymbol{Y}_{i(\gamma)} = (i^*\boldsymbol{Y})_\gamma \ .$$

If $\mathrm{G}\colon \boldsymbol{X} \to i^*\boldsymbol{Y}$ is a natural transformation, we let $(\psi\mathrm{G})\colon \mathbf{L}\boldsymbol{X} \to \boldsymbol{Y}$ be the natural transformation that on an object α of $\mathcal{C}$ is the composition

$$(\mathbf{L}\boldsymbol{X})_\alpha = \operatorname*{colim}_{(\beta,\sigma)\in \mathrm{Ob}\,(i\downarrow\alpha)} \boldsymbol{X}_\beta \xrightarrow{\operatorname{colim} \mathrm{G}} \operatorname*{colim}_{(\beta,\sigma)\in \mathrm{Ob}\,(i\downarrow\alpha)} (i^*\boldsymbol{Y})_\beta = \operatorname*{colim}_{(\beta,\sigma)\in \mathrm{Ob}\,(i\downarrow\alpha)} \boldsymbol{Y}_{i(\beta)} \to \boldsymbol{Y}_\alpha$$

where the last map in the composition is the natural map from the colimit. The compositions $\phi\psi$ and $\psi\phi$ are identity natural transformations, and so ϕ and ψ are isomorphisms. $\square$

THEOREM 11.9.4. *If $\mathcal{M}$ is a cofibrantly generated model category and $i\colon \mathcal{B} \to \mathcal{C}$ is a functor between small categories, then the adjoint functors $\mathbf{L}\colon \mathcal{M}^{\mathcal{B}} \rightleftarrows \mathcal{M}^{\mathcal{C}} :i^*$ of Theorem 11.9.3 are a Quillen pair.*

PROOF. Since fibrations and weak equivalences in both $\mathcal{M}^{\mathcal{B}}$ and $\mathcal{M}^{\mathcal{C}}$ are defined objectwise (see Theorem 11.6.1), the right adjoint i^* preserves both fibrations and trivial fibrations. The result now follows from Proposition 8.5.3. □

CHAPTER 12

Cellular Model Categories

A cellular model category is a cofibrantly generated model category (see Definition 11.1.2) in which the cell complexes (see Definition 11.1.4) are well behaved (see Definition 12.1.1). Most of the model categories with which I am acquainted are cellular model categories (but not all; see Example 12.1.7).

We define cellular model categories in Section 12.1. In Section 12.2 we show that the intersection of two subcomplexes of a cell complex in a cellular model category always exists, and in Section 12.3 we prove that the cell complexes in a cellular model category are *uniformly compact*, i.e., that there is a cardinal σ (called the "size of the cells"; see Definition 12.3.3) such that if τ is a cardinal and X is a cell complex of size τ, then X is $\sigma\tau$-compact (see Theorem 12.3.1).

In Section 12.4 we discuss smallness, and prove that every cofibrant object in a cellular model category is small relative to the class of all cofibrations (see Theorem 12.4.3). The main result of Section 12.5 is Proposition 12.5.3, which asserts that if a small object factorization (see Definition 10.5.19) is applied to a map between large enough cell complexes, then the resulting cell complex is no larger than those with which you started.

12.1. Cellular model categories

DEFINITION 12.1.1. A *cellular model category* is a cofibrantly generated (see Definition 11.1.2) model category $\mathcal{M}$ for which there are a set I of generating cofibrations and a set J of generating trivial cofibrations such that

(1) both the domains and the codomains of the elements of I are compact (see Definition 11.4.1),
(2) the domains of the elements of J are small relative to I (see Definition 10.5.12), and
(3) the cofibrations are effective monomorphisms (see Definition 10.9.1).

REMARK 12.1.2. Although the sets I and J in Definition 12.1.1 are not part of the structure of a cellular model category, we will generally assume that some specific sets I and J satisfying the conditions of Definition 12.1.1 have been chosen.

12.1.3. Examples of cellular model categories.

PROPOSITION 12.1.4. *The categories* SS, SS_*, Top, *and* Top_* *are cellular model categories.*

PROOF. This follows from Example 9.1.13, Example 9.1.14, Example 9.1.15, and Example 9.1.16. □

PROPOSITION 12.1.5. *If $\mathcal{M}$ is a cellular model category and $\mathcal{C}$ is a small category, then the diagram category $\mathcal{M}^{\mathcal{C}}$ with the model category structure of Theorem 11.6.1 is a cellular model category.*

PROOF. This follows from Theorem 11.6.1. □

PROPOSITION 12.1.6. *If $\mathcal{M}$ is a cellular model category and Z is an object of $\mathcal{M}$, then the overcategory $(\mathcal{M}\downarrow Z)$ (see Definition 11.8.1) is a cellular model category.*

PROOF. This follows from Theorem 7.6.5. □

We are indebted to C. Simpson for the following example.

EXAMPLE 12.1.7 (C. Simpson). We present here an example of a cofibrantly generated model category that fails to be a cellular model category. Let $\mathcal{M}$ be the category of sets, let the weak equivalences be the isomorphisms, and let both the cofibrations and the fibrations be all the maps in $\mathcal{M}$. We let I be the set containing the two maps $\emptyset \to *$ (where $\emptyset$ is the empty set and $*$ is the one point set) and $** \to *$ (where $**$ is the two point set). We let J be the set containing as its only element the identity map of the empty set. The cofibrantly generated model category $\mathcal{M}$ is not cellular because not all elements of I are monomorphisms.

12.1.8. Recognizing cellular model categories.

THEOREM 12.1.9. *If $\mathcal{M}$ is a model category, then $\mathcal{M}$ is a cellular model category if there are sets I and J of maps in $\mathcal{M}$ such that*

(1) *a map is a trivial fibration if and only if it has the right lifting property with respect to every element of I,*
(2) *a map is a fibration if and only if it has the right lifting property with respect to every element of J,*
(3) *the domains and codomains of the elements of I are compact relative to I,*
(4) *the domains of the elements of J are small relative to I, and*
(5) *relative I-cell complexes are effective monomorphisms.*

PROOF. Proposition 10.8.7 and Proposition 10.9.5 imply that I permits the small object argument (see Definition 10.5.15), and so I is a set of generating cofibrations for $\mathcal{M}$. Proposition 11.2.3 now implies that J is a set of generating trivial cofibrations for $\mathcal{M}$, and so the theorem follows from Proposition 11.2.10. □

12.2. Subcomplexes in cellular model categories

PROPOSITION 12.2.1. *If $\mathcal{M}$ is a cellular model category, then a subcomplex of a presented relative cell complex is entirely determined by its set of cells (see Definition 10.6.4).*

PROOF. This follows from Proposition 10.6.10 and Proposition 10.6.11. □

Thus, if $f\colon X \to Y$ is a presented relative cell complex, then the union of a set of subcomplexes of f is well defined. The intersection of a family of subcomplexes would also be well defined if it was known to exist, i.e., if it was known that the attaching maps of the cells factored as necessary to build the subcomplex. We will show in Theorem 12.2.6 that the intersection of any *two* subcomplexes does exist.

12.2.2. Intersections of subcomplexes. The main result of this section is Theorem 12.2.6, which asserts that the intersection of two subcomplexes of a presented cell complex always exists. We have not been able to determine whether an arbitrary intersection of subcomplexes must exist.

PROPOSITION 12.2.3. *Let $\mathcal{M}$ be a cellular model category and let X be a presented cell complex. If K and L are subcomplexes of X such that their intersection $K \cap L$ exists (see Remark 10.6.12), then the pushout square*

$$\begin{array}{ccc} K \cap L & \longrightarrow & K \\ \big\downarrow & & \big\downarrow{\scriptstyle v} \\ L & \xrightarrow[u]{} & K \cup L \end{array}$$

is a pullback square.

PROOF. If $f\colon W \to L$ and $g\colon W \to K$ are maps such that $vg = uf$, then we have the solid arrow diagram

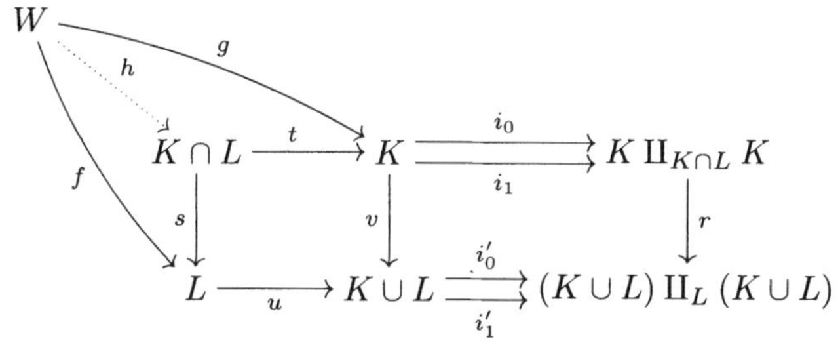

in which the left hand square commutes, $ri_0 = i_0'v$, and $ri_1 = i_1'v$. We now have $ri_0 g = i_0'vg = i_0'uf = i_1'uf = i_1'vg = ri_1 g$; since r is an inclusion of a subcomplex, it is a monomorphism (see Proposition 10.9.5), and so $i_0 g = i_1 g$. Since t is an inclusion of a subcomplex (and, thus, an effective monomorphism), this implies that there is a unique map $h\colon W \to K \cap L$ such that $th = g$. Since $ush = vth = vg = uf$ and u is an inclusion of a subcomplex (and, thus, a monomorphism; see Proposition 10.9.5), we have $sh = f$. □

THEOREM 12.2.4. *Let $\mathcal{M}$ be a cellular model category and let $\big(X, \emptyset = X_0 \to X_1 \to X_2 \to \cdots \to X_\beta \to \cdots \quad (\beta < \lambda), \{T^\beta, e^\beta, h^\beta\}_{\beta<\lambda}\big)$ be a presented cell complex. If $\{U^\beta\}_{\beta<\lambda}$ and $\{V^\beta\}_{\beta<\lambda}$ are subcomplexes of X (see Remark 10.6.12), then the sequence $\{\widetilde{T}^\beta\}_{\beta<\lambda}$ such that $\widetilde{T}^\beta = U^\beta \cap V^\beta$ for all $\beta < \lambda$ determines a subcomplex of X.*

PROOF. We must show that the sequence $\{\widetilde{T}^\beta\}_{\beta<\lambda}$ can be constructed by the inductive procedure of Proposition 10.6.11. Since Proposition 10.6.11 allows $\widetilde{T}^0$ to be any subset of T^0, the induction is begun.

Suppose now that β is an ordinal such that $\beta < \lambda$ and that the condition is satisfied for $\widetilde{T}^\gamma$ for all $\gamma < \beta$. We must show that if $i \in \widetilde{T}^\beta$ then $h_i^\beta\colon C_i \to X_\beta$ factors through $\widetilde{X}_\beta \to X_\beta$. Since $\widetilde{T}^\beta = U^\beta \cap V^\beta$, this follows from Proposition 12.2.3. □

DEFINITION 12.2.5. The subcomplex $\{\widetilde{T}^\beta\}_{\beta<\lambda}$ of Theorem 12.2.4 will be called the *intersection* of the subcomplexes $\{U^\beta\}_{\beta<\lambda}$ and $\{V^\beta\}_{\beta<\lambda}$.

THEOREM 12.2.6. *Let $\mathcal{M}$ be a cellular model category and let X be a cell complex. If K and L are subcomplexes (see Remark 10.6.8) of X (relative to some presentation of X), then the subcomplex $K \cap L$ of X exists.*

PROOF. This follows from Theorem 12.2.4. □

12.3. Compactness in cellular model categories

THEOREM 12.3.1 (Uniform compactness). *If $\mathcal{M}$ is a cellular model category then there is a cardinal σ such that if τ is a cardinal and X is a cell complex of size τ, then X is $\sigma\tau$-compact (see Definition 11.4.1).*

PROOF. Since the domains and codomains of the elements of I are compact, we can choose an infinite cardinal σ such that each of these domains and codomains is σ-compact (see Proposition 11.4.6).

If τ is a cardinal and X is a cell complex of size τ, then we can choose a presentation of X (see Definition 10.6.2), indexed by an ordinal λ whose cardinal is τ, that has no two cells with the same presentation ordinal (see Definition 10.6.4). Thus, we have a λ-sequence $\emptyset = X_0 \to X_1 \to X_2 \to \cdots \to X_\beta \to \cdots$ $(\beta < \lambda)$ whose colimit is X and such that every $X_{\beta+1}$ (for $\beta + 1 < \lambda$) is obtained as a pushout

$$\begin{array}{ccc} C_{\beta+1} & \longrightarrow & D_{\beta+1} \\ \downarrow & & \downarrow \\ X_\beta & \longrightarrow & X_{\beta+1} \end{array} \tag{12.3.2}$$

for some element $C_{\beta+1} \to D_{\beta+1}$ of I. If Y is a presented cell complex and $f\colon X \to Y$ is a map, then we must show that there is a subcomplex K of Y of size at most $\sigma\tau$ through which f factors. We will do this by showing (by induction on β) that for every $\beta < \lambda$ the composition $X_\beta \to X \to Y$ factors through a subcomplex K_β of Y of size at most $\sigma\tau$. The map f will then factor through the union of the $\{K_\beta\}_{\beta<\lambda}$ (since the inclusion of that union into Y is a monomorphism; see Proposition 10.9.5), which is of size at most $(\sigma\tau)\tau = \sigma\tau$.

The induction is begun by noting that $X_0 = \emptyset$ (the initial object of $\mathcal{M}$). If $\beta + 1 < \lambda$ and the composition $X_\beta \to X \to Y$ factors through a subcomplex K_β of Y of size at most $\sigma\tau$, then the composition of the attaching map $D_{\beta+1} \to X_{\beta+1} \to X \to Y$ (see Diagram 12.3.2) also factors through a subcomplex of size at most $\sigma\tau$, and (since σ is infinite) the union of these subcomplexes will be of size at most $\sigma\tau$ (see Proposition 10.1.13). Finally, if β is a limit ordinal such that $\beta < \lambda$ and for every $\alpha < \beta$ the composition $X_\alpha \to X \to Y$ factors through a subcomplex K_α of Y of size at most $\sigma\tau$, then the composition $X_\beta \to X \to Y$ factors through the union $\bigcup_{\alpha<\beta} K_\alpha$, which is of size at most $\sigma\tau$. □

DEFINITION 12.3.3. If $\mathcal{M}$ is a cellular model category, then the smallest cardinal σ satisfying the conclusion of Theorem 12.3.1 will be called the *size of the cells* of $\mathcal{M}$.

COROLLARY 12.3.4. *If $\mathcal{M}$ is a cellular model category and X is a cofibrant object of $\mathcal{M}$, then X is compact.*

PROOF. This follows from Theorem 12.3.1, Proposition 10.8.4, and Corollary 11.2.2. □

12.4. Smallness in cellular model categories

The main result of this section is Theorem 12.4.3, which asserts that all cofibrant objects in a cellular model category are small relative to the subcategory of all cofibrations.

LEMMA 12.4.1. *If $\mathcal{M}$ is a cellular model category with generating cofibrations I, then every cell complex (see Definition* 11.1.4*) is small relative to I.*

PROOF. This follows from Proposition 10.8.7 and Corollary 10.4.9. □

LEMMA 12.4.2. *If $\mathcal{M}$ is a cellular model category with generating cofibrations I, then every cofibrant object of $\mathcal{M}$ is small relative to I.*

PROOF. Corollary 11.2.2 implies that every cofibrant object of $\mathcal{M}$ is a retract of a cell complex, and so the result follows from Proposition 10.4.7 and Lemma 12.4.1. □

THEOREM 12.4.3. *If $\mathcal{M}$ is a cellular model category, then every cofibrant object is small relative to the subcategory of cofibrations.*

PROOF. This follows from Lemma 12.4.2 and Proposition 11.2.3. □

THEOREM 12.4.4. *If $\mathcal{M}$ is a cellular model category and J is a set of generating trivial cofibrations for $\mathcal{M}$ as in Definition* 12.1.1*, then the domains and the codomains of the elements of J are small relative to the subcategory of all cofibrations.*

PROOF. Proposition 11.2.3 implies that the domains are small relative to the subcategory of all cofibrations. Since every element of J is a cofibration (and thus a retract of a relative I-cell complex), Corollary 10.4.9, Proposition 10.4.7, and Proposition 10.8.7 imply that the codomains are small relative to the subcategory of all cofibrations. □

COROLLARY 12.4.5. *If $\mathcal{M}$ is a cellular model category and J is a set of generating trivial cofibrations for $\mathcal{M}$ as in Definition* 12.1.1*, then the domains and codomains of the elements of J are small relative to J.*

PROOF. Since every element of J is a cofibration, this follows from Theorem 12.4.4. □

PROPOSITION 12.4.6. *Let $\mathcal{M}$ be a cellular model category. If S is a set of cofibrations with cofibrant domains and J is a set of generating trivial cofibrations for $\mathcal{M}$ as in Definition* 12.1.1*, then there is a functorial factorization of every map $X \to Y$ as $X \xrightarrow{p} W \xrightarrow{q} Y$ where p is a relative $(S \cup J)$-cell complex and q is an $(S \cup J)$-injective.*

PROOF. Theorem 12.4.3 and Theorem 12.4.4 imply that the domains of the elements of $S \cup J$ are small relative to $S \cup J$, and so the result follows from Proposition 10.5.16. □

PROPOSITION 12.4.7. *Let $\mathcal{M}$ be a cellular model category, and let S be a set of inclusions of subcomplexes. If $X \to X'$ is the inclusion of a subcomplex and we apply a small object factorization using the set S and some ordinal λ (see*

Definition 10.5.19) to both of the maps $X \to *$ and $X' \to *$ to obtain the diagram

$$\begin{array}{ccccc} X & \longrightarrow & \mathrm{E}_S & \longrightarrow & * \\ \downarrow & & \downarrow & & \downarrow \\ X' & \longrightarrow & \mathrm{E}'_S & \longrightarrow & * \end{array}$$

then the map $\mathrm{E}_S \to \mathrm{E}'_S$ is the inclusion of a subcomplex.

PROOF. Using Proposition 10.9.5, one can check inductively that, at each stage in the construction of the factorization, the map $\mathrm{E}^\beta \to (\mathrm{E}^\beta)'$ is the inclusion of a subcomplex. □

12.5. Bounding the size of cell complexes

The main result of this section is Proposition 12.5.3, which asserts that if a small object factorization (see Definition 10.5.19) is applied to a map between "large enough" cell complexes, then the resulting cell complex is no larger than the ones with which you started.

PROPOSITION 12.5.1. *Let $\mathcal{M}$ be a cellular model category. If X is a cell complex (see Definition 11.1.4), then there is a cardinal η such that if ν is a cardinal, $\nu \geq 2$, and Y is a cell complex of size ν, then the set $\mathcal{M}(X,Y)$ has cardinal at most ν^η.*

PROOF. Let σ be the size of the cells of $\mathcal{M}$ (see Definition 12.3.3) and let τ be the size of X. The collection of isomorphism classes of cell complexes of size at most $\sigma\tau$ is a set, and so we can choose a set $\{Y_\alpha\}_{\alpha\in A}$ of representatives of those isomorphism classes. We let η be an infinite cardinal at least as large as the set $\left(\coprod_{\alpha\in A}\mathcal{M}(X,Y_\alpha)\right)\times(\sigma\tau)$.

Let ν be a cardinal such that $\nu \geq 2$ and let Y be a cell complex of size ν. Every map from X to Y must factor through a subcomplex of Y that is isomorphic to one of the Y_α (see Theorem 12.3.1). The set of such subcomplexes of Y has cardinal at most $\nu^{\sigma\tau} \leq \nu^\eta$ (see Proposition 10.6.10 and Lemma 10.1.16), and so the set $\mathcal{M}(X,Y)$ has cardinal at most $\eta\times(\nu^\eta) = \max(\eta,\nu^\eta) = \nu^\eta$. □

COROLLARY 12.5.2. *Let $\mathcal{M}$ be a cellular model category. If X is a cofibrant object then there is a cardinal η such that if ν is a cardinal, $\nu \geq 2$, and Y is a cell complex of size ν, then the set $\mathcal{M}(X,Y)$ has cardinal at most ν^η.*

PROOF. This follows from Proposition 12.5.1, Lemma 10.1.17, and Corollary 11.2.2. □

PROPOSITION 12.5.3. *Let $\mathcal{M}$ be a cellular model category with generating cofibrations I. If K is a set of relative I-cell complexes with cofibrant domains and κ is an infinite cardinal that is at least as large as each of the following cardinals:*

- *for each domain of an element of K, the cardinal η as in Corollary 12.5.2,*
- *for each codomain of an element of K, the cardinal η as in Corollary 12.5.2,*
- *for each relative I-cell complex in K, the cardinal of the set of cells in that relative I-cell complex, and*
- *the cardinal of the set K,*

then if $g\colon X \to Y$ is a map of cell complexes of size at most κ^κ (or if X is a cell complex of size at most κ^κ and Y is the terminal object of $\mathcal{M}$) and E_K is the object constructed by applying the small object factorization with the set K and an ordinal $\mu \le \kappa^\kappa$ to the map g (see Definition 10.5.19), then E_K is a cell complex of size at most κ^κ.

PROOF. Let μ be an ordinal such that $\mu \le \kappa^\kappa$, let $g\colon X \to Y$ be a map of cell complexes of size at most κ^κ, and let $X = X_0 \to X_1 \to X_2 \to \cdots \to X_\beta \to \cdots$ $(\beta < \mu)$ be the μ-sequence constructed by applying the small object factorization with the set K and the ordinal μ to g. We will show by transfinite induction that for $\beta < \mu$ the complex X_β has size at most κ^κ. Since $\mathrm{Succ}(\kappa^\kappa)$ (see Definition 10.1.10) is a regular cardinal (see Proposition 10.1.14), this will imply the proposition.

We begin the induction by noting that $X_0 = X$. If we now assume that β is an ordinal such that $\beta + 1 < \mu$ and that X_β has size at most κ^κ, then the domain of each element of K has at most $(\kappa^\kappa)^\kappa = \kappa^{(\kappa\times\kappa)} = \kappa^\kappa$ maps to X_β, the codomain has at most $(\kappa^\kappa)^\kappa = \kappa^{(\kappa\times\kappa)} = \kappa^\kappa$ maps to Y, and there are at most κ elements of K. Thus, $X_{\beta+1}$ is built from X_β by pushing out at most $(\kappa^\kappa) \times (\kappa^\kappa) \times \kappa = \kappa^\kappa$ maps, each of which attaches at most κ cells to X_β, and so $X_{\beta+1}$ has size at most κ^κ.

If β is a limit ordinal such that $\beta < \mu$, then X_β is a colimit of complexes of size at most κ^κ. Since $\beta < \mu \le \kappa^\kappa$, this implies that X_β is of size at most κ^κ. □

12.5.4. Natural cylinder objects.

DEFINITION 12.5.5. Let $\mathcal{M}$ be a cellular model category with generating cofibrations I and let ρ be the smallest regular cardinal such that the domains of the elements of I are ρ-small relative to I (see Definition 10.5.12). We define a *natural cylinder object* (see Definition 7.3.2) $X \amalg X \to \mathrm{Cyl}^{\mathcal{M}}(X) \to X$ on $\mathcal{M}$ by applying the small object factorization with the set I and the ordinal ρ to the fold map $1_X \amalg 1_X\colon X \amalg X \to X$ (see Definition 10.5.19).

DEFINITION 12.5.6. Let $\mathcal{M}$ be a cellular model category. If X is a cell complex, $H\colon \mathrm{Cyl}^{\mathcal{M}}(X) \to Y$ (see Definition 12.5.5) is a homotopy of maps from X to Y, and K is a subcomplex of X (see Definition 10.6.7), then we will use $H|_{\mathrm{Cyl}^{\mathcal{M}}(K)}\colon \mathrm{Cyl}^{\mathcal{M}}(K) \to Y$ to denote the composition $\mathrm{Cyl}^{\mathcal{M}}(K) \to \mathrm{Cyl}^{\mathcal{M}}(X) \xrightarrow{H} Y$, and we will call this composition the *restriction* of the homotopy H to the subcomplex K.

PROPOSITION 12.5.7. *If $\mathcal{M}$ is a cellular model category with generating cofibrations I and κ is an infinite cardinal that is at least as large as each of the following cardinals:*

- *for each domain of an element of I, the cardinal η as in Corollary 12.5.2,*
- *for each codomain of an element of I, the cardinal η as in Corollary 12.5.2,*
- *the cardinal ρ described in Definition 12.5.5, and*
- *the cardinal of the set I,*

and if X is a cell complex of size at most κ^κ, then the natural cylinder object $\mathrm{Cyl}^{\mathcal{M}}(X)$ (see Definition 12.5.5) is of size at most κ^κ.

PROOF. This follows from Proposition 12.5.3. □

CHAPTER 13

Proper Model Categories

A model category is *left proper* if weak equivalences are preserved by pushing them out along cofibrations, and it is *right proper* if they are preserved by pulling them back along fibrations (see Definition 13.1.1). Many model categories that come up in practice are left proper, right proper, or proper (i.e., both left proper and right proper), and even more model categories have homotopy theories equivalent to the homotopy theory of a proper model category (see, e.g., [**59**]).

In Section 13.1 we define properness and show that our categories of topological spaces and of simplicial sets are proper model categories. In Section 13.2 we prove a result relating lifting in left or right proper model categories and cofibrant or fibrant approximations that will be important for our localization results. In Sections 13.3 and 13.4 we discuss homotopy pullbacks and homotopy fibers in a right proper model category, and in Section 13.5 we discuss homotopy pushouts in a left proper model category.

13.1. Properness

DEFINITION 13.1.1. Let $\mathcal{M}$ be a model category.

(1) The model category $\mathcal{M}$ will be called *left proper* if every pushout of a weak equivalence along a cofibration (see Definition 7.2.10) is a weak equivalence.
(2) The model category $\mathcal{M}$ will be called *right proper* if every pullback of a weak equivalence along a fibration (see Definition 7.2.10) is a weak equivalence.
(3) The model category $\mathcal{M}$ will be called *proper* if it is both left proper and right proper.

The following proposition of C. L. Reedy shows that for weak equivalences between cofibrant objects, it follows from the definition of a model category that a pushout along a cofibration must be a weak equivalence (and, dually, that for weak equivalences between fibrant objects, a pullback along a fibration must be a weak equivalence).

PROPOSITION 13.1.2 (C. L. Reedy, [**57**]). *Let $\mathcal{M}$ be a model category.*

(1) *Every pushout of a weak equivalence between cofibrant objects along a cofibration (see Definition 7.2.10) is a weak equivalence.*
(2) *Every pullback of a weak equivalence between fibrant objects along a fibration (see Definition 7.2.10) is a weak equivalence.*

PROOF. We will prove part 1; the proof of part 2 is dual.

If we have a pushout diagram

$$\begin{array}{ccc} A & \xrightarrow{i} & C \\ {\scriptstyle f}\downarrow & & \downarrow{\scriptstyle g} \\ B & \xrightarrow[j]{} & D \end{array}$$

in which f is a weak equivalence, A and B are cofibrant, and i is a cofibration, then we must show that g is a weak equivalence. Since C and D are cofibrant (see Proposition 7.2.12), Theorem 7.8.6 implies that it is sufficient to show that if Z is a fibrant object of $\mathcal{M}$ then g induces an isomorphism of homotopy classes of maps $g^*\colon \pi(D, Z) \approx \pi(C, Z)$.

To see that g^* is an epimorphism, let $s\colon C \to Z$ be a map. Corollary 7.7.4 implies that there is a map $t\colon B \to Z$ such that $tf \simeq si$. Since i is a cofibration, Proposition 7.3.10 and Theorem 7.4.9 imply that there is a map $s'\colon C \to Z$ such that $s' \simeq s$ and $s'i = tf$. The maps s' and t combine to define $u\colon D \to Z$ such that $ug = s'$, and so $ug \simeq s$, and g^* is an epimorphism.

To see that g^* is a monomorphism, let u and u' be maps from D to Z such that $ug \simeq u'g$. Proposition 7.4.7 and Theorem 7.4.9 imply that there is a path object $Z \xrightarrow{s} \mathrm{Path}(Z) \xrightarrow{p_0 \times p_1} Z \times Z$ for Z and a map $v\colon C \to \mathrm{Path}(Z)$ such that $p_0 v = ug$ and $p_1 v = u'g$. Thus, we have the diagram

$$\begin{array}{ccccc} A & \xrightarrow{i} & C & \xrightarrow{v} & \mathrm{Path}(Z) \\ {\scriptstyle f}\downarrow & & {\scriptstyle g}\downarrow & & \downarrow{\scriptstyle p_0 \times p_1} \\ B & \xrightarrow[j]{} & D & \xrightarrow[u \times u']{} & Z \times Z \end{array} .$$

In the category $(\mathcal{M} \downarrow Z \times Z)$ of object of $\mathcal{M}$ over $Z \times Z$ (see Theorem 7.6.5), the object $\mathrm{Path}(Z)$ is fibrant (see Definition 7.3.2), and so Corollary 7.7.4 implies that there is a map $w\colon B \to \mathrm{Path}(Z)$ in $(\mathcal{M} \downarrow Z \times Z)$ such that $wf \simeq vi$ in $(\mathcal{M} \downarrow Z \times Z)$. Proposition 7.3.10 implies that there is a map $v'\colon C \to \mathrm{Path}(Z)$ in $(\mathcal{M} \downarrow Z \times Z)$ such that $v' \simeq v$ in $(\mathcal{M} \downarrow Z \times Z)$ and $v'i = wf$, and the pair (v', w) induces a map $H\colon D \to \mathrm{Path}(Z)$ such that $p_0 H = u$ and $p_1 H = u'$, i.e., a right homotopy from u to u'. □

COROLLARY 13.1.3. *Let $\mathcal{M}$ be a model category.*

(1) *If every object of $\mathcal{M}$ is cofibrant, then $\mathcal{M}$ is left proper.*
(2) *If every object of $\mathcal{M}$ is fibrant, then $\mathcal{M}$ is right proper.*
(3) *If every object of $\mathcal{M}$ is both cofibrant and fibrant, then $\mathcal{M}$ is proper.*

PROOF. This follows from Proposition 13.1.2. □

COROLLARY 13.1.4. *The categories* SS *and* SS$_*$ *(see Notation 7.10.5) are both left proper.*

PROOF. This follows from Corollary 13.1.3. □

COROLLARY 13.1.5. *The categories* Top *and* Top$_*$ *(see Notation 7.10.5) are both right proper.*

PROOF. This follows from Corollary 13.1.3. □

We will show in Theorem 13.1.10 that Top and Top_* are proper and in Theorem 13.1.13 that SS and SS_* are proper.

13.1.6. Topological spaces and simplicial sets.

LEMMA 13.1.7. *Let $f\colon X \to Y$ be a map of path connected topological spaces. If f induces an isomorphism of fundamental groups $f_*\colon \pi_1(X, x_0) \approx \pi_1\big(Y, f(x_0)\big)$ for some point $x_0 \in X$ and an isomorphism of homology $f_*\colon \mathrm{H}_*(X; f^*A) \approx \mathrm{H}_*(Y; A)$ for every local coefficient system A on Y, then f is a weak equivalence.*

PROOF. It is sufficient to show that the induced map of total singular complexes is a weak equivalence. Since this is a map of connected simplicial sets inducing an isomorphism of fundamental groups, it is sufficient to show that it induces isomorphisms of all higher homotopy groups, and for this it is sufficient to show that the induced map of universal covers $\widetilde{\operatorname{Sing} f}\colon \widetilde{\operatorname{Sing} X} \to \widetilde{\operatorname{Sing} Y}$ induces an isomorphism of all homology groups. Since the homology groups of the universal cover $\mathrm{H}_*(\widetilde{\operatorname{Sing} X})$ are naturally isomorphic to the local coefficient homology groups $\mathrm{H}_*\big(\operatorname{Sing} X; \mathbb{Z}[\pi_1 X]\big)$, this follows from our assumptions. □

THEOREM 13.1.8. *A map of topological spaces $f\colon X \to Y$ is a weak equivalence if and only if it induces an isomorphism of the sets of path components $f_*\colon \pi_0 X \approx \pi_0 Y$ and, for each path component of X and the corresponding path component of Y, isomorphisms of fundamental groups and of homology with all local coefficient systems.*

PROOF. The conditions are clearly necessary, and the converse follows from Lemma 13.1.7. □

PROPOSITION 13.1.9. *Let $f\colon X \to Y$ be a weak equivalence of topological spaces. If $n \geq 0$ and $\alpha\colon \mathrm{S}^n \to X$ is a map, then the induced map $\hat{f}\colon X \cup_\alpha \mathrm{D}^{n+1} \to Y \cup_{f\alpha} \mathrm{D}^{n+1}$ is a weak equivalence.*

PROOF. We will use Theorem 13.1.8. It follows immediately that $\hat{f}$ induces an isomorphism on the set of path components.

If $n = 0$ or $n = 1$, then the van Kampen theorem implies that $\hat{f}$ induces an isomorphism on the fundamental group of each path component. If $n > 1$, then the fundamental groups of the components of X and Y were unchanged when the cells were attached.

To see that $\hat{f}$ induces an isomorphism of homology with arbitrary local coefficients, we let

$$\begin{aligned} \mathrm{T}^{n+1} &= \{x \in \mathbb{R}^{n+1} \mid 0 < |x| \leq 1\} \\ \widetilde{X} &= X \cup_f \mathrm{T}^{n+1} \\ \widehat{X} &= X \cup_f \mathrm{D}^{n+1} \end{aligned}$$

and let $\widetilde{Y}$ and $\widehat{Y}$ be the corresponding constructions for Y. Since X is a deformation retract of $\widetilde{X}$ and Y is a deformation retract of $\widetilde{Y}$, the induced map $\tilde{f}\colon \widetilde{X} \to \widetilde{Y}$ is a weak equivalence. If B^{n+1} is the interior of D^{n+1}, then the subsets $\widetilde{X}$ and B^{n+1} of $\widehat{X}$ are an excisive pair, and so a Mayer-Vietoris argument shows that $\hat{f}$ induces an isomorphism of homology with arbitrary local coefficients. □

THEOREM 13.1.10. *The categories* Top *and* Top_* *(see Notation* 7.10.5*) are both left proper.*

PROOF. Let $f: X \to Y$ be a weak equivalence of topological spaces, let $s: X \to W$ is a cofibration, and let the square

$$\begin{array}{ccc} X & \xrightarrow{s} & W \\ {\scriptstyle f}\downarrow & & \downarrow{\scriptstyle g} \\ Y & \longrightarrow & Z \end{array}$$

is a pushout; we must show that g is a weak equivalence. The cofibration s must be a retract of a relative cell complex $t: X \to U$ (see Proposition 11.2.1, Example 11.1.8, and Example 11.1.9). If

$$\begin{array}{ccc} X & \xrightarrow{t} & U \\ {\scriptstyle f}\downarrow & & \downarrow{\scriptstyle h} \\ Y & \longrightarrow & V \end{array}$$

is a pushout, then g is a retract of h, and so it is sufficient to show that h is a weak equivalence. If we write t as a transfinite composition of maps, each of which attaches a single cell, then a transfinite induction using Proposition 13.1.9 and Proposition 10.7.4 implies that h is a weak equivalence. □

THEOREM 13.1.11. *The categories* Top *and* Top_* *(see Notation* 7.10.5*) are proper model categories.*

PROOF. Right properness follows from Corollary 13.1.3 and left properness follows from Theorem 13.1.10. □

PROPOSITION 13.1.12. *The geometric realization functor commutes with finite limits.*

PROOF. See [**38**, page 49]. □

THEOREM 13.1.13. *The categories* SS *and* SS_* *(see Notation* 7.10.5*) are proper model categories.*

PROOF. Left properness follows from Corollary 13.1.3. Right properness follows from the right properness of Top and Top_* (see Theorem 13.1.11) and the facts that

(1) the geometric realization functor commutes with pullbacks (see Proposition 13.1.12) and
(2) the geometric realization of a fibration of simplicial sets is a fibration of topological spaces (see [**53**]).

□

THEOREM 13.1.14. *Let $\mathcal{C}$ be a small category and let $\mathcal{M}$ be a cofibrantly generated model category. If $\mathcal{M}$ is left proper, right proper, or proper, then the model category structure on $\mathcal{M}^{\mathcal{C}}$ of Theorem* 11.6.1 *is, respectively, left proper, right proper, or proper.*

PROOF. Pullbacks in $\mathcal{M}^{\mathcal{C}}$ are constructed objectwise. Since fibrations in $\mathcal{M}^{\mathcal{C}}$ are objectwise fibrations and weak equivalences in $\mathcal{M}^{\mathcal{C}}$ are objectwise weak equivalences, if $\mathcal{M}$ is right proper then the pullback of a weak equivalence along a fibration is an objectwise weak equivalence, and so $\mathcal{M}^{\mathcal{C}}$ is right proper.

Pushouts in $\mathcal{M}^{\mathcal{C}}$ are also constructed objectwise. Since Proposition 11.6.3 implies that a cofibration in $\mathcal{M}^{\mathcal{C}}$ is an objectwise cofibration, if $\mathcal{M}$ is left proper then the pushout of an objectwise weak equivalence along a cofibration is an objectwise weak equivalence, and so $\mathcal{M}^{\mathcal{C}}$ is left proper. □

13.2. Properness and lifting

We are indebted to D. M. Kan for the following proposition.

PROPOSITION 13.2.1. *Let $\mathcal{M}$ be a model category.*

1. *Let $\mathcal{M}$ be left proper, let $g\colon A \to B$ be a cofibration, let $p\colon X \to Y$ be a fibration, and let $\tilde{g}\colon \widetilde{A} \to \widetilde{B}$ be a cofibrant approximation (see Definition 8.1.22) to g such that $\tilde{g}$ is a cofibration. If p has the right lifting property with respect to $\tilde{g}$, then p has the right lifting property with respect to g.*
2. *Let $\mathcal{M}$ be right proper, let $g\colon A \to B$ be a cofibration, let $p\colon X \to Y$ be a fibration, and let $\hat{p}\colon \widehat{X} \to \widehat{Y}$ be a fibrant approximation (see Definition 8.1.22) to p such that $\hat{p}$ is a fibration. If g has the left lifting property with respect to $\hat{p}$, then g has the left lifting property with respect to p.*

PROOF. We will prove part 2; the proof of part 1 is dual.

We have the diagram

$$\begin{CD} A @>>> X @>{i_X}>> \widehat{X} \\ @V{g}VV @V{p}VV @VV{\hat{p}}V \\ B @>>> Y @>>{i_Y}> \widehat{Y} \end{CD}$$

in which both i_X and i_Y are weak equivalences. If we let P be the pullback $Y \times_{\widehat{Y}} \widehat{X}$, then we have the diagram

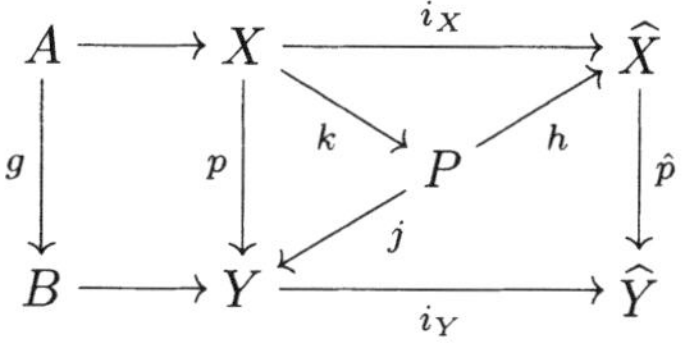

and, since g has the left lifting property with respect to $\hat{p}$, it also has the left lifting property with respect to j (see Lemma 7.2.11).

If we now consider the category $(A \downarrow \mathcal{M} \downarrow Y)$ of objects of $\mathcal{M}$ under A and over Y, then B, X, and P are objects in this category. Since g has the left lifting property with respect to j, we know that there is a map in this category from B to P, and we must show that there is a map in this category from B to X.

The category of objects under A and over Y is a model category in which a map is a cofibration, fibration, or weak equivalence if and only if it is one in $\mathcal{M}$ (see Theorem 7.6.5). Since j is a pullback of the fibration $\hat{p}$, it is also a fibration, and so

X and P are fibrant objects in our category, and B is a cofibrant object. If we knew that k was a weak equivalence, then the result would follow from Corollary 7.7.5.

Since i_Y is a weak equivalence, $\hat{p}$ is a fibration, and $\mathcal{M}$ is a *right proper* model category, the map h is also a weak equivalence. Since $i_X = hk$ and both i_X and h are weak equivalences, k is also a weak equivalence, and the proof is complete. □

COROLLARY 13.2.2. *Let $\mathcal{M}$ be a simplicial model category.*

(1) *Let $\mathcal{M}$ be left proper, let $g\colon A \to B$ be a cofibration, let $p\colon X \to Y$ be a fibration, and let $\tilde{g}\colon \widetilde{A} \to \widetilde{B}$ be a cofibrant approximation (see Definition 8.1.22) to g such that $\tilde{g}$ is a cofibration. If p has the homotopy right lifting property with respect to $\tilde{g}$ (see Definition 9.4.2), then p has the homotopy right lifting property with respect to g.*

(2) *Let $\mathcal{M}$ be right proper, let $g\colon A \to B$ be a cofibration, let $p\colon X \to Y$ be a fibration, and let $\hat{p}\colon \widehat{X} \to \widehat{Y}$ be a fibrant approximation (see Definition 8.1.22) to p such that $\hat{p}$ is a fibration. If g has the homotopy left lifting property with respect to $\hat{p}$ (see Definition 9.4.2), then g has the homotopy left lifting property with respect to p.*

PROOF. This follows from Proposition 13.2.1 and Lemma 9.4.7. □

13.3. Homotopy pullbacks and homotopy fiber squares

If all objects in a model category $\mathcal{M}$ were fibrant, then we would define homotopy pullbacks and homotopy fibers in terms of the homotopy limit functor (see Definition 19.1.5). Unfortunately, homotopy limits are homotopy invariant only for diagrams of fibrant objects (see Theorem 19.4.2). However, in a *right proper* model category (see Definition 13.1.1), we can define a homotopy pullback functor (see Definition 13.3.2) that is always homotopy invariant (see Proposition 13.3.4) and that is naturally weakly equivalent to the homotopy limit when all the objects in the diagram are fibrant (see Proposition 19.5.3).

13.3.1. Homotopy pullbacks. If $\mathcal{M}$ is a right proper model category (see Definition 13.1.1), then the homotopy pullback of the diagram $X \xrightarrow{g} Z \xleftarrow{h} Y$ is constructed by replacing g and h by fibrations and then taking a pullback (see Definition 13.3.2). In order to have a well defined functor, we need to choose a fixed functor to convert our maps into fibrations. We will show, however, that any other factorization into a weak equivalence followed by a fibration yields an object naturally weakly equivalent to the homotopy pullback and that, in fact, only one of the maps must be converted to a fibration (see Proposition 13.3.7). Thus, if either of the maps is already a fibration, then the pullback is naturally weakly equivalent to the homotopy pullback (see Corollary 13.3.8).

DEFINITION 13.3.2. Let $\mathcal{M}$ be a right proper model category and let E be an arbitrary but fixed functorial factorization of every map $g\colon X \to Y$ into $X \xrightarrow{i_g} \mathrm{E}(g) \xrightarrow{p_g} Y$, where i_g is a trivial cofibration and p_g is a fibration. The *homotopy pullback* of the diagram $X \xrightarrow{g} Z \xleftarrow{h} Y$ is defined to be the pullback of the diagram $\mathrm{E}(g) \xrightarrow{p_g} Z \xleftarrow{p_h} \mathrm{E}(h)$.

LEMMA 13.3.3. *Let $\mathcal{M}$ be a right proper model category. If $g\colon X \to Y$ is a weak equivalence and $h\colon W \to Z$ is a fibration, then, for any map $k\colon Y \to Z$, the* *natural map from the pullback of the diagram* $X \xrightarrow{kg} Z \xleftarrow{h} W$ *to the pullback of the diagram* $Y \xrightarrow{k} Z \xleftarrow{h} W$ *is a weak equivalence.*

PROOF. We have the commutative diagram

$$\begin{array}{ccccc} X \times_Z W & \longrightarrow & Y \times_Z W & \longrightarrow & W \\ \big\downarrow & & \big\downarrow & & \big\downarrow{\scriptstyle h} \\ X & \xrightarrow[g]{} & Y & \xrightarrow[k]{} & Z \end{array}$$

in which the vertical maps are all fibrations. Since g is a weak equivalence, the result follows from Proposition 7.2.14. $\square$

PROPOSITION 13.3.4 (Homotopy invariance of the homotopy pullback). *Let* $\mathcal{M}$ *be a right proper model category. If we have the diagram*

$$\begin{array}{ccccc} X & \xrightarrow{g} & Z & \xleftarrow{h} & Y \\ \big\downarrow & & \big\downarrow & & \big\downarrow \\ \widetilde{X} & \xrightarrow[\tilde{g}]{} & \widetilde{Z} & \xleftarrow[\tilde{h}]{} & \widetilde{Y} \end{array}$$

in which the vertical maps are weak equivalences, then the induced map of homotopy pullbacks

$$\mathrm{E}(g) \times_Z \mathrm{E}(h) \to \mathrm{E}(\tilde{g}) \times_{\widetilde{Z}} \mathrm{E}(\tilde{h})$$

is a weak equivalence.

PROOF. It is sufficient to show that if g, h, $\tilde{g}$, and $\tilde{h}$ are fibrations, then the map of pullbacks $X \times_Z Y \to \widetilde{X} \times_{\widetilde{Z}} \widetilde{Y}$ is a weak equivalence. This map equals the composition

$$X \times_Z Y \to (\widetilde{X} \times_{\widetilde{Z}} Z) \times_Z Y \approx \widetilde{X} \times_{\widetilde{Z}} Y \to \widetilde{X} \times_{\widetilde{Z}} \widetilde{Y}.$$

Since $\mathcal{M}$ is a right proper model category, the map $X \to \widetilde{X} \times_{\widetilde{Z}} Z$ is a weak equivalence, and Lemma 13.3.3 implies that the last map in the composition is a weak equivalence. $\square$

COROLLARY 13.3.5. *Let* $\mathcal{M}$ *be a right proper model category. If* $k\colon W \to X$ *is a weak equivalence, then the homotopy pullback of the diagram* $X \xrightarrow{g} Z \xleftarrow{h} Y$ *is naturally weakly equivalent to the homotopy pullback of the diagram* $W \xrightarrow{gk} Z \xleftarrow{h} Y$.

PROOF. We have the commutative diagram

$$\begin{array}{ccccc} W & \xrightarrow{gk} & Z & \xleftarrow{h} & Y \\ {\scriptstyle k}\big\downarrow & & \big\| & & \big\| \\ X & \xrightarrow[g]{} & Z & \xleftarrow[h]{} & Y \end{array}$$

in which the vertical maps are weak equivalences, and so the result follows from Proposition 13.3.4. $\square$

COROLLARY 13.3.6. *Let $\mathcal{M}$ be a right proper model category. If the maps $r, s\colon X \to Z$ are left homotopic (see Definition 7.3.2), right homotopic, or (if $\mathcal{M}$ is a simplicial model category) simplicially homotopic (see Definition 9.5.2), then the homotopy pullback of the diagram $X \xrightarrow{r} Z \xleftarrow{h} Y$ is weakly equivalent to the homotopy pullback of the diagram $X \xrightarrow{s} Z \xleftarrow{h} Y$.*

PROOF. We will prove this in the case that r and s are left homotopic; the proof in the case that they are right homotopic is similar, and either of these cases implies the corollary in the case that they are simplicially homotopic, since maps that are simplicially homotopic are both left and right homotopic (see Proposition 9.5.24).

If r and s are left homotopic, there is a diagram

$$X \underset{i_1}{\overset{i_0}{\rightrightarrows}} C \xrightarrow{H} Z$$

such that $Hi_0 = r$, $Hi_1 = s$, and both i_0 and i_1 are weak equivalences. The corollary now follows from Corollary 13.3.5. □

PROPOSITION 13.3.7. *Let $\mathcal{M}$ be a right proper model category. If $X \xrightarrow{j_g} W_g \xrightarrow{q_g} Z$ and $Y \xrightarrow{j_h} W_h \xrightarrow{q_h} Z$ are factorizations of, respectively, $g\colon X \to Z$ and $h\colon Y \to Z$, j_g and j_h are weak equivalences, and q_g and q_h are fibrations, then the homotopy pullback of the diagram $X \xrightarrow{g} Z \xleftarrow{h} Y$ is naturally weakly equivalent to each of $W_g \times_Z W_h$, $W_g \times_Z Y$, and $X \times_Z W_h$.*

PROOF. If E is the natural factorization used in Definition 13.3.2, then Lemma 13.3.3 implies that the homotopy pullback $\mathrm{E}(g) \times_Z \mathrm{E}(h)$ is naturally weakly equivalent to both $\mathrm{E}(g) \times_Z Y$ and $X \times_Z \mathrm{E}(h)$. Lemma 13.3.3 implies that these are naturally weakly equivalent to $\mathrm{E}(g) \times_Z W_h$ and $W_g \times_Z \mathrm{E}(h)$ respectively, and that these are naturally weakly equivalent to $X \times_Z W_h$ and $W_g \times_Z Y$, respectively. Lemma 13.3.3 implies that both of these are naturally weakly equivalent to $W_g \times_Z W_h$.

□

COROLLARY 13.3.8. *Let $\mathcal{M}$ be a right proper model category. If at least one of the maps $g\colon X \to Z$ and $h\colon Y \to Z$ is a fibration, then the pullback $X \times_Z Y$ is naturally weakly equivalent to the homotopy pullback of the diagram $X \xrightarrow{g} Z \xleftarrow{h} Y$.*

PROOF. This follows from Proposition 13.3.7. □

In Proposition 19.5.3, we show that if $\mathcal{M}$ is a right proper model category and X, Y, and Z are fibrant, then the homotopy pullback of the diagram $X \to Z \leftarrow Y$ is naturally weakly equivalent to the homotopy limit of that diagram (see Definition 19.1.5).

PROPOSITION 13.3.9. *Let $\mathcal{M}$ be a right proper model category. If the vertical maps in the diagram*

$$\begin{array}{ccccc} X & \longrightarrow & Z & \longleftarrow & Y \\ \downarrow & & \downarrow & & \downarrow \\ \widetilde{X} & \longrightarrow & \widetilde{Z} & \longleftarrow & \widetilde{Y} \end{array}$$

are weak equivalences and at least one map in each row is a fibration, then the map of pullbacks $X \times_Z Y \to \widetilde{X} \times_{\widetilde{Z}} \widetilde{Y}$ *is a weak equivalence.*

PROOF. This follows from Corollary 13.3.8 and Proposition 13.3.4. □

PROPOSITION 13.3.10. *Let* $\mathcal{M}$ *be a right proper model category. If we have a diagram* $X \xrightarrow{g} Z \xleftarrow{h} Y$ *in which at least one of* g *and* h *is a fibration and if* $\hat{h}\colon \widehat{Y} \to \widehat{Z}$ *is a fibrant approximation to* h*, then the pullback of* h *along* g *has a fibrant approximation that is a pullback of* $\hat{h}$*.*

PROOF. We have the diagram

$$\begin{array}{ccccc} W & \longrightarrow & Y & \xrightarrow{i_Y} & \widehat{Y} \\ {\scriptstyle k}\downarrow & & {\scriptstyle h}\downarrow & & \downarrow{\scriptstyle \hat{h}} \\ X & \xrightarrow[g]{} & Z & \xrightarrow[i_Z]{} & \widehat{Z} \end{array}$$

in which W is the pullback $X \times_Z Y$ and i_Y and i_Z are weak equivalences, and we must show that there is a pullback of $\hat{h}$ that is a fibrant approximation to k. If we factor the composition $i_Z g\colon X \to \widehat{Z}$ as $X \xrightarrow{i_X} \widehat{X} \xrightarrow{\hat{g}} \widehat{Z}$ where i_X is a trivial cofibration and $\hat{g}$ is a fibration, then we can let $\widehat{W} = \widehat{X} \times_{\widehat{Z}} \widehat{Y}$ and we have the diagram

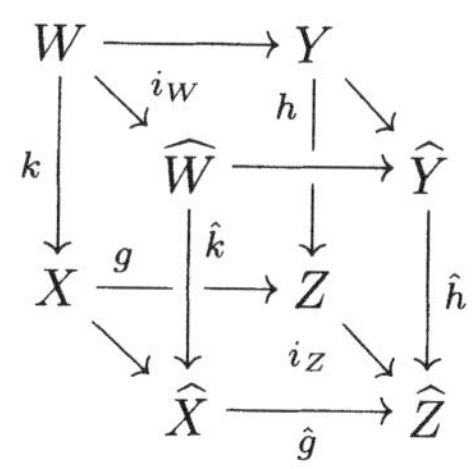

in which the front and back squares are pullbacks. Proposition 13.3.9 now implies that i_W is a weak equivalence, and so the pullback $\hat{k}$ of $\hat{h}$ is a fibrant approximation to k. □

13.3.11. Homotopy fiber squares.

DEFINITION 13.3.12. If $\mathcal{M}$ is a right proper model category, then a square

$$\begin{array}{ccc} A & \longrightarrow & C \\ \downarrow & & \downarrow \\ B & \longrightarrow & D \end{array}$$

will be called a *homotopy fiber square* if the natural map from A to the homotopy pullback (see Definition 13.3.2) of the diagram $B \to D \leftarrow C$ is a weak equivalence.

PROPOSITION 13.3.13. *If $\mathcal{M}$ is a right proper model category and we have the diagram*

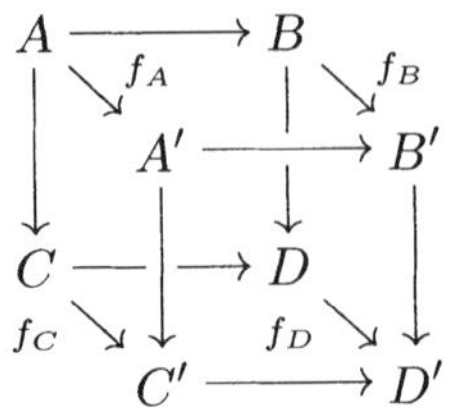

in which f_A, f_B, f_C, and f_D are weak equivalences, then the front square is a homotopy fiber square if and only if the back square is a homotopy fiber square.

PROOF. If P is the homotopy pullback of the diagram $C \to D \leftarrow B$ and P' is the homotopy pullback of the diagram $C' \to D' \leftarrow B'$, then we have the diagram

$$\begin{array}{ccc} A & \longrightarrow & P \\ {\scriptstyle f_A}\downarrow & & \downarrow{\scriptstyle f_P} \\ A' & \longrightarrow & P' \end{array}$$

and Proposition 13.3.4 implies that f_P is a weak equivalence. Since f_A is a weak equivalence, this implies that the top map is a weak equivalence if and only if the bottom map is a weak equivalence. □

PROPOSITION 13.3.14. *Let $\mathcal{M}$ be a right proper model category. If the front and back squares of the diagram*

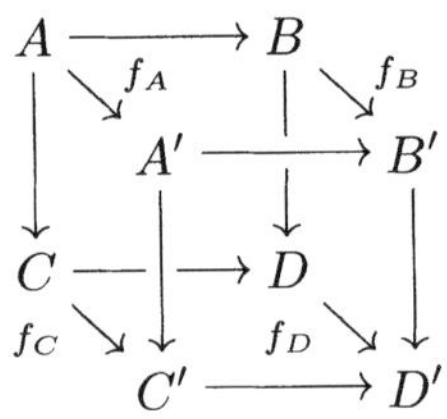

are homotopy fiber squares and if f_B, f_C, and f_D are weak equivalences, then f_A is a weak equivalence.

PROOF. This follows from Proposition 13.3.4. □

PROPOSITION 13.3.15. *Let $\mathcal{M}$ be a right proper model category. If the right hand square in the diagram*

$$\begin{array}{ccccc} A & \longrightarrow & B & \longrightarrow & C \\ \downarrow & & \downarrow & & \downarrow \\ D & \longrightarrow & E & \longrightarrow & F \end{array}$$

is a homotopy fiber square, then the left hand square is a homotopy fiber square if and only if the combined square is a homotopy fiber square.

PROOF. Factor $C \to F$ as $C \xrightarrow{i} G \xrightarrow{p} F$ where i is a trivial cofibration and p is a fibration, and let $P = E \times_F G$ and $P' = D \times_F G$. We now have the diagram

$$\begin{array}{ccccc} A & \longrightarrow & B & \longrightarrow & C \\ {\scriptstyle k}\downarrow & & \cong\downarrow{\scriptstyle j} & & \cong\downarrow{\scriptstyle i} \\ P' & \longrightarrow & P & \longrightarrow & G \\ \downarrow & & \downarrow{\scriptstyle q} & & \downarrow{\scriptstyle p} \\ D & \longrightarrow & E & \longrightarrow & F \end{array}$$

and Proposition 13.3.7 implies that j is a weak equivalence. Proposition 7.2.14 implies that P' is the pullback $D \times_E P$, and so Proposition 13.3.7 implies that k is a weak equivalence if and only if the (original) left hand square is a homotopy fiber square. Since Proposition 13.3.7 implies that k is a weak equivalence if and only if the (original) combined square is a homotopy fiber square, the proof is complete. □

13.4. Homotopy fibers

The *homotopy fiber* of the map $X \to Y$ over a point (see Definition 13.4.1) of Y will be defined so that it is a fibrant object weakly equivalent to the homotopy pullback of the diagram $X \to Y \leftarrow *$ (where "$*$" denotes the terminal object of $\mathcal{M}$) (see Definition 13.4.3 and Remark 13.4.5).

DEFINITION 13.4.1. If $\mathcal{M}$ is a model category and Z is an object of $\mathcal{M}$, then by a *point of* Z we will mean a map $* \to Z$ (where "$*$" is the terminal object of $\mathcal{M}$).

DEFINITION 13.4.2. If $\mathcal{M}$ is a model category, $g\colon Y \to Z$ is a map, and $z\colon * \to Z$ is a point of Z (see Definition 13.4.1), then the *fiber* of g over z is the pullback of the diagram $* \xrightarrow{z} Z \xleftarrow{g} Y$.

DEFINITION 13.4.3. Let $\mathcal{M}$ be a right proper model category. If $g\colon Y \to Z$ is a map and $z\colon * \to Z$ is a point of Z, then the *homotopy fiber* $\mathrm{HFib}_z(g)$ of g over z is the pullback of the diagram $* \xrightarrow{z} Z \xleftarrow{p_g} \mathrm{E}(Y)$ (see Definition 13.3.2).

PROPOSITION 13.4.4. *If $\mathcal{M}$ is a right proper model category, $g\colon Y \to Z$ is a map in $\mathcal{M}$, and $z\colon * \to Z$ is a point of Z, then the homotopy fiber of g over Z is a fibrant object of $\mathcal{M}$ that is naturally weakly equivalent to the homotopy pullback of the diagram $* \xrightarrow{z} Z \xleftarrow{g} Y$.*

PROOF. This follows from Proposition 7.2.12 and Proposition 13.3.7. □

REMARK 13.4.5. The homotopy fiber of the map $g\colon Y \to Z$ over a point $z\colon * \to Z$ was not defined to be the homotopy pullback of the diagram $* \xrightarrow{z} Z \xleftarrow{g} Y$ because that homotopy pullback need not be a fibrant object of $\mathcal{M}$.

PROPOSITION 13.4.6. *Let $\mathcal{M}$ be a right proper model category. If $g\colon Y \to Z$ is a fibration and $z\colon * \to Z$ is a point of Z, then the fiber of g over z is naturally weakly equivalent to the homotopy fiber of g over z.*

PROOF. This follows from Proposition 13.4.4 and Corollary 13.3.8. □

PROPOSITION 13.4.7. *Let $\mathcal{M}$ be a right proper model category. If $g\colon Y \to Z$ is a map and $z\colon * \to Z$ and $z'\colon * \to Z$ are points of Z that are (either left or right) homotopic, then the homotopy fiber of g over z is weakly equivalent to the homotopy fiber of g over z'.*

PROOF. This follows from Proposition 13.4.4 and Corollary 13.3.6. □

COROLLARY 13.4.8. *If $h\colon Y \to Z$ is a map in Spc and z and z' are points in the same path component of Z, then the homotopy fiber of h over z is weakly equivalent to the homotopy fiber of h over z'.*

PROOF. This follows from Proposition 13.4.7. □

PROPOSITION 13.4.9. *Let $\mathcal{M}$ be a right proper model category. If Z is an object of $\mathcal{M}$, $z\colon * \to Z$ is a point of Z, and $* \to P \to Z$ is a factorization of z into a weak equivalence followed by a fibration, then the homotopy fiber of any map $h\colon Y \to Z$ over z is naturally weakly equivalent to $P \times_Z Y$.*

PROOF. This follows from Proposition 13.3.7. □

PROPOSITION 13.4.10. *If $h\colon Y \to Z$ is a map in Top and z is a point of Z, then the total singular complex of the homotopy fiber of h over z is naturally homotopy equivalent to the corresponding homotopy fiber of $(\operatorname{Sing} h)\colon \operatorname{Sing} Y \to \operatorname{Sing} Z$.*

PROOF. If E is the factorization in Top of Definition 13.3.2 and $i_z\colon * \to Z$ is the constant map to z, then $\operatorname{Sing}(*) \to \operatorname{Sing} \mathrm{E}(i_z) \to \operatorname{Sing} Z$ is a factorization of $\operatorname{Sing}(*) \to \operatorname{Sing} Z$ into a weak equivalence followed by a fibration. Since the total singular complex functor commutes with pullbacks and all the simplicial sets involved are fibrant, the result now follows from Proposition 13.4.9. □

PROPOSITION 13.4.11. *If $h\colon Y \to Z$ is a map in SS and z is a vertex of Z, then the geometric realization of the homotopy fiber of h over z is naturally weakly equivalent to the corresponding homotopy fiber of $|h|\colon |Y| \to |Z|$.*

PROOF. Since the geometric realization functor commutes with pullbacks (see [**38**, page 49]), this is similar to the proof of Proposition 13.4.10. □

13.5. Homotopy pushouts and homotopy cofiber squares

Most of the definitions and results of this section are dual to those of Section 13.3.

13.5.1. Homotopy pushouts.

DEFINITION 13.5.2. Let $\mathcal{M}$ be a left proper model category and let E be an arbitrary but fixed functorial factorization of every map $g\colon X \to Y$ into $X \xrightarrow{i_g} E(g) \xrightarrow{p_g} Y$, where i_g is a cofibration and p_g is a trivial fibration. The *homotopy pushout* of the diagram $X \xleftarrow{g} Z \xrightarrow{h} Y$ is defined to be the pushout of the diagram $E(g) \xleftarrow{i_g} Z \xrightarrow{i_h} E(h)$.

PROPOSITION 13.5.3 (Homotopy invariance of the homotopy pushout). *Let $\mathcal{M}$ be a left proper model category. If we have the diagram*

$$\begin{array}{ccccc} X & \xleftarrow{g} & Z & \xrightarrow{h} & Y \\ \downarrow & & \downarrow & & \downarrow \\ \widetilde{X} & \xleftarrow[\tilde{g}]{} & \widetilde{Z} & \xrightarrow[\tilde{h}]{} & \widetilde{Y} \end{array}$$

in which the vertical maps are weak equivalences, then the induced map of homotopy pushouts $E(g) \amalg_Z E(h) \to E(\tilde{g}) \amalg_Z E(\tilde{h})$ is a weak equivalence.

PROOF. This follows from Proposition 13.3.4 and Proposition 7.1.9 (see Remark 7.1.10). □

PROPOSITION 13.5.4. *Let $\mathcal{M}$ be a left proper model category. If the vertical maps in the diagram*

$$\begin{array}{ccccc} Z & \longleftarrow & X & \longrightarrow & Y \\ \downarrow & & \downarrow & & \downarrow \\ \widetilde{Z} & \longleftarrow & \widetilde{X} & \longrightarrow & \widetilde{Y} \end{array}$$

are weak equivalences and at least one map in each row is a cofibration, then the induced map of pushouts $Z \amalg_X Y \to \widetilde{Z} \amalg_{\widetilde{X}} \widetilde{Y}$ is a weak equivalence.

PROOF. This follows from Proposition 13.3.9 and Proposition 7.1.9 (see Remark 7.1.10). □

PROPOSITION 13.5.5. *If we have a pushout diagram in $\mathrm{Top}_{(*)}$*

$$\begin{array}{ccc} A & \xrightarrow{i} & B \\ \downarrow & & \downarrow \\ C & \longrightarrow & D \end{array}$$

in which the map i is a cofibration, then the natural map of simplicial sets

$$(\operatorname{Sing} C) \amalg_{(\operatorname{Sing} A)} (\operatorname{Sing} B) \to \operatorname{Sing} D$$

is a weak equivalence.

PROOF. Since left adjoints commute with pushouts, there is a natural homeomorphism $\left|(\operatorname{Sing} C) \amalg_{(\operatorname{Sing} A)} (\operatorname{Sing} B)\right| \approx \left|\operatorname{Sing} C\right| \amalg_{|\operatorname{Sing} A|} \left|\operatorname{Sing} B\right|$, and so it is sufficient to show that the map $\left|\operatorname{Sing} C\right| \amalg_{|\operatorname{Sing} A|} \left|\operatorname{Sing} B\right| \to \left|\operatorname{Sing} D\right|$ is a weak equivalence. We have the diagram

$$\begin{array}{ccccc} \left|\operatorname{Sing} C\right| & \longleftarrow & \left|\operatorname{Sing} A\right| & \longrightarrow & \left|\operatorname{Sing} B\right| \\ \downarrow & & \downarrow & & \downarrow \\ C & \longleftarrow & A & \longrightarrow & B \end{array}$$

in which both of the maps pointing to the right are cofibrations (since both the geometric realization and total singular complex functors preserve cofibrations), and so Proposition 13.5.4 implies that the map $\left|\operatorname{Sing} C\right| \amalg_{|\operatorname{Sing} A|} \left|\operatorname{Sing} B\right| \to D$ is a weak equivalence. Since this map factors through the weak equivalence $|\operatorname{Sing} D| \to D$, the result follows from the "two out of three" axiom for weak equivalences. $\square$

PROPOSITION 13.5.6. *Let $\mathcal{M}$ be a left proper model category. If we have a diagram $Y \xleftarrow{g} X \xrightarrow{h} W$ in which at least one of g and h is a cofibration and if $\tilde{g}\colon \widetilde{X} \to \widetilde{Y}$ is a cofibrant approximation to g, then the pushout of g along h has a cofibrant approximation that is a pushout of $\tilde{g}$.*

PROOF. This follows from Proposition 13.3.10 and Proposition 7.1.9 (see Remark 7.1.10). $\square$

13.5.7. Homotopy cofiber squares.

DEFINITION 13.5.8. If $\mathcal{M}$ is a left proper model category, then a square

$$\begin{array}{ccc} A & \longrightarrow & C \\ \downarrow & & \downarrow \\ B & \longrightarrow & D \end{array}$$

will be called a *homotopy cofiber square* if the natural map to D from the homotopy pushout (see Definition 13.5.2) of the diagram $B \leftarrow A \to C$ is a weak equivalence.

PROPOSITION 13.5.9. *If $\mathcal{M}$ is a left proper model category and we have the diagram*

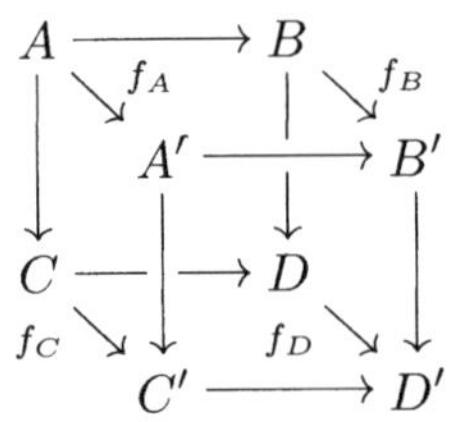

in which f_A, f_B, f_C, and f_D are weak equivalences, then the front square is a homotopy cofiber square if and only if the back square is a homotopy cofiber square.

PROOF. This follows from Proposition 13.3.13 and Proposition 7.1.9 (see Remark 7.1.10). $\square$

PROPOSITION 13.5.10. *Let $\mathcal{M}$ be a left proper model category. If the front and back squares of the diagram*

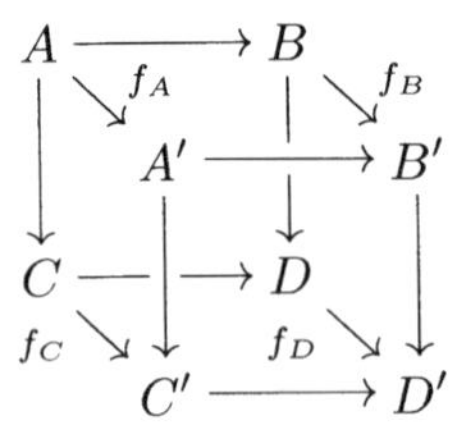

are homotopy cofiber squares and if f_A, f_B, and f_C are weak equivalences, then f_D is a weak equivalence.

PROOF. This follows from Proposition 13.3.14 and Proposition 7.1.9 (see Remark 7.1.10). $\square$

CHAPTER 14

The Classifying Space of a Small Category

The *classifying space* (or *nerve*) of a small category $\mathcal{C}$ is a simplicial set $B\mathcal{C}$ in which

- the vertices of $B\mathcal{C}$ are the objects of $\mathcal{C}$,
- the 1-simplices of $B\mathcal{C}$ are the morphisms of $\mathcal{C}$, and
- the n-simplices of $B\mathcal{C}$ for $n \geq 2$ are the strings of n composable morphisms in $\mathcal{C}$.

We will often want to know whether a category "has a contractible classifying space". If $\mathcal{C}$ is not small then the class of objects is not a set and so there cannot exist a simplicial set $B\mathcal{C}$ except possibly in some higher universe, but we are still able to describe what it means to say that "$\mathcal{C}$ has a contractible classifying space" by considering the classifying spaces of the small subcategories of $\mathcal{C}$ (see Definition 14.3.1). Our main use for this will be to prove the "essential uniqueness" of a construction that requires making choices: We build a category $\mathcal{C}$ whose objects are the possible outcomes and whose morphisms are equivalences between them. The assertion that $\mathcal{C}$ has a contractible classifying space then implies that any two outcomes are connected by an "essentially unique" zig-zag of equivalences (see Theorem 14.4.5).

We define the classifying space of a small category in Section 14.1. In Section 14.2 we define what it means for a functor between small categories to be *left (or right) cofinal* (see Definition 14.2.1). Our definition is in terms of whether the classifying spaces of the overcategories (or undercategories) are non-empty and connected, and we show in Theorem 14.2.5 that this is the "correct" notion for discussing limits (or colimits) of diagrams indexed by these categories. (We define the analogous notions of *homotopy left (or right) cofinal* in Definition 19.6.1 in terms of whether those classifying spaces are contractible, and we show in Theorem 19.6.13 that this is the "correct" notion for discussing homotopy limits and homotopy colimits.)

In Section 14.3 we discuss what it means to say that a category has a *contractible classifying space.* If $\mathcal{C}$ is a *small* category, then this just means that the simplicial set $B\mathcal{C}$ is contractible. For categories $\mathcal{C}$ that may not be small, we define this in terms of the classifying spaces of the small subcategories of $\mathcal{C}$ (see Definition 14.3.1). In Section 14.4 we define *equivalent zig-zags* in a category, and we say that there is an *essentially unique zig-zag* connecting two objects if the objects are connected by a zig-zag, any two of which are equivalent. We then show that if $\mathcal{C}$ has a contractible classifying space, then any pair of objects of $\mathcal{C}$ are connected by an essentially unique zig-zag in $\mathcal{C}$.

In Section 14.5 we discuss the situation in which we have categories $\mathcal{K}$ and $\mathcal{L}$ and our interest is in the functors from $\mathcal{K}$ to $\mathcal{L}$. If $\mathcal{K}$ is not small then there is no "category of all functors from $\mathcal{K}$ to $\mathcal{L}$ and all natural transformations between them", and so we consider collections of functors from $\mathcal{K}$ to $\mathcal{L}$ and natural transformations between them that do form categories. The main result is a sufficient condition for every small category of functors over a fixed functor to be contained in one with a contractible classifying space (see Theorem 14.5.4). Section 14.6 contains uniqueness results for cofibrant and fibrant approximations as our first application of contractible classifying spaces. Similar results for resolutions and homotopy function complexes will follow in Chapter 16 and Chapter 17.

In Sections 14.7 and 14.8 we discuss diagrams of classifying spaces of overcategories and undercategories. In Section 14.7 we describe the $\mathcal{D}^{\mathrm{op}}$-diagrams of opposites of undercategories and the $\mathcal{D}$-diagram of overcategories defined by a functor of small categories $\mathrm{F}\colon \mathcal{C} \to \mathcal{D}$. If we take the classifying spaces of the opposites of undercategories (or of the overcategories), then we obtain $\mathcal{D}^{\mathrm{op}}$-diagrams (or $\mathcal{D}$-diagrams) of simplicial sets, and we show in Section 14.8 that they are free cell complexes (see Proposition 14.8.5). Thus, they are cofibrant objects in the model category of $\mathcal{D}^{\mathrm{op}}$-diagrams (or $\mathcal{D}$-diagrams) of simplicial sets. These diagrams will be used in Chapters 18 and 19 to define homotopy colimit and homotopy limit functors.

14.1. The classifying space of a small category

DEFINITION 14.1.1. If $\mathcal{C}$ is a small category, then the *classifying space* of $\mathcal{C}$ (also called the *nerve* of $\mathcal{C}$) is the simplicial set $\mathrm{B}\mathcal{C}$ in which an n-simplex σ is a diagram in $\mathcal{C}$ of the form

$$\alpha_0 \xrightarrow{\sigma_0} \alpha_1 \xrightarrow{\sigma_1} \cdots \xrightarrow{\sigma_{n-1}} \alpha_n$$

and the face and degeneracy maps are defined by

(14.1.2)

$$d_i\sigma = \begin{cases} \alpha_1 \xrightarrow{\sigma_1} \alpha_2 \xrightarrow{\sigma_2} \cdots \xrightarrow{\sigma_{n-1}} \alpha_n & \text{if } i = 0 \\ \alpha_0 \xrightarrow{\sigma_0} \cdots \xrightarrow{\sigma_{i-2}} \alpha_{i-1} \xrightarrow{\sigma_i\sigma_{i-1}} \alpha_{i+1} \xrightarrow{\sigma_{i+1}} \cdots \xrightarrow{\sigma_{n-1}} \alpha_n & \text{if } 0 < i < n \\ \alpha_0 \xrightarrow{\sigma_0} \alpha_1 \xrightarrow{\sigma_1} \cdots \xrightarrow{\sigma_{n-2}} \alpha_{n-1} & \text{if } i = n \end{cases}$$

$$s_i\sigma = \alpha_0 \xrightarrow{\sigma_0} \cdots \xrightarrow{\sigma_{i-1}} \alpha_i \xrightarrow{1_{\alpha_i}} \alpha_i \xrightarrow{\sigma_i} \alpha_{i+1} \xrightarrow{\sigma_{i+1}} \cdots \xrightarrow{\sigma_{n-1}} \alpha_n \ .$$

If $\mathrm{F}\colon \mathcal{C} \to \mathcal{D}$ is a functor between small categories, then F induces a map of simplicial sets $\mathrm{BF}\colon \mathrm{B}\mathcal{C} \to \mathrm{B}\mathcal{D}$ defined by

$$\mathrm{BF}(\alpha_0 \xrightarrow{\sigma_0} \alpha_1 \xrightarrow{\sigma_1} \cdots \xrightarrow{\sigma_{n-1}} \alpha_n) = \mathrm{F}\alpha_0 \xrightarrow{\mathrm{F}\sigma_0} \mathrm{F}\alpha_1 \xrightarrow{\mathrm{F}\sigma_1} \cdots \xrightarrow{\mathrm{F}\sigma_{n-1}} \mathrm{F}\alpha_n \ .$$

EXAMPLE 14.1.3. For every integer $n \geq 0$ let $[n]$ denote the category with objects $\{0, 1, 2, \ldots, n\}$ and with a single morphism from i to j when $i \leq j$. There is a natural isomorphism of simplicial sets $\mathrm{B}[n] \approx \Delta[n]$ that takes the k-simplex $i_0 \to i_1 \to i_2 \to \cdots \to i_k$ of $\mathrm{B}[n]$ to the simplex $[i_0, i_1, i_2, \ldots, i_k]$ of $\Delta[n]$.

Let $\mathcal{O}$ denote the category with objects the $[n]$ for $n \geq 0$ and with $\mathcal{O}([n], [k])$ the functors from $[n]$ to $[k]$. $\mathcal{O}$ is a skeletal subcategory of the category of finite ordered sets, and we have a functor $\mathrm{B}\colon \mathcal{O} \to \mathrm{SS}$ whose image is isomorphic to the subcategory of SS consisting of the standard simplices $\Delta[n]$ for $n \geq 0$ and the standard maps between them.

EXAMPLE 14.1.4. Let G be a discrete group. If we consider G to be a category with one object and with morphisms equal to the group G, then $\mathrm{B}G$ is the standard classifying space of the group G, i.e., $\pi_1\mathrm{B}G \approx G$ and $\pi_i\mathrm{B}G \approx 0$ for $i \neq 1$.

PROPOSITION 14.1.5. *If $\mathcal{C}$ and $\mathcal{D}$ are small categories, then there is a natural isomorphism of simplicial sets* $\mathrm{B}(\mathcal{C} \times \mathcal{D}) \approx \mathrm{B}\mathcal{C} \times \mathrm{B}\mathcal{D}$.

PROOF. This follows directly from the definitions. □

Although there is an obvious one to one correspondence between the simplices of $\mathrm{B}\mathcal{C}$ and the simplices of $\mathrm{B}\mathcal{C}^{\mathrm{op}}$, this does not define a map of simplicial sets because it does not commute with the face and degeneracy operators. It does, however, define a homeomorphism between the geometric realizations of these simplicial sets.

PROPOSITION 14.1.6. *If $\mathcal{C}$ is a small category, then there is a natural homeomorphism of topological spaces* $|\mathrm{B}\mathcal{C}| \approx |\mathrm{B}\mathcal{C}^{\mathrm{op}}|$.

PROOF. We define $\phi\colon |\mathrm{B}\mathcal{C}| \to |\mathrm{B}\mathcal{C}^{\mathrm{op}}|$ by letting ϕ take the realization of the simplex $\alpha_0 \xrightarrow{\sigma_0} \alpha_1 \xrightarrow{\sigma_1} \cdots \xrightarrow{\sigma_{n-1}} \alpha_n$ of $\mathrm{B}\mathcal{C}$ to the realization of the simplex $\alpha_n \xleftarrow{\sigma_{n-1}} \alpha_{n-1} \xleftarrow{\sigma_{n-2}} \cdots \xleftarrow{\sigma_0} \alpha_0$ of $\mathrm{B}\mathcal{C}^{\mathrm{op}}$, reversing the orientation of the simplex. This commutes with the realizations of the face and degeneracy operators and so we have a map $|\mathrm{B}\mathcal{C}| \to |\mathrm{B}\mathcal{C}^{\mathrm{op}}|$ that has an obvious inverse. □

EXAMPLE 14.1.7. Let $\mathcal{C}$ and $\mathcal{D}$ be small categories, let $\mathrm{F}\colon \mathcal{C} \to \mathcal{D}$ be a functor, and let α be an object of $\mathcal{C}$.

(1) If $n \geq 0$ then an n-simplex σ of $\mathrm{B}(\mathrm{F}\downarrow\alpha)$ (see Definition 11.8.1) is a pair

$$\big((\beta_0 \xrightarrow{\sigma_0} \beta_1 \xrightarrow{\sigma_1} \cdots \xrightarrow{\sigma_{n-1}} \beta_n), \tau\colon \mathrm{F}\beta_n \to \alpha\big)$$

where $\beta_0 \xrightarrow{\sigma_0} \beta_1 \xrightarrow{\sigma_1} \cdots \xrightarrow{\sigma_{n-1}} \beta_n$ is a string of composable maps in $\mathcal{C}$ and $\tau\colon \mathrm{F}\beta_n \to \alpha$ is a map in $\mathcal{D}$. The face and degeneracy maps on the simplex σ are

$$d_i\sigma = \begin{cases} \big((\beta_1 \xrightarrow{\sigma_1} \beta_2 \xrightarrow{\sigma_2} \cdots \xrightarrow{\sigma_{n-1}} \beta_n), \tau\colon \mathrm{F}\beta_n \to \alpha\big) & \text{if } i = 0 \\ \big((\beta_0 \xrightarrow{\sigma_0} \cdots \xrightarrow{\sigma_{i-2}} \beta_{i-1} \xrightarrow{\sigma_i\sigma_{i-1}} \beta_{i+1} \\ \qquad \xrightarrow{\sigma_{i+1}} \cdots \xrightarrow{\sigma_{n-1}} \beta_n), \tau\colon \mathrm{F}\beta_n \to \alpha\big) & \text{if } 0 < i < n \\ \big((\beta_0 \xrightarrow{\sigma_0} \beta_1 \xrightarrow{\sigma_1} \cdots \xrightarrow{\sigma_{n-2}} \beta_{n-1}), \tau \circ (\mathrm{F}\sigma_{n-1})\colon \mathrm{F}\beta_{n-1} \to \alpha\big) & \text{if } i = n \end{cases}$$

$$s_i\sigma = \big((\beta_0 \xrightarrow{\sigma_0} \cdots \xrightarrow{\sigma_{i-1}} \beta_i \xrightarrow{1_{\beta_i}} \beta_i \xrightarrow{\sigma_i} \beta_{i+1} \xrightarrow{\sigma_{i+1}} \cdots \xrightarrow{\sigma_{n-1}} \beta_n), \tau\colon \mathrm{F}\beta_n \to \alpha\big)\ .$$

(2) If $n \geq 0$ then an n-simplex σ of $\mathrm{B}(\alpha\downarrow\mathrm{F})^{\mathrm{op}}$ (see Definition 11.8.3) is a pair

$$\big((\beta_0 \xleftarrow{\sigma_0} \beta_1 \xleftarrow{\sigma_1} \cdots \xleftarrow{\sigma_{n-1}} \beta_n), \tau\colon \alpha \to \mathrm{F}\beta_n\big)$$

where $\beta_0 \xleftarrow{\sigma_0} \beta_1 \xleftarrow{\sigma_1} \cdots \xleftarrow{\sigma_{n-1}} \beta_n$ is a string of composable maps in $\mathcal{C}$ and $\tau\colon \alpha \to \mathrm{F}\beta_n$ is a map in $\mathcal{D}$. The face and degeneracy maps on the simplex

σ are

$$d_i\sigma = \begin{cases} \bigl((\beta_1 \xleftarrow{\sigma_1} \beta_2 \xleftarrow{\sigma_2} \cdots \xleftarrow{\sigma_{n-1}} \beta_n), \tau\colon \alpha \to \mathrm{F}\beta_n\bigr) & \text{if } i = 0 \\ \bigl((\beta_0 \xleftarrow{\sigma_0} \cdots \xleftarrow{\sigma_{i-2}} \beta_{i-1} \xleftarrow{\sigma_{i-1}\sigma_i} \beta_{i+1} \\ \qquad \xleftarrow{\sigma_{i+1}} \cdots \xleftarrow{\sigma_{n-1}} \beta_n), \tau\colon \alpha \to \mathrm{F}\beta_n\bigr) & \text{if } 0 < i < n \\ \bigl((\beta_0 \xleftarrow{\sigma_0} \beta_1 \xleftarrow{\sigma_1} \cdots \xleftarrow{\sigma_{n-2}} \beta_{n-1}), (\mathrm{F}\sigma_{n-1}) \circ \tau\colon \alpha \to \mathrm{F}\beta_{n-1}\bigr) & \text{if } i = n \end{cases}$$

$$s_i\sigma = \bigl((\beta_0 \xleftarrow{\sigma_0} \cdots \xleftarrow{\sigma_{i-1}} \beta_i \xleftarrow{1_{\beta_i}} \beta_i \xleftarrow{\sigma_i} \beta_{i+1} \xleftarrow{\sigma_{i+1}} \cdots \xleftarrow{\sigma_{n-1}} \beta_n), \tau\colon \alpha \to \mathrm{F}\beta_n\bigr) .$$

EXAMPLE 14.1.8. Let $\mathcal{C}$ and $\mathcal{D}$ be small categories, let $\mathrm{F}\colon \mathcal{C} \to \mathcal{D}$ be a functor, and let α be an object of $\mathcal{C}$.

(1) The map of simplicial sets $\mathrm{BF}_*\colon \mathrm{B}(\mathcal{C}\downarrow\alpha) \to \mathrm{B}(\mathrm{F}\downarrow\mathrm{F}\alpha)$ (see Example 11.8.2) takes the n-simplex $\bigl((\beta_0 \xrightarrow{\sigma_0} \beta_1 \xrightarrow{\sigma_1} \cdots \xrightarrow{\sigma_{n-1}} \beta_n), \tau\colon \beta_n \to \alpha\bigr)$ of $\mathrm{B}(\mathcal{C}\downarrow\alpha)$ to the simplex $\bigl((\beta_0 \xrightarrow{\sigma_0} \beta_1 \xrightarrow{\sigma_1} \cdots \xrightarrow{\sigma_{n-1}} \beta_n), \mathrm{F}\tau\colon \mathrm{F}\beta_n \to \mathrm{F}\alpha\bigr)$ of $\mathrm{B}(\mathrm{F}\downarrow\mathrm{F}\alpha)$.

(2) The map of simplicial sets $\mathrm{BF}_*\colon \mathrm{B}(\alpha\downarrow\mathcal{C})^{\mathrm{op}} \to \mathrm{B}(\mathrm{F}\alpha\downarrow\mathrm{F})^{\mathrm{op}}$ (see Example 11.8.4) takes the n-simplex $\bigl((\beta_0 \xleftarrow{\sigma_0} \beta_1 \xleftarrow{\sigma_1} \cdots \xleftarrow{\sigma_{n-1}} \beta_n), \tau\colon \alpha \to \beta_n\bigr)$ of $\mathrm{B}(\mathcal{C}\downarrow\alpha)$ to the simplex $\bigl((\beta_0 \xleftarrow{\sigma_0} \beta_1 \xleftarrow{\sigma_1} \cdots \xleftarrow{\sigma_{n-1}} \beta_n), \mathrm{F}\tau\colon \mathrm{F}\alpha \to \mathrm{F}\beta_n\bigr)$ of $\mathrm{B}(\mathrm{F}\downarrow\mathrm{F}\alpha)$.

14.2. Cofinal functors

If $\mathcal{C}$ and $\mathcal{D}$ are small categories and $\mathrm{F}\colon \mathcal{C} \to \mathcal{D}$ is a functor, then for every cocomplete and complete category $\mathcal{M}$ and every diagram $\boldsymbol{X}\colon \mathcal{D} \to \mathcal{M}$ there is an induced $\mathcal{C}$-diagram $\mathrm{F}^*\boldsymbol{X}$ and natural maps $\operatorname{colim}_{\mathcal{C}} \mathrm{F}^*\boldsymbol{X} \to \operatorname{colim}_{\mathcal{D}} \boldsymbol{X}$ and $\lim_{\mathcal{D}} \boldsymbol{X} \to \lim_{\mathcal{C}} \mathrm{F}^*\boldsymbol{X}$. The main result of this section is Theorem 14.2.5, which characterizes those functors F for which that natural map of colimits or that natural map of limits is always an isomorphism.

DEFINITION 14.2.1. Let $\mathcal{C}$ and $\mathcal{D}$ be small categories and let $\mathrm{F}\colon \mathcal{C} \to \mathcal{D}$ be a functor.

- The functor F is *left cofinal* (or *initial*) if for every object α of $\mathcal{D}$ the classifying space $\mathrm{B}(\mathrm{F}\downarrow\alpha)$ of the overcategory $(\mathrm{F}\downarrow\alpha)$ (see Definition 11.8.1) is non-empty and connected. If in addition F is an inclusion of a subcategory, then we will say that $\mathcal{C}$ is a *left cofinal subcategory* (or *initial subcategory*) of $\mathcal{D}$.
- The functor F is *right cofinal* (or *terminal*) if for every object α of $\mathcal{D}$ the classifying space $\mathrm{B}(\alpha\downarrow\mathrm{F})$ of the undercategory $(\alpha\downarrow\mathrm{F})$ (see Definition 11.8.3) is non-empty and connected. If in addition F is an inclusion of a subcategory, then we will say that $\mathcal{C}$ is a *right cofinal subcategory* (or *terminal subcategory*) of $\mathcal{D}$.

There are differing uses of the above terms in the literature; see Remark 19.6.2.

REMARK 14.2.2. We will show in Theorem 14.2.5 that *left cofinal* and *right cofinal* are the "correct" notions when considering colimits and limits of functors. In Definition 19.6.1 we will define a functor between small categories to be *homotopy left cofinal* or *homotopy right cofinal* if the classifying spaces of the overcategories (or undercategories) are *contractible*, rather than just non-empty and connected. We show in Theorem 19.6.13 that these are the "correct" notions when considering homotopy colimits and homotopy limits.

LEMMA 14.2.3. *Let $\mathcal{C}$ be a small category. If α is an object of $\mathcal{C}$ and $\mathbf{F}^{\alpha}_{*}$ is the free $\mathcal{C}$-diagram of sets generated at α (see Definition 11.5.7), then $\operatorname{colim}_{\mathcal{C}} \mathbf{F}^{\alpha}_{*}$ is a set with one element.*

PROOF. If β is an object of $\mathcal{C}$ and $h \in \mathbf{F}^{\alpha}_{*}(\beta) = \mathcal{C}(\alpha, \beta)$, then $\big(\mathbf{F}^{\alpha}_{*}(h)\big)(1_{\alpha}) = h \circ 1_{\alpha} = h$. That is, $\mathbf{F}^{\alpha}_{*}(h)$ takes $1_{\alpha} \in \mathbf{F}^{\alpha}_{*}(\alpha) = \mathcal{C}(\alpha,\alpha)$ to $h \in \mathbf{F}^{\alpha}_{*}(\beta)$, and so h and 1_{α} represent the same element of $\operatorname{colim}_{\mathcal{C}} \mathbf{F}^{\alpha}_{*}$. □

LEMMA 14.2.4. *Let $\mathcal{C}$ and $\mathcal{D}$ be small categories and let $\mathrm{G}\colon \mathcal{C} \to \mathcal{D}$ be a functor. If α is an object of $\mathcal{D}$ and $\mathbf{F}^{\alpha}_{*}$ is the free $\mathcal{D}$-diagram of sets generated at α (see Definition 11.5.7), then there is a natural one to one correspondence between $\operatorname{colim}_{\mathcal{C}} \mathrm{G}^{*}\mathbf{F}^{\alpha}_{*}$ and the components of $\mathrm{B}(\alpha \downarrow \mathrm{G})$.*

PROOF. There is a natural one to one correspondence between the vertices of $\mathrm{B}(\alpha\downarrow \mathrm{G})$ and the set $\coprod_{\sigma\in \mathrm{Ob}(\mathcal{C})} \mathbf{F}^{\alpha}_{*}\big(\mathrm{G}(\sigma)\big) = \coprod_{\sigma\in\mathrm{Ob}(\mathcal{C})} (\mathrm{G}^{*}\mathbf{F}^{\alpha}_{*})(\sigma)$. Under this correspondence, there is a 1-simplex from the vertex $f\colon \alpha \to \mathrm{F}(\sigma)$ to the vertex $g\colon \alpha \to \mathrm{G}(\tau)$ if and only if there is a map $h\colon \sigma \to \tau$ in $\mathcal{C}$ such that $\mathrm{F}(h)\circ f = g$. Thus, two vertices of $\mathrm{B}(\alpha\downarrow \mathrm{F})$ are in the same component of $\mathrm{B}(\alpha \downarrow \mathrm{G})$ if and only if they represent the same element of $\operatorname{colim}_{\mathcal{C}} \mathrm{G}^{*}\mathbf{F}^{\alpha}_{*}$. □

THEOREM 14.2.5. *Let $\mathcal{C}$ and $\mathcal{D}$ be small categories and let $\mathrm{G}\colon \mathcal{C} \to \mathcal{D}$ be a functor.*

(1) *The functor G is right cofinal (see Definition 14.2.1) if and only if for every cocomplete category $\mathcal{M}$ and every diagram $\boldsymbol{X}\colon \mathcal{D} \to \mathcal{M}$ the natural map $\operatorname{colim}_{\mathcal{C}} \mathrm{G}^{*}\boldsymbol{X} \to \operatorname{colim}_{\mathcal{D}} \boldsymbol{X}$ is an isomorphism.*
(2) *The functor G is left cofinal (see Definition 14.2.1) if and only if for every complete category $\mathcal{M}$ and every diagram $\boldsymbol{X}\colon \mathcal{D} \to \mathcal{M}$ the natural map $\lim_{\mathcal{D}} \boldsymbol{X} \to \lim_{\mathcal{C}} \mathrm{G}^{*}\boldsymbol{X}$ is an isomorphism.*

Proof of part 1: Let G be right cofinal, let $\mathcal{M}$ be a cocomplete category, and let $\boldsymbol{X}\colon \mathcal{D} \to \mathcal{M}$ be a diagram; we will define a map $\psi\colon \operatorname{colim}_{\mathcal{D}} \boldsymbol{X} \to \operatorname{colim}_{\mathcal{C}} \mathrm{G}^{*}\boldsymbol{X}$ that is an inverse to the natural map $\phi\colon \operatorname{colim}_{\mathcal{C}} \mathrm{G}^{*}\boldsymbol{X} \to \operatorname{colim}_{\mathcal{D}} \boldsymbol{X}$. If α is an object of $\mathcal{D}$, we can choose an object σ of $\mathcal{C}$ and a map $f\colon \alpha \to \mathrm{G}(\sigma)$ (since $\mathrm{B}(\alpha\downarrow \mathrm{G})$ is nonempty) and define a map $\boldsymbol{X}_{\alpha} \to \operatorname{colim}_{\mathcal{C}} \mathrm{G}^{*}\boldsymbol{X}$ as the composition $\boldsymbol{X}_{\alpha} \xrightarrow{f_{*}} \boldsymbol{X}_{\mathrm{G}(\sigma)} = (\mathrm{G}^{*}\boldsymbol{X})_{\sigma} \to \operatorname{colim}_{\mathcal{C}} \mathrm{G}^{*}\boldsymbol{X}$. If $h\colon \sigma \to \tau$ is a map in $\mathcal{C}$ then the map $(\mathrm{G}^{*}\boldsymbol{X})_{\sigma} \to \operatorname{colim}_{\mathcal{C}} \mathrm{G}^{*}\boldsymbol{X}$ equals the composition $(\mathrm{G}^{*}\boldsymbol{X})_{\sigma} \xrightarrow{h_{*}} (\mathrm{G}^{*}\boldsymbol{X})_{\tau} \to \operatorname{colim}_{\mathcal{C}} \mathrm{G}^{*}\boldsymbol{X}$. Since $\mathrm{B}(\alpha\downarrow G)$ is connected, this implies that our map $\boldsymbol{X}_{\sigma} \to \operatorname{colim}_{\mathcal{C}} G^{*}\boldsymbol{X}$ is independent of the choices. If $k\colon \alpha \to \beta$ is a map in $\mathcal{D}$, then it also implies that the map $\boldsymbol{X}_{\alpha} \to \operatorname{colim}_{\mathcal{C}} \mathrm{G}^{*}\boldsymbol{X}$ equals the composition $\boldsymbol{X}_{\alpha} \xrightarrow{k_{*}} \boldsymbol{X}_{\beta} \to \operatorname{colim}_{\mathcal{C}} \mathrm{G}^{*}\boldsymbol{X}$, and so we have an induced map $\psi\colon \operatorname{colim}_{\mathcal{D}} \boldsymbol{X} \to \operatorname{colim}_{\mathcal{C}} \mathrm{G}^{*}\boldsymbol{X}$. The composition $\boldsymbol{X}_{\alpha} \to \operatorname{colim}_{\mathcal{D}} \boldsymbol{X} \xrightarrow{\psi} \operatorname{colim}_{\mathcal{C}} \mathrm{G}^{*}\boldsymbol{X} \xrightarrow{\phi} \operatorname{colim}_{\mathcal{D}} \boldsymbol{X}$ equals the map $\boldsymbol{X}_{\alpha} \to \operatorname{colim}_{\mathcal{D}} \boldsymbol{X}$, and so $\phi\psi = 1_{\operatorname{colim}_{\mathcal{D}} \boldsymbol{X}}$. Similarly, the composition $(\mathrm{G}^{*}\boldsymbol{X})_{\sigma} \to \operatorname{colim}_{\mathcal{C}} \mathrm{G}^{*}\boldsymbol{X} \xrightarrow{\phi} \operatorname{colim}_{\mathcal{D}} \boldsymbol{X} \xrightarrow{\psi} \operatorname{colim}_{\mathcal{C}} \mathrm{G}^{*}\boldsymbol{X}$ equals the map $(\mathrm{G}^{*}\boldsymbol{X})_{\sigma} \to \operatorname{colim}_{\mathcal{C}} \mathrm{G}^{*}\boldsymbol{X}$, and so the composition $\psi\phi = 1_{\operatorname{colim}_{\mathcal{C}} \mathrm{G}^{*}\boldsymbol{X}}$.

If G is not right cofinal, then we can choose an object α of $\mathcal{D}$ such that $\mathrm{B}(\alpha\downarrow G)$ is either empty or not connected. If we let $\mathbf{F}^{\alpha}_{*}$ be the free $\mathcal{D}$-diagram of sets generated at α (see Definition 11.5.7), then $\operatorname{colim}_{\mathcal{D}} \mathbf{F}^{\alpha}_{*}$ is a set with one element (see Lemma 14.2.3) but $\operatorname{colim}_{\mathcal{C}} \mathrm{G}^* \mathbf{F}^{\alpha}_{*}$ has as many elements as the number of components of $\mathrm{B}(\alpha \downarrow \mathrm{G})$ (see Lemma 14.2.4).

Proof of part 2: Proposition 11.8.5 implies that the functor G is left cofinal if and only if $\mathrm{B}(\alpha \downarrow \mathrm{F}^{\mathrm{op}})^{\mathrm{op}}$ is nonempty and connected for every object α of $\mathcal{D}$, and Proposition 14.1.6 implies that this is true if and only if $\mathrm{B}(\alpha \downarrow \mathrm{F}^{\mathrm{op}})$ is nonempty and connected for every object α of $\mathcal{D}$. Part 1 implies that this is true if and only if for every cocomplete category $\mathcal{M}^{\mathrm{op}}$ and every diagram $\boldsymbol{X}^{\mathrm{op}}\colon \mathcal{D}^{\mathrm{op}} \to \mathcal{M}^{\mathrm{op}}$ the natural map $\operatorname{colim}_{\mathcal{C}^{\mathrm{op}}} (\mathrm{G}^{\mathrm{op}})^* \boldsymbol{X}^{\mathrm{op}} \to \operatorname{colim}_{\mathcal{D}^{\mathrm{op}}} \boldsymbol{X}^{\mathrm{op}}$ is an isomorphism, and this is true if and only if for every complete category $\mathcal{M}$ and every diagram $\boldsymbol{X}\colon \mathcal{D} \to \mathcal{M}$ the natural map $\lim_{\mathcal{D}} \boldsymbol{X} \to \lim_{\mathcal{C}} \mathrm{G}^* \boldsymbol{X}$ is an isomorphism.

14.3. Contractible classifying spaces

If $\mathcal{C}$ is a small category, then its classifying space $\mathrm{B}\mathcal{C}$ (see Definition 14.1.1) exists and is a simplicial set, and it makes sense to ask whether $\mathrm{B}\mathcal{C}$ is contractible. If $\mathcal{C}$ is not small, though, then there is no simplicial set $\mathrm{B}\mathcal{C}$ unless we are working in a universe $\mathcal{U}$ from which we can pass to a higher universe $\mathcal{U}'$ and construct the simplicial set $\mathrm{B}\mathcal{C}$ in $\mathcal{U}'$. Definition 14.3.1 allows us to say what we mean by "$\mathcal{C}$ has a contractible classifying space" without assuming that there exists a simplicial set $\mathrm{B}\mathcal{C}$. Proposition 14.3.3 shows that this definition is equivalent to our intuitive notion of having a simplicial set $\mathrm{B}\mathcal{C}$ that is contractible.

DEFINITION 14.3.1. If $\mathcal{C}$ is a category (that is not necessarily small), then we will say that $\mathcal{C}$ *has a contractible classifying space* if for every small subcategory $\mathcal{D}$ of $\mathcal{C}$ there is a small subcategory $\mathcal{D}'$ of $\mathcal{C}$ such that $\mathcal{D} \subset \mathcal{D}'$ and $\mathrm{B}\mathcal{D}'$ is a contractible simplicial set.

REMARK 14.3.2. If $\mathcal{C}$ is a small category, then Definition 14.3.1 is equivalent to the assertion that $\mathrm{B}\mathcal{C}$ is a contractible simplicial set. However, if a category $\mathcal{C}$ is not small, then there is no simplicial set $\mathrm{B}\mathcal{C}$ unless we are working in a universe from which we can pass to a higher universe in which the class of objects of $\mathcal{C}$ is a set (see, e.g., [**60**, page 17]), so that $\mathrm{B}\mathcal{C}$ is a simplicial set in that higher universe. For this situation, see Proposition 14.3.3.

PROPOSITION 14.3.3. *If $\mathcal{C}$ is a category in a universe $\mathcal{U}$ and $\mathcal{U}'$ is a higher universe in which $\mathrm{B}\mathcal{C}$ is a simplicial set, then $\mathrm{B}\mathcal{C}$ is contractible in the sense of Definition 14.3.1 if and only if $\mathrm{B}\mathcal{C}$ is a contractible simplicial set in $\mathcal{U}'$.*

PROOF. If $\mathrm{B}\mathcal{C}$ is contractible in the sense of Definition 14.3.1 and if $f\colon \mathrm{S}^n \to |\mathrm{B}\mathcal{C}|$ is a map of topological spaces in $\mathcal{U}'$, then the image of f factors through the image of $|\mathrm{B}\mathcal{D}| \to |\mathrm{B}\mathcal{C}|$ for some small (in $\mathcal{U}$) subcategory $\mathcal{D}$ of $\mathcal{C}$. There is then a small (in $\mathcal{U}$) subcategory $\mathcal{D}'$ of $\mathcal{C}$ such that $\mathcal{D} \subset \mathcal{D}'$ and $\mathrm{B}\mathcal{D}'$ is contractible, and so the map f is nullhomotopic.

Conversely, let $\mathrm{B}\mathcal{C}$ be contractible in $\mathcal{U}'$ and let $\mathcal{D}$ be a small (in $\mathcal{U}$) subcategory of $\mathcal{C}$. We will inductively define a sequence $\mathcal{D} = \mathcal{D}_0 \subset \mathcal{D}_1 \subset \mathcal{D}_2 \subset \cdots$ of small (in $\mathcal{U}$) subcategories of $\mathcal{C}$ such that for every $k \geq 0$, $n \geq 0$, and map $\mathrm{S}^k \to |\mathrm{B}\mathcal{D}_n|$, the composition $\mathrm{S}^k \to |\mathrm{B}\mathcal{D}_n| \subset |\mathrm{B}\mathcal{D}_{n+1}|$ is nullhomotopic. If we then let $\mathcal{D}' = \bigcup_{n\geq 0} \mathcal{D}_n$, then $\mathrm{B}\mathcal{D}'$ will be contractible (because any map from a sphere to $|\mathrm{B}\mathcal{D}'|$ must factor through $|\mathrm{B}\mathcal{D}_n|$ for some $n \geq 0$ and will thus be nullhomotopic in $|\mathrm{B}\mathcal{D}_{n+1}|$).

If $n \geq 0$ and we have defined $\mathcal{D}_n$, then for every $k \geq 0$ there is a set (in $\mathcal{U}$) of maps $f\colon S^k \to |B\mathcal{D}_n|$, and for each of these maps there is a nullhomotopy of the composition $S^k \to |B\mathcal{D}_n| \subset |B\mathcal{C}|$. The image of each of these nullhomotopies is contained in some finite subcomplex of $|B\mathcal{C}|$ and the union of all of these finite subcomplexes of $|B\mathcal{C}|$ is a small subcomplex of $|B\mathcal{C}|$. We can thus find a small subcategory $\mathcal{D}_{n+1}$ of $\mathcal{C}$ such that $B\mathcal{D}_{n+1}$ contains both $B\mathcal{D}_n$ and the image of all of the nullhomotopies. □

PROPOSITION 14.3.4. *Let $\mathcal{K}$ and $\mathcal{L}$ be categories. If both $B\mathcal{K}$ and $B\mathcal{L}$ are contractible (see Definition 14.3.1), then for every small subcategory $\mathcal{D}$ of $\mathcal{K} \times \mathcal{L}$ there are small subcategories $\mathcal{D}'_{\mathcal{K}}$ of $\mathcal{K}$ and $\mathcal{D}'_{\mathcal{L}}$ of $\mathcal{L}$ such that*

(1) *$B\mathcal{D}'_{\mathcal{K}}$ and $B\mathcal{D}'_{\mathcal{L}}$ are contractible and*
(2) *$\mathcal{D} \subset \mathcal{D}'_{\mathcal{K}} \times \mathcal{D}'_{\mathcal{L}}$.*

PROOF. If $\mathcal{D}$ is a small subcategory of $\mathcal{K} \times \mathcal{L}$, let $\mathcal{D}_{\mathcal{K}}$ be the image of $\mathcal{D}$ under $\mathrm{pr}_{\mathcal{K}}\colon \mathcal{K} \times \mathcal{L} \to \mathcal{K}$ and let $\mathcal{D}_{\mathcal{L}}$ be the image of $\mathcal{D}$ under $\mathrm{pr}_{\mathcal{L}}\colon \mathcal{K} \times \mathcal{L} \to \mathcal{L}$. Both $\mathcal{D}_{\mathcal{K}}$ and $\mathcal{D}_{\mathcal{L}}$ are small, and so there exist small subcategories $\mathcal{D}'_{\mathcal{K}}$ of $\mathcal{K}$ and $\mathcal{D}'_{\mathcal{L}}$ of $\mathcal{L}$ such that $\mathcal{D}_{\mathcal{K}} \subset \mathcal{D}'_{\mathcal{K}}$, $\mathcal{D}_{\mathcal{L}} \subset \mathcal{D}'_{\mathcal{L}}$, and both $B\mathcal{D}'_{\mathcal{K}}$ and $B\mathcal{D}'_{\mathcal{L}}$ are contractible. □

PROPOSITION 14.3.5. *Let $\mathcal{K}$ and $\mathcal{L}$ be categories. If both $B\mathcal{K}$ and $B\mathcal{L}$ are contractible (see Definition 14.3.1), then $B(\mathcal{K} \times \mathcal{L})$ is contractible.*

PROOF. This follows from Proposition 14.3.4 and Proposition 14.1.5. □

14.3.6. Homotopic maps of classifying spaces.

PROPOSITION 14.3.7. *If $\mathcal{C}$ is a small category then there is a natural isomorphism*

$$B(\mathcal{C} \times [1]) \approx (B\mathcal{C}) \times \Delta[1]$$

(where $[1]$ is the category of Example 14.1.3).

PROOF. This follows from Proposition 14.1.5 and Example 14.1.3. □

LEMMA 14.3.8. *Let $[1]$ be the category of Example 14.1.3. If $\mathcal{C}$ is a small category, then*

- *the objects of the category $\mathcal{C} \times [1]$ consist of two objects $(\alpha, 0)$ and $(\alpha, 1)$ for every object α of $\mathcal{C}$, and*
- *the morphisms of $\mathcal{C} \times [1]$ consist of three morphisms $(\sigma, 1_0)$, $(\sigma, 1_1)$, and $(\sigma, 0 \to 1)$ for every morphism σ of $\mathcal{C}$ (where $0 \to 1$ is the unique non-identity map of $[1]$).*

PROOF. This follows directly from the definitions. □

LEMMA 14.3.9. *Let $\mathcal{C}$ and $\mathcal{D}$ be categories, let $F, G\colon \mathcal{C} \to \mathcal{D}$ be functors from $\mathcal{C}$ to $\mathcal{D}$, and let $[1]$ be the category of Example 14.1.3. If $i_0\colon \mathcal{C} \to \mathcal{C} \times [1]$ is the functor that takes an object α to $(\alpha, 0)$ and a morphism g to $(g, 1_0)$ and $i_1\colon \mathcal{C} \to \mathcal{C} \times [1]$ is the functor that takes an object α to $(\alpha, 1)$ and a morphism g to $(g, 1_1)$, then there is a natural transformation $\phi\colon F \to G$ if and only if there is a functor $\Phi\colon \mathcal{C} \times [1] \to \mathcal{D}$ such that $\Phi i_0 = F$ and $\Phi i_1 = G$.*

PROOF. If $\phi\colon F \to G$ is a natural transformation, then we define Φ by letting

- $\Phi(\alpha, 0) = F(\alpha)$ and $\Phi(\alpha, 1) = G(\alpha)$ for every object α of $\mathcal{C}$, and

- $\Phi(\sigma, 1_0) = \mathrm{F}(\sigma)$, $\Phi(\sigma, 1_1) = \mathrm{G}(\sigma)$, and $\Phi(\sigma, 0 \to 1) = \phi(\sigma)$ for every morphism σ of $\mathcal{C}$ (see Lemma 14.3.8).

Conversely, if $\Phi\colon \mathcal{C} \times [1] \to \mathcal{D}$ is a functor such that $\Phi i_0 = \mathrm{F}$ and $\Phi i_1 = \mathrm{G}$, define a natural transformation $\phi\colon \mathrm{F} \to \mathrm{G}$ by letting $\phi(\alpha) = \Phi(\alpha, 0 \to 1)$ for every object α of $\mathcal{C}$. □

PROPOSITION 14.3.10. *Let $\mathcal{C}$ and $\mathcal{D}$ be small categories and let $\mathrm{F}, \mathrm{G}\colon \mathcal{C} \to \mathcal{D}$ be functors from $\mathcal{C}$ to $\mathcal{D}$. If there is a natural transformation from F to G, then the induced maps of classifying spaces $\mathrm{BF}, \mathrm{BG}\colon \mathrm{B}\mathcal{C} \to \mathrm{B}\mathcal{D}$ are homotopic.*

PROOF. This follows from Lemma 14.3.9 and Proposition 14.3.7. □

DEFINITION 14.3.11. If $\mathcal{C}$ and $\mathcal{D}$ are categories and α is an object of $\mathcal{D}$, then the *constant functor from $\mathcal{C}$ to $\mathcal{D}$ at α* is the functor that takes every object of $\mathcal{C}$ to α and every morphism of $\mathcal{C}$ to 1_α.

COROLLARY 14.3.12. *Let $\mathcal{C}$ be a category and let α be an object of $\mathcal{C}$.*

1. *If there is a natural transformation from the identity functor of $\mathcal{C}$ to the constant functor from $\mathcal{C}$ to $\mathcal{C}$ at α (see Definition 14.3.11), then $\mathrm{B}\mathcal{C}$ is contractible (see Definition 14.3.1).*
2. *If there is a natural transformation from the constant functor from $\mathcal{C}$ to $\mathcal{C}$ at α (see Definition 14.3.11) to the identity functor of $\mathcal{C}$, then $\mathrm{B}\mathcal{C}$ is contractible (see Definition 14.3.1).*

PROOF. We will prove part 1; the proof of part 2 is similar.

If $\mathcal{D}$ is a small subcategory of $\mathcal{C}$, let $\mathcal{D}'$ be the subcategory of $\mathcal{C}$ consisting of $\mathcal{D}$, the object α, and the maps $\{\phi(\beta) \mid \beta \in \mathrm{Ob}(\mathcal{D})\}$. $\mathcal{D}'$ is also small, and Proposition 14.3.10 implies that $\mathrm{B}\mathcal{D}'$ is contractible. □

PROPOSITION 14.3.13. *Let $\mathcal{C}$ be a category.*

1. *If α is an initial object of $\mathcal{C}$, then there is a natural transformation from the constant functor at α (see Definition 14.3.11) to the identity functor of $\mathcal{C}$.*
2. *If α is a terminal object of $\mathcal{C}$, then there is a natural transformation from the identity functor of $\mathcal{C}$ to the constant functor at α (see Definition 14.3.11).*

PROOF. We will prove part 1; the proof of part 2 is dual.

For every object β of $\mathcal{C}$, let $\phi(\beta)$ be the unique map from α to β. Since α is an initial object of $\mathcal{C}$, it follows that ϕ is a natural transformation, and so the result follows from Corollary 14.3.12. □

PROPOSITION 14.3.14. *If the small category $\mathcal{C}$ has either a terminal or an initial object, then $\mathrm{B}\mathcal{C}$ is contractible (see Definition 14.3.1.)*

PROOF. This follows from Corollary 14.3.12 and Proposition 14.3.13. □

14.4. Uniqueness of weak equivalences

DEFINITION 14.4.1. Let $\mathcal{C}$ be a category. If X and Y are objects of $\mathcal{C}$, then two zig-zags (see Definition 7.9.1) in $\mathcal{C}$ from X to Y are *equivalent* if one can be changed into the other by a finite sequence of the following transformations and their inverses:

(1) If two consecutive arrows in a zig-zag point in the same direction, compose them; i.e.,

$$X \to W_1 \leftarrow \cdots \to W_{k-1} \xrightarrow{\alpha_k} W_k \xrightarrow{\alpha_{k+1}} W_{k+1} \leftarrow \cdots \to Y$$

is equivalent to

$$X \to W_1 \leftarrow \cdots \to W_{k-1} \xrightarrow{\alpha_{k+1}\alpha_k} W_{k+1} \leftarrow \cdots \to Y$$

and

$$X \to W_1 \leftarrow \cdots \to W_{k-1} \xleftarrow{\beta_k} W'_k \xleftarrow{\beta_{k+1}} W_{k+1} \leftarrow \cdots \to Y$$

is equivalent to

$$X \to W_1 \leftarrow \cdots \to W_{k_1} \xleftarrow{\beta_k\beta_{k+1}} W_{k+1} \leftarrow \cdots \to Y\ .$$

(2) If an arrow is immediately followed by the same arrow pointing in the opposite direction, remove the pair; i.e., both

$$X \to W_1 \leftarrow \cdots \to W_{k-1} \xrightarrow{\alpha_k} W_k \xleftarrow{\alpha_k} W_{k-1} \leftarrow \cdots \to Y$$

and

$$X \to W_1 \leftarrow \cdots \to W_{k-1} \xleftarrow{\beta_k} W'_k \xrightarrow{\beta_k} W_{k-1} \leftarrow \cdots \to Y$$

are equivalent to

$$X \to W_1 \leftarrow \cdots W_{k-2} \to W_{k-1} \leftarrow W_{k+2} \cdots \to Y\ .$$

DEFINITION 14.4.2. Let $\mathcal{C}$ be a category and let X and Y be objects of $\mathcal{C}$. If any two zig-zags in $\mathcal{C}$ from X to Y are equivalent (see Definition 14.4.1), then we will say that there is an *essentially unique zig-zag* in $\mathcal{C}$ from X to Y.

PROPOSITION 14.4.3. *Let $\mathcal{M}$ be a model category and let $\mathcal{K}$ be a small and full subcategory of $\mathcal{M}$. If X, Y, and Z are objects in $\mathcal{K}$, then composition of zig-zags of weak equivalences (see Definition 7.9.1) passes to equivalence classes of zig-zags of weak equivalences in $\mathcal{K}$ (see Definition 14.4.1) to define the composition of an equivalence class of zig-zags of weak equivalences from X to Y in $\mathcal{K}$ with an equivalence class of zig-zags of weak equivalences from Y to Z in $\mathcal{K}$.*

PROOF. This follows directory from the definitions. □

PROPOSITION 14.4.4. *Let $\mathcal{K}$ be a small category. If X and Y are objects of $\mathcal{K}$, then the set of equivalence classes of zig-zags in $\mathcal{K}$ from X to Y is isomorphic to the set of maps from X to Y in the edge path groupoid of $\mathrm{B}\mathcal{K}$.*

PROOF. This almost follows directly from the definitions. If α and β are composable maps in $\mathcal{K}$, then there is a 2-simplex in $\mathrm{B}\mathcal{K}$ with faces α, β, and $\beta \circ \alpha$, and so the definition of the edge path groupoid contains the relation $(\beta\circ\alpha)^{-1}\cdot\beta = \alpha^{-1}$, which is not part of the definition of equivalence of zig-zags. However, this relation is a consequence of the definition of equivalence of zig-zags because $(\beta\circ\alpha)^{-1}\cdot\beta = \alpha^{-1}\cdot\beta^{-1}\cdot\beta = \alpha^{-1}$ (and similarly for the other relations derived from the 2-simplex). □

THEOREM 14.4.5. *Let $\mathcal{C}$ be a category and let X and Y be objects of $\mathcal{C}$. If $\mathrm{B}\mathcal{C}$ is contractible (see Definition 14.3.1), then there is an essentially unique zig-zag in $\mathcal{C}$ from X to Y (see Definition 14.4.2).*

PROOF. This follows from Proposition 14.4.4. □

14.4.6. Homotopy equivalences.

PROPOSITION 14.4.7. *Let $\mathcal{M}$ be a model category, let $\mathcal{C}$ be a category, and let $F\colon \mathcal{C} \to \mathcal{M}$ be a functor such that $F(\alpha)$ is a weak equivalence between cofibrant-fibrant objects of $\mathcal{M}$ for every map α of $\mathcal{C}$. If*

$$X \xrightarrow{\alpha_1} W_1 \xleftarrow{\alpha_2} W_2 \xrightarrow{\alpha_3} \cdots \xleftarrow{\alpha_{n-1}} W_{n-1} \xrightarrow{\alpha_n} Y$$

is a zig-zag in $\mathcal{C}$ from X to Y, then we can choose a homotopy inverse g_k to $F(\alpha_k)$ for each α_k that points to the left and let $g_k = F(\alpha_k)$ for each α_k that points to the right and the composition $g_n g_{n-1} \cdots g_1\colon X \to Y$ will be a homotopy equivalence whose homotopy class is independent of the choices made.

PROOF. This follows from Theorem 7.5.10. □

DEFINITION 14.4.8. If $\mathcal{M}$ is a model category, $\mathcal{C}$ is a category, $F\colon \mathcal{C} \to \mathcal{M}$ is a functor such that $F(\alpha)$ is a weak equivalence between cofibrant-fibrant objects of $\mathcal{M}$ for every map α of $\mathcal{C}$, and if

$$X \xrightarrow{\alpha_1} W_1 \xleftarrow{\alpha_2} W_2 \xrightarrow{\alpha_3} \cdots \xleftarrow{\alpha_{n-1}} W_{n-1} \xrightarrow{\alpha_n} Y$$

is a zig-zag in $\mathcal{C}$ from X to Y, then any homotopy equivalence from $F(X)$ to $F(Y)$ that is homotopic to one obtained from the zig-zag as in Proposition 14.4.7 will be called a homotopy equivalence *determined* by that zig-zag.

LEMMA 14.4.9. *Let $\mathcal{M}$ be a model category, let $\mathcal{C}$ be a category, and let $F\colon \mathcal{C} \to \mathcal{M}$ be a functor such that $F(\alpha)$ is a weak equivalence between cofibrant-fibrant objects of $\mathcal{M}$ for every map α of $\mathcal{C}$. If X and Y are objects of $\mathcal{C}$ and we have two equivalent zig-zags (see Definition 14.4.1) from X to Y, then those zig-zags determine (see Definition 14.4.8) the same homotopy class of homotopy equivalences from $F(X)$ to $F(Y)$.*

PROOF. This follows immediately from the definitions. □

PROPOSITION 14.4.10. *Let $\mathcal{M}$ be a model category, let $\mathcal{C}$ be a category, and let $F\colon \mathcal{C} \to \mathcal{M}$ be a functor such that $F(\alpha)$ is a weak equivalence between cofibrant-fibrant objects of $\mathcal{M}$ for every map α of $\mathcal{C}$. If there is an essentially unique zig-zag (see Definition 14.4.2) in $\mathcal{C}$ from X to Y then the zig-zags from X to Y determine (see Definition 14.4.8) a homotopy class of homotopy equivalences from $F(X)$ to $F(Y)$.*

PROOF. This follows from Lemma 14.4.9. □

PROPOSITION 14.4.11. *Let $\mathcal{M}$ be a model category, let $\mathcal{C}$ be a category, and let $F\colon \mathcal{C} \to \mathcal{M}$ be a functor such that $F(\alpha)$ is a weak equivalence between cofibrant-fibrant objects of $\mathcal{M}$ for every map α of $\mathcal{C}$. If there is an essentially unique zig-zag (see Definition 14.4.2) in $\mathcal{C}$ between any two objects of $\mathcal{C}$, then for any two objects X and Y of $\mathcal{C}$ those zig-zags define a homotopy class of homotopy equivalences $h_{XY}\colon F(X) \to F(Y)$ such that if X, Y, and Z are objects of $\mathcal{C}$, then $h_{YZ}h_{XY} = h_{XZ}$.*

PROOF. This follows from Proposition 14.4.10. □

14.5. Categories of functors

In this section, we consider a situation in which we have categories $\mathcal{K}$ and $\mathcal{L}$ and a functor $\boldsymbol{X}: \mathcal{K} \to \mathcal{L}$. We would like to discuss what would be called "subcategories of the overcategory $(\mathcal{L}^{\mathcal{K}} \downarrow \boldsymbol{X})$" (see Definition 11.8.1), where $\mathcal{L}^{\mathcal{K}}$ is the "category of functors from $\mathcal{K}$ to $\mathcal{L}$", but we cannot do this because *the collection of functors from $\mathcal{K}$ to $\mathcal{L}$ and all natural transformations between them might not form a category.* This is because if $\mathcal{K}$ is not small, then there might be a proper class of natural transformations between any pair of functors from $\mathcal{K}$ to $\mathcal{L}$.

There are two ways to deal with this situation. The first is to work in a universe that allows us to pass, temporarily, to a "higher universe" (see, e.g., [**60**, page 17]) in which $\mathcal{K}$ and $\mathcal{L}$ are small and so there is a category of "all functors from $\mathcal{K}$ to $\mathcal{L}$ and all natural transformations between them". We could pass to such a higher universe, prove theorems there, and then state the implications of those theorems for our original universe. The second is to work with subcollections of those functors and natural transformations, restricting ourselves to those subcollections that do form categories, and draw our results from the relationships between those categories of functors and natural transformations. This second method is the one that we shall use (see Definition 14.5.2 and Remark 14.5.3). We are indebted to D. Dugger for suggesting this approach.

DEFINITION 14.5.1. Let $\mathcal{K}$ and $\mathcal{L}$ be categories, let $\mathcal{W}$ be a class of maps in $\mathcal{L}$, and let $\boldsymbol{X}: \mathcal{K} \to \mathcal{L}$ be a functor.

(1) A *functor over $\boldsymbol{X}$ relative to $\mathcal{W}$* is a pair $(\widetilde{\boldsymbol{X}}, i)$ in which $\widetilde{\boldsymbol{X}}: \mathcal{K} \to \mathcal{L}$ is a functor and $i: \widetilde{\boldsymbol{X}} \to \boldsymbol{X}$ is a natural transformation such that $i_\alpha: \widetilde{\boldsymbol{X}}_\alpha \to \boldsymbol{X}_\alpha$ is in $\mathcal{W}$ for every object α of $\mathcal{K}$.
(2) A *functor under $\boldsymbol{X}$ relative to $\mathcal{W}$* is a pair $(\widehat{\boldsymbol{X}}, j)$ in which $\widehat{\boldsymbol{X}}: \mathcal{K} \to \mathcal{L}$ is a functor and $j: \boldsymbol{X} \to \widehat{\boldsymbol{X}}$ is a natural transformation such that $j_\alpha: \boldsymbol{X}_\alpha \to \widehat{\boldsymbol{X}}_\alpha$ is in $\mathcal{W}$ for every object α of $\mathcal{K}$.

DEFINITION 14.5.2. Let $\mathcal{K}$ and $\mathcal{L}$ be categories, let $\mathcal{W}$ be a class of maps in $\mathcal{L}$, and let $\boldsymbol{X}: \mathcal{K} \to \mathcal{L}$ be a functor.

(1) A *category of functors over $\boldsymbol{X}$ relative to $\mathcal{W}$* is a category $\mathcal{C}$ such that
   (a) every object of $\mathcal{C}$ is a functor over $\boldsymbol{X}$ relative to $\mathcal{W}$ (see Definition 14.5.1),
   (b) if $(\widetilde{\boldsymbol{X}}, i)$ and $(\widetilde{\boldsymbol{X}}', i')$ are objects of $\mathcal{C}$, then $\mathcal{C}\big((\widetilde{\boldsymbol{X}}, i), (\widetilde{\boldsymbol{X}}', i')\big)$ is a set of natural transformations $\phi: \widetilde{\boldsymbol{X}} \to \widetilde{\boldsymbol{X}}'$ such that $i'\phi = i$,
   (c) if $(\widetilde{\boldsymbol{X}}, i)$ is an object of $\mathcal{C}$, then the identity natural transformation of $\widetilde{\boldsymbol{X}}$ is an element of $\mathcal{C}\big((\widetilde{\boldsymbol{X}}, i), (\widetilde{\boldsymbol{X}}, i)\big)$, and
   (d) composition of morphisms in $\mathcal{C}$ is defined by composition of natural transformations.
(2) A *category of functors under $\boldsymbol{X}$ relative to $\mathcal{W}$* is a category $\mathcal{C}$ such that
   (a) every object of $\mathcal{C}$ is a functor under $\boldsymbol{X}$ relative to $\mathcal{W}$ (see Definition 14.5.1),
   (b) if $(\widehat{\boldsymbol{X}}, j)$ and $(\widehat{\boldsymbol{X}}', j')$ are objects of $\mathcal{C}$, then $\mathcal{C}\big((\widehat{\boldsymbol{X}}, j), (\widehat{\boldsymbol{X}}', j')\big)$ is a set of natural transformations $\phi: \widehat{\boldsymbol{X}} \to \widehat{\boldsymbol{X}}'$ such that $\phi j = j'$,
   (c) if $(\widehat{\boldsymbol{X}}, j)$ is an object of $\mathcal{C}$, then the identity natural transformation of $\widehat{\boldsymbol{X}}$ is an element of $\mathcal{C}\big((\widehat{\boldsymbol{X}}, j), (\widehat{\boldsymbol{X}}, j)\big)$, and

(d) composition of morphisms in $\mathcal{C}$ is defined by composition of natural transformations.

REMARK 14.5.3. If $\mathcal{K}$ is a *small* category and $\mathcal{L}$ is a category, then there is a category $\mathcal{L}^{\mathcal{K}}$ with objects all the functors from $\mathcal{K}$ to $\mathcal{L}$ and morphisms all natural transformations between such functors. In this case, a category of functors over $\boldsymbol{X}$ is just a subcategory of $(\mathcal{L}^{\mathcal{K}} \downarrow \boldsymbol{X})$ (see Definition 11.8.1) and a category of functors under $\boldsymbol{X}$ is just a subcategory of $(\boldsymbol{X} \downarrow \mathcal{L}^{\mathcal{K}})$ (see Definition 11.8.3). If $\mathcal{K}$ is not small, however, then the collection of natural transformations between two functors may not be a set, and so there may not be a "category of all functors and all natural transformations" except in a universe higher than the one in which we work.

We are indebted to D. M. Kan for the following theorem.

THEOREM 14.5.4. *Let $\mathcal{K}$ and $\mathcal{L}$ be categories, let $\boldsymbol{X}\colon \mathcal{K} \to \mathcal{L}$ be a functor, and let $\mathcal{W}$ be a class of maps in $\mathcal{L}$ that is closed under composition.*

(1) *If there is an augmented functor (see Definition 8.1.12) (F, p) on $\mathcal{L}$ such that $p_Y\colon \mathrm{F}Y \to Y$ is in $\mathcal{W}$ for every object Y of $\mathcal{L}$, then for every small category $\mathcal{D}$ of functors over $\boldsymbol{X}$ relative to $\mathcal{W}$ (see Definition 14.5.2) there is a small category $\mathcal{D}'$ of functors over $\boldsymbol{X}$ relative to $\mathcal{W}$ such that $\mathcal{D} \subset \mathcal{D}'$ and $\mathrm{B}\mathcal{D}'$ is contractible.*
(2) *If there is a coaugmented functor (see Definition 8.1.12) (G, q) on $\mathcal{L}$ such that $q_Y\colon Y \to \mathrm{G}Y$ is in $\mathcal{W}$ for every object Y of $\mathcal{L}$, then for every small category $\mathcal{D}$ of functors under $\boldsymbol{X}$ relative to $\mathcal{W}$ (see Definition 14.5.2) there is a small category $\mathcal{D}'$ of functors under $\boldsymbol{X}$ relative to $\mathcal{W}$ such that $\mathcal{D} \subset \mathcal{D}'$ and $\mathrm{B}\mathcal{D}'$ is contractible.*

PROOF. We will prove part 1; the proof of part 2 is dual.

Let $\mathcal{D}$ be a small category of functors over $\boldsymbol{X}$ relative to $\mathcal{W}$. If we let $\mathcal{D}'$ be the category generated by

- $\mathcal{D}$,
- the objects $\{(\mathrm{F}\widetilde{\boldsymbol{X}}, i \circ p_{\widetilde{\boldsymbol{X}}}) \mid (\widetilde{\boldsymbol{X}}, i) \in \mathrm{Ob}(\mathcal{D})\}$,
- the object $(\mathrm{F}\boldsymbol{X}, p_{\boldsymbol{X}})$,
- the maps $\{\mathrm{F}(g) \mid g \text{ is a map in } \mathcal{D}\}$,
- the maps $\{\mathrm{F}(i)\colon (\mathrm{F}\widetilde{\boldsymbol{X}}, i \circ p_{\widetilde{\boldsymbol{X}}}) \to (\mathrm{F}\boldsymbol{X}, p_{\boldsymbol{X}}) \mid (\widetilde{\boldsymbol{X}}, i) \in \mathrm{Ob}(\mathcal{D})\}$, and
- the maps $\{p_{\widetilde{\boldsymbol{X}}} \mid (\widetilde{\boldsymbol{X}}, i) \in \mathrm{Ob}(\mathcal{D})\}$,

then $\mathcal{D}'$ is a small category and $\mathcal{D} \subset \mathcal{D}'$. We will show that $\mathrm{B}\mathcal{D}'$ is contractible by showing that there is a subcategory $\widetilde{\mathcal{D}} \subset \mathcal{D}'$ such that

(1) $\widetilde{\mathcal{D}}$ has a terminal object (and so $\mathrm{B}\widetilde{\mathcal{D}}$ is contractible; see Proposition 14.3.14), and
(2) $\mathrm{B}\widetilde{\mathcal{D}}$ is a deformation retract of $\mathrm{B}\mathcal{D}'$.

Let $\widetilde{\mathcal{D}}$ be the subcategory of $\mathcal{D}'$ generated by

- the objects $\{(\mathrm{F}\widetilde{\boldsymbol{X}}, i \circ p_{\widetilde{\boldsymbol{X}}}) \mid (\widetilde{\boldsymbol{X}}, i) \in \mathrm{Ob}(\mathcal{D})\}$,
- the object $(\mathrm{F}\boldsymbol{X}, p_{\boldsymbol{X}})$,
- the maps $\{\mathrm{F}(g) \mid g \text{ is a map in } \mathcal{D}\}$, and
- the maps $\{\mathrm{F}(i)\colon (\mathrm{F}\widetilde{\boldsymbol{X}}, i \circ p_{\widetilde{\boldsymbol{X}}}) \to (\mathrm{F}\boldsymbol{X}, p_{\boldsymbol{X}}) \mid (\widetilde{\boldsymbol{X}}, i) \in \mathrm{Ob}(\mathcal{D})\}$.

The object $(\mathrm{F}\boldsymbol{X}, p_{\boldsymbol{X}})$ is a terminal object of $\widetilde{\mathcal{D}}$, and so it remains only to show that $\mathrm{B}\widetilde{\mathcal{D}}$ is a deformation retract of $\mathrm{B}\mathcal{D}'$.

We define a retraction $\widetilde{\mathrm{F}}\colon \mathcal{D}' \to \widetilde{\mathcal{D}}$ by letting

$$\widetilde{\mathrm{F}}(\widetilde{\boldsymbol{X}}, i) = (\mathrm{F}\widetilde{\boldsymbol{X}}, i \circ p_{\widetilde{\boldsymbol{X}}}) \text{ for } (\widetilde{\boldsymbol{X}}, i) \in \mathrm{Ob}(\mathcal{D}),$$

$$\widetilde{\mathrm{F}}(g\colon (\widetilde{\boldsymbol{X}}, i) \to (\widetilde{\boldsymbol{X}}', i')) = \mathrm{F}(g)\colon (\mathrm{F}\widetilde{\boldsymbol{X}}, i \circ p_{\widetilde{\boldsymbol{X}}}) \to (\mathrm{F}\widetilde{\boldsymbol{X}}', i' \circ p_{\widetilde{\boldsymbol{X}}'}) \text{ for } g \text{ a map in } \mathcal{D},$$

$$\widetilde{\mathrm{F}}(p_{\widetilde{\boldsymbol{X}}}\colon (\mathrm{F}\widetilde{\boldsymbol{X}}, i \circ p_{\widetilde{\boldsymbol{X}}}) \to (\widetilde{\boldsymbol{X}}, i)) = 1_{(\mathrm{F}\widetilde{\boldsymbol{X}}, i \circ p_{\widetilde{\boldsymbol{X}}})} \text{ for } (\widetilde{\boldsymbol{X}}, i) \in \mathrm{Ob}(\mathcal{D}),$$

and by letting $\widetilde{\mathrm{F}}$ be the identity on the subcategory $\widetilde{\mathcal{D}}$. Proposition 14.3.10 implies that it is sufficient to construct a natural transformation ϕ from $\widetilde{\mathrm{F}}$ to $1_{\mathcal{D}'}$. We do this by letting $\phi(\widetilde{\boldsymbol{X}}, i) = p_{\widetilde{\boldsymbol{X}}}$ for $(\widetilde{\boldsymbol{X}}, i) \in \mathrm{Ob}(\mathcal{D})$ and by letting ϕ take every object of $\widetilde{\mathcal{D}}$ to the identity map of that object. □

THEOREM 14.5.5. *Let $\mathcal{K}$ and $\mathcal{L}$ be categories, let $\boldsymbol{X}\colon \mathcal{K} \to \mathcal{L}$ be a functor, and let $\mathcal{W}$ be a class of maps in $\mathcal{L}$ that is closed under composition.*

(1) *If there is an augmented functor (see Definition 8.1.12) (F, p) on $\mathcal{L}$ such that $p_Y\colon \mathrm{F}Y \to Y$ is in $\mathcal{W}$ for every object Y of $\mathcal{L}$ and if $(\widetilde{\boldsymbol{X}}\colon \mathcal{K} \to \mathcal{L}, i\colon \widetilde{\boldsymbol{X}} \to \boldsymbol{X})$ and $(\widetilde{\boldsymbol{X}}'\colon \mathcal{K} \to \mathcal{L}, i'\colon \widetilde{\boldsymbol{X}}' \to \boldsymbol{X})$ are functors over $\boldsymbol{X}$ relative to $\mathcal{W}$ (see Definition 14.5.1), then there is an essentially unique zig-zag (see Definition 14.4.2) of natural transformations of functors over $\boldsymbol{X}$ relative to $\mathcal{W}$ from $\widetilde{\boldsymbol{X}}$ to $\widetilde{\boldsymbol{X}}'$.*
(2) *If there is a coaugmented functor (see Definition 8.1.12) (G, q) on $\mathcal{L}$ such that $q_Y\colon Y \to \mathrm{G}Y$ is in $\mathcal{W}$ for every object Y of $\mathcal{L}$ and if $(\widehat{\boldsymbol{X}}\colon \mathcal{K} \to \mathcal{L}, j\colon \boldsymbol{X} \to \widehat{\boldsymbol{X}})$ and $(\widehat{\boldsymbol{X}}'\colon \mathcal{K} \to \mathcal{L}, j'\colon \boldsymbol{X} \to \widehat{\boldsymbol{X}}')$ are functors under $\boldsymbol{X}$ relative to $\mathcal{W}$ (see Definition 14.5.1), then there is an essentially unique zig-zag (see Definition 14.4.2) of natural transformations of functors under $\boldsymbol{X}$ relative to $\mathcal{W}$ from $\widehat{\boldsymbol{X}}$ to $\widehat{\boldsymbol{X}}'$.*

PROOF. This follows from Proposition 14.5.7 and Theorem 14.5.4. □

THEOREM 14.5.6. *Let $\mathcal{C}$ be a category and let $\mathcal{W}$ be a class of maps in $\mathcal{C}$ that is closed under composition.*

(1) *Let X be an object of $\mathcal{C}$ and let $\widetilde{\mathcal{C}}_X$ be the full subcategory of $(\mathcal{C} \downarrow X)$ determined by the objects $\widetilde{X} \to X$ that are in $\mathcal{W}$. If there is an augmented functor (see Definition 8.1.12) (F, i) on $\mathcal{C}$ such that $i_Y\colon \mathrm{F}Y \to Y$ is in $\mathcal{W}$ for every object Y of $\mathcal{C}$, then $\mathrm{B}\widetilde{\mathcal{C}}_X$ is contractible (see Definition 14.3.1).*
(2) *Let X be an object of $\mathcal{C}$ and let $\widehat{\mathcal{C}}_X$ be the full subcategory of $(X \downarrow \mathcal{C})$ determined by the objects $X \to \widehat{X}$ that are in $\mathcal{W}$. If there is a coaugmented functor (see Definition 8.1.12) (G, j) on $\mathcal{C}$ such that $j_Y\colon Y \to \mathrm{G}Y$ is in $\mathcal{W}$ for every object Y of $\mathcal{C}$, then $\mathrm{B}\widehat{\mathcal{C}}_X$ is contractible.*

PROOF. If we let $\mathcal{K}$ be the category with one object and one (identity) map, then this follows from Theorem 14.5.4. □

PROPOSITION 14.5.7. *Let $\mathcal{K}$ and $\mathcal{L}$ be categories, let $\mathcal{W}$ be a class of maps in $\mathcal{L}$, and let $\boldsymbol{X}\colon \mathcal{K} \to \mathcal{L}$ be a functor.*

(1) *If every small category $\mathcal{D}$ of functors over $\boldsymbol{X}$ relative to $\mathcal{W}$ (see Definition 14.5.2) is contained in a small category $\mathcal{D}'$ of functors over $\boldsymbol{X}$ relative to $\mathcal{W}$ such that $\mathrm{B}\mathcal{D}'$ is contractible, then there is an essentially unique zig-zag (see Definition 14.4.2) of natural transformations of functors over $\boldsymbol{X}$ relative to $\mathcal{W}$ connecting any two functors over $\boldsymbol{X}$.*

(2) *If every small category $\mathcal{D}$ of functors under $\boldsymbol{X}$ relative to $\mathcal{W}$ (see Definition 14.5.2) is contained in a small category $\mathcal{D}'$ of functors under $\boldsymbol{X}$ relative to $\mathcal{W}$ such that $\mathrm{B}\mathcal{D}'$ is contractible, then there is an essentially unique zig-zag (see Definition 14.4.2) of natural transformations of functors under $\boldsymbol{X}$ relative to $\mathcal{W}$ connecting any two functors under $\boldsymbol{X}$.*

PROOF. This follows from Theorem 14.4.5. □

14.6. Cofibrant approximations and fibrant approximations

DEFINITION 14.6.1. Let $\mathcal{M}$ be a model category.

(1) If X is an object of $\mathcal{M}$, we let $\mathrm{CofAp}(X)$ denote the category whose objects are cofibrant approximations to X (see Definition 8.1.2) and whose morphisms are maps of cofibrant approximations (see Definition 8.1.4).
(2) If X is an object of $\mathcal{M}$, we let $\mathrm{FibAp}(X)$ denote the category whose objects are fibrant approximations to X and whose morphisms are maps of fibrant approximations.

THEOREM 14.6.2. *Let $\mathcal{M}$ be a model category.*

(1) *If X is an object of $\mathcal{M}$, then $\mathrm{B}\,\mathrm{CofAp}(X)$ is contractible (see Definition 14.3.1).*
(2) *If X is an object of $\mathcal{M}$, then $\mathrm{B}\,\mathrm{FibAp}(X)$ is contractible (see Definition 14.3.1).*

PROOF. We will prove part 1; the proof of part 2 is dual.

If we let $\mathcal{W}$ be the class of weak equivalences in $\mathcal{M}$ with cofibrant domain, then the result follows from Theorem 14.5.6 and Proposition 8.1.17. □

PROPOSITION 14.6.3. *Let $\mathcal{M}$ be a model category and let X be an object of $\mathcal{M}$.*

(1) *If $(\widetilde{X}, i)$ and $(\widetilde{X}', i')$ are cofibrant approximations to X, then there is an essentially unique zig-zag (see Definition 14.4.2) of weak equivalences of cofibrant approximations to X from $(\widetilde{X}, i)$ to $(\widetilde{X}', i')$.*
(2) *If $(\widehat{X}, j)$ and $(\widehat{X}', j')$ are fibrant approximations to X, then there is an essentially unique zig-zag (see Definition 14.4.2) of weak equivalences of fibrant approximations to X from $(\widehat{X}, j)$ to $(\widehat{X}', j')$.*

PROOF. This follows from Theorem 14.4.5 and Theorem 14.6.2. □

DEFINITION 14.6.4. Let $\mathcal{M}$ be a model category.

(1) If $g\colon X \to Y$ is a map in $\mathcal{M}$, we let $\mathrm{CofAp}(g)$ denote the category whose objects are cofibrant approximations to g (see Definition 8.1.22) and whose morphisms are maps of cofibrant approximations (see Definition 8.1.26).
(2) If $g\colon X \to Y$ is a map in $\mathcal{M}$, we let $\mathrm{FibAp}(g)$ denote the category whose objects are fibrant approximations to g and whose morphisms are maps of fibrant approximations.

PROPOSITION 14.6.5. *Let $\mathcal{M}$ be a model category.*

(1) *If $g\colon X \to Y$ is a map in $\mathcal{M}$, then $\mathrm{B}\,\mathrm{CofAp}(g)$ is contractible (see Definition 14.3.1).*
(2) *If $g\colon X \to Y$ is a map in $\mathcal{M}$, then $\mathrm{B}\,\mathrm{FibAp}(g)$ is contractible.*

PROOF. We will prove part 1; the proof of part 2 is dual.

Let $\mathcal{K}$ be the category of maps in $\mathcal{M}$, in which a map from $f\colon A \to B$ to $f'\colon A' \to B'$ is a commutative square

$$\begin{array}{ccc} A & \xrightarrow{s} & A' \\ {\scriptstyle f}\downarrow & & \downarrow{\scriptstyle f'} \\ B & \xrightarrow[t]{} & B' \end{array} ,$$

and let $\mathcal{W}$ be the class of maps for which both s and t are weak equivalences with cofibrant domains. If (F, i) is a functorial cofibrant approximation on $\mathcal{M}$, then (F, i) defines a functorial cofibrant approximation on $\mathcal{K}$, and so the result follows from Theorem 14.5.6. □

PROPOSITION 14.6.6. *Let $\mathcal{M}$ be a model category and let $f\colon X \to Y$ be a map in $\mathcal{M}$.*

(1) *If $\big((\widetilde{X}, i_X), (\widetilde{Y}, i_Y), \tilde{g}\colon \widetilde{X} \to \widetilde{Y}\big)$ and $\big((\widetilde{X}', i'_X), (\widetilde{Y}', i'_Y), \tilde{g}'\colon \widetilde{X}' \to \widetilde{Y}'\big)$ are cofibrant approximations to g, then they are connected by an essentially unique zig-zag (see Definition 14.4.2) of weak equivalences of cofibrant approximations to g.*

(2) *If $\big((\widehat{X}, j_X), (\widehat{Y}, j_Y), \hat{g}\colon \widehat{X} \to \widehat{Y}\big)$ and $\big((\widehat{X}', j'_X), (\widehat{Y}', j'_Y), \hat{g}'\colon \widehat{X}' \to \widehat{Y}'\big)$ are fibrant approximations to g, then they are connected by an essentially unique zig-zag (see Definition 14.4.2) of weak equivalences of fibrant approximations to g.*

PROOF. This follows from Theorem 14.4.5 and Proposition 14.6.5. □

DEFINITION 14.6.7. Let $\mathcal{M}$ be a model category and let $\mathcal{K}$ be a subcategory of $\mathcal{M}$.

(1) A *category of functorial cofibrant approximations on $\mathcal{K}$* is category of functors from $\mathcal{K}$ to $\mathcal{M}$ over the inclusion functor with respect to those maps in $\mathcal{M}$ that are weak equivalences with cofibrant domains (see Definition 14.5.2).

(2) A *category of functorial fibrant approximations on $\mathcal{K}$* is category of functors from $\mathcal{K}$ to $\mathcal{M}$ under the inclusion functor with respect to those maps in $\mathcal{M}$ that are weak equivalences with fibrant codomains (see Definition 14.5.2).

THEOREM 14.6.8. *Let $\mathcal{M}$ be a model category and let $\mathcal{K}$ be a subcategory of $\mathcal{M}$.*

(1) *For every small category $\mathcal{D}$ of functorial cofibrant approximations on $\mathcal{K}$ (see Definition 14.6.7) there is a small category $\mathcal{D}'$ of functorial cofibrant approximations on $\mathcal{K}$ such that $\mathcal{D} \subset \mathcal{D}'$ and $\mathrm{B}\mathcal{D}'$ is contractible.*

(2) *For every small category $\mathcal{D}$ of functorial fibrant approximations on $\mathcal{K}$ (see Definition 14.6.7) there is a small category $\mathcal{D}'$ of functorial fibrant approximations on $\mathcal{K}$ such that $\mathcal{D} \subset \mathcal{D}'$ and $\mathrm{B}\mathcal{D}'$ is contractible.*

PROOF. This follows from Theorem 14.5.4 and Proposition 8.1.17. □

THEOREM 14.6.9. *Let $\mathcal{M}$ be a model category and let $\mathcal{K}$ be a subcategory of $\mathcal{M}$.*

(1) *If* $(\widetilde{\boldsymbol{X}}, i)$ *and* $(\widetilde{\boldsymbol{X}'}, i')$ *are functorial cofibrant approximations on* $\mathcal{K}$, *then there is an essentially unique zig-zag of weak equivalences of functorial cofibrant approximations on* $\mathcal{K}$ *from* $(\widetilde{\boldsymbol{X}}, i)$ *to* $(\widetilde{\boldsymbol{X}'}, i')$.

(2) *If* $(\widehat{\boldsymbol{X}}, j)$ *and* $(\widehat{\boldsymbol{X}'}, j')$ *are functorial fibrant approximations on* $\mathcal{K}$, *then there is an essentially unique zig-zag of weak equivalences of functorial fibrant approximations on* $\mathcal{K}$ *from* $(\widehat{\boldsymbol{X}}, j)$ *to* $(\widehat{\boldsymbol{X}'}, j')$.

PROOF. This follows from Proposition 14.5.7 and Theorem 14.6.8. □

14.7. Diagrams of undercategories and overcategories

In this section, for every small category $\mathcal{C}$ we define a natural $\mathcal{C}^{\mathrm{op}}$-diagram of simplicial sets $\mathrm{B}(-\downarrow \mathcal{C})^{\mathrm{op}}$ that will be used to define the homotopy colimit of a $\mathcal{C}$-diagram in a model category (see Definition 18.1.2 and Definition 19.1.2) and a natural $\mathcal{C}$-diagram of simplicial sets $\mathrm{B}(\mathcal{C}\downarrow -)$ that will be used to define the homotopy limit of a $\mathcal{C}$-diagram in a model category (see Definition 18.1.8 and Definition 19.1.5). We also derive a relation between them (see Corollary 14.7.13) that we will use to obtain a relation between the homotopy colimit and the homotopy limit functors (see Theorem 18.1.10). We will show in Proposition 14.8.5 that these diagrams are free cell complexes (see Definition 11.5.35).

14.7.1. Diagrams of undercategories.

DEFINITION 14.7.2. If $\mathcal{C}$ and $\mathcal{D}$ are small categories and $\mathrm{F}\colon \mathcal{C} \to \mathcal{D}$ is a functor, then for every object α of $\mathcal{D}$ we have the category $(\alpha\downarrow \mathrm{F})^{\mathrm{op}}$, the opposite of the category of objects of $\mathcal{C}$ under α (see Definition 11.8.3). If $\sigma\colon \alpha \to \alpha'$ is a map in $\mathcal{D}$, then σ induces a functor $\sigma^*\colon (\alpha'\downarrow \mathrm{F})^{\mathrm{op}} \to (\alpha\downarrow \mathrm{F})^{\mathrm{op}}$, defined on objects by

$$\sigma^*(\alpha' \xrightarrow{\tau} \mathrm{F}\beta) = \alpha \xrightarrow{\tau\sigma} \mathrm{F}\beta.$$

If we take the classifying space (see Definition 14.1.1) of each undercategory, we obtain the $\mathcal{D}^{\mathrm{op}}$*-diagram of simplicial sets* $\mathrm{B}(-\downarrow \mathrm{F})^{\mathrm{op}}$ that on the object α of $\mathcal{D}$ takes the value $\mathrm{B}(\alpha\downarrow \mathrm{F})^{\mathrm{op}}$. Thus, an n-simplex of $\mathrm{B}(-\downarrow \mathrm{F})^{\mathrm{op}}(\alpha) = \mathrm{B}(\alpha\downarrow \mathrm{F})^{\mathrm{op}}$ is a pair

$$\big((\beta_0 \xleftarrow{\sigma_0} \beta_1 \xleftarrow{\sigma_1} \cdots \xleftarrow{\sigma_{n-1}} \beta_n), \tau\colon \alpha \to \mathrm{F}\beta_n\big)$$

where $\beta_0 \xleftarrow{\sigma_0} \beta_1 \xleftarrow{\sigma_1} \cdots \xleftarrow{\sigma_{n-1}} \beta_n$ is a string of composable maps in $\mathcal{C}$ and $\tau\colon \alpha \to \mathrm{F}\beta_n$ is a map in $\mathcal{D}$, with face and degeneracy maps defined as in (14.1.2).

As in Definition 11.8.3, if $\mathcal{C} = \mathcal{D}$ and F is the identity functor, then we use $\mathrm{B}(-\downarrow \mathcal{C})^{\mathrm{op}}$ to denote the diagram of classifying spaces of the opposites of the undercategories, and an n-simplex of $B(-\downarrow \mathcal{C})^{\mathrm{op}}(\alpha) = \mathrm{B}(\alpha\downarrow \mathcal{C})^{\mathrm{op}}$ is a commutative diagram in $\mathcal{C}$

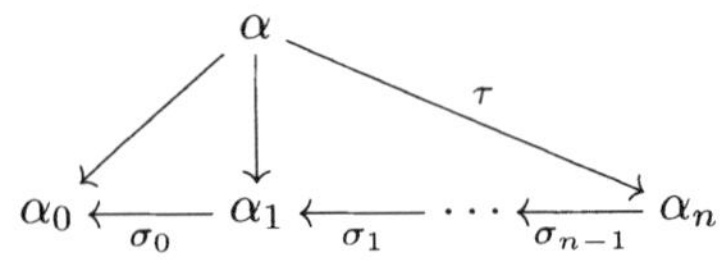

with face and degeneracy maps defined as in (14.1.2).

LEMMA 14.7.3. *If* $\mathcal{C}$ *and* $\mathcal{D}$ *are small categories and* $\mathrm{F}\colon \mathcal{C} \to \mathcal{D}$ *is a functor, then for every object* α *of* $\mathcal{C}$ *there is a map of simplicial sets* $\mathrm{F}_*\colon \mathrm{B}(\alpha\downarrow \mathcal{C})^{\mathrm{op}} \to \mathrm{B}(\mathrm{F}\alpha\downarrow \mathrm{F})^{\mathrm{op}}$ *that takes the simplex*

$$\big((\alpha_0 \xleftarrow{\sigma_0} \alpha_1 \xleftarrow{\sigma_1} \cdots \xleftarrow{\sigma_{n-1}} \alpha_n), \tau\colon \alpha \to \alpha_n\big)$$

of $B(\alpha\downarrow\mathcal{C})^{op}$ *to the simplex*

$$\bigl((\alpha_0 \xleftarrow{\sigma_0} \alpha_1 \xleftarrow{\sigma_1} \cdots \xleftarrow{\sigma_{n-1}} \alpha_n), F\tau\colon F\alpha \to F\alpha_n\bigr)$$

of $B(F\alpha\downarrow F)$.

PROOF. This follows directly from the definitions. $\square$

LEMMA 14.7.4. *If* $\mathcal{C}$ *is a small category and* α *is an object of* $\mathcal{C}$, *then* $B(\alpha\downarrow\mathcal{C})^{op}$ *is contractible.*

PROOF. This follows from Proposition 14.3.14, since $(\alpha\downarrow\mathcal{C})^{op}$ has the terminal object $1_\alpha\colon \alpha\to\alpha$. $\square$

The $\mathcal{C}^{op}$-diagram $B(-\downarrow\mathcal{C})^{op}$ will be used to define the homotopy colimit functor (see Definition 18.1.2 and Definition 19.1.2). Lemma 14.7.4 implies that, in the model category of $\mathcal{C}^{op}$-diagrams of simplicial sets (see Theorem 11.6.1), the $\mathcal{C}^{op}$-diagram $B(-\downarrow\mathcal{C})^{op}$ is weakly equivalent to the constant diagram at a point. We will show in Corollary 14.8.8 that $B(-\downarrow\mathcal{C})^{op}$ is also a free cell complex, and so $B(-\downarrow\mathcal{C})^{op}$ is a cofibrant approximation to the constant diagram at a point (see Definition 8.1.2). This will imply (in Theorem 19.4.7) that if we use a different cofibrant approximation to the constant diagram at a point in the definition of the homotopy colimit of a diagram, then for objectwise cofibrant diagrams we will get a functor naturally weakly equivalent to the homotopy colimit functor.

PROPOSITION 14.7.5. *If* $\mathcal{C}$ *and* $\mathcal{D}$ *are small categories and* $F\colon \mathcal{C}\to\mathcal{D}$ *is a functor, then the colimit of the* $\mathcal{D}^{op}$*-diagram of classifying spaces of undercategories* $\operatorname{colim}_{\mathcal{D}^{op}} B(-\downarrow F)$ *is naturally isomorphic to* $B\mathcal{C}$.

PROOF. We define a map $\operatorname{colim}_{\mathcal{D}^{op}} B(-\downarrow F)\to B\mathcal{C}$ by taking the simplex $(\beta_0\to\beta_1\to\cdots\to\beta_n, \sigma\colon \alpha\to F\beta_0)$ of $B(\alpha\downarrow F)$ to the simplex $\beta_0\to\beta_1\to\cdots\to\beta_n$ of $B\mathcal{C}$. This map is onto because the simplex $\beta_0\to\beta_1\to\cdots\to\beta_n$ of $B\mathcal{C}$ is in the image of $(\beta_0\to\beta_1\to\cdots\to\beta_n, 1_{F\beta_0}\colon F\beta_0\to F\beta_0)$, and it is one to one because the simplex $(\beta_0\to\beta_1\to\cdots\to\beta_n,\sigma\colon\alpha\to F\beta_0)$ of $B(\alpha\downarrow F)$ is identified with the simplex $(\beta_0\to\beta_1\to\cdots\to\beta_n, 1_{F\beta_0}\colon F\beta_0\to F\beta_0)$ of $B(F\beta_0\downarrow F)$ in $\operatorname{colim} B(-\downarrow F)$. $\square$

REMARK 14.7.6. We will show in Proposition 14.8.5 that the $\mathcal{D}^{op}$-diagram $B(-\downarrow F)$ is also a free cell complex (see Definition 11.5.35). It will then follow from Proposition 18.9.4 that the natural map $\operatorname{hocolim} B(-\downarrow F)\to\operatorname{colim} B(-\downarrow F)$ is a weak equivalence, and so $\operatorname{hocolim} B(-\downarrow F)$ is naturally weakly equivalent to $B\mathcal{C}$ (see Proposition 18.9.5).

14.7.7. Diagrams of overcategories.

DEFINITION 14.7.8. If $\mathcal{C}$ and $\mathcal{D}$ are small categories and $F\colon\mathcal{C}\to\mathcal{D}$ is a functor, then for every object α of $\mathcal{D}$ we have the category $(F\downarrow\alpha)$, the category of objects of $\mathcal{C}$ over α (see Definition 11.8.1). If $\sigma\colon\alpha\to\alpha'$ is a map in $\mathcal{D}$, then σ induces a functor $\sigma_*\colon (F\downarrow\alpha)\to(F\downarrow\alpha')$, defined on objects by

$$\sigma_*(F\beta\xrightarrow{\tau}\alpha) = F\beta\xrightarrow{\sigma\tau}\alpha'.$$

If we take the classifying space of each overcategory (see Definition 14.1.1), we obtain the $\mathcal{D}$-*diagram of simplicial sets* $B(F\downarrow -)$ that on the object α of $\mathcal{D}$ takes the value $\mathrm{B}(\mathrm{F}\downarrow\alpha)$. Thus, an n-simplex of $\mathrm{B}(\mathrm{F}\downarrow -)(\alpha) = \mathrm{B}(\mathrm{F}\downarrow\alpha)$ is a pair

$$\bigl((\beta_0 \xrightarrow{\sigma_0} \beta_1 \xrightarrow{\sigma_1} \cdots \xrightarrow{\sigma_{n-1}} \beta_n), \tau\colon \mathrm{F}\beta_n \to \alpha\bigr)$$

where $\beta_0 \xrightarrow{\sigma_0} \beta_1 \xrightarrow{\sigma_1} \cdots \xrightarrow{\sigma_{n-1}} \beta_n$ is a string of composable maps in $\mathcal{C}$ and $\tau\colon \mathrm{F}\beta_n \to \alpha$ is a map in $\mathcal{D}$, with face and degeneracy maps defined as in (14.1.2).

As in Definition 14.7.2, if $\mathcal{C} = \mathcal{D}$ and F is the identity functor, then we use $\mathrm{B}(\mathcal{C}\downarrow -)$ to denote the diagram of overcategories, and an n-simplex of $\mathrm{B}(\mathcal{C}\downarrow -)(\alpha) = \mathrm{B}(\mathcal{C}\downarrow\alpha)$ is a commutative diagram in $\mathcal{C}$

$$\begin{array}{ccccccc} \alpha_0 & \xrightarrow{\sigma_0} & \alpha_1 & \xrightarrow{\sigma_1} & \cdots & \xrightarrow{\sigma_{n-1}} & \alpha_n \\ & \searrow & \downarrow & & & \swarrow_{\tau} & \\ & & \alpha & & & & \end{array}$$

with face and degeneracy maps defined as in (14.1.2).

LEMMA 14.7.9. *If $\mathcal{C}$ and $\mathcal{D}$ are small categories and $\mathrm{F}\colon \mathcal{C} \to \mathcal{D}$ is a functor, then for every object α of $\mathcal{C}$ there is a map of simplicial sets $\mathrm{F}_*\colon \mathrm{B}(\mathcal{C}\downarrow\alpha) \to \mathrm{B}(\mathrm{F}\downarrow\mathrm{F}\alpha)$ that takes the simplex*

$$\bigl((\alpha_0 \xrightarrow{\sigma_0} \alpha_1 \xrightarrow{\sigma_1} \cdots \xrightarrow{\sigma_{n-1}} \alpha_n), \tau\colon \alpha_n \to \alpha\bigr)$$

of $\mathrm{B}(\alpha\downarrow\mathcal{C})^{\mathrm{op}}$ to the simplex

$$\bigl((\alpha_0 \xrightarrow{\sigma_0} \alpha_1 \xrightarrow{\sigma_1} \cdots \xrightarrow{\sigma_{n-1}} \alpha_n), \mathrm{F}\tau\colon \mathrm{F}\alpha_n \to \mathrm{F}\alpha\bigr)$$

of $\mathrm{B}(\mathrm{F}\alpha\downarrow\mathrm{F})$.

PROOF. This follows directly from the definitions. □

LEMMA 14.7.10. *If $\mathcal{C}$ is a small category and α is an object of $\mathcal{C}$, then $\mathrm{B}(\mathcal{C}\downarrow\alpha)$ is contractible.*

PROOF. This follows from Proposition 14.3.14, since $(\mathcal{C}\downarrow\alpha)$ has the terminal object $1_\alpha\colon \alpha \to \alpha$. □

The $\mathcal{C}$-diagram $\mathrm{B}(\mathcal{C}\downarrow -)$ will be used to define the homotopy limit functor (see Definition 18.1.8 and Definition 19.1.5). Lemma 14.7.10 implies that in the model category of $\mathcal{C}$-diagrams of simplicial sets (see Theorem 11.6.1), the $\mathcal{C}$-diagram $\mathrm{B}(\mathcal{C}\downarrow -)$ is weakly equivalent to the constant diagram at a point. We will show in Corollary 14.8.8 that $\mathrm{B}(\mathcal{C}\downarrow -)$ is also a free cell complex, and so $\mathrm{B}(\mathcal{C}\downarrow -)$ is a cofibrant approximation to the constant diagram at a point (see Definition 8.1.2). This will imply (in Theorem 19.4.7) that if we use a different cofibrant approximation to the constant diagram at a point in the definition of the homotopy limit of a diagram, then, for objectwise fibrant diagrams, we will get a functor naturally weakly equivalent to the homotopy limit functor.

14.7.11. Relations.

PROPOSITION 14.7.12. *If $\mathcal{C}$ is a small category, then the isomorphism $(\alpha\downarrow\mathcal{C})^{\mathrm{op}} \approx (\mathcal{C}^{\mathrm{op}}\downarrow\alpha)$ of Corollary 11.8.7 is natural in the object α of $\mathcal{C}$.*

PROOF. This follows directly from the definitions. □

COROLLARY 14.7.13. *If $\mathcal{C}$ is a small category, then there is a natural isomorphism of $\mathcal{C}^{\mathrm{op}}$-diagrams of simplicial sets*

$$\mathrm{B}(-\downarrow \mathcal{C})^{\mathrm{op}} \approx \mathrm{B}(\mathcal{C}^{\mathrm{op}}\downarrow -) .$$

PROOF. This follows from Proposition 14.7.12. □

14.8. Free cell complexes of simplicial sets

In this section, we characterize those diagrams of simplicial sets that are free cell complexes (see Theorem 14.8.4). Our main application of this will be to the diagrams of opposites of undercategories and of overcategories (see Proposition 14.8.5 and Corollary 14.8.8), which will be used to define the homotopy colimit and homotopy limit functors (see Definition 18.1.2, Definition 18.1.8, Definition 19.1.2, and Definition 19.1.5).

PROPOSITION 14.8.1. *If $\mathcal{C}$ is a small category and $\boldsymbol{X}$ is a $\mathcal{C}$-diagram of simplicial sets, then $\boldsymbol{X}$ is a free cell complex if and only if there is a sequence $\boldsymbol{S} = \{\boldsymbol{S}^0, \boldsymbol{S}^1, \boldsymbol{S}^2, \ldots\}$ of $\mathcal{C}^{\mathrm{disc}}$-diagrams of sets (where $\mathcal{C}^{\mathrm{disc}}$ is the discrete category with objects equal to the objects of $\mathcal{C}$) such that*

1. *for $n \geq 0$ and α an object of $\mathcal{C}$, the set $\boldsymbol{S}^n_\alpha$ is a subset of the set of n-simplices of $\boldsymbol{X}_\alpha$,*
2. *for $0 \leq i \leq n$ and α an object of $\mathcal{C}$, we have $s_i(\boldsymbol{S}^n_\alpha) \subset \boldsymbol{S}^{n+1}_\alpha$ (i.e., $\boldsymbol{S}$ is closed under degeneracies), and*
3. *for $n \geq 0$ the natural map $\mathbf{F}(\boldsymbol{S}^n) \to \boldsymbol{X}_n$ (see Theorem 11.5.20) is an isomorphism of $\mathcal{C}$-diagrams of sets (where $\boldsymbol{X}_n$ is the $\mathcal{C}$-diagram of n-simplices of $\boldsymbol{X}_\alpha$ at every object α of $\mathcal{C}$).*

PROOF. We first assume that there is a sequence $\{\boldsymbol{S}^0, \boldsymbol{S}^1, \boldsymbol{S}^2, \ldots\}$ of $\mathcal{C}^{\mathrm{disc}}$-diagrams of sets satisfying conditions 1 through 3, and we will show that the n-skeleton $\boldsymbol{X}^n$ of $\boldsymbol{X}$ can be obtained from the $(n-1)$-skeleton $\boldsymbol{X}^{n-1}$ of $\boldsymbol{X}$ as a pushout of a coproduct of free cells. Proposition 10.2.14 will then imply that $\boldsymbol{X}$ is a free cell complex.

We begin by noting that $\boldsymbol{X}^0 = \Delta[0] \otimes \mathbf{F}(\boldsymbol{S}^0)$ (see Definition 11.5.19 and Definition 11.5.24). We now assume that n is a positive integer. If α is an object of $\mathcal{C}$, let $\widetilde{\boldsymbol{S}}^n_\alpha \subset \boldsymbol{S}^n_\alpha$ be the subset of nondegenerate simplices. If $\sigma \in \widetilde{\boldsymbol{S}}^n_\alpha$, then all faces of σ are contained in $\boldsymbol{X}^{n-1}_\alpha$, and so σ defines a map $\partial\sigma\colon \partial\Delta[n] \to \boldsymbol{X}^{n-1}_\alpha$. Proposition 11.5.26 implies that this defines a map of $\mathcal{C}$-diagrams $\partial\sigma \otimes \mathbf{F}^\alpha_*\colon \partial\Delta[n] \otimes \mathbf{F}^\alpha_* \to \boldsymbol{X}^{n-1}$, and we can take the coproduct of these to obtain

$$\coprod_{\sigma\in\widetilde{\boldsymbol{S}}^n_\alpha} \partial\sigma \otimes \mathbf{F}^\alpha_*\colon \coprod_{\sigma\in\widetilde{\boldsymbol{S}}^n_\alpha} \partial\Delta[n] \otimes \mathbf{F}^\alpha_* = \partial\Delta[n] \otimes \mathbf{F}^\alpha_{\widetilde{\boldsymbol{S}}^n_\alpha} \to \boldsymbol{X}^{n-1} .$$

If we combine these for all objects α of $\mathcal{C}$, we obtain the map

$$\coprod_{\alpha\in\mathrm{Ob}(\mathcal{C})} \partial\Delta[n] \otimes \mathbf{F}^\alpha_{\widetilde{\boldsymbol{S}}^n_\alpha} = \partial\Delta[n] \otimes \mathbf{F}(\widetilde{\boldsymbol{S}}^n) \to \boldsymbol{X}^{n-1}$$

(see Definition 11.5.19), and condition (3) implies that the square

$$\begin{array}{ccc} \partial\Delta[n] \otimes \mathbf{F}(\widetilde{\boldsymbol{S}}^n) & \longrightarrow & \boldsymbol{X}^{n-1} \\ \big\downarrow & & \big\downarrow \\ \Delta[n] \otimes \mathbf{F}(\widetilde{\boldsymbol{S}}^n) & \longrightarrow & \boldsymbol{X}^n \end{array}$$

is a pushout, which completes the first direction of the proof.

We now assume that $\boldsymbol{X}$ is a free cell complex. If γ is an ordinal and

$$\emptyset \to \boldsymbol{X}_1 \to \boldsymbol{X}_2 \to \cdots \to \boldsymbol{X}_\beta \to \cdots \qquad (\beta < \gamma)$$

is a presentation of $\boldsymbol{X}$ as a transfinite composition of pushouts of free cells, then for every $\beta < \gamma$ there is an integer $n \geq 0$, an object α_β of $\mathcal{C}$, and a pushout diagram

$$\begin{array}{ccc} \partial\Delta[n] \otimes \mathbf{F}_*^{\alpha_\beta} & \longrightarrow & \boldsymbol{X}_\beta \\ \big\downarrow & & \big\downarrow \\ \Delta[n] \otimes \mathbf{F}_*^{\alpha_\beta} & \longrightarrow & \boldsymbol{X}_{\beta+1} \end{array} .$$

For every object α of $\mathcal{C}$ let $\boldsymbol{S}_\alpha$ be the union over all β for which $\alpha_\beta = \alpha$ of the images in $\boldsymbol{X}$ of $\delta_n \otimes 1_{\alpha_\beta}$ and its degeneracies (where δ_n is the nondegenerate n-simplex of $\Delta[n]$ and we mean the image under the composition $\delta_n \otimes 1_\alpha \subset \Delta[n] \otimes \mathbf{F}_*^{\alpha_\beta}(\alpha_\beta) \to (\boldsymbol{X}_{\beta+1})_\alpha$). Let $\boldsymbol{S}^n_\alpha$ be the set of n-simplices in $\boldsymbol{S}_\alpha$. Since for every $0 \leq \beta < \gamma$ the diagram $\boldsymbol{X}$ is enlarged by adding the free diagram of simplices generated by the images of $\delta_n \otimes 1_{\alpha_\beta}$ and its degeneracies, it follows that the sets $\boldsymbol{S}^n$ satisfy conditions (1) through (3). □

DEFINITION 14.8.2. If $\mathcal{C}$ is a small category and $\boldsymbol{X}$ is a $\mathcal{C}$-diagram of simplicial sets that is a free cell complex, then a sequence $\{\boldsymbol{S}^0, \boldsymbol{S}^1, \boldsymbol{S}^2, \ldots\}$ as in Proposition 14.8.1 will be called a *basis* for $\boldsymbol{X}$, and an element of an $\boldsymbol{S}^n_\alpha$ will be called a *generator* of the free cell complex $\boldsymbol{X}$. We will use $\boldsymbol{S}$ to denote the sequence $\{\boldsymbol{S}^0, \boldsymbol{S}^1, \boldsymbol{S}^2, \ldots\}$. We will let $\widetilde{\boldsymbol{S}}^n_\alpha \subset \boldsymbol{S}^n_\alpha$ be the subset of nondegenerate simplices, and we will call an element of an $\widetilde{\boldsymbol{S}}^n_\alpha$ a *nondegenerate generator* of $\boldsymbol{X}$. An element of an $\boldsymbol{S}^n_\alpha - \widetilde{\boldsymbol{S}}^n_\alpha$ will be called a *degenerate generator*.

REMARK 14.8.3. The reader should note the similarity between the *free cell complexes* among diagrams of simplicial sets and the *free simplicial groups* among simplicial groups (see, e.g., [**46**, Section 5]). Since a $\mathcal{C}$-diagram of simplicial sets is equivalently a simplicial object in the category of $\mathcal{C}$-diagrams of sets, we are comparing the definitions of free simplicial groups and free simplicial $\mathcal{C}$-diagrams of sets. This similarity can be made more precise by noting that a group is an algebra over the "underlying set of the free group" triple on the category of sets (see, e.g., [**3**, page 339] or [**48**, pages 176–177]), while a $\mathcal{C}$-diagram of sets is an algebra over the "underlying $\mathcal{C}^{\mathrm{disc}}$-diagram of sets on the free $\mathcal{C}$-diagram of sets" triple on the category of $\mathcal{C}^{\mathrm{disc}}$-diagrams of sets. The sequence $\boldsymbol{S}$ in Proposition 14.8.1 is the analogue for $\mathcal{C}$-diagrams of simplicial sets of a basis of a free simplicial group (see Definition 14.8.2). Free cell complexes are also free objects in the category of simplicial $\mathcal{C}$-diagrams of sets in the sense of [**45**, Definition 5.1].

THEOREM 14.8.4. *Let $\mathcal{C}$ be a small category and let $\boldsymbol{X}$ be a $\mathcal{C}$-diagram of simplicial sets. If $\boldsymbol{S} = \{\boldsymbol{S}^0, \boldsymbol{S}^1, \boldsymbol{S}^2, \ldots\}$ is a sequence of $\mathcal{C}^{\mathrm{disc}}$-diagrams of sets, then $\boldsymbol{X}$ is a free cell complex with basis $\boldsymbol{S}$ (see Definition 14.8.2) if and only if:*

1. *for $n \geq 0$ and α an object of $\mathcal{C}$, the set $\boldsymbol{S}^n_\alpha$ is a subset of the set of n-simplices of $\boldsymbol{X}_\alpha$,*
2. *for $0 \leq i \leq n$ and α an object of $\mathcal{C}$, we have $s_i(\boldsymbol{S}^n_\alpha) \subset \boldsymbol{S}^{n+1}_\alpha$ (i.e., $\boldsymbol{S}$ is closed under degeneracies), and*
3. *if $n \geq 0$, β is an object of $\mathcal{C}$, and τ is an n-simplex of $\boldsymbol{X}_\beta$, then there exist an object α of $\mathcal{C}$, an element σ of $\boldsymbol{S}^n_\alpha$, and a map $\gamma\colon \alpha \to \beta$ in $\mathcal{C}$ such that $\boldsymbol{X}_\gamma(\sigma) = \tau$, and such a triple (α, σ, γ) is unique.*

PROOF. This follows directly from Proposition 14.8.1 and Definition 11.5.19. $\square$

PROPOSITION 14.8.5. *Let $\mathcal{C}$ and $\mathcal{D}$ be small categories and let $\mathrm{F}\colon \mathcal{C} \to \mathcal{D}$ be a functor.*

1. *The $\mathcal{D}^{\mathrm{op}}$-diagram of simplicial sets $\mathrm{B}(-\downarrow \mathrm{F})^{\mathrm{op}}$ (see Definition 14.7.2) is a free cell complex with a basis (see Definition 14.8.2) consisting of the simplices of the form*

$$\big((\beta_0 \xleftarrow{\sigma_0} \beta_1 \xleftarrow{\sigma_1} \cdots \xleftarrow{\sigma_{n-1}} \beta_n), 1_{\mathrm{F}\beta_n} : \mathrm{F}\beta_n \to \mathrm{F}\beta_n\big) \tag{14.8.6}$$

   (see Definition 14.7.2).
2. *The $\mathcal{D}^{\mathrm{op}}$-diagram of simplicial sets $\mathrm{B}(-\downarrow \mathrm{F})$ is a free cell complex with a basis consisting of the simplices of the form*

$$\big((\beta_0 \xrightarrow{\sigma_0} \beta_1 \xrightarrow{\sigma_1} \cdots \xrightarrow{\sigma_{n-1}} \beta_n), 1_{\mathrm{F}\beta_n} : \mathrm{F}\beta_n \to \mathrm{F}\beta_n\big)\ .$$

3. *The $\mathcal{D}$-diagram of simplicial sets $\mathrm{B}(\mathrm{F}\downarrow -)$ (see Definition 14.7.8) is a free cell complex with a basis consisting of the simplices of the form*

$$\big((\beta_0 \xrightarrow{\sigma_0} \beta_1 \xrightarrow{\sigma_1} \cdots \xrightarrow{\sigma_{n-1}} \beta_n), 1_{\mathrm{F}\beta_n} : \mathrm{F}\beta_n \to \mathrm{F}\beta_n\big) \tag{14.8.7}$$

   (see Definition 14.7.8).

PROOF. We will prove part 1; the proof of the other parts are similar.

For every object α of $\mathcal{D}$ and every $n \geq 0$, let $\boldsymbol{S}^n_\alpha$ be the set of n-simplices of the form (14.8.6) for which $\mathrm{F}\beta_n = \alpha$; the result now follows from Theorem 14.8.4. $\square$

COROLLARY 14.8.8. *If $\mathcal{C}$ is a small category, then*

- *the $\mathcal{C}^{\mathrm{op}}$-diagram of simplicial sets $\mathrm{B}(-\downarrow \mathcal{C})^{\mathrm{op}}$ (see Definition 14.7.2) and*
- *the $\mathcal{C}$-diagram of simplicial sets $\mathrm{B}(\mathcal{C}\downarrow -)$ (see Definition 14.7.8)*

are both free cell complexes.

PROOF. This follows from Proposition 14.8.5. $\square$

PROPOSITION 14.8.9. *Let $\mathcal{C}$ be a small category.*

1. *The $\mathcal{C}^{\mathrm{op}}$-diagram of simplicial sets $\mathrm{B}(-\downarrow \mathcal{C})$ (see Definition 14.7.2) is a cofibrant approximation (see Theorem 11.6.1) to the constant $\mathcal{C}^{\mathrm{op}}$-diagram at a point.*
2. *The $\mathcal{C}$-diagram of simplicial sets $\mathrm{B}(\mathcal{C}\downarrow -)$ (see Definition 14.7.8) is a cofibrant approximation (see Theorem 11.6.1) to the constant diagram at a point.*

PROOF. This follows from Corollary 14.8.8, Proposition 11.6.2, Lemma 14.7.4, and Lemma 14.7.9. □

COROLLARY 14.8.10. *Let $\mathcal{C}$ and $\mathcal{D}$ be small categories and let $F\colon \mathcal{C} \to \mathcal{D}$ be a functor.*

(1) (a) *There is a basis for the free cell complex $B(-\downarrow F)^{op}$ in $SS^{\mathcal{D}^{op}}$ consisting of the simplices of the form*

$$((\beta_0 \xleftarrow{\sigma_0} \beta_1 \xleftarrow{\sigma_1} \cdots \xleftarrow{\sigma_{n-1}} \beta_n), 1_{F\beta_n}\colon F\beta_n \to F\beta_n) \tag{14.8.11}$$

(where $(\beta_0 \xleftarrow{\sigma_0} \beta_1 \xleftarrow{\sigma_1} \cdots \xleftarrow{\sigma_{n-1}} \beta_n)$ is a string of n composable maps in $\mathcal{C}$). The simplex of (14.8.11) is a nondegenerate generator when none of the maps $\sigma_0, \sigma_1, \ldots, \sigma_{n-1}$ is an identity map.

(b) *There is a basis for the free cell complex $B(-\downarrow \mathcal{C})^{op}$ in $SS^{\mathcal{C}^{op}}$ consisting of the simplices of the form*

$$((\beta_0 \xleftarrow{\sigma_0} \beta_1 \xleftarrow{\sigma_1} \cdots \xleftarrow{\sigma_{n-1}} \beta_n), 1_{\beta_n}\colon \beta_n \to \beta_n) \tag{14.8.12}$$

(where $(\beta_0 \xleftarrow{\sigma_0} \beta_1 \xleftarrow{\sigma_1} \cdots \xleftarrow{\sigma_{n-1}} \beta_n)$ is a string of n composable maps in $\mathcal{C}$). The simplex of (14.8.12) is a nondegenerate generator when none of the maps $\sigma_0, \sigma_1, \ldots, \sigma_{n-1}$ is an identity map.

(c) *The maps of simplicial sets F_* of Lemma 14.7.3 induce a natural one to one correspondence F_* from the set of simplices in the basis of $B(-\downarrow F)^{op}$ to the set of simplices in the basis of $B(-\downarrow \mathcal{C})^{op}$ that takes the simplex of (14.8.11) to the simplex of (14.8.12). This one to one correspondence restricts to a one to one correspondence between the sets of nondegenerate generators. Furthermore, if γ is the simplex of (14.8.11) and $i < n$, then $d_i F_*(\gamma) = F_* d_i(\gamma)$, and if γ is a nondegenerate generator then there is a nondegenerate $(n-1)$-dimensional generator η such that $d_n(\gamma) = \sigma_{n-1}^*(\eta)$ and $F_* d_n(\gamma) = (F\sigma_{n-1})^*(F_*(\eta)) = d_n F_*(\gamma)$*

(2) (a) *There is a basis for the free cell complex $B(F\downarrow -)$ in $SS^{\mathcal{D}}$ consisting of the simplices of the form*

$$((\beta_0 \xrightarrow{\sigma_0} \beta_1 \xrightarrow{\sigma_1} \cdots \xrightarrow{\sigma_{n-1}} \beta_n), 1_{F\beta_n}\colon F\beta_n \to F\beta_n) \tag{14.8.13}$$

(where $(\beta_0 \xrightarrow{\sigma_0} \beta_1 \xrightarrow{\sigma_1} \cdots \xrightarrow{\sigma_{n-1}} \beta_n)$ is a string of n composable maps in $\mathcal{C}$). The simplex of (14.8.13) is a nondegenerate generator when none of the maps $\sigma_0, \sigma_1, \ldots, \sigma_{n-1}$ is an identity map.

(b) *There is a basis for the free cell complex $B(\mathcal{C}\downarrow -)$ in $SS^{\mathcal{C}}$ consisting of the simplices of the form*

$$((\beta_0 \xrightarrow{\sigma_0} \beta_1 \xrightarrow{\sigma_1} \cdots \xrightarrow{\sigma_{n-1}} \beta_n), 1_{\beta_n}\colon \beta_n \to \beta_n) \tag{14.8.14}$$

(where $(\beta_0 \xrightarrow{\sigma_0} \beta_1 \xrightarrow{\sigma_1} \cdots \xrightarrow{\sigma_{n-1}} \beta_n)$ is a string of n composable maps in $\mathcal{C}$). The simplex of (14.8.14) is a nondegenerate generator when none of the maps $\sigma_0, \sigma_1, \ldots, \sigma_{n-1}$ is an identity map.

(c) *The maps of simplicial sets F_* of Lemma 14.7.9 induce a natural one to one correspondence F_* from the set of simplices in the basis of $B(F\downarrow -)$ to the set of simplices in the basis of $B(\mathcal{C}\downarrow -)$ that takes the simplex of (14.8.13) to the simplex of (14.8.14). This one to one correspondence restricts to a one to one correspondence* *between the sets of nondegenerate generators. Furthermore, if γ is the simplex of (14.8.13) and $i < n$, then $d_i\mathrm{F}_*(\gamma) = \mathrm{F}_* d_i(\gamma)$, and if γ is a nondegenerate generator then there is a nondegenerate $(n-1)$-dimensional generator η such that $d_n(\gamma) = (\sigma_{n-1})_*(\eta)$ and $\mathrm{F}_* d_n(\gamma) = (\mathrm{F}\sigma_{n-1})_*(\mathrm{F}_*(\eta)) = d_n \mathrm{F}_*(\gamma)$*

PROOF. This follows from Proposition 14.8.5 and Example 14.1.8. □

PROPOSITION 14.8.15. *Let $\mathcal{C}$ be a small category, let $\boldsymbol{X}$ be a $\mathcal{C}$-diagram of simplicial sets that is a free cell complex with basis $\{\boldsymbol{S}^0, \boldsymbol{S}^1, \boldsymbol{S}^2, \ldots\}$ (see Definition 14.8.2), and for every $n \geq 0$ let $\boldsymbol{X}^n$ be the $\mathcal{C}$-diagram of n-skeletons of $\boldsymbol{X}$ (i.e., let $\boldsymbol{X}^n_\alpha$ be the n-skeleton of $\boldsymbol{X}_\alpha$ for every object α of $\mathcal{C}$). If $\boldsymbol{Y}$ is a $\mathcal{C}$-diagram of simplicial sets and $g\colon \boldsymbol{X}^n \to \boldsymbol{Y}$ is a map of $\mathcal{C}$-diagrams, then extensions of g to the $(n+1)$-skeleton of $\boldsymbol{X}$ correspond to maps of $\mathcal{C}^{\mathrm{disc}}$-diagrams $h\colon \widetilde{\boldsymbol{S}}^{n+1} \to \boldsymbol{Y}_{n+1}$ such that $d_i h_\alpha = g_\alpha d_i$ for every object α of $\mathcal{C}$ and every $0 \leq i \leq n+1$.*

PROOF. This follows from Theorem 14.8.4. □

PROPOSITION 14.8.16. *Let $\mathcal{C}$ be a small category.*

1. *If $\boldsymbol{X}$ is a $\mathcal{C}$-diagram of unpointed simplicial sets that is a free cell complex, then $\boldsymbol{X}^+\colon \mathcal{C} \to \mathrm{SS}_*$ (defined by $\boldsymbol{X}^+_\alpha = (\boldsymbol{X}_\alpha)^+$ for every object α of $\mathcal{C}$) is a free cell complex of pointed simplicial sets.*
2. *If $\boldsymbol{X}$ is a $\mathcal{C}$-diagram of (pointed or unpointed) simplicial sets that is a free cell complex, then $|\boldsymbol{X}|\colon \mathcal{C} \to \mathrm{Top}_{(*)}$ (defined by $|\boldsymbol{X}|_\alpha = |\boldsymbol{X}_\alpha|$) is a free cell complex of (pointed or unpointed) topological spaces.*

PROOF. This follows from the definition of free cell complex and the facts that if

$$\begin{array}{ccc} \partial\Delta \otimes \mathbf{F}^\alpha_* & \longrightarrow & \boldsymbol{X}_\beta \\ \downarrow & & \downarrow \\ \Delta \otimes \mathbf{F}^\alpha_* & \longrightarrow & \boldsymbol{X}_{\beta+1} \end{array} \tag{14.8.17}$$

is a pushout of $\mathcal{C}$-diagrams of (pointed or unpointed) simplicial sets, then

$$\begin{array}{ccc} \partial\Delta \otimes (\mathbf{F}^\alpha_*)^+ & \longrightarrow & \boldsymbol{X}^+_\beta \\ \downarrow & & \downarrow \\ \Delta \otimes (\mathbf{F}^\alpha_*)^+ & \longrightarrow & \boldsymbol{X}^+_{\beta+1} \end{array}$$

is a pushout of $\mathcal{C}$-diagrams of pointed simplicial sets, and if Diagram 14.8.17 is a pushout of $\mathcal{C}$-diagrams of unpointed simplicial sets, then

$$\begin{array}{ccc} |\partial\Delta| \otimes \mathbf{F}^\alpha_* & \longrightarrow & |\boldsymbol{X}_\beta| \\ \downarrow & & \downarrow \\ |\Delta| \otimes \mathbf{F}^\alpha_* & \longrightarrow & |\boldsymbol{X}_{\beta+1}| \end{array}$$

is a pushout of $\mathcal{C}$-diagrams of (pointed or unpointed) topological spaces. □

CHAPTER 15

The Reedy Model Category Structure

The model category structures on a category of simplicial objects in a model category and a category of cosimplicial objects in a model category of [**14**, Chapter X], [**57**, Section 1], [**13**, Theorem B.6], and [**34**, Section 2.4] have a common generalization: the Reedy model category structure on a category of diagrams in a model category indexed by a Reedy category. A Reedy category (see Definition 15.1.2) is D. M. Kan's generalization of both the indexing category for simplicial objects and the indexing category for cosimplicial objects (see Definition 15.1.8). The Reedy model category structure will be defined for diagrams in a model category indexed by a Reedy category (see Definition 15.3.3). The main examples of Reedy categories are the cosimplicial and simplicial indexing categories (see Definition 15.1.8) and, more generally, the category of simplices of a simplicial set (see Definition 15.1.16) and its opposite.

If $\mathcal{C}$ is a Reedy category and $\mathcal{M}$ is a cofibrantly generated model category, then we have already defined a model category structure on $\mathcal{M}^{\mathcal{C}}$, the category of $\mathcal{C}$-diagrams in $\mathcal{M}$ (see Theorem 11.6.1). The Reedy model category structure on $\mathcal{M}^{\mathcal{C}}$ has the same weak equivalences as that model category structure but has more cofibrations (see Proposition 15.6.3 and Corollary 15.9.11) and, thus, fewer fibrations (see Proposition 7.2.3).

Most of the definitions and results of this chapter are due to D. M. Kan.

In Section 15.1 we define Reedy categories and describe the main examples. In Section 15.2 we describe how to construct a diagram indexed by a Reedy category inductively over the filtrations of the Reedy category. This leads us to the definition of the latching and matching categories of a Reedy category (see Definition 15.2.3) and the latching and matching objects of a diagram indexed by a Reedy category (see Definition 15.2.5). We also show how maps between diagrams indexed by a Reedy category are naturally analyzed inductively over the filtrations (see Section 15.2.11).

In Section 15.3 we define the Reedy model category structure and prove that it is a model category. In Section 15.4 we show that if $\mathrm{F}\colon \mathcal{M} \rightleftarrows \mathcal{N} :\mathrm{U}$ is a Quillen pair and $\mathcal{C}$ is a Reedy category, then the induced functors $\mathrm{F}^{\mathcal{C}}\colon \mathcal{M}^{\mathcal{C}} \rightleftarrows \mathcal{N}^{\mathcal{C}} :\mathrm{U}^{\mathcal{C}}$ form a Quillen pair and that $\mathrm{F}^{\mathcal{C}}$ and $\mathrm{U}^{\mathcal{C}}$ are Quillen equivalences if F and U are Quillen equivalences. In Section 15.5 we discuss diagrams indexed by a product of Reedy categories, and show that the various possible model category structures are the same. In Section 15.6 we discuss diagrams in a cofibrantly generated model category indexed by a Reedy category, and we show that the two possible model category structures are different, although they are Quillen equivalent. We also show that the Reedy model category structure is also cofibrantly generated. In Section 15.7 we show that the model category of diagrams in a cellular model category indexed by a Reedy category is a cellular model category.

We discuss the model category of bisimplicial sets (i.e., simplicial simplicial sets) in Section 15.8 and the model category of cosimplicial simplicial sets in Section 15.9. In Section 15.10 we discuss *Reedy categories with fibrant constants*, which are Reedy categories for which an objectwise weak equivalence of cofibrant diagrams induces a weak equivalence of their colimits, and *Reedy categories with cofibrant constants*, which have the dual property. We will show in Theorem 19.9.1 that if $\mathcal{C}$ is a Reedy category with fibrant constants and $\boldsymbol{X}$ is a cofibrant $\mathcal{C}$-diagram in a model category, then the homotopy colimit of $\boldsymbol{X}$ is naturally weakly equivalent to the colimit of $\boldsymbol{X}$ (with a dual statement for Reedy categories with cofibrant constants).

In Section 15.11 we discuss bisimplicial sets. We define the realization of a bisimplicial set and show that it is isomorphic to the diagonal, and we show that a levelwise weak equivalence of bisimplicial sets induces a weak equivalence of their realizations.

15.1. Reedy categories

15.1.1. Reedy categories.

DEFINITION 15.1.2. A *Reedy Category* is a small category $\mathcal{C}$ together with two subcategories $\overrightarrow{\mathcal{C}}$ (the *direct subcategory*) and $\overleftarrow{\mathcal{C}}$ (the *inverse subcategory*), both of which contain all the objects of $\mathcal{C}$, in which every object can be assigned a nonnegative integer (called its *degree*) such that

(1) Every non-identity morphism of $\overrightarrow{\mathcal{C}}$ raises degree.
(2) Every non-identity morphism of $\overleftarrow{\mathcal{C}}$ lowers degree.
(3) Every morphism g in $\mathcal{C}$ has a unique factorization $g = \overrightarrow{g}\overleftarrow{g}$ where $\overrightarrow{g}$ is in $\overrightarrow{\mathcal{C}}$ and $\overleftarrow{g}$ is in $\overleftarrow{\mathcal{C}}$.

REMARK 15.1.3. According to Definition 15.1.2, a Reedy category consists of a category and two subcategories, subject to certain conditions. The function that assigns to each object its degree is not a part of the structure, but we will often implicitly assume that a degree function has been chosen.

REMARK 15.1.4. It is possible to use a more general definition of a Reedy category in which the degree function takes values that are *ordinal*, rather than just positive integers. All of the results of this chapter would go through, although some arguments would have to be rephrased slightly to be correct for limit ordinals.

PROPOSITION 15.1.5. *If $\mathcal{C}$ is a Reedy category, then $\mathcal{C}^{\mathrm{op}}$ is a Reedy category in which $\overrightarrow{\mathcal{C}^{\mathrm{op}}} = (\overleftarrow{\mathcal{C}})^{\mathrm{op}}$ and $\overleftarrow{\mathcal{C}^{\mathrm{op}}} = (\overrightarrow{\mathcal{C}})^{\mathrm{op}}$.*

PROOF. A degree function for $\mathcal{C}$ will serve as a degree function for $\mathcal{C}^{\mathrm{op}}$. □

PROPOSITION 15.1.6. *If $\mathcal{C}$ and $\mathcal{D}$ are Reedy categories, then $\mathcal{C} \times \mathcal{D}$ is a Reedy category with $\overrightarrow{\mathcal{C} \times \mathcal{D}} = \overrightarrow{\mathcal{C}} \times \overrightarrow{\mathcal{D}}$ and $\overleftarrow{\mathcal{C} \times \mathcal{D}} = \overleftarrow{\mathcal{C}} \times \overleftarrow{\mathcal{D}}$.*

PROOF. If we have chosen degree functions for $\mathcal{C}$ and $\mathcal{D}$, we define a degree function for $\mathcal{C} \times \mathcal{D}$ by $\deg(X \times Y) = \deg X + \deg Y$. The existence and uniqueness of the required factorization of maps in $\mathcal{C} \times \mathcal{D}$ follows from that of $\mathcal{C}$ and $\mathcal{D}$. □

15.1.7. Examples: The simplicial and cosimplicial indexing categories.

DEFINITION 15.1.8 (The cosimplicial and simplicial indexing categories). If n is a nonnegative integer, we let $[n]$ denote the ordered set $(0, 1, 2, \ldots, n)$. The category

$\mathbf{\Delta}$ is the category with objects the $[n]$ for $n \geq 0$ and with morphisms $\mathbf{\Delta}([n],[k])$ the weakly monotone functions $[n] \to [k]$, i.e., the functions $\sigma\colon [n] \to [k]$ such that $\sigma(i) \leq \sigma(j)$ for $0 \leq i \leq j \leq n$.

(1) The *cosimplicial indexing category* is the category $\mathbf{\Delta}$.
(2) The *simplicial indexing category* is the category $\mathbf{\Delta}^{\mathrm{op}}$.

REMARK 15.1.9. The cosimplicial indexing category $\mathbf{\Delta}$ (see Definition 15.1.8) is a skeletal subcategory of the category whose objects are the finite ordered sets and whose morphisms are the weakly monotone maps.

DEFINITION 15.1.10. Let $\mathcal{M}$ be a category.

(1) A *simplicial object in* $\mathcal{M}$ is a functor $\mathbf{\Delta}^{\mathrm{op}} \to \mathcal{M}$.
(2) A *cosimplicial object in* $\mathcal{M}$ is a functor $\mathbf{\Delta} \to \mathcal{M}$.

NOTATION 15.1.11. Let $\mathcal{M}$ be a category.

(1) If $\boldsymbol{X}$ is a simplicial object in $\mathcal{M}$, we will usually denote the object $\boldsymbol{X}_{[n]}$ by $\boldsymbol{X}_n$.
(2) If $\boldsymbol{X}$ is a cosimplicial object in $\mathcal{M}$, we will usually denote the object $\boldsymbol{X}_{[n]}$ by $\boldsymbol{X}^n$.

EXAMPLE 15.1.12. The cosimplicial indexing category $\mathbf{\Delta}$ (see Definition 15.1.8) is a Reedy category in which the object $[n]$ has degree n, the direct subcategory consists of the injective maps, and the inverse subcategory consists of the surjective maps.

EXAMPLE 15.1.13. The simplicial indexing category $\mathbf{\Delta}^{\mathrm{op}}$ (see Definition 15.1.8) is a Reedy category in which the object $[n]$ has degree n, the direct subcategory consists of the opposites of the surjective maps, and the inverse subcategory consists of the opposites of the injective maps (see Proposition 15.1.5).

15.1.14. Example: The category of simplices of a simplicial set. If X is a simplicial set, we will define a category ΔX whose objects are the simplices of X and whose morphisms from the simplex σ to the simplex τ are the simplicial operators that take τ to σ (see Definition 15.1.16). Note the reversal of direction: If $d_i\tau = \sigma$, then d_i corresponds to a morphism that takes σ to τ. This is because a simplicial set is a functor $\mathbf{\Delta}^{\mathrm{op}} \to \mathrm{Set}$, while ΔX is defined as an overcategory using a covariant functor $\mathbf{\Delta} \to \mathrm{SS}$. A diagram indexed by ΔX is a sort of generalized cosimplicial object, and a diagram indexed by $\Delta^{\mathrm{op}} X$ is a sort of generalized simplicial object (see Example 15.1.18 and Definition 15.1.10).

DEFINITION 15.1.15. The *cosimplicial standard simplex* is the cosimplicial simplicial set $\Delta\colon \mathbf{\Delta} \to \mathrm{SS}$ (see Definition 15.1.10) that takes the object $[n]$ of $\mathbf{\Delta}$ to the standard n-simplex $\Delta[n]$. The simplicial set $\Delta[n]$ has as k-simplices the weakly monotone functions $[k] \to [n]$, i.e., $\Delta[n]_k = \mathbf{\Delta}([k],[n])$.

DEFINITION 15.1.16. Let $\mathbf{\Delta}$ be the cosimplicial indexing category (see Definition 15.1.8), and let $\Delta\colon \mathbf{\Delta} \to \mathrm{SS}$ be the cosimplicial standard simplex (see Definition 15.1.15).

(1) If K is a simplicial set, then ΔK, the *category of simplices of* K, is defined to be the overcategory $(\Delta \downarrow K)$ (see Definition 11.8.1). Thus, ΔK is the category with objects the simplicial maps $\Delta[n] \to K$ (for some $n \geq 0$)

and with morphisms from $\sigma\colon \Delta[n] \to K$ to $\tau\colon \Delta[k] \to K$ the commutative triangles of simplicial maps

$$\begin{array}{ccc} \Delta[n] & \longrightarrow & \Delta[k] \\ & {\scriptstyle\sigma}\searrow \quad \swarrow{\scriptstyle\tau} & \\ & K\,. & \end{array}$$

(2) if K is a simplicial set, then $\Delta^{\mathrm{op}}K$ is defined to be $(\Delta K)^{\mathrm{op}}$, the opposite of the category of simplices of K.

PROPOSITION 15.1.17. *If K is a simplicial set, then there is a natural isomorphism of sets $\mathrm{Ob}(\Delta K) \approx \coprod_{n\geq 0} K_n$. If τ is an n-simplex (for some $n > 0$), k is an integer satisfying $0 \leq k \leq n$, and $d_k\tau = \sigma$, then d_k corresponds under this isomorphism to a morphism from $\chi_\sigma\colon \Delta[n-1] \to K$ to $\chi_\tau\colon \Delta[n] \to K$ (where the characteristic map χ_τ of an n-simplex τ of K is the unique map $\Delta[n] \to K$ that takes the non-degenerate n-simplex of $\Delta[n]$ to τ; see Example 11.5.16).*

PROOF. This follows from the one to one correspondence between n-simplices of K and maps of simplicial sets $\Delta[n] \to K$ (see Example 11.5.16). □

EXAMPLE 15.1.18. If K is the one point simplicial set (i.e., $K_n = *$ for all $n \geq 0$), then ΔK is the cosimplicial indexing category $\mathbf{\Delta}$ (see Definition 15.1.8).

EXAMPLE 15.1.19. If X is a simplicial set, then the category ΔX of simplices of X (see Definition 15.1.16) is a Reedy category in which the degree of an object is the dimension of the simplex of X to which it corresponds, the direct subcategory consists of the morphisms corresponding to iterated face maps in X, and the inverse subcategory consists of the morphisms corresponding to iterated degeneracy maps of X. Note that Example 15.1.12 is a special case of this example (see Example 15.1.18).

PROPOSITION 15.1.20. *If K is a simplicial set and $\mathrm{G}\colon \Delta K \to \mathrm{SS}$ is the ΔK-diagram of simplicial sets that takes the object $\sigma\colon \Delta[n] \to K$ of ΔK to $\Delta[n]$, then there is a natural isomorphism $\operatorname{colim}_{\Delta K} \mathrm{G} \approx K$.*

PROOF. The objects $\sigma\colon \Delta[n] \to K$ of ΔK come with natural maps $\mathrm{G}(\sigma) \to K$ that commute with the structure maps of G, and so there is a natural map $\operatorname{colim}_{\Delta K} \mathrm{G} \to K$. Since every n-simplex σ of K defines an object $\chi_\sigma\colon \Delta[n] \to K$ of ΔK for which the image of the natural map $\mathrm{G}(\chi_\sigma) \to K$ contains σ, the map $\operatorname{colim}_{\Delta K} \mathrm{G} \to K$ is surjective.

To show that the map $\operatorname{colim}_{\Delta K} \mathrm{G} \to K$ is injective, assume that there are objects $\sigma\colon \Delta[m] \to K$ and $\tau\colon \Delta[n] \to K$ of ΔK together with a k-simplex η of $\Delta[m]$ and a k-simplex μ of $\Delta[n]$ such that the image in K of η under $\mathrm{G}(\sigma) \to K$ equals the image in K of μ under $\mathrm{G}(\tau) \to K$. Example 11.5.16 implies that there is a commutative square in SS

$$\begin{array}{ccc} \Delta[k] & \xrightarrow{\chi_\mu} & \Delta[n] \\ {\scriptstyle\chi_\eta}\downarrow & & \downarrow{\scriptstyle\tau} \\ \Delta[m] & \xrightarrow[\sigma]{} & K \end{array}$$

which we can regard as a diagram in ΔK. This diagram in ΔK implies that the image of η in $\operatorname{colim}_{\Delta K} \mathrm{G}$ equals the image of μ in $\operatorname{colim}_{\Delta K} \mathrm{G}$, and so the natural surjection $\operatorname{colim}_{\Delta K} \mathrm{G} \to K$ is a natural isomorphism. □

15.1.21. Filtrations. Most arguments involving diagrams indexed by a Reedy category are done inductively on the degree of the object in the Reedy category. We define the filtrations of a Reedy category in order to facilitate such arguments.

DEFINITION 15.1.22. If $\mathcal{C}$ is a Reedy category with a degree function (see Remark 15.1.3) and n is a nonnegative integer, the *n-filtration* $\mathrm{F}^n\mathcal{C}$ is the full subcategory of $\mathcal{C}$ whose objects are the objects of $\mathcal{C}$ of degree less than or equal to n.

EXAMPLE 15.1.23. If $\mathcal{C}$ is a Reedy category, then the 0-filtration of $\mathcal{C}$ is a category with no non-identity maps.

EXAMPLE 15.1.24. If X is a simplicial set and $\mathcal{C} = \Delta X$ (see Example 15.1.19), then the n-filtration $\mathrm{F}^n\Delta X$ of ΔX is *not* the same as $\Delta(X^n)$, the category of simplices of the n-skeleton of the simplicial set X. This is because $\mathrm{F}^n\Delta X$ has no objects of degree greater than n, while $\Delta(X^n)$ has among its objects the high dimensional simplices of X that are degeneracies of simplices of dimension less than or equal to n.

PROPOSITION 15.1.25. *If $\mathcal{C}$ is a Reedy category, then each of its filtrations $\mathrm{F}^n\mathcal{C}$ (see Definition 15.1.22) is a Reedy category with $\overrightarrow{\mathrm{F}^n\mathcal{C}} = \overrightarrow{\mathcal{C}} \cap (\mathrm{F}^n\mathcal{C})$ and $\overleftarrow{\mathrm{F}^n\mathcal{C}} = \overleftarrow{\mathcal{C}} \cap (\mathrm{F}^n\mathcal{C})$, and $\mathcal{C}$ equals the union of the increasing sequence of subcategories $\mathrm{F}^0\mathcal{C} \subset \mathrm{F}^1\mathcal{C} \subset \mathrm{F}^2\mathcal{C} \subset \cdots$.*

PROOF. This follows directly from the definitions. □

15.2. Diagrams indexed by a Reedy category

Diagrams indexed by a Reedy category and maps of such diagrams are most naturally analyzed inductively on the degree of the object. In this section, we assume that we have a Reedy category with a degree function (see Remark 15.1.3), and we describe how to define a diagram indexed by the Reedy category by defining it inductively over the filtrations of the Reedy category (see Definition 15.1.22 and Proposition 15.1.25). In Remark 15.2.10, we summarize this description in terms of the latching objects and matching objects of the diagram, which we define in Definition 15.2.5. In Section 15.2.11, we will describe how to define a map between two such diagrams. We will use this analysis in Section 15.3 to define a model category structure on a category of diagrams in a model category indexed by a Reedy category.

Since the 0-filtration of a Reedy category (see Definition 15.1.22) contains no non-identity maps, we can define a diagram $\boldsymbol{X}\colon \mathrm{F}^0\mathcal{C} \to \mathcal{M}$ by choosing an object $\boldsymbol{X}_\alpha$ of $\mathcal{M}$ for each object α of $\mathcal{C}$ of degree 0.

Suppose that we have a diagram $\boldsymbol{X}\colon \mathrm{F}^{n-1}\mathcal{C} \to \mathcal{M}$ indexed by the $(n-1)$-filtration of a Reedy category $\mathcal{C}$, and we wish to extend it to a diagram $\boldsymbol{X}\colon \mathrm{F}^n\mathcal{C} \to \mathcal{M}$. We begin by choosing an object $\boldsymbol{X}_\alpha$ in $\mathcal{M}$ for each object α of $\mathcal{C}$ of degree

n. For each object β of $\mathrm{F}^{n-1}\mathcal{C}$ and map $\beta \to \alpha$ in $\mathrm{F}^n\mathcal{C}$, we must choose a map $\boldsymbol{X}_\beta \to \boldsymbol{X}_\alpha$ in $\mathcal{M}$. We must do this so that if $\beta \to \beta'$ is a map in $\mathrm{F}^{n-1}\mathcal{C}$ and

$$\begin{array}{ccc} \beta & \longrightarrow & \beta' \\ & \searrow \quad \swarrow & \\ & \alpha & \end{array}$$

is a commutative triangle in $\mathrm{F}^n\mathcal{C}$, then the triangle in $\mathcal{M}$

$$\begin{array}{ccc} \boldsymbol{X}_\beta & \longrightarrow & \boldsymbol{X}_{\beta'} \\ & \searrow \quad \swarrow & \\ & \boldsymbol{X}_\alpha & \end{array}$$

commutes. If $\mathrm{I}^n \colon \mathrm{F}^{n-1}\mathcal{C} \to \mathrm{F}^n\mathcal{C}$ is the inclusion functor, then this is equivalent to choosing a map $\operatorname{colim}_{(\mathrm{I}^n\downarrow\alpha)} \boldsymbol{X} \to \boldsymbol{X}_\alpha$ (see Definition 11.8.1). (The object $\operatorname{colim}_{(\mathrm{I}^n\downarrow\alpha)} \boldsymbol{X}$ is the value on α of the left Kan extension of $\boldsymbol{X} \colon \mathrm{F}^{n-1}\mathcal{C} \to \mathcal{M}$ along the inclusion $\mathrm{F}^{n-1}\mathcal{C} \to \mathrm{F}^n\mathcal{C}$ (see Remark 8.4.3 and [**47**, page 232–236]).) We will show in Proposition 15.2.8 that this colimit is actually independent of the choice of degree function (see Remark 15.1.3).

Similarly, for each object γ of $\mathrm{F}^{n-1}\mathcal{C}$ and map $\alpha \to \gamma$ in $\mathrm{F}^n\mathcal{C}$, we must choose a map $\boldsymbol{X}_\alpha \to \boldsymbol{X}_\gamma$ such that if $\gamma \to \gamma'$ is a map in $\mathrm{F}^{n-1}\mathcal{C}$ and

$$\begin{array}{ccc} & \alpha & \\ & \swarrow \quad \searrow & \\ \gamma & \longrightarrow & \gamma' \end{array}$$

is a commutative triangle in $\mathrm{F}^n\mathcal{C}$, then the triangle in $\mathcal{M}$

$$\begin{array}{ccc} & \boldsymbol{X}_\alpha & \\ & \swarrow \quad \searrow & \\ \boldsymbol{X}_\gamma & \longrightarrow & \boldsymbol{X}_{\gamma'} \end{array}$$

commutes. This is equivalent to choosing a map $\boldsymbol{X}_\alpha \to \lim_{(\alpha\downarrow\mathrm{I}^n)} \boldsymbol{X}$ (see Definition 11.8.3). (The object $\lim_{(\alpha\downarrow\mathrm{I}^n)} \boldsymbol{X}$ is the value on α of the right Kan extension of $\boldsymbol{X} \colon \mathrm{F}^{n-1}\mathcal{C} \to \mathcal{M}$ along the inclusion $\mathrm{F}^{n-1}\mathcal{C} \to \mathrm{F}^n\mathcal{C}$ (see Remark 8.4.3 and [**47**, page 232–236]).) We will show in Proposition 15.2.8 that this limit is actually independent of the choice of degree function.

The maps $\operatorname{colim}_{(\mathrm{I}^n\downarrow\alpha)} \boldsymbol{X} \to \boldsymbol{X}_\alpha$ and $\boldsymbol{X}_\alpha \to \lim_{(\alpha\downarrow\mathrm{I}^n)} \boldsymbol{X}$ cannot be totally arbitrary. If $\beta \to \gamma$ is a map in $\mathrm{F}^{n-1}\mathcal{C}$ and

$$\begin{array}{ccc} & \alpha & \\ & \nearrow \quad \searrow & \\ \beta & \longrightarrow & \gamma \end{array}$$

is a commutative triangle in $\mathrm{F}^n\mathcal{C}$, then the triangle in $\mathcal{M}$

$$\begin{array}{ccc} & \boldsymbol{X}_\alpha & \\ \nearrow & & \searrow \\ \boldsymbol{X}_\beta & \longrightarrow & \boldsymbol{X}_\gamma \end{array}$$

must commute. This is equivalent to requiring that the composition

$$\operatorname*{colim}_{(\mathrm{I}^n\downarrow\alpha)} \boldsymbol{X} \to \boldsymbol{X}_\alpha \to \lim_{(\alpha\downarrow\mathrm{I}^n)} \boldsymbol{X}$$

be a factorization of the natural map

$$\operatorname*{colim}_{(\mathrm{I}^n\downarrow\alpha)} \boldsymbol{X} \to \lim_{(\alpha\downarrow\mathrm{I}^n)} \boldsymbol{X} \ .$$

We will now show that the definition of a Reedy category implies that this last condition suffices to construct an extension of $\boldsymbol{X}$ from $\mathrm{F}^{n-1}\mathcal{C}$ to $\mathrm{F}^n\mathcal{C}$.

THEOREM 15.2.1. *Let $\mathcal{C}$ be a Reedy category, let $\mathcal{M}$ be a category closed under limits and colimits, let n be a positive integer, and let $\boldsymbol{X}\colon \mathrm{F}^{n-1}\mathcal{C} \to \mathcal{M}$ be a diagram. If for every object α of $\mathcal{C}$ of degree n we choose an object $\boldsymbol{X}_\alpha$ of $\mathcal{M}$ and a factorization $\operatorname{colim}_{(\mathrm{I}^n\downarrow\alpha)} \boldsymbol{X} \to \boldsymbol{X}_\alpha \to \lim_{(\alpha\downarrow\mathrm{I}^n)} \boldsymbol{X}$ of the natural map $\operatorname{colim}_{(\mathrm{I}^n\downarrow\alpha)} \boldsymbol{X} \to \lim_{(\alpha\downarrow\mathrm{I}^n)} \boldsymbol{X}$, then this uniquely determines an extension $\boldsymbol{X}\colon \mathrm{F}^n\mathcal{C} \to \mathcal{M}$ of the diagram $\boldsymbol{X}$.*

PROOF. The discussion above explains why our choices determine everything except the maps $\boldsymbol{X}_\alpha \to \boldsymbol{X}_{\alpha'}$ for a map $\alpha \to \alpha'$ in $\mathrm{F}^n\mathcal{C}$ between objects of degree n. Given such a map, if $\alpha \xrightarrow{\overleftarrow{g}} \beta \xrightarrow{\overrightarrow{g}} \alpha'$ is the factorization described in Definition 15.1.2, then we must define $\boldsymbol{X}_\alpha \to \boldsymbol{X}_{\alpha'}$ to be the composition $\boldsymbol{X}_\alpha \to \boldsymbol{X}_\beta \to \boldsymbol{X}_{\alpha'}$. It remains only to show that, if $\alpha \to \alpha' \to \alpha''$ are composable maps in $\mathrm{F}^n\mathcal{C}$ between objects of degree n, then the triangle

$$\begin{array}{ccc} & \boldsymbol{X}_\alpha & \\ \swarrow & & \searrow \\ \boldsymbol{X}_{\alpha'} & \longrightarrow & \boldsymbol{X}_{\alpha''} \end{array}$$

commutes.

Let $\alpha \xrightarrow{\overleftarrow{g}} \beta \xrightarrow{\overrightarrow{g}} \alpha'$ and $\alpha' \xrightarrow{\overleftarrow{h}} \beta' \xrightarrow{\overrightarrow{h}} \alpha''$ be the factorization of Definition 15.1.2 applied to $\alpha \to \alpha'$ and $\alpha' \to \alpha''$, respectively. If the factorization of Definition 15.1.2 applied to $\overleftarrow{h}\,\overrightarrow{g}\,\overleftarrow{g}\colon \alpha \to \beta'$ is $\alpha \xrightarrow{\overleftarrow{k}} \beta'' \xrightarrow{\overrightarrow{k}} \beta'$, then we have the commutative diagram

$$\begin{array}{ccccc} & & \alpha & & \\ & {\scriptstyle\overleftarrow{g}}\swarrow & \downarrow{\scriptstyle\overleftarrow{k}} & & \\ & \beta & \beta'' & & \\ {\scriptstyle\overrightarrow{g}}\swarrow & & \downarrow{\scriptstyle\overrightarrow{k}} & & \\ \alpha' & \xrightarrow{\overleftarrow{h}} & \beta' & \xrightarrow{\overrightarrow{h}} & \alpha'' \ . \end{array}$$

Since $\overrightarrow{h}\,\overrightarrow{k}\,\overleftarrow{k} = \overrightarrow{h}\,\overleftarrow{h}\,\overrightarrow{g}\,\overleftarrow{g}$ and $\overrightarrow{h}\,\overrightarrow{k}$ is in $\overrightarrow{\mathcal{C}}$, the factorization $\alpha \xrightarrow{\overleftarrow{k}} \beta'' \xrightarrow{\overrightarrow{h}\,\overrightarrow{k}} \alpha''$ must be the factorization of $\alpha \to \alpha''$ described in Definition 15.1.2. Thus, it is sufficient to show that the composition $\boldsymbol{X}_\alpha \xrightarrow{\overleftarrow{k}_*} \boldsymbol{X}_{\beta''} \xrightarrow{\overrightarrow{k}_*} \boldsymbol{X}_{\beta'}$ equals the composition $\boldsymbol{X}_\alpha \xrightarrow{\overleftarrow{g}_*} \boldsymbol{X}_\beta \xrightarrow{\overrightarrow{g}_*} \boldsymbol{X}_{\alpha'} \xrightarrow{\overleftarrow{h}_*} \boldsymbol{X}_{\beta'}$. Since both of the maps $\boldsymbol{X}_\alpha \to \boldsymbol{X}_{\beta'}$ and $\boldsymbol{X}_\alpha \to \boldsymbol{X}_\beta$ are defined as the composition of our map $\boldsymbol{X}_\alpha \to \lim_{(\alpha\downarrow \mathrm{I}^n)} \boldsymbol{X}$ with a projection from the limit, the first of these maps equals the composition

$$\boldsymbol{X}_\alpha \to \lim_{(\alpha\downarrow \mathrm{I}^n)} \boldsymbol{X} \to \boldsymbol{X}_{\beta''} \xrightarrow{\overrightarrow{k}_*} \boldsymbol{X}_{\beta'}$$

while the second equals the composition

$$\boldsymbol{X}_\alpha \to \lim_{(\alpha\downarrow \mathrm{I}^n)} \boldsymbol{X} \to \boldsymbol{X}_\beta \xrightarrow{\overrightarrow{g}_*} \boldsymbol{X}_{\alpha'} \xrightarrow{\overleftarrow{h}_*} \boldsymbol{X}_{\beta'} \ .$$

The universal property of the limit implies that these are equal. □

15.2.2. Latching objects and matching objects. In this section, we show that the colimits and limits used in Section 15.2 to construct diagrams indexed by a Reedy category (which will also be used in Section 15.2.11 to construct maps of such diagrams) are independent of the choice of degree function (see Remark 15.1.3) and have a particularly convenient form. These colimits and limits are the *latching objects* and *matching objects* (see Definition 15.2.5). We continue to assume that we have chosen a degree function for our Reedy category (see Remark 15.1.3).

DEFINITION 15.2.3. Let $\mathcal{C}$ be a Reedy category and let α be an object of $\mathcal{C}$.

(1) The *latching category* $\partial(\overrightarrow{\mathcal{C}} \downarrow \alpha)$ of $\mathcal{C}$ at α is the full subcategory of $(\overrightarrow{\mathcal{C}} \downarrow \alpha)$ containing all the objects except the identity map of α.
(2) The *matching category* $\partial(\alpha \downarrow \overleftarrow{\mathcal{C}})$ of $\mathcal{C}$ at α is the full subcategory of $(\alpha \downarrow \overleftarrow{\mathcal{C}})$ containing all the objects except the identity map of α.

PROPOSITION 15.2.4. *Let $\mathcal{C}$ be a Reedy category and let α be an object of $\mathcal{C}$.*

(1) *The opposite of the latching category of $\mathcal{C}$ at α is naturally isomorphic to the matching category of $\mathcal{C}^{\mathrm{op}}$ at α (see Proposition 15.1.5).*
(2) *The opposite of the matching category of $\mathcal{C}$ at α is naturally isomorphic to the latching category of $\mathcal{C}^{\mathrm{op}}$ at α.*

PROOF. This follows from Corollary 11.8.7. □

DEFINITION 15.2.5. Let $\mathcal{C}$ be a Reedy category, let $\mathcal{M}$ be a model category, let $\boldsymbol{X}$ be a $\mathcal{C}$-diagram in $\mathcal{M}$, and let α be an object of $\mathcal{C}$. We use $\boldsymbol{X}$ to denote also the induced $\partial(\overrightarrow{\mathcal{C}} \downarrow \alpha)$-diagram (defined on objects by $\boldsymbol{X}_{(\beta\to\alpha)} = \boldsymbol{X}_\beta$) and the induced $\partial(\alpha \downarrow \overleftarrow{\mathcal{C}})$-diagram (defined on objects by $\boldsymbol{X}_{(\alpha\to\beta)} = \boldsymbol{X}_\beta$).

(1) The *latching object* of $\boldsymbol{X}$ at α is $\mathrm{L}_\alpha \boldsymbol{X} = \operatorname{colim}_{\partial(\overrightarrow{\mathcal{C}}\downarrow\alpha)} \boldsymbol{X}$ and the *latching map* of $\boldsymbol{X}$ at α is the natural map $\mathrm{L}_\alpha \boldsymbol{X} \to \boldsymbol{X}_\alpha$. We will sometimes use $\mathrm{L}^{\mathcal{C}}_\alpha \boldsymbol{X}$ to denote $\mathrm{L}_\alpha \boldsymbol{X}$ when we want to emphasize the indexing category $\mathcal{C}$.
(2) The *matching object* of $\boldsymbol{X}$ at α is $\mathrm{M}_\alpha \boldsymbol{X} = \lim_{\partial(\alpha\downarrow\overleftarrow{\mathcal{C}})} \boldsymbol{X}$ and the *matching map* of $\boldsymbol{X}$ at α is the natural map $\boldsymbol{X}_\alpha \to \mathrm{M}_\alpha \boldsymbol{X}$. We will sometimes use $\mathrm{M}^{\mathcal{C}}_\alpha \boldsymbol{X}$ to denote $\mathrm{M}_\alpha \boldsymbol{X}$ when we want to emphasize the indexing category $\mathcal{C}$.

Matching objects were first defined for cosimplicial simplicial sets, in [**14**, page 274], where they were called *matching spaces*. The following proposition shows that the definition used there agrees with Definition 15.2.5.

PROPOSITION 15.2.6. *Let $\mathcal{M}$ be a model category.*

1. *If $\boldsymbol{X}$ is a simplicial object in $\mathcal{M}$ (see Definition 15.1.10) and $n \geq 2$, then the latching object of $\boldsymbol{X}$ at $[n]$ is the colimit of the diagram obtained by restricting $\boldsymbol{X}$ (see Definition 15.2.5) to the full subcategory of $\partial(\overrightarrow{\boldsymbol{\Delta}^{\mathrm{op}}} \downarrow [n])$ (see Definition 15.1.10) with objects the maps $[k] \to [n]$ with $k = n-1$ or $k = n-2$.*
2. *If $\boldsymbol{X}$ is a cosimplicial object in $\mathcal{M}$ and $n \geq 2$, then the matching object of $\boldsymbol{X}$ at $[n]$ is the limit of the diagram obtained by restricting $\boldsymbol{X}$ to the full subcategory of $\partial([n] \downarrow \overleftarrow{\boldsymbol{\Delta}})$ with objects the maps $[n] \to [k]$ with $k = n-1$ or $k = n-2$.*

PROOF. We will prove part 1; the proof of part 2 is dual.

Let $\mathcal{D}$ denote the full subcategory of $\partial(\overrightarrow{\boldsymbol{\Delta}^{\mathrm{op}}} \downarrow [n])$ with objects the maps $[k] \to [n]$ with $k = n-1$ or $k = n-2$; we will show that the inclusion $\mathcal{D} \to \partial(\overrightarrow{\boldsymbol{\Delta}^{\mathrm{op}}} \downarrow [n])$ is right cofinal (see Theorem 14.2.5).

If $k = n-1$ or $k = n-2$ then the identity map of $[k]$ is an initial object of $([k] \downarrow \mathcal{D})$, and so $\mathrm{B}([k] \downarrow \mathcal{D})$ is connected. If $k < n-2$, then there are morphisms from the object $s_{i_1} s_{i_2} \cdots s_{i_{n-k-1}} \colon [k] \to [n-1]$ to each of the objects

$$\begin{aligned} s_0 s_{i_1} s_{i_2} \cdots s_{i_{n-k-1}} &\colon [k] \to [n], \\ s_1 s_{i_1} s_{i_2} \cdots s_{i_{n-k-1}} &\colon [k] \to [n], \ldots, \text{ and} \\ s_{n-1} s_{i_1} s_{i_2} \cdots s_{i_{n-k-1}} &\colon [k] \to [n], \end{aligned}$$

and so it is sufficient to show that the object $s_{i_1} s_{i_2} \cdots s_{i_{n-k-1}}$ is connected to the object $(s_0)^{n-k} \colon [k] \to [n]$. Since $s_{i_1} s_{i_2} \cdots s_{i_{n-k-1}}$ is connected to

$$s_0 (s_{i_1} s_{i_2} \cdots s_{i_{n-k-1}}) = s_{i_1+1} s_{i_2+1} \cdots s_{i_{n-k-1}+1} s_0$$

which is connected to

$$s_0 s_{i_2+1} \cdots s_{i_{n-k-1}+1} s_0 = s_{i_2+2} s_{i_3+2} \cdots s_{i_{n-k-1}+2} (s_0)^2$$

which is connected to ... which is connected to

$$s_0 s_{i_{n-k-1}+n-k-2} (s_0^{n-k-2}) = s_{i_{n-k-1}+n-k-1} (s_0)^{n-k-1}$$

which is connected to $(s_0)^{n-k}$, the proof is complete. □

DEFINITION 15.2.7. If $\mathcal{C}$ is a Reedy category and α is an object of $\mathcal{C}$ of degree n, then

1. $\partial(\alpha \downarrow \mathrm{F}^n\mathcal{C})$ is the full subcategory of $(\alpha \downarrow \mathrm{F}^n\mathcal{C})$ with objects the maps $\alpha \xrightarrow{g} \beta$ for which there is a factorization $\alpha \xrightarrow{\overleftarrow{g}} \gamma \xrightarrow{\overrightarrow{g}} \beta$ with $\overleftarrow{g} \in \overleftarrow{\mathcal{C}}$, $\overrightarrow{g} \in \overrightarrow{\mathcal{C}}$, and $\overleftarrow{g} \neq 1_\alpha$, and
2. $\partial(\mathrm{F}^n\mathcal{C} \downarrow \alpha)$ is the full subcategory of $(\mathrm{F}^n\mathcal{C} \downarrow \alpha)$ with objects the maps $\beta \xrightarrow{g} \alpha$ for which there is a factorization $\beta \xrightarrow{\overleftarrow{g}} \gamma \xrightarrow{\overrightarrow{g}} \alpha$ with $\overleftarrow{g} \in \overleftarrow{\mathcal{C}}$, $\overrightarrow{g} \in \overrightarrow{\mathcal{C}}$, and $\overrightarrow{g} \neq 1_\alpha$.

The objects $\operatorname{colim}_{(\mathrm{I}^n\downarrow\alpha)} \boldsymbol{X}$ and $\lim_{(\alpha\downarrow\mathrm{I}^n)} \boldsymbol{X}$ (where $\mathrm{I}^n\colon \mathrm{F}^{n-1}\mathcal{C} \to \mathrm{F}^n\mathcal{C}$ is the inclusion functor and $\boldsymbol{X}$ is a diagram defined on $\mathrm{F}^{n-1}\mathcal{C}$) were used in Section 15.2 to construct diagrams indexed by a Reedy category. The objects $\operatorname{colim}_{\partial(\mathrm{F}^n\mathcal{C}\downarrow\alpha)} \boldsymbol{X}$ and $\lim_{\partial(\alpha\downarrow\mathrm{F}^n\mathcal{C})} \boldsymbol{X}$ (where $\boldsymbol{X}$ is a diagram defined on $\mathrm{F}^n\mathcal{C}$) will be used in Section 15.2.11 to analyze maps between such diagrams. Corollary 15.2.9 shows that all of these colimits are latching objects of $\boldsymbol{X}$ and all of these limits are matching objects of $\boldsymbol{X}$.

PROPOSITION 15.2.8. *Let $\mathcal{C}$ be a Reedy category, let α be an object of $\mathcal{C}$ of degree n, and let $\mathrm{I}^n\colon \mathrm{F}^{n-1}\mathcal{C} \to \mathrm{F}^n\mathcal{C}$ be the inclusion functor.*

1. *The latching category $\partial(\overrightarrow{\mathcal{C}}\downarrow\alpha)$ is a right cofinal subcategory (see Definition 19.6.1) of both $(\mathrm{I}^n\downarrow\alpha)$ and $\partial(\mathrm{F}^n\mathcal{C}\downarrow\alpha)$ (see Definition 15.2.7).*
2. *The matching category $\partial(\alpha\downarrow\overleftarrow{\mathcal{C}})$ is a left cofinal subcategory of both $(\alpha\downarrow\mathrm{I}^n)$ and $\partial(\alpha\downarrow\mathrm{F}^n\mathcal{C})$ (see Definition 15.2.7).*

PROOF. We will prove part 1; the proof of part 2 is dual.

Let $\mathrm{J}^n\colon \partial(\overrightarrow{\mathcal{C}}\downarrow\alpha) \to (\mathrm{I}^n\downarrow\alpha)$ be the inclusion functor. If $g\colon \beta\to\alpha$ is an object of $(\mathrm{I}^n\downarrow\alpha)$, then we can factor it as $\beta \xrightarrow{\overleftarrow{g}} \beta' \xrightarrow{\overrightarrow{g}} \alpha$ where $\overleftarrow{g} \in \overleftarrow{\mathcal{C}}$ and $\overrightarrow{g} \in \overrightarrow{\mathcal{C}}$. This gives us the object

$$\begin{array}{ccc} \beta & \xrightarrow{\;\overleftarrow{g}\;} & \beta' \\ & \searrow & \swarrow{\scriptstyle \overrightarrow{g}} \\ & \alpha & \end{array}$$

of $((\beta\to\alpha)\downarrow\mathrm{J}^n)$; we will show that there is a map from this object to every other object of $((\beta\to\alpha)\downarrow\mathrm{J}^n)$, which will imply that $((\beta\to\alpha)\downarrow\mathrm{J}^n)$ is nonempty and connected (see Definition 14.2.1).

Any object of $((\beta\to\alpha)\downarrow\mathrm{J}^n)$ is of the form $\beta \xrightarrow{h} \gamma \xrightarrow{\overrightarrow{k}} \alpha$ where $\overrightarrow{k} \in \overrightarrow{\mathcal{C}}$, $\overrightarrow{k} \neq 1_\alpha$, and the composition $\overrightarrow{k}h$ equals our map $\beta\to\alpha$. If we factor h as $h = \overrightarrow{h}\overleftarrow{h}$, then the uniqueness of the factorization in Definition 15.1.2 implies that $\overleftarrow{h} = \overleftarrow{g}$ and $\overrightarrow{k}\overrightarrow{h} = \overrightarrow{g}$, i.e., $\overrightarrow{h}$ is a map from $\beta \xrightarrow{\overleftarrow{g}} \beta' \xrightarrow{\overrightarrow{g}} \alpha$ to $\beta \xrightarrow{h} \gamma \xrightarrow{\overrightarrow{k}} \alpha$.

The proof that $\partial(\overrightarrow{\mathcal{C}}\downarrow\alpha)$ is right cofinal in $\partial(\mathrm{F}^n\mathcal{C}\downarrow\alpha)$ is identical to this. □

COROLLARY 15.2.9. *Let $\mathcal{C}$ be a Reedy category, let $\mathcal{M}$ be a model category, let α be an object of $\mathcal{C}$ of degree n, and let $\boldsymbol{X}$ be a $\mathcal{C}$-diagram in $\mathcal{M}$. If $\mathrm{I}^n\colon \mathrm{F}^{n-1}\mathcal{C} \to \mathrm{F}^n\mathcal{C}$ is the inclusion functor, then there are natural isomorphisms*

$$\operatorname*{colim}_{(\mathrm{I}^n\downarrow\alpha)} \boldsymbol{X} \approx \mathrm{L}_\alpha\boldsymbol{X} \approx \operatorname*{colim}_{\partial(\mathrm{F}^n\mathcal{C}\downarrow\alpha)} \boldsymbol{X} \quad\text{and}\quad \lim_{(\alpha\downarrow\mathrm{I}^n)} \boldsymbol{X} \approx \mathrm{M}_\alpha\boldsymbol{X} \approx \lim_{\partial(\alpha\downarrow\mathrm{F}^n\mathcal{C})} \boldsymbol{X}$$

(see Definition 15.2.7).

PROOF. This follows from Proposition 15.2.8 and Theorem 14.2.5. □

REMARK 15.2.10. In light of Definition 15.2.5 and Corollary 15.2.9, the discussion in Section 15.2 can be summarized as follows: If $\mathcal{C}$ is a Reedy category, $\mathcal{M}$ is a model category, $\boldsymbol{X}\colon \mathrm{F}^{n-1}\mathcal{C} \to \mathcal{M}$ is a diagram indexed by the $(n-1)$-filtration of $\mathcal{C}$, and α is an object of $\mathcal{C}$ of degree n, then there is a natural map $\mathrm{L}_\alpha\boldsymbol{X} \to \mathrm{M}_\alpha\boldsymbol{X}$ from the latching object of $\boldsymbol{X}$ at α to the matching object of $\boldsymbol{X}$ at α. Extending $\boldsymbol{X}$ to a diagram $\mathrm{F}^n\mathcal{C} \to \mathcal{M}$ is equivalent to choosing, for every object α of degree n, an object X_α and a factorization $\mathrm{L}_\alpha X \to X_\alpha \to \mathrm{M}_\alpha X$ of that natural map, and this can be done independently for each of the objects of degree n.

15.2.11. Maps between diagrams. Maps between diagrams indexed by a Reedy category are most naturally defined inductively over the filtrations of the Reedy category (see Definition 15.1.22). We assume that we have chosen a degree function for our Reedy category (see Remark 15.1.3).

Let $\mathcal{C}$ be a Reedy category, let $\mathcal{M}$ be a model category, and let X and Y be $\mathcal{C}$-diagrams in $\mathcal{M}$. Since the 0-filtration (see Definition 15.1.22) of a Reedy category contains no non-identity maps, a map $f\colon X|_{\mathrm{F}^0\mathcal{C}} \to Y|_{\mathrm{F}^0\mathcal{C}}$ is determined by choosing a map $X_\alpha \to Y_\alpha$ for every object α of degree 0.

Suppose that $f\colon X|_{\mathrm{F}^{n-1}\mathcal{C}} \to Y|_{\mathrm{F}^{n-1}\mathcal{C}}$ is a map of the restrictions of the diagrams to the $(n-1)$-filtration of $\mathcal{C}$. For every object α of $\mathcal{C}$ of degree n we have the solid arrow diagram

$$\begin{array}{ccccc} \operatorname*{colim}_{(\mathrm{I}^n\downarrow\alpha)} X & \longrightarrow & X_\alpha & \longrightarrow & \lim_{(\alpha\downarrow\mathrm{I}^n)} X \\ \downarrow & & \vdots & & \downarrow \\ \operatorname*{colim}_{(\mathrm{I}^n\downarrow\alpha)} Y & \longrightarrow & Y_\alpha & \longrightarrow & \lim_{(\alpha\downarrow\mathrm{I}^n)} Y \end{array}$$

(where $\mathrm{I}^n\colon \mathrm{F}^{n-1}\mathcal{C} \to \mathrm{F}^n\mathcal{C}$ is the inclusion functor) and Corollary 15.2.9 implies that this diagram is isomorphic to the diagram

$$\begin{array}{ccccc} \operatorname*{colim}_{\partial(\mathrm{F}^n\mathcal{C}\downarrow\alpha)} X & \longrightarrow & X_\alpha & \longrightarrow & \lim_{\partial(\alpha\downarrow\mathrm{F}^n\mathcal{C})} X \\ \downarrow & & \vdots & & \downarrow \\ \operatorname*{colim}_{\partial(\mathrm{F}^n\mathcal{C}\downarrow\alpha)} Y & \longrightarrow & Y_\alpha & \longrightarrow & \lim_{\partial(\alpha\downarrow\mathrm{F}^n\mathcal{C})} Y \end{array}.$$

Thus, extensions of f to the n-filtration of $\mathcal{C}$ correspond to a choice, for every object α of degree n, of a dotted arrow that makes both squares commute. Corollary 15.2.9 implies that this diagram is also isomorphic to the diagram

$$\begin{array}{ccccc} \mathrm{L}_\alpha X & \longrightarrow & X_\alpha & \longrightarrow & \mathrm{M}_\alpha X \\ \downarrow & & \vdots & & \downarrow \\ \mathrm{L}_\alpha Y & \longrightarrow & Y_\alpha & \longrightarrow & \mathrm{M}_\alpha Y \end{array}.$$

Thus, if A, B, X, and Y are $\mathcal{C}$-diagrams in $\mathcal{M}$ and we have a diagram

$$\begin{array}{ccc} A & \longrightarrow & X \\ \downarrow & \overset{h}{\nearrow} & \downarrow \\ B & \longrightarrow & Y \end{array} \tag{15.2.12}$$

in which the dotted arrow h is defined only on the restriction of B to the $(n-1)$-filtration of $\mathcal{C}$, then for every object α of $\mathcal{C}$ of degree n we have an induced solid arrow diagram

$$\begin{array}{ccc} \boldsymbol{A}_\alpha \amalg_{\mathrm{L}_\alpha \boldsymbol{A}} \mathrm{L}_\alpha \boldsymbol{B} & \longrightarrow & \boldsymbol{X}_\alpha \\ \big\downarrow & \nearrow & \big\downarrow \\ \boldsymbol{B}_\alpha & \longrightarrow & \boldsymbol{Y}_\alpha \times_{\mathrm{M}_\alpha \boldsymbol{Y}} \mathrm{M}_\alpha \boldsymbol{X} \end{array}$$

and there is a map $\boldsymbol{B}_\alpha \to \boldsymbol{X}_\alpha$ for every object α of degree n that makes both triangles commute if and only if h can be extended over the restriction of $\boldsymbol{B}$ to the n-filtration of $\mathcal{C}$ so that both triangles in Diagram 15.2.12 commute. This is the motivation for the definitions of the relative latching map and relative matching map (see Definition 15.3.2) and their appearance in the definitions of *Reedy cofibration* and *Reedy fibration* (see Definition 15.3.3).

15.3. The Reedy model category structure

If $\mathcal{C}$ is a Reedy category and $\mathcal{M}$ is a model category, we will define a model category structure on $\mathcal{M}^{\mathcal{C}}$, the category of $\mathcal{C}$-diagrams in $\mathcal{M}$, called the *Reedy model category structure.* If $\mathcal{M}$ is a *simplicial* model category, then we will show that the simplicial structure of Definition 11.7.1 and Definition 11.7.2 makes the Reedy model category structure on $\mathcal{M}^{\mathcal{C}}$ a simplicial model category.

If $\mathcal{M}$ is a cofibrantly generated model category, then the Reedy model category structure will have the same weak equivalences as the model category structure of Theorem 11.6.1, but it will have a larger class of cofibrations (see Proposition 15.6.3). Thus, free cell complexes and their retracts will be cofibrant in the Reedy model category structure, as will some diagrams that are not retracts of free cell complexes.

15.3.1. Statement of the theorem.

DEFINITION 15.3.2. Let $\mathcal{C}$ be a Reedy category, let $\mathcal{M}$ be a model category, let $\boldsymbol{X}$ and $\boldsymbol{Y}$ be $\mathcal{C}$-diagrams in $\mathcal{M}$, and let $f\colon \boldsymbol{X} \to \boldsymbol{Y}$ be a map of $\mathcal{C}$-diagrams.

(1) If α is an object of $\mathcal{C}$, then the *relative latching map of f at α* is the map $\boldsymbol{X}_\alpha \amalg_{\mathrm{L}_\alpha \boldsymbol{X}} \mathrm{L}_\alpha \boldsymbol{Y} \to \boldsymbol{Y}_\alpha$ (see Definition 15.2.5).
(2) If α is an object of $\mathcal{C}$, then the *relative matching map of f at α* is the map $\boldsymbol{X}_\alpha \to \boldsymbol{Y}_\alpha \times_{\mathrm{M}_\alpha \boldsymbol{Y}} \mathrm{M}_\alpha \boldsymbol{X}$.

DEFINITION 15.3.3. Let $\mathcal{C}$ be a Reedy category, let $\mathcal{M}$ be a model category, and let $\boldsymbol{X}, \boldsymbol{Y}\colon \mathcal{C} \to \mathcal{M}$ be $\mathcal{C}$-diagrams in $\mathcal{M}$.

(1) A map of diagrams $f\colon \boldsymbol{X} \to \boldsymbol{Y}$ is a *Reedy weak equivalence* if, for every object α of $\mathcal{C}$, the map $f_\alpha\colon \boldsymbol{X}_\alpha \to \boldsymbol{Y}_\alpha$ is a weak equivalence in $\mathcal{M}$.
(2) A map of diagrams $f\colon \boldsymbol{X} \to \boldsymbol{Y}$ is a *Reedy cofibration* if, for every object α of $\mathcal{C}$, the relative latching map (see Definition 15.3.2)
$$\boldsymbol{X}_\alpha \amalg_{\mathrm{L}_\alpha \boldsymbol{X}} \mathrm{L}_\alpha \boldsymbol{Y} \to \boldsymbol{Y}_\alpha$$
is a cofibration in $\mathcal{M}$.
(3) A map of diagrams $f\colon \boldsymbol{X} \to \boldsymbol{Y}$ is a *Reedy fibration* if, for every object α of $\mathcal{C}$, the relative matching map (see Definition 15.3.2)
$$\boldsymbol{X}_\alpha \to \boldsymbol{Y}_\alpha \times_{\mathrm{M}_\alpha \boldsymbol{Y}} \mathrm{M}_\alpha \boldsymbol{X}$$
is a fibration in $\mathcal{M}$.

THEOREM 15.3.4 (D. M. Kan). *Let $\mathcal{C}$ be a Reedy category and let $\mathcal{M}$ be a model category.*

(1) *The category $\mathcal{M}^{\mathcal{C}}$ of $\mathcal{C}$-diagrams in $\mathcal{M}$ with the Reedy weak equivalences, Reedy cofibrations, and Reedy fibrations (see Definition 15.3.3) is a model category.*
(2) *If $\mathcal{M}$ is a left proper, right proper, or proper model category (see Definition 13.1.1), then the model category of part 1 is, respectively, left proper, right proper, or proper.*
(3) *If $\mathcal{M}$ is a simplicial model category (see Definition 9.1.6), then the model category of part 1 with the simplicial structure defined in Definition 11.7.1 and Definition 11.7.2 is a simplicial model category.*

The proof of Theorem 15.3.4 is in Section 15.3.16.

EXAMPLE 15.3.5. If $\mathcal{M}$ is a model category, then the category $\mathcal{M}^{\mathbf{\Delta}^{\mathrm{op}}}$ of simplicial objects in $\mathcal{M}$ has a model category structure from the Reedy category structure of $\mathbf{\Delta}^{\mathrm{op}}$ (see Definition 15.1.10 and Proposition 15.1.5).

EXAMPLE 15.3.6. If $\mathcal{M}$ is a model category, then the category $\mathcal{M}^{\mathbf{\Delta}}$ of cosimplicial objects in $\mathcal{M}$ has a model category structure from the Reedy category structure of $\mathbf{\Delta}$.

LEMMA 15.3.7. *Let $\mathcal{C}$ be a Reedy category, let $\mathcal{M}$ be a model category, and let $\boldsymbol{X}$ be a $\mathcal{C}$-diagram in $\mathcal{M}$.*

(1) *If $\boldsymbol{X}$ is Reedy cofibrant then for every object α of $\mathcal{C}$ the restriction of $\boldsymbol{X}$ to $\partial(\overrightarrow{\mathcal{C}}\downarrow\alpha)$ (see Definition 15.2.3) is Reedy cofibrant.*
(2) *If $\boldsymbol{X}$ is Reedy fibrant then for every object α of $\mathcal{C}$ the restriction of $\boldsymbol{X}$ to $\partial(\alpha\downarrow\overleftarrow{\mathcal{C}})$ (see Definition 15.2.3) is Reedy fibrant.*

PROOF. We will prove part 1; the proof of part 2 is similar.

If $\beta\to\alpha$ is an object of $\partial(\overrightarrow{\mathcal{C}}\downarrow\alpha)$, then the latching category of β in $\mathcal{C}$ equals the latching category of β in $\partial(\overrightarrow{\mathcal{C}}\downarrow\alpha)$, and so the latching map of the restriction of $\boldsymbol{X}$ at $\beta\to\alpha$ equals the latching map of $\boldsymbol{X}$ at β. $\square$

15.3.8. Trivial cofibrations and trivial fibrations. In order to prove Theorem 15.3.4, we need to identify those maps of diagrams that are both Reedy cofibrations and Reedy weak equivalences and those maps that are both Reedy fibrations and Reedy weak equivalences. In this section, we will show that f is both a Reedy cofibration and a Reedy weak equivalence if and only if each of the maps $\boldsymbol{X}_\alpha \amalg_{\mathrm{L}_\alpha\boldsymbol{X}} \mathrm{L}_\alpha\boldsymbol{Y}\to\boldsymbol{Y}_\alpha$ is a trivial cofibration in $\mathcal{M}$, and that f is both a Reedy fibration and a Reedy weak equivalence if and only if each of the maps $\boldsymbol{X}_\alpha\to\boldsymbol{Y}_\alpha\times_{\mathrm{M}_\alpha\boldsymbol{Y}}\mathrm{M}_\alpha\boldsymbol{X}$ is a trivial fibration in $\mathcal{M}$ (see Theorem 15.3.15). We will use this theorem in Section 15.3.16 to prove Theorem 15.3.4.

LEMMA 15.3.9. *Let $\mathcal{C}$ be a Reedy category, let $\mathcal{M}$ be a model category, let $f\colon\boldsymbol{X}\to\boldsymbol{Y}$ be a map of $\mathcal{C}$-diagrams in $\mathcal{M}$, let α be an object of $\mathcal{C}$, and let S be a class of maps in $\mathcal{M}$.*

(1) *If for every object β of $\mathcal{C}$ whose degree is less than that of α the relative latching map*

$$\boldsymbol{X}_\beta \amalg_{\mathrm{L}_\beta\boldsymbol{X}} \mathrm{L}_\beta\boldsymbol{Y}\to\boldsymbol{Y}_\beta$$

has the left lifting property (see Definition 7.2.1) with respect to every element of S, then the induced map of latching objects $\mathrm{L}_\alpha\boldsymbol{X} \to \mathrm{L}_\alpha\boldsymbol{Y}$ has the left lifting property with respect to every element of S.

(2) *If for every object β of $\mathcal{C}$ whose degree is less than that of α the relative matching map*

$$\boldsymbol{X}_\beta \to \boldsymbol{Y}_\beta \times_{\mathrm{M}_\beta\boldsymbol{Y}} \mathrm{M}_\beta\boldsymbol{X}$$

has the right lifting property (see Definition 7.2.1) with respect to every element of S, then the induced map of matching objects $\mathrm{M}_\alpha\boldsymbol{X} \to \mathrm{M}_\alpha\boldsymbol{Y}$ has the right lifting property with respect to every element of S.

PROOF. We will prove part 1; the proof of part 2 is dual. We assume that we have chosen a degree function for $\mathcal{C}$ (see Remark 15.1.3).

There is a filtration of the category $\partial(\overrightarrow{\mathcal{C}}\downarrow\alpha)$ in which $\mathrm{F}^k\partial(\overrightarrow{\mathcal{C}}\downarrow\alpha)$ is the full subcategory of $\partial(\overrightarrow{\mathcal{C}}\downarrow\alpha)$ whose objects are the maps $\beta\to\alpha$ in $\overrightarrow{\mathcal{C}}$ such that the degree of β is less than or equal to k. Thus, $\mathrm{F}^0\partial(\overrightarrow{\mathcal{C}}\downarrow\alpha)$ has no non-identity maps, and $\mathrm{F}^{\deg(\alpha)-1}\partial(\overrightarrow{\mathcal{C}}\downarrow\alpha) = \partial(\overrightarrow{\mathcal{C}}\downarrow\alpha)$. If $E\to B$ is an element of S and we have the solid arrow diagram

$$\begin{array}{ccc} \mathrm{L}_\alpha\boldsymbol{X} & \longrightarrow & E \\ \downarrow & \overset{h}{\nearrow} & \downarrow \\ \mathrm{L}_\alpha\boldsymbol{Y} & \longrightarrow & B \end{array}$$

then we will define the map h by defining it inductively over $\mathrm{colim}_{\mathrm{F}^k\partial(\overrightarrow{\mathcal{C}}\downarrow\alpha)}\boldsymbol{Y}$.

For objects $\beta\to\alpha$ of $(\overrightarrow{\mathcal{C}}\downarrow\alpha)$ such that β is of degree zero, the latching objects $\mathrm{L}_\beta\boldsymbol{X}$ and $\mathrm{L}_\beta\boldsymbol{Y}$ are the initial object of $\mathcal{M}$, and so the map $\boldsymbol{X}_\beta\to\boldsymbol{Y}_\beta$ equals the relative latching map $\boldsymbol{X}_\beta \amalg_{\mathrm{L}_\beta\boldsymbol{X}} \mathrm{L}_\beta\boldsymbol{Y} \to \boldsymbol{Y}_\beta$, which we have assumed has the left lifting property with respect to $E\to B$. Thus, there exists a dotted arrow h that makes both triangles commute in the diagram

$$\begin{array}{ccc} \boldsymbol{X}_\beta & \longrightarrow & E \\ \downarrow & \overset{h}{\nearrow} & \downarrow \\ \boldsymbol{Y}_\beta & \longrightarrow & B \end{array}$$

Since $\mathrm{F}^0\partial(\overrightarrow{\mathcal{C}}\downarrow\alpha)$ has no non-identity maps, this defines h on $\mathrm{F}^0\partial(\overrightarrow{\mathcal{C}}\downarrow\alpha)$.

For the inductive step, we assume that $0<k<\deg(\alpha)$ and that the map has been defined on $\mathrm{colim}_{\mathrm{F}^{k-1}\partial(\overrightarrow{\mathcal{C}}\downarrow\alpha)}\boldsymbol{Y}$. Let $\beta\to\alpha$ be an object of $\partial(\overrightarrow{\mathcal{C}}\downarrow\alpha)$ such that β is of degree k. The map $\beta\to\alpha$ defines a functor $\partial(\overrightarrow{\mathcal{C}}\downarrow\beta)\to\mathrm{F}^{k-1}\partial(\overrightarrow{\mathcal{C}}\downarrow\alpha)$ which, defines the map h on $\mathrm{L}_\beta\boldsymbol{Y}$. Thus, we have the commutative diagram

$$\begin{array}{ccc} \boldsymbol{X}_\beta \amalg_{\mathrm{L}_\beta\boldsymbol{X}} \mathrm{L}_\beta\boldsymbol{Y} & \longrightarrow & E \\ \downarrow & & \downarrow \\ \boldsymbol{Y}_\beta & \longrightarrow & B \end{array}$$

and the vertical map on the left is assumed to have the left lifting property with respect to $E\to B$. This implies that the map h can be defined on $\boldsymbol{Y}_\beta$, and the discussion in Section 15.2.11 explains why this can be done independently for the various objects of degree k. This completes the induction, and the proof. □

LEMMA 15.3.10. *Let $\mathcal{C}$ be a Reedy category, let $\mathcal{M}$ be a model category, let $f\colon \boldsymbol{X} \to \boldsymbol{Y}$ be a map of $\mathcal{C}$-diagrams in $\mathcal{M}$, and let S be a class of maps in $\mathcal{M}$.*

(1) *If for every object α of $\mathcal{C}$ the relative latching map*

$$\boldsymbol{X}_\alpha \amalg_{\mathrm{L}_\alpha \boldsymbol{X}} \mathrm{L}_\alpha \boldsymbol{Y} \to \boldsymbol{Y}_\alpha$$

has the left lifting property with respect to every element of S, then for every object α of $\mathcal{C}$ the map $f_\alpha\colon \boldsymbol{X}_\alpha \to \boldsymbol{Y}_\alpha$ has the left lifting property with respect to every element of S.

(2) *If for every object α of $\mathcal{C}$ the relative matching map*

$$\boldsymbol{X}_\alpha \to \boldsymbol{Y}_\alpha \times_{\mathrm{M}_\alpha \boldsymbol{Y}} \mathrm{M}_\alpha \boldsymbol{X}$$

has the right lifting property with respect to every element of S, then for every object α of $\mathcal{C}$ the map $f_\alpha\colon \boldsymbol{X}_\alpha \to \boldsymbol{Y}_\alpha$ has the right lifting property with respect to every element of S.

PROOF. We will prove part 1; the proof of part 2 is dual.

The map $f_\alpha\colon \boldsymbol{X}_\alpha \to \boldsymbol{Y}_\alpha$ equals the composition $\boldsymbol{X}_\alpha \to \boldsymbol{X}_\alpha \amalg_{\mathrm{L}_\alpha \boldsymbol{X}} \mathrm{L}_\alpha \boldsymbol{Y} \to \boldsymbol{Y}_\alpha$. Since the first of these maps is the pushout of $\mathrm{L}_\alpha \boldsymbol{X} \to \mathrm{L}_\alpha \boldsymbol{Y}$ along $\mathrm{L}_\alpha \boldsymbol{X} \to \boldsymbol{X}_\alpha$, the result follows from Lemma 15.3.9 and Lemma 7.2.11. □

PROPOSITION 15.3.11. *Let $\mathcal{C}$ be a Reedy category, let $\mathcal{M}$ be a model category, and let $f\colon \boldsymbol{X} \to \boldsymbol{Y}$ be a map of $\mathcal{C}$-diagrams in $\mathcal{M}$.*

(1) *If f is a Reedy cofibration, then for every object α of $\mathcal{C}$ both the map $f_\alpha\colon \boldsymbol{X}_\alpha \to \boldsymbol{Y}_\alpha$ and the induced map of latching objects $\mathrm{L}_\alpha \boldsymbol{X} \to \mathrm{L}_\alpha \boldsymbol{Y}$ are cofibrations in $\mathcal{M}$.*

(2) *If f is a Reedy fibration, then for every object α of $\mathcal{C}$ both the map $f_\alpha\colon \boldsymbol{X}_\alpha \to \boldsymbol{Y}_\alpha$ and the induced map of matching objects $\mathrm{M}_\alpha \boldsymbol{X} \to \mathrm{M}_\alpha \boldsymbol{Y}$ are fibrations in $\mathcal{M}$.*

PROOF. This follows from Lemma 15.3.9, Lemma 15.3.10, and Proposition 7.2.3. □

COROLLARY 15.3.12. *Let $\mathcal{C}$ be a Reedy category, let $\mathcal{M}$ be a model category, and let $\boldsymbol{X}$ be a $\mathcal{C}$-diagram in $\mathcal{M}$.*

(1) *If $\boldsymbol{X}$ is Reedy cofibrant, then for every object α of $\mathcal{C}$ both the object $\boldsymbol{X}_\alpha$ and the latching object $\mathrm{L}_\alpha \boldsymbol{X}$ are cofibrant objects of $\mathcal{M}$.*

(2) *If $\boldsymbol{X}$ is Reedy fibrant, then for every object α of $\mathcal{C}$ both the object $\boldsymbol{X}_\alpha$ and the matching object $\mathrm{M}_\alpha \boldsymbol{X}$ are fibrant objects of $\mathcal{M}$.*

PROOF. This follows from Proposition 15.3.11. □

PROPOSITION 15.3.13. *Let $\mathcal{C}$ be a Reedy category, let $\mathcal{M}$ be a model category, and let $f\colon \boldsymbol{X} \to \boldsymbol{Y}$ be a map of $\mathcal{C}$-diagrams in $\mathcal{M}$.*

(1) *If for every object α of $\mathcal{C}$ the relative latching map $\boldsymbol{X}_\alpha \amalg_{\mathrm{L}_\alpha \boldsymbol{X}} \mathrm{L}_\alpha \boldsymbol{Y} \to \boldsymbol{Y}_\alpha$ is a trivial cofibration, then for every object α of $\mathcal{C}$ both the map $f_\alpha\colon \boldsymbol{X}_\alpha \to \boldsymbol{Y}_\alpha$ and the induced map of latching objects $\mathrm{L}_\alpha \boldsymbol{X} \to \mathrm{L}_\alpha \boldsymbol{Y}$ are trivial cofibrations.*

(2) *If for every object α of $\mathcal{C}$ the relative matching map $\boldsymbol{X}_\alpha \to \boldsymbol{Y}_\alpha \times_{\mathrm{M}_\alpha \boldsymbol{Y}} \mathrm{M}_\alpha \boldsymbol{X}$ is a trivial fibration, then for every object α of $\mathcal{C}$ both the map $f_\alpha\colon \boldsymbol{X}_\alpha \to \boldsymbol{Y}_\alpha$ and the induced map of matching objects $\mathrm{M}_\alpha \boldsymbol{X} \to \mathrm{M}_\alpha \boldsymbol{Y}$ are trivial fibrations.*

PROOF. This follows from Lemma 15.3.9, Lemma 15.3.10, and Proposition 7.2.3. □

PROPOSITION 15.3.14. *Let $\mathcal{C}$ be a Reedy category, let $\mathcal{M}$ be a model category, and let $f\colon \boldsymbol{X} \to \boldsymbol{Y}$ be a map of $\mathcal{C}$-diagrams in $\mathcal{M}$.*

(1) *If f is both a Reedy cofibration and a Reedy weak equivalence, then for every object α of $\mathcal{C}$ the map $f_\alpha\colon \boldsymbol{X}_\alpha \to \boldsymbol{Y}_\alpha$, the induced map of latching objects $\mathrm{L}_\alpha \boldsymbol{X} \to \mathrm{L}_\alpha \boldsymbol{Y}$, and the relative latching map $\boldsymbol{X}_\alpha \amalg_{\mathrm{L}_\alpha \boldsymbol{X}} \mathrm{L}_\alpha \boldsymbol{Y} \to \boldsymbol{Y}_\alpha$ are trivial cofibrations.*
(2) *If f is both a Reedy fibration and a Reedy weak equivalence, then for every object α of $\mathcal{C}$ the map $f_\alpha\colon \boldsymbol{X}_\alpha \to \boldsymbol{Y}_\alpha$, the induced map of matching objects $\mathrm{M}_\alpha \boldsymbol{X} \to \mathrm{M}_\alpha \boldsymbol{Y}$, and the relative matching map $\boldsymbol{X}_\alpha \to \boldsymbol{Y}_\alpha \times_{\mathrm{M}_\alpha \boldsymbol{Y}} \mathrm{M}_\alpha \boldsymbol{X}$ are trivial fibrations.*

PROOF. We will prove part 1; the proof of part 2 is dual. We assume that we have chosen a degree function for $\mathcal{C}$ (see Remark 15.1.3).

Proposition 15.3.11 implies that f_α is a cofibration for every object α in $\mathcal{C}$. Since f is a Reedy weak equivalence, this implies that f_α is a trivial cofibration for every object α of $\mathcal{C}$.

We will prove that the maps $\mathrm{L}_\alpha \boldsymbol{X} \to \mathrm{L}_\alpha \boldsymbol{Y}$ and $\boldsymbol{X}_\alpha \amalg_{\mathrm{L}_\alpha \boldsymbol{X}} \mathrm{L}_\alpha \boldsymbol{Y} \to \boldsymbol{Y}_\alpha$ are trivial cofibrations for every object α of $\mathcal{C}$ by induction on the degree of α. If $\mathrm{L}_\alpha \boldsymbol{X} \to \mathrm{L}_\alpha \boldsymbol{Y}$ is a trivial cofibration in $\mathcal{M}$ for some particular object α of $\mathcal{C}$, then, since $\boldsymbol{X}_\alpha \to \boldsymbol{X}_\alpha \amalg_{\mathrm{L}_\alpha \boldsymbol{X}} \mathrm{L}_\alpha \boldsymbol{Y}$ is a pushout of $\mathrm{L}_\alpha \boldsymbol{X} \to \mathrm{L}_\alpha \boldsymbol{Y}$, this map is also a trivial cofibration. Since the weak equivalence $f_\alpha\colon \boldsymbol{X}_\alpha \to \boldsymbol{Y}_\alpha$ equals the composition $\boldsymbol{X}_\alpha \to \boldsymbol{X}_\alpha \amalg_{\mathrm{L}_\alpha \boldsymbol{X}} \mathrm{L}_\alpha \boldsymbol{Y} \to \boldsymbol{Y}_\alpha$, this implies that the cofibration $\boldsymbol{X}_\alpha \amalg_{\mathrm{L}_\alpha \boldsymbol{X}} \mathrm{L}_\alpha \boldsymbol{Y} \to \boldsymbol{Y}_\alpha$ is actually a trivial cofibration.

If α is of degree 0, then $\mathrm{L}_\alpha \boldsymbol{X}$ and $\mathrm{L}_\alpha \boldsymbol{Y}$ are both the initial object of $\mathcal{M}$, and so $\mathrm{L}_\alpha \boldsymbol{X} \to \mathrm{L}_\alpha \boldsymbol{Y}$ is the identity map, which is certainly a trivial cofibration.

We now assume that n is a positive integer, $\mathrm{L}_\beta \boldsymbol{X} \to \mathrm{L}_\beta \boldsymbol{Y}$ is a trivial cofibration for all objects β of degree less than n, and α is an object of degree n. The discussion above explains why our inductive hypothesis implies that $\boldsymbol{X}_\beta \amalg_{\mathrm{L}_\beta \boldsymbol{X}} \mathrm{L}_\beta \boldsymbol{Y} \to \boldsymbol{Y}_\beta$ is a trivial cofibration for all objects β of degree less than n, and so Lemma 15.3.9 and Proposition 7.2.3 imply that $\mathrm{L}_\alpha \boldsymbol{X} \to \mathrm{L}_\alpha \boldsymbol{Y}$ is a trivial cofibration. □

THEOREM 15.3.15. *Let $\mathcal{C}$ be a Reedy category, let $\mathcal{M}$ be a model category, and let $f\colon \boldsymbol{X} \to \boldsymbol{Y}$ be a map of $\mathcal{C}$-diagrams in $\mathcal{M}$.*

(1) *The map f is both a Reedy cofibration and a Reedy weak equivalence if and only if for every object α of $\mathcal{C}$ the relative latching map $\boldsymbol{X}_\alpha \amalg_{\mathrm{L}_\alpha \boldsymbol{X}} \mathrm{L}_\alpha \boldsymbol{Y} \to \boldsymbol{Y}_\alpha$ is a trivial cofibration in $\mathcal{M}$.*
(2) *The map f is both a Reedy fibration and a Reedy weak equivalence if and only if for every object α of $\mathcal{C}$ the relative matching map $\boldsymbol{X}_\alpha \to \boldsymbol{Y}_\alpha \times_{\mathrm{M}_\alpha \boldsymbol{Y}} \mathrm{M}_\alpha \boldsymbol{X}$ is a trivial fibration in $\mathcal{M}$.*

PROOF. This follows from Proposition 15.3.13 and Proposition 15.3.14. □

15.3.16. Proof of Theorem 15.3.4. For part 1, we must show that axioms M1 through M5 of Definition 7.1.3 are satisfied. Axioms M1 and M2 follow from the fact that limits, colimits, and weak equivalences of diagrams are all defined objectwise.

Axiom M3 follows from the observation that if the map $g\colon \boldsymbol{X} \to \boldsymbol{Y}$ is a retract of the map $h\colon \boldsymbol{W} \to \boldsymbol{Z}$, then for every object α of $\mathcal{C}$ the relative latching map $\boldsymbol{X}_\alpha \amalg_{\mathrm{L}_\alpha \boldsymbol{X}} \mathrm{L}_\alpha \boldsymbol{Y} \to \boldsymbol{Y}_\alpha$ is a retract of the relative latching map $\boldsymbol{W}_\alpha \amalg_{\mathrm{L}_\alpha \boldsymbol{W}} \mathrm{L}_\alpha \boldsymbol{Z} \to \boldsymbol{Z}_\alpha$ and the relative matching map $\boldsymbol{X}_\alpha \to \boldsymbol{Y}_\alpha \times_{\mathrm{M}_\alpha \boldsymbol{Y}} \mathrm{M}_\alpha \boldsymbol{X}$ is a retract of the relative matching map $\boldsymbol{W}_\alpha \to \boldsymbol{Z}_\alpha \times_{\mathrm{M}_\alpha \boldsymbol{Z}} \mathrm{M}_\alpha \boldsymbol{W}$.

If we choose a degree function for $\mathcal{C}$ (see Remark 15.1.3), then the maps required by axiom M4 are constructed inductively on the degree of the objects of $\mathcal{C}$, using Theorem 15.3.15 (see the discussion in Section 15.2.11).

The factorizations required by axiom M5 are also constructed inductively on the degree of the objects of $\mathcal{C}$. For axiom M5 part 1, if $g\colon \boldsymbol{X} \to \boldsymbol{Y}$ is a map in $\mathcal{M}^{\mathcal{C}}$, then, for every object α of degree zero of $\mathcal{C}$, we have a functorial factorization of g_α in $\mathcal{M}$ as $\boldsymbol{X}_\alpha \xrightarrow{i} \boldsymbol{Z}_\alpha \xrightarrow{h} \boldsymbol{Y}_\alpha$ with i a cofibration and h a trivial fibration. If we now assume that g has been factored on all objects of degree less than n and that α is an object of degree n, then we have an induced map $\boldsymbol{X}_\alpha \amalg_{\mathrm{L}_\alpha \boldsymbol{X}} \mathrm{L}_\alpha \boldsymbol{Z} \to \boldsymbol{Y}_\alpha \times_{\mathrm{M}_\alpha \boldsymbol{Y}} \mathrm{M}_\alpha \boldsymbol{Z}$. We can factor this map (functorially) in $\mathcal{M}$ as

$$\boldsymbol{X}_\alpha \amalg_{\mathrm{L}_\alpha \boldsymbol{X}} \mathrm{L}_\alpha \boldsymbol{Z} \xrightarrow{i} \boldsymbol{Z}_\alpha \xrightarrow{h} \boldsymbol{Y}_\alpha \times_{\mathrm{M}_\alpha \boldsymbol{Y}} \mathrm{M}_\alpha \boldsymbol{Z}$$

with i a cofibration and h a trivial fibration to obtain $\boldsymbol{Z}_\alpha$. This completes the construction, and Theorem 15.3.15 implies that it has the required properties. The proof for axiom M5 part 2 is similar, and so $\mathcal{M}^{\mathcal{C}}$ is a model category, and the proof of part 1 is complete.

For part 2, Proposition 15.3.11 implies that a Reedy cofibration is an objectwise cofibration and a Reedy fibration is an objectwise fibration. Since weak equivalences are defined objectwise and both pushouts and pullbacks are constructed objectwise, the conditions of Definition 13.1.1 follow if they hold in $\mathcal{M}$.

For part 3, if $\mathcal{M}$ is a simplicial model category, then axiom M6 of Definition 9.1.6 follows because the constructions are all done objectwise and $\mathcal{M}$ is a simplicial model category, and so it remains only to show that axiom M7 follows as well. Proposition 9.3.7 implies that it is sufficient to show that if $i\colon \boldsymbol{A} \to \boldsymbol{B}$ is a Reedy cofibration and $j\colon K \to L$ is a cofibration of simplicial sets, then $\boldsymbol{A} \otimes L \amalg_{\boldsymbol{A} \otimes K} \boldsymbol{B} \otimes K \to \boldsymbol{B} \otimes L$ is a Reedy cofibration that is also a weak equivalence if either i or j is a weak equivalence. Thus, we must show that for every object α of $\mathcal{C}$ the map

$$(\boldsymbol{A} \otimes L \amalg_{\boldsymbol{A} \otimes K} \boldsymbol{B} \otimes K)_\alpha \amalg_{\mathrm{L}_\alpha (\boldsymbol{A} \otimes L \amalg_{\boldsymbol{A} \otimes K} \boldsymbol{B} \otimes K)} \mathrm{L}_\alpha (\boldsymbol{B} \otimes L) \to (\boldsymbol{B} \otimes L)_\alpha$$

is a cofibration in $\mathcal{M}$ that is also a weak equivalence if either i or j is a weak equivalence. Since each latching object is a colimit, Lemma 9.2.1 implies that this map is isomorphic to the map

$$(\boldsymbol{B}_\alpha \otimes K) \amalg_{(\mathrm{L}_\alpha \boldsymbol{B} \amalg_{\mathrm{L}_\alpha \boldsymbol{A}} \boldsymbol{A}_\alpha) \otimes K} \big((\mathrm{L}_\alpha \boldsymbol{B} \amalg_{\mathrm{L}_\alpha \boldsymbol{A}} \boldsymbol{A}_\alpha) \otimes L\big) \to \boldsymbol{B}_\alpha \otimes L.$$

Since $i\colon A \to B$ is a Reedy cofibration and $\mathcal{M}$ is a simplicial model category, this map is a cofibration that is a weak equivalence if either i or j is a weak equivalence, and so the proof is complete.

15.4. Quillen functors

PROPOSITION 15.4.1. *Let $\mathcal{C}$ be a Reedy category and let $\mathcal{M}$ and $\mathcal{N}$ be model categories.*

1. *If $\mathrm{F}\colon \mathcal{M} \rightleftarrows \mathcal{N} :\mathrm{U}$ is a Quillen pair (see Definition 8.5.2), then the induced functors $\mathrm{F}^{\mathcal{C}}\colon \mathcal{M}^{\mathcal{C}} \rightleftarrows \mathcal{N}^{\mathcal{C}} :\mathrm{U}^{\mathcal{C}}$ form a Quillen pair.*
2. *If (F, U) is a pair of Quillen equivalences, then so is the induced pair $(\mathrm{F}^{\mathcal{C}}, \mathrm{U}^{\mathcal{C}})$.*

PROOF. The induced functors $\mathrm{F}^{\mathcal{C}}$ and $\mathrm{U}^{\mathcal{C}}$ are adjoint (see Lemma 11.6.4), and so for part 1 it is sufficient to show that $\mathrm{F}^{\mathcal{C}}$ preserves both cofibrations and trivial cofibrations (see Proposition 8.5.3). If $f\colon \boldsymbol{A} \to \boldsymbol{B}$ is a cofibration or a trivial cofibration in $\mathcal{M}^{\mathcal{C}}$, then for every object α of $\mathcal{C}$ the relative latching map $\mathrm{L}_\alpha \boldsymbol{B} \amalg_{\mathrm{L}_\alpha \boldsymbol{A}} \boldsymbol{A}_\alpha \to \boldsymbol{B}_\alpha$ is, respectively, a cofibration or a trivial cofibration in $\mathcal{M}$ (see Theorem 15.3.15). Since the latching objects $\mathrm{L}_\alpha \boldsymbol{A}$ and $\mathrm{L}_\alpha \boldsymbol{B}$ are defined as colimits (see Definition 15.2.5) and left adjoints commute with colimits, the relative latching map $\mathrm{L}_\alpha \mathrm{F}\boldsymbol{B} \amalg_{\mathrm{L}_\alpha \mathrm{F}\boldsymbol{A}} \mathrm{F}\boldsymbol{A}_\alpha \to \mathrm{F}\boldsymbol{B}_\alpha$ is isomorphic to the map $\mathrm{F}(\mathrm{L}_\alpha \boldsymbol{B} \amalg_{\mathrm{L}_\alpha \boldsymbol{A}} \boldsymbol{A}_\alpha) \to \mathrm{F}\boldsymbol{B}_\alpha$, and is thus, respectively, a cofibration or a trivial cofibration in $\mathcal{N}$. Thus, $\mathrm{F}^{\mathcal{C}}$ is a left Quillen functor. Part 2 follows immediately, since weak equivalences in $\mathcal{M}^{\mathcal{C}}$ and $\mathcal{N}^{\mathcal{C}}$ are defined objectwise in $\mathcal{C}$. □

COROLLARY 15.4.2. *Let $\mathcal{C}$ be a Reedy category, let $\mathcal{M}$ and $\mathcal{N}$ be model categories, and let $\mathrm{F}\colon \mathcal{M} \rightleftarrows \mathcal{N} :\mathrm{U}$ be a Quillen pair.*

1. *If $\boldsymbol{B}\colon \mathcal{C} \to \mathcal{M}$ is a cofibrant $\mathcal{C}$-diagram in $\mathcal{M}$, then $\mathrm{F}\boldsymbol{B}\colon \mathcal{C} \to \mathcal{N}$ is a cofibrant $\mathcal{C}$-diagram in $\mathcal{N}$.*
2. *If $\boldsymbol{X}\colon \mathcal{C} \to \mathcal{N}$ is a fibrant $\mathcal{C}$-diagram in $\mathcal{N}$, then $\mathrm{U}\boldsymbol{X}\colon \mathcal{C} \to \mathcal{M}$ is a fibrant $\mathcal{C}$-diagram in $\mathcal{M}$.*

PROOF. This follows from Proposition 15.4.1. □

15.5. Products of Reedy categories

In this section, we show that if $\mathcal{C}$ and $\mathcal{D}$ are Reedy categories and $\mathcal{M}$ is a model category, then the three possible Reedy model category structures on $\mathcal{M}^{\mathcal{C}\times\mathcal{D}}$ are all the same (see Theorem 15.5.2).

LEMMA 15.5.1. *Let $\mathcal{C}$ and $\mathcal{D}$ be Reedy categories, let $\mathcal{M}$ be a model category, and let $\boldsymbol{X}$ be a $\mathcal{C}\times\mathcal{D}$-diagram in $\mathcal{M}$.*

1. *For every object (α,β) of $\mathcal{C}\times\mathcal{D}$ (see Proposition 15.1.6) the latching object $\mathrm{L}^{\mathcal{C}\times\mathcal{D}}_{(\alpha,\beta)}\boldsymbol{X}$ is naturally isomorphic to the pushout $\mathrm{L}^{\mathcal{C}}_\alpha \boldsymbol{X}_{(-,\beta)} \amalg_{\mathrm{L}^{\mathcal{C}}_\alpha \mathrm{L}^{\mathcal{D}}_\beta \boldsymbol{X}} \mathrm{L}^{\mathcal{D}}_\beta \boldsymbol{X}_{(\alpha,-)}$.*
2. *For every object (α,β) of $\mathcal{C}\times\mathcal{D}$ the matching object $\mathrm{M}^{\mathcal{C}\times\mathcal{D}}_{(\alpha,\beta)}\boldsymbol{X}$ is naturally isomorphic to the pullback $\mathrm{M}^{\mathcal{C}}_\alpha \boldsymbol{X}_{(-,\beta)} \times_{\mathrm{M}^{\mathcal{C}}_\alpha \mathrm{M}^{\mathcal{D}}_\beta \boldsymbol{X}} \mathrm{M}^{\mathcal{D}}_\beta \boldsymbol{X}_{(\alpha,-)}$.*

PROOF. We will prove part 1; the proof of part 2 is similar.

We begin by defining a map $\mathrm{L}^{\mathcal{C}}_\alpha \boldsymbol{X}_{(-,\beta)} \to \mathrm{L}^{\mathcal{C}\times\mathcal{D}}_{(\alpha,\beta)}\boldsymbol{X}$. We define an embedding of $\partial(\overrightarrow{\mathcal{C}}\downarrow\alpha)$ in $\partial(\overrightarrow{\mathcal{C}\times\mathcal{D}}\downarrow(\alpha,\beta))$ by taking the object $f\colon \alpha' \to \alpha$ of $\partial(\overrightarrow{\mathcal{C}}\downarrow\alpha)$ to $(f, 1_\beta)$ in $\partial(\overrightarrow{\mathcal{C}\times\mathcal{D}}\downarrow(\alpha,\beta))$. This defines a map $\mathrm{L}^{\mathcal{C}}_\alpha \boldsymbol{X}_{(-,\beta)} \to \mathrm{L}^{\mathcal{C}\times\mathcal{D}}_{(\alpha,\beta)}\boldsymbol{X}$. Similarly, we have an embedding of $\partial(\overrightarrow{\mathcal{D}}\downarrow\beta)$ in $\partial(\overrightarrow{\mathcal{C}\times\mathcal{D}}\downarrow(\alpha,\beta))$ that defines a map $\mathrm{L}^{\mathcal{D}}_\beta \boldsymbol{X}_{(\alpha,-)} \to$

$\mathrm{L}^{\mathcal{C}\times\mathcal{D}}_{(\alpha,\beta)}\boldsymbol{X}$. We have natural isomorphisms

$$\begin{aligned}
\mathrm{L}^{\mathcal{C}}_{\alpha}\mathrm{L}^{\mathcal{D}}_{\beta}\boldsymbol{X} &= \operatorname*{colim}_{(\alpha'\to\alpha)\in\mathrm{Ob}\,\partial(\overrightarrow{\mathcal{C}}\downarrow\alpha)} \big(\mathrm{L}^{\mathcal{D}}_{\beta}\boldsymbol{X}_{(\alpha',-)}\big)\\
&\approx \operatorname*{colim}_{(\alpha'\to\alpha)\in\mathrm{Ob}\,\partial(\overrightarrow{\mathcal{C}}\downarrow\alpha)} \Big(\operatorname*{colim}_{(\beta'\to\beta)\in\mathrm{Ob}\,\partial(\overrightarrow{\mathcal{D}}\downarrow\beta)} \boldsymbol{X}_{(\alpha',\beta')}\Big)\\
&\approx \operatorname*{colim}_{(\alpha'\to\alpha,\beta'\to\beta)\in\mathrm{Ob}(\partial(\overrightarrow{\mathcal{C}}\downarrow\alpha)\times\partial(\overrightarrow{\mathcal{D}}\downarrow\beta))} \boldsymbol{X}_{(\alpha'\beta')}
\end{aligned}$$

and natural maps $\mathrm{L}^{\mathcal{C}}_{\alpha}\boldsymbol{X}_{(-,\beta)} \leftarrow \mathrm{L}^{\mathcal{C}}_{\alpha}\mathrm{L}^{\mathcal{D}}_{\beta}\boldsymbol{X} \to \mathrm{L}^{\mathcal{D}}_{\beta}\boldsymbol{X}_{(\alpha,-)}$ such that the composition $\mathrm{L}^{\mathcal{C}}_{\alpha}\mathrm{L}^{\mathcal{D}}_{\beta}\boldsymbol{X} \to \mathrm{L}^{\mathcal{C}}_{\alpha}\boldsymbol{X}_{(-,\beta)} \to \mathrm{L}^{\mathcal{C}\times\mathcal{D}}_{(\alpha,\beta)}$ equals the composition $\mathrm{L}^{\mathcal{C}}_{\alpha}\mathrm{L}^{\mathcal{D}}_{\beta}\boldsymbol{X} \to \mathrm{L}^{\mathcal{D}}_{\beta}\boldsymbol{X}_{(\alpha,-)} \to \mathrm{L}^{\mathcal{C}\times\mathcal{D}}_{(\alpha,\beta)}$. Thus, we have a well defined map $\mathrm{L}^{\mathcal{C}}_{\alpha}\boldsymbol{X}_{(-,\beta)} \amalg_{\mathrm{L}^{\mathcal{C}}_{\alpha}\mathrm{L}^{\mathcal{D}}_{\beta}\boldsymbol{X}} \mathrm{L}^{\mathcal{D}}_{\beta}\boldsymbol{X}_{(\alpha,-)} \to \mathrm{L}^{\mathcal{C}\times\mathcal{D}}_{(\alpha,\beta)}\boldsymbol{X}$.

The latching object $\mathrm{L}^{\mathcal{C}\times\mathcal{D}}_{(\alpha,\beta)}\boldsymbol{X}$ is a colimit indexed by pairs $(f,g)\in\mathrm{Ob}\big(\partial(\overrightarrow{\mathcal{C}}\downarrow\alpha)\times\partial(\overrightarrow{\mathcal{D}}\downarrow\beta)\big)$ in which at least one of f and g is not an identity map. The pushout $\mathrm{L}^{\mathcal{C}}_{\alpha}\boldsymbol{X}_{(-,\beta)} \amalg_{\mathrm{L}^{\mathcal{C}}_{\alpha}\mathrm{L}^{\mathcal{D}}_{\beta}\boldsymbol{X}} \mathrm{L}^{\mathcal{D}}_{\beta}\boldsymbol{X}_{(\alpha,-)}$ is the same coproduct with the indexing category partitioned into three subcategories according to whether $f\neq 1_\alpha$ and $g=1_\beta$, $f\neq 1_\alpha$ and $g\neq 1_\beta$, or $f=1_\alpha$ and $g\neq 1_\beta$, and so our map is an isomorphism. □

THEOREM 15.5.2. *If $\mathcal{C}$ and $\mathcal{D}$ are Reedy categories and $\mathcal{M}$ is a model category, then the category $\mathcal{M}^{\mathcal{C}\times\mathcal{D}}$ of $(\mathcal{C}\times\mathcal{D})$-diagrams in $\mathcal{M}$ has the same model category structure when viewed as either*

1. *diagrams in $\mathcal{M}$ indexed by the Reedy category $(\mathcal{C}\times\mathcal{D})$ (see Proposition 15.1.6),*
2. *the category $(\mathcal{M}^{\mathcal{D}})^{\mathcal{C}}$, i.e., diagrams in $\mathcal{M}^{\mathcal{D}}$ indexed by the Reedy category $\mathcal{C}$, or*
3. *the category $(\mathcal{M}^{\mathcal{C}})^{\mathcal{D}}$, i.e., diagrams in $\mathcal{M}^{\mathcal{C}}$ indexed by the Reedy category $\mathcal{D}$.*

PROOF. We will prove that the model category structure of 1 equals that of 2; the proof that the model category structure of 1 equals that of 3 is similar.

Since the weak equivalences of both $\mathcal{M}^{\mathcal{C}\times\mathcal{D}}$ and $(\mathcal{M}^{\mathcal{D}})^{\mathcal{C}}$ are defined objectwise, these two model categories have he same weak equivalences. Thus, Proposition 7.2.3 implies that it is sufficient to show that they have the same cofibrations.

A map $f\colon \boldsymbol{X}\to\boldsymbol{Y}$ is a cofibration in $(\mathcal{M}^{\mathcal{D}})^{\mathcal{C}}$ if and only if, for every object α of $\mathcal{C}$, the relative latching map $\boldsymbol{X}_\alpha \amalg_{\mathrm{L}^{\mathcal{C}}_{\alpha}\boldsymbol{X}} \mathrm{L}^{\mathcal{C}}_{\alpha}\boldsymbol{Y} \to \boldsymbol{Y}_\alpha$ is a cofibration in $\mathcal{M}^{\mathcal{D}}$. This is the case if and only if, for every object β of $\mathcal{D}$, the relative latching map

$$(\boldsymbol{X}_\alpha \amalg_{\mathrm{L}^{\mathcal{C}}_{\alpha}\boldsymbol{X}} \mathrm{L}^{\mathcal{C}}_{\alpha}\boldsymbol{Y})_\beta \amalg_{\mathrm{L}^{\mathcal{D}}_{\beta}(\boldsymbol{X}_\alpha \amalg_{\mathrm{L}^{\mathcal{C}}_{\alpha}\boldsymbol{X}} \mathrm{L}^{\mathcal{C}}_{\alpha}\boldsymbol{Y})} \mathrm{L}^{\mathcal{D}}_{\beta}\boldsymbol{Y}_\alpha \to \boldsymbol{Y}_{(\alpha,\beta)} \tag{15.5.3}$$

is a cofibration in $\mathcal{M}$. Since colimits commute, the domain of this map is isomorphic to

$$(\boldsymbol{X}_{(\alpha,\beta)} \amalg_{\mathrm{L}^{\mathcal{C}}_{\alpha}\boldsymbol{X}_\beta} \mathrm{L}^{\mathcal{C}}_{\alpha}\boldsymbol{Y}_\beta) \amalg_{(\mathrm{L}^{\mathcal{D}}_{\beta}\boldsymbol{X}_\alpha \amalg_{\mathrm{L}^{\mathcal{D}}_{\beta}\mathrm{L}^{\mathcal{C}}_{\alpha}\boldsymbol{X}} \mathrm{L}^{\mathcal{D}}_{\beta}\mathrm{L}^{\mathcal{C}}_{\alpha}\boldsymbol{Y})} \mathrm{L}^{\mathcal{D}}_{\beta}\boldsymbol{Y}_\alpha\ ,$$

which is the colimit of the diagram

$$\begin{array}{ccccc}
\boldsymbol{X}_{(\alpha,\beta)} & \longleftarrow & \mathrm{L}_\alpha^{\mathcal{C}}\boldsymbol{X}_\beta & \longrightarrow & \mathrm{L}_\alpha^{\mathcal{C}}\boldsymbol{Y}_\beta \\
\uparrow & & \uparrow & & \uparrow \\
\mathrm{L}_\beta^{\mathcal{D}} & \longleftarrow & \mathrm{L}_\beta^{\mathcal{D}}\mathrm{L}_\alpha^{\mathcal{C}}\boldsymbol{X} & \longrightarrow & \mathrm{L}_\beta^{\mathcal{D}}\mathrm{L}_\alpha^{\mathcal{C}}\boldsymbol{Y} \\
& \searrow & \downarrow & \swarrow & \\
& & \mathrm{L}_\beta^{\mathcal{D}}\boldsymbol{Y}_\alpha & & .
\end{array}$$

This is isomorphic to the pushout

$$\boldsymbol{X}_{(\alpha,\beta)} \amalg_{(\mathrm{L}_\beta^{\mathcal{D}}\boldsymbol{X}_\alpha \amalg_{\mathrm{L}_\beta^{\mathcal{D}}\mathrm{L}_\alpha^{\mathcal{C}}\boldsymbol{X}} \mathrm{L}_\alpha^{\mathcal{C}}\boldsymbol{X}_\beta)} (\mathrm{L}_\beta^{\mathcal{D}}\boldsymbol{Y}_\alpha \amalg_{\mathrm{L}_\beta^{\mathcal{D}}\mathrm{L}_\alpha^{\mathcal{C}}\boldsymbol{Y}} \mathrm{L}_\alpha^{\mathcal{C}}\boldsymbol{Y}_\beta) \ ,$$

and so Lemma 15.5.1 implies that the map (15.5.3) is isomorphic to the map $\boldsymbol{X}_{(\alpha,\beta)} \amalg_{\mathrm{L}_{(\alpha,\beta)}^{\mathcal{C}\times\mathcal{D}}\boldsymbol{X}} \mathrm{L}_{(\alpha,\beta)}^{\mathcal{C}\times\mathcal{D}}\boldsymbol{Y} \to \boldsymbol{Y}_{(\alpha,\beta)}$, which is the relative latching map of f at (α,β) in $\mathcal{M}^{\mathcal{C}\times\mathcal{D}}$. Thus, the class of cofibrations of $(\mathcal{M}^{\mathcal{D}})^{\mathcal{C}}$ equals that of $\mathcal{M}^{\mathcal{C}\times\mathcal{D}}$. □

15.6. Reedy diagrams in a cofibrantly generated model category

If $\mathcal{C}$ is a Reedy category and $\mathcal{M}$ is a cofibrantly generated model category (see Definition 11.1.2), then we have two model category structures on $\mathcal{M}^{\mathcal{C}}$, the category of $\mathcal{C}$-diagrams in $\mathcal{M}$: The first is constructed using the cofibrantly generated model category structure on $\mathcal{M}$ (see Theorem 11.6.1), and the second is constructed using the Reedy category structure on $\mathcal{C}$ (see Theorem 15.3.4). Although these two model category structures have the same class of weak equivalences, they are not, in general, equal (see Example 15.6.2).

We begin by showing (in Section 15.6.1) that although these two model category structures are not, in general, equal, they are always Quillen equivalent (see Theorem 15.6.4). In Section 15.6.22 we will show that the Reedy model category structure on $\mathcal{M}^{\mathcal{C}}$ is nearly always cofibrantly generated (see Theorem 15.6.27).

15.6.1. Two model category structures. We begin with an example that shows that if $\mathcal{M}$ is a cofibrantly generated model category and $\mathcal{C}$ is a Reedy category, then the two model category structures on $\mathcal{M}^{\mathcal{C}}$ (see Theorem 11.6.1 and Theorem 15.3.4) are not, in general, equal. In Theorem 15.6.4 we will show that these two model category structures are always Quillen equivalent.

EXAMPLE 15.6.2. Let $\mathcal{C}$ be $\boldsymbol{\Delta}^{\mathrm{op}}$, the simplicial indexing category (see Example 15.1.13), and let $\mathcal{M}$ be the standard model category of simplicial sets. If $\boldsymbol{X}$ is a simplicial object in $\mathcal{M}$, then $\boldsymbol{X}$ is fibrant in the cofibrantly generated model category structure on $\mathcal{M}^{\boldsymbol{\Delta}^{\mathrm{op}}}$ whenever $\boldsymbol{X}_n$ is a fibrant simplicial set for all $n \geq 0$. However, for $\boldsymbol{X}$ to be fibrant in the Reedy structure the map $\boldsymbol{X}_1 \xrightarrow{d_0 \times d_1} \boldsymbol{X}_0 \times \boldsymbol{X}_0$ must be a fibration, which is a strictly stronger requirement. (For example, let Z be a nontrivial fibrant simplicial set and let $\boldsymbol{X}$ be the constant simplicial object at Z. The map $\boldsymbol{X}_1 \xrightarrow{d_0 \times d_1} \boldsymbol{X}_0 \times \boldsymbol{X}_0$ is then the diagonal map $Z \to Z \times Z$, which is not, in general, a fibration.)

PROPOSITION 15.6.3. *Let $\mathcal{C}$ be a Reedy category, let $\mathcal{M}$ be a cofibrantly generated model category (see Definition 11.1.2), and let $\boldsymbol{X}$ and $\boldsymbol{Y}$ be $\mathcal{C}$-diagrams in $\mathcal{M}$.*

1. *If the map $f\colon \boldsymbol{X} \to \boldsymbol{Y}$ is a Reedy fibration (see Definition 15.3.3) then it is also a fibration in the cofibrantly generated model category structure on $\mathcal{M}^{\mathcal{C}}$ (see Theorem 11.6.1).*
2. *If the map $f\colon \boldsymbol{X} \to \boldsymbol{Y}$ is a cofibration in the cofibrantly generated model category structure on $\mathcal{M}^{\mathcal{C}}$ then it is a Reedy cofibration.*

PROOF. Part 1 follows from Proposition 15.3.11.

Part 2 follows from part 1 and Proposition 7.2.3, since the weak equivalences are the same in both model category structures. □

THEOREM 15.6.4. *If $\mathcal{C}$ is a Reedy category and $\mathcal{M}$ is a cofibrantly generated model category, then the identity functor of $\mathcal{M}^{\mathcal{C}}$ is a left Quillen equivalence (see Definition 8.5.20) from the cofibrantly generated model category structure (see Theorem 11.6.1) to the Reedy model category structure (see Theorem 15.3.4) and a right Quillen equivalence in the opposite direction.*

PROOF. This follows from Proposition 15.6.3. □

COROLLARY 15.6.5. *If $\mathcal{C}$ is a Reedy category, $\mathcal{M}$ is a cofibrantly generated model category, $\boldsymbol{X}$ and $\boldsymbol{Y}$ are $\mathcal{C}$-diagrams in $\mathcal{M}$, and $f\colon \boldsymbol{X} \to \boldsymbol{Y}$ is a relative free cell complex (see Definition 11.5.35), then f is a Reedy cofibration.*

PROOF. This follows from Theorem 11.6.1 and Proposition 15.6.3. □

COROLLARY 15.6.6. *Let $\mathcal{C}$ be a Reedy category, let $\mathcal{M}$ be a cofibrantly generated model category, and let $\boldsymbol{X}$ be a $\mathcal{C}$-diagram in $\mathcal{M}$. If $\boldsymbol{X}$ is a free cell complex (see Definition 11.5.35) then $\boldsymbol{X}$ is Reedy cofibrant.*

PROOF. This follows from Corollary 15.6.5. □

COROLLARY 15.6.7. *If $\mathcal{C}$ is a Reedy category then the $\mathcal{C}^{\mathrm{op}}$-diagram of simplicial sets $\mathrm{B}(-\downarrow \mathcal{C})^{\mathrm{op}}$ and the $\mathcal{C}$-diagram of simplicial sets $\mathrm{B}(\mathcal{C}\downarrow -)$ (see Section 14.7) are Reedy cofibrant diagrams.*

PROOF. This follows from Corollary 14.8.8 and Corollary 15.6.6. □

COROLLARY 15.6.8. *Let $\mathcal{C}$ be a Reedy category.*

1. *The $\mathcal{C}^{\mathrm{op}}$-diagram of simplicial sets $\mathrm{B}(-\downarrow \mathcal{C})^{\mathrm{op}}$ is a Reedy cofibrant approximation to the constant $\mathcal{C}^{\mathrm{op}}$-diagram at a point.*
2. *The $\mathcal{C}$-diagram of simplicial sets $\mathrm{B}(\mathcal{C}\downarrow -)$ is a Reedy cofibrant approximation to the constant $\mathcal{C}$-diagram at a point.*

PROOF. Corollary 15.6.7 implies that these diagrams are Reedy cofibrant, and Lemma 14.7.4 and Lemma 14.7.10 imply that for every object α of $\mathcal{C}$ the maps from $\mathrm{B}(\alpha\downarrow \mathcal{C})^{\mathrm{op}}$ and $\mathrm{B}(\mathcal{C}\downarrow \alpha)$ to a point are weak equivalences. □

15.6.9. An adjoint to the matching object. If $\mathcal{C}$ is a Reedy category, α is an object of $\mathcal{C}$, and $\mathcal{M}$ is a category, then we construct in this section a left adjoint to the matching object functor $\mathrm{M}_\alpha\colon \mathcal{M}^{\mathcal{C}} \to \mathcal{M}$ (see Proposition 15.6.20). This will be used in Section 15.6.22 to show that the Reedy model category structure is cofibrantly generated if $\mathcal{M}$ is almost any cofibrantly generated model category (see Theorem 15.6.27).

DEFINITION 15.6.10. If $\mathcal{C}$ is a Reedy category and α and β are objects of $\mathcal{C}$ then the *boundary* $\partial\mathcal{C}(\alpha,\beta)$ of $\mathcal{C}(\alpha,\beta)$ is the set of maps $g\colon \alpha \to \beta$ for which there is a factorization $g = \overrightarrow{g}\overleftarrow{g}$ with $\overrightarrow{g} \in \overrightarrow{\mathcal{C}}$, $\overleftarrow{g} \in \overleftarrow{\mathcal{C}}$, and $\overleftarrow{g} \neq 1_\alpha$. That is, $\partial\mathcal{C}(\alpha,\beta)$ is the set of maps from α to β that factor through an object of degree less than that of α.

LEMMA 15.6.11. *Let $\mathcal{C}$ be a Reedy category. If α, β, and γ are objects of $\mathcal{C}$, $g \in \partial\mathcal{C}(\alpha,\beta)$, and $h\colon \beta \to \gamma$ is any map, then $hg \in \partial\mathcal{C}(\alpha,\gamma)$.*

PROOF. Let $g = \overrightarrow{g}\overleftarrow{g}$ with $\overrightarrow{g} \in \overrightarrow{\mathcal{C}}$, $\overleftarrow{g} \in \overleftarrow{\mathcal{C}}$, and $\overleftarrow{g} \neq 1_\alpha$. The composition $(h\overrightarrow{g})$ has a factorization $(h\overrightarrow{g}) = \overrightarrow{k}\overleftarrow{k}$ with $\overrightarrow{k} \in \overrightarrow{\mathcal{C}}$ and $\overleftarrow{k} \in \overleftarrow{\mathcal{C}}$, and so the composition (hg) has the factorization $(hg) = \overrightarrow{k}(\overleftarrow{k}\overleftarrow{g})$ with $\overrightarrow{k} \in \overrightarrow{\mathcal{C}}$ and $(\overleftarrow{k}\overleftarrow{g}) \in \overleftarrow{\mathcal{C}}$, and $(\overleftarrow{k}\overleftarrow{g}) \neq 1_\alpha$ because $\overleftarrow{g} \neq 1_\alpha$. □

PROPOSITION 15.6.12. *If $\mathcal{C}$ is a Reedy category and α is an object of $\mathcal{C}$ then there is a sub-diagram of $\mathbf{F}^\alpha_*$, the free $\mathcal{C}$-diagram of sets generated at α (see Definition* 11.5.7*), that on an object β of $\mathcal{C}$ equals $\partial\mathcal{C}(\alpha,\beta)$.*

PROOF. This follows from Lemma 15.6.11. □

DEFINITION 15.6.13. If $\mathcal{C}$ is a Reedy category and α is an object of $\mathcal{C}$ then the $\mathcal{C}$-diagram of sets described in Proposition 15.6.12 will be called the *boundary* $\boldsymbol{\partial}\mathbf{F}^\alpha_*$ of $\mathbf{F}^\alpha_*$.

EXAMPLE 15.6.14. If $\mathcal{C} = \boldsymbol{\Delta}^{\mathrm{op}}$ (the simplicial indexing category (see Definition 15.1.8)) and $n \geq 0$, then $\mathbf{F}^{[n]}_*$ is the simplicial set $\Delta[n]$ (see Example 11.5.15) and $\boldsymbol{\partial}\mathbf{F}^{[n]}_*$ is what is commonly called $\partial\Delta[n]$.

PROPOSITION 15.6.15. *If $\mathcal{C}$ is a Reedy category and α is an object of $\mathcal{C}$ then $\boldsymbol{\partial}\mathbf{F}^\alpha_*$ (see Definition* 15.6.13*) is naturally isomorphic to the colimit of the $\partial(\alpha\downarrow\overleftarrow{\mathcal{C}})^{\mathrm{op}}$-diagram (see Definition* 15.2.3*) of $\mathcal{C}$-diagrams of sets that takes the object $g\colon \alpha \to \beta$ of $\partial(\alpha\downarrow\overleftarrow{\mathcal{C}})^{\mathrm{op}}$ to the diagram $\mathbf{F}^\beta_*$ (see Definition* 11.5.7*) and the morphism*

(15.6.16)

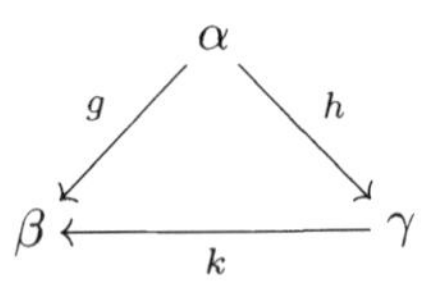

from $g\colon \alpha \to \beta$ to $h\colon \alpha \to \gamma$ in $\partial(\alpha\downarrow\overleftarrow{\mathcal{C}})^{\mathrm{op}}$ to the map of diagrams $k^*\colon \mathbf{F}^\beta_* \to \mathbf{F}^\gamma_*$ determined (see Proposition 11.5.8*) by the element $k\colon \gamma \to \beta$ of $\mathbf{F}^\gamma_*(\beta)$.*

PROOF. If $g\colon \alpha \to \beta$ is an object of $\partial(\alpha\downarrow\overleftarrow{\mathcal{C}})^{\mathrm{op}}$, then composition with g defines a map of diagrams $g^*\colon \mathbf{F}^\beta_* \to \mathbf{F}^\alpha_*$ whose image is contained in $\boldsymbol{\partial}\mathbf{F}^\alpha_*$ (see Lemma 15.6.11). For each morphism (15.6.16) of $\partial(\alpha\downarrow\overleftarrow{\mathcal{C}})^{\mathrm{op}}$, the map $k^*\colon \mathbf{F}^\beta_* \to \mathbf{F}^\gamma_*$ is defined by composition with k. Since $kh = g$, we have $k^*h^* = g^*$, and so we have a well defined map of diagrams

$$\operatorname*{colim}_{\partial(\alpha\downarrow\overleftarrow{\mathcal{C}})^{\mathrm{op}}} \mathbf{F}^{-}_* \to \boldsymbol{\partial}\mathbf{F}^\alpha_* \,. \tag{15.6.17}$$

The map (15.6.17) is surjective because if $g\colon \alpha \to \beta$ is an element of $\boldsymbol{\partial}\mathbf{F}^{\alpha}_{*}(\beta)$, then we can factor g as $\alpha \xrightarrow{h} \gamma \xrightarrow{k} \beta$ with $h \in \overleftarrow{\mathcal{C}}$, $k \in \overrightarrow{\mathcal{C}}$, and $h \neq 1_{\alpha}$, and g is in the image of k under $h^{*}\colon \mathbf{F}^{\gamma}_{*} \to \boldsymbol{\partial}\mathbf{F}^{\alpha}_{*}$.

To see that the map (15.6.17) is injective, suppose that $g\colon \alpha \to \beta$ and $h\colon \alpha \to \gamma$ are objects of $\partial(\alpha \downarrow \overleftarrow{\mathcal{C}})^{\mathrm{op}}$ and $s \in \mathbf{F}^{\beta}_{*}(\delta)$ and $t \in \mathbf{F}^{\gamma}_{*}(\delta)$ are elements such that $g^{*}(s) = h^{*}(t)$. We then have $sg = th$. If we factor $s\colon \beta \to \delta$ as $\beta \xrightarrow{\overleftarrow{s}} \eta \xrightarrow{\overrightarrow{s}} \delta$ with $\overleftarrow{s} \in \overleftarrow{\mathcal{C}}$ and $\overrightarrow{s} \in \overrightarrow{\mathcal{C}}$ and factor $t\colon \gamma \to \delta$ as $\gamma \xrightarrow{\overleftarrow{t}} \mu \xrightarrow{\overrightarrow{t}} \delta$ with $\overleftarrow{t} \in \overleftarrow{\mathcal{C}}$ and $\overrightarrow{t} \in \overrightarrow{\mathcal{C}}$, then $\overrightarrow{s}(\overleftarrow{s}g) = sg = th = \overrightarrow{t}(\overleftarrow{t}h)$, and so $(\overrightarrow{s})(\overleftarrow{s}g)$ and $(\overrightarrow{t})(\overleftarrow{t}h)$ are two factorizations of the same map into a map in $\overleftarrow{\mathcal{C}}$ followed by a map in $\overrightarrow{\mathcal{C}}$. By the uniqueness of such factorizations, we must have $\overrightarrow{s} = \overrightarrow{t}$ and $\overleftarrow{s}g = \overleftarrow{t}h$ (and, of course, $\eta = \mu$), and so we have the diagram

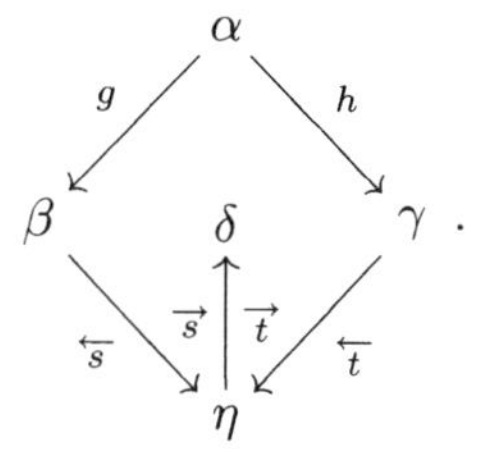

Thus $(\overleftarrow{t})^{*}\colon \mathbf{F}^{\eta}_{*} \to \mathbf{F}^{\gamma}_{*}$ takes $\overrightarrow{s} = \overrightarrow{t}$ to $\overrightarrow{t}\overleftarrow{t} = t$ and $(\overleftarrow{s})^{*}\colon \mathbf{F}^{\eta}_{*} \to \mathbf{F}^{\beta}_{*}$ takes $\overrightarrow{s} = \overrightarrow{t}$ to $\overrightarrow{s}\overleftarrow{s} = s$, and so s and t represent the same element of $\operatorname{colim}_{\partial(\alpha\downarrow\overleftarrow{\mathcal{C}})^{\mathrm{op}}} \mathbf{F}^{-}_{*}$, and the map (15.6.17) is injective. □

DEFINITION 15.6.18. If $\mathcal{C}$ is a Reedy category, $\mathcal{M}$ is a cocomplete category, X is an object of $\mathcal{M}$, and α is an object of $\mathcal{C}$, then the *boundary* $\boldsymbol{\partial}\mathbf{F}^{\alpha}_{X}$ of the free diagram on X generated at α is the $\mathcal{C}$-diagram $\boldsymbol{\partial}\mathbf{F}^{\alpha}_{X} = X \otimes \boldsymbol{\partial}\mathbf{F}^{\alpha}_{*}$ (see Definition 15.6.13, Definition 11.5.25, Definition 11.5.24, and Definition 11.5.7).

PROPOSITION 15.6.19. *Let $\mathcal{M}$ be a cocomplete category and let $\mathcal{C}$ be a Reedy category. If X is an object of $\mathcal{M}$ and α is an object of $\mathcal{C}$, then $\boldsymbol{\partial}\mathbf{F}^{\alpha}_{X}$ (see Definition 15.6.18) is naturally isomorphic to the colimit of the $\partial(\overrightarrow{\mathcal{C}} \downarrow \alpha)^{\mathrm{op}}$-diagram of $\mathcal{C}$-diagrams in $\mathcal{M}$ that takes the object $g\colon \alpha \to \beta$ of $\partial(\overrightarrow{\mathcal{C}} \downarrow \alpha)^{\mathrm{op}}$ to the diagram $\mathbf{F}^{\beta}_{X}$ (see Definition 11.5.25) and the morphism*

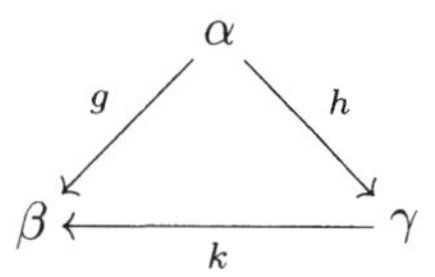

from $g\colon \alpha \to \beta$ to $h\colon \alpha \to \gamma$ in $\partial(\overrightarrow{\mathcal{C}} \downarrow \alpha)^{\mathrm{op}}$ to the map of diagrams $k^{*}\colon \mathbf{F}^{\beta}_{X} \to \mathbf{F}^{\gamma}_{X}$ determined (see Proposition 11.5.8) be the element $k\colon \gamma \to \beta$ of $\mathbf{F}^{\gamma}_{*}(\beta)$.

PROOF. This follows from Proposition 15.6.15. □

PROPOSITION 15.6.20. *Let $\mathcal{C}$ be a Reedy category and let $\mathcal{M}$ be a model category. For every object α of $\mathcal{C}$ the functor $\mathcal{M} \to \mathcal{M}^{\mathcal{C}}$ that takes the object X of $\mathcal{M}$ to the $\mathcal{C}$-diagram $\boldsymbol{\partial}\mathbf{F}^{\alpha}_{X}$ (see Definition 15.6.18) is left adjoint to the matching object* *functor* $\mathrm{M}_\alpha\colon \mathcal{M}^{\mathcal{C}} \to \mathcal{M}$ *(see Definition 15.2.5), i.e., for every* $\mathcal{C}$*-diagram* $\boldsymbol{Y}$ *there is a natural isomorphism of sets*

$$\mathcal{M}^{\mathcal{C}}(\boldsymbol{\partial}\mathbf{F}^\alpha_X, \boldsymbol{Y}) \approx \mathcal{M}(X, \mathrm{M}_\alpha \boldsymbol{Y}) \ .$$

PROOF. We have natural isomorphisms

$$\begin{aligned}
\mathcal{M}^{\mathcal{C}}(\boldsymbol{\partial}\mathbf{F}^\alpha_X, \boldsymbol{Y}) &= \mathcal{M}^{\mathcal{C}}\Big(\operatorname*{colim}_{(\alpha\to\beta)\in \mathrm{Ob}\,\partial(\overrightarrow{\mathcal{C}}\downarrow\alpha)^{\mathrm{op}}} \mathbf{F}^\beta_X, \boldsymbol{Y}\Big) \qquad \text{(see Proposition 15.6.19)}\\
&\approx \lim_{(\alpha\to\beta)\in \mathrm{Ob}\,\partial(\overrightarrow{\mathcal{C}}\downarrow\alpha)} \mathcal{M}^{\mathcal{C}}(\mathbf{F}^\beta_X, \boldsymbol{Y})\\
&\approx \lim_{(\alpha\to\beta)\in \mathrm{Ob}\,\partial(\overrightarrow{\mathcal{C}}\downarrow\alpha)} \mathcal{M}(X, \boldsymbol{Y}_\beta)\\
&\approx \mathcal{M}(X, \lim_{(\alpha\to\beta)\in \mathrm{Ob}\,\partial(\overrightarrow{\mathcal{C}}\downarrow\alpha)} \boldsymbol{Y}_\beta)\\
&= \mathcal{M}(X, \mathrm{M}_\alpha \boldsymbol{Y}) \ .
\end{aligned}$$

□

COROLLARY 15.6.21. *Let* $\mathcal{C}$ *be a Reedy category and let* $\mathcal{M}$ *be a model category. If* $A \to B$ *is a map in* $\mathcal{M}$ *and* $\boldsymbol{X} \to \boldsymbol{Y}$ *is a map of* $\mathcal{C}$*-diagrams in* $\mathcal{M}$*, then for every object* α *of* $\mathcal{C}$ *the following are equivalent:*

(1) *The dotted arrow exists in every solid arrow diagram of the form*

$$\begin{array}{ccc}
A & \longrightarrow & \boldsymbol{X}_\alpha \\
\downarrow & \nearrow & \downarrow \\
B & \longrightarrow & \boldsymbol{Y}_\alpha \times_{\mathrm{M}_\alpha \boldsymbol{Y}} \mathrm{M}_\alpha \boldsymbol{X} \ .
\end{array}$$

(2) *The dotted arrow exists in every solid arrow diagram of the form*

$$\begin{array}{ccc}
\mathbf{F}^\alpha_A \amalg_{\boldsymbol{\partial}\mathbf{F}^\alpha_A} \boldsymbol{\partial}\mathbf{F}^\alpha_B & \longrightarrow & \boldsymbol{X} \\
\downarrow & \nearrow & \downarrow \\
\mathbf{F}^\alpha_B & \longrightarrow & \boldsymbol{Y} \ .
\end{array}$$

PROOF. This follows from Proposition 15.6.20. □

15.6.22. Cofibrant generation of the Reedy model category structure. In this section, we show that if $\mathcal{C}$ is a Reedy category and $\mathcal{M}$ is a cofibrantly generated model category in which both the domains and the codomains of the elements of I are small relative to I and both the domains and the codomains of the elements of J are small relative to J, then the Reedy model category structure on $\mathcal{M}^{\mathcal{C}}$ (see Theorem 15.3.4) is cofibrantly generated (see Theorem 15.6.27). Although this seems to be a restriction on the class of cofibrantly generated model categories to which our results apply, it includes all cofibrantly generated model categories of which I am aware.

DEFINITION 15.6.23. If $\mathcal{C}$ is a Reedy category, $\mathcal{M}$ is a model category, and K is a set of maps in $\mathcal{M}$, then $\mathbf{RF}^{\mathcal{C}}_K$ will denote the set of maps in $\mathcal{M}^{\mathcal{C}}$ of the form

$$\mathbf{F}^\alpha_{A_k} \amalg_{\boldsymbol{\partial}\mathbf{F}^\alpha_{A_k}} \boldsymbol{\partial}\mathbf{F}^\alpha_{B_k} \to \mathbf{F}^\alpha_{B_k}$$

(see Definition 11.5.25 and Definition 15.6.13) for α an object of $\mathcal{C}$ and $A_k \to B_k$ an element of K.

PROPOSITION 15.6.24. *If $\mathcal{C}$ is a Reedy category and $\mathcal{M}$ is a cofibrantly generated model category with generating cofibrations I and generating trivial cofibrations J, then a map of $\mathcal{C}$-diagrams in $\mathcal{M}$ is a Reedy fibration if and only if it has the right lifting property with respect to every element of $\mathbf{RF}_J^{\mathcal{C}}$ (see Definition 15.6.23) and it is a Reedy trivial fibration if and only if it has the right lifting property with respect to every element of $\mathbf{RF}_I^{\mathcal{C}}$.*

PROOF. This follows from Corollary 15.6.21 and Theorem 15.3.15. $\square$

LEMMA 15.6.25. *If $\mathcal{C}$ is a Reedy category, $\mathcal{M}$ is a cocomplete category, and I is a set of maps in $\mathcal{M}$, then for every object β of $\mathcal{C}$ and every element $\mathbf{F}_A^\alpha \amalg_{\partial \mathbf{F}_A^\alpha} \partial \mathbf{F}_B^\alpha \to \mathbf{F}_B^\alpha$ of $\mathbf{RF}_I^{\mathcal{C}}$ (see Definition 15.6.23) the map*

$$\mathbf{F}_A^\alpha \amalg_{\partial \mathbf{F}_A^\alpha} \partial \mathbf{F}_B^\alpha(\beta) \to \mathbf{F}_B^\alpha(\beta)$$

is a relative I-cell complex.

PROOF. If $A \to B$ is a map in $\mathcal{M}$ and α and β are objects of $\mathcal{C}$, then the map $\partial \mathbf{F}_A^\alpha(\beta) \to \mathbf{F}_A^\alpha(\beta)$ is the inclusion of a summand. Thus, the pushout $\mathbf{F}_A^\alpha(\beta) \amalg_{\partial \mathbf{F}_A^\alpha(\beta)} \partial \mathbf{F}_B^\alpha(\beta)$ is isomorphic to the coproduct $\left(\coprod_{(\mathbf{F}_*^\alpha(\beta) - \partial \mathbf{F}_*^\alpha(\beta))} A\right) \amalg \left(\coprod_{\partial \mathbf{F}_*^\alpha(\beta)} B\right)$ and the map $\mathbf{F}_A^\alpha(\beta) \amalg_{\partial \mathbf{F}_A^\alpha(\beta)} \partial \mathbf{F}_B^\alpha(\beta) \to \mathbf{F}_B^\alpha(\beta)$ is isomorphic to the coproduct of the identity map of $\coprod_{\partial \mathbf{F}_B^\alpha(\beta)} B$ with the map $\coprod_{(\mathbf{F}_*^\alpha(\beta) - \partial \mathbf{F}_*^\alpha(\beta))} A \to \coprod_{(\mathbf{F}_*^\alpha(\beta) - \partial \mathbf{F}_*^\alpha(\beta))} B$. Proposition 10.2.7 now implies that the map $\mathbf{F}_A^\alpha(\beta) \amalg_{\partial \mathbf{F}_A^\alpha(\beta)} \partial \mathbf{F}_B^\alpha(\beta) \to \mathbf{F}_B^\alpha(\beta)$ is a transfinite composition of pushouts of the map $A \to B$. $\square$

LEMMA 15.6.26. *Let $\mathcal{C}$ be a Reedy category and let $\mathcal{M}$ be a cocomplete category. If I and K are sets of maps in $\mathcal{M}$ such that the domains and codomains of the elements of K are small relative to I, then the domains and codomains of the elements of $\mathbf{RF}_K^{\mathcal{C}}$ (see Definition 15.6.23) are small relative to $\mathbf{RF}_I^{\mathcal{C}}$.*

PROOF. Proposition 10.5.13, Proposition 11.5.26, and Lemma 15.6.25 imply that if A is a domain or a codomain of an element of K and α is an object of $\mathcal{C}$ then $\mathbf{F}_A^\alpha$ is small relative to $\mathbf{RF}_I^{\mathcal{C}}$. Proposition 10.4.8 and Proposition 15.6.15 now imply that the domains and codomains of the elements of $\mathbf{RF}_K^{\mathcal{C}}$ are small relative to $\mathbf{RF}_I^{\mathcal{C}}$. $\square$

The next theorem may seem to be weak in that it applies only to those cofibrantly generated model categories $\mathcal{M}$ for which there are a set I of generating cofibrations whose domains and codomains are small relative to I and a set J of generating trivial cofibrations whose domains and codomains are small relative to J. However, this is a property shared by every cofibrantly generated model category of which I am aware.

THEOREM 15.6.27. *Let $\mathcal{M}$ be a cofibrantly generated model category for which there are a set I of generating cofibrations whose domains and codomains are small relative to I (see Definition 10.5.12) and a set J of generating trivial cofibrations whose domains and codomains are small relative to J. If $\mathcal{C}$ is a Reedy category then the Reedy model category structure on $\mathcal{M}^{\mathcal{C}}$ is cofibrantly generated with generating cofibrations $\mathbf{RF}_I^{\mathcal{C}}$ (see Definition 15.6.23) and generating trivial cofibrations $\mathbf{RF}_J^{\mathcal{C}}$.*

PROOF. This follows from Proposition 15.6.24, Lemma 15.6.25, and Lemma 15.6.26. □

15.7. Reedy diagrams in a cellular model category

In this section we show that if $\mathcal{C}$ is a Reedy category and $\mathcal{M}$ is a cellular model category (see Definition 12.1.1), then the Reedy model category structure on $\mathcal{M}^{\mathcal{C}}$ (see Theorem 15.3.4) is a cellular model category (see Theorem 15.7.6).

PROPOSITION 15.7.1. *If $\mathcal{M}$ is a cellular model category and $\mathcal{C}$ is a Reedy category, then the cofibrations of the Reedy model category structure on $\mathcal{M}^{\mathcal{C}}$ (see Theorem 15.3.4) are effective monomorphisms.*

PROOF. Let $f\colon \boldsymbol{X} \to \boldsymbol{Y}$ be a Reedy cofibration in $\mathcal{M}^{\mathcal{C}}$. Proposition 15.3.11 implies that for every object α of $\mathcal{C}$ the map $f_\alpha\colon \boldsymbol{X}_\alpha \to \boldsymbol{Y}_\alpha$ is a cofibration in $\mathcal{M}$, and so f_α is an effective monomorphism. Thus, for every object α of $\mathcal{C}$ the map $f_\alpha\colon \boldsymbol{X}_\alpha \to \boldsymbol{Y}_\alpha$ is the equalizer of the natural inclusions $\boldsymbol{Y}_\alpha \rightrightarrows \boldsymbol{Y}_\alpha \amalg_{\boldsymbol{X}_\alpha} \boldsymbol{Y}_\alpha$. Since the pushout $\boldsymbol{Y} \amalg_{\boldsymbol{X}} \boldsymbol{Y}$ on an object α of $\mathcal{C}$ is $\boldsymbol{Y}_\alpha \amalg_{\boldsymbol{X}_\alpha} \boldsymbol{Y}_\alpha$, this implies that the map $f\colon \boldsymbol{X} \to \boldsymbol{Y}$ is the equalizer of the natural inclusion $\boldsymbol{Y} \rightrightarrows \boldsymbol{Y} \amalg_{\boldsymbol{X}} \boldsymbol{Y}$. □

COROLLARY 15.7.2. *If $\mathcal{M}$ is a cellular model category and $\mathcal{C}$ is a Reedy category, then the cofibrations of the Reedy model category structure on $\mathcal{M}^{\mathcal{C}}$ (see Theorem 15.3.4) are monomorphisms.*

PROOF. This follows from Proposition 15.7.1 and Proposition 10.9.5. □

COROLLARY 15.7.3. *If $\mathcal{M}$ is a cellular model category with generating cofibrations I and $\mathcal{C}$ is a Reedy category then the relative $\mathbf{RF}^{\mathcal{C}}_I$-cell complexes (see Definition 15.6.23) are monomorphisms.*

PROOF. This follows from Corollary 15.7.2 and Theorem 15.6.27. □

LEMMA 15.7.4. *Let $\mathcal{M}$ be a cellular model category with generating cofibrations I and let $\mathcal{C}$ be a Reedy category. If β is an object of $\mathcal{C}$ and W is an object of $\mathcal{M}$ that is compact relative to I, then $\mathbf{F}^{\beta}_{W}$ (see Definition 11.5.25) is compact relative to $\mathbf{RF}^{\mathcal{C}}_I$.*

PROOF. Proposition 11.5.26 and Proposition 11.4.9 imply that it is sufficient to show that for every element $A \to B$ of I and every object α of $\mathcal{C}$ the map $(\mathbf{F}^{\alpha}_{A} \amalg_{\partial\mathbf{F}^{\alpha}_{A}} \partial\mathbf{F}^{\alpha}_{B})(\beta) \to \mathbf{F}^{\alpha}_{B}(\beta)$ is a cofibration whose domain is compact relative to I. Lemma 15.6.25 implies that that map is a cofibration, and (since $(\mathbf{F}^{\alpha}_{A} \amalg_{\partial\mathbf{F}^{\alpha}_{A}} \partial\mathbf{F}^{\alpha}_{B})(\beta) = \mathbf{F}^{\alpha}_{A}(\beta) \amalg_{\partial\mathbf{F}^{\alpha}_{A}(\beta)} \partial\mathbf{F}^{\alpha}_{B}(\beta)$ and both A and B are compact relative to I) Proposition 10.8.8 implies that its domain is compact relative to I. □

PROPOSITION 15.7.5. *If $\mathcal{M}$ is a cellular model category with generating cofibrations I and $\mathcal{C}$ is a Reedy category, then the domains and codomains of the elements of $\mathbf{RF}^{\mathcal{C}}_I$ (see Definition 15.6.23) are compact relative to $\mathbf{RF}^{\mathcal{C}}_I$.*

PROOF. Since the domains and codomains of the elements of I are compact relative to I, this follows from Proposition 15.6.15, Proposition 10.8.8, Corollary 15.7.3, and Lemma 15.7.4. □

THEOREM 15.7.6. *If $\mathcal{M}$ is a cellular model category and $\mathcal{C}$ is a Reedy category then the Reedy model category structure on $\mathcal{M}^{\mathcal{C}}$ (see Theorem 15.3.4) is a cellular model category.*

PROOF. This follows from Theorem 15.6.27, Proposition 15.7.5, Lemma 15.6.26, Corollary 12.4.5, and Proposition 15.7.1. □

15.8. Bisimplicial sets

There are two possible Reedy model category structures on the category of bisimplicial sets, obtained by viewing a bisimplicial set as a simplicial object in the category of simplicial sets in two different ways. We show in Proposition 15.8.1 that these model category structures have different classes of weak equivalences. (This does not contradict Theorem 15.5.2 because the category of simplicial sets is not a Reedy model category structure obtained from some model category of sets.) We also show (in Theorem 15.8.7) that in either of these model category structures the cofibrations are the monomorphisms.

PROPOSITION 15.8.1. *The two possible Reedy model category structures on the category of bisimplicial sets (obtained by viewing a bisimplicial set as either a horizontal simplicial object in the category of vertical simplicial sets or as a vertical simplicial object in the category of horizontal simplicial sets) are not the same.*

PROOF. We will show that these two model category structures have different classes of weak equivalences. Let $\boldsymbol{X}$ be the bisimplicial set such that $\boldsymbol{X}_{n,*} = \Delta[1]$, with all horizontal face and degeneracy maps equal to the identity, let $\boldsymbol{Y}$ be the bisimplicial set such that $\boldsymbol{Y}_{n,k}$ is a single point for all $n \geq 0$ and $k \geq 0$, and let $f\colon \boldsymbol{X} \to \boldsymbol{Y}$ be the unique map from $\boldsymbol{X}$ to $\boldsymbol{Y}$.

As a map of horizontal simplicial objects in the category of vertical simplicial sets, f is a weak equivalence, because for every $n \geq 0$ the map of simplicial sets $\boldsymbol{X}_{n,*} \to \boldsymbol{Y}_{n,*}$ is the map $\Delta[1] \to \Delta[0]$, which is a weak equivalence of simplicial sets. However, as a map of vertical simplicial objects in the category of horizontal simplicial sets, f is not a weak equivalence, because (for example) the map of simplicial sets $f_{*,0}\colon \boldsymbol{X}_{*,0} \to \boldsymbol{Y}_{*,0}$ is the map $(\Delta[0] \amalg \Delta[0]) \to \Delta[0]$, which is not a weak equivalence of simplicial sets. □

LEMMA 15.8.2. *Let X be a simplicial set, let $n \geq 0$, and let σ and τ be elements of X_n for which there are iterated degeneracy operators $s_{i_1}s_{i_2}\cdots s_{i_k}$ and $s_{j_1}s_{j_2}\cdots s_{j_k}$ such that $s_{i_1}s_{i_2}\cdots s_{i_k}(\sigma) = s_{j_1}s_{j_2}\cdots s_{j_k}(\tau)$. If σ is nondegenerate, then so is τ.*

PROOF. If $\tau = s_m\nu$ for some $0 \leq m \leq n-1$, then

$$\begin{aligned}
\sigma &= d_{i_k}\cdots d_{i_2}d_{i_1}s_{i_1}s_{i_2}\cdots s_{i_k}\sigma \\
&= d_{i_k}\cdots d_{i_2}d_{i_1}s_{j_1}s_{j_2}\cdots s_{j_k}\tau \\
&= d_{i_k}\cdots d_{i_2}d_{i_1}s_{j_1}s_{j_2}\cdots s_{j_k}s_m\nu \ ,
\end{aligned}$$

and this last expression for σ has k face operators and $(k+1)$-degeneracy operators. The simplicial identities would then imply that σ was degenerate, which was assumed not to be the case. □

LEMMA 15.8.3. *If $\mathcal{M}$ is a category and $\boldsymbol{X}$ is a simplicial object in $\mathcal{M}$, then every iterated degeneracy operator $\boldsymbol{X}_n \to \boldsymbol{X}_{n+k}$ in $\boldsymbol{X}$ has a unique expression in the form $s_{i_1} s_{i_2} \cdots s_{i_k}$ with $i_1 > i_2 > \cdots > i_k$.*

PROOF. Such an iterated degeneracy operator corresponds to an epimorphism $\alpha\colon [n+k] \to [n]$ in $\boldsymbol{\Delta}$ (see Definition 15.1.8), and the set $\{i_1, i_2, \ldots, i_k\}$ is the set of integers i in $[n+k]$ such that $\alpha(i+1) = \alpha(i)$. □

LEMMA 15.8.4. *If X is a simplicial set and μ is a degenerate simplex of X, then there is a unique nondegenerate simplex ν of X and a unique iterated degeneracy operator α such that $\alpha(\nu) = \mu$.*

PROOF. Lemma 15.8.2 implies that it is sufficient to show that

(1) if $n \geq 0$ and σ and τ are nondegenerate n-simplices such that some degeneracy of σ equals some (possibly different) degeneracy of τ, then $\sigma = \tau$, and
(2) if σ is a nondegenerate simplex and α and β are iterated degeneracy operators such that $\alpha(\sigma) = \beta(\tau)$, then $\alpha = \beta$.

For assertion 1, let k be the smallest positive integer for which there are iterated degeneracy operators $s_{i_1} s_{i_2} \cdots s_{i_k}$ with $i_1 > i_2 > \cdots > i_k$ and $s_{j_1} s_{j_2} \cdots s_{j_k}$ with $j_1 > j_2 > \cdots > j_k$ (see Lemma 15.8.3) such that $s_{i_1} s_{i_2} \cdots s_{i_k}(\sigma) = s_{j_1} s_{j_2} \cdots s_{j_k}(\tau)$. If we apply the face operator d_{i_1} to both sides of this equation, we obtain

$$s_{i_2} s_{i_3} \cdots s_{i_k}(\sigma) = d_{i_1} s_{j_1} s_{j_2} \cdots s_{j_k}(\tau) \ ,$$

and the simplicial identities imply that the right hand side is either a $(k-1)$-fold iterated degeneracy of τ or a k-fold iterated degeneracy of a face of τ. Lemma 15.8.2 implies that it cannot be the latter, and so our assumption that k was the smallest positive integer of its type implies that $k = 1$, i.e., $s_{i_1}\sigma = s_{j_1}\tau$. If $i_1 > j_1$, then $\sigma = d_{i_1+1} s_{i_1} \sigma = d_{i_1+1} s_j \tau = s_j d_{i_1} \tau$, which is impossible because σ is nondegenerate. Similarly, we cannot have $i_1 < j_1$. Thus, $i_1 = j_1$, and so $\sigma = \tau$ (because degeneracy operators have left inverses).

For assertion 2, let k be the smallest positive integer for which there are iterated degeneracy operators $s_{i_1} s_{i_2} \cdots s_{i_k}$ with $i_1 > i_2 > \cdots > i_k$ and $s_{j_1} s_{j_2} \cdots s_{j_k}$ with $j_1 > j_2 > \cdots > j_k$ such that $s_{i_1} s_{i_2} \cdots s_{i_k}(\sigma) = s_{j_1} s_{j_2} \cdots s_{j_k}(\sigma)$ (see Lemma 15.8.3). Because k is the smallest such integer and degeneracy operators have left inverses, we must have $i_1 \neq j_1$. If $i_1 > j_1$, then we can apply d_{i_1+1} to obtain

$$\begin{aligned} s_{i_2} s_{i_3} \cdots s_{i_k}(\sigma) &= d_{i_1+1} s_{j_1} s_{j_2} \cdots s_{j_k}(\sigma) \\ &= s_{j_1} s_{j_2} \cdots s_{j_k} d_{i_1+1-k}(\sigma) \end{aligned}$$

which contradicts Lemma 15.8.2. Similarly, we cannot have $i_1 < j_1$. Thus, $i_1 = j_1$, and so $s_{i_2} s_{i_3} \cdots s_{i_k}(\sigma) = s_{j_2} s_{j_3} \cdots s_{j_k}(\sigma)$, which implies that $k = 1$ (or else we have contradicted our assumption that k is the smallest positive integer of its type). □

LEMMA 15.8.5. *Let X be a simplicial set. If $n \geq 0$ and σ is an n-simplex of X, then σ is nondegenerate if and only if no two of the simplices $s_0\sigma$, $s_1\sigma$, …, $s_n\sigma$ are equal.*

PROOF. If σ is degenerate, then $\sigma = s_i \tau$ for some $0 \leq i < n$ and some $(n-1)$-simplex τ, and so $s_{i+1}\sigma = s_{i+1} s_i \tau = s_i s_i \tau = s_i \sigma$.

Conversely, if $s_i\sigma = s_j\sigma$ for $0 \leq i < j \leq n$, then $\sigma = \partial_i s_i \sigma = \partial_i s_j \sigma = s_{j-1} \partial_i \sigma$, and so σ is degenerate. □

PROPOSITION 15.8.6. *If $\mathbf{\Delta}^{\mathrm{op}}$ is the indexing category for simplicial sets (see Example 15.1.13 and Definition 15.1.10) and $\boldsymbol{X}$ is an object of $\mathrm{SS}^{\mathbf{\Delta}^{\mathrm{op}}}$ (i.e., a bisimplicial set, which we view as a "horizontal" simplicial object in the category of "vertical" simplicial sets, so that the object of degree n is the simplicial set $\boldsymbol{X}_{n,*}$), then for $n \geq 0$ the latching object $\mathrm{L}_n\boldsymbol{X}$ of $\boldsymbol{X}$ at n (see Definition 15.2.5) is naturally isomorphic to the subcomplex of $\boldsymbol{X}_{n,*}$ consisting of those simplices that are in the image of a horizontal degeneracy operator.*

PROOF. For every $k \geq 0$ there is a natural map from the set $(\mathrm{L}_n\boldsymbol{X})_k$ of k-simplices of $\mathrm{L}_n\boldsymbol{X}$ onto the set of "horizontally" degenerate k-simplices of the "horizontal" simplicial set $\boldsymbol{X}_{*,k}$; we must show that this map is one to one. Lemma 15.8.4 implies that for every horizontally degenerate simplex $\sigma \in \boldsymbol{X}_{n,k}$ there is a unique horizontally nondegenerate simplex $\tau \in \boldsymbol{X}_{m,k}$ (for some $m < n$) and a unique horizontal iterated degeneracy operator α such that $\alpha(\tau) = \sigma$, and that if μ is any other simplex for which there is an iterated horizontal degeneracy operator β such that $\beta(\mu) = \sigma$ then there is an iterated horizontal degeneracy operator γ such that $\gamma(\tau) = \mu$ and $\gamma\beta = \alpha$. □

THEOREM 15.8.7. *In the Reedy model category structure on $\mathrm{SS}^{\mathbf{\Delta}^{\mathrm{op}}}$ (the category of simplicial simplicial sets; see Example 15.1.13 and Definition 15.1.10) a map $f\colon \boldsymbol{X} \to \boldsymbol{Y}$ of simplicial simplicial sets is a cofibration if and only if it is a monomorphism.*

PROOF. Proposition 15.3.11 implies that if f is a Reedy cofibration, then for every $n \geq 0$ the map $f_n\colon \boldsymbol{X}_n \to \boldsymbol{Y}_n$ is a cofibration (i.e., a monomorphism) of simplicial sets.

Conversely, assume that $f_n\colon \boldsymbol{X}_n \to \boldsymbol{Y}_n$ is a monomorphism of simplicial sets for every $n \geq 0$. Proposition 15.8.6 implies that each latching object $\mathrm{L}_n\boldsymbol{X}$ is a subcomplex of $\boldsymbol{X}_n$ and that each $\mathrm{L}_n\boldsymbol{Y}$ is a subcomplex of $\boldsymbol{Y}_n$, and Lemma 15.8.5 implies that the intersection of $\boldsymbol{X}_n$ and $\mathrm{L}_n\boldsymbol{Y}$ in $\boldsymbol{Y}_n$ is $\mathrm{L}_n\boldsymbol{X}$. Thus, $\boldsymbol{X}_n \amalg_{\mathrm{L}_n\boldsymbol{X}} \mathrm{L}_n\boldsymbol{Y} \to \boldsymbol{Y}_n$ is an inclusion of simplicial sets, and so $f\colon \boldsymbol{X} \to \boldsymbol{Y}$ is a Reedy cofibration. □

COROLLARY 15.8.8. *A simplicial object in the category of simplicial sets is always Reedy cofibrant.*

PROOF. This follows from Theorem 15.8.7. □

15.9. Cosimplicial simplicial sets

Lemma 15.8.4 implies that the cofibrations in the Reedy model category structure on simplicial simplicial sets are the monomorphisms (see Theorem 15.8.7) and that every object is cofibrant (see Corollary 15.8.8). The precise analogue of Lemma 15.8.4 for cosimplicial simplicial sets is false, because there can be simplices of codegree zero whose images under the coface operators d^0 and d^1 coincide. Thus, we define the *maximal augmentation* (see Definition 15.9.2) of a cosimplicial simplicial set to be the subspace of simplices of codegree 0 with this property, and we establish Lemma 15.9.5 as our replacement for Lemma 15.8.4. We then show in Theorem 15.9.9 and Corollary 15.9.10 that, except for the special attention required by the maximal augmentation, the situation for cosimplicial simplicial sets is as convenient as that for simplicial simplicial sets.

LEMMA 15.9.1. *If* $\mathcal{M}$ *is a category and* $\boldsymbol{X}$ *is a cosimplicial object in* $\mathcal{M}$*, then every iterated coface operator* $\boldsymbol{X}^n \to \boldsymbol{X}^{n+k}$ *in* $\boldsymbol{X}$ *has a unique expression in the form* $d^{i_1}d^{i_2}\cdots d^{i_k}$ *with* $i_1 > i_2 > \cdots > i_k$.

PROOF. Such an iterated coface operator corresponds to a monomorphism $[n] \to [n+k]$ in $\boldsymbol{\Delta}$, the cosimplicial indexing category (see Definition 15.1.8), and the set $\{i_1, i_2, \ldots, i_k\}$ is the complement of the image of $[n]$ in $[n+k]$. □

DEFINITION 15.9.2. If $\boldsymbol{X}$ is a cosimplicial simplicial set, then the *maximal augmentation* of $\boldsymbol{X}$ is the simplicial set that is the equalizer of the coface operators d^0 and d^1 from $\boldsymbol{X}^0$ to $\boldsymbol{X}^1$. That is, an n-simplex of the maximal augmentation of $\boldsymbol{X}$ is an n-simplex σ of $\boldsymbol{X}^0$ such that $d^0\sigma = d^1\sigma$.

If $\boldsymbol{X}$ is a cosimplicial set, then by the *maximal augmentation* of $\boldsymbol{X}$ we mean the maximal augmentation of the cosimplicial discrete simplicial set determined by $\boldsymbol{X}$.

DEFINITION 15.9.3. If $\boldsymbol{X}$ is a cosimplicial set, $n \geq 0$, and $\sigma \in \boldsymbol{X}^n$, then we will say that σ *is a coface* if σ is in the image of some coface operator, and that σ *is a non-coface* if σ is not in the image of any coface operator.

LEMMA 15.9.4. *Let* $\boldsymbol{X}$ *be a cosimplicial set, let* $n \geq 0$*, and let* σ *and* τ *be elements of* $\boldsymbol{X}^n$ *for which there are iterated coface operators* $d^{i_1}d^{i_2}\cdots d^{i_k}$ *and* $d^{j_1}d^{j_2}\cdots d^{j_k}$ *such that* $d^{i_1}d^{i_2}\cdots d^{i_k}\sigma = d^{j_1}d^{j_2}\cdots d^{j_k}\tau$. *If* σ *is a non-coface (see Definition* 15.9.3*), then so is* τ.

PROOF. If $\tau = d^m\nu$, then

$$\begin{aligned}\sigma &= s^{i_k-1}\cdots s^{i_2-1}s^{i_1-1}d^{i_1}d^{i_2}\cdots d^{i_k}(\sigma)\\ &= s^{i_k-1}\cdots s^{i_2-1}s^{i_1-1}d^{j_1}d^{j_2}\cdots d^{j_k}(\tau)\\ &= s^{i_k-1}\cdots s^{i_2-1}s^{i_1-1}d^{j_1}d^{j_2}\cdots d^{j_k}d^m(\nu)\ ,\end{aligned}$$

and this last expression for σ has k codegeneracy operators and $(k+1)$ coface operators. The cosimplicial identities would then imply that σ was a coface, which was assumed not to be the case. □

LEMMA 15.9.5. *If* $\boldsymbol{X}$ *is a cosimplicial set,* $n \geq 0$*, and* μ *is an element of* $\boldsymbol{X}^n$ *that is not the image of an element of the maximal augmentation (see Definition* 15.9.2*) under an iterated coface operator, then there is a unique non-coface* σ *(see Definition* 15.9.3*) and a unique iterated coface operator* α *such that* $\alpha(\sigma) = \mu$.

PROOF. Lemma 15.9.4 implies that it is sufficient to show that

(1) if $m \geq 0$ and σ and τ are non-coface elements of $\boldsymbol{X}^m$ for which some iterated coface of σ equals some (possibly different) iterated coface of τ, then either $\sigma = \tau$ or one of σ and τ is an element of the maximal augmentation, and
(2) if σ is not a coface and not an element of the maximal augmentation and α and β are iterated coface operators such that $\alpha(\sigma) = \beta(\sigma)$, then $\alpha = \beta$.

For assertion 1, let k be the smallest positive integer for which there are iterated coface operators $d^{i_1}d^{i_2}\cdots d^{i_k}$ and $d^{j_1}d^{j_2}\cdots d^{j_k}$ with $i_1 > i_2 > \cdots > i_k$ and $j_1 > j_2 > \cdots > j_k$ (see Lemma 15.9.1) such that $d^{i_1}d^{i_2}\cdots d^{i_k}(\sigma) = d^{j_1}d^{j_2}\cdots d^{j_k}(\tau)$. If we apply the operator s^{i_1-1} to both sides of this equation, the we obtain $d^{i_2}d^{i_3}\cdots d^{i_k}(\sigma) = s^{i_1-1}d^{j_1}d^{j_2}\cdots d^{j_k}(\tau)$ and the cosimplicial identities imply that the right hand side is either a $(k-1)$-fold iterated coface of τ or a k-fold iterated coface of a codegeneracy of τ. Lemma 15.9.4 implies that it cannot be the latter, and so our assumption that k was the smallest positive integer of its type implies that $k=1$, i.e., $d^{i_1}(\sigma) = d^{j_1}(\tau)$. If $i_1 = j_1$, then $\sigma = \tau$ (because coface operators have left inverses) and we are done. If $i_1 \neq j_1$, then we assume that $i_1 > j_1$ (the case $i_1 < j_1$ is similar).

- If s^{i_1} is defined on $d^{i_1}(\sigma)$, then $\sigma = s^{i_1}d^{i_1}(\sigma) = s^{i_1}d^{j_1}(\tau) = d^{j_1}s^{i_1-1}(\tau)$, which contradicts Lemma 15.9.4.
- If $i_1 > j_1 + 1$, then $\sigma = s^{i_1-1}d^{i_1}(\sigma) = s^{i_1-1}d^{j_1}(\tau) = d^{j_1}s^{i_1-2}(\tau)$, which contradicts Lemma 15.9.4.
- If $j_1 > 0$, then $\tau = s^{j_1-1}d^{j_1}(\tau) = s^{j_1-1}d^{i_1}(\sigma) = d^{i_1-1}s^{j_1-1}(\sigma)$, which contradicts Lemma 15.9.4.

Thus, we must have $j_1 = 0$, $i_1 = 1$, and $m = 0$; that is, $\sigma, \tau \in \boldsymbol{X}^0$ and $d^1\sigma = d^0\tau$. Thus, $\sigma = s_0 d^1\sigma = s^0 d^0\tau = \tau$, i.e., $d^0\sigma = d^1\sigma$, and so σ is an element of the maximal augmentation.

For assertion 2, let k be the smallest positive integer for which there are iterated coface operators $d^{i_1}d^{i_2}\cdots d^{i_k}$ with $i_1 > i_2 > \cdots > i_k$ and $d^{j_1}d^{j_2}\cdots d^{j_k}$ with $j_1 > j_2 > \cdots > j_k$ such that $d^{i_1}d^{i_2}\cdots d^{i_k}(\sigma) = d^{j_1}d^{j_2}\cdots d^{j_k}(\sigma)$ (see Lemma 15.9.1). Applying s^{i_1-1} to both sides of this equation, we obtain

$$d^{i_2}d^{i_3}\cdots d^{i_k}(\sigma) = s^{i_1-1}d^{j_1}d^{j_2}\cdots d^{j_k}(\sigma) \ .$$

The right hand side of this equation is either a $(k-1)$-fold coface of σ or a k-fold coface of a codegeneracy of σ. The latter would contradict Lemma 15.9.4, and so we must have $k=1$, i.e., $d^{i_1}(\sigma) = d^{j_1}(\sigma)$, and we must show that $i_1 = j_1$. If not, then we'll assume that $i_1 > j_1$ (the other case is similar). If $i_1 > j_1 + 1$, then $\sigma = s^{i_1-1}d^{i_1}(\sigma) = s^{i_1-1}d^{j_1}(\sigma) = d^{j_1}s^{i_1-2}(\sigma)$, which contradicts out assumption about σ. Thus, $i_1 = j_1 + 1$. Similarly, $j_1 = 0$ and $i_1 = n+1$, i.e., $\sigma \in \boldsymbol{X}^0$, $i_1 = 1$, and $j_1 = 0$. This would imply that σ is an element of the maximal augmentation, which was assumed not to be the case, and so we are done. ☐

PROPOSITION 15.9.6. *Let $\boldsymbol{X}$ be a cosimplicial simplicial set (i.e., an object of $\mathrm{SS}^{\boldsymbol{\Delta}}$).*

1. *If $n \geq 2$, then the latching object $\mathrm{L}_n\boldsymbol{X}$ of $\boldsymbol{X}$ at codegree n is naturally isomorphic to the subcomplex of $\boldsymbol{X}^n$ consisting of those simplices that are in the image of a coface operator.*
2. *The latching object $\mathrm{L}_1\boldsymbol{X}$ of $\boldsymbol{X}$ at codegree 1 is naturally isomorphic to the pushout $\mathrm{C}_1\boldsymbol{X} \amalg_{\mathrm{N}_1\boldsymbol{X}} \mathrm{C}_1\boldsymbol{X}$ where $\mathrm{C}_1\boldsymbol{X}$ is the subspace of $\boldsymbol{X}^1$ consisting of the simplices that are cofaces (see Definition 15.9.3) and $\mathrm{N}_1\boldsymbol{X}$ is the subspace of $\mathrm{C}_1\boldsymbol{X}$ consisting of those cofaces that are not in the image of the maximal augmentation (see Definition 15.9.2) under a coface operator.*
3. *The latching object $\mathrm{L}_0\boldsymbol{X}$ of $\boldsymbol{X}$ at codegree 0 is the empty simplicial set.*

PROOF. This follows from Lemma 15.9.5. ☐

LEMMA 15.9.7. *Let $\boldsymbol{X}$ be a cosimplicial set. If $n \geq 1$ and σ is an element of $\boldsymbol{X}^n$, then σ is a non-coface (see Definition 15.9.3) if and only if no two of the elements $d^0\sigma$, $d^1\sigma$, ..., $d^{n+1}\sigma$ of $\boldsymbol{X}^{n+1}$ are equal.*

PROOF. If σ is a coface, then $\sigma = d^i\tau$ for some $\tau \in \boldsymbol{X}^{n-1}$ and $0 \leq i \leq n$, and so $d^{i+1}\sigma = d^{i+1}d^i\tau = d^i d^i\tau = d^i\sigma$.

Conversely, suppose $d^i\sigma = d^j\sigma$ for $0 \le i < j \le n+1$. There are three (non-exclusive) cases:

- If $j < n+1$, then $\sigma = s^j d^j \sigma = s^j d^i \sigma = d^i s^{j-1}\sigma$.
- If $i < j-1$, then $\sigma = s^{j-1} d^j \sigma = s^{j-1} d^i \sigma = d^i s^j \sigma$.
- If $j = n+1$ and $i = j-1$, then $i = n$; since $n \ge 1$, $\sigma = s^{i-1} d^i \sigma = s^{i-1} d^j \sigma = d^{j-1} s^{i-1} \sigma$.

Thus, in each case, σ is a coface. □

REMARK 15.9.8. The assertion of Lemma 15.9.7 for $n = 0$ is false; this is why we needed Definition 15.9.2.

THEOREM 15.9.9. *In the Reedy model category structure on* $\mathrm{SS}^{\mathbf{\Delta}}$ *(the category of cosimplicial simplicial sets; see Definition* 15.1.8 *and Example* 15.1.12*), a map* $f\colon \boldsymbol{X} \to \boldsymbol{Y}$ *of cosimplicial simplicial sets is a cofibration if and only if it is a monomorphism that takes the maximal augmentation (see Definition* 15.9.2*) of* $\boldsymbol{X}$ *isomorphically onto that of* $\boldsymbol{Y}$*.*

PROOF. If f is a cofibration, then Proposition 15.3.11 implies that $f^n\colon \boldsymbol{X}^n \to \boldsymbol{Y}^n$ is a monomorphism for every $n \ge 0$. Since the relative latching map $\boldsymbol{X}^1 \amalg_{\mathrm{L}_1 \boldsymbol{X}} \mathrm{L}_1 \boldsymbol{Y} \to \boldsymbol{Y}^1$ is a monomorphism, Proposition 15.9.6 implies that f must map the maximal augmentation of $\boldsymbol{X}$ isomorphically onto that of $\boldsymbol{Y}$.

Conversely, if f is a monomorphism, then Proposition 15.9.6 and Lemma 15.9.7 imply that for $n \ne 1$, the intersection of $\boldsymbol{X}^n$ and $\mathrm{L}_n\boldsymbol{Y}$ in $\boldsymbol{Y}^n$ is $\mathrm{L}_n\boldsymbol{X}$, and so the relative latching map $\boldsymbol{X}^n \amalg_{\mathrm{L}_n \boldsymbol{X}} \mathrm{L}_n \boldsymbol{Y} \to \boldsymbol{Y}^n$ is a monomorphism. If f takes the maximal augmentation of $\boldsymbol{X}$ isomorphically onto that of $\boldsymbol{Y}$, then Proposition 15.9.6 implies that the relative latching map $\boldsymbol{X}^1 \amalg_{\mathrm{L}_1 \boldsymbol{Y}} \mathrm{L}_1 \boldsymbol{Y} \to \boldsymbol{Y}^1$ is a monomorphism, and so f is a cofibration. □

COROLLARY 15.9.10. *A cosimplicial simplicial set is Reedy cofibrant if and only if its maximal augmentation (see Definition* 15.9.2*) is empty.*

PROOF. This follows from Theorem 15.9.9. □

COROLLARY 15.9.11. *The cosimplicial standard simplex (see Definition* 15.1.15*) is a Reedy cofibrant cosimplicial set.*

PROOF. This follows from Corollary 15.9.10. □

COROLLARY 15.9.12. *The cosimplicial standard simplex (see Definition* 15.1.15*) is a Reedy cofibrant approximation (see Definition* 8.1.2*) to the constant* $\mathbf{\Delta}$*-diagram at a point.*

PROOF. Since for every $n \ge 0$ the map from $\Delta[n]$ to a point is a weak equivalence, this follows from Corollary 15.9.11. □

15.10. Cofibrant constants and fibrant constants

Some Reedy categories $\mathcal{C}$ have the property that, for every model category $\mathcal{M}$, the colimit of an objectwise weak equivalence of Reedy cofibrant $\mathcal{C}$-diagrams in $\mathcal{M}$ is a weak equivalence of cofibrant objects. These are the Reedy categories with *fibrant constants* (see Definition 15.10.1 and Theorem 15.10.9). Dually, a Reedy category $\mathcal{C}$ with *cofibrant constants* has the property that, for every model category $\mathcal{M}$, the limit of an objectwise weak equivalence of Reedy fibrant $\mathcal{C}$-diagrams in $\mathcal{M}$ is a weak equivalence of fibrant objects.

DEFINITION 15.10.1. Let $\mathcal{C}$ be a Reedy category.

(1) We will say that $\mathcal{C}$ has *cofibrant constants* if for every model category $\mathcal{M}$ and every cofibrant object B of $\mathcal{M}$ the constant $\mathcal{C}$-diagram at B is cofibrant in the Reedy model category structure on $\mathcal{M}^{\mathcal{C}}$.
(2) We will say that $\mathcal{C}$ has *fibrant constants* if for every model category $\mathcal{M}$ and every fibrant object X of $\mathcal{M}$ the constant $\mathcal{C}$-diagram at X is fibrant in the Reedy model category structure on $\mathcal{M}^{\mathcal{C}}$.

PROPOSITION 15.10.2. *Let $\mathcal{C}$ be a Reedy category.*

(1) *The Reedy category $\mathcal{C}$ has cofibrant constants (see Definition 15.10.1) if and only if for every object α of $\mathcal{C}$ the latching category (see Definition 15.2.3) of $\mathcal{C}$ at α is either connected or empty.*
(2) *The Reedy category $\mathcal{C}$ has fibrant constants if and only if for every object α of $\mathcal{C}$ the matching category of $\mathcal{C}$ at α is either connected or empty.*

PROOF. We will prove part 1; the proof of part 2 is dual.

If $\mathcal{M}$ is a model category, B is an object of $\mathcal{M}$, $\boldsymbol{X}\colon \mathcal{C} \to \mathcal{M}$ is the constant diagram at B, and α is an object of $\mathcal{C}$, then the latching object of $\boldsymbol{X}$ at α (see Definition 15.2.5) is the colimit of a diagram in which every map is the identity map of B, and so it is isomorphic to a coproduct, indexed by the components of $\partial(\overrightarrow{\mathcal{C}} \downarrow \alpha)$, of copies of B. Thus, if $\partial(\overrightarrow{\mathcal{C}} \downarrow \alpha)$ is either connected or empty for every object α of $\mathcal{C}$, then the latching map of $\boldsymbol{X}$ at α is either the identity map of B or the map $\emptyset \to B$ (where "$\emptyset$" is the initial object of $\mathcal{M}$) for every object α of $\mathcal{C}$, and so if B is cofibrant then so is $\boldsymbol{X}$.

Conversely, if there is an object α of $\mathcal{C}$ such that $\partial(\overrightarrow{\mathcal{C}} \downarrow \alpha)$ has more than one component, B is a nonempty simplicial set, and $\boldsymbol{X}\colon \mathcal{C} \to \mathrm{SS}$ is the constant diagram at B, then the latching map of $\boldsymbol{X}$ at $\mathcal{C}$ will not be a monomorphism. □

PROPOSITION 15.10.3. *If $\mathcal{C}$ is a Reedy category, then $\mathcal{C}$ has cofibrant constants if and only if $\mathcal{C}^{\mathrm{op}}$ has fibrant constants.*

PROOF. This follows from Proposition 15.10.2 and Proposition 15.2.4. □

PROPOSITION 15.10.4. *If $\mathcal{C}$ is the category of simplices of a simplicial set (see Definition 15.1.16), then*

(1) *the category $\mathcal{C}$ has fibrant constants (see Definition 15.10.1), and*
(2) *the category $\mathcal{C}^{\mathrm{op}}$ has cofibrant constants.*

PROOF. Proposition 15.10.3 implies that it is sufficient to prove part 1. Let K be a simplicial set such that $\mathcal{C} = \Delta K$. If σ is a nondegenerate simplex of K, then $\partial(\alpha \downarrow \overleftarrow{\mathcal{C}})$ is empty. If σ is a degenerate simplex of K, then there is a unique nondegenerate simplex τ such that σ is the image of τ under a degeneracy operator (see Lemma 15.8.4), and the map $\sigma \to \tau$ is a terminal object of $\partial(\alpha \downarrow \overleftarrow{\mathcal{C}})$. The proposition now follows from Proposition 15.10.2. □

COROLLARY 15.10.5. *The cosimplicial indexing category (see Definition 15.1.8) is a Reedy category with fibrant constants and the simplicial indexing category is a Reedy category with cofibrant constants.*

PROOF. This follows from Proposition 15.10.4 and Example 15.1.18. □

PROPOSITION 15.10.6. *Let $\mathcal{C}$ be a Reedy category and let α be an object of $\mathcal{C}$.*

(1) *The latching category $\partial(\overrightarrow{\mathcal{C}} \downarrow \alpha)$ of $\mathcal{C}$ at α is a Reedy category with fibrant constants in which $\overrightarrow{\partial(\overrightarrow{\mathcal{C}} \downarrow \alpha)} = \partial(\overrightarrow{\mathcal{C}} \downarrow \alpha)$ and $\overleftarrow{\partial(\overrightarrow{\mathcal{C}} \downarrow \alpha)}$ has only identity maps.*

(2) *The matching category $\partial(\alpha \downarrow \overleftarrow{\mathcal{C}})$ of $\mathcal{C}$ at α is a Reedy category with cofibrant constants in which $\overleftarrow{\partial(\alpha \downarrow \overleftarrow{\mathcal{C}})} = \partial(\alpha \downarrow \overleftarrow{\mathcal{C}})$ and $\overrightarrow{\partial(\alpha \downarrow \overleftarrow{\mathcal{C}})}$ has only identity maps.*

PROOF. We will prove part 1; the proof of part 2 is dual.

The restriction of a degree function for $\mathcal{C}$ yields a degree function for $\partial(\overrightarrow{\mathcal{C}} \downarrow \alpha)$, and so $\partial(\overrightarrow{\mathcal{C}} \downarrow \alpha)$ is a Reedy category. Since the matching category at every object of $\partial(\overrightarrow{\mathcal{C}} \downarrow \alpha)$ is empty, the result follows from Proposition 15.10.2. □

PROPOSITION 15.10.7. *Let $\mathcal{C}$ be a Reedy category.*

(1) *The Reedy category $\mathcal{C}$ has cofibrant constants if and only if the $\mathcal{C}$-diagram of simplicial sets that is a single point at every object of $\mathcal{C}$ is a Reedy cofibrant diagram.*

(2) *The Reedy category $\mathcal{C}$ has fibrant constants if and only if the $\mathcal{C}^{\mathrm{op}}$-diagram of simplicial sets that is a single point at every object of $\mathcal{C}^{\mathrm{op}}$ is a Reedy cofibrant diagram.*

PROOF. We will prove part 1; part 2 will then follow from part 1 and Proposition 15.10.3.

Definition 15.10.1 implies one direction of part 1. Conversely. if the constant diagram at a point is cofibrant, then every latching object is the domain of a cofibration with codomain a single point. Thus, every latching object is either empty or a single point, and so every latching category is either empty or connected. The result now follows from Proposition 15.10.2. □

THEOREM 15.10.8. *Let $\mathcal{C}$ be a Reedy category.*

(1) *The Reedy category $\mathcal{C}$ has cofibrant constants if and only if, for every model category $\mathcal{M}$, the constant diagram functor $\mathcal{M} \to \mathcal{M}^{\mathcal{C}}$ and the limit functor $\mathcal{M}^{\mathcal{C}} \to \mathcal{M}$ are a Quillen pair (see Definition 8.5.2).*

(2) *The Reedy category $\mathcal{C}$ has fibrant constants if and only if, for every model category $\mathcal{M}$, the colimit functor $\mathcal{M}^{\mathcal{C}} \to \mathcal{M}$ and the constant diagram functor $\mathcal{M} \to \mathcal{M}^{\mathcal{C}}$ are a Quillen pair (see Definition 8.5.2).*

PROOF. We will prove part 1; part 2 will then follow from Proposition 15.10.3, Proposition 15.10.2, and Proposition 15.2.4.

The colimit and constant diagram functors are an adjoint pair for all categories $\mathcal{M}$ and small categories $\mathcal{C}$. Proposition 15.10.7 implies that if the constant diagram functor is a left Quillen functor, then $\mathcal{C}$ has cofibrant constants. For the converse, Proposition 15.10.2 implies that if $\mathcal{C}$ has cofibrant constants, $i\colon A \to B$ is a cofibration in a model category $\mathcal{M}$, and α is an object of $\mathcal{C}$, then the relative latching map (see Definition 15.3.2) at α of the induced map of constant $\mathcal{C}$-diagrams is either the identity map of B or is isomorphic to the map i, and is thus a cofibration. □

THEOREM 15.10.9. *Let $\mathcal{M}$ be a model category and let $\mathcal{C}$ be a Reedy category.*

(1) *If $\mathcal{C}$ has cofibrant constants, then the limit functor $\lim_{\mathcal{C}}\colon \mathcal{M}^{\mathcal{C}} \to \mathcal{M}$ takes Reedy fibrant diagrams to fibrant objects of $\mathcal{M}$ and takes objectwise weak equivalences between Reedy fibrant diagrams to weak equivalences between fibrant objects of $\mathcal{M}$.*

(2) *If $\mathcal{C}$ has fibrant constants, then the colimit functor $\operatorname{colim}_{\mathcal{C}}\colon \mathcal{M}^{\mathcal{C}} \to \mathcal{M}$ takes Reedy cofibrant diagrams to cofibrant objects of $\mathcal{M}$ and takes objectwise weak equivalences between Reedy cofibrant diagrams to weak equivalences between cofibrant objects of $\mathcal{M}$.*

PROOF. This follows from Theorem 15.10.8 and Corollary 7.7.2. □

PROPOSITION 15.10.10. *Let $\mathcal{M}$ be a model category and let*

(15.10.11)

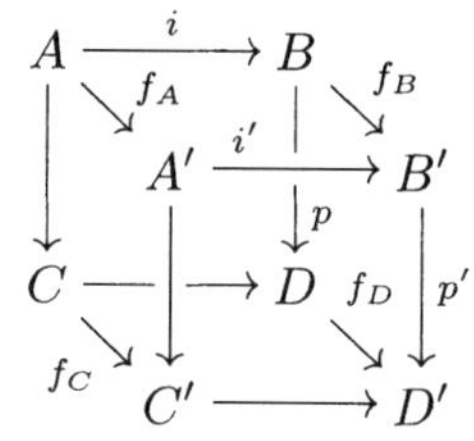

be a diagram in $\mathcal{M}$.

(1) *If all of the objects of Diagram 15.10.11 are cofibrant, the front and back squares are pushouts, i and i' are cofibrations, and all of f_A, f_B, and f_C are weak equivalences, then f_D is a weak equivalence.*

(2) *If all of the objects of Diagram 15.10.11 are fibrant, the front and back squares are pullbacks, p and p' are fibrations, and all of f_B, f_C,and f_D are weak equivalences, then f_A is a weak equivalence.*

PROOF. We will prove part 1; the proof of part 2 is dual.

Let $\mathcal{C}$ be the Reedy category with three objects and two non-identity maps $\gamma \leftarrow \alpha \to \beta$, in which we let $\deg(\alpha) = 2$, $\deg(\beta) = 3$, and $\deg(\gamma) = 1$. The Reedy category $\mathcal{C}$ has fibrant constants (see Proposition 15.10.2), and we have an objectwise weak equivalence of Reedy cofibrant $\mathcal{C}$-diagrams in $\mathcal{M}$

$$\begin{array}{ccccc} C & \longleftarrow & A & \xrightarrow{\ i\ } & B \\ {\scriptstyle f_C}\downarrow & & \downarrow{\scriptstyle f_A} & & \downarrow{\scriptstyle f_B} \\ C' & \longleftarrow & A' & \xrightarrow[\ i'\ ]{} & B' \end{array} .$$

Theorem 15.10.9 now implies that the induced map of pushouts $f_D\colon D \to D'$ is a weak equivalence. □

PROPOSITION 15.10.12. *Let $\mathcal{M}$ be a model category.*

(1) *If we have a map of sequences in $\mathcal{M}$*

$$\begin{array}{ccccccc} X_0 & \xrightarrow{\ i_0\ } & X_1 & \xrightarrow{\ i_1\ } & X_2 & \xrightarrow{\ i_2\ } & \cdots \\ \downarrow{\scriptstyle f_0} & & \downarrow{\scriptstyle f_1} & & \downarrow{\scriptstyle f_2} & & \\ Y_0 & \xrightarrow[\ j_0\ ]{} & Y_1 & \xrightarrow[\ j_1\ ]{} & Y_2 & \xrightarrow[\ j_2\ ]{} & \cdots \end{array}$$

in which all of the objects are cofibrant, the maps i_n and j_n are cofibrations for all $n \geq 0$, and the maps f_n are weak equivalences for all $n \geq 0$, then the induced map of colimits $\operatorname{colim} f_n\colon \operatorname{colim} X_n \to \operatorname{colim} Y_n$ is a weak equivalence.

(2) *If we have a map of towers in $\mathcal{M}$*

$$\begin{array}{ccccccc} \cdots \xrightarrow{p_3} & X_2 & \xrightarrow{p_2} & X_1 & \xrightarrow{p_1} & X_0 \\ & \downarrow{\scriptstyle f_2} & & \downarrow{\scriptstyle f_1} & & \downarrow{\scriptstyle f_0} \\ \cdots \xrightarrow[q_3]{} & Y_2 & \xrightarrow[q_2]{} & Y_1 & \xrightarrow[q_1]{} & Y_0 \end{array}$$

in which all of the objects are fibrant, the maps p_n and q_n are fibrations for all $n > 0$, and the maps f_n are weak equivalences for all $n \geq 0$, then the induced map of limits $\lim f_n\colon \lim X_n \to \lim Y_n$ is a weak equivalence.

PROOF. This follows from Theorem 15.10.9. □

15.11. The realization of a bisimplicial set

DEFINITION 15.11.1. If $\boldsymbol{X}$ is a simplicial object in the category of simplicial sets (i.e., a bisimplicial set), then the *realization* $|\boldsymbol{X}|$ of $\boldsymbol{X}$ is the simplicial set built from $\boldsymbol{X}$ and the cosimplicial standard simplex (see Definition 15.1.15) as the coequalizer of the diagram

$$\coprod_{(\alpha\colon [n]\to[k])\in\boldsymbol{\Delta}^{\mathrm{op}}} \boldsymbol{X}_n \times \Delta[k] \overset{\phi}{\underset{\psi}{\rightrightarrows}} \coprod_{n\geq 0} \boldsymbol{X}_n \times \Delta[n]$$

where the map ϕ on the summand $\alpha\colon [n] \to [k]$ is $1_{\boldsymbol{X}_n} \times \alpha^*\colon \boldsymbol{X}_n \times \Delta[k] \to \boldsymbol{X}_n \times \Delta[n]$ composed with the natural injection into the coproduct and the map ψ on the summand $\alpha\colon [n] \to [k]$ is $\alpha_* \times 1_{\Delta[k]}\colon \boldsymbol{X}_n \times \Delta[k] \to \boldsymbol{X}_k \times \Delta[k]$ composed with the natural injection into the coproduct.

REMARK 15.11.2. The realization of a bisimplicial set is an example of a tensor product of functors; see Definition 18.3.2.

DEFINITION 15.11.3. If $\boldsymbol{X}$ is a bisimplicial set, its *diagonal* is the simplicial set $\operatorname{diag}\boldsymbol{X}$ for which $(\operatorname{diag}\boldsymbol{X})_n = \boldsymbol{X}_{n,n}$ for $n \geq 0$, $d_i\colon (\operatorname{diag}\boldsymbol{X})_n \to (\operatorname{diag}\boldsymbol{X})_{n-1}$ is $d_i^h d_i^v\colon \boldsymbol{X}_{n,n} \to \boldsymbol{X}_{n-1,n-1}$ for $0 \leq i \leq n$, and $s_i\colon (\operatorname{diag}\boldsymbol{X})_n \to (\operatorname{diag}\boldsymbol{X})_{n+1}$ is $s_i^h s_i^v\colon \boldsymbol{X}_{n,n} \to \boldsymbol{X}_{n+1,n+1}$ for $0 \leq i \leq n$.

PROPOSITION 15.11.4. *If X is a simplicial set and $\mathrm{cs}_* X$ denotes the constant simplicial object at X (i.e., the bisimplicial set such that $(\mathrm{cs}_* X)_{i,j} = X_j$ and such that all horizontal face and degeneracy maps are the identity), then there is a natural isomorphism $X \approx \operatorname{diag}(\mathrm{cs}_* X)$.*

PROOF. This follows directly from the definitions. □

DEFINITION 15.11.5. If $\boldsymbol{X}$ is a simplicial object in the category of simplicial sets, then there is a natural map $\mathrm{cs}_*(\boldsymbol{X}_0) \to \boldsymbol{X}$ from the constant simplicial simplicial set at $\boldsymbol{X}_0$ to $\boldsymbol{X}$ that on $\big(\mathrm{cs}_*(\boldsymbol{X}_0)\big)_n = \boldsymbol{X}_0$ is the map $(s_0)^n\colon \boldsymbol{X}_0 \to \boldsymbol{X}_n$. We will call the composition $\boldsymbol{X}_0 \approx \operatorname{diag}\mathrm{cs}_*(\boldsymbol{X}_0) \to \operatorname{diag}\boldsymbol{X}$ (see Proposition 15.11.4) *the natural map* $\boldsymbol{X}_0 \to \operatorname{diag}\boldsymbol{X}$.

THEOREM 15.11.6. *If $\boldsymbol{X}$ is a bisimplicial set, then its realization $|\boldsymbol{X}|$ is naturally isomorphic to its diagonal* $\operatorname{diag}\boldsymbol{X}$.

PROOF. We first define a map $f\colon (\operatorname{diag}\boldsymbol{X}) \to |\boldsymbol{X}|$. If $\sigma \in (\operatorname{diag}\boldsymbol{X})_n = \boldsymbol{X}_{n,n}$, we let $f(\sigma)$ be the image of $(\sigma, 1_{[n]}) \in \boldsymbol{X}_n \times \Delta[n]$ in $|\boldsymbol{X}|$, where $1_{[n]}$ is the nondegenerate n-simplex of $\Delta[n]$ (see Definition 15.11.1). If $\alpha \in \boldsymbol{\Delta}^{\mathrm{op}}([n],[k])$ is a simplicial operator, then $1_{\boldsymbol{X}_n} \times \alpha^*\colon \boldsymbol{X}_n \times \Delta[k] \to \boldsymbol{X}_n \times \Delta[n]$ takes $(\sigma, 1_{[k]})$ to $(\sigma, 1_{[n]})$ and $\alpha_* \times 1_{\Delta[k]}\colon \boldsymbol{X}_n \times \Delta[k] \to \boldsymbol{X}_k \times \Delta[k]$ takes $(\sigma, 1_{[k]})$ to $(\alpha(\sigma), 1_{[k]})$ in $|\boldsymbol{X}|$. Thus, $f\bigl(\alpha(\sigma)\bigr) = \alpha\bigl(f(\sigma)\bigr)$, and so f is a map of simplicial sets.

We now define a map $g\colon |\boldsymbol{X}| \to (\operatorname{diag}\boldsymbol{X})$. Since $\Delta[n]$ is the free $\boldsymbol{\Delta}^{\mathrm{op}}$-diagram generated at $[n]$ (see Example 11.5.15), a k-simplex of $\Delta[n]$ is a simplicial operator $\alpha\colon [n] \to [k]$, and so a k-simplex of $\boldsymbol{X}_n \times \Delta[n]$ is of the form (σ, α) for $\sigma \in \boldsymbol{X}_{n,k}$ and $\alpha \in \boldsymbol{\Delta}^{\mathrm{op}}([n],[k])$. We define $g_n\colon \boldsymbol{X}_n \times \Delta[n] \to (\operatorname{diag}\boldsymbol{X})$ by letting $g_n(\sigma,\alpha) = \alpha^h_*(\sigma) \in \boldsymbol{X}_{k,k}$. If $\beta \in \boldsymbol{\Delta}^{\mathrm{op}}([k],[m])$ is a simplicial operator, then $\beta_*(\sigma,\alpha) = \bigl(\beta^v_*(\sigma), \beta\circ\alpha\bigr)$, and so $g_n\bigl(\beta_*(\sigma,\alpha)\bigr) = g_n\bigl(\beta^v_*(\sigma), \beta\alpha\bigr) = \beta^h_*\alpha^h_*\beta^v_*(\sigma) = \beta^h_*\beta^v_*\alpha^h_*(\sigma) = \beta^h_*\beta^v_* g_n(\sigma,\alpha) = \beta_* g_n(\sigma,\alpha)$, so g_n is a map of simplicial sets. To see that the g_n define a map on $|\boldsymbol{X}|$, let $\alpha \in \boldsymbol{\Delta}^{\mathrm{op}}([n],[k])$ (see Definition 15.11.1), and let (σ,β) be a simplex of $\boldsymbol{X}_n \times \Delta[k]$; then $g_k\phi(\sigma,\beta) = g_k\bigl(\sigma, \alpha^*(\beta)\bigr) = g_k(\sigma, \beta\alpha) = (\beta\alpha)^h_*(\sigma) = \beta^h_*\alpha^h_*(\sigma) = g_n\bigl(\alpha^h_*(\sigma), \beta\bigr) = g_n\psi(\sigma,\beta)$. Thus, the g_n combine to define $g\colon |\boldsymbol{X}| \to (\operatorname{diag}\boldsymbol{X})$.

We first show that $gf = 1_{(\operatorname{diag}\boldsymbol{X})}$. If $\sigma \in (\operatorname{diag}\boldsymbol{X})_n = \boldsymbol{X}_{n,n}$, then $gf(\sigma) = g_n(\sigma, 1_n) = (1_{[n]})_*(\sigma) = \sigma$.

We now show that $fg = 1_{|\boldsymbol{X}|}$. If $\sigma \in \boldsymbol{X}_{n,n}$, then $fg(\sigma, 1_{[n]}) = f\bigl((1_{[n]})_*(\sigma)\bigr) = f(\sigma) = (\sigma, 1_{[n]})$ in $|\boldsymbol{X}|$, and so it is sufficient to show that every simplex of $|\boldsymbol{X}|$ is equivalent to one of the form $(\sigma, 1_{[n]})$. If (σ,α) is a k-simplex of $\boldsymbol{X}_n \times \Delta[n]$, then $\sigma \in \boldsymbol{X}_{n,k}$ and $\alpha \in \boldsymbol{\Delta}^{\mathrm{op}}([n],[k])$, and so $(\sigma, 1_{[k]})$ is a k-simplex of $\boldsymbol{X}_n \times \Delta[k]$. We have $\phi(\sigma, 1_{[k]}) = \bigl(\sigma, \alpha^*(1_{[k]})\bigr) = (\sigma,\alpha)$ and $\psi(\sigma, 1_{[k]}) = \bigl(\alpha(\sigma), 1_{[k]}\bigr)$ and $\alpha(\sigma) \in \boldsymbol{X}_{k,k}$, and so the simplex of $|\boldsymbol{X}|$ represented by (σ,α) is also represented by $\bigl(\sigma(\alpha), 1_{[k]}\bigr)$. □

THEOREM 15.11.7 (A. K. Bousfield and E. M. Friedlander, [**13**]). *If $f\colon \boldsymbol{X} \to \boldsymbol{Y}$ is a map of bisimplicial sets such that*

1. *as a map of horizontal simplicial objects in the category of vertical simplicial sets (i.e., $(\boldsymbol{X}_n)_k = \boldsymbol{X}_{n,k}$), f is a Reedy fibration, and*
2. *as a map of vertical simplicial objects in the category of horizontal simplicial sets (i.e., $(\boldsymbol{X}_n)_k = \boldsymbol{X}_{k,n}$), f is an objectwise fibration (i.e., every induced map $\boldsymbol{X}_{*,n} \to \boldsymbol{Y}_{*,n}$ is a fibration of simplicial sets),*

then the induced map of diagonals $\operatorname{diag} f\colon (\operatorname{diag}\boldsymbol{X}) \to (\operatorname{diag}\boldsymbol{Y})$ *is a fibration of simplicial sets.*

PROOF. This is [**13**, Lemma B.9]. □

DEFINITION 15.11.8. If $\boldsymbol{X}$ is a bisimplicial set, i.e., an object of $\mathrm{SS}^{\boldsymbol{\Delta}^{\mathrm{op}}}$, and Y is a simplicial set, then $\operatorname{Map}(\boldsymbol{X}, Y)$ is the cosimplicial simplicial set given by $\operatorname{Map}(\boldsymbol{X}, Y)^n = \operatorname{Map}(\boldsymbol{X}_n, Y)$, with coface and codegeneracy maps induced by the face and degeneracy maps of $\boldsymbol{X}$.

THEOREM 15.11.9. *If $\boldsymbol{X}\colon \boldsymbol{\Delta}^{\mathrm{op}} \to \mathrm{SS}$ is a bisimplicial set, $\boldsymbol{Y}\colon \boldsymbol{\Delta} \to \mathrm{SS}$ is a cosimplicial simplicial set, and Z is a simplicial set, then there is a natural isomorphism of simplicial sets*

$$\mathrm{Map}(\boldsymbol{X} \otimes_{\boldsymbol{\Delta}} \boldsymbol{Y}, Z) \approx \mathrm{Map}\big(\boldsymbol{Y}, \mathrm{Map}(\boldsymbol{X}, Z)\big).$$

PROOF. We have the coequalizer diagram of simplicial sets

$$\coprod_{(\sigma\colon [n]\to[m])\in\boldsymbol{\Delta}} \boldsymbol{X}_m \times \boldsymbol{Y}^n \underset{\psi}{\overset{\phi}{\rightrightarrows}} \coprod_{n\geq 0} \boldsymbol{X}_n \times \boldsymbol{Y}^n \to \boldsymbol{X} \otimes_{\boldsymbol{\Delta}} \boldsymbol{Y}.$$

Since the functor $- \times \Delta[k]\colon \mathrm{SS} \to \mathrm{SS}$ is a left adjoint, the diagram

$$\coprod_{(\sigma\colon [n]\to[m])\in\boldsymbol{\Delta}} \boldsymbol{X}_m \times \boldsymbol{Y}^n \times \Delta[k] \rightrightarrows \coprod_{n\geq 0} \boldsymbol{X}_n \times \boldsymbol{Y}^n \times \Delta[k] \to (\boldsymbol{X} \otimes_{\boldsymbol{\Delta}} \boldsymbol{Y}) \times \Delta[k]$$

is also a coequalizer diagram, and so we have the equalizer diagram

$$\begin{aligned} \mathrm{SS}\big((\boldsymbol{X} \otimes_{\boldsymbol{\Delta}} \boldsymbol{Y}) \times \Delta[k], Z\big) \to \prod_{n\geq 0} &\mathrm{SS}(\boldsymbol{X}_n \times \boldsymbol{Y}^n \times \Delta[k], Z) \\ &\rightrightarrows \prod_{(\sigma\colon [n]\to[m])\in\boldsymbol{\Delta}} \mathrm{SS}(\boldsymbol{X}_m \times \boldsymbol{Y}^n \times \Delta[k], Z) \end{aligned}$$

which is isomorphic to the diagram

$$\begin{aligned} \mathrm{SS}\big((\boldsymbol{X} \otimes_{\boldsymbol{\Delta}} \boldsymbol{Y}) \times \Delta[k], Z\big) \to \prod_{n\geq 0} &\mathrm{SS}\big(\boldsymbol{Y}^n \times \Delta[k], \mathrm{Map}(\boldsymbol{X}_n, Z)\big) \\ &\rightrightarrows \prod_{(\sigma\colon [n]\to[m])\in\boldsymbol{\Delta}} \mathrm{SS}\big(\boldsymbol{Y}^n \times \Delta[k], \mathrm{Map}(\boldsymbol{X}_m, Z)\big). \end{aligned}$$

This implies that the diagram

$$\begin{aligned} \mathrm{Map}(\boldsymbol{X} \otimes_{\boldsymbol{\Delta}} \boldsymbol{Y}, Z) \to \prod_{n\geq 0} &\mathrm{Map}\big(\boldsymbol{Y}^n, \mathrm{Map}(\boldsymbol{X}_n, Z)\big) \\ &\rightrightarrows \prod_{(\sigma\colon [n]\to[m])\in\boldsymbol{\Delta}} \mathrm{Map}\big(\boldsymbol{Y}^n, \mathrm{Map}(\boldsymbol{X}_m, Z)\big) \end{aligned}$$

is an equalizer diagram, from which the result follows. □

LEMMA 15.11.10. *Let $\mathcal{C}$ be a Reedy category and let $\mathcal{M}$ be a simplicial model category. If $\boldsymbol{X}$ is a Reedy cofibrant $\mathcal{C}$-diagram in $\mathcal{M}$ and Y is a fibrant object of $\mathcal{M}$, then the $\mathcal{C}^{\mathrm{op}}$-diagram of simplicial sets $\mathrm{Map}(\boldsymbol{X}, Y)$ is Reedy fibrant (see Proposition 15.1.5).*

PROOF. If α is an object of $\mathcal{C}$ and $\mathrm{L}_\alpha \boldsymbol{X}$ is the latching object of $\boldsymbol{X}$ at α (see Definition 15.2.5), then Proposition 15.2.4 implies that there are natural isomorphisms

$$\begin{aligned} \mathrm{Map}(\mathrm{L}_\alpha \boldsymbol{X}, Y) &= \mathrm{Map}\Big(\operatorname*{colim}_{\partial(\overrightarrow{\mathcal{C}}\downarrow\alpha)} \boldsymbol{X}, Y\Big) \\ &\approx \lim_{\partial(\overrightarrow{\mathcal{C}}\downarrow\alpha)^{\mathrm{op}}} \mathrm{Map}(\boldsymbol{X}, Y) \\ &\approx \lim_{\partial(\alpha\downarrow\overleftarrow{\mathcal{C}^{\mathrm{op}}})} \mathrm{Map}(\boldsymbol{X}, Y) \\ &= \mathrm{M}_\alpha \mathrm{Map}(\boldsymbol{X}, Y) \ , \end{aligned}$$

i.e., $\mathrm{Map}(\mathrm{L}_\alpha \boldsymbol{X}, Y)$ is naturally isomorphic to the matching object at α of the $\mathcal{C}^{\mathrm{op}}$-diagram $\mathrm{Map}(\boldsymbol{X}, Y)$. Since the latching map $\mathrm{L}_\alpha \boldsymbol{X} \to \boldsymbol{X}_\alpha$ is a cofibration and Y is fibrant, Proposition 9.3.1 implies that the matching map $\mathrm{Map}(\boldsymbol{X}_\alpha, Y) \to \mathrm{M}_\alpha \mathrm{Map}(\boldsymbol{X}, Y)$ is a fibration, and so $\mathrm{Map}(\boldsymbol{X}, Y)$ is a Reedy fibrant $\mathcal{C}^{\mathrm{op}}$-diagram of simplicial sets. □

THEOREM 15.11.11. *If $f\colon \boldsymbol{X} \to \boldsymbol{Y}$ is a map of bisimplicial sets such that $f_n\colon \boldsymbol{X}_n \to \boldsymbol{Y}_n$ is a weak equivalence of simplicial sets for every $n \geq 0$, then the induced map of realizations $|f|\colon |\boldsymbol{X}| \to |\boldsymbol{Y}|$ is a weak equivalence of simplicial sets.*

PROOF. It is sufficient to show that if Z is a fibrant simplicial set, then the induced map $|f|^*\colon \mathrm{Map}(|\boldsymbol{Y}|, Z) \to \mathrm{Map}(|\boldsymbol{X}|, Z)$ is a weak equivalence (see Corollary 9.7.5).

Corollary 15.8.8 implies that $\boldsymbol{X}$ and $\boldsymbol{Y}$ are Reedy cofibrant. Since Z is fibrant, Lemma 15.11.10 implies that the map $\mathrm{Map}(\boldsymbol{Y}, Z) \to \mathrm{Map}(\boldsymbol{X}, Z)$ is a map of Reedy fibrant cosimplicial simplicial sets, and Corollary 9.3.3 implies that it is a Reedy weak equivalence of cosimplicial simplicial sets. Since Δ (see Definition 15.1.15) is a cofibrant cosimplicial simplicial set (see Corollary 15.9.11), the map $\mathrm{Map}\big(\Delta, \mathrm{Map}(\boldsymbol{Y}, Z)\big) \to \mathrm{Map}\big(\Delta, \mathrm{Map}(\boldsymbol{X}, Z)\big)$ is a weak equivalence of simplicial sets (see Corollary 9.3.3 and Theorem 15.3.4). This is isomorphic to the map $\mathrm{Map}(\boldsymbol{Y} \otimes_{\boldsymbol{\Delta}} \Delta, Z) \to \mathrm{Map}(\boldsymbol{X} \otimes_{\boldsymbol{\Delta}} \Delta, Z)$ (see Theorem 15.11.9), which is the definition of the map $\mathrm{Map}(|\boldsymbol{Y}|, Z) \to \mathrm{Map}(|\boldsymbol{X}|, Z)$ (see Definition 15.11.1). □

COROLLARY 15.11.12. *If $\boldsymbol{X}$ is a simplicial object in the category of simplicial sets in which all the face and degeneracy operators are weak equivalences, then the natural map $\boldsymbol{X}_0 \to |\boldsymbol{X}|$ (defined as the composition $\boldsymbol{X}_0 \to \mathrm{diag}\, \boldsymbol{X} \approx |\boldsymbol{X}|$; see Definition 15.11.5) is a weak equivalence.*

PROOF. This follows from Theorem 15.11.11. □

CHAPTER 16

Cosimplicial and Simplicial Resolutions

If $\mathcal{M}$ is a simplicial model category and $W \to X$ is a cofibrant approximation to X, then the cosimplicial object $\widetilde{\boldsymbol{X}}$ in which $\widetilde{\boldsymbol{X}}^n = W \otimes \Delta[n]$ is a cosimplicial resolution of X. Dually, if $\mathcal{M}$ is a simplicial model category and $Y \to Z$ is a fibrant approximation to Y, then the simplicial object $\widehat{\boldsymbol{Y}}$ in which $\widehat{\boldsymbol{Y}}_n = Z^{\Delta[n]}$ is a simplicial resolution of Y. In this chapter, we define cosimplicial and simplicial resolutions in an arbitrary model category (see Definition 16.1.2), and establish a number of their technical properties. The constructions of this chapter will be used in Chapter 17 to define homotopy function complexes between objects in a model category (see Definition 17.4.1) and in Chapter 19 to define homotopy colimits and homotopy limits of diagrams in model categories (see Definition 19.1.2 and Definition 19.1.5).

In Section 16.1 we define cosimplicial and simplicial resolutions of objects and maps, and we establish existence and uniqueness theorems. In Section 16.2 we show that left Quillen functors preserve cosimplicial resolutions of cofibrant objects and that right Quillen functors preserve simplicial resolutions of fibrant objects.

In Section 16.3 we define the *realization* $\boldsymbol{X} \otimes K$ of a cosimplicial object $\boldsymbol{X}$ in $\mathcal{M}$ and a simplicial set K. This is an object of $\mathcal{M}$ that is the colimit of a diagram of the $\boldsymbol{X}_n$ indexed by the simplices of the simplicial set K. If $\mathcal{M} = \mathrm{Top}$ and $\boldsymbol{X}^n = |\Delta[n]|$, then $\boldsymbol{X} \otimes K$ is the geometric realization of K (see Example 16.3.5). Dually, we also define the *corealization* $\boldsymbol{Y}^K$ of a simplicial object $\boldsymbol{Y}$ in $\mathcal{M}$ and a simplicial set K. This is an object of $\mathcal{M}$ that is the limit of a diagram of the $\boldsymbol{Y}_n$ indexed by the simplices of K.

If $\boldsymbol{X}$ is a simplicial object in $\mathcal{M}$ and Y is an object of $\mathcal{M}$, then there is a simplicial set $\mathcal{M}(\boldsymbol{X}, Y)$ in which $\mathcal{M}(\boldsymbol{X}, Y)_n = \mathcal{M}(\boldsymbol{X}^n, Y)$ (with face and degeneracy operators induced by the coface and codegeneracy operators of $\boldsymbol{X}$). Dually, if X is an object of $\mathcal{M}$ and $\boldsymbol{Y}$ is a simplicial object in $\mathcal{M}$, then there is a simplicial set $\mathcal{M}(X, \boldsymbol{Y})$ in which $\mathcal{M}(X, \boldsymbol{Y})_n = \mathcal{M}(X, \boldsymbol{Y}_n)$, and if $\boldsymbol{X}$ is a cosimplicial object in $\mathcal{M}$ and $\boldsymbol{Y}$ is a simplicial object in $\mathcal{M}$ then there is a bisimplicial set $\mathcal{M}(\boldsymbol{X}, \boldsymbol{Y})$ in which $\mathcal{M}(\boldsymbol{X}, \boldsymbol{Y})_{n,k} = \mathcal{M}(\boldsymbol{X}^k, \boldsymbol{Y}_n)$ (see Notation 16.4.1). We will use these constructions in Chapter 17 to define homotopy function complexes (see Definition 17.1.1, Definition 17.2.1, and Definition 17.3.1). In Section 16.4 we establish some adjointness properties for these constructions, and we use these in Section 16.5 to prove several *homotopy lifting extension theorems* (see Theorem 16.5.2, Theorem 16.5.13, and Theorem 16.5.18).

If $\mathcal{M}$ is a simplicial model category and X is an object of $\mathcal{M}$, then the cosimplicial object $\widetilde{\boldsymbol{X}}$ in which $\widetilde{\boldsymbol{X}}^n = X \otimes \Delta[n]$ will not be a cosimplicial resolution of X unless X is cofibrant. In Section 16.6 we define cosimplicial and simplicial *frames* on an object to describe this situation, and we show that a cosimplicial resolution of an object is exactly a cosimplicial frame on a cofibrant approximation to that object (with dual definitions and results for *simplicial frames.*) These will be used in Chapter 19 to define homotopy colimit and homotopy limit functors.

If $\mathcal{C}$ is a Reedy category, $\boldsymbol{X}$ is a $\mathcal{C}$-diagram in a model category $\mathcal{M}$, and $\widetilde{\boldsymbol{X}}$ is a natural cosimplicial frame on $\boldsymbol{X}$, then even if $\boldsymbol{X}$ is a Reedy cofibrant diagram it need not be true that $\widetilde{\boldsymbol{X}}$ is a Reedy cofibrant diagram of cosimplicial objects in $\mathcal{M}$. (A dual statement applies to simplicial frames.) In Section 16.7 we define a *Reedy frame* on a diagram (see Definition 16.7.8) as one in which this difficulty does not arise, and we obtain existence and uniqueness results (see Theorem 16.7.6, Proposition 16.7.11, and Theorem 16.7.14). These will be used in Chapter 19 to discuss homotopy colimits of Reedy cofibrant diagrams and homotopy limits of Reedy fibrant diagrams (see Theorem 19.9.1).

16.1. Resolutions

NOTATION 16.1.1. Let $\mathcal{M}$ be a model category.

- The category of cosimplicial objects in $\mathcal{M}$ will be denoted $\mathcal{M}^{\boldsymbol{\Delta}}$.
- The category of simplicial objects in $\mathcal{M}$ will be denoted $\mathcal{M}^{\boldsymbol{\Delta}^{\mathrm{op}}}$.
- If X is an object of $\mathcal{M}$, then
  - the constant cosimplicial object at X will be denoted cc_*X, and
  - the constant simplicial object at X will be denoted cs_*X.

DEFINITION 16.1.2. Let $\mathcal{M}$ be a model category and let X be an object of $\mathcal{M}$.

(1)
- A *cosimplicial resolution* of X is a cofibrant approximation (see Definition 8.1.2) $\widetilde{\boldsymbol{X}} \to \mathrm{cc}_*X$ to cc_*X (see Notation 16.1.1) in the Reedy model category structure (see Definition 15.3.3) on $\mathcal{M}^{\boldsymbol{\Delta}}$.
- A *fibrant cosimplicial resolution* is a cosimplicial resolution in which the weak equivalence $\widetilde{\boldsymbol{X}} \to \mathrm{cc}_*X$ is a Reedy trivial fibration.

We will sometimes use the term *cosimplicial resolution* to refer to the object $\widetilde{\boldsymbol{X}}$ without explicitly mentioning the weak equivalence $\widetilde{\boldsymbol{X}} \to \mathrm{cc}_*X$.

(2)
- A *simplicial resolution* of X is a fibrant approximation $\mathrm{cs}_*X \to \widehat{\boldsymbol{X}}$ to cs_*X in the Reedy model category structure on $\mathcal{M}^{\boldsymbol{\Delta}^{\mathrm{op}}}$.
- A *cofibrant simplicial resolution* is a simplicial resolution in which the weak equivalence $\mathrm{cs}_*X \to \widehat{\boldsymbol{X}}$ is a Reedy trivial cofibration.

We will sometimes use the term *simplicial resolution* to refer to the object $\widehat{\boldsymbol{X}}$ without explicitly mentioning the weak equivalence $\mathrm{cs}_*X \to \widehat{\boldsymbol{X}}$.

PROPOSITION 16.1.3. *Let $\mathcal{M}$ be a simplicial model category.*

1. *If X is an object of $\mathcal{M}$ and $W \to X$ is a cofibrant approximation to X, then the cosimplicial object $\widetilde{\boldsymbol{W}}$ in which $\widetilde{\boldsymbol{W}}^n = W \otimes \Delta[n]$ is a cosimplicial resolution of X.*
2. *If Y is an object of $\mathcal{M}$ and $Y \to Z$ is a fibrant approximation to Y, then the simplicial object $\widehat{\boldsymbol{Z}}$ in which $\widehat{\boldsymbol{Z}}_n = Z^{\Delta[n]}$ is a simplicial resolution of Y.*

PROOF. We will prove part 1; the proof of part 2 is similar.

Since all of the inclusions $\Delta[0] \to \Delta[n]$ are trivial cofibrations and W is cofibrant, all of the maps $W \approx W \otimes \Delta[0] \to W \otimes \Delta[n]$ are trivial cofibrations (see Proposition 9.3.9). Thus, $\widetilde{\boldsymbol{W}}$ is weakly equivalent to cc_*X. Since each $\partial\Delta[n] \to \Delta[n]$ is a cofibration and W is cofibrant, each latching map $W \otimes \partial\Delta[n] \to W \otimes \Delta[n]$ (see Lemma 9.2.1) is a cofibration, and so $\widetilde{\boldsymbol{W}}$ is cofibrant. □

COROLLARY 16.1.4. *Let $\mathcal{M}$ be a simplicial model category.*

(1) *If X is a cofibrant object of $\mathcal{M}$, then the cosimplicial object $\widetilde{\boldsymbol{X}}$ in which $\widetilde{\boldsymbol{X}}^n = X \otimes \Delta[n]$ is a cosimplicial resolution of X.*
(2) *If Y is a fibrant object of $\mathcal{M}$, then the simplicial object $\widehat{\boldsymbol{Y}}$ in which $\widehat{\boldsymbol{Y}}_n = Y^{\Delta[n]}$ is a simplicial resolution of Y.*

PROOF. This follows from Proposition 16.1.3. □

The next two propositions show that if X is an object in a model category and $\widetilde{\boldsymbol{X}} \to \mathrm{cc}_* X$ is a cosimplicial resolution of X, then $\widetilde{\boldsymbol{X}}^0 \to X$ is a cofibrant approximation to X and $\widetilde{\boldsymbol{X}}^1$ is a cylinder object for $\widetilde{\boldsymbol{X}}^0$. Thus, a cosimplicial resolution of X is a sort of collection of "higher cylinder objects" for a cofibrant approximation to X. Dually, a simplicial resolution is a sort of collection of "higher path objects" for a fibrant approximation to X.

PROPOSITION 16.1.5. *Let $\mathcal{M}$ be a model category and let X be an object of $\mathcal{M}$.*

(1) *If $\widetilde{\boldsymbol{X}} \to \mathrm{cc}_* X$ is a cosimplicial resolution of X (see Definition 16.1.2), then $\widetilde{\boldsymbol{X}}^0 \to X$ is a cofibrant approximation to X. If $\widetilde{\boldsymbol{X}} \to \mathrm{cc}_* X$ is a fibrant cosimplicial resolution of X, then $\widetilde{\boldsymbol{X}}^0 \to X$ is a fibrant cofibrant approximation to X.*
(2) *If $\mathrm{cs}_* X \to \widehat{\boldsymbol{X}}$ is a simplicial resolution of X, then $X \to \widehat{\boldsymbol{X}}_0$ is a fibrant approximation to X. If $\mathrm{cs}_* X \to \widehat{\boldsymbol{X}}$ is a cofibrant simplicial resolution of X, then $X \to \widehat{\boldsymbol{X}}_0$ is a cofibrant fibrant approximation to X.*

PROOF. This follows from Proposition 15.3.11. □

PROPOSITION 16.1.6. *Let $\mathcal{M}$ be a model category.*

(1) *If $\widetilde{\boldsymbol{X}}$ is a cosimplicial resolution in $\mathcal{M}$, then $\widetilde{\boldsymbol{X}}^0 \amalg \widetilde{\boldsymbol{X}}^0 \xrightarrow{d^0 \amalg d^1} \widetilde{\boldsymbol{X}}^1 \xrightarrow{s^0} \widetilde{\boldsymbol{X}}^0$ is a cylinder object (see Definition 7.3.2) for $\widetilde{\boldsymbol{X}}^0$.*
(2) *If $\widehat{\boldsymbol{X}}$ is a simplicial resolution in $\mathcal{M}$, then $\widehat{\boldsymbol{X}}_0 \xrightarrow{s_0} \widehat{\boldsymbol{X}}_1 \xrightarrow{d_0 \times d_1} \widehat{\boldsymbol{X}}_0 \times \widehat{\boldsymbol{X}}_0$ is a path object for $\widehat{\boldsymbol{X}}_0$.*

PROOF. This follows directory from the definitions. □

16.1.7. Existence of functorial resolutions.

DEFINITION 16.1.8. Let $\mathcal{M}$ be a model category and let $\mathcal{K}$ be a subcategory of $\mathcal{M}$.

(1) A *functorial cosimplicial resolution* on $\mathcal{K}$ is a pair (F, i) in which $\mathrm{F}\colon \mathcal{K} \to \mathcal{M}^{\boldsymbol{\Delta}}$ is a functor and i is a natural transformation such that $i_X \colon \mathrm{F}X \to \mathrm{cc}_* X$ is a cosimplicial resolution of X for every object X of $\mathcal{K}$.
(2) A *functorial simplicial resolution* on $\mathcal{K}$ is a pair (G, j) in which $\mathrm{G}\colon \mathcal{K} \to \mathcal{M}^{\boldsymbol{\Delta}^{\mathrm{op}}}$ is a functor and j is a natural transformation such that $j_X \colon \mathrm{cs}_* X \to \mathrm{G}X$ is a simplicial resolution of X for every object X of $\mathcal{K}$.

PROPOSITION 16.1.9. *Let $\mathcal{M}$ be a model category.*

(1) *There is a functorial cosimplicial resolution* (F, i) *on* $\mathcal{M}$ *such that for every object* X *of* $\mathcal{M}$ *the cosimplicial resolution* $i_X\colon \mathrm{F}X \to \mathrm{cc}_* X$ *is a fibrant cosimplicial resolution of* X

(2) *There is a functorial simplicial resolution* (G, j) *on* $\mathcal{M}$ *such that for every object* X *of* $\mathcal{M}$ *the simplicial resolution* $j_X\colon \mathrm{cs}_* X \to \mathrm{G}X$ *is a cofibrant simplicial resolution of* X

PROOF. This follows from Proposition 8.1.17. □

16.1.10. Uniqueness of resolutions.

DEFINITION 16.1.11. Let $\mathcal{M}$ be a model category and let X be an object of $\mathcal{M}$.

(1) If $\widetilde{\boldsymbol{X}} \xrightarrow{i} \mathrm{cc}_* X$ and $\widetilde{\boldsymbol{X}}' \xrightarrow{i'} \mathrm{cc}_* X$ are cosimplicial resolutions of X, then a *map of cosimplicial resolutions* from $(\widetilde{\boldsymbol{X}}, i)$ to $(\widetilde{\boldsymbol{X}}', i')$ is a map $g\colon \widetilde{\boldsymbol{X}} \to \widetilde{\boldsymbol{X}}'$ such that $i'g = i$.

(2) If $\mathrm{cs}_* X \xrightarrow{j} \widehat{\boldsymbol{X}}$ and $\mathrm{cs}_* X \xrightarrow{j'} \widehat{\boldsymbol{X}}'$ are simplicial resolutions of X, then a *map of simplicial resolutions* from $(\widehat{\boldsymbol{X}}, j)$ to $(\widehat{\boldsymbol{X}}', j')$ is a map $g\colon \widehat{\boldsymbol{X}} \to \widehat{\boldsymbol{X}}'$ such that $gj = j'$.

LEMMA 16.1.12. *Let* $\mathcal{M}$ *be a model category and let* X *be an object of* $\mathcal{M}$.

(1) *If* $(\widetilde{\boldsymbol{X}}, i)$ *and* $(\widetilde{\boldsymbol{X}}', i')$ *are cosimplicial resolutions of* X *and* $g\colon \widetilde{\boldsymbol{X}} \to \widetilde{\boldsymbol{X}}'$ *is a map of cosimplicial resolutions, then* g *is a Reedy weak equivalence.*

(2) *If* $(\widehat{\boldsymbol{X}}, j)$ *and* $(\widehat{\boldsymbol{X}}', j')$ *are simplicial resolutions of* X *and* $g\colon \widehat{\boldsymbol{X}} \to \widehat{\boldsymbol{X}}'$ *is a map of simplicial resolutions, then* g *is a Reedy weak equivalence.*

PROOF. This follows from Lemma 8.1.5. □

PROPOSITION 16.1.13. *Let* $\mathcal{M}$ *be a model category and let* X *be an object of* $\mathcal{M}$.

(1) *If* $\widetilde{\boldsymbol{X}} \to \mathrm{cc}_* X$ *is cosimplicial resolution of* X *and* $\widetilde{\boldsymbol{X}}' \to \mathrm{cc}_* X$ *is a fibrant cosimplicial resolution of* X, *then there is a map* $\widetilde{\boldsymbol{X}} \to \widetilde{\boldsymbol{X}}'$ *of cosimplicial resolutions, unique up to homotopy over* $\mathrm{cc}_* X$, *and any such map is a weak equivalence.*

(2) *If* $\mathrm{cs}_* X \to \widehat{\boldsymbol{X}}$ *is a simplicial resolution of* X *and* $\mathrm{cs}_* X \to \widehat{\boldsymbol{X}}'$ *is a cofibrant simplicial resolution of* X, *then there is a map* $\widehat{\boldsymbol{X}}' \to \widehat{\boldsymbol{X}}$ *of simplicial resolutions, unique up to homotopy under* $\mathrm{cs}_* X$, *and any such map is a weak equivalence.*

PROOF. This follows from Proposition 8.1.7. □

DEFINITION 16.1.14. Let $\mathcal{M}$ be a model category and let X be an object of $\mathcal{M}$.

(1) The category $\mathrm{CRes}(X)$ is the category whose objects are cosimplicial resolutions of X and whose morphisms are maps of cosimplicial resolutions.

(2) The category $\mathrm{SRes}(X)$ is the category whose objects are simplicial resolutions of X and whose morphisms are maps of simplicial resolutions.

PROPOSITION 16.1.15. *Let* $\mathcal{M}$ *be a model category and let* X *be an object of* $\mathcal{M}$.

(1) *The classifying space* $\mathrm{B}\,\mathrm{CRes}(X)$ *of the category of cosimplicial resolutions of* X *(see Definition 16.1.14) is contractible (see Definition 14.3.1).*

(2) *The classifying space* $\mathrm{B}\,\mathrm{SRes}(X)$ *of the category of simplicial resolutions of* X *(see Definition 16.1.14) is contractible (see Definition 14.3.1).*

PROOF. This follows from Theorem 14.6.2. □

PROPOSITION 16.1.16. *Let* $\mathcal{M}$ *be a model category and let* $\mathcal{K}$ *be a subcategory of* $\mathcal{M}$.

(1) *For every small category* $\mathcal{D}$ *of functorial cosimplicial resolutions on* $\mathcal{K}$ *(see Definition 14.6.7) there is a small category* $\mathcal{D}'$ *of functorial cosimplicial resolutions on* $\mathcal{K}$ *such that* $\mathcal{D} \subset \mathcal{D}'$ *and* $\mathrm{B}\mathcal{D}'$ *is contractible.*
(2) *For every small category* $\mathcal{D}$ *of functorial simplicial resolutions on* $\mathcal{K}$ *(see Definition 14.6.7) there is a small category* $\mathcal{D}'$ *of functorial simplicial resolutions on* $\mathcal{K}$ *such that* $\mathcal{D} \subset \mathcal{D}'$ *and* $\mathrm{B}\mathcal{D}'$ *is contractible.*

PROOF. This follows from Theorem 14.6.8. □

PROPOSITION 16.1.17. *Let* $\mathcal{M}$ *be a model category and let* X *be an object of* $\mathcal{M}$.

(1) *Any two cosimplicial resolutions of* X *are connected by an essentially unique zig-zag (see Definition 14.4.2) of weak equivalences.*
(2) *Any two simplicial resolutions of* X *are connected by an essentially unique zig-zag (see Definition 14.4.2) of weak equivalences.*

PROOF. This follows from Proposition 16.1.15. □

PROPOSITION 16.1.18. *Let* $\mathcal{M}$ *be a model category and let* $\mathcal{K}$ *be a subcategory of* $\mathcal{M}$.

(1) *Any two functorial cosimplicial resolutions on* $\mathcal{K}$ *are connected by an essentially unique zig-zag (see Definition 14.4.2) of weak equivalences.*
(2) *Any two functorial simplicial resolutions on* $\mathcal{K}$ *are connected by an essentially unique zig-zag (see Definition 14.4.2) of weak equivalences.*

PROOF. This follows from Proposition 16.1.16. □

16.1.19. Resolutions of maps.

DEFINITION 16.1.20. Let $\mathcal{M}$ be a model category, and let $g\colon X \to Y$ be a map in $\mathcal{M}$.

(1) A *cosimplicial resolution of* g consists of a cosimplicial resolution $\widetilde{\boldsymbol{X}} \to \mathrm{cc}_*X$ of X, a cosimplicial resolution $\widetilde{\boldsymbol{Y}} \to \mathrm{cc}_*Y$ of Y, and a map $\tilde{g}\colon \widetilde{\boldsymbol{X}} \to \widetilde{\boldsymbol{Y}}$ that makes the square

$$\begin{array}{ccc} \widetilde{\boldsymbol{X}} & \xrightarrow{\tilde{g}} & \widetilde{\boldsymbol{Y}} \\ \downarrow & & \downarrow \\ \mathrm{cc}_*X & \longrightarrow & \mathrm{cc}_*Y \end{array}$$

commute.

(2) A *simplicial resolution of* g consists of a simplicial resolution $\mathrm{cs}_* X \to \widehat{\boldsymbol{X}}$ of X, a simplicial resolution $\mathrm{cs}_* Y \to \widehat{\boldsymbol{Y}}$ of Y, and a map $\hat{g}\colon \widehat{\boldsymbol{X}} \to \widehat{\boldsymbol{Y}}$ that makes the square

$$\begin{array}{ccc} \mathrm{cs}_* X & \longrightarrow & \mathrm{cs}_* Y \\ \downarrow & & \downarrow \\ \widehat{\boldsymbol{X}} & \xrightarrow[\hat{g}]{} & \widehat{\boldsymbol{Y}} \end{array}$$

commute.

Remark 16.1.21. The effect of Definition 16.1.2 and Definition 16.1.20 is that

- a cosimplicial resolution of an object or map in a model category is exactly a Reedy cofibrant approximation to a constant cosimplicial object or map, and
- a simplicial resolution of an object or map in a model category is exactly a Reedy fibrant approximation to a constant simplicial object or map.

This is the explanation of the terminology "fibrant cosimplicial resolution" and "cofibrant simplicial resolution".

Proposition 16.1.22. *Let* $\mathcal{M}$ *be a model category and let* $g\colon X \to Y$ *be a map in* $\mathcal{M}$.

(1) *There exists a natural cosimplicial resolution* $\tilde{g}\colon \widetilde{\boldsymbol{X}} \to \widetilde{\boldsymbol{Y}}$ *of* g *such that* $\widetilde{\boldsymbol{X}}$ *and* $\widetilde{\boldsymbol{Y}}$ *are fibrant cosimplicial resolutions of, respectively,* X *and* Y, *and* $\tilde{g}$ *is a Reedy cofibration.*
(2) *There exists a natural simplicial resolution* $\hat{g}\colon \widehat{\boldsymbol{X}} \to \widehat{\boldsymbol{Y}}$ *of* g *such that* $\widehat{\boldsymbol{X}}$ *and* $\widehat{\boldsymbol{Y}}$ *are cofibrant simplicial resolutions of, respectively,* X *and* Y, *and* $\hat{g}$ *is a Reedy fibration.*

Proof. This follows from Proposition 8.1.23. □

Proposition 16.1.23. *Let* $\mathcal{M}$ *be a model category and let* $g\colon X \to Y$ *be a map in* $\mathcal{M}$.

(1) *If* $\widetilde{\boldsymbol{X}} \to \mathrm{cc}_* X$ *is a cosimplicial resolution of* X *and* $\widetilde{\boldsymbol{Y}} \to \mathrm{cc}_* Y$ *is a fibrant cosimplicial resolution of* Y, *then there exists a resolution* $\tilde{g}\colon \widetilde{\boldsymbol{X}} \to \widetilde{\boldsymbol{Y}}$ *of* g, *and* $\tilde{g}$ *is unique up to homotopy in* $(\mathcal{M}^{\boldsymbol{\Delta}} \downarrow \mathrm{cc}_* Y)$.
(2) *If* $\mathrm{cs}_* Y \to \widehat{\boldsymbol{Y}}$ *is a simplicial resolution of* Y *and* $\mathrm{cs}_* X \to \widehat{\boldsymbol{X}}$ *is a cofibrant simplicial resolution of* X, *then there exists a resolution* $\hat{g}\colon \widehat{\boldsymbol{X}} \to \widehat{\boldsymbol{Y}}$ *of* g, *and* $\hat{g}$ *is unique up to homotopy in* $(\mathrm{cs}_* X \downarrow \mathcal{M}^{\boldsymbol{\Delta}^{\mathrm{op}}})$.

Proof. This follows from Proposition 8.1.25. □

Proposition 16.1.24. *If* $\mathcal{M}$ *is a model category and* $g\colon X \to Y$ *is a weak equivalence in* $\mathcal{M}$, *then every cosimplicial resolution of* g *and every simplicial resolution of* g *are Reedy weak equivalences.*

Proof. This follows from the "two out of three" axiom for weak equivalences. □

16.1.25. Recognizing resolutions.

DEFINITION 16.1.26. Let $\mathcal{M}$ be a model category.

(1) If $\widetilde{\boldsymbol{X}}$ is a cosimplicial object in $\mathcal{M}$, then we will say that $\widetilde{\boldsymbol{X}}$ is a *cosimplicial resolution* if there is an object X in $\mathcal{M}$ and a map $\widetilde{\boldsymbol{X}} \to \mathrm{cc}_*X$ that is a cosimplicial resolution of X (see Definition 16.1.2).
(2) If $\widehat{\boldsymbol{Y}}$ is a simplicial object in $\mathcal{M}$, then we will say that $\widehat{\boldsymbol{Y}}$ is a *simplicial resolution* if there is an object Y in $\mathcal{M}$ and a map $\mathrm{cs}_*Y \to \widehat{\boldsymbol{Y}}$ that is a simplicial resolution of Y.

PROPOSITION 16.1.27. *Let $\mathcal{M}$ be a model category.*

(1) *If $\boldsymbol{X}$ is a cosimplicial object in $\mathcal{M}$, then $\boldsymbol{X}$ is a cosimplicial resolution (see Definition 16.1.26) if and only if $\boldsymbol{X}$ is Reedy cofibrant and all of the coface and codegeneracy operators of $\boldsymbol{X}$ are weak equivalences.*
(2) *If $\boldsymbol{Y}$ is a simplicial object in $\mathcal{M}$, then $\boldsymbol{Y}$ is a simplicial resolution if and only if $\boldsymbol{Y}$ is Reedy fibrant and all of the face and degeneracy operators of $\boldsymbol{Y}$ are weak equivalences.*

PROOF. We will prove part 1; the proof of part 2 is dual.

If $\boldsymbol{X}$ is a cosimplicial resolution, then it follows directly from the definitions that $\boldsymbol{X}$ is Reedy cofibrant and all of the coface and codegeneracy operators of $\boldsymbol{X}$ are weak equivalences. For the converse, the map $\boldsymbol{X} \to \mathrm{cc}_*\boldsymbol{X}^0$ defined on $\boldsymbol{X}^n$ as any n-fold iterated coface map is a cosimplicial resolution of $\boldsymbol{X}^0$. □

LEMMA 16.1.28. *Let $\mathcal{M}$ be a model category.*

(1) *If $i\colon \boldsymbol{A} \to \boldsymbol{B}$ is a weak equivalence of cosimplicial resolutions in $\mathcal{M}$, then there is a natural factorization of i as $\boldsymbol{A} \xrightarrow{q} \boldsymbol{C} \xrightarrow{r} \boldsymbol{B}$ such that $\boldsymbol{C}$ is a cosimplicial resolution in $\mathcal{M}$, q is a Reedy trivial cofibration, and r has a right inverse that is a Reedy trivial cofibration.*
(2) *If $p\colon \boldsymbol{X} \to \boldsymbol{Y}$ is a weak equivalence of simplicial resolutions in $\mathcal{M}$, then there is a natural factorization of p as $\boldsymbol{X} \xrightarrow{q} \boldsymbol{Z} \xrightarrow{r} \boldsymbol{Y}$ such that $\boldsymbol{Z}$ is a simplicial resolution in $\mathcal{M}$, r is a Reedy trivial fibration, and q has a left inverse that is a Reedy trivial fibration.*

PROOF. This follows from Lemma 7.7.1 and Proposition 16.1.27. □

16.2. Quillen functors and resolutions

PROPOSITION 16.2.1. *Let $\mathcal{M}$ and $\mathcal{N}$ be model categories and let $\mathrm{F}\colon \mathcal{M} \rightleftarrows \mathcal{N} :\mathrm{U}$ be a Quillen pair (see Definition 8.5.2).*

(1) *If X is a cofibrant object of $\mathcal{M}$ and $\widetilde{\boldsymbol{X}} \to \mathrm{cc}_*X$ is a cosimplicial resolution of X (see Definition 16.1.2), then $\mathrm{F}\widetilde{\boldsymbol{X}} \to \mathrm{cc}_*\mathrm{F}X$ is a cosimplicial resolution of $\mathrm{F}X$.*
(2) *If Y is a fibrant object of $\mathcal{N}$ and $\mathrm{cs}_*Y \to \widehat{\boldsymbol{Y}}$ is a simplicial resolution of Y, then $\mathrm{cs}_*\mathrm{U}Y \to \mathrm{U}\widehat{\boldsymbol{Y}}$ is a simplicial resolution of $\mathrm{U}Y$.*

PROOF. We will prove part 1; the proof of part 2 is dual.

Corollary 15.4.2 implies that $\mathrm{F}\widetilde{\boldsymbol{X}}$ is Reedy cofibrant. Since X and $\widetilde{\boldsymbol{X}}^n$ for all $n \geq 0$ are cofibrant, Proposition 8.5.7 implies that $\mathrm{F}\widetilde{\boldsymbol{X}} \to \mathrm{cc}_*\mathrm{F}X$ is a Reedy weak equivalence. □

COROLLARY 16.2.2. *Let $\mathcal{C}$ be a Reedy category, let $\mathcal{M}$ and $\mathcal{N}$ be small categories, and let* $\mathrm{F}\colon \mathcal{M} \rightleftarrows \mathcal{N} :\mathrm{U}$ *be a Quillen pair.*

1. *If $i\colon A \to B$ is a map of cofibrant objects in $\mathcal{M}$ and $\tilde{\imath}\colon \widetilde{\boldsymbol{A}} \to \widetilde{\boldsymbol{B}}$ is a cosimplicial resolution of i such that $\tilde{\imath}$ is a Reedy cofibration, then $\mathrm{F}\tilde{\imath}\colon \mathrm{F}\widetilde{\boldsymbol{A}} \to \mathrm{F}\widetilde{\boldsymbol{B}}$ is a cosimplicial resolution of $\mathrm{F}i$ and $\mathrm{F}\tilde{\imath}$ is a Reedy cofibration.*
2. *If $p\colon X \to Y$ is a map of fibrant objects in $\mathcal{N}$ and $\hat{p}\colon \widehat{\boldsymbol{X}} \to \widehat{\boldsymbol{Y}}$ is a simplicial resolution of p such that $\hat{p}$ is a Reedy fibration, then $\mathrm{U}\hat{p}$ is a simplicial resolution of $\mathrm{U}p$ and $\mathrm{U}\hat{p}$ is a Reedy fibration.*

PROOF. This follows from Proposition 15.4.1 and Proposition 16.2.1. □

16.3. Realizations

This section contains a number of technical results needed for the homotopy lifting extension theorems of Section 16.5.

DEFINITION 16.3.1. Let $\mathcal{M}$ be a model category.

1. If $\boldsymbol{X}$ is a cosimplicial object in $\mathcal{M}$ and K is a simplicial set, then $\boldsymbol{X} \otimes K$ is defined to be the object of $\mathcal{M}$ that is the colimit of the (ΔK)-diagram in $\mathcal{M}$ (see Definition 15.1.16) that takes the object $\Delta[n] \to K$ of $\Delta K = (\Delta \downarrow K)$ to $\boldsymbol{X}^n$ and takes the commutative triangle

$$\begin{array}{ccc} \Delta[n] & \xrightarrow{\ \alpha\ } & \Delta[k] \\ & \searrow \quad \swarrow & \\ & K & \end{array} \tag{16.3.2}$$

   to the map $\alpha_*\colon \boldsymbol{X}^n \to \boldsymbol{X}^k$.
2. If $\boldsymbol{Y}$ is a simplicial object in $\mathcal{M}$ and K is a simplicial set, then $\boldsymbol{Y}^K$ is defined to be the object of $\mathcal{M}$ that is the limit of the $(\Delta^{\mathrm{op}}K)$-diagram in $\mathcal{M}$ (see Definition 15.1.16) that takes the object $\Delta[n] \to K$ of $\Delta^{\mathrm{op}}K = (\Delta \downarrow K)^{\mathrm{op}}$ to $\boldsymbol{Y}_n$ and takes the commutative triangle (16.3.2) to the map $\alpha^*\colon \boldsymbol{Y}_k \to \boldsymbol{Y}_n$.

PROPOSITION 16.3.3. *If $\mathcal{M}$ is a model category, then the constructions of Definition 16.3.1 are natural in $\boldsymbol{X}$, $\boldsymbol{Y}$ and K.*

PROOF. This follows directly from the definitions. □

PROPOSITION 16.3.4. *If $\mathcal{M} = \mathrm{SS}$, the cosimplicial object $\boldsymbol{X}$ is the cosimplicial standard simplex (see Definition 15.1.15), and K is a simplicial set, then $\boldsymbol{X} \otimes K$ is naturally isomorphic to K.*

PROOF. This is a restatement of Proposition 15.1.20. □

EXAMPLE 16.3.5. If $\mathcal{M} = \mathrm{Top}$, the cosimplicial object $\boldsymbol{X}$ is the geometric realization of the cosimplicial standard simplex (i.e., $\boldsymbol{X}^n = \bigl|\Delta[n]\bigr|$), and K is a simplicial set, then $\boldsymbol{X} \otimes K$ is the usual geometric realization of K.

LEMMA 16.3.6. *Let $\mathcal{M}$ be a model category.*

1. *If $\boldsymbol{B}$ is a cosimplicial object in $\mathcal{M}$ and $n \geq 0$, then $\boldsymbol{B} \otimes \Delta[n]$ is naturally isomorphic to $\boldsymbol{B}^n$.*

(2) *If $\boldsymbol{X}$ is a simplicial object in $\mathcal{M}$ and $n \geq 0$, then $\boldsymbol{X}^{\Delta[n]}$ is naturally isomorphic to $\boldsymbol{X}_n$.*

PROOF. The nondegenerate n-simplex of $\Delta[n]$ is a terminal object of $\Delta(\Delta[n])$ and an initial object of $\Delta^{\mathrm{op}}(\Delta[n])$. □

LEMMA 16.3.7. *Let $\mathcal{M}$ be a model category.*

(1) *If $\boldsymbol{B}$ is a cosimplicial object in $\mathcal{M}$ and $n \geq 0$, then $\boldsymbol{B} \otimes \partial\Delta[n]$ is naturally isomorphic to $\mathrm{L}_n\boldsymbol{B}$, the latching object of $\boldsymbol{B}$ at $[n]$ (see Definition 15.2.5).*
(2) *If $\boldsymbol{X}$ is a simplicial object in $\mathcal{M}$ and $n \geq 0$, then $\boldsymbol{X}^{\partial\Delta[n]}$ is naturally isomorphic to $\mathrm{M}_n\boldsymbol{X}$, the matching object of $\boldsymbol{X}$ at $[n]$.*

PROOF. We will prove part 1; the proof of part 2 is dual.

If $n \geq 0$, then the latching object of $\boldsymbol{B}$ at n is

$$\mathrm{L}_n\boldsymbol{B} = \operatorname*{colim}_{\partial(\overrightarrow{\boldsymbol{\Delta}}\downarrow[n])} \boldsymbol{B} \approx \operatorname*{colim}_{\substack{k<n \\ \boldsymbol{\Delta}([k],[n])}} \boldsymbol{B}$$

(see Corollary 15.2.9). Since $\boldsymbol{\Delta}([k],[n])$ is naturally isomorphic to the set of k-simplices of $\Delta[n]$, this is the colimit of the diagram with one copy of $\boldsymbol{B}^k$ for every k-simplex of $\Delta[n]$ for $k < n$. The result now follows from Definition 16.3.1. □

PROPOSITION 16.3.8. *Let $\mathcal{M}$ be a model category.*

(1) *If $\boldsymbol{B}$ is a cosimplicial object in $\mathcal{M}$ and $n \geq 0$, then the latching map (see Definition 15.2.5) of $\boldsymbol{B}$ at $[n]$ is naturally isomorphic to the map $\boldsymbol{B} \otimes \partial\Delta[n] \to \boldsymbol{B} \otimes \Delta[n]$.*
(2) *If $\boldsymbol{X}$ is a simplicial object in $\mathcal{M}$ and $n \geq 0$, then the matching map of $\boldsymbol{X}$ at $[n]$ is naturally isomorphic to the map $\boldsymbol{X}^{\Delta[n]} \to \boldsymbol{X}^{\partial\Delta[n]}$.*

PROOF. This follows from Lemma 16.3.6, Lemma 16.3.7, and the proof of Lemma 16.3.7. □

COROLLARY 16.3.9. *Let $\mathcal{M}$ be a model category.*

(1) *If $\boldsymbol{B}$ is a Reedy cofibrant cosimplicial object in $\mathcal{M}$ and $n \geq 0$, then both $\boldsymbol{B} \otimes \partial\Delta[n]$ and $\boldsymbol{B} \otimes \Delta[n]$ are cofibrant objects of $\mathcal{M}$.*
(2) *If $\boldsymbol{X}$ is a Reedy fibrant simplicial object in $\mathcal{M}$ and $n \geq 0$, then both $\boldsymbol{X}^{\Delta[n]}$ and $\boldsymbol{X}^{\partial\Delta[n]}$ are fibrant objects of $\mathcal{M}$.*

PROOF. This follows from Proposition 16.3.8 and Corollary 15.3.12. □

PROPOSITION 16.3.10. *Let $\mathcal{M}$ be a model category.*

(1) *If $\boldsymbol{A} \to \boldsymbol{B}$ is a Reedy cofibration of cosimplicial objects in $\mathcal{M}$ and $n \geq 0$, then the induced map $\boldsymbol{A} \otimes \Delta[n] \amalg_{\boldsymbol{A}\otimes\partial\Delta[n]} \boldsymbol{B} \otimes \partial\Delta[n] \to \boldsymbol{B} \otimes \Delta[n]$ is a cofibration in $\mathcal{M}$ that is a trivial cofibration if $\boldsymbol{A} \to \boldsymbol{B}$ is a Reedy trivial cofibration.*
(2) *If $\boldsymbol{X} \to \boldsymbol{Y}$ is a Reedy fibration of simplicial objects in $\mathcal{M}$ and $n \geq 0$, then the induced map $\boldsymbol{X}^{\Delta[n]} \to \boldsymbol{Y}^{\Delta[n]} \times_{\boldsymbol{Y}^{\partial\Delta[n]}} \boldsymbol{X}^{\partial\Delta[n]}$ is a fibration in $\mathcal{M}$ that is a trivial fibration if $\boldsymbol{X} \to \boldsymbol{Y}$ is a Reedy trivial fibration.*

PROOF. This follows from Proposition 16.3.8 and Theorem 15.3.15. □

COROLLARY 16.3.11. *Let $\mathcal{M}$ be a model category.*

(1) *If* $\boldsymbol{A} \to \boldsymbol{B}$ *is a Reedy cofibration of Reedy cofibrant cosimplicial objects in* $\mathcal{M}$ *and* $n \geq 0$, *then the induced map* $\boldsymbol{A} \otimes \Delta[n] \amalg_{\boldsymbol{A} \otimes \partial\Delta[n]} \boldsymbol{B} \otimes \partial\Delta[n] \to \boldsymbol{B} \otimes \Delta[n]$ *is a cofibration between cofibrant objects in* $\mathcal{M}$ *that is a trivial cofibration if* $\boldsymbol{A} \to \boldsymbol{B}$ *is a Reedy trivial cofibration.*

(2) *If* $\boldsymbol{X} \to \boldsymbol{Y}$ *is a Reedy fibration of Reedy fibrant simplicial objects in* $\mathcal{M}$ *and* $n \geq 0$, *then the induced map* $\boldsymbol{X}^{\Delta[n]} \to \boldsymbol{Y}^{\Delta[n]} \times_{\boldsymbol{Y}^{\partial\Delta[n]}} \boldsymbol{X}^{\partial\Delta[n]}$ *is a fibration between fibrant objects in* $\mathcal{M}$ *that is a trivial fibration if* $\boldsymbol{X} \to \boldsymbol{Y}$ *is a Reedy trivial fibration.*

PROOF. We will prove part 1; the proof of part 2 is dual.

Proposition 16.3.8 and Corollary 16.3.9 imply that $\boldsymbol{A} \otimes \Delta[n] \amalg_{\boldsymbol{A} \otimes \partial\Delta[n]} \boldsymbol{B} \otimes \partial\Delta[n]$ and $\boldsymbol{B} \otimes \Delta[n]$ are cofibrant objects, and so the result follows from Proposition 16.3.10. □

PROPOSITION 16.3.12. *Let* $\mathcal{M}$ *be a model category.*

(1) *If* $\boldsymbol{X}$ *is a Reedy cofibrant cosimplicial object in* $\mathcal{M}$ *and* K *is a simplicial set, then the* (ΔK)*-diagram in* $\mathcal{M}$ *whose colimit is defined to be* $\boldsymbol{X} \otimes K$ *(see Definition* 16.3.1*) is a Reedy cofibrant diagram (see Example* 15.1.19*).*

(2) *If* $\boldsymbol{Y}$ *is a Reedy fibrant simplicial object in* $\mathcal{M}$ *and* K *is a simplicial set, then the* $(\Delta^{\mathrm{op}} K)$*-diagram in* $\mathcal{M}$ *whose limit is defined to be* $\boldsymbol{Y}^K$ *(see Definition* 16.3.1*) is a Reedy fibrant diagram.*

PROOF. We will prove part 1; the proof of part 2 is similar.

If $\sigma\colon \Delta[n] \to K$ is an object of ΔK, then the latching category $\partial(\overrightarrow{\Delta K} \downarrow \alpha)$ of ΔK at σ has an object for each $k < n$ and each iterated coface operator $\alpha\colon \Delta[k] \to \Delta[n]$. Thus, the latching category of ΔK at σ is isomorphic to the latching category of $\boldsymbol{\Delta}$ at $[n]$, and the latching map of our diagram at σ is isomorphic to the latching map of $\boldsymbol{X}$ at $[n]$. Since $\boldsymbol{X}$ is Reedy cofibrant, so is our diagram. □

16.4. Adjointness

This section contains technical results for various simplicial sets constructed from cosimplicial objects and simplicial objects in a category (see Notation 16.4.1). These constructions will be used in Chapter 17 to define homotopy function complexes between objects in a model category (see Definition 17.1.1, Definition 17.2.1, and Definition 17.3.1).

NOTATION 16.4.1. Let $\mathcal{M}$ be a model category.

(1) If $\boldsymbol{X}$ is a cosimplicial object in $\mathcal{M}$ and Y is an object of $\mathcal{M}$, then $\mathcal{M}(\boldsymbol{X}, Y)$ will denote the simplicial set, natural in both $\boldsymbol{X}$ and Y, defined by $\mathcal{M}(\boldsymbol{X}, Y)_n = \mathcal{M}(\boldsymbol{X}^n, Y)$, with face and degeneracy maps induced by the coface and codegeneracy maps in $\boldsymbol{X}$.

(2) If X is an object of $\mathcal{M}$ and $\boldsymbol{Y}$ is a simplicial object in $\mathcal{M}$, then $\mathcal{M}(X, \boldsymbol{Y})$ will denote the simplicial set, natural in both X and $\boldsymbol{Y}$, defined by $\mathcal{M}(X, \boldsymbol{Y})_n = \mathcal{M}(X, \boldsymbol{Y}_n)$, with face and degeneracy maps induced by those in $\boldsymbol{Y}$.

(3) If $\boldsymbol{X}$ is a cosimplicial object in $\mathcal{M}$ and $\boldsymbol{Y}$ is a simplicial object in $\mathcal{M}$, then $\mathcal{M}(\boldsymbol{X}, \boldsymbol{Y})$ will denote the bisimplicial set, natural in both $\boldsymbol{X}$ and $\boldsymbol{Y}$, defined by $\mathcal{M}(\boldsymbol{X}, \boldsymbol{Y})_{n,k} = \mathcal{M}(\boldsymbol{X}^k, \boldsymbol{Y}_n)$, with face and degeneracy maps

induced by the coface and codegeneracy maps in $\boldsymbol{X}$ and the face and degeneracy maps in $\boldsymbol{Y}$.

(4) If $\boldsymbol{X}$ is a cosimplicial object in $\mathcal{M}$ and $\boldsymbol{Y}$ is a simplicial object in $\mathcal{M}$, then $\operatorname{diag}\mathcal{M}(\boldsymbol{X},\boldsymbol{Y})$ will denote the simplicial set, natural in both $\boldsymbol{X}$ and $\boldsymbol{Y}$, defined by $\big(\operatorname{diag}\mathcal{M}(\boldsymbol{X},\boldsymbol{Y})\big)_n = \mathcal{M}(\boldsymbol{X}^n,\boldsymbol{Y}_n)$, with face and degeneracy maps induced by the coface and codegeneracy maps in $\boldsymbol{X}$ and the face and degeneracy maps in $\boldsymbol{Y}$.

THEOREM 16.4.2. *Let $\mathcal{M}$ be a model category.*

(1) *If $\boldsymbol{A}$ is a cosimplicial object in $\mathcal{M}$, X is an object of $\mathcal{M}$, and K is a simplicial set, then there is a natural isomorphism of sets*

$$\mathrm{SS}\big(K,\mathcal{M}(\boldsymbol{A},X)\big) \approx \mathcal{M}(\boldsymbol{A}\otimes K, X)$$

(see Notation 16.4.1 *and Definition* 16.3.1*).*

(2) *If B is an object of $\mathcal{M}$, $\boldsymbol{Y}$ is a simplicial object in $\mathcal{M}$, and K is a simplicial set, then there is a natural isomorphism of sets*

$$\mathrm{SS}\big(K,\mathcal{M}(B,\boldsymbol{Y})\big) \approx \mathcal{M}(B,\boldsymbol{Y}^K)$$

(see Notation 16.4.1 *and Definition* 16.3.1*).*

PROOF. We will prove part 1; the proof of part 2 is similar.

Since $\boldsymbol{A}\otimes K$ is the colimit of a (ΔK)-diagram, a map in $\mathcal{M}$ from $\boldsymbol{A}\otimes K$ to X corresponds to a coherent set of maps from each object in the diagram to X. Thus, a map $\boldsymbol{A}\otimes K \to X$ is defined by a map $\boldsymbol{A}^n \to X$ for each n-simplex of K that commute with the simplicial operators. This is also a description of a map of simplicial sets from K to $\mathcal{M}(\boldsymbol{A},X)$. □

PROPOSITION 16.4.3. *Let $\mathcal{M}$ be a model category.*

(1) *If $\boldsymbol{A}$ is a cosimplicial object in $\mathcal{M}$, $\mathcal{C}$ is a small category, and $\boldsymbol{K}\colon \mathcal{C}\to \mathrm{SS}$ is a $\mathcal{C}$-diagram of simplicial sets, then the natural map* $\operatorname{colim}_{\mathcal{C}}(\boldsymbol{A}\otimes\boldsymbol{K}) \to \boldsymbol{A}\otimes(\operatorname{colim}_{\mathcal{C}}\boldsymbol{K})$ *is an isomorphism.*

(2) *If $\boldsymbol{X}$ is a simplicial object in $\mathcal{M}$, $\mathcal{C}$ is a small category, and $\boldsymbol{K}\colon \mathcal{C}\to\mathrm{SS}$ is a $\mathcal{C}$-diagram of simplicial sets, then the natural map* $\boldsymbol{X}^{(\operatorname{colim}_{\mathcal{C}}\boldsymbol{K})} \to \lim_{\mathcal{C}^{\mathrm{op}}}(\boldsymbol{X}^{\boldsymbol{K}})$ *is an isomorphism.*

PROOF. This follows from the adjointness relations of Theorem 16.4.2. □

LEMMA 16.4.4. *Let $\mathcal{M}$ be a model category, and let (K,L) be a pair of simplicial sets.*

(1) *If $\boldsymbol{A}$ is a Reedy cofibrant cosimplicial object in $\mathcal{M}$, then the map $\boldsymbol{A}\otimes L\to \boldsymbol{A}\otimes K$ is a cofibration in $\mathcal{M}$.*

(2) *If $\boldsymbol{X}$ is a Reedy fibrant simplicial object in $\mathcal{M}$, then the map $\boldsymbol{X}^K\to\boldsymbol{X}^L$ is a fibration in $\mathcal{M}$.*

PROOF. Since an inclusion $L\to K$ of simplicial sets is a transfinite composition of pushouts of the maps $\partial\Delta[n]\to\Delta[n]$ for $n\ge 0$ (see Proposition 10.2.18), the map $\boldsymbol{A}\otimes L\to\boldsymbol{A}\otimes K$ is a transfinite composition of pushouts of the maps $\boldsymbol{A}\otimes\partial\Delta[n]\to \boldsymbol{A}\otimes\Delta[n]$ for $n\ge 0$, and so part 1 follows from Proposition 16.4.3, Proposition 16.3.8, and Proposition 10.3.4. The proof of part 2 is similar. □

PROPOSITION 16.4.5. *Let $\mathcal{M}$ be a model category.*

(1) *If $i\colon \boldsymbol{A} \to \boldsymbol{B}$ is a map of cosimplicial objects in $\mathcal{M}$, $p\colon X \to Y$ is a map in $\mathcal{M}$, and (K, L) is a pair of simplicial sets, then the following are equivalent:*

(a) *The dotted arrow exists in every solid arrow diagram of the form*

$$\begin{array}{ccc} L & \longrightarrow & \mathcal{M}(\boldsymbol{B}, X) \\ \big\downarrow & \nearrow & \big\downarrow \\ K & \longrightarrow & \mathcal{M}(\boldsymbol{A}, X) \times_{\mathcal{M}(\boldsymbol{A}, Y)} \mathcal{M}(\boldsymbol{B}, Y) \end{array} .$$

(b) *The dotted arrow exists in every solid arrow diagram of the form*

$$\begin{array}{ccc} \boldsymbol{A} \otimes K \amalg_{\boldsymbol{A} \otimes L} \boldsymbol{B} \otimes L & \longrightarrow & X \\ \big\downarrow & \nearrow & \big\downarrow \\ \boldsymbol{B} \otimes K & \longrightarrow & Y \end{array} .$$

(2) *If $i\colon A \to B$ is a map in $\mathcal{M}$, $p\colon \boldsymbol{X} \to \boldsymbol{Y}$ is a map of simplicial objects in $\mathcal{M}$, and (K, L) is a pair of simplicial sets, then the following are equivalent:*

(a) *The dotted arrow exists in every solid arrow diagram of the form*

$$\begin{array}{ccc} L & \longrightarrow & \mathcal{M}(B, \boldsymbol{X}) \\ \big\downarrow & \nearrow & \big\downarrow \\ K & \longrightarrow & \mathcal{M}(A, \boldsymbol{X}) \times_{\mathcal{M}(A, \boldsymbol{Y})} \mathcal{M}(B, \boldsymbol{Y}) \end{array} .$$

(b) *The dotted arrow exists in every solid arrow diagram of the form*

$$\begin{array}{ccc} A & \longrightarrow & \boldsymbol{X}^K \\ \big\downarrow & \nearrow & \big\downarrow \\ B & \longrightarrow & \boldsymbol{X}^L \times_{\boldsymbol{Y}^L} \boldsymbol{Y}^K \end{array} .$$

PROOF. This follows from Theorem 16.4.2. □

PROPOSITION 16.4.6 (Partial homotopy lifting extension theorem). *Let $\mathcal{M}$ be a model category.*

(1) *If $i\colon \boldsymbol{A} \to \boldsymbol{B}$ is a Reedy cofibration of cosimplicial objects in $\mathcal{M}$, $p\colon X \to Y$ is a fibration in $\mathcal{M}$, and at least one of i and p is also a weak equivalence, then the map of simplicial sets*

$$\mathcal{M}(\boldsymbol{B}, X) \to \mathcal{M}(\boldsymbol{A}, X) \times_{\mathcal{M}(\boldsymbol{A}, Y)} \mathcal{M}(\boldsymbol{B}, Y)$$

is a trivial fibration.

(2) *If $i\colon A \to B$ is a cofibration in $\mathcal{M}$, $p\colon \boldsymbol{X} \to \boldsymbol{Y}$ is a Reedy fibration of simplicial objects in $\mathcal{M}$, and at least one of i and p is also a weak equivalence, then the map of simplicial sets*

$$\mathcal{M}(B, \boldsymbol{X}) \to \mathcal{M}(A, \boldsymbol{X}) \times_{\mathcal{M}(A, \boldsymbol{Y})} \mathcal{M}(B, \boldsymbol{Y})$$

is a trivial fibration.

PROOF. A map of simplicial sets is a trivial fibration if and only if it has the right lifting property with respect to the maps $\partial\Delta[n] \to \Delta[n]$ for $n \geq 0$, and so the result follows from Proposition 16.4.5 and Proposition 16.3.10. □

Proposition 16.4.6 may seem to be incomplete in that it does not assert the full homotopy lifting extension theorem. We will show in Theorem 16.5.2 that if the cosimplicial and simplicial objects are assumed to be cosimplicial and simplicial *resolutions* (see Definition 16.1.26) then the full homotopy lifting extension theorem does hold. Example 16.4.7 shows that it does not hold without the assumption that the cosimplicial or simplicial objects are resolutions.

EXAMPLE 16.4.7. We present here an example of a model category $\mathcal{M}$, a Reedy cofibrant cosimplicial object $\boldsymbol{B}$ in $\mathcal{M}$, and a fibration $p\colon X \to Y$ in $\mathcal{M}$ such that the map of simplicial sets $\mathcal{M}(\boldsymbol{B}, X) \to \mathcal{M}(\boldsymbol{B}, Y)$ is not a fibration. (This implies that the partial homotopy lifting extension theorem of Proposition 16.4.6 is the strongest result possible without assuming that the cosimplicial objects $\boldsymbol{A}$ and $\boldsymbol{B}$ are cosimplicial resolutions; see also Theorem 16.5.2.)

Let $\mathcal{M}$ be the category SS_* of pointed simplicial sets. Let $\boldsymbol{B}$ be the cosimplicial object in $\mathcal{M}$ that is the free diagram on S^1 generated at $[1]$ (see Definition 11.5.25 and Definition 15.1.8), so that $\boldsymbol{B}^n = \bigvee_{\boldsymbol{\Delta}([1],[n])} \mathrm{S}^1$ (where $\boldsymbol{\Delta}([1],[n])$ is the set of 1-simplices of $\Delta[n]$). Corollary 15.6.6 implies that $\boldsymbol{B}$ is a Reedy cofibrant cosimplicial object.

Let $p\colon X \to Y$ be any fibration of fibrant pointed simplicial sets for which the induced homomorphism of fundamental groups $p_*\colon \pi_1 X \to \pi_1 Y$ is *not* surjective. We will show that the map of simplicial sets $\mathcal{M}(\boldsymbol{B}, X) \to \mathcal{M}(\boldsymbol{B}, Y)$ is not a fibration.

$\boldsymbol{B}^1$ is the wedge of three copies of S^1 (indexed by $[0,0]$, $[1,1]$, and $[0,1]$), $\boldsymbol{B}^0$ is a single copy of S^1, and the maps $d^0, d^1\colon \boldsymbol{B}^0 \to \boldsymbol{B}^1$ take the S^1 in $\boldsymbol{B}^0$ to the summand indexed by, respectively, $[0,0]$ and $[1,1]$. Thus, we can define a 1-simplex of $\mathcal{M}(\boldsymbol{B}, Y)$ by sending the summands of $\boldsymbol{B}^1$ corresponding to $[0,0]$ and $[1,1]$ to the basepoint of Y and sending the summand S^1 of $\boldsymbol{B}^1$ corresponding to $[0,1]$ to some 1-simplex of Y that represents an element of $\pi_1 Y$ that is not in the image of $p_*\colon \pi_1 X \to \pi_1 Y$. If we define a 0-simplex of $\mathcal{M}(\boldsymbol{B}, X)$ by sending $\boldsymbol{B}^0$ to the basepoint of X, then we have a solid arrow diagram

$$\begin{array}{ccc} \Delta[0] & \longrightarrow & \mathcal{M}(\boldsymbol{B}, X) \\ \big\downarrow & \nearrow & \big\downarrow \\ \Delta[1] & \longrightarrow & \mathcal{M}(\boldsymbol{B}, Y) \end{array}$$

for which there is no dotted arrow making the triangles commute.

LEMMA 16.4.8. *Let $\mathcal{M}$ be a model category.*

(1) *If $\boldsymbol{A} \to \boldsymbol{B}$ is a Reedy cofibration of cosimplicial objects in $\mathcal{M}$, $n \geq 1$, and $n \geq k \geq 0$, then the induced map $\boldsymbol{A} \otimes \Delta[n] \amalg_{\boldsymbol{A} \otimes \Lambda[n,k]} \boldsymbol{B} \otimes \Lambda[n,k] \to \boldsymbol{B} \otimes \Delta[n]$ is a cofibration.*

(2) *If $\boldsymbol{X} \to \boldsymbol{Y}$ is a Reedy fibration of simplicial objects in $\mathcal{M}$, $n \geq 1$, and $n \geq k \geq 0$, then the induced map $\boldsymbol{X}^{\Delta[n]} \to \boldsymbol{Y}^{\Delta[n]} \times_{\boldsymbol{Y}^{\Lambda[n,k]}} \boldsymbol{X}^{\Lambda[n,k]}$ is a fibration.*

PROOF. We will prove part 1; the proof of part 2 is similar.

We have the diagram

$$\begin{array}{ccc} \boldsymbol{A}\otimes\Delta[n-1]\amalg_{\boldsymbol{A}\otimes\partial\Delta[n-1]}\boldsymbol{B}\otimes\partial\Delta[n-1] & \longrightarrow & \boldsymbol{A}\otimes\Delta[n]\amalg_{\boldsymbol{A}\otimes\Lambda[n,k]}\boldsymbol{B}\otimes\Lambda[n,k] \\ \downarrow & & \downarrow \\ \boldsymbol{B}\otimes\Delta[n-1] & \longrightarrow & \boldsymbol{A}\otimes\Delta[n]\amalg_{\boldsymbol{A}\otimes\partial\Delta[n]}\boldsymbol{B}\otimes\partial\Delta[n] \\ & & \downarrow \\ & & \boldsymbol{B}\otimes\Delta[n] \end{array}$$

in which the square is a pushout, and so Proposition 16.3.10 implies that all of the vertical maps are cofibrations. Our map is thus the composition of two cofibrations. □

16.4.9. Resolutions.

LEMMA 16.4.10. *If $n \geq 1$ and $n \geq k \geq 0$, then there is a finite sequence of inclusions of simplicial sets*

$$\Delta[0] = K_0 \to K_1 \to K_2 \to \cdots \to K_p = \Lambda[n,k]$$

where each map $K_i \to K_{i+1}$ for $i < p$ is constructed as a pushout

$$\begin{array}{ccc} \Lambda[m_i, l_i] & \longrightarrow & K_i \\ \downarrow & & \downarrow \\ \Delta[m_i] & \longrightarrow & K_{i+1} \end{array}$$

with $m_i < n$.

PROOF. We let $\Delta[0] = K_0$ be vertex k of $\Delta[n]$. We can then add in all the 1-simplices of $\Lambda[n,k]$ that contain that vertex, followed by the 2-simplices of $\Lambda[n,k]$ that contain that vertex, etc., until we've added in all of $\Lambda[n,k]$. □

LEMMA 16.4.11. *Let $\mathcal{M}$ be a model category.*

(1) *If $\boldsymbol{A}$ is a cosimplicial resolution in $\mathcal{M}$, $n \geq 1$, and $n \geq k \geq 0$, then the natural map $\boldsymbol{A}\otimes\Lambda[n,k] \to \boldsymbol{A}\otimes\Delta[n]$ is a trivial cofibration.*
(2) *If $\boldsymbol{X}$ is a simplicial resolution in $\mathcal{M}$, $n \geq 1$, and $n \geq k \geq 0$, then the natural map $\boldsymbol{X}^{\Delta[n]} \to \boldsymbol{X}^{\Lambda[n,k]}$ is a trivial fibration.*

PROOF. We will prove part 1; the proof of part 2 is similar.

We will prove the lemma by induction on n. If $n = 1$, then the result follows from Lemma 16.3.6, Proposition 16.1.27, and Lemma 16.4.4.

We now assume that $\boldsymbol{A}\otimes\Lambda[m,l] \to \boldsymbol{A}\otimes\Delta[m]$ is a trivial cofibration for $l \leq m < n$. Lemma 16.4.10 implies that there is a finite sequence of maps in $\mathcal{M}$

$$\boldsymbol{A}\otimes\Delta[0] = \boldsymbol{A}\otimes K_0 \to \boldsymbol{A}\otimes K_1 \to \boldsymbol{A}\otimes K_2 \to \cdots \to \boldsymbol{A}\otimes K_p = \boldsymbol{A}\otimes\Lambda[n,k]$$

where each $\boldsymbol{A}\otimes K_i \to \boldsymbol{A}\otimes K_{i+1}$ for $i < p$ is constructed as a pushout

$$\begin{array}{ccc} \boldsymbol{A}\otimes\Lambda[m_i, l_i] & \longrightarrow & \boldsymbol{A}\otimes K_i \\ \downarrow & & \downarrow \\ \boldsymbol{A}\otimes\Delta[m_i] & \longrightarrow & \boldsymbol{A}\otimes K_{i+1} \end{array}$$

with $m_i < n$. The induction hypothesis implies that each of these maps is a trivial cofibration, and so $\boldsymbol{A} \otimes \Delta[0] \to \boldsymbol{A} \otimes \Lambda[n,k]$ is a trivial cofibration. Since $\boldsymbol{A} \otimes \Delta[0] \to \boldsymbol{A} \otimes \Delta[n]$ is a weak equivalence, the "two out of three" property of weak equivalences implies that $\boldsymbol{A} \otimes \Lambda[n,k] \to \boldsymbol{A} \otimes \Delta[n]$ is a weak equivalence and Lemma 16.4.4 implies that it is a cofibration. □

PROPOSITION 16.4.12. *Let $\mathcal{M}$ be a model category.*

(1) *If $\boldsymbol{A} \to \boldsymbol{B}$ is a Reedy cofibration of cosimplicial resolutions in $\mathcal{M}$, $n \geq 1$, and $n \geq k \geq 0$, then the map $\boldsymbol{A} \otimes \Delta[n] \amalg_{\boldsymbol{A} \otimes \Lambda[n,k]} \boldsymbol{B} \otimes \Lambda[n,k] \to \boldsymbol{B} \otimes \Delta[n]$ is a trivial cofibration.*
(2) *If $\boldsymbol{X} \to \boldsymbol{Y}$ is a Reedy fibration of simplicial resolutions in $\mathcal{M}$, $n > 1$, and $n \geq k \geq 0$, then the map $\boldsymbol{X}^{\Delta[n]} \to \boldsymbol{X}^{\Lambda[n,k]} \times_{\boldsymbol{Y}^{\Lambda[n,k]}} \boldsymbol{Y}^{\Delta[n]}$ is a trivial fibration.*

PROOF. We will prove part 1; the proof of part 2 is similar.

Lemma 16.4.8 implies that our map is a cofibration, and so it remains only to show that it is a weak equivalence. Lemma 16.4.11 implies that $\boldsymbol{A} \otimes \Lambda[n,k] \to \boldsymbol{A} \otimes \Delta[n]$ is a trivial cofibration. Since the diagram

$$\begin{array}{ccc} \boldsymbol{A} \otimes \Lambda[n,k] & \longrightarrow & \boldsymbol{B} \otimes \Lambda[n,k] \\ \downarrow & & \downarrow \\ \boldsymbol{A} \otimes \Delta[n] & \longrightarrow & \boldsymbol{A} \otimes \Delta[n] \amalg_{\boldsymbol{A} \otimes \Lambda[n,k]} \boldsymbol{B} \otimes \Lambda[n,k] \end{array}$$

is a pushout, the map $\boldsymbol{B} \otimes \Lambda[n,k] \to \boldsymbol{A} \otimes \Delta[n] \amalg_{\boldsymbol{A} \otimes \Lambda[n,k]} \boldsymbol{B} \otimes \Lambda[n,k]$ is also a trivial cofibration. Since Lemma 16.4.11 implies that the map $\boldsymbol{B} \otimes \Lambda[n,k] \to \boldsymbol{B} \otimes \Delta[n]$ is a weak equivalence, the result follows from the "two out of three" property of weak equivalences. □

16.5. Homotopy lifting extension theorems

This section contains several versions of the homotopy lifting extension theorem (Theorem 16.5.2, Theorem 16.5.13, and Theorem 16.5.18). These will be used in Chapters 17 and 19 to obtain homotopy invariance results for homotopy function complexes (see Definition 17.1.1, Definition 17.2.1, and Definition 17.3.1), homotopy colimits (see Definition 19.1.2), and homotopy limits (see Definition 19.1.5).

16.5.1. One-sided constructions.

THEOREM 16.5.2 (The one-sided homotopy lifting extension theorem). *Let $\mathcal{M}$ be a model category.*

(1) *If $i\colon \boldsymbol{A} \to \boldsymbol{B}$ is a Reedy cofibration of cosimplicial resolutions in $\mathcal{M}$ and $p\colon X \to Y$ is a fibration in $\mathcal{M}$, then the map of simplicial sets*
$$\mathcal{M}(\boldsymbol{B}, X) \to \mathcal{M}(\boldsymbol{A}, X) \times_{\mathcal{M}(\boldsymbol{A}, Y)} \mathcal{M}(\boldsymbol{B}, Y)$$
is a fibration that is a trivial fibration if at least one of i and p is also a weak equivalence.
(2) *If $i\colon A \to B$ is a cofibration in $\mathcal{M}$ and $p\colon \boldsymbol{X} \to \boldsymbol{Y}$ is a Reedy fibration of simplicial resolutions in $\mathcal{M}$, then the map of simplicial sets*
$$\mathcal{M}(B, \boldsymbol{X}) \to \mathcal{M}(A, \boldsymbol{X}) \times_{\mathcal{M}(A, \boldsymbol{Y})} \mathcal{M}(B, \boldsymbol{Y})$$

is a fibration that is a trivial fibration if at least one of i and p is also a weak equivalence.

PROOF. A map of simplicial sets is a fibration if and only if it has the right lifting property with respect to the maps $\Lambda[n,k] \to \Delta[n]$ for $n > 0$ and $n \geq k \geq 0$, and so the result follows from Proposition 16.4.5, Proposition 16.4.12, and Proposition 16.4.6. □

COROLLARY 16.5.3. *Let $\mathcal{M}$ be a model category.*

(1) *If $\boldsymbol{B}$ is a cosimplicial resolution in $\mathcal{M}$ and X is a fibrant object of $\mathcal{M}$, then $\mathcal{M}(\boldsymbol{B}, X)$ is a fibrant simplicial set.*
(2) *If B is a cofibrant object of $\mathcal{M}$ and $\boldsymbol{X}$ is a simplicial resolution in $\mathcal{M}$, then $\mathcal{M}(B, \boldsymbol{X})$ is a fibrant simplicial set.*

PROOF. This follows from Theorem 16.5.2. □

COROLLARY 16.5.4. *Let $\mathcal{M}$ be a model category.*

(1) *If $i\colon \boldsymbol{A} \to \boldsymbol{B}$ is a Reedy cofibration of cosimplicial resolutions in $\mathcal{M}$ and X is a fibrant object in $\mathcal{M}$, then the map $i^*\colon \mathcal{M}(\boldsymbol{B}, X) \to \mathcal{M}(\boldsymbol{A}, X)$ is a fibration of simplicial sets that is a trivial fibration if i is a Reedy trivial cofibration.*
(2) *If $\boldsymbol{B}$ is a cosimplicial resolution in $\mathcal{M}$ and $p\colon X \to Y$ is a fibration in $\mathcal{M}$, then the map $p_*\colon \mathcal{M}(\boldsymbol{B}, X) \to \mathcal{M}(\boldsymbol{B}, Y)$ is a fibration of simplicial sets that is a trivial fibration if p is a trivial fibration.*
(3) *If $i\colon A \to B$ is a cofibration in $\mathcal{M}$ and $\boldsymbol{X}$ is a simplicial resolution in $\mathcal{M}$, then the map $i^*\colon \mathcal{M}(B, \boldsymbol{X}) \to \mathcal{M}(A, \boldsymbol{X})$ is a fibration of simplicial sets that is a trivial fibration if i is a trivial cofibration.*
(4) *If B is a cofibrant object of $\mathcal{M}$ and $p\colon \boldsymbol{X} \to \boldsymbol{Y}$ is a Reedy fibration of simplicial resolutions in $\mathcal{M}$, then the map $p_*\colon \mathcal{M}(B, \boldsymbol{X}) \to \mathcal{M}(B, \boldsymbol{Y})$ is a fibration of simplicial sets that is a trivial fibration if p is a Reedy trivial cofibration.*

PROOF. This follows from Theorem 16.5.2. □

COROLLARY 16.5.5. *Let $\mathcal{M}$ be a model category.*

(1) *If $i\colon \boldsymbol{A} \to \boldsymbol{B}$ is a Reedy weak equivalence of cosimplicial resolutions in $\mathcal{M}$ and X is a fibrant object of $\mathcal{M}$, then the map $i^*\colon \mathcal{M}(\boldsymbol{B}, X) \to \mathcal{M}(\boldsymbol{A}, X)$ is a weak equivalence of fibrant simplicial sets.*
(2) *If $\boldsymbol{B}$ is a cosimplicial resolution in $\mathcal{M}$ and $p\colon X \to Y$ is a weak equivalence of fibrant objects of $\mathcal{M}$, then the map $p_*\colon \mathcal{M}(\boldsymbol{B}, X) \to \mathcal{M}(\boldsymbol{B}, Y)$ is a weak equivalence of fibrant simplicial sets.*
(3) *If $i\colon A \to B$ is a weak equivalence of cofibrant objects of $\mathcal{M}$ and $\boldsymbol{X}$ is a simplicial resolution in $\mathcal{M}$, then the map $i^*\colon \mathcal{M}(B, \boldsymbol{X}) \to \mathcal{M}(A, \boldsymbol{Y})$ is a weak equivalence of fibrant simplicial sets.*
(4) *If B is a cofibrant object of $\mathcal{M}$ and $p\colon \boldsymbol{X} \to \boldsymbol{Y}$ is a Reedy weak equivalence of simplicial resolutions in $\mathcal{M}$, then the map $p_*\colon \mathcal{M}(B, \boldsymbol{X}) \to \mathcal{M}(B, \boldsymbol{Y})$ is a weak equivalence of fibrant simplicial sets.*

PROOF. This follows from Corollary 16.5.4, Corollary 7.7.2, and Corollary 16.5.3. □

PROPOSITION 16.5.6. *Let $\mathcal{M}$ be a model category.*

(1) *If* $i\colon \boldsymbol{A} \to \boldsymbol{B}$ *is a Reedy cofibration of cosimplicial resolutions in* $\mathcal{M}$ *and* $j\colon L \to K$ *is a cofibration of simplicial sets, then the map* $\boldsymbol{A} \otimes K \amalg_{\boldsymbol{A}\otimes L} \boldsymbol{B} \otimes L \to \boldsymbol{B} \otimes K$ *is a cofibration in* $\mathcal{M}$ *that is a trivial cofibration if either* i *or* j *is a weak equivalence.*

(2) *If* $p\colon \boldsymbol{X} \to \boldsymbol{Y}$ *is a Reedy fibration of simplicial resolutions in* $\mathcal{M}$ *and* $j\colon L \to K$ *is a cofibration of simplicial sets, then the map* $\boldsymbol{X}^K \to \boldsymbol{X}^L \times_{\boldsymbol{Y}^L} \boldsymbol{Y}^K$ *is a fibration in* $\mathcal{M}$ *that is a trivial fibration if either* p *or* j *is a weak equivalence.*

PROOF. This follows from Proposition 7.2.3, Proposition 16.4.5, and Theorem 16.5.2. □

THEOREM 16.5.7. *Let* $\mathcal{M}$ *be a model category.*

(1) *If* $i\colon \boldsymbol{A} \to \boldsymbol{B}$ *is a Reedy cofibration of cosimplicial resolutions in* $\mathcal{M}$ *and* $j\colon L \to K$ *is an inclusion of simplicial sets, then the pushout corner map* $\boldsymbol{A} \otimes K \amalg_{\boldsymbol{A}\otimes L} \boldsymbol{B} \otimes L \to \boldsymbol{B} \otimes K$ *is a cofibration in* $\mathcal{M}$ *that is a trivial cofibration if at least one of* i *and* p *is a weak equivalence.*

(2) *If* $j\colon L \to K$ *is an inclusion of simplicial sets and* $p\colon \boldsymbol{X} \to \boldsymbol{Y}$ *is a Reedy fibration of simplicial resolutions in* $\mathcal{M}$*, then the pullback corner map* $\boldsymbol{X}^K \to \boldsymbol{X}^L \times_{\boldsymbol{Y}^L} \boldsymbol{Y}^K$ *is a fibration in* $\mathcal{M}$ *that is a trivial fibration if at least one of* j *and* p *is a weak equivalence.*

PROOF. This follows from Proposition 7.2.3, Proposition 16.4.5, and Theorem 16.5.2. □

16.5.8. Two-sided constructions. The main theorems of this section are the bisimplicial homotopy lifting extension theorem (Theorem 16.5.13) and the two-sided homotopy lifting extension theorem (Theorem 16.5.18).

LEMMA 16.5.9. *Let* $\mathcal{C}$ *and* $\mathcal{D}$ *be Reedy categories, let* $\mathcal{M}$ *be a complete and cocomplete category, let* $\boldsymbol{X}$ *be a* $\mathcal{C}$*-diagram in* $\mathcal{M}$*, and let* $\boldsymbol{Y}$ *be a* $\mathcal{D}$*-diagram in* $\mathcal{M}$*.*

(1) *If we view* $\mathcal{M}(\boldsymbol{X},\boldsymbol{Y})$ *as a* $\mathcal{C}^{\mathrm{op}}$*-diagram in the category of* $\mathcal{D}$*-diagrams of sets (see Proposition 15.1.5), then for every object* α *of* $\mathcal{C}$ *there is a natural isomorphism of* $\mathcal{D}$*-diagrams of sets* $\mathrm{M}_\alpha\mathcal{M}(\boldsymbol{X},\boldsymbol{Y}) \approx \mathcal{M}(\mathrm{L}_\alpha\boldsymbol{X},\boldsymbol{Y})$*.*

(2) *If we view* $\mathcal{M}(\boldsymbol{X},\boldsymbol{Y})$ *as a* $\mathcal{D}$*-diagram in the category of* $\mathcal{C}^{\mathrm{op}}$*-diagrams of sets, then for every object* α *of* $\mathcal{D}$ *there is a natural isomorphism of* $\mathcal{C}^{\mathrm{op}}$*-diagrams of sets* $\mathrm{M}_\alpha\mathcal{M}(\boldsymbol{X},\boldsymbol{Y}) \approx \mathcal{M}(\boldsymbol{X},\mathrm{M}_\alpha\boldsymbol{Y})$*.*

PROOF. We will prove part 1; the proof of part 2 is similar.

Proposition 15.2.4 implies that we have natural isomorphisms

$$\begin{aligned}
\mathrm{M}_\alpha\mathcal{M}(\boldsymbol{X},\boldsymbol{Y}) &= \lim_{(\alpha\to\beta)\in \mathrm{Ob}\,\partial(\alpha\downarrow\overleftarrow{\mathcal{C}^{\mathrm{op}}})} \mathcal{M}(\boldsymbol{X}_\beta,\boldsymbol{Y}) \\
&\approx \lim_{(\beta\to\alpha)\in \mathrm{Ob}\,\partial(\overrightarrow{\mathcal{C}}\downarrow\alpha)} \mathcal{M}(\boldsymbol{X}_\beta,\boldsymbol{Y}) \\
&\approx \mathcal{M}\Big(\operatorname*{colim}_{(\beta\to\alpha)\in \mathrm{Ob}\,\partial(\overrightarrow{\mathcal{C}}\downarrow\alpha)} \boldsymbol{X}_\beta,\boldsymbol{Y}\Big) \\
&= \mathcal{M}(\mathrm{L}_\alpha\boldsymbol{X},\boldsymbol{Y}) \ .
\end{aligned}$$

□

DEFINITION 16.5.10. Let $\mathcal{M}$ be a model category. If $\boldsymbol{B}$ is a cosimplicial object in $\mathcal{M}$ and $\boldsymbol{X}$ is a simplicial object in $\mathcal{M}$, then the bisimplicial set $\mathcal{M}(\boldsymbol{B},\boldsymbol{X})$ (for which $\mathcal{M}(\boldsymbol{B},\boldsymbol{X})_{n,k} = \mathcal{M}(\boldsymbol{B}^k,\boldsymbol{X}_n))$ can be considered a simplicial object in the category of simplicial objects in $\mathcal{M}$ in two ways. We define the *horizontal simplicial object* to be the one whose object in degree n is $\mathcal{M}(\boldsymbol{B},\boldsymbol{X})_{n,*} = \mathcal{M}(\boldsymbol{B},\boldsymbol{X}_n)$ and the *vertical simplicial object* to be the one whose object in degree k is $\mathcal{M}(\boldsymbol{B},\boldsymbol{X})_{*,k} = \mathcal{M}(\boldsymbol{B}^k,\boldsymbol{X})$.

LEMMA 16.5.11. *Let $\mathcal{M}$ be a model category, let $\boldsymbol{B}$ be a cosimplicial object in $\mathcal{M}$, and let $\boldsymbol{X}$ be a simplicial object in $\mathcal{M}$.*

(1) *If we view $\mathcal{M}(\boldsymbol{B},\boldsymbol{X})$ as a horizontal simplicial object in the category of simplicial sets (see Definition 16.5.10), then for every $n \ge 0$ there is a natural isomorphism of simplicial sets (see Definition 15.2.5) $\mathrm{M}_n\mathcal{M}(\boldsymbol{B},\boldsymbol{X}) \approx \mathcal{M}(\boldsymbol{B},\mathrm{M}_n\boldsymbol{X})$.*
(2) *If we view $\mathcal{M}(\boldsymbol{B},\boldsymbol{X})$ as a vertical simplicial object in the category of simplicial sets (see Definition 16.5.10), then for every $n \ge 0$ there is a natural isomorphism of simplicial sets $\mathrm{M}_n\mathcal{M}(\boldsymbol{B},\boldsymbol{X}) \approx \mathcal{M}(\mathrm{L}_n\boldsymbol{B},\boldsymbol{X})$.*

PROOF. This follows from Lemma 16.5.9, letting $\mathcal{C}$ be the cosimplicial indexing category and letting $\mathcal{D}$ be the simplicial indexing category (see Definition 15.1.8). □

LEMMA 16.5.12. *Let $\mathcal{M}$ be a model category, let $\boldsymbol{A} \to \boldsymbol{B}$ be a map of cosimplicial objects in $\mathcal{M}$, and let $\boldsymbol{X} \to \boldsymbol{Y}$ be a map of simplicial objects in $\mathcal{M}$.*

(1) *If all bisimplicial sets are viewed as horizontal simplicial objects, then for every $n \ge 0$ there is a natural isomorphism of simplicial sets (see Definition 15.2.5)*

$$\mathrm{M}_n\big(\mathcal{M}(\boldsymbol{A},\boldsymbol{X}) \times_{\mathcal{M}(\boldsymbol{A},\boldsymbol{Y})} \mathcal{M}(\boldsymbol{B},\boldsymbol{Y})\big) \approx \mathcal{M}(\boldsymbol{A},\mathrm{M}_n\boldsymbol{X}) \times_{\mathcal{M}(\boldsymbol{A},\mathrm{M}_n\boldsymbol{Y})} \mathcal{M}(\boldsymbol{B},\mathrm{M}_n\boldsymbol{Y}) .$$

(2) *If all bisimplicial sets are viewed as vertical simplicial objects, then for every $n \ge 0$ there is a natural isomorphism of simplicial sets*

$$\mathrm{M}_n\big(\mathcal{M}(\boldsymbol{A},\boldsymbol{X}) \times_{\mathcal{M}(\boldsymbol{A},\boldsymbol{Y})} \mathcal{M}(\boldsymbol{B},\boldsymbol{Y})\big) \approx \mathcal{M}(\mathrm{L}_n\boldsymbol{A},\boldsymbol{X}) \times_{\mathcal{M}(\mathrm{L}_n\boldsymbol{A},\boldsymbol{Y})} \mathcal{M}(\mathrm{L}_n\boldsymbol{B},\boldsymbol{Y}) .$$

PROOF. This follows from Lemma 16.5.11. □

THEOREM 16.5.13 (The bisimplicial homotopy lifting extension theorem). *Let $\mathcal{M}$ be a model category. If $i\colon \boldsymbol{A} \to \boldsymbol{B}$ is a Reedy cofibration of cosimplicial resolutions in $\mathcal{M}$ and $p\colon \boldsymbol{X} \to \boldsymbol{Y}$ is a Reedy fibration of simplicial resolutions in $\mathcal{M}$, then for both the horizontal and the vertical simplicial object structures (see Definition 16.5.10), the induced map of bisimplicial sets*

$$\mathcal{M}(\boldsymbol{B},\boldsymbol{X}) \to \mathcal{M}(\boldsymbol{A},\boldsymbol{X}) \times_{\mathcal{M}(\boldsymbol{A},\boldsymbol{Y})} \mathcal{M}(\boldsymbol{B},\boldsymbol{Y})$$

is a Reedy fibration of simplicial objects that is a Reedy trivial fibration if at least one of i and p is a weak equivalence.

PROOF. We will prove this for the horizontal structure; the proof for the vertical structure is similar.

Theorem 15.3.15 implies that it is sufficient to show that for every $n \ge 0$ the map

$$\begin{aligned}&\mathcal{M}(\boldsymbol{B},\boldsymbol{X})_n\\ &\quad\to \big(\mathcal{M}(\boldsymbol{A},\boldsymbol{X}) \times_{\mathcal{M}(\boldsymbol{A},\boldsymbol{Y})} \mathcal{M}(\boldsymbol{B},\boldsymbol{Y})\big)_n \times_{\mathrm{M}_n(\mathcal{M}(\boldsymbol{A},\boldsymbol{X})\times_{\mathcal{M}(\boldsymbol{A},\boldsymbol{Y})}\mathcal{M}(\boldsymbol{B},\boldsymbol{Y}))} \mathrm{M}_n\mathcal{M}(\boldsymbol{B},\boldsymbol{X})\end{aligned}$$

is a fibration of simplicial sets that is a trivial fibration if either of i and p is a weak equivalence. Lemma 16.5.11 and Lemma 16.5.12 imply that this map is isomorphic to the map

$$\begin{aligned}&\mathcal{M}(\boldsymbol{B}, \boldsymbol{X}_n)\\ &\to \big(\mathcal{M}(\boldsymbol{A}, \boldsymbol{X}_n) \times_{\mathcal{M}(\boldsymbol{A}, \boldsymbol{Y}_n)} \mathcal{M}(\boldsymbol{B}, \boldsymbol{Y}_n)\big) \times_{(\mathcal{M}(\boldsymbol{A}, \mathrm{M}_n \boldsymbol{X}) \times_{\mathcal{M}(\boldsymbol{A}, \mathrm{M}_n \boldsymbol{Y})} \mathcal{M}(\boldsymbol{B}, \mathrm{M}_n \boldsymbol{Y}))} \mathcal{M}(\boldsymbol{B}, \mathrm{M}_n \boldsymbol{X})\end{aligned}$$

The codomain of this map is the limit of the diagram

$$\begin{array}{ccccc} \mathcal{M}(\boldsymbol{A}, \boldsymbol{X}_n) & \longrightarrow & \mathcal{M}(\boldsymbol{A}, \boldsymbol{Y}_n) & \longleftarrow & \mathcal{M}(\boldsymbol{B}, \boldsymbol{Y}_n) \\ \downarrow & & \downarrow & & \downarrow \\ \mathcal{M}(\boldsymbol{A}, \mathrm{M}_n \boldsymbol{X}) & \longrightarrow & \mathcal{M}(\boldsymbol{A}, \mathrm{M}_n \boldsymbol{Y}) & \longleftarrow & \mathcal{M}(\boldsymbol{B}, \mathrm{M}_n \boldsymbol{Y}) \\ & \nwarrow & \uparrow & \nearrow & \\ & & \mathcal{M}(\boldsymbol{B}, \mathrm{M}_n \boldsymbol{X}) & & \end{array}$$

and so our map is isomorphic to the map

$$\begin{aligned}&\mathcal{M}(\boldsymbol{B}, \boldsymbol{X}_n)\\ &\to \mathcal{M}(\boldsymbol{A}, \boldsymbol{X}_n) \times_{(\mathcal{M}(\boldsymbol{A}, \boldsymbol{Y}_n) \times_{\mathcal{M}(\boldsymbol{A}, \mathrm{M}_n \boldsymbol{Y})} \mathcal{M}(\boldsymbol{A}, \mathrm{M}_n \boldsymbol{X}))} \big(\mathcal{M}(\boldsymbol{B}, \boldsymbol{Y}_n) \times_{\mathcal{M}(\boldsymbol{B}, \mathrm{M}_n \boldsymbol{Y})} \mathcal{M}(\boldsymbol{B}, \mathrm{M}_n \boldsymbol{X})\big)\end{aligned}$$

Since p is a Reedy fibration, the map $\boldsymbol{X}_n \to \boldsymbol{Y}_n \times_{\mathrm{M}_n \boldsymbol{Y}} \mathrm{M}_n \boldsymbol{X}$ is a fibration of simplicial sets, and so the result now follows from Theorem 16.5.2 and Theorem 15.3.15. □

COROLLARY 16.5.14. *If $\mathcal{M}$ is a model category, $\boldsymbol{B}$ is a cosimplicial resolution in $\mathcal{M}$, and $\boldsymbol{X}$ is a simplicial resolution in $\mathcal{M}$, then $\mathcal{M}(\boldsymbol{B}, \boldsymbol{X})$ is a Reedy fibrant simplicial object in both the horizontal and vertical simplicial object structures (see Definition 16.5.10).*

PROOF. This follows from Theorem 16.5.13. □

COROLLARY 16.5.15. *Let $\mathcal{M}$ be a model category, let $\boldsymbol{B}$ be a cosimplicial resolution in $\mathcal{M}$, and let $\boldsymbol{X}$ be a simplicial resolution in $\mathcal{M}$.*

1. *If we consider the bisimplicial set $\mathcal{M}(\boldsymbol{B}, \boldsymbol{X})$ as a horizontal simplicial object (see Definition 16.5.10) in the category of simplicial sets (so that in simplicial degree n we have the simplicial set $\mathcal{M}(\boldsymbol{B}, \boldsymbol{X}_n)$), then $\mathcal{M}(\boldsymbol{B}, \boldsymbol{X})$ is a simplicial resolution of the simplicial set $\mathcal{M}(\boldsymbol{B}, \boldsymbol{X}_0)$.*
2. *If we consider the bisimplicial set $\mathcal{M}(\boldsymbol{B}, \boldsymbol{X})$ as a vertical simplicial object in the category of simplicial sets (so that in simplicial degree n we have the simplicial set $\mathcal{M}(\boldsymbol{B}^n, \boldsymbol{X})$), then $\mathcal{M}(\boldsymbol{B}, \boldsymbol{X})$ is a simplicial resolution of the simplicial set $\mathcal{M}(\boldsymbol{B}^0, \boldsymbol{X})$.*

PROOF. Corollary 16.5.14 implies that $\mathcal{M}(\boldsymbol{B}, \boldsymbol{X})$ is a Reedy fibrant simplicial object, and Corollary 16.5.5 implies that, for every $n > 0$, the natural maps $\mathcal{M}(\boldsymbol{B}, \boldsymbol{X}_0) \to \mathcal{M}(\boldsymbol{B}, \boldsymbol{X}_n)$ and $\mathcal{M}(\boldsymbol{B}_0, \boldsymbol{X}) \to \mathcal{M}(\boldsymbol{B}_n, \boldsymbol{X})$ are weak equivalences. □

COROLLARY 16.5.16. *Let $\mathcal{M}$ be a model category, let B be a cofibrant object of $\mathcal{M}$ with cosimplicial resolution $\widetilde{\boldsymbol{B}}$, and let X be a fibrant object of $\mathcal{M}$ with simplicial resolution $\widehat{\boldsymbol{X}}$.*

(1) *If we consider the bisimplicial set* $\mathcal{M}(\widetilde{\boldsymbol{B}}, \widehat{\boldsymbol{X}})$ *as a horizontal simplicial object (see Definition 16.5.10) in the category of simplicial sets (so that in simplicial degree* n *we have the simplicial set* $\mathcal{M}(\widetilde{\boldsymbol{B}}, \widehat{\boldsymbol{X}}_n)$*), then* $\mathcal{M}(\widetilde{\boldsymbol{B}}, \widehat{\boldsymbol{X}})$ *is a simplicial resolution of the simplicial set* $\mathcal{M}(\widetilde{\boldsymbol{B}}, X)$.

(2) *If we consider the bisimplicial set* $\mathcal{M}(\widetilde{\boldsymbol{B}}, \widehat{\boldsymbol{X}})$ *as a vertical simplicial object in the category of simplicial sets (so that in simplicial degree* n *we have the simplicial set* $\mathcal{M}(\widetilde{\boldsymbol{B}}^n, \widehat{\boldsymbol{X}})$*), then* $\mathcal{M}(\widetilde{\boldsymbol{B}}, \widehat{\boldsymbol{X}})$ *is a simplicial resolution of the simplicial set* $\mathcal{M}(B, \widehat{\boldsymbol{X}})$.

PROOF. This follows from Corollary 16.5.15 and Corollary 16.5.5. □

COROLLARY 16.5.17. *Let* $\mathcal{M}$ *be a model category.*

(1) *If* $i\colon \boldsymbol{A} \to \boldsymbol{B}$ *is a Reedy cofibration of cosimplicial resolutions in* $\mathcal{M}$ *and* $\boldsymbol{X}$ *is a simplicial resolution in* $\mathcal{M}$ *then, for both the horizontal and vertical simplicial object structures (see Definition 16.5.10), the induced map* $i^*\colon \mathcal{M}(\boldsymbol{B}, \boldsymbol{X}) \to \mathcal{M}(\boldsymbol{A}, \boldsymbol{X})$ *is a Reedy fibration of Reedy fibrant simplicial objects that is a Reedy trivial fibration if* i *is a Reedy trivial cofibration.*

(2) *If* $\boldsymbol{B}$ *is a cosimplicial resolution in* $\mathcal{M}$ *and* $p\colon \boldsymbol{X} \to \boldsymbol{Y}$ *is a Reedy fibration of simplicial resolutions in* $\mathcal{M}$ *then, for both the horizontal and vertical simplicial object structures, the induced map* $p_*\colon \mathcal{M}(\boldsymbol{B}, \boldsymbol{X}) \to \mathcal{M}(\boldsymbol{B}, \boldsymbol{Y})$ *is a Reedy fibration of Reedy fibrant simplicial objects that is a Reedy trivial fibration if* p *is a Reedy trivial fibration.*

PROOF. This follows from Theorem 16.5.13. □

THEOREM 16.5.18 (The two-sided homotopy lifting extension theorem). *Let* $\mathcal{M}$ *be a model category. If* $i\colon \boldsymbol{A} \to \boldsymbol{B}$ *is a Reedy cofibration of cosimplicial resolutions in* $\mathcal{M}$ *and* $p\colon \boldsymbol{X} \to \boldsymbol{Y}$ *is a Reedy fibration of simplicial resolutions in* $\mathcal{M}$*, then the induced map of simplicial sets*

$$\operatorname{diag} \mathcal{M}(\boldsymbol{B}, \boldsymbol{X}) \to \operatorname{diag} \mathcal{M}(\boldsymbol{A}, \boldsymbol{X}) \times_{\operatorname{diag} \mathcal{M}(\boldsymbol{A}, \boldsymbol{Y})} \operatorname{diag} \mathcal{M}(\boldsymbol{B}, \boldsymbol{Y})$$

is a fibration of fibrant simplicial sets that is a trivial fibration if at least one of i *and* p *is a weak equivalence.*

PROOF. The result follows from Theorem 16.5.13, Proposition 15.3.11, Theorem 15.11.7, Proposition 15.3.13, Theorem 15.11.11, and Theorem 15.11.6. □

COROLLARY 16.5.19. *If* $\mathcal{M}$ *is a model category,* $\boldsymbol{B}$ *is a cosimplicial resolution in* $\mathcal{M}$*, and* $\boldsymbol{X}$ *is a simplicial resolution in* $\mathcal{M}$*, then* $\operatorname{diag} \mathcal{M}(\boldsymbol{B}, \boldsymbol{X})$ *is a fibrant simplicial set.*

PROOF. This follows from Theorem 16.5.18. □

COROLLARY 16.5.20. *Let* $\mathcal{M}$ *be a model category.*

(1) *If* $i\colon \boldsymbol{A} \to \boldsymbol{B}$ *is a Reedy cofibration of cosimplicial resolutions in* $\mathcal{M}$ *and* $\boldsymbol{X}$ *is a simplicial resolution in* $\mathcal{M}$*, then the induced map*

$$\operatorname{diag} i^*\colon \operatorname{diag} \mathcal{M}(\boldsymbol{B}, \boldsymbol{X}) \to \operatorname{diag} \mathcal{M}(\boldsymbol{A}, \boldsymbol{X})$$

is a fibration of fibrant simplicial sets that is a trivial fibration if i *is a Reedy trivial cofibration.*

(2) *If $\boldsymbol{B}$ is a cosimplicial resolution in $\mathcal{M}$ and $p\colon \boldsymbol{X} \to \boldsymbol{Y}$ is a Reedy fibration of simplicial resolutions in $\mathcal{M}$, then the induced map*

$$\operatorname{diag} p_*\colon \operatorname{diag} \mathcal{M}(\boldsymbol{B}, \boldsymbol{X}) \to \operatorname{diag} \mathcal{M}(\boldsymbol{B}, \boldsymbol{Y})$$

is a fibration of fibrant simplicial sets that is a trivial fibration if p is a Reedy trivial fibration.

PROOF. This follows from Theorem 16.5.18 and Corollary 16.5.19. □

COROLLARY 16.5.21. *Let $\mathcal{M}$ be a model category.*

(1) *If $i\colon \boldsymbol{A} \to \boldsymbol{B}$ is a Reedy weak equivalence of cosimplicial resolutions in $\mathcal{M}$ and $\boldsymbol{X}$ is a simplicial resolution in $\mathcal{M}$, then the induced map*

$$\operatorname{diag} i^*\colon \operatorname{diag} \mathcal{M}(\boldsymbol{B}, \boldsymbol{X}) \to \operatorname{diag} \mathcal{M}(\boldsymbol{A}, \boldsymbol{X})$$

is a weak equivalence of fibrant simplicial sets.

(2) *If $\boldsymbol{B}$ is a cosimplicial resolution in $\mathcal{M}$ and $p\colon \boldsymbol{X} \to \boldsymbol{Y}$ is a Reedy weak equivalence of simplicial resolutions in $\mathcal{M}$, then the induced map*

$$\operatorname{diag} p_*\colon \operatorname{diag} \mathcal{M}(\boldsymbol{B}, \boldsymbol{X}) \to \operatorname{diag} \mathcal{M}(\boldsymbol{B}, \boldsymbol{Y})$$

is a weak equivalence of fibrant simplicial sets.

PROOF. This follows from Corollary 16.5.20 and Corollary 7.7.2. □

16.6. Frames

Proposition 16.1.5 shows how a cosimplicial resolution of an object in a model category yields a cofibrant approximation to that object (and how a simplicial resolution yields a fibrant approximation). Frames (see Definition 16.6.1) allow us to discuss the reverse operation (see Proposition 16.6.7). Frames will also be used to define the homotopy colimit and homotopy limit functors (see Definition 19.1.2 and Definition 19.1.5).

DEFINITION 16.6.1. Let $\mathcal{M}$ be a model category and let X be an object of $\mathcal{M}$.

- A *cosimplicial frame* on X is a cosimplicial object $\widetilde{\boldsymbol{X}}$ in $\mathcal{M}$ together with a weak equivalence $\widetilde{\boldsymbol{X}} \to \mathrm{cc}_* X$ (see Notation 16.1.1) in the Reedy model category structure (see Definition 15.3.3) on $\mathcal{M}^{\boldsymbol{\Delta}}$ such that
  (1) the induced map $\widetilde{\boldsymbol{X}}^0 \to X$ is an isomorphism, and
  (2) if X is a cofibrant object of $\mathcal{M}$, then $\widetilde{\boldsymbol{X}}$ is a cofibrant object of $\mathcal{M}^{\boldsymbol{\Delta}}$.
  We will sometimes refer to $\widetilde{\boldsymbol{X}}$ as a cosimplicial frame on X, without explicitly mentioning the map $\widetilde{\boldsymbol{X}} \to \mathrm{cc}_* X$.
- A *simplicial frame* on X is a simplicial object $\widehat{\boldsymbol{X}}$ in $\mathcal{M}$ together with a weak equivalence $\mathrm{cs}_* X \to \widehat{\boldsymbol{X}}$ in the Reedy model category structure on $\mathcal{M}^{\boldsymbol{\Delta}^{\mathrm{op}}}$ such that
  (1) the induced map $X \to \widehat{\boldsymbol{X}}_0$ is an isomorphism, and
  (2) if X is a fibrant object of $\mathcal{M}$, then $\widehat{\boldsymbol{X}}$ is a fibrant object of $\mathcal{M}^{\boldsymbol{\Delta}^{\mathrm{op}}}$.
  We will sometimes refer to $\widehat{\boldsymbol{X}}$ as a simplicial frame on X, without explicitly mentioning the map $\mathrm{cs}_* X \to \widehat{\boldsymbol{X}}$.

REMARK 16.6.2. Note that Definition 16.6.1 does not require cosimplicial frames on non-cofibrant objects to be cofibrant or simplicial frames on non-fibrant objects to be fibrant. This was done in order to make Proposition 16.6.4 true.

PROPOSITION 16.6.3. *Let $\mathcal{M}$ be a model category and let X be an object of $\mathcal{M}$.*

(1) *If X is cofibrant, then any cosimplicial frame on X is a cosimplicial resolution of X.*
(2) *If X is fibrant, then any simplicial frame on X is a simplicial resolution of X.*

PROOF. This follows directly from the definitions. □

PROPOSITION 16.6.4. *If $\mathcal{M}$ is a simplicial model category and X is an object of $\mathcal{M}$, then*

- *the cosimplicial object $\widetilde{\boldsymbol{X}}$ in which $\widetilde{\boldsymbol{X}}^n = X \otimes \Delta[n]$ is a cosimplicial frame on X, and*
- *the simplicial object $\widehat{\boldsymbol{Y}}$ in which $\widehat{\boldsymbol{Y}}_n = X^{\Delta[n]}$ is a simplicial frame on X.*

PROOF. This follows from Proposition 9.5.20 and Proposition 16.1.3. □

DEFINITION 16.6.5. If $\mathcal{M}$ is a simplicial model category and X is an object of $\mathcal{M}$, then the cosimplicial frame on X of Proposition 16.6.4 will be called the *standard cosimplicial frame on* X, and the simplicial frame on X of Proposition 16.6.4 will be called the *standard simplicial frame on* X.

PROPOSITION 16.6.6. *Let $\mathcal{M}$ be a simplicial model category.*

(1) *If X is an object of $\mathcal{M}$, $\widetilde{\boldsymbol{X}}$ is the standard cosimplicial frame on X (see Proposition 16.6.23), and K is a simplicial set, then $\widetilde{\boldsymbol{X}} \otimes K$ is naturally isomorphic to $X \otimes K$.*
(2) *If X is an object of $\mathcal{M}$, $\widehat{\boldsymbol{X}}$ is the standard simplicial frame on X, and K is a simplicial set, then $\widehat{\boldsymbol{X}}^K$ is naturally isomorphic to X^K.*

PROOF. This follows from Proposition 16.4.3 and Proposition 15.1.20. □

PROPOSITION 16.6.7. *Let $\mathcal{M}$ be a model category.*

(1) *If X is an object of $\mathcal{M}$, $\widetilde{X} \to X$ is a cofibrant approximation to X, and $\widetilde{\boldsymbol{X}}' \to \mathrm{cc}_*\widetilde{X}$ is a cosimplicial frame on $\widetilde{X}$, then the induced map $\widetilde{\boldsymbol{X}}' \to \mathrm{cc}_*X$ is a cosimplicial resolution of X, and every cosimplicial resolution of X can be constructed in this way.*
(2) *If X is an object of $\mathcal{M}$, $X \to \widehat{X}$ is a fibrant approximation to X, and $\mathrm{cs}_*\widehat{X} \to \widehat{\boldsymbol{X}}'$ is a simplicial frame on $\widehat{X}$, then the induced map $\mathrm{cs}_*X \to \widehat{\boldsymbol{X}}'$ is a simplicial resolution of X, and every simplicial resolution of X can be constructed in this way.*

PROOF. This follows from Proposition 16.1.5. □

PROPOSITION 16.6.8. *Let $\mathcal{M}$ be a model category.*

(1) *There is an augmented functor (F, i) on $\mathcal{M}^{\boldsymbol{\Delta}}$ (see Definition 8.1.12) such that*
    (a) *$i_{\boldsymbol{X}}\colon \mathrm{F}\boldsymbol{X} \to \boldsymbol{X}$ is a Reedy trivial fibration for every object $\boldsymbol{X}$ of $\mathcal{M}^{\boldsymbol{\Delta}}$,*
    (b) *$(i_{\boldsymbol{X}})^0\colon (\mathrm{F}\boldsymbol{X})^0 \to \boldsymbol{X}^0$ is an isomorphism for every object $\boldsymbol{X}$ of $\mathcal{M}^{\boldsymbol{\Delta}}$, and*
    (c) *if $\boldsymbol{X}^0$ is cofibrant in $\mathcal{M}$, then $\mathrm{F}\boldsymbol{X}$ is Reedy cofibrant.*
(2) *There is a coaugmented functor (G, j) on $\mathcal{M}^{\boldsymbol{\Delta}^{\mathrm{op}}}$ (see Definition 8.1.12) such that*

(a) $j_{\boldsymbol{X}}\colon \boldsymbol{X} \to \mathrm{G}\boldsymbol{X}$ *is a Reedy trivial cofibration for every object* $\boldsymbol{X}$ *of* $\mathcal{M}^{\boldsymbol{\Delta}^{\mathrm{op}}}$,

(b) $(j_{\boldsymbol{X}})_0\colon \boldsymbol{X}_0 \to (\mathrm{G}\boldsymbol{X})_0$ *is an isomorphism for every object* $\boldsymbol{X}$ *of* $\mathcal{M}^{\boldsymbol{\Delta}^{\mathrm{op}}}$, *and*

(c) *if* $\boldsymbol{X}_0$ *is fibrant in* $\mathcal{M}$, *then* $\mathrm{G}\boldsymbol{X}$ *is Reedy fibrant.*

PROOF. We will construct $\mathrm{F}\boldsymbol{X}$ and the map $\mathrm{F}\boldsymbol{X} \to \boldsymbol{X}$ inductively, and we begin by letting $(\mathrm{F}\boldsymbol{X})^0 = \boldsymbol{X}^0$. If $n > 0$ and we have constructed $\mathrm{F}\boldsymbol{X} \to \boldsymbol{X}$ in degrees less than n, then we have the induced map $\mathrm{L}_n(\mathrm{F}\boldsymbol{X}) \to \boldsymbol{X}^n \times_{\mathrm{M}_n\boldsymbol{X}} \mathrm{M}_n(\mathrm{F}\boldsymbol{X})$. We can factor this map functorially in $\mathcal{M}$ as

$$\mathrm{L}_n(\mathrm{F}\boldsymbol{X}) \xrightarrow{\;i\;} (\mathrm{F}\boldsymbol{X})^n \xrightarrow{\;p\;} \boldsymbol{X}^n \times_{\mathrm{M}_n\boldsymbol{X}} \mathrm{M}_n(\mathrm{F}\boldsymbol{X})$$

with i a cofibration and p a trivial fibration. This completes the construction, and Theorem 15.3.15 implies that the map $\mathrm{F}\boldsymbol{X} \to \boldsymbol{X}$ is always a Reedy trivial fibration. If $\boldsymbol{X}^0$ is cofibrant, then $\mathrm{L}_n(\mathrm{F}\boldsymbol{X}) \to (\mathrm{F}\boldsymbol{X})^n$ is a cofibration for all $n \geq 0$, and so $\mathrm{F}\boldsymbol{X}$ is Reedy cofibrant. □

THEOREM 16.6.9. *If* $\mathcal{M}$ *is a model category then there exists a functorial cosimplicial frame on* $\mathcal{M}$ *and a functorial simplicial frame on* $\mathcal{M}$.

PROOF. This follows from Proposition 16.6.8. □

THEOREM 16.6.10. *Let* $\mathcal{M}$ *be a model category and let* $\mathcal{K}$ *be a subcategory of* $\mathcal{M}$.

(1) *Any two functorial cosimplicial frames on* $\mathcal{K}$ *are connected by an essentially unique zig-zag (see Definition 14.4.2) of weak equivalences of functorial cosimplicial frames on* $\mathcal{K}$.

(2) *Any two functorial simplicial frames on* $\mathcal{K}$ *are connected by an essentially unique zig-zag (see Definition 14.4.2) of weak equivalences of functorial simplicial frames on* $\mathcal{K}$.

PROOF. This follows from Theorem 14.5.5 and Proposition 16.6.8. □

16.6.11. Frames on maps.

DEFINITION 16.6.12. Let $\mathcal{M}$ be a model category and let $g\colon X \to Y$ be a map in $\mathcal{M}$.

(1) A *cosimplicial frame on* g consists of a cosimplicial frame $\widetilde{\boldsymbol{X}} \to \mathrm{cc}_*X$ on X, a cosimplicial frame $\widetilde{\boldsymbol{Y}} \to \mathrm{cc}_*Y$ on Y, and a map $\tilde{g}\colon \widetilde{\boldsymbol{X}} \to \widetilde{\boldsymbol{Y}}$ that makes the square

$$\begin{array}{ccc} \widetilde{\boldsymbol{X}} & \xrightarrow{\;\tilde{g}\;} & \widetilde{\boldsymbol{Y}} \\ \downarrow & & \downarrow \\ \mathrm{cc}_*X & \longrightarrow & \mathrm{cc}_*Y \end{array}$$

commute.

(2) A *simplicial frame on* g consists of a simplicial frame $\mathrm{cs}_*X \to \widehat{\boldsymbol{X}}$ on X, a simplicial frame $\mathrm{cs}_*Y \to \widehat{\boldsymbol{Y}}$ on Y, and a map $\hat{g}\colon \widehat{\boldsymbol{X}} \to \widehat{\boldsymbol{Y}}$ that makes the

square

$$\begin{array}{ccc} \mathrm{cs}_* X & \longrightarrow & \mathrm{cs}_* Y \\ \downarrow & & \downarrow \\ \widehat{\boldsymbol{X}} & \xrightarrow[\hat{g}]{} & \widehat{\boldsymbol{Y}} \end{array}$$

commute.

EXAMPLE 16.6.13. Let $\mathcal{M}$ be a simplicial model category.

(1) Let $i\colon A \to B$ be a map in $\mathcal{M}$, let $\widetilde{\boldsymbol{A}}$ and $\widetilde{\boldsymbol{B}}$ be the cosimplicial objects in $\mathcal{M}$ such that $\widetilde{\boldsymbol{A}}^n = A \otimes \Delta[n]$ and $\widetilde{\boldsymbol{B}}^n = B \otimes \Delta[n]$, and let $\tilde{\imath}\colon \widetilde{\boldsymbol{A}} \to \widetilde{\boldsymbol{B}}$ be the obvious map. Proposition 16.6.4 implies that $\tilde{\imath}$ is a cosimplicial frame on i, and Proposition 9.3.8 implies that $\tilde{\imath}$ is a Reedy cofibration if i is a cofibration in $\mathcal{M}$.

(2) Let $p\colon X \to Y$ be a map in $\mathcal{M}$, let $\widehat{\boldsymbol{X}}$ and $\widehat{\boldsymbol{Y}}$ be the simplicial objects in $\mathcal{M}$ such that $\widehat{\boldsymbol{X}}_n = X^{\Delta[n]}$ and $\widehat{\boldsymbol{Y}}_n = Y^{\Delta[n]}$, and let $\hat{p}\colon \widehat{\boldsymbol{X}} \to \widehat{\boldsymbol{Y}}$ be the obvious map. Proposition 16.6.4 implies that $\hat{p}$ is a simplicial frame on p, and Proposition 9.3.8 implies that $\hat{p}$ is a Reedy fibration if p is a fibration in $\mathcal{M}$.

PROPOSITION 16.6.14. *Let $\mathcal{M}$ be a model category and let $g\colon X \to Y$ be a map in $\mathcal{M}$.*

(1) *There is a natural cosimplicial frame $\tilde{g}\colon \widetilde{\boldsymbol{X}} \to \widetilde{\boldsymbol{Y}}$ on g that is a Reedy cofibration if g is a cofibration.*

(2) *There is a natural simplicial frame $\hat{g}\colon \widehat{\boldsymbol{X}} \to \widehat{\boldsymbol{Y}}$ on g that is a Reedy fibration if g is a fibration.*

PROOF. We will prove part 1; the proof of part 2 is dual.

We begin by constructing a natural cosimplicial frame $\widetilde{\boldsymbol{X}} \to \mathrm{cc}_* X$ on X as in the proof of Theorem 16.6.9.

We will define $\widetilde{\boldsymbol{Y}}$ and $\tilde{g}$ inductively. We let $\widetilde{\boldsymbol{Y}}^0 = Y$. If $n > 0$ and we have constructed $\widetilde{\boldsymbol{Y}}$ and $\tilde{g}$ in degrees less than n, then we have the induced map $\mathrm{L}_n\widetilde{\boldsymbol{Y}} \amalg_{\mathrm{L}_n\widetilde{\boldsymbol{X}}} \widetilde{\boldsymbol{X}}^n \to (\mathrm{cc}_* Y)^n \times_{\mathrm{M}_n \mathrm{cc}_* Y} \mathrm{M}_n\widetilde{\boldsymbol{Y}}$. We factor this map functorially in $\mathcal{M}$ as

$$\mathrm{L}_n\widetilde{\boldsymbol{Y}} \amalg_{\mathrm{L}_n\widetilde{\boldsymbol{X}}} \widetilde{\boldsymbol{X}}^n \xrightarrow{i} \widetilde{\boldsymbol{Y}}^n \xrightarrow{p} (\mathrm{cc}_* Y)^n \times_{\mathrm{M}_n \mathrm{cc}_* Y} \mathrm{M}_n\widetilde{\boldsymbol{Y}}$$

with i a cofibration and p a trivial fibration. This completes the construction, and Theorem 15.3.15 implies that the map $\widetilde{\boldsymbol{Y}} \to \mathrm{cc}_* Y$ is always a Reedy trivial fibration. Since $\mathrm{L}_n\widetilde{\boldsymbol{X}} \to \widetilde{\boldsymbol{X}}^n$ was constructed to be a cofibration for all $n > 0$, and $\mathrm{L}_n\widetilde{\boldsymbol{Y}} \to \mathrm{L}_n\widetilde{\boldsymbol{Y}} \amalg_{\mathrm{L}_n\widetilde{\boldsymbol{X}}} \widetilde{\boldsymbol{X}}^n$ is a pushout of that cofibration, the composition $\mathrm{L}_n\widetilde{\boldsymbol{Y}} \to \mathrm{L}_n\widetilde{\boldsymbol{Y}} \amalg_{\mathrm{L}_n\widetilde{\boldsymbol{X}}} \widetilde{\boldsymbol{X}}^n \to \widetilde{\boldsymbol{Y}}^n$ is a cofibration for all $n > 0$. Thus, if Y is cofibrant, then $\widetilde{\boldsymbol{Y}}$ is Reedy cofibrant. Finally, if g is a cofibration, then $\mathrm{L}_n\widetilde{\boldsymbol{Y}} \amalg_{\mathrm{L}_n\widetilde{\boldsymbol{X}}} \widetilde{\boldsymbol{X}}^n \to \widetilde{\boldsymbol{Y}}^n$ is a cofibration for all $n \geq 0$, and so $\tilde{g}$ is a Reedy cofibration. □

16.6.15. Uniqueness of frames.

DEFINITION 16.6.16. Let $\mathcal{M}$ be a model category and let X be an object of $\mathcal{M}$.

(1) The category $\mathrm{CosFr}(X)$ is the category whose objects are cosimplicial frames on X and whose morphisms are maps of cosimplicial frames on X.

(2) The category $\mathrm{SimpFr}(X)$ is the category whose objects are simplicial frames on X and whose morphisms are maps of simplicial frames on X.

THEOREM 16.6.17. *Let $\mathcal{M}$ be a model category.*

(1) *If X is an object of $\mathcal{M}$, then the category $\mathrm{CosFr}(X)$ of cosimplicial frames on X (see Definition 16.6.16) has a contractible classifying space (see Definition 14.3.1).*
(2) *If X is an object of $\mathcal{M}$, then the category $\mathrm{SimpFr}(X)$ of simplicial frames on X (see Definition 16.6.16) has a contractible classifying space (see Definition 14.3.1).*

PROOF. We will prove part 1; the proof of part 2 is dual.

Let $\mathcal{W}$ be the class of Reedy weak equivalences $\boldsymbol{X} \to \boldsymbol{Y}$ in $\mathcal{M}^{\boldsymbol{\Delta}}$ such that $\boldsymbol{X}^0 \to \boldsymbol{Y}^0$ is an isomorphism and such that if $\boldsymbol{Y}^0$ is cofibrant in $\mathcal{M}$ then $\boldsymbol{X}$ is Reedy cofibrant. The result now follows from Theorem 14.5.6 and Proposition 16.6.8. □

THEOREM 16.6.18. *Let $\mathcal{M}$ be a model category and let X be an object of $\mathcal{M}$.*

(1) *Any two cosimplicial frames on X are connected by an essentially unique zig-zag (see Definition 14.4.2) of weak equivalences of cosimplicial frames on X.*
(2) *Any two simplicial frames on X are connected by an essentially unique zig-zag (see Definition 14.4.2) of weak equivalences of simplicial frames on X.*

PROOF. This follows from Theorem 16.6.17 and Theorem 14.4.5. □

PROPOSITION 16.6.19. *Let $\mathcal{M}$ and $\mathcal{N}$ be model categories and let $\mathrm{F}: \mathcal{M} \rightleftarrows \mathcal{N} : \mathrm{U}$ be a Quillen pair (see Definition 8.5.2).*

(1) *If X is a cofibrant object of $\mathcal{M}$ and $\widetilde{\boldsymbol{X}} \to \mathrm{cc}_* X$ is a cosimplicial frame on X, then $\mathrm{F}\widetilde{\boldsymbol{X}} \to \mathrm{cc}_* \mathrm{F}X$ is a cosimplicial frame on $\mathrm{F}X$.*
(2) *If Y is a fibrant object of $\mathcal{N}$ and $\mathrm{cs}_* Y \to \widehat{\boldsymbol{Y}}$ is a simplicial frame on Y, then $\mathrm{cs}_* \mathrm{U}Y \to \mathrm{U}\widehat{\boldsymbol{Y}}$ is a simplicial frame on $\mathrm{U}Y$.*

PROOF. This follows from Proposition 16.2.1 and Proposition 16.6.7. □

16.6.20. Framed model categories.

DEFINITION 16.6.21. A *framed model category* is a model category $\mathcal{M}$ together with

(1) a functorial cosimplicial frame (see Definition 16.6.1) $\widetilde{\boldsymbol{X}}$ on every object X in $\mathcal{M}$, and
(2) a functorial simplicial frame $\widehat{\boldsymbol{X}}$ on every object X in $\mathcal{M}$.

PROPOSITION 16.6.22. *If $\mathcal{M}$ is a model category, then there exists a framed model category structure on $\mathcal{M}$.*

PROOF. This follows from Theorem 16.6.9. □

PROPOSITION 16.6.23. *If $\mathcal{M}$ is a simplicial model category, then there is a natural framing on $\mathcal{M}$ (called the* standard framing*) defined on objects X in $\mathcal{M}$ by $\widetilde{\boldsymbol{X}}^n = X \otimes \Delta[n]$ and $\widehat{\boldsymbol{X}}_n = X^{\Delta[n]}$.*

PROOF. This follows from Proposition 16.6.4. □

REMARK 16.6.24. If $\mathcal{M}$ is a simplicial model category and we make reference to $\mathcal{M}$ in a context that calls for a framed model category, then we will consider $\mathcal{M}$ as a framed model category using the standard framing of Proposition 16.6.23.

16.7. Reedy frames

If $\mathcal{M}$ is a framed model category, $\mathcal{C}$ is a small category, and $\boldsymbol{X}$ is a $\mathcal{C}$-diagram in $\mathcal{M}$, then the framing defines a $\mathcal{C}$-diagram $\widehat{\boldsymbol{X}}$ in $\mathcal{M}^{\boldsymbol{\Delta}}$, i.e., a $\mathcal{C}$-diagram of cosimplicial objects in $\mathcal{M}$. If $\mathcal{C}$ is a Reedy category and $\boldsymbol{X}$ is Reedy cofibrant, though, there is no reason to expect $\widehat{\boldsymbol{X}}$ to be Reedy cofibrant. Thus, we define a *Reedy frame* on a diagram (see Definition 16.7.8), and we show that Reedy frames always exist (see Proposition 16.7.11). We also show that any two frames on a diagram (see Definition 16.7.2) are connected by an essentially unique zig-zag of equivalences (see Theorem 16.7.6), so that a frame on a diagram defined by a framed model category structure can always be replaced by a Reedy frame.

16.7.1. Frames on diagrams.

DEFINITION 16.7.2. Let $\mathcal{M}$ be a model category, let $\mathcal{C}$ be a small category, and let $\boldsymbol{X}$ be a $\mathcal{C}$-diagram in $\mathcal{M}$.

(1) A *cosimplicial frame on* $\boldsymbol{X}$ is a diagram $\widetilde{\boldsymbol{X}}\colon \mathcal{C} \to \mathcal{M}^{\boldsymbol{\Delta}}$ of cosimplicial objects in $\mathcal{M}$ together with a map of diagrams $i\colon \widetilde{\boldsymbol{X}} \to \mathrm{cc}_*\boldsymbol{X}$ to the diagram of constant cosimplicial objects such that, for every object α in $\mathcal{C}$, the map $i_\alpha\colon \widetilde{\boldsymbol{X}}_\alpha \to \mathrm{cc}_*\boldsymbol{X}_\alpha$ is a cosimplicial frame on $\boldsymbol{X}_\alpha$ (see Definition 16.6.1).
(2) A *simplicial frame on* $\boldsymbol{X}$ is a diagram $\widehat{\boldsymbol{X}}\colon \mathcal{C} \to \mathcal{M}^{\boldsymbol{\Delta}^{\mathrm{op}}}$ of simplicial objects in $\mathcal{M}$ together with a map of diagrams $j\colon \mathrm{cs}_*\boldsymbol{X} \to \widehat{\boldsymbol{X}}$ from the diagram of constant simplicial objects such that, for every object α in $\mathcal{C}$, the map $j_\alpha\colon \mathrm{cs}_*\boldsymbol{Z}_\alpha \to \widehat{\boldsymbol{X}}_\alpha$ is a simplicial frame on $\boldsymbol{X}_\alpha$.

EXAMPLE 16.7.3. Let $\mathcal{M}$ be a framed model category (see Definition 16.6.21). If $\mathcal{C}$ is a small category and $\boldsymbol{X}$ is a $\mathcal{C}$-diagram in $\mathcal{M}$, then the framing on $\mathcal{M}$ defines a natural cosimplicial frame $\widetilde{\boldsymbol{X}}\colon \mathcal{C} \to \mathcal{M}$ on $\boldsymbol{X}$ and a natural simplicial frame $\widehat{\boldsymbol{X}}\colon \mathcal{C} \to \mathcal{M}$ on $\boldsymbol{X}$.

DEFINITION 16.7.4. Let $\mathcal{M}$ be a model category, let $\mathcal{C}$ be a small category, and let $\boldsymbol{X}$ be a $\mathcal{C}$-diagram in $\mathcal{M}$.

(1) The category $\mathrm{CosFr}(\boldsymbol{X})$is the category whose objects are cosimplicial frames on $\boldsymbol{X}$ and whose morphisms are maps of cosimplicial frames on $\boldsymbol{X}$.
(2) The category $\mathrm{SimpFr}(\boldsymbol{X})$is the category whose objects are simplicial frames on $\boldsymbol{X}$ and whose morphisms are maps of simplicial frames on $\boldsymbol{X}$.

THEOREM 16.7.5. *Let $\mathcal{M}$ be a model category, let $\mathcal{C}$ be a small category, and let $\boldsymbol{X}$ be a $\mathcal{C}$-diagram in $\mathcal{M}$.*

(1) *The category $\mathrm{CosFr}(\boldsymbol{X})$ of cosimplicial frames on $\boldsymbol{X}$ has a contractible classifying space (see Definition 14.3.1).*
(2) *The category $\mathrm{SimpFr}(\boldsymbol{X})$ of simplicial frames on $\boldsymbol{X}$ has a contractible classifying space.*

PROOF. We will prove part 1; the proof of part 2 is dual.

Let $\mathcal{W}$ be the class of maps of $\mathcal{C}$-diagrams $\widetilde{\boldsymbol{X}} \to \boldsymbol{X}$ in $\mathcal{M}^{\boldsymbol{\Delta}}$ such that for every object α of $\mathcal{C}$

(1) the map $\widetilde{\boldsymbol{X}}_\alpha \to \boldsymbol{X}_\alpha$ is a Reedy weak equivalence,
(2) $(\widetilde{\boldsymbol{X}}_\alpha)^0 \to (\boldsymbol{X}_\alpha)^0$ is an isomorphism, and
(3) if $(\boldsymbol{X}_\alpha)^0$ is cofibrant in $\mathcal{M}$ then $\widetilde{\boldsymbol{X}}_\alpha$ is Reedy cofibrant.

The result now follows from Theorem 14.5.6 and Proposition 16.6.8. □

THEOREM 16.7.6. *Let $\mathcal{M}$ be a model category, let $\mathcal{C}$ be a small category, and let $\boldsymbol{X}$ be a $\mathcal{C}$-diagram in $\mathcal{M}$.*

(1) *Any two cosimplicial frames on $\boldsymbol{X}$ are connected by an essentially unique zig-zag of maps of cosimplicial frames on $\boldsymbol{X}$.*
(2) *Any two simplicial frames on $\boldsymbol{X}$ are connected by an essentially unique zig-zag of maps of simplicial frames on $\boldsymbol{X}$.*

PROOF. This follows from Theorem 14.4.5 and Theorem 16.7.5. □

16.7.7. Reedy frames. The notion of a *Reedy frame* (see Definition 16.7.8) on a diagram will be used in our discussion of homotopy limits and homotopy colimits of diagrams indexed by a Reedy category (see Section 19.9).

DEFINITION 16.7.8. Let $\mathcal{M}$ be a model category, let $\mathcal{C}$ be a Reedy category (see Definition 15.1.2), and let $\boldsymbol{X}$ be a $\mathcal{C}$-diagram in $\mathcal{M}$.

(1) A *Reedy cosimplicial frame on* $\boldsymbol{X}$ is a cosimplicial frame $\widetilde{\boldsymbol{X}}\colon \mathcal{C} \to \mathcal{M}^{\boldsymbol{\Delta}}$ on $\boldsymbol{X}$ (see Definition 16.7.2) such that if $\boldsymbol{X}$ is a Reedy cofibrant diagram in $\mathcal{M}$ (see Definition 15.3.3) then $\widetilde{\boldsymbol{X}}$ is a Reedy cofibrant diagram in $\mathcal{M}^{\boldsymbol{\Delta}}$.
(2) A *Reedy simplicial frame on* $\boldsymbol{X}$ is a simplicial frame $\widehat{\boldsymbol{X}}\colon \mathcal{C} \to \mathcal{M}^{\boldsymbol{\Delta}^{\mathrm{op}}}$ on $\boldsymbol{X}$ such that if $\boldsymbol{X}$ is a Reedy fibrant diagram in $\mathcal{M}$ then $\widehat{\boldsymbol{X}}$ is a Reedy fibrant diagram in $\mathcal{M}^{\boldsymbol{\Delta}^{\mathrm{op}}}$

PROPOSITION 16.7.9. *Let $\mathcal{M}$ be a simplicial model category, let $\mathcal{C}$ be a Reedy category, and let $\boldsymbol{X}$ be a $\mathcal{C}$-diagram in $\mathcal{M}$.*

(1) *The cosimplicial frame on $\boldsymbol{X}$ defined by the standard frame on $\mathcal{M}$ (see Definition 16.6.5) is a Reedy cosimplicial frame on $\boldsymbol{X}$.*
(2) *The simplicial frame on $\boldsymbol{X}$ defined by the standard frame on $\mathcal{M}$ is a Reedy simplicial frame on $\boldsymbol{X}$.*

PROOF. We will prove part 1; the proof of part 2 is dual.

Let $\boldsymbol{X}$ be Reedy cofibrant, and let $\widetilde{\boldsymbol{X}}\colon \mathcal{C} \to \mathcal{M}^{\boldsymbol{\Delta}}$ be the cosimplicial frame on $\boldsymbol{X}$ defined by the standard frame on $\mathcal{M}$. For every object α in $\mathcal{C}$, let $\mathrm{L}_\alpha^{\mathcal{M}}\boldsymbol{X} \to \boldsymbol{X}_\alpha$ denote the latching map of $\boldsymbol{X}$ in $\mathcal{M}$, and let $\mathrm{L}_\alpha^{\mathcal{M}^{\boldsymbol{\Delta}}}\widetilde{\boldsymbol{X}} \to \widetilde{\boldsymbol{X}}$ denote the latching map of $\widetilde{\boldsymbol{X}}$ in $\mathcal{M}^{\boldsymbol{\Delta}}$. For every object α in $\mathcal{C}$, $\mathrm{L}_\alpha^{\mathcal{M}}\boldsymbol{X} \to \boldsymbol{X}_\alpha$ is a cofibration in $\mathcal{M}$, and we must show that $\mathrm{L}_\alpha^{\mathcal{M}^{\boldsymbol{\Delta}}}\widetilde{\boldsymbol{X}} \to \widetilde{\boldsymbol{X}}_\alpha$ is a cofibration in $\mathcal{M}^{\boldsymbol{\Delta}}$. Thus, Proposition 16.3.8 implies that we must show that for every $n \geq 0$ the relative latching map

$$\widetilde{\boldsymbol{X}}_\alpha \otimes \partial\Delta[n] \amalg_{(\mathrm{L}_\alpha^{\mathcal{M}^{\boldsymbol{\Delta}}}\widetilde{\boldsymbol{X}})\otimes\partial\Delta[n]} (\mathrm{L}_\alpha^{\mathcal{M}^{\boldsymbol{\Delta}}}\widetilde{\boldsymbol{X}}) \otimes \Delta[n] \to \widetilde{\boldsymbol{X}}_\alpha \otimes \Delta[n] \tag{16.7.10}$$

(see Proposition 16.3.8) is a cofibration in $\mathcal{M}$. Since the latching object $\mathrm{L}_\alpha^{\mathcal{M}^{\boldsymbol{\Delta}}}\widetilde{\boldsymbol{X}}$ is defined as a colimit (see Definition 15.2.5), Proposition 16.6.6 and Lemma 9.2.1

imply that the map (16.7.10) is isomorphic to the map

$$\boldsymbol{X}_\alpha \otimes \partial\Delta[n] \amalg_{(\mathrm{L}_\alpha^{\mathcal{M}} \boldsymbol{X})\otimes\partial\Delta[n]} (\mathrm{L}_\alpha^{\mathcal{M}} \boldsymbol{X}) \otimes \Delta[n] \to \boldsymbol{X}_\alpha \otimes \Delta[n] \ .$$

Since $\mathrm{L}_\alpha^{\mathcal{M}} \boldsymbol{X} \to \boldsymbol{X}_\alpha$ is a cofibration in the simplicial model category $\mathcal{M}$, Proposition 9.3.8 implies that this is a cofibration. □

PROPOSITION 16.7.11. *If $\mathcal{M}$ is a model category and $\mathcal{C}$ is a Reedy category, then*

1. *there is a functorial Reedy cosimplicial frame on every $\mathcal{C}$-diagram in $\mathcal{M}$, and*
2. *there is a functorial Reedy simplicial frame on every $\mathcal{C}$-diagram in $\mathcal{M}$.*

PROOF. We will prove part 1; the proof of part 2 is similar.

Theorem 16.6.9 implies that we can choose a functorial cosimplicial frame $\widetilde{\boldsymbol{X}}$ on every object $\boldsymbol{X}$ of the model category $\mathcal{M}^{\mathcal{C}}$. The definition of a frame on an object implies that if $\boldsymbol{X}$ is Reedy cofibrant, then $\widetilde{\boldsymbol{X}}$ is a cofibrant object of $(\mathcal{M}^{\mathcal{C}})^{\boldsymbol{\Delta}}$, and Theorem 15.5.2 implies that this is equivalent to the assertion that $\widetilde{\boldsymbol{X}}$ is cofibrant in $(\mathcal{M}^{\boldsymbol{\Delta}})^{\mathcal{C}}$. □

DEFINITION 16.7.12. Let $\mathcal{M}$ be a model category, let $\mathcal{C}$ be a Reedy category, and let $\boldsymbol{X}$ be a $\mathcal{C}$-diagram in $\mathcal{M}$.

1. The category $\mathrm{ReCosFr}(X)$ is the category whose objects are Reedy cosimplicial frames on $\boldsymbol{X}$ and whose morphisms are maps of cosimplicial frames on $\boldsymbol{X}$.
2. The category $\mathrm{ReSimpFr}(X)$ is the category whose objects are Reedy simplicial frames on $\boldsymbol{X}$ and whose morphisms are maps of simplicial frames on $\boldsymbol{X}$.

THEOREM 16.7.13. *Let $\mathcal{M}$ be a model category, let $\mathcal{C}$ be a Reedy category, and let $\boldsymbol{X}$ be a $\mathcal{C}$-diagram in $\mathcal{M}$.*

1. *The classifying space of the category of Reedy cosimplicial frames on $\boldsymbol{X}$ is contractible (see Definition 14.3.1).*
2. *The classifying space of the category of Reedy simplicial frames on $\boldsymbol{X}$ is contractible.*

PROOF. We will prove part 1; the proof of part 2 is dual.

Let $\mathcal{W}$ be the class of maps $\widetilde{\boldsymbol{X}} \to \boldsymbol{X}$ of $\mathcal{C}$-diagrams in $\mathcal{M}^{\boldsymbol{\Delta}}$ such that

1. for every object α of $\mathcal{C}$ the map $\widetilde{\boldsymbol{X}}_\alpha \to \boldsymbol{X}_\alpha$ is a Reedy weak equivalence,
2. for every object α of $\mathcal{C}$ the map $(\widetilde{\boldsymbol{X}}_\alpha)^0 \to (\boldsymbol{X}_\alpha)^0$ is an isomorphism, and
3. if $\boldsymbol{X}^0$ is a Reedy cofibrant diagram in $\mathcal{M}$ then $\widetilde{\boldsymbol{X}}$ is a Reedy cofibrant diagram in $\mathcal{M}^{\boldsymbol{\Delta}}$.

The result now follows from Theorem 14.5.6 and Proposition 16.7.11. □

THEOREM 16.7.14. *Let $\mathcal{M}$ be a model category, let $\mathcal{C}$ be a Reedy category, and let $\boldsymbol{X}$ be a $\mathcal{C}$-diagram in $\mathcal{M}$.*

1. *Any two Reedy cosimplicial frames on $\boldsymbol{X}$ are connected by an essentially unique zig-zag of maps of Reedy cosimplicial frames on $\boldsymbol{X}$.*
2. *Any two Reedy simplicial frames on $\boldsymbol{X}$ are connected by an essentially unique zig-zag of maps of Reedy simplicial frames on $\boldsymbol{X}$.*

PROOF. This follows from Theorem 14.4.5 and Theorem 16.7.13. □

DEFINITION 16.7.15. A *Reedy framed diagram category* consists of

(1) a Reedy category $\mathcal{C}$,
(2) a model category $\mathcal{M}$,
(3) a choice of a functorial Reedy cosimplicial frame on every $\mathcal{C}$-diagram in $\mathcal{M}$, and
(4) a choice of a functorial Reedy simplicial frame on every $\mathcal{C}$-diagram in $\mathcal{M}$

(see Proposition 16.7.11). We will often denote a Reedy framed diagram category by $(\mathcal{C}, \mathcal{M})$, without making explicit reference to the choice of functorial Reedy cosimplicial frame or the choice of functorial Reedy simplicial frame.

PROPOSITION 16.7.16. *Let $\mathcal{C}$ be a Reedy category, let $\mathcal{M}$ be a model category, and let $\boldsymbol{X}$ be a $\mathcal{C}$-diagram in $\mathcal{M}$.*

(1) *If $\boldsymbol{X}$ is Reedy cofibrant and $\widetilde{\boldsymbol{X}} \colon \mathcal{C} \to \mathcal{M}^{\boldsymbol{\Delta}}$ is a Reedy cosimplicial frame on $\boldsymbol{X}$, then for every object α of $\mathcal{C}$ the latching object $\mathrm{L}_\alpha \widetilde{\boldsymbol{X}} = \operatorname{colim}_{\partial(\overrightarrow{\mathcal{C}} \downarrow \alpha)} \widetilde{\boldsymbol{X}}$ of $\widetilde{\boldsymbol{X}}$ at α is a cosimplicial frame on $\mathrm{L}_\alpha \boldsymbol{X}$.*
(2) *If $\boldsymbol{X}$ is Reedy fibrant and $\widehat{\boldsymbol{X}} \colon \mathcal{C} \to \mathcal{M}^{\boldsymbol{\Delta}^{\mathrm{op}}}$ is a Reedy simplicial frame on $\boldsymbol{X}$, then for every object α of $\mathcal{C}$ the matching object $\mathrm{M}_\alpha \widehat{\boldsymbol{X}} = \lim_{\partial(\alpha \downarrow \overleftarrow{\mathcal{C}})} \widehat{\boldsymbol{X}}$ of $\widehat{\boldsymbol{X}}$ at α is a simplicial frame on $\mathrm{M}_\alpha \boldsymbol{X}$.*

PROOF. We will prove part 1; the proof of part 2 is dual.

Lemma 15.3.7 and Theorem 15.10.9 imply that $\mathrm{L}_\alpha \widetilde{\boldsymbol{X}}$ is a cofibrant cosimplicial object. Corollary 15.3.12 implies that $\boldsymbol{X}_\beta$ is cofibrant in $\mathcal{M}$ for every object $\beta \to \alpha$ of $\partial(\overrightarrow{\mathcal{C}} \downarrow \alpha)$, and so every coface and codegeneracy operator of $\mathrm{L}_\alpha \widetilde{\boldsymbol{X}}$ is a colimit of an objectwise weak equivalence between cofibrant objects. Theorem 15.10.9 thus implies that every coface and codegeneracy operator of $\mathrm{L}_\alpha \widetilde{\boldsymbol{X}}$ is a weak equivalence. □

CHAPTER 17

Homotopy Function Complexes

In this chapter, we define *homotopy function complexes* between objects in a model category. A homotopy function complex between a pair of objects is a simplicial set that plays the role of the "space of functions" between those objects, and its set of components is isomorphic to the set of maps in the homotopy category between those objects.

In a simplicial model category, a homotopy function complex between a cofibrant object and a fibrant object is weakly equivalent to the simplicial mapping space between those objects, and in the category of topological spaces it is weakly equivalent to the total singular complex of the topological space of continuous functions between them. Homotopy function complexes are defined for all model categories, though, and for a simplicial model category they give the "correct" function space even between objects that may not be cofibrant or fibrant (which is not true of the space of maps obtained from the simplicial structure).

If $\mathcal{M}$ is a model category and X and Y are objects of $\mathcal{M}$, then there are three varieties of homotopy function complexes from X to Y:

- *Left homotopy function complexes*, obtained by resolving the first object (see Definition 17.1.1),
- *Right homotopy function complexes*, obtained by resolving the second object (see Definition 17.2.1), and
- *Two-sided homotopy function complexes*, obtained by resolving both objects (see Definition 17.3.1).

We can work with any one of these three varieties, or work with all three combined.

Although constructing a homotopy function complex requires making choices, there is an essentially unique zig-zag (see Definition 14.4.2) of change of homotopy function complex maps (see Definition 17.4.7) connecting any two homotopy function complexes between a pair of objects (see Theorem 17.1.11, Theorem 17.2.11, Theorem 17.3.9, and Theorem 17.4.14). Since every change of homotopy function complex map is a weak equivalence of fibrant simplicial sets (see Theorem 17.4.8), this implies that there is a distinguished homotopy class of homotopy equivalences connecting any two homotopy function complexes between a pair of objects, and the composition of two of these distinguished homotopy classes of homotopy equivalences is another (see Theorem 17.5.30).

Homotopy function complexes are actually only a part of the larger theory of the *simplicial localization* of W. G. Dwyer and D. M. Kan ([**33, 31, 32**]). Dwyer and Kan start with a category $\mathcal{C}$ and a subcategory $\mathcal{W}$ of $\mathcal{C}$, the maps of which are called "weak equivalences". They then construct the *simplicial localization* $\mathrm{sL}_{\mathcal{W}}\mathcal{C}$ of $\mathcal{C}$ with respect to $\mathcal{W}$, which is a simplicial category, i.e., a category enriched over simplicial sets. (If $\mathcal{C}$ is not assumed to be small, then the simplicial localization may exist only in a higher universe; see, e.g., [**60**, page 17].) The simplicial localization of $\mathcal{C}$ with respect to $\mathcal{W}$ is the derived functor of the localization of $\mathcal{C}$ with respect to $\mathcal{W}$ (which also may exist only in a higher universe if $\mathcal{C}$ is not small; see Remark 8.3.3). The simplicial localization thus constructs composable function complexes between objects in the category, and the sets of components of these function complexes are the sets of maps in the localization of $\mathcal{C}$ with respect to $\mathcal{W}$.

Dwyer and Kan show that if $\mathcal{M}$ is a simplicial model category and $\mathcal{W}$ is its subcategory of weak equivalences, then when X is cofibrant and Y is fibrant the simplicial set $\mathrm{Map}(X,Y)$ that is part of the simplicial structure of $\mathcal{M}$ is naturally weakly equivalent to $\mathrm{sL}_{\mathcal{W}}\mathcal{M}(X,Y)$. They show that a weak equivalence $Y \to Z$ in $\mathcal{M}$ always induces a weak equivalence $\mathrm{sL}_{\mathcal{W}}\mathcal{M}(X,Y) \cong \mathrm{sL}_{\mathcal{W}}\mathcal{M}(X,Z)$, while the map $\mathrm{Map}(X,Y) \to \mathrm{Map}(X,Z)$ is guaranteed to be a weak equivalence only when X is cofibrant and both Y and Z are fibrant (and a similar statement is true for weak equivalences of the first argument). Thus, the simplicial set $\mathrm{sL}_{\mathcal{W}}\mathcal{M}(X,Y)$ is the "correct" function complex of maps from X to Y, even for simplicial model categories.

Dwyer and Kan also show that if $\mathcal{M}$ is a model category and $\mathcal{W}$ is the subcategory of weak equivalences in $\mathcal{M}$, then the simplicial sets $\mathrm{sL}_{\mathcal{W}}\mathcal{M}(X,Y)$ can be computed (up to weak equivalence) using resolutions (see Definition 16.1.2) in the model category $\mathcal{M}$ (see [**32**, Section 4]), with no need to consider higher universes. In this chapter, we define a *homotopy function complex* to be a simplicial set obtained from the Dwyer-Kan construction using resolutions in the model category $\mathcal{M}$ (see Definition 17.4.1). We present a self-contained development of the properties of these homotopy function complexes, with no explicit reference to the more general construction of the simplicial localization of Dwyer and Kan.

We define *left homotopy function complexes* in Section 17.1, *right homotopy function complexes* in Section 17.2, and *two-sided homotopy function complexes* in Section 17.3, proving existence and uniqueness theorems for each of these. In Section 17.4 we discuss homotopy function complexes in general (left, right, and two-sided). We define *left to two-sided change of homotopy function complex maps* and *right to two-sided change of homotopy function complex maps*, and we prove a uniqueness theorem for homotopy function complexes. We also show that a Quillen pair induces isomorphisms of homotopy function complexes for cofibrant domains and fibrant codomains.

In Section 17.5 we discuss functorial homotopy function complexes. We prove existence and uniqueness theorems for functorial left homotopy function complexes, functorial right homotopy function complexes, functorial two-sided homotopy function complexes, and for all functorial homotopy function complexes combined. In Section 17.6 we show that (left or right) homotopic maps induced homotopic maps of homotopy function complexes. In Section 17.7 we show that the set of components of a homotopy function complex between a pair of objects is isomorphic to the set of maps in the homotopy category between those objects, and that weak equivalences can be detected as maps that induce weak equivalences of homotopy function complexes.

In Section 17.8 we discuss *homotopy orthogonal maps*, which is the generalization for homotopy function complexes of homotopy lifting-extension pairs in a simplicial model category, and in Section 17.9 we use homotopy function complexes to obtain some results on colimits of λ-sequences of weak equivalences.

17.1. Left homotopy function complexes

DEFINITION 17.1.1. If $\mathcal{M}$ is a model category and X and Y are objects of $\mathcal{M}$, then a *left homotopy function complex* from X to Y is a triple

$$(\widetilde{\boldsymbol{X}}, \widehat{Y}, \mathcal{M}(\widetilde{\boldsymbol{X}}, \widehat{Y}))$$

where

- $\widetilde{\boldsymbol{X}}$ is a cosimplicial resolution of X (see Definition 16.1.2),
- $\widehat{Y}$ is a fibrant approximation to Y (see Definition 8.1.2), and
- $\mathcal{M}(\widetilde{\boldsymbol{X}}, \widehat{Y})$ is the simplicial set of Notation 16.4.1.

The left homotopy function complex $(\widetilde{\boldsymbol{X}}, \widehat{Y}, \mathcal{M}(\widetilde{\boldsymbol{X}}, \widehat{Y}))$ is thus entirely determined by the cosimplicial resolution $\widetilde{\boldsymbol{X}}$ and the fibrant approximation $\widehat{Y}$, but we will commonly refer to the simplicial set $\mathcal{M}(\widetilde{\boldsymbol{X}}, \widehat{Y})$ that is a part of the left homotopy function complex as though it were the left homotopy function complex (see also Notation 17.4.2). Strictly speaking, though, a left homotopy function complex from X to Y can be identified with an object of the undercategory $\big((\mathrm{cc}_* X, Y) \downarrow (\mathcal{M}^{\boldsymbol{\Delta}})^{\mathrm{op}} \times \mathcal{M}\big)$ (see Notation 16.1.1 and Definition 11.8.3)).

REMARK 17.1.2. If we embed $\mathcal{M}$ in $\mathcal{M}^{\boldsymbol{\Delta}^{\mathrm{op}}}$ as the subcategory of constant simplicial objects, then a left homotopy function complex from X to Y can be identified with an object of the undercategory $\big((\mathrm{cc}_* X, \mathrm{cs}_* Y) \downarrow (\mathcal{M}^{\boldsymbol{\Delta}})^{\mathrm{op}} \times \mathcal{M}^{\boldsymbol{\Delta}^{\mathrm{op}}}\big)$ (see Notation 16.1.1), in which case the simplicial set $\mathcal{M}(\widetilde{\boldsymbol{X}}, \widehat{Y})$ is naturally isomorphic to $\operatorname{diag} \mathcal{M}(\widetilde{\boldsymbol{X}}, \mathrm{cc}_* \widehat{Y})$.

PROPOSITION 17.1.3. *If $\mathcal{M}$ is a model category and X and Y are objects of $\mathcal{M}$, then a left homotopy function complex from X to Y is a fibrant simplicial set.*

PROOF. This follows from Corollary 16.5.3. □

EXAMPLE 17.1.4. If $\mathcal{M}$ is a simplicial model category, X is a cofibrant object of $\mathcal{M}$, and Y is a fibrant object of $\mathcal{M}$, then Corollary 16.1.4 implies that $\operatorname{Map}(X, Y)$ (i.e., the simplicial set that is part of the simplicial structure of $\mathcal{M}$) is a left homotopy function complex from X to Y.

DEFINITION 17.1.5. A *change of left homotopy function complex map*

$$(\widetilde{\boldsymbol{X}}, \widehat{Y}, \mathcal{M}(\widetilde{\boldsymbol{X}}, \widehat{Y})) \to (\widetilde{\boldsymbol{X}}', \widehat{Y}', \mathcal{M}(\widetilde{\boldsymbol{X}}', \widehat{Y}'))$$

is a triple (f, g, h) where

- $f\colon \widetilde{\boldsymbol{X}}' \to \widetilde{\boldsymbol{X}}$ is a map of cosimplicial resolutions of X,
- $g\colon \widehat{Y} \to \widehat{Y}'$ is a map of fibrant approximations to Y (see Definition 8.1.4), and
- $h\colon \mathcal{M}(\widetilde{\boldsymbol{X}}, \widehat{Y}) \to \mathcal{M}(\widetilde{\boldsymbol{X}}', \widehat{Y}')$ is the map of simplicial sets induced by f and g.

The change of left homotopy function complex map is thus entirely determined by f and g, but we will commonly refer to the map of simplicial sets h as though it were the change of left homotopy function complex map. Strictly speaking, though, a change of left homotopy function complex map can be identified with a map in the undercategory $\big((\mathrm{cc}_* X, \mathrm{cs}_* Y) \downarrow (\mathcal{M}^{\boldsymbol{\Delta}})^{\mathrm{op}} \times \mathcal{M}^{\boldsymbol{\Delta}^{\mathrm{op}}}\big)$ (see Remark 17.1.2).

PROPOSITION 17.1.6. *If $\mathcal{M}$ is a model category and X and Y are objects of $\mathcal{M}$, then a change of left homotopy function complex map is a weak equivalence of fibrant simplicial sets.*

PROOF. This follows from Corollary 16.5.5, Lemma 16.1.12, and Lemma 8.1.5. □

DEFINITION 17.1.7. If $\mathcal{M}$ is a model category and X and Y are objects of $\mathcal{M}$, then the *category of left homotopy function complexes* from X to Y is the category $\mathrm{LHFC}(X, Y)$ whose objects are left homotopy function complexes from X to Y and whose maps are change of left homotopy function complex maps.

PROPOSITION 17.1.8. *If $\mathcal{M}$ is a model category and X and Y are objects of $\mathcal{M}$, then the category $\mathrm{LHFC}(X, Y)$ (see Definition 17.1.7) can be identified with a full subcategory of $\big((\mathrm{cc}_* X, \mathrm{cs}_* Y) \downarrow (\mathcal{M}^{\mathbf{\Delta}})^{\mathrm{op}} \times \mathcal{M}^{\mathbf{\Delta}^{\mathrm{op}}}\big)$ (see Notation 16.1.1 and Definition 11.8.3).*

PROOF. This follows from Remark 17.1.2. □

PROPOSITION 17.1.9. *If $\mathcal{M}$ is a model category and X and Y are objects of $\mathcal{M}$, then the category $\mathrm{LHFC}(X, Y)$ (see Definition 17.1.7) is naturally isomorphic to $\mathrm{CRes}(X)^{\mathrm{op}} \times \mathrm{FibAp}(Y)$ (see Definition 16.1.14).*

PROOF. This follows directly from the definitions. □

PROPOSITION 17.1.10. *If $\mathcal{M}$ is a model category and X and Y are objects of $\mathcal{M}$, then the classifying space $\mathrm{B}\,\mathrm{LHFC}(X, Y)$ of the category of left homotopy function complexes from X to Y (see Definition 17.1.7) is contractible.*

PROOF. This follows from Proposition 14.3.5, Proposition 17.1.9, Proposition 14.1.5, Theorem 14.6.2, and Proposition 16.1.15. □

THEOREM 17.1.11. *If $\mathcal{M}$ is a model category and X and Y are objects of $\mathcal{M}$, then any two left homotopy function complexes from X to Y are connected by an essentially unique (see Definition 14.4.2) zig-zag of change of left homotopy function complex maps.*

PROOF. This follows from Theorem 14.4.5 and Proposition 17.1.10. □

17.2. Right homotopy function complexes

DEFINITION 17.2.1. If $\mathcal{M}$ is a model category and X and Y are objects of $\mathcal{M}$, then a *right homotopy function complex* from X to Y is a triple

$$(\widetilde{X}, \widehat{\boldsymbol{Y}}, \mathcal{M}(\widetilde{X}, \widehat{\boldsymbol{Y}}))$$

where

- $\widetilde{X}$ is a cofibrant approximation to X (see Definition 8.1.2),
- $\widehat{\boldsymbol{Y}}$ is a simplicial resolution of Y (see Definition 16.1.2), and
- $\mathcal{M}(\widetilde{X}, \widehat{\boldsymbol{Y}})$ is the simplicial set of Notation 16.4.1.

The right homotopy function complex $(\widetilde{X}, \widehat{\boldsymbol{Y}}, \mathcal{M}(\widetilde{X}, \widehat{\boldsymbol{Y}}))$ is thus entirely determined by the cofibrant approximation $\widetilde{X}$ and the simplicial resolution $\widehat{\boldsymbol{Y}}$, but we will commonly refer to the simplicial set $\mathcal{M}(\widetilde{X}, \widehat{\boldsymbol{Y}})$ that is a part of the right homotopy function complex as though it were the right homotopy function complex (see also Notation 17.4.2). Strictly speaking, though, a right homotopy function complex from X to Y can be identified with an object of the undercategory $\big((X, \mathrm{cs}_* Y) \downarrow \mathcal{M}^{\mathrm{op}} \times \mathcal{M}^{\boldsymbol{\Delta}^{\mathrm{op}}}\big)$ (see Notation 16.1.1 and Definition 11.8.3).

REMARK 17.2.2. If we embed $\mathcal{M}$ in $\mathcal{M}^{\boldsymbol{\Delta}}$ as the subcategory of constant cosimplicial objects, then a right homotopy function complex from X to Y can be identified with an object of the undercategory $\big((\mathrm{cc}_* X, \mathrm{cs}_* Y) \downarrow (\mathcal{M}^{\boldsymbol{\Delta}})^{\mathrm{op}} \times \mathcal{M}^{\boldsymbol{\Delta}^{\mathrm{op}}}\big)$ (see Notation 16.1.1), in which case the simplicial set $\mathcal{M}(\widetilde{X}, \widehat{\boldsymbol{Y}})$ is naturally isomorphic to $\operatorname{diag} \mathcal{M}(\mathrm{cc}_* \widetilde{X}, \widehat{\boldsymbol{Y}})$.

PROPOSITION 17.2.3. *If $\mathcal{M}$ is a model category and X and Y are objects of $\mathcal{M}$, then a right homotopy function complex from X to Y is a fibrant simplicial sets.*

PROOF. This follows from Corollary 16.5.3. □

EXAMPLE 17.2.4. If $\mathcal{M}$ is a simplicial model category, X is a cofibrant object of $\mathcal{M}$, and Y is a fibrant object of $\mathcal{M}$, then Corollary 16.1.4 implies that $\mathrm{Map}(X, Y)$ (i.e., the simplicial set that is part of the simplicial structure of $\mathcal{M}$) is a right homotopy function complex from X to Y.

DEFINITION 17.2.5. A *change of right homotopy function complex map*

$$(\widetilde{X}, \widehat{\boldsymbol{Y}}, \mathcal{M}(\widetilde{X}, \widehat{\boldsymbol{Y}})) \to (\widetilde{X}', \widehat{\boldsymbol{Y}}', \mathcal{M}(\widetilde{X}', \widehat{\boldsymbol{Y}}'))$$

is a triple (f, g, h) where

- $f\colon \widetilde{X}' \to \widetilde{X}$ is a map of cofibrant approximations to X (see Definition 8.1.4),
- $g\colon \widehat{\boldsymbol{Y}} \to \widehat{\boldsymbol{Y}}'$ is a map of simplicial resolutions of Y (see Definition 16.1.11), and
- $h\colon \mathcal{M}(\widetilde{X}, \widehat{\boldsymbol{Y}}) \to \mathcal{M}(\widetilde{X}', \widehat{\boldsymbol{Y}}')$ is the map of simplicial sets induced by f and g.

The change of right homotopy function complex map is thus entirely determined by f and g, but we will commonly refer to the map of simplicial sets h as though it were the change of right homotopy function complex map. Strictly speaking, though, a change of right homotopy function complex map can be identified with a map in the undercategory $\big((\mathrm{cc}_* X, \mathrm{cs}_* Y) \downarrow (\mathcal{M}^{\boldsymbol{\Delta}})^{\mathrm{op}} \times \mathcal{M}^{\boldsymbol{\Delta}^{\mathrm{op}}}\big)$ (see Remark 17.2.2).

PROPOSITION 17.2.6. *If $\mathcal{M}$ is a model category and X and Y are objects of $\mathcal{M}$, then a change of right homotopy function complex map is a weak equivalence of fibrant simplicial sets.*

PROOF. This follows from Corollary 16.5.5, Lemma 16.1.12, and Lemma 8.1.5. □

DEFINITION 17.2.7. If $\mathcal{M}$ is a model category and X and Y are objects of $\mathcal{M}$, then the *category of right homotopy function complexes* from X to Y is the category $\mathrm{RHFC}(X, Y)$ whose objects are right homotopy function complexes from X to Y and whose maps are change of right homotopy function complex maps.

PROPOSITION 17.2.8. *If $\mathcal{M}$ is a model category and X and Y are objects of $\mathcal{M}$, then the category* $\mathrm{RHFC}(X, Y)$ *(see Definition 17.2.7) can be identified with* *a full subcategory of* $\big((\mathrm{cc}_*X, \mathrm{cs}_*Y)\downarrow(\mathcal{M}^{\boldsymbol{\Delta}})^{\mathrm{op}}\times\mathcal{M}^{\boldsymbol{\Delta}^{\mathrm{op}}}\big)$ *(see Notation* 16.1.1 *and Definition* 11.8.3*).*

PROOF. This follows from Remark 17.2.2. □

PROPOSITION 17.2.9. *If* $\mathcal{M}$ *is a model category and* X *and* Y *are objects of* $\mathcal{M}$*, then the category* $\mathrm{RHFC}(X, Y)$ *(see Definition* 17.2.7*) is naturally isomorphic to* $\mathrm{CofAp}(X)^{\mathrm{op}}\times\mathrm{SRes}(Y)$ *(see Definition* 16.1.14*).*

PROOF. This follows directly from the definitions. □

PROPOSITION 17.2.10. *If* $\mathcal{M}$ *is a model category and* X *and* Y *are objects of* $\mathcal{M}$*, then the classifying space* $\mathrm{B}\,\mathrm{RHFC}(X, Y)$ *of the category of right homotopy function complexes from* X *to* Y *(see Definition* 17.2.7*) is contractible.*

PROOF. This follows from Proposition 14.3.5, Proposition 17.2.9, Proposition 14.1.5, Theorem 14.6.2, and Proposition 16.1.15. □

THEOREM 17.2.11. *If* $\mathcal{M}$ *is a model category and* X *and* Y *are objects of* $\mathcal{M}$*, then any two right homotopy function complexes from* X *to* Y *are connected by an essentially unique (see Definition* 14.4.2*) zig-zag of change of right homotopy function complex maps.*

PROOF. This follows from Theorem 14.4.5 and Proposition 17.2.10. □

17.3. Two-sided homotopy function complexes

DEFINITION 17.3.1. If $\mathcal{M}$ is a model category and X and Y are objects of $\mathcal{M}$, then a *two-sided homotopy function complex* from X to Y is a triple

$$\big(\widetilde{\boldsymbol{X}}, \widehat{\boldsymbol{Y}}, \operatorname{diag}\mathcal{M}(\widetilde{\boldsymbol{X}}, \widehat{\boldsymbol{Y}})\big)$$

where

- $\widetilde{\boldsymbol{X}}$ is a cosimplicial resolution of X (see Definition 16.1.2),
- $\widehat{\boldsymbol{Y}}$ is a simplicial resolution of Y (see Definition 16.1.2), and
- $\operatorname{diag}\mathcal{M}(\widetilde{\boldsymbol{X}}, \widehat{\boldsymbol{Y}})$ is the simplicial set of Notation 16.4.1.

The two-sided homotopy function complex $\big(\widetilde{\boldsymbol{X}}, \widehat{\boldsymbol{Y}}, \operatorname{diag}\mathcal{M}(\widetilde{\boldsymbol{X}}, \widehat{\boldsymbol{Y}})\big)$ is thus entirely determined by the cosimplicial resolution $\widetilde{\boldsymbol{X}}$ and the simplicial resolution $\widehat{\boldsymbol{Y}}$, but we will commonly refer to the simplicial set $\operatorname{diag}\mathcal{M}(\widetilde{\boldsymbol{X}}, \widehat{\boldsymbol{Y}})$ that is a part of the two-sided homotopy function complex as though it were the two-sided homotopy function complex (see also Notation 17.4.2). Strictly speaking, though, a two-sided homotopy function complex from X to Y can be identified with an object of the undercategory $\big((\mathrm{cc}_*X, \mathrm{cs}_*Y)\downarrow(\mathcal{M}^{\boldsymbol{\Delta}})^{\mathrm{op}}\times\mathcal{M}^{\boldsymbol{\Delta}^{\mathrm{op}}}\big)$ (see Notation 16.1.1).

PROPOSITION 17.3.2. *If* $\mathcal{M}$ *is a model category and* X *and* Y *are objects of* $\mathcal{M}$*, then a two-sided homotopy function complex from* X *to* Y *is a fibrant simplicial set.*

PROOF. This follows from Corollary 16.5.19. □

DEFINITION 17.3.3. A *change of two-sided homotopy function complex map*

$$\big(\widetilde{\boldsymbol{X}}, \widehat{\boldsymbol{Y}}, \operatorname{diag}\mathcal{M}(\widetilde{\boldsymbol{X}}, \widehat{\boldsymbol{Y}})\big)\to\big(\widetilde{\boldsymbol{X}}', \widehat{\boldsymbol{Y}}', \operatorname{diag}\mathcal{M}(\widetilde{\boldsymbol{X}}', \widehat{\boldsymbol{Y}}')\big)$$

is a triple (f, g, h) where

- $f\colon \widetilde{\boldsymbol{X}}' \to \widetilde{\boldsymbol{X}}$ is a map of cosimplicial resolutions of X,
- $g\colon \widehat{\boldsymbol{Y}} \to \widehat{\boldsymbol{Y}}'$ is a map of simplicial resolutions of Y (see Definition 16.1.11), and
- $h\colon \operatorname{diag} \mathcal{M}(\widetilde{\boldsymbol{X}}, \widehat{\boldsymbol{Y}}) \to \operatorname{diag} \mathcal{M}(\widetilde{\boldsymbol{X}}', \widehat{\boldsymbol{Y}}')$ is the map of simplicial sets induced by f and g.

The change of two-sided homotopy function complex map is thus entirely determined by f and g, but we will commonly refer to the map of simplicial sets h as though it were the change of two-sided homotopy function complex map. Strictly speaking, though, a change of two-sided homotopy function complex map can be identified with a map in the undercategory $\big((\mathrm{cc}_* X, \mathrm{cs}_* Y) \downarrow (\mathcal{M}^{\boldsymbol{\Delta}})^{\mathrm{op}} \times \mathcal{M}^{\boldsymbol{\Delta}^{\mathrm{op}}}\big)$.

PROPOSITION 17.3.4. *If $\mathcal{M}$ is a model category and X and Y are objects of $\mathcal{M}$, then a change of two-sided homotopy function complex map is a weak equivalence of fibrant simplicial sets.*

PROOF. This follows from Corollary 16.5.21 and Lemma 16.1.12. □

DEFINITION 17.3.5. If $\mathcal{M}$ is a model category and X and Y are objects of $\mathcal{M}$, then the *category of two-sided homotopy function complexes* from X to Y is the category $\mathrm{TSHFC}(X,Y)$ whose objects are two-sided homotopy function complexes from X to Y and whose maps are change of two-sided homotopy function complex maps.

PROPOSITION 17.3.6. *If $\mathcal{M}$ is a model category and X and Y are objects of $\mathcal{M}$, then the category $\mathrm{TSHFC}(X,Y)$ (see Definition 17.3.5) can be identified with a full subcategory of $\big((\mathrm{cc}_* X, \mathrm{cs}_* Y) \downarrow (\mathcal{M}^{\boldsymbol{\Delta}})^{\mathrm{op}} \times \mathcal{M}^{\boldsymbol{\Delta}^{\mathrm{op}}}\big)$ (see Notation 16.1.1 and Definition 11.8.3).*

PROOF. This follows directly from the definitions. □

PROPOSITION 17.3.7. *If $\mathcal{M}$ is a model category and X and Y are objects of $\mathcal{M}$, then the category $\mathrm{TSHFC}(X,Y)$ (see Definition 17.3.5) is naturally isomorphic to $\mathrm{CRes}(X)^{\mathrm{op}} \times \mathrm{SRes}(Y)$ (see Definition 16.1.14).*

PROOF. This follows directly from the definitions. □

PROPOSITION 17.3.8. *If $\mathcal{M}$ is a model category and X and Y are objects of $\mathcal{M}$, then the classifying space $\mathrm{B}\,\mathrm{TSHFC}(X,Y)$ of the category of two-sided homotopy function complexes from X to Y (see Definition 17.3.5) is contractible.*

PROOF. This follows from Proposition 14.3.5, Proposition 17.3.7, Proposition 14.1.5, and Proposition 16.1.15. □

THEOREM 17.3.9. *If $\mathcal{M}$ is a model category and X and Y are objects of $\mathcal{M}$, then any two two-sided homotopy function complexes from X to Y are connected by an essentially unique (see Definition 14.4.2) zig-zag of change of two-sided homotopy function complex maps.*

PROOF. This follows from Theorem 14.4.5 and Proposition 17.3.8. □

17.4. Homotopy function complexes

DEFINITION 17.4.1. A *homotopy function complex* from X to Y is either

- a left homotopy function complex from X to Y (see Definition 17.1.1),
- a right homotopy function complex from X to Y (see Definition 17.2.1), or
- a two-sided homotopy function complex from X to Y (see Definition 17.3.1).

Every homotopy function complex from X to Y can be identified with an object of the undercategory $\big((\mathrm{cc}_*X, \mathrm{cs}_*Y)\downarrow(\mathcal{M}^{\mathbf{\Delta}})^{\mathrm{op}}\times\mathcal{M}^{\mathbf{\Delta}^{\mathrm{op}}}\big)$ (see Remark 17.1.2, Remark 17.2.2, and Definition 17.3.1), but we will commonly refer to the simplicial set $\operatorname{diag}\mathcal{M}(\widetilde{\boldsymbol{X}}, \widehat{\boldsymbol{Y}})$ that is a part of the homotopy function complex $\big((\mathrm{cc}_*X, \mathrm{cs}_*Y)\to(\widetilde{\boldsymbol{X}}, \widehat{\boldsymbol{Y}})\big)$ as though it were the homotopy function complex.

NOTATION 17.4.2. If $\mathcal{M}$ is a model category and X and Y are objects of $\mathcal{M}$, then we will use the notation $\operatorname{map}(X,Y)$ to denote a simplicial set that is some unspecified homotopy function complex from X to Y (see Definition 17.4.1). Thus, $\operatorname{map}(X,Y)$ will denote either

- the simplicial set $\mathcal{M}(\widetilde{\boldsymbol{X}}, \widehat{Y})$ that is part of a left homotopy function complex $\big(\widetilde{\boldsymbol{X}}, \widehat{Y}, \mathcal{M}(\widetilde{\boldsymbol{X}}, \widehat{Y})\big)$,
- the simplicial set $\mathcal{M}(\widetilde{X}, \widehat{\boldsymbol{Y}})$ that is part of a right homotopy function complex $\big(\widetilde{X}, \widehat{\boldsymbol{Y}}, \mathcal{M}(\widetilde{X}, \widehat{\boldsymbol{Y}})\big)$, or
- the simplicial set $\operatorname{diag}\mathcal{M}(\widetilde{\boldsymbol{X}}, \widehat{\boldsymbol{Y}})$ that is part of a two-sided homotopy function complex $\big(\widetilde{\boldsymbol{X}}, \widehat{\boldsymbol{Y}}, \operatorname{diag}\mathcal{M}(\widetilde{\boldsymbol{X}}, \widehat{\boldsymbol{Y}})\big)$.

PROPOSITION 17.4.3. *If $\mathcal{M}$ is a model category and X and Y are objects of $\mathcal{M}$, then a homotopy function complex from X to Y is a fibrant simplicial set.*

PROOF. This follows from Proposition 17.1.3, Proposition 17.2.3, and Proposition 17.3.2. □

PROPOSITION 17.4.4. *Let $\mathcal{M}$ be a model category and let X and Y be objects of $\mathcal{M}$. If $\big(\widetilde{\boldsymbol{X}}, \widehat{\boldsymbol{Y}}, \operatorname{diag}\mathcal{M}(\widetilde{\boldsymbol{X}}, \widehat{\boldsymbol{Y}})\big)$ is a two-sided homotopy function complex from X to Y, then*

1. *there is a left homotopy function complex $\big(\widetilde{\boldsymbol{X}}, \widehat{\boldsymbol{Y}}_0, \mathcal{M}(\widetilde{\boldsymbol{X}}, \widehat{\boldsymbol{Y}}_0)\big)$ obtained by replacing the simplicial resolution $\widehat{\boldsymbol{Y}}$ with the fibrant approximation $\widehat{\boldsymbol{Y}}_0$, and*
2. *there is a right homotopy function complex $\big(\widetilde{\boldsymbol{X}}^0, \widehat{\boldsymbol{Y}}, \mathcal{M}(\widetilde{\boldsymbol{X}}^0, \widehat{\boldsymbol{Y}})\big)$ obtained by replacing the cosimplicial resolution $\widetilde{\boldsymbol{X}}$ with the cofibrant approximation $\widetilde{\boldsymbol{X}}^0$.*

PROOF. This follows from Proposition 16.1.5. □

DEFINITION 17.4.5. Let $\mathcal{M}$ be a model category and let X and Y be objects of $\mathcal{M}$. If $\big(\widetilde{\boldsymbol{X}}, \widehat{\boldsymbol{Y}}, \operatorname{diag}\mathcal{M}(\widetilde{\boldsymbol{X}}, \widehat{\boldsymbol{Y}})\big)$ is a two-sided homotopy function complex from X to Y, then

1. We will let
$$\mathrm{LTS}(\widetilde{\boldsymbol{X}}, \widehat{\boldsymbol{Y}})\colon \big(\widetilde{\boldsymbol{X}}, \mathrm{cs}_*\widehat{\boldsymbol{Y}}_0, \operatorname{diag}\mathcal{M}(\widetilde{\boldsymbol{X}}, \mathrm{cs}_*\widehat{\boldsymbol{Y}}_0)\big)\to\big(\widetilde{\boldsymbol{X}}, \widehat{\boldsymbol{Y}}, \operatorname{diag}\mathcal{M}(\widetilde{\boldsymbol{X}}, \widehat{\boldsymbol{Y}})\big)$$
denote the natural map in $\big((\mathrm{cc}_*X, \mathrm{cs}_*Y)\downarrow(\mathcal{M}^{\mathbf{\Delta}})^{\mathrm{op}}\times\mathcal{M}^{\mathbf{\Delta}^{\mathrm{op}}}\big)$ defined by the identity of $\widetilde{\boldsymbol{X}}$ and the natural map $\mathrm{cs}_*\widehat{\boldsymbol{Y}}_0\to\widehat{\boldsymbol{Y}}$ (see Notation 16.1.1 and

Remark 17.1.2), and we will call such a map a *left to two-sided change of homotopy function complex map.* Although this map is, strictly speaking, a map in the undercategory $\big((\mathrm{cc}_* X, \mathrm{cs}_* Y)\downarrow(\mathcal{M}^{\mathbf{\Delta}})^{\mathrm{op}} \times \mathcal{M}^{\mathbf{\Delta}^{\mathrm{op}}}\big)$, we will commonly refer to the map of simplicial sets $\operatorname{diag}\mathcal{M}(\widetilde{\boldsymbol{X}}, \mathrm{cs}_*\widehat{\boldsymbol{Y}}_0) \to \operatorname{diag}\mathcal{M}(\widetilde{\boldsymbol{X}}, \widehat{\boldsymbol{Y}})$ that is a part of the left to two-sided change of homotopy function complex map as though it were the left to two-sided change of homotopy function complex map.

(2) We will let

$$\mathrm{RTS}(\widetilde{\boldsymbol{X}}, \widehat{\boldsymbol{Y}})\colon \big(\mathrm{cc}_*\widetilde{\boldsymbol{X}}^0, \widehat{\boldsymbol{Y}}, \operatorname{diag}\mathcal{M}(\mathrm{cc}_*\widetilde{\boldsymbol{X}}^0, \widehat{\boldsymbol{Y}})\big) \to \big(\widetilde{\boldsymbol{X}}, \widehat{\boldsymbol{Y}}, \operatorname{diag}\mathcal{M}(\widetilde{\boldsymbol{X}}, \widehat{\boldsymbol{Y}})\big)$$

denote the natural map in $\big((\mathrm{cc}_* X, \mathrm{cs}_* Y)\downarrow(\mathcal{M}^{\mathbf{\Delta}})^{\mathrm{op}} \times \mathcal{M}^{\mathbf{\Delta}^{\mathrm{op}}}\big)$ defined by the natural map $\widetilde{\boldsymbol{X}} \to \mathrm{cc}_*\widetilde{\boldsymbol{X}}^0$ and the identity of $\widehat{\boldsymbol{Y}}$ (see Notation 16.1.1 and Remark 17.2.2), and we will call such a map a *right to two-sided change of homotopy function complex map.* Although this map is, strictly speaking, a map in the undercategory $\big((\mathrm{cc}_* X, \mathrm{cs}_* Y)\downarrow(\mathcal{M}^{\mathbf{\Delta}})^{\mathrm{op}} \times \mathcal{M}^{\mathbf{\Delta}^{\mathrm{op}}}\big)$ we will commonly refer to the map of simplicial sets $\operatorname{diag}\mathcal{M}(\mathrm{cc}_*\widetilde{\boldsymbol{X}}^0, \widehat{\boldsymbol{Y}}) \to \operatorname{diag}\mathcal{M}(\widetilde{\boldsymbol{X}}, \widehat{\boldsymbol{Y}})$ that is a part of the right to two-sided change of homotopy function complex map as though it were the right to two-sided change of homotopy function complex map.

PROPOSITION 17.4.6. *Let $\mathcal{M}$ be a model category. If $\widetilde{\boldsymbol{X}}$ is a cosimplicial resolution in $\mathcal{M}$ and $\widehat{\boldsymbol{Y}}$ is a simplicial resolution in $\mathcal{M}$, then the left to two-sided change of homotopy function complex map $\mathcal{M}(\widetilde{\boldsymbol{X}}, \widehat{\boldsymbol{Y}}_0) \approx \operatorname{diag}\mathcal{M}(\widetilde{\boldsymbol{X}}, \mathrm{cs}_*\widehat{\boldsymbol{Y}}_0) \to \operatorname{diag}\mathcal{M}(\widetilde{\boldsymbol{X}}, \widehat{\boldsymbol{Y}})$ (see Definition 17.4.5) and the right to two-sided change of homotopy function complex map $\mathcal{M}(\widetilde{\boldsymbol{X}}^0, \widehat{\boldsymbol{Y}}) \approx \operatorname{diag}\mathcal{M}(\mathrm{cc}_*\widetilde{\boldsymbol{X}}^0, \widehat{\boldsymbol{Y}}) \to \operatorname{diag}\mathcal{M}(\widetilde{\boldsymbol{X}}, \widehat{\boldsymbol{Y}})$ are weak equivalences of fibrant simplicial sets.*

PROOF. Corollary 16.5.5 implies that the bisimplicial set $\mathcal{M}(\widetilde{\boldsymbol{X}}, \widehat{\boldsymbol{Y}})$ satisfies the hypotheses of Corollary 15.11.12, and so Theorem 15.11.6 implies that the natural map $\mathcal{M}(\widetilde{\boldsymbol{X}}^0, \widehat{\boldsymbol{Y}}) \to \operatorname{diag}\mathcal{M}(\widetilde{\boldsymbol{X}}, \widehat{\boldsymbol{Y}})$ is a weak equivalence. Similarly, reversing the indices of the bisimplicial set $\mathcal{M}(\widetilde{\boldsymbol{X}}, \widehat{\boldsymbol{Y}})$, the natural map $\mathcal{M}(\widetilde{\boldsymbol{X}}, \widehat{\boldsymbol{Y}}_0) \to \operatorname{diag}\mathcal{M}(\widetilde{\boldsymbol{X}}, \widehat{\boldsymbol{Y}})$ is a weak equivalence. □

DEFINITION 17.4.7. If $\mathcal{M}$ is a model category and X and Y are objects of $\mathcal{M}$, then a *change of homotopy function complex map* is a finite composition (in $\big((\mathrm{cc}_* X, \mathrm{cs}_* Y)\downarrow(\mathcal{M}^{\mathbf{\Delta}})^{\mathrm{op}} \times \mathcal{M}^{\mathbf{\Delta}^{\mathrm{op}}}\big)$; see Notation 16.1.1) of

- change of left homotopy function complex maps (see Definition 17.1.5),
- change of right homotopy function complex maps (see Definition 17.2.5),
- change of two-sided homotopy function complex maps (see Definition 17.3.3),
- left to two-sided change of homotopy function complex maps (see Definition 17.4.5), and
- right to two-sided change of homotopy function complex maps (see Definition 17.4.5).

Although these maps are, strictly speaking, maps in the undercategory

$$\big((\mathrm{cc}_* X, \mathrm{cs}_* Y)\downarrow(\mathcal{M}^{\mathbf{\Delta}})^{\mathrm{op}} \times \mathcal{M}^{\mathbf{\Delta}^{\mathrm{op}}}\big)\ ,$$

we will commonly refer to the corresponding maps of simplicial sets as though they were the change of homotopy function complex maps.

THEOREM 17.4.8. *If $\mathcal{M}$ is a model category and X and Y are objects of $\mathcal{M}$, then a change of homotopy function complex map (see Definition 17.4.1) is a weak equivalence of fibrant simplicial sets.*

PROOF. This follows from Proposition 17.1.6, Proposition 17.2.6, Proposition 17.3.4, and Proposition 17.4.6. □

DEFINITION 17.4.9. If $\mathcal{M}$ is a model category and X and Y are objects of $\mathcal{M}$, then the *category of homotopy function complexes* from X to Y is the category $\mathrm{HFC}(X,Y)$ whose objects are homotopy function complexes from X to Y and whose maps are change of homotopy function complex maps.

PROPOSITION 17.4.10. *If $\mathcal{M}$ is a model category and X and Y are objects of $\mathcal{M}$, then the category $\mathrm{HFC}(X,Y)$ (see Definition 17.4.9) can be identified with a subcategory of $\big((\mathrm{cc}_*X, \mathrm{cs}_*Y)\downarrow (\mathcal{M}^{\boldsymbol{\Delta}})^{\mathrm{op}} \times \mathcal{M}^{\boldsymbol{\Delta}^{\mathrm{op}}}\big)$ (see Notation 16.1.1 and Definition 11.8.3).*

PROOF. This follows from Proposition 17.1.8, Proposition 17.2.8, and Proposition 17.3.6. □

DEFINITION 17.4.11. Let $\mathcal{M}$ be a model category and let X and Y be objects of $\mathcal{M}$. We will call an object $(\boldsymbol{X},\boldsymbol{Y})$ of $(\mathcal{M}^{\boldsymbol{\Delta}})^{\mathrm{op}} \times \mathcal{M}^{\boldsymbol{\Delta}^{\mathrm{op}}}$

(1) a *left resolving pair* if $\boldsymbol{X}$ is a cosimplicial resolution (see Definition 16.1.26) and $\boldsymbol{Y}$ is isomorphic to cs_*Y (see Notation 16.1.1) for some fibrant object Y of $\mathcal{M}$,
(2) a *right resolving pair* if $\boldsymbol{X}$ is isomorphic to cc_*X (see Notation 16.1.1) for some cofibrant object X of $\mathcal{M}$ and $\boldsymbol{Y}$ is a simplicial resolution (see Definition 16.1.26), or
(3) a *two-sided resolving pair* if $\boldsymbol{X}$ is a cosimplicial resolution and $\boldsymbol{Y}$ is a simplicial resolution (see Definition 16.1.26).

An object $(\boldsymbol{X},\boldsymbol{Y})$ of $(\mathcal{M}^{\boldsymbol{\Delta}})^{\mathrm{op}} \times \mathcal{M}^{\boldsymbol{\Delta}^{\mathrm{op}}}$ will be called a *resolving pair* if it is either a left resolving pair, a right resolving pair, or a two-sided resolving pair.

REMARK 17.4.12. The conditions of Definition 17.4.11 are not mutually exclusive. For example, in the category of simplicial simplicial sets, the constant simplicial object at the one point simplicial set is both a simplicial resolution and a constant simplicial object at a fibrant simplicial set.

THEOREM 17.4.13. *If $\mathcal{M}$ is a model category and X and Y are objects of $\mathcal{M}$, then the classifying space $\mathrm{B}\,\mathrm{HFC}(X,Y)$ of homotopy function complexes from X to Y (see Definition 17.4.9) is contractible (see Definition 14.3.1).*

PROOF. We view $\mathrm{HFC}(X,Y)$ as a subcategory of the undercategory

$$\big((\mathrm{cc}_*X, \mathrm{cs}_*Y)\downarrow (\mathcal{M}^{\boldsymbol{\Delta}})^{\mathrm{op}} \times \mathcal{M}^{\boldsymbol{\Delta}^{\mathrm{op}}}\big)$$

(see Remark 17.1.2, Remark 17.2.2, and Definition 17.3.1), and we let $\mathcal{W}$ be the class of maps $h = (f^{\mathrm{op}}, g)$ in $(\mathcal{M}^{\boldsymbol{\Delta}})^{\mathrm{op}} \times \mathcal{M}^{\boldsymbol{\Delta}^{\mathrm{op}}}$ such that

(1) both f and g are degreewise weak equivalences and
(2) the codomain of h is a resolving pair (see Definition 17.4.11).

The result now follows from Theorem 14.5.6 and Proposition 8.1.17. □

THEOREM 17.4.14. *Let $\mathcal{M}$ be a model category. If X and Y are objects of $\mathcal{M}$, then any two homotopy function complexes from X to Y are connected by an essentially unique (see Definition 14.4.2) zig-zag of change of homotopy function complex maps.*

PROOF. This follows from Theorem 14.4.5 and Theorem 17.4.13. □

17.4.15. Quillen functors and homotopy function complexes.

PROPOSITION 17.4.16. *Let $\mathcal{M}$ and $\mathcal{N}$ be small categories, let $\mathrm{F}\colon \mathcal{M} \rightleftarrows \mathcal{N} : \mathrm{U}$ be a Quillen pair (see Definition 8.5.2), let X be a cofibrant object of $\mathcal{M}$ and let Y be a fibrant object of $\mathcal{N}$.*

1. *If $\widetilde{\boldsymbol{X}}$ is a cosimplicial resolution of X then $\mathrm{F}\widetilde{\boldsymbol{X}}$ is a cosimplicial resolution of $\mathrm{F}X$ and the adjointness isomorphism defines a natural isomorphism of left homotopy function complexes $\mathcal{N}(\mathrm{F}\widetilde{\boldsymbol{X}}, Y) \approx \mathcal{M}(\widetilde{\boldsymbol{X}}, \mathrm{U}Y)$.*
2. *If $\widehat{\boldsymbol{Y}}$ is a simplicial resolution of Y then $\mathrm{U}\widehat{\boldsymbol{Y}}$ is a simplicial resolution of $\mathrm{U}Y$ and the adjointness isomorphism defines a natural isomorphism of right homotopy function complexes $\mathcal{N}(\mathrm{F}X, \widehat{\boldsymbol{Y}}) \approx \mathcal{M}(X, \mathrm{U}\widehat{\boldsymbol{Y}})$.*
3. *If $\widetilde{\boldsymbol{X}}$ is a cosimplicial resolution of X and $\widehat{\boldsymbol{Y}}$ is a simplicial resolution of Y, then the adjointness isomorphism defines a natural isomorphism of two-sided homotopy function complexes $\operatorname{diag}\mathcal{N}(\mathrm{F}\widetilde{\boldsymbol{X}}, \widehat{\boldsymbol{Y}}) \approx \operatorname{diag}\mathcal{M}(\widetilde{\boldsymbol{X}}, \mathrm{U}\widehat{\boldsymbol{Y}})$.*

PROOF. This follows from Proposition 16.2.1. □

17.5. Functorial homotopy function complexes

17.5.1. Functorial left homotopy function complexes.

DEFINITION 17.5.2. If $\mathcal{M}$ is a model category and $\mathcal{K}$ is a subcategory of $\mathcal{M}^{\mathrm{op}} \times \mathcal{M}$, then a *functorial left homotopy function complex* on $\mathcal{K}$ is a pair (F, ϕ) where F is a functor $\mathrm{F}\colon \mathcal{K} \to (\mathcal{M}^{\boldsymbol{\Delta}})^{\mathrm{op}} \times \mathcal{M}^{\boldsymbol{\Delta}^{\mathrm{op}}}$ and ϕ is a natural transformation $\phi\colon (\mathrm{cc}_*X, \mathrm{cs}_*Y) \to \mathrm{F}(X, Y)$ (see Remark 17.1.2) such that

1. $\phi(X, Y)$ is a left homotopy function complex from X to Y for every object (X, Y) of $\mathcal{K}$ (see Proposition 17.1.8) and
2. F takes maps of $\mathcal{K}$ into compositions of maps of left homotopy function complexes.

DEFINITION 17.5.3. Let $\mathcal{M}$ be a model category and let $\mathcal{K}$ be a subcategory of $\mathcal{M}^{\mathrm{op}} \times \mathcal{M}$. If (F, ϕ) and (F', ϕ') are functorial left homotopy function complexes on $\mathcal{K}$, then a *change of functorial left homotopy function complex map* from (F, ϕ) to (F', ϕ') is a natural transformation $g\colon \mathrm{F} \to \mathrm{F}'$ such that $g(X, Y)\colon \mathrm{F}(X, Y) \to \mathrm{F}'(X, Y)$ is a change of left homotopy function complex map for every object (X, Y) of $\mathcal{K}$.

PROPOSITION 17.5.4. *If $\mathcal{M}$ is a model category, then there exists a functorial left homotopy function complex defined on all of $\mathcal{M}^{\mathrm{op}} \times \mathcal{M}$.*

PROOF. This follows from Proposition 16.1.9 and Proposition 8.1.17. □

DEFINITION 17.5.5. If $\mathcal{M}$ is a model category and $\mathcal{K}$ is a subcategory of $\mathcal{M}^{\mathrm{op}} \times \mathcal{M}$, then a *category of functorial left homotopy function complexes on $\mathcal{K}$* is a category of functors from $\mathcal{K}$ to $(\mathcal{M}^{\boldsymbol{\Delta}})^{\mathrm{op}} \times \mathcal{M}^{\boldsymbol{\Delta}^{\mathrm{op}}}$ under the "constant object" functor (that takes (X, Y) to $(\mathrm{cc}_* X, \mathrm{cs}_* Y)$) with respect to those maps in $(\mathcal{M}^{\boldsymbol{\Delta}})^{\mathrm{op}} \times \mathcal{M}^{\boldsymbol{\Delta}^{\mathrm{op}}}$ that are componentwise Reedy weak equivalences with codomain a left resolving pair (see Definition 17.4.11).

THEOREM 17.5.6. *If $\mathcal{M}$ is a model category and $\mathcal{K}$ is a subcategory of $\mathcal{M}^{\mathrm{op}} \times \mathcal{M}$, then for every small category $\mathcal{D}$ of functorial left homotopy function complexes on $\mathcal{K}$ there is a small category $\mathcal{D}'$ of functorial left homotopy function complexes on $\mathcal{K}$ such that $\mathcal{D} \subset \mathcal{D}'$ and $\mathrm{B}\mathcal{D}'$ is contractible.*

PROOF. This follows from Theorem 14.5.4, Proposition 16.1.9, and Proposition 8.1.17. □

THEOREM 17.5.7. *If $\mathcal{M}$ is a model category, $\mathcal{K}$ is a subcategory of $\mathcal{M}^{\mathrm{op}} \times \mathcal{M}$, and (F, ϕ) and (F', ϕ') are functorial left homotopy function complexes on $\mathcal{K}$, then there is an essentially unique zig-zag (see Definition 14.4.2) of change of functorial left homotopy function complex maps from (F, ϕ) to (F', ϕ').*

PROOF. This follows from Proposition 14.5.7 and Theorem 17.5.6. □

17.5.8. Functorial right homotopy function complexes.

DEFINITION 17.5.9. If $\mathcal{M}$ is a model category and $\mathcal{K}$ is a subcategory of $\mathcal{M}^{\mathrm{op}} \times \mathcal{M}$, then a *functorial right homotopy function complex* on a subcategory $\mathcal{K}$ of $\mathcal{M}^{\mathrm{op}} \times \mathcal{M}$ is a pair (F, ϕ) where F is a functor $\mathrm{F}\colon \mathcal{K} \to (\mathcal{M}^{\boldsymbol{\Delta}})^{\mathrm{op}} \times \mathcal{M}^{\boldsymbol{\Delta}^{\mathrm{op}}}$ and ϕ is a natural transformation $\phi\colon (\mathrm{cc}_* X, \mathrm{cs}_* Y) \to \mathrm{F}(X, Y)$ such that

(1) $\phi(X, Y)$ is a right homotopy function complex from X to Y for every object (X, Y) of $\mathcal{K}$ (see Proposition 17.2.8) and
(2) F takes maps of $\mathcal{K}$ into compositions of maps of right homotopy function complexes.

DEFINITION 17.5.10. Let $\mathcal{M}$ be a model category and let $\mathcal{K}$ be a subcategory of $\mathcal{M}^{\mathrm{op}} \times \mathcal{M}$. If (F, ϕ) and (F', ϕ') are functorial right homotopy function complexes on $\mathcal{K}$, then a *change of functorial right homotopy function complex map* from (F, ϕ) to (F', ϕ') is a natural transformation $g\colon \mathrm{F} \to \mathrm{F}'$ such that $g(X, Y)\colon \mathrm{F}(X, Y) \to \mathrm{F}'(X, Y)$ is a change of right homotopy function complex map for every object (X, Y) of $\mathcal{K}$.

PROPOSITION 17.5.11. *If $\mathcal{M}$ is a model category, then there exists a functorial right homotopy function complex defined on all of $\mathcal{M}^{\mathrm{op}} \times \mathcal{M}$.*

PROOF. This follows from Proposition 16.1.9 and Proposition 8.1.17. □

DEFINITION 17.5.12. If $\mathcal{M}$ is a model category and $\mathcal{K}$ is a subcategory of $\mathcal{M}^{\mathrm{op}} \times \mathcal{M}$, then a *category of functorial right homotopy function complexes on* $\mathcal{K}$ is a category of functors from $\mathcal{K}$ to $(\mathcal{M}^{\boldsymbol{\Delta}})^{\mathrm{op}} \times \mathcal{M}^{\boldsymbol{\Delta}^{\mathrm{op}}}$ under the "constant object" functor (that takes (X, Y) to $(\mathrm{cc}_* X, \mathrm{cs}_* Y)$) with respect to those maps in $(\mathcal{M}^{\boldsymbol{\Delta}})^{\mathrm{op}} \times \mathcal{M}^{\boldsymbol{\Delta}^{\mathrm{op}}}$ that are componentwise Reedy weak equivalences with codomain a right resolving pair (see Definition 17.4.11).

THEOREM 17.5.13. *If $\mathcal{M}$ is a model category and $\mathcal{K}$ is a subcategory of $\mathcal{M}^{\mathrm{op}} \times \mathcal{M}$, then for every small category $\mathcal{D}$ of functorial right homotopy function complexes on $\mathcal{K}$ there is a small category $\mathcal{D}'$ of functorial right homotopy function complexes on $\mathcal{K}$ such that $\mathcal{D} \subset \mathcal{D}'$ and $\mathrm{B}\mathcal{D}'$ is contractible.*

PROOF. This follows from Theorem 14.5.4, Proposition 16.1.9, and Proposition 8.1.17. □

THEOREM 17.5.14. *If $\mathcal{M}$ is a model category, $\mathcal{K}$ is a subcategory of $\mathcal{M}^{\mathrm{op}} \times \mathcal{M}$, and (F, ϕ) and (F', ϕ') are functorial right homotopy function complexes on $\mathcal{K}$, then there is an essentially unique zig-zag (see Definition 14.4.2) of change of functorial right homotopy function complex maps from (F, ϕ) to (F', ϕ').*

PROOF. This follows from Proposition 14.5.7 and Theorem 17.5.13. □

17.5.15. Functorial two-sided homotopy function complexes.

DEFINITION 17.5.16. If $\mathcal{M}$ is a model category and $\mathcal{K}$ is a subcategory of $\mathcal{M}^{\mathrm{op}} \times \mathcal{M}$, then a *functorial two-sided homotopy function complex* on a subcategory $\mathcal{K}$ of $\mathcal{M}^{\mathrm{op}} \times \mathcal{M}$ is a pair (F, ϕ) where F is a functor $\mathrm{F}\colon \mathcal{K} \to (\mathcal{M}^{\mathbf{\Delta}})^{\mathrm{op}} \times \mathcal{M}^{\mathbf{\Delta}^{\mathrm{op}}}$ and ϕ is a natural transformation $\phi\colon (\mathrm{cc}_* X, \mathrm{cs}_* Y) \to \mathrm{F}(X, Y)$ such that

(1) $\phi(X, Y)$ is a two-sided homotopy function complex from X to Y for every object (X, Y) of $\mathcal{K}$ (see Proposition 17.3.6) and
(2) F takes maps of $\mathcal{K}$ into compositions of maps of two-sided homotopy function complexes.

DEFINITION 17.5.17. Let $\mathcal{M}$ be a model category and let $\mathcal{K}$ be a subcategory of $\mathcal{M}^{\mathrm{op}} \times \mathcal{M}$. If (F, ϕ) and (F', ϕ') are functorial two-sided homotopy function complexes on $\mathcal{K}$, then a *change of functorial two-sided homotopy function complex map* from (F, ϕ) to (F', ϕ') is a natural transformation $g\colon \mathrm{F} \to \mathrm{F}'$ such that $g(X, Y)\colon \mathrm{F}(X, Y) \to \mathrm{F}'(X, Y)$ is a change of two-sided homotopy function complex map for every object (X, Y) of $\mathcal{K}$.

PROPOSITION 17.5.18. *If $\mathcal{M}$ is a model category, then there exists a functorial two-sided homotopy function complex defined on all of $\mathcal{M}^{\mathrm{op}} \times \mathcal{M}$.*

PROOF. This follows from Proposition 16.1.9. □

DEFINITION 17.5.19. If $\mathcal{M}$ is a model category and $\mathcal{K}$ is a subcategory of $\mathcal{M}^{\mathrm{op}} \times \mathcal{M}$, then a *category of functorial two-sided homotopy function complexes on $\mathcal{K}$* is a category of functors from $\mathcal{K}$ to $(\mathcal{M}^{\mathbf{\Delta}})^{\mathrm{op}} \times \mathcal{M}^{\mathbf{\Delta}^{\mathrm{op}}}$ under the "constant object" functor (that takes (X, Y) to $(\mathrm{cc}_* X, \mathrm{cs}_* Y)$) with respect to those maps in $(\mathcal{M}^{\mathbf{\Delta}})^{\mathrm{op}} \times \mathcal{M}^{\mathbf{\Delta}^{\mathrm{op}}}$ that are componentwise Reedy weak equivalences with codomain a two-sided resolving pair (see Definition 17.4.11).

THEOREM 17.5.20. *If $\mathcal{M}$ is a model category and $\mathcal{K}$ is a subcategory of $\mathcal{M}^{\mathrm{op}} \times \mathcal{M}$, then for every small category $\mathcal{D}$ of functorial two-sided homotopy function complexes on $\mathcal{K}$ there is a small category $\mathcal{D}'$ of functorial two-sided homotopy function complexes on $\mathcal{K}$ such that $\mathcal{D} \subset \mathcal{D}'$ and $\mathrm{B}\mathcal{D}'$ is contractible.*

PROOF. This follows from Theorem 14.5.4 and Proposition 16.1.9. □

THEOREM 17.5.21. *If $\mathcal{M}$ is a model category, $\mathcal{K}$ is a subcategory of $\mathcal{M}^{\mathrm{op}} \times \mathcal{M}$, and (F, ϕ) and (F', ϕ') are functorial two-sided homotopy function complexes on $\mathcal{K}$, then there is an essentially unique zig-zag (see Definition 14.4.2) of change of functorial two-sided homotopy function complex maps from (F, ϕ) to (F', ϕ').*

PROOF. This follows from Proposition 14.5.7 and Theorem 17.5.20. □

17.5.22. Functorial homotopy function complexes.

DEFINITION 17.5.23. If $\mathcal{M}$ is a model category and $\mathcal{K}$ is a subcategory of $\mathcal{M}^{\mathrm{op}} \times \mathcal{M}$, then a *functorial homotopy function complex* on a subcategory $\mathcal{K}$ of $\mathcal{M}^{\mathrm{op}} \times \mathcal{M}$ is either

- a functorial left homotopy function complex on $\mathcal{K}$,
- a functorial right homotopy function complex on $\mathcal{K}$, or
- a functorial two-sided homotopy function complex on $\mathcal{K}$.

PROPOSITION 17.5.24. *Let $\mathcal{M}$ be a model category and let $\mathcal{K}$ be a subcategory of $\mathcal{M}^{\mathrm{op}} \times \mathcal{M}$. If (F, ϕ) is a functorial two-sided homotopy function complex on $\mathcal{K}$ (see Definition 17.5.16) such that $\mathrm{F}(X,Y) = \big(\widetilde{\boldsymbol{F}}(X,Y), \widehat{\boldsymbol{F}}(X,Y)\big)$ for every object (X,Y) of $\mathcal{K}$, then*

1. *there is a functorial left homotopy function complex $(\widetilde{\boldsymbol{F}}, \widehat{\boldsymbol{F}}_0)$ obtained by replacing the functorial simplicial resolution $\widehat{\boldsymbol{F}}$ with the functorial fibrant approximation $\widehat{\boldsymbol{F}}_0$, and*
2. *there is a functorial right homotopy function complex $(\widetilde{\boldsymbol{F}}^0, \widehat{\boldsymbol{F}})$ obtained by replacing the functorial cosimplicial resolution $\widetilde{\boldsymbol{F}}$ with the functorial cofibrant approximation $\widetilde{\boldsymbol{F}}^0$.*

PROOF. This follows from Proposition 17.4.4. □

DEFINITION 17.5.25. Let $\mathcal{M}$ be a model category, let $\mathcal{K}$ be a subcategory of $\mathcal{M}^{\mathrm{op}} \times \mathcal{M}$, and let $\mathrm{F} = \big(\widetilde{\boldsymbol{F}}, \widehat{\boldsymbol{F}}, \operatorname{diag} \mathcal{M}(\widetilde{\boldsymbol{F}}, \widehat{\boldsymbol{F}})\big)$ be a functorial two-sided homotopy function complex on $\mathcal{K}$.

1. We will let
$$\mathrm{LTS}(\mathrm{F})\colon \big(\widetilde{\boldsymbol{F}}, \mathrm{cs}_*\widehat{\boldsymbol{F}}_0, \operatorname{diag} \mathcal{M}(\widetilde{\boldsymbol{F}}, \mathrm{cs}_*\widehat{\boldsymbol{F}}_0)\big) \to \big(\widetilde{\boldsymbol{F}}, \widehat{\boldsymbol{F}}, \operatorname{diag} \mathcal{M}(\widetilde{\boldsymbol{F}}, \widehat{\boldsymbol{F}})\big)$$
denote the natural map defined by the identity of $\widetilde{\boldsymbol{F}}$ and the natural transformation $\mathrm{cs}_*\widehat{\boldsymbol{F}}_0 \to \widehat{\boldsymbol{F}}$, and we will call such a map a *functorial left to two-sided change of homotopy function complex map*. We will commonly refer to the natural map of simplicial sets $\operatorname{diag} \mathcal{M}(\widetilde{\boldsymbol{F}}, \mathrm{cs}_*\widehat{\boldsymbol{F}}_0) \to \operatorname{diag} \mathcal{M}(\widetilde{\boldsymbol{F}}, \widehat{\boldsymbol{F}})$ as though it were the functorial left to two-sided change of homotopy function complex map.
2. We will let
$$\mathrm{RTS}(\mathrm{F})\colon \big(\mathrm{cc}_*\widetilde{\boldsymbol{F}}^0, \widehat{\boldsymbol{F}}, \operatorname{diag} \mathcal{M}(\mathrm{cc}_*\widetilde{\boldsymbol{F}}^0, \widehat{\boldsymbol{F}})\big) \to \big(\widetilde{\boldsymbol{F}}, \widehat{\boldsymbol{F}}, \operatorname{diag} \mathcal{M}(\widetilde{\boldsymbol{F}}, \widehat{\boldsymbol{F}})\big)$$
denote the natural map defined by the natural transformation $\widetilde{\boldsymbol{F}} \to \mathrm{cc}_*\widetilde{\boldsymbol{F}}^0$ and the identity of $\widehat{\boldsymbol{F}}$, and we will call such a map a *functorial right to two-sided change of homotopy function complex map*. We will commonly refer to the natural map of simplicial sets $\operatorname{diag} \mathcal{M}(\mathrm{cs}_*\widetilde{\boldsymbol{F}}^0, \widehat{\boldsymbol{F}}) \to \operatorname{diag} \mathcal{M}(\widetilde{\boldsymbol{F}}, \widehat{\boldsymbol{F}})$ as though it wee the functorial left to two-sided change of homotopy function complex map.

DEFINITION 17.5.26. If $\mathcal{M}$ is a model category and $\mathcal{K}$ is a subcategory of $\mathcal{M}^{\mathrm{op}} \times \mathcal{M}$, then a *functorial change of homotopy function complex map* on $\mathcal{K}$ is a finite composition of

1. functorial change of left homotopy function complex maps (see Definition 17.5.3),

(2) functorial change of right homotopy function complex maps (see Definition 17.5.10),
(3) functorial change of two-sided homotopy function complex maps (see Definition 17.5.17),
(4) functorial left to two-sided change of homotopy function complex maps (see Definition 17.5.25), and
(5) functorial right to two-sided change of homotopy function complex maps (see Definition 17.5.25).

We will commonly refer to the natural map of simplicial sets that is a part of the change of functorial homotopy function complex map as though it were the functorial change of homotopy function complex map.

DEFINITION 17.5.27. If $\mathcal{M}$ is a model category and $\mathcal{K}$ is a subcategory of $\mathcal{M}^{\mathrm{op}} \times \mathcal{M}$, then a *category of functorial homotopy function complexes on* $\mathcal{K}$ is a category of functors from $\mathcal{K}$ to $(\mathcal{M}^{\mathbf{\Delta}})^{\mathrm{op}} \times \mathcal{M}^{\mathbf{\Delta}^{\mathrm{op}}}$ under the "constant object" functor (that takes (X, Y) to $(\mathrm{cc}_* X, \mathrm{cs}_* Y)$) with respect to those maps in $(\mathcal{M}^{\mathbf{\Delta}})^{\mathrm{op}} \times \mathcal{M}^{\mathbf{\Delta}^{\mathrm{op}}}$ that are componentwise Reedy weak equivalences with codomain a resolving pair (see Definition 17.4.11).

THEOREM 17.5.28. *If $\mathcal{M}$ is a model category and $\mathcal{K}$ is a subcategory of $\mathcal{M}^{\mathrm{op}} \times \mathcal{M}$, then for every small category $\mathcal{D}$ of functorial homotopy function complexes on $\mathcal{K}$ there is a small category $\mathcal{D}'$ of functorial homotopy function complexes on $\mathcal{K}$ such that $\mathcal{D} \subset \mathcal{D}'$ and $\mathrm{B}\mathcal{D}'$ is contractible.*

PROOF. This follows from Theorem 14.5.4 and Proposition 16.1.9. □

THEOREM 17.5.29. *If $\mathcal{M}$ is a model category, $\mathcal{K}$ is a subcategory of $\mathcal{M}^{\mathrm{op}} \times \mathcal{M}$, and (F, ϕ) and (F', ϕ') are functorial homotopy function complexes on $\mathcal{K}$, then there is an essentially unique zig-zag (see Definition 14.4.2) of change of functorial homotopy function complex maps from (F, ϕ) to (F', ϕ').*

PROOF. This follows from Proposition 14.5.7 and Theorem 17.5.28. □

THEOREM 17.5.30. *Let $\mathcal{M}$ be a model category and let $\mathcal{K}$ be a subcategory of $\mathcal{M}^{\mathrm{op}} \times \mathcal{M}$.*

(1) *For every ordered pair of functorial left homotopy function complexes $\mathrm{map}_1(X, Y)$ and $\mathrm{map}_2(X, Y)$ defined on $\mathcal{K}$ there is a homotopy equivalence $h_{1,2}\colon \mathrm{map}_1(X, Y) \to \mathrm{map}_2(X, Y)$, defined up to homotopy and natural up to homotopy, such that if $\mathrm{map}_3(X, Y)$ is a third one and $h_{1,3}\colon \mathrm{map}_1(X, Y) \to \mathrm{map}_3(X, Y)$ and $h_{2,3}\colon \mathrm{map}_2(X, Y) \to \mathrm{map}_3(X, Y)$ are the corresponding homotopy equivalences, then $h_{2,3}h_{1,2} \simeq h_{1,3}$.*
(2) *For every ordered pair of functorial right homotopy function complexes $\mathrm{map}_1(X, Y)$ and $\mathrm{map}_2(X, Y)$ defined on $\mathcal{K}$ there is a homotopy equivalence $h_{1,2}\colon \mathrm{map}_1(X, Y) \to \mathrm{map}_2(X, Y)$, defined up to homotopy and natural up to homotopy, such that if $\mathrm{map}_3(X, Y)$ is a third one and $h_{1,3}\colon \mathrm{map}_1(X, Y) \to \mathrm{map}_3(X, Y)$ and $h_{2,3}\colon \mathrm{map}_2(X, Y) \to \mathrm{map}_3(X, Y)$ are the corresponding homotopy equivalences, then $h_{2,3}h_{1,2} \simeq h_{1,3}$.*
(3) *For every ordered pair of functorial two-sided homotopy function complexes $\mathrm{map}_1(X, Y)$ and $\mathrm{map}_2(X, Y)$ defined on $\mathcal{K}$ there is a homotopy equivalence $h_{1,2}\colon \mathrm{map}_1(X, Y) \to \mathrm{map}_2(X, Y)$, defined up to homotopy and natural up to homotopy, such that if $\mathrm{map}_3(X, Y)$ is a third one and*

$h_{1,3}\colon \mathrm{map}_1(X,Y) \to \mathrm{map}_3(X,Y)$ *and* $h_{2,3}\colon \mathrm{map}_2(X,Y) \to \mathrm{map}_3(X,Y)$ *are the corresponding homotopy equivalences, then* $h_{2,3}h_{1,2} \simeq h_{1,3}$.

(4) *For every ordered pair of functorial homotopy function complexes (see Definition 17.4.1)* $\mathrm{map}_1(X,Y)$ *and* $\mathrm{map}_2(X,Y)$ *defined on* $\mathcal{K}$ *there is a homotopy equivalence* $h_{1,2}\colon \mathrm{map}_1(X,Y) \to \mathrm{map}_2(X,Y)$, *defined up to homotopy and natural up to homotopy, such that if* $\mathrm{map}_3(X,Y)$ *is a third one and* $h_{1,3}\colon \mathrm{map}_1(X,Y) \to \mathrm{map}_3(X,Y)$ *and* $h_{2,3}\colon \mathrm{map}_2(X,Y) \to \mathrm{map}_3(X,Y)$ *are the corresponding homotopy equivalences, then* $h_{2,3}h_{1,2} \simeq h_{1,3}$.

PROOF. This follows from Proposition 14.4.11 and Theorem 17.5.29. □

THEOREM 17.5.31. *Let* $\mathcal{M}$ *be a model category.*

(1) *If* B *is an object of* $\mathcal{M}$ *and* $g\colon X \to Y$ *is a map for which there is some map of homotopy function complexes* $g_*\colon \mathrm{map}(B,X) \to \mathrm{map}(B,Y)$ *(see Notation 17.4.2) induced by* g *that is a weak equivalence, then every such map of homotopy function complexes induced by* g *is a weak equivalence.*
(2) *If* X *is an object of* $\mathcal{M}$ *and* $f\colon A \to B$ *is a map for which there is some map of homotopy function complexes (see Definition 17.6.2)* $f^*\colon \mathrm{map}(B,X) \to \mathrm{map}(A,X)$ *(see Notation 17.4.2) induced by* f *that is a weak equivalence, then every such map of homotopy function complexes induced by* f *is a weak equivalence.*

PROOF. This follows from Theorem 17.5.30, Proposition 7.7.6, and the "two out of three" axiom (see Definition 7.1.3). □

17.6. Homotopic maps of homotopy function complexes

17.6.1. Induced maps of homotopy function complexes.

DEFINITION 17.6.2. Let $\mathcal{M}$ be a model category, let W, X, Y, and Z be objects of $\mathcal{M}$, and let $g\colon X \to Y$ be a map.

(1) A *map of left homotopy function complexes induced by* g will mean either
    (a) the map $\hat{g}_*\colon \mathcal{M}(\widetilde{\boldsymbol{W}}, \widehat{X}) \to \mathcal{M}(\widetilde{\boldsymbol{W}}, \widehat{Y})$ where $\widetilde{\boldsymbol{W}}$ is a cosimplicial resolution of W and $\hat{g}\colon \widehat{X} \to \widehat{Y}$ is a fibrant approximation to g (see Definition 8.1.22), or
    (b) the map $\tilde{g}^*\colon \mathcal{M}(\widetilde{\boldsymbol{Y}}, \widehat{Z}) \to \mathcal{M}(\widetilde{\boldsymbol{X}}, \widehat{Z})$ where $\tilde{g}\colon \widetilde{\boldsymbol{X}} \to \widetilde{\boldsymbol{Y}}$ is a cosimplicial resolution of g (see Definition 16.1.20), and $\widehat{Z}$ is a fibrant approximation to Z.
(2) A *map of right homotopy function complexes induced by* g will mean either
    (a) the map $\hat{g}_*\colon \mathcal{M}(\widetilde{W}, \widehat{\boldsymbol{X}}) \to \mathcal{M}(\widetilde{W}, \widehat{\boldsymbol{Y}})$ where $\widetilde{W}$ is a cofibrant approximation to W and $\hat{g}\colon \widehat{\boldsymbol{X}} \to \widehat{\boldsymbol{Y}}$ is a simplicial resolution of g, or
    (b) the map $\tilde{g}^*\colon \mathcal{M}(\widetilde{Y}, \widehat{\boldsymbol{Z}}) \to \mathcal{M}(\widetilde{X}, \widehat{\boldsymbol{Z}})$ where $\tilde{g}$ is a cofibrant approximation to g (see Definition 8.1.22) and $\widehat{\boldsymbol{Z}}$ is a simplicial resolution of Z.
(3) A *map of two-sided homotopy function complexes induced by* g will mean either
    (a) the map $\operatorname{diag}\hat{g}_*\colon \operatorname{diag}\mathcal{M}(\widetilde{\boldsymbol{W}}, \widehat{\boldsymbol{X}}) \to \operatorname{diag}\mathcal{M}(\widetilde{\boldsymbol{W}}, \widehat{\boldsymbol{Y}})$ where $\widetilde{\boldsymbol{W}}$ is a cosimplicial resolution of W and $\hat{g}\colon \widehat{\boldsymbol{X}} \to \widehat{\boldsymbol{Y}}$ is a simplicial resolution of g, or

(b) the map $\operatorname{diag} \tilde{g}^*\colon \operatorname{diag} \mathcal{M}(\widetilde{\boldsymbol{Y}}, \widehat{\boldsymbol{Z}}) \to \operatorname{diag} \mathcal{M}(\widetilde{\boldsymbol{X}}, \widehat{\boldsymbol{Z}})$ where $\tilde{g}\colon \widetilde{\boldsymbol{X}} \to \widetilde{\boldsymbol{Y}}$ is a cosimplicial resolution of g and $\widehat{\boldsymbol{Z}}$ is a simplicial resolution of Z.

(4) A *map of homotopy function complexes induced by* g will mean either a map of left homotopy function complexes induced by g, a map of right homotopy function complexes induced by g, or a map of two-sided homotopy function complexes induced by g.

THEOREM 17.6.3. *Let $\mathcal{M}$ be a model category and let W, X, Y, and Z be objects in $\mathcal{M}$. If $g\colon X \to Y$ is a weak equivalence, then*

(1) *any map of homotopy function complexes $g_*\colon \operatorname{map}(W, X) \to \operatorname{map}(W, Y)$ induced by g (see Notation 17.4.2 and Definition 17.6.2) is a weak equivalence of fibrant simplicial sets, and*
(2) *any map of homotopy function complexes $g^*\colon \operatorname{map}(Y, Z) \to \operatorname{map}(X, Z)$ induced by g is a weak equivalence of fibrant simplicial sets.*

PROOF. This follows from Corollary 16.5.5, Corollary 16.5.21, and Proposition 16.1.24. □

17.6.4. Homotopic maps of homotopy function complexes. The main result of this section is Theorem 17.6.7, which implies that if $\mathcal{M}$ is a model category and if f and g are maps in $\mathcal{M}$ that are either left homotopic or right homotopic, then any maps of homotopy function complexes induced by f and g are homotopic maps of fibrant simplicial sets.

LEMMA 17.6.5. *If $\mathcal{M}$ is a model category and $f, g\colon X \to Y$ are left homotopic, right homotopic, or homotopic, then both the induced maps of constant cosimplicial objects $\mathrm{cc}_* f, \mathrm{cc}_* g\colon \mathrm{cc}_* X \to \mathrm{cc}_* Y$ and the induced maps of constant simplicial objects $\mathrm{cs}_* f, \mathrm{cs}_* g\colon \mathrm{cs}_* X \to \mathrm{cs}_* Y$ are, respectively, left homotopic, right homotopic, or homotopic.*

PROOF. The constant cosimplicial and constant simplicial objects obtained from either a cylinder object for X or a path object for Y satisfy the conditions of Proposition 7.3.5. □

PROPOSITION 17.6.6. *Let $\mathcal{M}$ be a model category.*

(1) (a) *If $\widetilde{\boldsymbol{B}}$ is a cosimplicial resolution in $\mathcal{M}$ and $\hat{f}, \hat{g}\colon \widehat{X} \to \widehat{Y}$ are left homotopic, right homotopic, or homotopic maps of fibrant objects in $\mathcal{M}$, then the induced maps of left homotopy function complexes $\hat{f}_*, \hat{g}_*\colon \mathcal{M}(\widetilde{\boldsymbol{B}}, \widehat{X}) \to \mathcal{M}(\widetilde{\boldsymbol{B}}, \widehat{Y})$ are homotopic.*
(b) *If $\tilde{f}, \tilde{g}\colon \widetilde{\boldsymbol{A}} \to \widetilde{\boldsymbol{B}}$ are left homotopic, right homotopic, or homotopic maps of cosimplicial resolutions in $\mathcal{M}$ and $\widehat{X}$ is a fibrant object of $\mathcal{M}$, then the induced maps of left homotopy function complexes $\tilde{f}^*, \tilde{g}^*\colon \mathcal{M}(\widetilde{\boldsymbol{B}}, \widehat{X}) \to \mathcal{M}(\widetilde{\boldsymbol{A}}, \widehat{X})$ are homotopic.*

(2) (a) *If $\widetilde{B}$ is a cofibrant object of $\mathcal{M}$ and $\hat{f}, \hat{g}\colon \widehat{\boldsymbol{X}} \to \widehat{\boldsymbol{Y}}$ are left homotopic, right homotopic, or homotopic maps of simplicial resolutions in $\mathcal{M}$, then the induced maps of right homotopy function complexes $\hat{f}_*, \hat{g}_*\colon \mathcal{M}(\widetilde{B}, \widehat{\boldsymbol{X}}) \to \mathcal{M}(\widetilde{B}, \widehat{\boldsymbol{Y}})$ are homotopic.*
(b) *If $\tilde{f}, \tilde{g}\colon \widetilde{A} \to \widetilde{B}$ are left homotopic, right homotopic, or homotopic maps of cofibrant objects in $\mathcal{M}$ and $\widehat{\boldsymbol{X}}$ is a simplicial resolution in*

$\mathcal{M}$, then the induced maps of right homotopy function complexes $\tilde{f}^*, \tilde{g}^* \colon \mathcal{M}(\widetilde{B}, \widehat{\boldsymbol{X}}) \to \mathcal{M}(\widetilde{A}, \widehat{\boldsymbol{X}})$ are homotopic.

(3) (a) *If $\widetilde{\boldsymbol{B}}$ is a cosimplicial resolution in $\mathcal{M}$ and $\hat{f}, \hat{g} \colon \widehat{\boldsymbol{X}} \to \widehat{\boldsymbol{Y}}$ are left homotopic, right homotopic, or homotopic maps of simplicial resolutions in $\mathcal{M}$, then the induced maps of two-sided homotopy function complexes $\operatorname{diag} \hat{f}_*, \operatorname{diag} \hat{g}_* \colon \operatorname{diag} \mathcal{M}(\widetilde{\boldsymbol{B}}, \widehat{\boldsymbol{X}}) \to \operatorname{diag} \mathcal{M}(\widetilde{\boldsymbol{B}}, \widehat{\boldsymbol{Y}})$ are homotopic.*

(b) *If $\tilde{f}, \tilde{g} \colon \widetilde{\boldsymbol{A}} \to \widetilde{\boldsymbol{B}}$ are left homotopic, right homotopic, or homotopic maps of cosimplicial resolutions in $\mathcal{M}$ and $\widehat{\boldsymbol{X}}$ is a simplicial resolution in $\mathcal{M}$, then the induced maps of two-sided homotopy function complexes $\operatorname{diag} \tilde{f}^*, \operatorname{diag} \tilde{g}^* \colon \operatorname{diag} \mathcal{M}(\widetilde{\boldsymbol{B}}, \widehat{\boldsymbol{X}}) \to \operatorname{diag} \mathcal{M}(\widetilde{\boldsymbol{A}}, \widehat{\boldsymbol{X}})$ are homotopic.*

PROOF. We will prove part 1(a); the proofs of the other parts are similar.

If $\hat{f}$ and $\hat{g}$ are left homotopic, then Proposition 7.3.4 implies that there is a cylinder object $\widehat{X} \amalg \widehat{X} \to \operatorname{Cyl}(\widehat{X}) \xrightarrow{p} \widehat{X}$ for $\widehat{X}$ such that p is a trivial fibration and a left homotopy $H \colon \operatorname{Cyl}(\widehat{X}) \to \widehat{Y}$ from $\hat{f}$ to $\hat{g}$. Corollary 16.5.4 implies that the map $\mathcal{M}\big(\widetilde{\boldsymbol{B}}, \operatorname{Cyl}(\widehat{X})\big) \to \mathcal{M}(\widetilde{\boldsymbol{B}}, \widehat{X})$ is a weak equivalence, and so Proposition 7.3.5 implies that $\hat{f}_*$ and $\hat{g}_*$ are left homotopic. Proposition 17.1.3 and Theorem 7.4.9 now imply that $\hat{f}_*$ and $\hat{g}_*$ are homotopic.

If $\hat{f}$ and $\hat{g}$ are right homotopic and if $\widehat{Y} \to \operatorname{Path}(\widehat{Y}) \to \widehat{Y} \times \widehat{Y}$ is a path object for $\widehat{Y}$ and $H \colon \widehat{X} \to \operatorname{Path}(\widehat{Y})$ is a right homotopy from $\hat{f}$ to $\hat{g}$, then Corollary 16.5.5 implies that the map $\mathcal{M}(\widetilde{\boldsymbol{B}}, \widehat{Y}) \to \mathcal{M}\big(\widetilde{\boldsymbol{B}}, \operatorname{Path}(\widehat{Y})\big)$ is a weak equivalence. Thus, Proposition 7.3.5 implies that $\hat{f}_*$ and $\hat{g}_*$ are right homotopic. Proposition 17.1.3 and Theorem 7.4.9 now imply that $\hat{f}_*$ and $\hat{g}_*$ are homotopic. □

THEOREM 17.6.7. *Let $\mathcal{M}$ be a model category, and let W, X, Y, and Z be objects in $\mathcal{M}$.*

1. *If $f, g \colon X \to Y$ are left homotopic, right homotopic, or homotopic, and if $f_*, g_* \colon \operatorname{map}(W, X) \to \operatorname{map}(W, Y)$ (see Notation 17.4.2) are maps of homotopy function complexes induced by, respectively, f and g, then f_* and g_* are homotopic maps of fibrant simplicial sets.*
2. *If $f, g \colon X \to Y$ are left homotopic, right homotopic, or homotopic, and if $f^*, g^* \colon \operatorname{map}(Z, W) \to \operatorname{map}(Z, W)$ (see Notation 17.4.2) are maps of homotopy function complexes induced by, respectively, f and g, then f^* and g^* are homotopic maps of fibrant simplicial sets.*

PROOF. We will prove part 1 in the case in which f_* and g_* are maps of left homotopy function complexes; the proofs in the other cases (and of part 2) are similar.

Let $\widetilde{\boldsymbol{W}}$ be a cosimplicial resolution of W and let $\hat{f}, \hat{g} \colon \widehat{X} \to \widehat{Y}$ be fibrant approximations to, respectively, f and g, such that the maps f_* and g_* are, respectively, the maps $\hat{f}_* \colon \mathcal{M}(\widetilde{\boldsymbol{W}}, \widehat{X}) \to \mathcal{M}(\widetilde{\boldsymbol{W}}, \widehat{Y})$ and $\hat{g}_* \colon \mathcal{M}(\widetilde{\boldsymbol{W}}, \widehat{X}) \to \mathcal{M}(\widetilde{\boldsymbol{W}}, \widehat{Y})$. If we factor the weak equivalences $X \to \widehat{X}$ and $Y \to \widehat{Y}$ as, respectively, $X \xrightarrow{i_X} \widehat{X}' \xrightarrow{p_X} \widehat{X}$ and $Y \xrightarrow{i_Y} \widehat{Y}' \xrightarrow{p_Y} \widehat{Y}$ such that i_X and i_Y are trivial cofibrations and p_X and p_Y are fibrations, then the "two out of three" axiom implies that p_X and p_Y are trivial fibrations.

The dotted arrow exists in the solid arrow diagram

$$\begin{array}{ccccc} X & \xrightarrow{f} & Y & \xrightarrow{i_Y} & \widehat{Y}' \\ {\scriptstyle i_X}\downarrow & & \nearrow{\scriptstyle \hat{f}'} & & \downarrow{\scriptstyle p_Y} \\ \widehat{X}' & \xrightarrow[p_X]{} & \widehat{X} & \xrightarrow[\hat{f}]{} & \widehat{Y} \end{array}$$

and a similar diagram implies that the corresponding map $\hat{g}'\colon \widehat{X}' \to \widehat{Y}'$ exists. Thus, $\hat{f}'$ and $\hat{g}'$ are cofibrant fibrant approximations to, respectively, f and g, and we have the diagram

$$\begin{array}{ccc} \widehat{X}' & \overset{\hat{f}'}{\underset{\hat{g}'}{\rightrightarrows}} & \widehat{Y}' \\ {\scriptstyle p_X}\downarrow & & \downarrow{\scriptstyle p_Y} \\ \widehat{X} & \overset{\hat{f}}{\underset{\hat{g}}{\rightrightarrows}} & \widehat{Y} \end{array}$$

in which $p_Y\hat{f}' = \hat{f}p_X$ and $p_Y\hat{g}' = \hat{g}p_X$. Lemma 17.6.5 and Proposition 8.2.4 imply that if f and g are left homotopic, right homotopic, or homotopic, then $\hat{f}'$ and $\hat{g}'$ are, respectively, left homotopic, right homotopic, or homotopic. In any of these cases, Proposition 17.6.6 implies that the maps $\hat{f}'_*\colon \mathcal{M}(\widetilde{\boldsymbol{W}}, \widehat{X}') \to \mathcal{M}(\widetilde{\boldsymbol{W}}, \widehat{Y}')$ and $\hat{g}'_*\colon \mathcal{M}(\widetilde{\boldsymbol{W}}, \widehat{X}') \to \mathcal{M}(\widetilde{\boldsymbol{W}}, \widehat{Y}')$ are homotopic. Since p_X and p_Y are weak equivalences of fibrant objects, Corollary 16.5.5 implies that the maps $\mathcal{M}(\widetilde{\boldsymbol{W}}, \widehat{X}') \to \mathcal{M}(\widetilde{\boldsymbol{W}}, \widehat{X})$ and $\mathcal{M}(\widetilde{\boldsymbol{W}}, \widehat{Y}') \to \mathcal{M}(\widetilde{\boldsymbol{W}}, \widehat{Y})$ are homotopy equivalences of fibrant simplicial sets, and this implies that $\hat{f}_*\colon \mathcal{M}(\widetilde{\boldsymbol{W}}, \widehat{X}) \to \mathcal{M}(\widetilde{\boldsymbol{W}}, \widehat{Y})$ and $\hat{g}_*\colon \mathcal{M}(\widetilde{\boldsymbol{W}}, \widehat{X}) \to \mathcal{M}(\widetilde{\boldsymbol{W}}, \widehat{Y})$ are homotopic. The result now follows from Proposition 17.4.3. □

17.7. Homotopy classes of maps

PROPOSITION 17.7.1. *Let $\mathcal{M}$ be a model category.*

1. *If $\widetilde{\boldsymbol{B}}$ is a cosimplicial resolution in $\mathcal{M}$ and X is a fibrant object of $\mathcal{M}$, then the set $\pi_0\mathcal{M}(\widetilde{\boldsymbol{B}}, X)$ is naturally isomorphic to the set of homotopy classes of maps from $\widetilde{\boldsymbol{B}}^0$ to X.*
2. *If B is a cofibrant object of $\mathcal{M}$ and $\widehat{\boldsymbol{X}}$ is a simplicial resolution in $\mathcal{M}$, then the set $\pi_0\mathcal{M}(B, \widehat{\boldsymbol{X}})$ is naturally isomorphic to the set of homotopy classes of maps from B to $\widehat{\boldsymbol{X}}_0$.*

PROOF. We will prove part 1; the proof of part 2 is dual.

The set of vertices of $\mathcal{M}(\widetilde{\boldsymbol{B}}, X)$ is the set of maps from $\widetilde{\boldsymbol{B}}^0$ to X and Proposition 16.1.6 implies that if two vertices of $\mathcal{M}(\widetilde{\boldsymbol{B}}, X)$ represent the same element of $\pi_0\mathcal{M}(\widetilde{\boldsymbol{B}}, X)$ then those vertices (i.e., maps from $\widetilde{\boldsymbol{B}}_0$ to X) are homotopic. Finally, if two maps from $\widetilde{\boldsymbol{B}}^0$ to X are homotopic, then Proposition 7.4.7 and Proposition 16.1.6 imply that there is a 1-simplex of $\mathcal{M}(\widetilde{\boldsymbol{B}}, X)$ whose faces are those maps. □

THEOREM 17.7.2. *Let $\mathcal{M}$ be a model category. If X and Y are objects of $\mathcal{M}$ and $\operatorname{map}(X, Y)$ is a homotopy function complex from X to Y, then $\pi_0 \operatorname{map}(X, Y)$ is naturally isomorphic to the set of maps from X to Y in $\operatorname{Ho}\mathcal{M}$.*

PROOF. This follows from Proposition 17.7.1 and Proposition 16.1.5. □

LEMMA 17.7.3. *Let $\mathcal{M}$ be a model category.*

1. *If $\widetilde{\boldsymbol{B}}$ is a cosimplicial resolution in $\mathcal{M}$ and $p\colon X \to Y$ is a map of fibrant objects in $\mathcal{M}$ that induces a weak equivalence of simplicial sets $p_*\colon \mathcal{M}(\widetilde{\boldsymbol{B}}, X) \cong \mathcal{M}(\widetilde{\boldsymbol{B}}, Y)$, then p induces an isomorphism of the sets of homotopy classes of maps $p_*\colon \pi(\widetilde{\boldsymbol{B}}^0, X) \approx \pi(\widetilde{\boldsymbol{B}}^0, Y)$.*
2. *If $\widehat{\boldsymbol{X}}$ is a simplicial resolution in $\mathcal{M}$ and $i\colon A \to B$ is a map of cofibrant objects in $\mathcal{M}$ that induces a weak equivalence of simplicial sets $i^*\colon \mathcal{M}(B, \widehat{\boldsymbol{X}}) \cong \mathcal{M}(A, \widehat{\boldsymbol{X}})$, then i induces an isomorphism of the sets of homotopy classes of maps $i^*\colon \pi(B, \widehat{\boldsymbol{X}}_0) \approx \pi(A, \widehat{\boldsymbol{X}}_0)$.*

PROOF. This follows from Proposition 17.7.1. □

PROPOSITION 17.7.4. *Let $\mathcal{M}$ be a model category.*

1. *If B is cofibrant and $p\colon X \to Y$ is a map of fibrant objects that induces a weak equivalence of homotopy function complexes $p_*\colon \mathrm{map}(B, X) \to \mathrm{map}(B, Y)$ (see Notation 17.4.2), then p induces an isomorphism of the sets of homotopy classes of maps $p_*\colon \pi(B, X) \approx \pi(B, Y)$.*
2. *If X is fibrant and $i\colon A \to B$ is a map of cofibrant objects that induces a weak equivalence of homotopy function complexes $i^*\colon \mathrm{map}(B, X) \to \mathrm{map}(A, X)$ (see Notation 17.4.2), then i induces an isomorphism of the sets of homotopy classes of maps $i^*\colon \pi(B, X) \approx \pi(A, X)$.*

PROOF. We will prove part 1; the proof of part 2 is dual.

If $\widetilde{\boldsymbol{B}}$ is a cosimplicial resolution of B, then p induces a weak equivalence $p_*\colon \mathcal{M}(\widetilde{\boldsymbol{B}}, X) \to \mathcal{M}(\widetilde{\boldsymbol{B}}, Y)$ (see Theorem 17.5.31), and so Lemma 17.7.3 implies that p induces an isomorphism $p_*\colon \pi(\widetilde{\boldsymbol{B}}^0, X) \approx \pi(\widetilde{\boldsymbol{B}}^0, Y)$. Since $\widetilde{\boldsymbol{B}}^0 \to B$ is a weak equivalence of cofibrant objects, the result now follows from Corollary 7.7.4. □

COROLLARY 17.7.5. *Let $\mathcal{M}$ be a model category.*

1. *If B is cofibrant and $p\colon X \to Y$ is a fibration of fibrant objects for which the induced map of homotopy function complexes $p_*\colon \mathrm{map}(B, X) \to \mathrm{map}(B, Y)$ is a weak equivalence, then for every map $f\colon B \to Y$ there is a map $g\colon B \to X$, unique up to homotopy, such that $pg = f$.*
2. *If X is fibrant and $i\colon A \to B$ is a cofibration of cofibrant objects for which the induced map of homotopy function complexes $i^*\colon \mathrm{map}(B, X) \to \mathrm{map}(A, X)$ is a weak equivalence, then for every map $f\colon A \to X$ there is a map $g\colon B \to X$, unique up to homotopy, such that $gi = f$.*

PROOF. This follows from Proposition 17.7.4, Proposition 7.3.13, and Theorem 7.4.9. □

PROPOSITION 17.7.6. *If $\mathcal{M}$ is a model category, then a map $g\colon X \to Y$ is a weak equivalence if either of the following two conditions is satisfied:*

1. *The map g induces weak equivalences of homotopy function complexes*

$$g_*\colon \mathrm{map}(X, X) \cong \mathrm{map}(X, Y) \quad \text{and} \quad g_*\colon \mathrm{map}(Y, X) \cong \mathrm{map}(Y, Y)$$

(see Notation 17.4.2).

(2) *The map g induces weak equivalences of homotopy function complexes*

$$g^*\colon \operatorname{map}(Y,X) \cong \operatorname{map}(X,X) \quad \textit{and} \quad g^*\colon \operatorname{map}(Y,Y) \cong \operatorname{map}(X,Y)$$

(see Notation 17.4.2*).*

PROOF. We will prove this using condition 1; the proof using condition 2 is dual.

If $\tilde{g}\colon \widetilde{X} \to \widetilde{Y}$ is a cofibrant approximation to g, then Theorem 17.6.3 implies that $\tilde{g}$ induces weak equivalences of homotopy function complexes $\tilde{g}_*\colon \operatorname{map}(\widetilde{X},\widetilde{X}) \cong \operatorname{map}(\widetilde{X},\widetilde{Y})$ and $\tilde{g}_*\colon \operatorname{map}(\widetilde{Y},\widetilde{X}) \cong \operatorname{map}(\widetilde{Y},\widetilde{Y})$. If $\hat{g}\colon \widehat{X} \to \widehat{Y}$ is a cofibrant fibrant approximation to $\tilde{g}$, then $\hat{g}$ is a map of cofibrant-fibrant objects, and Theorem 17.6.3 implies that $\hat{g}$ induces weak equivalences of homotopy function complexes $\hat{g}_*\colon \operatorname{map}(\widehat{X},\widehat{X}) \cong \operatorname{map}(\widehat{X},\widehat{Y})$ and $\hat{g}_*\colon \operatorname{map}(\widehat{Y},\widehat{X}) \cong \operatorname{map}(\widehat{Y},\widehat{Y})$. Proposition 17.7.4 now implies that $\hat{g}$ induces isomorphisms of the sets of homotopy classes of maps $\hat{g}_*\colon \pi(\widehat{X},\widehat{X}) \approx \pi(\widehat{X},\widehat{Y})$ and $\hat{g}_*\colon \pi(\widehat{Y},\widehat{X}) \approx \pi(\widehat{Y},\widehat{Y})$, and so Proposition 7.5.12 implies that $\hat{g}$ is a homotopy equivalence. Thus, $\hat{g}$ is a weak equivalence (see Theorem 7.8.5), and so $\tilde{g}$ is a weak equivalence, and so g is a weak equivalence. □

THEOREM 17.7.7. *If $\mathcal{M}$ is a model category and $g\colon X \to Y$ is a map in $\mathcal{M}$, then the following are equivalent:*

(1) *The map g is a weak equivalence.*
(2) *For every object W in $\mathcal{M}$ the map g induces a weak equivalence of homotopy function complexes $g_*\colon \operatorname{map}(W,X) \cong \operatorname{map}(W,Y)$.*
(3) *For every cofibrant object W in $\mathcal{M}$ the map g induces a weak equivalence of homotopy function complexes $g_*\colon \operatorname{map}(W,X) \cong \operatorname{map}(W,Y)$.*
(4) *For every object Z in $\mathcal{M}$ the map g induces a weak equivalence of homotopy function complexes $g^*\colon \operatorname{map}(Y,Z) \cong \operatorname{map}(X,Z)$.*
(5) *For every fibrant object Z in $\mathcal{M}$ the map g induces a weak equivalence of homotopy function complexes $g^*\colon \operatorname{map}(Y,Z) \cong \operatorname{map}(X,Z)$.*

PROOF. This follows from Theorem 17.6.3, Proposition 17.7.6, and Proposition 8.1.17. □

17.8. Homotopy orthogonal maps

If $\mathcal{M}$ is a simplicial model category and if $i\colon A \to B$ and $p\colon X \to Y$ are maps in $\mathcal{M}$ such that either

(1) i is a trivial cofibration and p is a fibration, or
(2) i is a cofibration and p is a trivial fibration,

then the map of function complexes $\operatorname{Map}(B,X) \to \operatorname{Map}(A,X) \times_{\operatorname{Map}(A,Y)} \operatorname{Map}(B,Y)$ is a trivial fibration (see Definition 9.1.6). If at least one of A and B is cofibrant then at least one of the maps $\operatorname{Map}(A,X) \to \operatorname{Map}(A,Y)$ and $\operatorname{Map}(B,Y) \to \operatorname{Map}(A,Y)$ is a fibration, and so the pullback $\operatorname{Map}(A,X) \times_{\operatorname{Map}(A,Y)} \operatorname{Map}(B,Y)$ is weakly equivalent to the homotopy pullback (see Corollary 13.3.8 and Theorem 13.1.13). If both A and B are cofibrant and both X and Y are fibrant, then these function complexes are homotopy function complexes (see Example 17.1.4 and Example 17.2.4), and in this case the "orthogonality" condition is equivalent to saying that the square

$$\begin{array}{ccc} \mathrm{Map}(B,X) & \longrightarrow & \mathrm{Map}(B,Y) \\ \downarrow & & \downarrow \\ \mathrm{Map}(A,X) & \longrightarrow & \mathrm{Map}(A,Y) \end{array}$$

is a homotopy fiber square (see Definition 13.3.12).

DEFINITION 17.8.1. If $\mathcal{M}$ is a model category and $i\colon A \to B$ and $p\colon X \to Y$ are maps in $\mathcal{M}$, then we will say that

(1) *(i,p) is a homotopy orthogonal pair,*
(2) *i is left homotopy orthogonal to p, and*
(3) *p is right homotopy orthogonal to i*

if there is a homotopy function complex $\mathrm{map}(-,-)$ on $\mathcal{M}$ (see Notation 17.4.2) such that the square

$$\begin{array}{ccc} \mathrm{map}(B,X) & \longrightarrow & \mathrm{map}(B,Y) \\ \downarrow & & \downarrow \\ \mathrm{map}(A,X) & \longrightarrow & \mathrm{map}(A,Y) \end{array}$$

is a homotopy fiber square (see Definition 13.3.12). (We will show in Proposition 17.8.2 that if this is true for any one homotopy function complex, then it is true for every homotopy function complex.)

PROPOSITION 17.8.2. *Let $\mathcal{M}$ be a model category, and let $i\colon A \to B$ and $p\colon X \to Y$ be maps in $\mathcal{M}$. If there is some homotopy function complex $\mathrm{map}(-,-)$ (see Notation 17.4.2) such that the square*

$$\begin{array}{ccc} \mathrm{map}(B,X) & \longrightarrow & \mathrm{map}(B,Y) \\ \downarrow & & \downarrow \\ \mathrm{map}(A,X) & \longrightarrow & \mathrm{map}(A,Y) \end{array} \qquad (17.8.3)$$

is a homotopy fiber square of simplicial sets (see Definition 13.3.12), then Diagram 17.8.3 for any other homotopy function complex is also a homotopy fiber square.

PROOF. If $\mathrm{map}_1(-,-)$ and $\mathrm{map}_2(-,-)$ are homotopy function complexes on $\mathcal{M}$, then Theorem 17.5.30 implies that there is a homotopy equivalence $\mathrm{map}_1(-,-) \cong \mathrm{map}_2(-,-)$ that is natural up to homotopy. If we can alter these homotopy equivalences by homotopies to get maps from Diagram 17.8.3 for map_1 to Diagram 17.8.3 for map_2 that commute on the nose, then the result will follow from Proposition 13.3.13. If the maps $\mathrm{map}_2(A,X) \to \mathrm{map}_2(A,Y)$, $\mathrm{map}_2(B,Y) \to \mathrm{map}_2(A,Y)$, and $\mathrm{map}_2(B,X) \to \mathrm{map}_2(A,X) \times_{\mathrm{map}_2(A,Y)} \mathrm{map}_2(B,Y)$ are fibrations, then we can use the homotopy lifting property (see Proposition 7.3.11) to alter the homotopy equivalences from map_1 to map_2 in our diagrams by homotopies so that we do get a map of diagrams. Thus, it is sufficient to show that for any homotopy function complex, Diagram 17.8.3 maps to one with fibrations as described. We will do this for left homotopy function complexes; the proofs for right and two-sided homotopy function complexes are similar.

If map is a left homotopy function complex defined by the cosimplicial resolution $\tilde{\imath}\colon \widetilde{\boldsymbol{A}} \to \widetilde{\boldsymbol{B}}$ to i and the fibrant approximation $\hat{p}\colon \widehat{X} \to \widehat{Y}$ to p, then we can factor $\tilde{\imath}$ into a cofibration followed by a trivial fibration $\widetilde{\boldsymbol{A}} \to \widetilde{\boldsymbol{B}}' \to \widetilde{\boldsymbol{B}}$ and factor $\hat{p}$ into a trivial cofibration followed by a fibration $\widehat{X} \to \widehat{X}' \to \widehat{Y}$. This yields a diagram

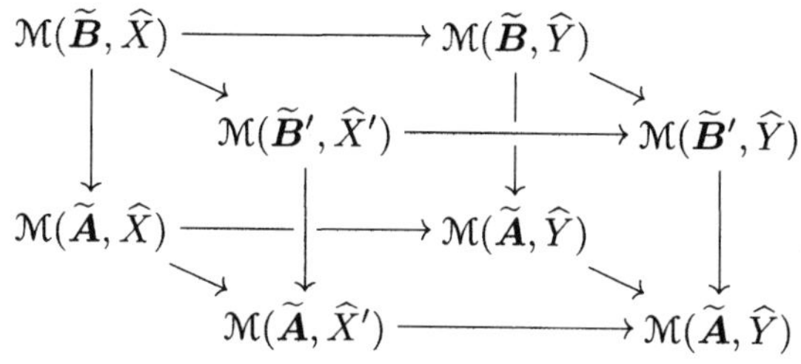

in which all four maps from the back square to the front square are weak equivalences (see Corollary 16.5.5), and Corollary 16.5.4 and Theorem 16.5.2 imply that the front square has the fibrations required. □

THEOREM 17.8.4. *Let $\mathcal{M}$ be a model category. If $i\colon A \to B$ and $p\colon X \to Y$ are maps in $\mathcal{M}$, then the following are equivalent:*

(1) *(i, p) is a homotopy orthogonal pair.*
(2) *For some cosimplicial resolution $\tilde{\imath}\colon \widetilde{\boldsymbol{A}} \to \widetilde{\boldsymbol{B}}$ of i such that $\tilde{\imath}$ is a Reedy cofibration and some fibrant approximation $\hat{p}\colon \widehat{X} \to \widehat{Y}$ to p such that $\hat{p}$ is a fibration, the map of simplicial sets*
$$\mathcal{M}(\widetilde{\boldsymbol{B}}, \widehat{X}) \to \mathcal{M}(\widetilde{\boldsymbol{A}}, \widehat{X}) \times_{\mathcal{M}(\widetilde{\boldsymbol{A}}, \widehat{Y})} \mathcal{M}(\widetilde{\boldsymbol{B}}, \widehat{Y})$$
is a trivial fibration.
(3) *For every cosimplicial resolution $\tilde{\imath}\colon \widetilde{\boldsymbol{A}} \to \widetilde{\boldsymbol{B}}$ of i such that $\tilde{\imath}$ is a Reedy cofibration and every fibrant approximation $\hat{p}\colon \widehat{X} \to \widehat{Y}$ to p such that $\hat{p}$ is a fibration, the map of simplicial sets*
$$\mathcal{M}(\widetilde{\boldsymbol{B}}, \widehat{X}) \to \mathcal{M}(\widetilde{\boldsymbol{A}}, \widehat{X}) \times_{\mathcal{M}(\widetilde{\boldsymbol{A}}, \widehat{Y})} \mathcal{M}(\widetilde{\boldsymbol{B}}, \widehat{Y})$$
is a trivial fibration.
(4) *For some cofibrant approximation $\tilde{\imath}\colon \widetilde{A} \to \widetilde{B}$ to i such that $\tilde{\imath}$ is a cofibration and some simplicial resolution $\hat{p}\colon \widehat{\boldsymbol{X}} \to \widehat{\boldsymbol{Y}}$ of p such that $\hat{p}$ is a Reedy fibration, the map of simplicial sets*
$$\mathcal{M}(\widetilde{B}, \widehat{\boldsymbol{X}}) \to \mathcal{M}(\widetilde{A}, \widehat{\boldsymbol{X}}) \times_{\mathcal{M}(\widetilde{A}, \widehat{\boldsymbol{Y}})} \mathcal{M}(\widetilde{B}, \widehat{\boldsymbol{Y}})$$
is a trivial fibration.
(5) *For every cofibrant approximation $\tilde{\imath}\colon \widetilde{A} \to \widetilde{B}$ to i such that $\tilde{\imath}$ is a cofibration and every simplicial resolution $\hat{p}\colon \widehat{\boldsymbol{X}} \to \widehat{\boldsymbol{Y}}$ of p such that $\hat{p}$ is a Reedy fibration, the map of simplicial sets*
$$\mathcal{M}(\widetilde{B}, \widehat{\boldsymbol{X}}) \to \mathcal{M}(\widetilde{A}, \widehat{\boldsymbol{X}}) \times_{\mathcal{M}(\widetilde{A}, \widehat{\boldsymbol{Y}})} \mathcal{M}(\widetilde{B}, \widehat{\boldsymbol{Y}})$$
is a trivial fibration.
(6) *For some cosimplicial resolution $\tilde{\imath}\colon \widetilde{\boldsymbol{A}} \to \widetilde{\boldsymbol{B}}$ of i such that $\tilde{\imath}$ is a Reedy cofibration and some simplicial resolution $\hat{p}\colon \widehat{\boldsymbol{X}} \to \widehat{\boldsymbol{Y}}$ of p such that $\hat{p}$ is a Reedy fibration, the map of simplicial sets*
$$\operatorname{diag} \mathcal{M}(\widetilde{\boldsymbol{B}}, \widehat{\boldsymbol{X}}) \to \operatorname{diag} \mathcal{M}(\widetilde{\boldsymbol{A}}, \widehat{\boldsymbol{X}}) \times_{\operatorname{diag} \mathcal{M}(\widetilde{\boldsymbol{A}}, \widehat{\boldsymbol{Y}})} \operatorname{diag} \mathcal{M}(\widetilde{\boldsymbol{B}}, \widehat{\boldsymbol{Y}})$$

is a trivial fibration.

(7) *For every cosimplicial resolution* $\tilde{\imath}\colon \widetilde{\boldsymbol{A}} \to \widetilde{\boldsymbol{B}}$ *of* i *such that* $\tilde{\imath}$ *is a Reedy cofibration and every simplicial resolution* $\hat{p}\colon \widehat{\boldsymbol{X}} \to \widehat{\boldsymbol{Y}}$ *of* p *such that* $\hat{p}$ *is a Reedy fibration, the map of simplicial sets*

$$\operatorname{diag} \mathcal{M}(\widetilde{\boldsymbol{B}}, \widehat{\boldsymbol{X}}) \to \operatorname{diag} \mathcal{M}(\widetilde{\boldsymbol{A}}, \widehat{\boldsymbol{X}}) \times_{\operatorname{diag} \mathcal{M}(\widetilde{\boldsymbol{A}}, \widehat{\boldsymbol{Y}})} \operatorname{diag} \mathcal{M}(\widetilde{\boldsymbol{B}}, \widehat{\boldsymbol{Y}})$$

is a trivial fibration.

PROOF. This follows from Proposition 17.8.2, Theorem 16.5.2, and Theorem 16.5.18. □

PROPOSITION 17.8.5. *Let* $\mathcal{M}$ *be a model category.*

(1) *If* $i\colon A \to B$ *is a map in* $\mathcal{M}$ *and* $p\colon X \to *$ *is the map to the terminal object of* $\mathcal{M}$, *then* (i,p) *is a homotopy orthogonal pair if and only if* i *induces a weak equivalence of homotopy function complexes* $i^*\colon \operatorname{map}(B,X) \cong \operatorname{map}(A,X)$ *(see Notation 17.4.2).*
(2) *If* $p\colon X \to Y$ *is a map in* $\mathcal{M}$ *and* $i\colon \emptyset \to B$ *is the map from the initial object of* $\mathcal{M}$, *then* (i,p) *is a homotopy orthogonal pair if and only if* p *induces a weak equivalence of homotopy function complexes* $p_*\colon \operatorname{map}(B,X) \cong \operatorname{map}(B,Y)$ *(see Notation 17.4.2).*

PROOF. This follows directly from the definitions. □

PROPOSITION 17.8.6. *Let* $\mathcal{M}$ *be a model category.*

(1) *If* $p\colon X \to Y$ *is a map in* $\mathcal{M}$ *and we have a square*

$$\begin{array}{ccc} A & \xrightarrow{\cong} & A' \\ {\scriptstyle i}\downarrow & & \downarrow{\scriptstyle i'} \\ B & \xrightarrow[\cong]{} & B' \end{array}$$

in which the horizontal maps are weak equivalences, then (i,p) *is a homotopy orthogonal pair if and only if* (i',p) *is one.*

(2) *If* $i\colon A \to B$ *is a map in* $\mathcal{M}$ *and we have a square*

$$\begin{array}{ccc} X & \xrightarrow{\cong} & X' \\ {\scriptstyle p}\downarrow & & \downarrow{\scriptstyle p'} \\ Y & \xrightarrow[\cong]{} & Y' \end{array}$$

in which the horizontal maps are weak equivalences, then (i,p) *is a homotopy orthogonal pair if and only if* (i,p') *is one.*

PROOF. This follows from Proposition 13.3.13 and Theorem 17.6.3. □

PROPOSITION 17.8.7. *Let* $\mathcal{M}$ *be a model category and let* $i\colon A \to B$ *and* $p\colon X \to Y$ *be maps in* $\mathcal{M}$ *such that* (i,p) *is a homotopy orthogonal pair.*

(1) *If* $i'\colon A' \to B'$ *is a retract of* i *(see Definition 7.1.1), then* (i',p) *is a homotopy orthogonal pair.*
(2) *If* $p'\colon X' \to Y'$ *is a retract of* p, *then* (i,p') *is a homotopy orthogonal pair.*

PROOF. We will prove part 1; the proof of part 2 is dual.

Let $\hat{p}\colon \widehat{\mathbf{X}} \to \widehat{\mathbf{Y}}$ be a simplicial resolution of p such that $\hat{p}$ is a Reedy fibration (see Proposition 16.1.22). Proposition 8.1.23 implies that there are cofibrant approximations $\tilde{\imath}\colon \widetilde{A} \to \widetilde{B}$ to i and $\tilde{\imath}'\colon \widetilde{A}' \to \widetilde{B}'$ to i' such that $\tilde{\imath}$ and $\tilde{\imath}'$ are cofibrations and $\tilde{\imath}'$ is a retract of $\tilde{\imath}$. The map $\mathcal{M}(\widetilde{B}', \widehat{\mathbf{X}}) \to \mathcal{M}(\widetilde{A}', \widehat{\mathbf{X}}) \times_{\mathcal{M}(\widetilde{A}', \widehat{\mathbf{Y}})} \mathcal{M}(\widetilde{B}', \widehat{\mathbf{Y}})$ is thus a retract of the map $\mathcal{M}(\widetilde{B}, \widehat{\mathbf{X}}) \to \mathcal{M}(\widetilde{A}, \widehat{\mathbf{X}}) \times_{\mathcal{M}(\widetilde{A}, \widehat{\mathbf{Y}})} \mathcal{M}(\widetilde{B}, \widehat{\mathbf{Y}})$, and so the result follows from Theorem 17.8.4. □

PROPOSITION 17.8.8. *Let $\mathcal{M}$ be a model category. If $i\colon A \to B$ and $p\colon X \to Y$ are maps in $\mathcal{M}$, then the following are equivalent:*

(1) *(i, p) is a homotopy orthogonal pair.*

(2) *For some cosimplicial resolution $\tilde{\imath}\colon \widetilde{\mathbf{A}} \to \widetilde{\mathbf{B}}$ of i such that $\tilde{\imath}$ is a Reedy cofibration, some fibrant approximation $\hat{p}\colon \widehat{X} \to \widehat{Y}$ to p such that $\hat{p}$ is a fibration, and every $n \ge 0$, the dotted arrow exists in every solid arrow diagram of the form*

$$\begin{array}{ccc} \widetilde{\mathbf{A}} \otimes \Delta[n] \amalg_{\widetilde{\mathbf{A}} \otimes \partial\Delta[n]} \widetilde{\mathbf{B}} \otimes \partial\Delta[n] & \longrightarrow & \widehat{X} \\ \big\downarrow & \nearrow & \big\downarrow \\ \widetilde{\mathbf{B}} \otimes \Delta[n] & \longrightarrow & \widehat{Y} \end{array}$$

(3) *For every cosimplicial resolution $\tilde{\imath}\colon \widetilde{\mathbf{A}} \to \widetilde{\mathbf{B}}$ of i such that $\tilde{\imath}$ is a Reedy cofibration, every fibrant approximation $\hat{p}\colon \widehat{X} \to \widehat{Y}$ to p such that $\hat{p}$ is a fibration, and every $n \ge 0$, the dotted arrow exists in every solid arrow diagram of the form*

$$\begin{array}{ccc} \widetilde{\mathbf{A}} \otimes \Delta[n] \amalg_{\widetilde{\mathbf{A}} \otimes \partial\Delta[n]} \widetilde{\mathbf{B}} \otimes \partial\Delta[n] & \longrightarrow & \widehat{X} \\ \big\downarrow & \nearrow & \big\downarrow \\ \widetilde{\mathbf{B}} \otimes \Delta[n] & \longrightarrow & \widehat{Y} \end{array}$$

(4) *For some cofibrant approximation $\tilde{\imath}\colon \widetilde{A} \to \widetilde{B}$ to i such that $\tilde{\imath}$ is a cofibration, some simplicial resolution $\hat{p}\colon \widehat{\mathbf{X}} \to \widehat{\mathbf{Y}}$ to p such that $\hat{p}$ is a Reedy fibration, and every $n \ge 0$, the dotted arrow exists in every solid arrow diagram of the form*

$$\begin{array}{ccc} \widetilde{A} & \longrightarrow & \widehat{\mathbf{X}}^{\Delta[n]} \\ \big\downarrow & \nearrow & \big\downarrow \\ \widetilde{B} & \longrightarrow & \widehat{\mathbf{Y}}^{\Delta[n]} \times_{\widehat{\mathbf{Y}}^{\partial\Delta[n]}} \widehat{\mathbf{X}}^{\partial\Delta[n]} \end{array}$$

(5) *For every cofibrant approximation $\tilde{\imath}\colon \widetilde{A} \to \widetilde{B}$ to i such that $\tilde{\imath}$ is a cofibration, every simplicial resolution $\hat{p}\colon \widehat{\mathbf{X}} \to \widehat{\mathbf{Y}}$ to p such that $\hat{p}$ is a Reedy fibration, and every $n \ge 0$, the dotted arrow exists in every solid arrow*

diagram of the form

$$\begin{array}{ccc}
\widetilde{A} & \longrightarrow & \widehat{\boldsymbol{X}}^{\Delta[n]} \\
\big\downarrow & \nearrow & \big\downarrow \\
\widetilde{B} & \longrightarrow & \widehat{\boldsymbol{Y}}^{\Delta[n]} \times_{\widehat{\boldsymbol{Y}}^{\partial\Delta[n]}} \widehat{\boldsymbol{X}}^{\partial\Delta[n]}
\end{array}$$

PROOF. Since a map of simplicial sets is a trivial fibration if and only if it has the right lifting property with respect to the map $\partial\Delta[n] \to \Delta[n]$ for every $n \geq 0$, this follows from Theorem 17.8.4 and Proposition 16.4.5. $\square$

PROPOSITION 17.8.9. *Let $\mathcal{M}$ be a model category. If $i\colon A \to B$ is a cofibration between cofibrant objects, $p\colon X \to Y$ is a fibration between fibrant objects, and (i,p) is a homotopy orthogonal pair, then (i,p) is a lifting-extension pair (see Definition* 7.2.1*).*

PROOF. Proposition 16.6.14 implies that there is a cosimplicial frame $\tilde{\imath}\colon \widetilde{\boldsymbol{A}} \to \widetilde{\boldsymbol{B}}$ on i such that $\tilde{\imath}$ is a Reedy cofibration. Proposition 17.8.8 now implies that $\widetilde{\boldsymbol{A}} \otimes \Delta[0] \to \widetilde{\boldsymbol{B}} \otimes \Delta[0]$ has the left lifting property with respect to p, and Lemma 16.3.6 implies that $\widetilde{\boldsymbol{A}} \otimes \Delta[0] \to \widetilde{\boldsymbol{B}} \otimes \Delta[0]$ is isomorphic to the map i. $\square$

THEOREM 17.8.10. *If $\mathcal{M}$ is a model category and $g\colon X \to Y$ is a map in $\mathcal{M}$, then the following are equivalent:*

(1) *g is a weak equivalence.*
(2) *g is right homotopy orthogonal to every map in $\mathcal{M}$.*
(3) *For every cofibrant object W of $\mathcal{M}$, g is right homotopy orthogonal to the map $\emptyset \to W$ (where $\emptyset$ is the initial object of $\mathcal{M}$).*
(4) *g is left homotopy orthogonal to every map in $\mathcal{M}$.*
(5) *For every fibrant object Z of $\mathcal{M}$, g is left homotopy orthogonal to the map $Z \to *$ (where $*$ is the terminal object of $\mathcal{M}$).*

PROOF. We will prove that conditions 1, 2, and 3 are equivalent. The proof that conditions 1, 4, and 5 are equivalent is dual.

1 implies 2: If $i\colon A \to B$ is a map in $\mathcal{M}$, choose a cofibrant approximation $\tilde{\imath}\colon \widetilde{A} \to \widetilde{B}$ to i such that $\tilde{\imath}$ is a cofibration (see Proposition 8.1.23) and choose a simplicial resolution $\hat{g}\colon \widehat{\boldsymbol{X}} \to \widehat{\boldsymbol{Y}}$ such that $\hat{g}$ is a Reedy trivial fibration (see Proposition 16.1.22 and Proposition 16.1.24); the result now follows from Theorem 16.5.2.

2 implies 3: This is immediate.

3 implies 1: This follows from Proposition 17.8.5 and Theorem 17.7.7.

$\square$

PROPOSITION 17.8.11. *Let $\mathcal{M}$ be a model category.*

(1) *If $i\colon A \to B$ is a cofibration between cofibrant objects and $p\colon X \to Y$ is a map such that i is left homotopy orthogonal to p, then for every map $j\colon A \to C$ such that C is cofibrant the map $C \to B \amalg_A C$ is left homotopy orthogonal to p.*
(2) *If $p\colon X \to Y$ is a fibration between fibrant objects and $i\colon A \to B$ is a map such that p is right homotopy orthogonal to i, then for every map*

$W \to Y$ such that W is fibrant the map $W \times_Y X \to W$ is right homotopy orthogonal to i.

PROOF. We will prove part 1; the proof of part 2 is dual.

If we choose a simplicial resolution $\hat{p}\colon \widehat{\boldsymbol{X}} \to \widehat{\boldsymbol{Y}}$ of p such that $\hat{p}$ is a Reedy fibration (see Proposition 16.1.22), then Proposition 17.8.8 implies that i has the left lifting property with respect to the map $\widehat{\boldsymbol{X}}^{\Delta[n]} \to \widehat{\boldsymbol{Y}}^{\Delta[n]} \times_{\widehat{\boldsymbol{Y}}^{\partial\Delta[n]}} \widehat{\boldsymbol{X}}^{\partial\Delta[n]}$ for every $n \geq 0$. Since the map $C \to B \amalg_A C$ is also a cofibration between cofibrant objects, the result follows from Lemma 7.2.11 and Proposition 17.8.8. □

COROLLARY 17.8.12. *Let $\mathcal{M}$ be a model category.*

1. *If X is an object of $\mathcal{M}$ and $i\colon A \to B$ is a cofibration between cofibrant objects that induces a weak equivalence of homotopy function complexes $i^*\colon \operatorname{map}(B, X) \cong \operatorname{map}(A, X)$ (see Notation 17.4.2), then for every map $A \to C$ such that C is cofibrant the map $C \to B \amalg_A C$ also induces a weak equivalence of homotopy function complexes to X.*
2. *If B is an object of $\mathcal{M}$ and $p\colon X \to Y$ is a fibration between fibrant objects that induces a weak equivalence of homotopy function complexes $p_*\colon \operatorname{map}(B, X) \cong \operatorname{map}(B, Y)$ (see Notation 17.4.2), then for every map $W \to Y$ such that W is fibrant the map $W \times_Y X \to W$ also induces a weak equivalence of homotopy function complexes from B.*

PROOF. This follows from Proposition 17.8.5 and Proposition 17.8.11. □

PROPOSITION 17.8.13. *Let $\mathcal{M}$ be a model category.*

1. *If $i\colon A \to B$, $j\colon B \to C$, and $p\colon X \to Y$ are maps in $\mathcal{M}$ such that (i, p) is a homotopy orthogonal pair, then (j, p) is a homotopy orthogonal pair if and only if (ji, p) is one.*
2. *If $i\colon A \to B$, $p\colon X \to Y$, and $q\colon Y \to Z$ are maps in $\mathcal{M}$ such that (i, q) is a homotopy orthogonal pair, then (i, p) is a homotopy orthogonal pair if and only if (i, qp) is one.*

PROOF. This follows from Proposition 13.3.15. □

PROPOSITION 17.8.14. *Let $\mathcal{M}$ be a model category, and let $i\colon A \to B$ and $p\colon X \to Y$ be maps in $\mathcal{M}$ such that (i, p) is a homotopy orthogonal pair.*

1. *If $\tilde{\imath}\colon \widetilde{\boldsymbol{A}} \to \widetilde{\boldsymbol{B}}$ is a cosimplicial resolution of i such that $\tilde{\imath}$ is a Reedy cofibration, then for every $n \geq 0$ the pushout corner map $\widetilde{\boldsymbol{A}} \otimes \Delta[n] \amalg_{\widetilde{\boldsymbol{A}} \otimes \partial\Delta[n]} \widetilde{\boldsymbol{B}} \otimes \partial\Delta[n] \to \widetilde{\boldsymbol{B}} \otimes \Delta[n]$ is left homotopy orthogonal to p.*
2. *If $\hat{p}\colon \widehat{\boldsymbol{X}} \to \widehat{\boldsymbol{Y}}$ is a simplicial resolution of p such that $\hat{p}$ is a Reedy fibration, then for every $n \geq 0$ the pullback corner map $\widehat{\boldsymbol{X}}^{\Delta[n]} \to \widehat{\boldsymbol{Y}}^{\Delta[n]} \times_{\widehat{\boldsymbol{Y}}^{\partial\Delta[n]}} \widehat{\boldsymbol{X}}^{\partial\Delta[n]}$ is right homotopy orthogonal to i.*

PROOF. We will prove part 1; the proof of part 2 is dual.

Corollary 16.3.11implies that for every $n \geq 0$ the map $\sigma_n\colon \widetilde{\boldsymbol{A}} \otimes \Delta[n] \amalg_{\widetilde{\boldsymbol{A}} \otimes \partial\Delta[n]} \widetilde{\boldsymbol{B}} \otimes \partial\Delta[n] \to \widetilde{\boldsymbol{B}} \otimes \Delta[n]$ is a cofibration between cofibrant objects. Thus, Proposition 17.8.8 implies that if $\hat{p}\colon \widehat{\boldsymbol{X}} \to \widehat{\boldsymbol{Y}}$ is a simplicial resolution of p such that $\hat{p}$ is a Reedy fibration, then it is sufficient to show that σ_n has the left lifting property with respect to the map $\tau_k\colon \widehat{\boldsymbol{X}}^{\Delta[k]} \to \widehat{\boldsymbol{Y}}^{\Delta[k]} \times_{\widehat{\boldsymbol{Y}}^{\partial\Delta[k]}} \widehat{\boldsymbol{X}}^{\partial\Delta[k]}$ for every $k \geq 0$. We will do this by induction on n.

Lemma 16.3.6 and Proposition 17.8.6 imply that for every $n \geq 0$ the map $\widetilde{\boldsymbol{A}} \otimes \Delta[n] \to \widetilde{\boldsymbol{B}} \otimes \Delta[n]$ is left homotopy orthogonal to p. Since the map σ_0 is the map $\widetilde{\boldsymbol{A}} \otimes \Delta[0] \to \widetilde{\boldsymbol{B}} \otimes \Delta[0]$, the induction is begun.

We now assume that $n > 0$ and that the result is true for all lesser values of n. Lemma 15.3.9 now implies that $\mathrm{L}_n\widetilde{\boldsymbol{A}} \to \mathrm{L}_n\widetilde{\boldsymbol{B}}$ has the left lifting property with respect to τ_k for every $k \geq 0$. Proposition 15.3.11 and Corollary 15.3.12 imply that $\mathrm{L}_n\widetilde{\boldsymbol{A}} \to \mathrm{L}_n\widetilde{\boldsymbol{B}}$ is a cofibration between cofibrant objects and Lemma 16.3.7 implies that this map is isomorphic to the map $\widetilde{\boldsymbol{A}} \otimes \partial\Delta[n] \to \widetilde{\boldsymbol{B}} \otimes \partial\Delta[n]$. Since the map $\widetilde{\boldsymbol{A}} \otimes \Delta[n] \to \widetilde{\boldsymbol{A}} \otimes \Delta[n] \amalg_{\widetilde{\boldsymbol{A}} \otimes \partial\Delta[n]} \widetilde{\boldsymbol{B}} \otimes \partial\Delta[n]$ is the pushout of $\widetilde{\boldsymbol{A}} \otimes \partial\Delta[n] \to \widetilde{\boldsymbol{B}} \otimes \partial\Delta[n]$ along the map $\widetilde{\boldsymbol{A}} \otimes \partial\Delta[n] \to \widetilde{\boldsymbol{A}} \otimes \Delta[n]$, Proposition 17.8.11 implies that it is left homotopy orthogonal to p. Since the composition $\widetilde{\boldsymbol{A}} \otimes \Delta[n] \to \widetilde{\boldsymbol{A}} \otimes \Delta[n] \amalg_{\widetilde{\boldsymbol{A}} \otimes \partial\Delta[n]} \widetilde{\boldsymbol{B}} \otimes \partial\Delta[n] \to \widetilde{\boldsymbol{B}} \otimes \Delta[n]$ is also left homotopy orthogonal to p, Proposition 17.8.13 completes the inductive step. □

17.8.15. Properness.

PROPOSITION 17.8.16. *Let $\mathcal{M}$ be a left proper model category and let $i\colon A \to B$ and $p\colon X \to Y$ be maps in $\mathcal{M}$ such that (i, p) is a homotopy orthogonal pair (see Definition 17.8.1).*

(1) *If the diagram*

$$\begin{array}{ccc} A & \xrightarrow{j} & C \\ {\scriptstyle i}\downarrow & & \downarrow{\scriptstyle k} \\ B & \longrightarrow & D \end{array}$$

is a pushout and at least one of i and j is a cofibration, then (k, p) is a homotopy orthogonal pair.

(2) *If the diagram*

$$\begin{array}{ccc} W & \longrightarrow & X \\ {\scriptstyle r}\downarrow & & \downarrow{\scriptstyle p} \\ Z & \xrightarrow[q]{} & Y \end{array}$$

is a pullback and at least one of p and q is a fibration, then (i, r) is a homotopy orthogonal pair.

PROOF. We will prove part 1; the proof of part 2 is dual.

Let $\tilde{\imath}\colon \widetilde{A} \to \widetilde{B}$ be a cofibrant approximation to i such that $\tilde{\imath}$ is a cofibration (see Proposition 8.1.23). Proposition 17.8.6 implies that $\tilde{\imath}$ is left homotopy orthogonal to p, and Proposition 13.5.6 implies that k has a cofibrant approximation $\tilde{k}$ that is a pushout of $\tilde{\imath}$ (which must be a pushout of k along a map to a cofibrant object). Thus, Proposition 17.8.11 implies that $(\tilde{k}, p)$ is a homotopy orthogonal pair, and so Proposition 17.8.6 implies that (k, p) is a homotopy orthogonal pair. □

17.8.17. Cofibrantly generated model categories.

THEOREM 17.8.18. *Let $\mathcal{M}$ be a cofibrantly generated model category. If there is a set I of generating cofibrations for $\mathcal{M}$ such that either*

(1) *the domains of the elements of I are cofibrant, or*
(2) *$\mathcal{M}$ is left proper,*

then a map $g\colon X \to Y$ in $\mathcal{M}$ is a weak equivalence if and only if it is homotopy right orthogonal to every element of I.

PROOF. If g is a weak equivalence, then Theorem 17.8.10 implies that g is right homotopy orthogonal to every map in $\mathcal{M}$. Conversely, assume that g is right homotopy orthogonal to every element of I. We will show that if W is a cofibrant object of $\mathcal{M}$, then g is right homotopy orthogonal to the map $\emptyset \to W$ (where $\emptyset$ is the initial object of $\mathcal{M}$); the result will then follow from Theorem 17.8.10.

Since every cofibrant object of $\mathcal{M}$ is a retract of a cell complex (see Corollary 11.2.2), it is sufficient to assume that W is a cell complex (see Proposition 17.8.7). Thus, there is an ordinal λ and a λ-sequence $\emptyset \to W_1 \to W_2 \to \cdots \to W_\beta \to \cdots \quad (\beta < \lambda)$ such that $W = \operatorname{colim}_{\beta<\lambda} W_\beta$ and each map $W_\beta \to W_{\beta+1}$ for $\beta < \lambda$ is a pushout of an element of I. We will show by induction on β that $\emptyset \to W$ is left homotopy orthogonal to g.

The induction is begun because $\emptyset \to W_0$ is the identity map of $\emptyset$. If $\beta < \lambda$ and $\emptyset \to W_\beta$ is left homotopy orthogonal to g, then there is an element $A \to B$ of I and a pushout diagram

$$\begin{array}{ccc} A & \longrightarrow & B \\ \big\downarrow & & \big\downarrow \\ W_\beta & \underset{h}{\longrightarrow} & W_{\beta+1} \end{array} .$$

Either Proposition 17.8.11 or Proposition 17.8.16 implies that h is left homotopy orthogonal to g, and so Proposition 17.8.13 implies that $\emptyset \to W_{\beta+1}$ is left homotopy orthogonal to g.

Finally, let γ be a limit ordinal with $\gamma < \lambda$ and assume that $\emptyset \to W_\beta$ is left homotopy orthogonal to g for all $\beta < \gamma$. If $\hat{g}\colon \widehat{\boldsymbol{X}} \to \widehat{\boldsymbol{Y}}$ is a simplicial resolution of g such that $\hat{g}$ is a Reedy fibration, then we have a map of towers of simplicial sets

$$\begin{array}{ccccccccc} \mathcal{M}(\emptyset, \widehat{\boldsymbol{X}}) & \longleftarrow & \mathcal{M}(W_1, \widehat{\boldsymbol{X}}) & \longleftarrow & \mathcal{M}(W_2, \widehat{\boldsymbol{X}}) & \longleftarrow \cdots \longleftarrow & \mathcal{M}(W_\beta, \widehat{\boldsymbol{X}}) & \longleftarrow & \cdots \\ \big\downarrow & & \big\downarrow & & \big\downarrow & & \big\downarrow & & \\ \mathcal{M}(\emptyset, \widehat{\boldsymbol{Y}}) & \longleftarrow & \mathcal{M}(W_1, \widehat{\boldsymbol{Y}}) & \longleftarrow & \mathcal{M}(W_2, \widehat{\boldsymbol{Y}}) & \longleftarrow \cdots \longleftarrow & \mathcal{M}(W_\beta, \widehat{\boldsymbol{Y}}) & \longleftarrow & \cdots \end{array}$$

in which all the horizontal maps are fibrations and all the vertical maps are weak equivalences of fibrant simplicial sets. Thus, the induced map $\lim_{\beta<\gamma} \mathcal{M}(W_\beta, \widehat{\boldsymbol{X}}) \to \lim_{\beta<\gamma} \mathcal{M}(W_\beta, \widehat{\boldsymbol{Y}})$ is a weak equivalence, and so we have weak equivalences

$$\begin{aligned} \mathcal{M}(W_\gamma, \widehat{\boldsymbol{X}}) \approx \mathcal{M}(\operatorname*{colim}_{\beta<\gamma} W_\beta, \widehat{\boldsymbol{X}}) \approx \lim_{\beta<\gamma} \mathcal{M}(W_\beta, \widehat{\boldsymbol{X}}) \\ \cong \lim_{\beta<\gamma} \mathcal{M}(W_\beta, \widehat{\boldsymbol{Y}}) \approx \mathcal{M}(\operatorname*{colim}_{\beta<\gamma} W_\beta, \widehat{\boldsymbol{Y}}) \approx \mathcal{M}(W_\gamma, \widehat{\boldsymbol{Y}}) \end{aligned} ,$$

and the result follows from Proposition 17.8.5. □

17.9. Sequential colimits

PROPOSITION 17.9.1. *If $\mathcal{M}$ is a model category, λ is an ordinal, and*

$$\begin{array}{ccccccc} X_0 & \longrightarrow & X_1 & \longrightarrow & X_2 & \longrightarrow & \cdots \\ {\scriptstyle g_0}\downarrow & & {\scriptstyle g_1}\downarrow & & {\scriptstyle g_2}\downarrow & & \\ Y_0 & \longrightarrow & Y_1 & \longrightarrow & Y_2 & \longrightarrow & \cdots \end{array}$$

is a map of λ-sequences in $\mathcal{M}$ such that

(1) *each of the maps $g_\alpha\colon X_\alpha \to Y_\alpha$ for $\alpha < \lambda$ is a weak equivalence of cofibrant objects and*
(2) *each of the maps $X_\alpha \to X_{\alpha+1}$ and $Y_\alpha \to Y_{\alpha+1}$ for $\alpha < \lambda$ is a cofibration,*

then the induced map of colimits $(\operatorname{colim} g_\alpha)\colon \operatorname{colim} X_\alpha \to \operatorname{colim} Y_\alpha$ is a weak equivalence.

PROOF. If Z is an object of $\mathcal{M}$ and $\mathrm{cs}_* Z \to \widehat{\boldsymbol{Z}}$ is a simplicial resolution of Z, then Theorem 17.7.7 implies that it is sufficient to show that the map $\mathcal{M}(\operatorname{colim} Y_\alpha, \widehat{\boldsymbol{Z}}) \to \mathcal{M}(\operatorname{colim} X_\alpha, \widehat{\boldsymbol{Z}})$ is a weak equivalence of simplicial sets.

Corollary 16.5.5 implies that the map $g^*\colon \mathcal{M}(Y_\alpha, \widehat{\boldsymbol{Z}}) \to \mathcal{M}(X_\alpha, \widehat{\boldsymbol{Z}})$ is a weak equivalence of fibrant simplicial sets for every $\alpha < \lambda$, and so the diagram

$$\begin{array}{ccccccc} \cdots \longrightarrow & \mathcal{M}(Y_2, \widehat{\boldsymbol{Z}}) & \longrightarrow & \mathcal{M}(Y_1, \widehat{\boldsymbol{Z}}) & \longrightarrow & \mathcal{M}(Y_0, \widehat{\boldsymbol{Z}}) \\ & \downarrow & & \downarrow & & \downarrow \\ \cdots \longrightarrow & \mathcal{M}(X_2, \widehat{\boldsymbol{Z}}) & \longrightarrow & \mathcal{M}(X_1, \widehat{\boldsymbol{Z}}) & \longrightarrow & \mathcal{M}(X_0, \widehat{\boldsymbol{Z}}) \end{array}$$

is a weak equivalence of towers of fibrations of fibrant simplicial sets. Thus, the induced map $\lim \mathcal{M}(Y_\alpha, \widehat{\boldsymbol{Z}}) \to \lim \mathcal{M}(X_\alpha, \widehat{\boldsymbol{Z}})$ is a weak equivalence, and that map is isomorphic to the map $\mathcal{M}(\operatorname{colim} Y_\alpha, \widehat{\boldsymbol{Z}}) \to \mathcal{M}(\operatorname{colim} X_\alpha, \widehat{\boldsymbol{Z}})$. □

17.9.2. Properness. We are indebted to D. M. Kan for the following proposition.

PROPOSITION 17.9.3. *Let $\mathcal{M}$ be a left proper model category. If λ is an ordinal and*

$$\begin{array}{ccccccc} X_0 & \longrightarrow & X_1 & \longrightarrow & X_2 & \longrightarrow & \cdots \\ {\scriptstyle g_0}\downarrow & & {\scriptstyle g_1}\downarrow & & {\scriptstyle g_2}\downarrow & & \\ Y_0 & \longrightarrow & Y_1 & \longrightarrow & Y_2 & \longrightarrow & \cdots \end{array}$$

is a map of λ-sequences in $\mathcal{M}$ such that

(1) *each of the maps $g_\alpha\colon X_\alpha \to Y_\alpha$ for $\alpha < \lambda$ is a weak equivalence and*
(2) *each of the maps $X_\alpha \to X_{\alpha+1}$ and $Y_\alpha \to Y_{\alpha+1}$ for $\alpha < \lambda$ is a cofibration,*

then the induced map of colimits $(\operatorname{colim} g_\alpha)\colon \operatorname{colim} X_\alpha \to \operatorname{colim} Y_\alpha$ is a weak equivalence.

PROOF. We construct a λ-sequence $Z_0 \to Z_1 \to Z_2 \to \cdots$ intermediate between the given ones by letting Z_α be the pushout $Y_0 \amalg_{X_0} X_\alpha$ for every $\alpha < \beta$. Proposition 7.2.14 implies that $Z_\alpha \to Z_{\alpha+1}$ is a cofibration for every $\alpha < \lambda$, and we have maps of λ-sequences

$$\begin{array}{ccccccc} X_0 & \longrightarrow & X_1 & \longrightarrow & X_2 & \longrightarrow & \cdots \\ {\scriptstyle h_0}\downarrow & & {\scriptstyle h_1}\downarrow & & {\scriptstyle h_2}\downarrow & & \\ Z_0 & \longrightarrow & Z_1 & \longrightarrow & Z_2 & \longrightarrow & \cdots \\ {\scriptstyle k_0}\downarrow & & {\scriptstyle k_1}\downarrow & & {\scriptstyle k_2}\downarrow & & \\ Y_0 & \longrightarrow & Y_1 & \longrightarrow & Y_2 & \longrightarrow & \cdots \end{array}$$

such that

(1) each of the maps $Z_0 \to Z_\alpha$ (for $\alpha < \lambda$) is a cofibration,
(2) the map $k_0 \colon Z_0 \to Y_0$ is an isomorphism, and
(3) (since $\mathcal{M}$ is left proper) each of the maps $k_\alpha \colon Z_\alpha \to Y_\alpha$ (for $\alpha < \lambda$) is a weak equivalence.

Since left adjoints commute with colimits, $\operatorname{colim} Z_\alpha$ is isomorphic to the pushout $Y_0 \amalg_{X_0} (\operatorname{colim} X_\alpha)$ (see Lemma 7.6.6); thus, the map $\operatorname{colim} X_\alpha \to \operatorname{colim} Z_\alpha$ is a weak equivalence. Thus, it is sufficient to show that $\operatorname{colim} Z_\alpha \to \operatorname{colim} Y_\alpha$ is a weak equivalence. Since $k_0 \colon Z_0 \to Y_0$ is an isomorphism, each of the maps $k_\alpha \colon Z_\alpha \to Y_\alpha$ (for $\alpha < \lambda$) is a weak equivalence of cofibrant objects in the category $(Z_0 \downarrow \mathcal{M})$ of objects under Z_0 (see Theorem 7.6.5). Thus, Proposition 17.9.1 implies that the map $\operatorname{colim} Z_\alpha \to \operatorname{colim} Y_\alpha$ is a weak equivalence, and the proof is complete. $\square$

PROPOSITION 17.9.4. *Let $\mathcal{M}$ be a left proper model category. If λ is an ordinal and*

$$X_0 \to X_1 \to X_2 \to \cdots \to X_\beta \to \cdots \qquad (\beta < \lambda)$$

is a λ-sequence in $\mathcal{M}$ such that $X_\beta \to X_{\beta+1}$ is a cofibration for every $\beta < \lambda$, then there is a λ-sequence

$$\widetilde{X}_0 \to \widetilde{X}_1 \to \widetilde{X}_2 \to \cdots \to \widetilde{X}_\beta \to \cdots \qquad (\beta < \lambda)$$

and a map of λ-sequences

$$\begin{array}{ccccccccccc} \widetilde{X}_0 & \longrightarrow & \widetilde{X}_1 & \longrightarrow & \widetilde{X}_2 & \longrightarrow & \cdots & \longrightarrow & \widetilde{X}_\beta & \longrightarrow & \cdots \qquad (\beta < \lambda) \\ {\scriptstyle g_0}\downarrow & & {\scriptstyle g_1}\downarrow & & {\scriptstyle g_2}\downarrow & & & & {\scriptstyle g_\beta}\downarrow & & \\ X_0 & \longrightarrow & X_1 & \longrightarrow & X_2 & \longrightarrow & \cdots & \longrightarrow & X_\beta & \longrightarrow & \cdots \qquad (\beta < \lambda) \end{array}$$

such that

(1) *every $\widetilde{X}_\beta$ is cofibrant,*
(2) *every $g_\beta \colon \widetilde{X}_\beta \to X_\beta$ is a weak equivalence,*
(3) *every $\widetilde{X}_\beta \to \widetilde{X}_{\beta+1}$ is a cofibration, and*
(4) *the map $\operatorname{colim}_{\beta<\lambda} \widetilde{X}_\beta \to \operatorname{colim}_{\beta<\lambda} X_\beta$ is a weak equivalence.*

PROOF. We will define the $\widetilde{X}_\beta$ inductively. We begin by choosing a cofibrant approximation $g_0 \colon \widetilde{X}_0 \to X_0$ to X_0 (see Proposition 8.1.17). If $\beta + 1 < \lambda$ and we have defined $g_\beta \colon \widetilde{X}_\beta \to X_\beta$, then we factor the composition $\widetilde{X}_\beta \to X_\beta \to X_{\beta+1}$ into a cofibration followed by a trivial fibration, to obtain $\widetilde{X}_\beta \to \widetilde{X}_{\beta+1} \xrightarrow{g_{\beta+1}} X_{\beta+1}$. If $\beta < \lambda$ and β is a limit ordinal, then Proposition 17.9.3 implies that $\operatorname{colim}_{\alpha<\beta} \widetilde{X}_\alpha \to \operatorname{colim}_{\alpha<\beta} X_\alpha$ is a weak equivalence, and so we can construct the $\widetilde{X}_\beta$ as required.

Proposition 17.9.3 implies that the map $\operatorname{colim}_{\beta<\lambda} \widetilde{X}_\beta \to \operatorname{colim}_{\beta<\lambda} X_\beta$ is a weak equivalence, and so the proof is complete. □

CHAPTER 18

Homotopy Limits in Simplicial Model Categories

If $\mathcal{C}$ is a small category and $\mathcal{M}$ is a model category, then the colimit functor takes a $\mathcal{C}$-diagram $\boldsymbol{X}$ in $\mathcal{M}$ to an object $\operatorname{colim} \boldsymbol{X}$ of $\mathcal{M}$. For most diagrams, though, the colimit functor does not have good homotopy properties: If $f\colon \boldsymbol{X} \to \boldsymbol{Y}$ is an objectwise weak equivalence of $\mathcal{C}$-diagrams in $\mathcal{M}$, then it will not in general be true that $(\operatorname{colim} f)\colon \operatorname{colim} \boldsymbol{X} \to \operatorname{colim} \boldsymbol{Y}$ is a weak equivalence. The homotopy colimit functor is an attempt to repair this deficiency of the ordinary colimit.

There is a class of $\mathcal{C}$-diagrams in $\mathcal{M}$ for which the colimit functor does take objectwise weak equivalences of diagrams into weak equivalences. For example, if $\mathcal{C}$ is a Reedy category with fibrant constants (see Definition 15.10.1), then the Reedy cofibrant diagrams are in that class (see Theorem 15.10.9). It can be shown that our definition of the homotopy colimit of a diagram $\boldsymbol{X}$ (see Definition 18.1.2) is equivalent to constructing an objectwise weak equivalence $\widetilde{\boldsymbol{X}} \to \boldsymbol{X}$ such that $\widetilde{\boldsymbol{X}}$ is in this "special class" when $\boldsymbol{X}$ is objectwise cofibrant, and then defining $\operatorname{hocolim} \boldsymbol{X}$ to be $\operatorname{colim} \widetilde{\boldsymbol{X}}$. Thus, at the cost of replacing our original diagram with an objectwise weakly equivalent one, we obtain a version of the colimit functor that takes objectwise weak equivalences between objectwise cofibrant diagrams into weak equivalences in $\mathcal{M}$. It can be shown that, although there may not be a model category structure on the category of all $\mathcal{C}$-diagrams in $\mathcal{M}$, the localization of that category of diagrams with respect to the objectwise weak equivalences does exist (see Remark 8.3.3), and that the homotopy colimit functor represents the total left derived functor (see Definition 8.4.7) of the colimit functor. For this, see [**30**].

The definition that we use (see Definition 18.1.2) is homotopy invariant only for objectwise cofibrant diagrams. To obtain a definition that is homotopy invariant for all diagrams, we could first functorially take a cofibrant approximation to each object in the diagram and then apply Definition 18.1.2. Our definition provides simpler formulas, though, and it is the standard definition that is already in wide use.

All of the above remarks can be dualized to describe the homotopy limit functor as a replacement for the ordinary limit. The formula that we give below for the homotopy limit (see Definition 18.1.8) is homotopy invariant only for objectwise fibrant diagrams, and it can also be made completely homotopy invariant by first functorially taking a fibrant approximation to each object in the diagram.

The standard reference for homotopy colimits and homotopy limits of diagrams of simplicial sets, total spaces of cosimplicial simplicial sets, and realizations of bisimplicial sets is [**14**, Chapters X through XII], and our definitions are essentially the ones used there (but see Remark 18.1.11). The reference [**19**] gives a useful discussion of the idea of a *free diagram* (see Definition 11.5.35), and [**35**] gives a careful development of the homotopy colimit of certain small diagrams in a model category.

In this chapter we restrict ourselves to diagrams in a simplicial model category. This includes most of the examples of interest and makes for simpler formulas. In Chapter 19 we will work with general model categories, providing definitions and results that specialize to those in this chapter for simplicial model categories with the standard framing (see Proposition 16.6.23).

In Section 18.1 we define the homotopy colimit and homotopy limit of a diagram in a simplicial model category and give several examples. In Section 18.2 we show that the homotopy limit of a diagram of spaces can be described as a space of maps between diagrams. In Section 18.3 we discuss *coends* and *ends*, which are constructions that generalize the definitions of the homotopy colimit and the homotopy limit. If $\mathcal{C}$ is a small category, $\mathcal{M}$ is a simplicial model category, $\boldsymbol{X}$ is a $\mathcal{C}$-diagram in $\mathcal{M}$, and $\boldsymbol{K}$ is a $\mathcal{C}^{\mathrm{op}}$-diagram of simplicial sets, then the coend $\boldsymbol{X} \otimes_{\mathcal{C}} \boldsymbol{K}$ (also called the *tensor product* of the functors $\boldsymbol{X}$ and $\boldsymbol{K}$) reduces to the homotopy colimit of $\boldsymbol{X}$ when $\boldsymbol{K}$ is the $\mathcal{C}^{\mathrm{op}}$-diagram $\mathrm{B}(-\downarrow \mathcal{C})^{\mathrm{op}}$. Dually, if $\boldsymbol{K}$ is a $\mathcal{C}$-diagram of simplicial sets, then the end $\hom^{\mathcal{C}}(\boldsymbol{K}, \boldsymbol{X})$ (also called the *hom* of the functors $\boldsymbol{K}$ and $\boldsymbol{X}$) reduces to the homotopy limit of $\boldsymbol{X}$ when $\boldsymbol{K}$ is the $\mathcal{C}$-diagram $\mathrm{B}(\mathcal{C}\downarrow -)$. We also establish adjointness properties for ends and coends, and we use those adjointness properties in Section 18.4 together with the homotopy lifting extension theorem (see Remark 9.1.7) to obtain homotopy invariance results for the pushout corner map for coends and the pullback corner map for ends (see Theorem 18.4.1). We also establish a homotopy lifting extension theorem for diagram indexed by a Reedy category. In Section 18.5 we obtain homotopy invariance results for the homotopy colimit and homotopy limit functors.

In Section 18.6 we discuss realizations of simplicial objects and total objects of cosimplicial objects, and establish homotopy invariance results. In Section 18.7 we discuss the Bousfield-Kan maps from the homotopy colimit of a simplicial object to its realization and from the total object of a cosimplicial object to its homotopy limit. In Section 18.8 we compare the homotopy colimit of a diagram of pointed spaces with the homotopy colimit of the diagram of unpointed spaces obtained by forgetting the basepoints, and in Section 18.9 we discuss diagrams of simplicial sets.

18.1. Homotopy colimits and homotopy limits

In this section we define the homotopy colimit and homotopy limit of a diagram in a simplicial model category, and give several examples.

18.1.1. Homotopy colimits.

DEFINITION 18.1.2. Let $\mathcal{M}$ be a simplicial model category and let $\mathcal{C}$ be a small category. If $\boldsymbol{X}$ is a $\mathcal{C}$-diagram in $\mathcal{M}$ (see Definition 11.5.2), then the *homotopy colimit* $\operatorname{hocolim} \boldsymbol{X}$ of $\boldsymbol{X}$ is defined to be the coequalizer of the maps

$$\coprod_{(\sigma\colon \alpha\to\alpha')\in\mathcal{C}} \boldsymbol{X}_\alpha \otimes \mathrm{B}(\alpha'\downarrow\mathcal{C})^{\mathrm{op}} \underset{\psi}{\overset{\phi}{\rightrightarrows}} \coprod_{\alpha\in\mathrm{Ob}(\mathcal{C})} \boldsymbol{X}_\alpha \otimes \mathrm{B}(\alpha\downarrow\mathcal{C})^{\mathrm{op}} .$$

(see Definition 14.1.1 and Definition 11.8.3) where the map ϕ on the summand $\sigma\colon \alpha \to \alpha'$ is the composition of the map

$$\sigma_* \otimes 1_{\mathrm{B}(\alpha'\downarrow\mathcal{C})}\colon \boldsymbol{X}_\alpha \otimes \mathrm{B}(\alpha'\downarrow\mathcal{C})^{\mathrm{op}} \longrightarrow \boldsymbol{X}_{\alpha'} \otimes \mathrm{B}(\alpha'\downarrow\mathcal{C})^{\mathrm{op}}$$

with the natural injection into the coproduct, and the map ψ on the summand $\sigma\colon \alpha \to \alpha'$ is the composition of the map

$$1_{\boldsymbol{X}_\alpha} \otimes \mathrm{B}(\sigma^*)\colon \boldsymbol{X}_\alpha \otimes \mathrm{B}(\alpha' \downarrow \mathcal{C})^{\mathrm{op}} \longrightarrow \boldsymbol{X}_\alpha \otimes \mathrm{B}(\alpha \downarrow \mathcal{C})^{\mathrm{op}}$$

(where $\sigma^*\colon (\alpha' \downarrow \mathcal{C})^{\mathrm{op}} \to (\alpha \downarrow \mathcal{C})^{\mathrm{op}}$; see Definition 14.7.2) with the natural injection into the coproduct.

For a discussion of the relation of our definition of the homotopy colimit to that of [**14**], see Remark 18.1.11.

REMARK 18.1.3. If $\mathrm{Spc}_{(*)}$ is one of our categories of spaces (see Notation 7.10.5), $\mathcal{C}$ is a small category, and $\boldsymbol{X}$ is a $\mathcal{C}$-diagram in $\mathrm{Spc}_{(*)}$, then

$$\boldsymbol{X}_\alpha \otimes \mathrm{B}(\alpha \downarrow \mathcal{C})^{\mathrm{op}} \approx \begin{cases} \boldsymbol{X}_\alpha \times \mathrm{B}(\alpha \downarrow \mathcal{C})^{\mathrm{op}} & \text{if } \mathrm{Spc}_{(*)} = \mathrm{SS} \\ \boldsymbol{X}_\alpha \wedge \left(\mathrm{B}(\alpha \downarrow \mathcal{C})^{\mathrm{op}}\right)^+ & \text{if } \mathrm{Spc}_{(*)} = \mathrm{SS}_* \\ \boldsymbol{X}_\alpha \times \left|\mathrm{B}(\alpha \downarrow \mathcal{C})^{\mathrm{op}}\right| & \text{if } \mathrm{Spc}_{(*)} = \mathrm{Top} \\ \boldsymbol{X}_\alpha \wedge \left|\mathrm{B}(\alpha \downarrow \mathcal{C})^{\mathrm{op}}\right|^+ & \text{if } \mathrm{Spc}_{(*)} = \mathrm{Top}_* \end{cases}$$

(see Example 9.1.13, Example 9.1.14, Example 9.1.15, and Example 9.1.16).

EXAMPLE 18.1.4. If $\mathrm{Spc}_{(*)}$ is one of our categories of spaces (see Notation 7.10.5) and $g\colon X \to Y$ is a map in $\mathrm{Spc}_{(*)}$, then the homotopy colimit of the diagram consisting of just the map g is the mapping cylinder of g.

EXAMPLE 18.1.5. If $\mathrm{Spc}_{(*)}$ is one of our categories of spaces (see Notation 7.10.5) and $Z \xleftarrow{h} X \xrightarrow{g} Y$ is a diagram in $\mathrm{Spc}_{(*)}$, then the homotopy colimit of this diagram is the double mapping cylinder of g and h.

PROPOSITION 18.1.6. *If $\mathcal{C}$ is a small category and $\boldsymbol{P}$ is the diagram of simplicial sets in which $\boldsymbol{P}_\alpha$ is a single point for every object α of $\mathcal{C}$ (i.e., $\boldsymbol{P}$ is the constant diagram at a point), then there is a natural isomorphism* $\operatorname{hocolim} \boldsymbol{P} \approx \mathrm{B}\mathcal{C}^{\mathrm{op}}$.

PROOF. Definition 18.1.2 defines $\operatorname{hocolim} \boldsymbol{P}$ to be the coequalizer of the maps

$$\coprod_{(\sigma\colon \alpha \to \alpha') \in \mathcal{C}} \mathrm{B}(\alpha' \downarrow \mathcal{C})^{\mathrm{op}} \overset{\phi}{\underset{\psi}{\rightrightarrows}} \coprod_{\alpha \in \mathrm{Ob}(\mathcal{C})} \mathrm{B}(\alpha \downarrow \mathcal{C})^{\mathrm{op}}$$

where the map ϕ is induced by the identity map on each summand and the map ψ on the summand $\sigma\colon \alpha \to \alpha'$ is the composition of the map $\mathrm{B}(\sigma^*)\colon \mathrm{B}(\alpha' \downarrow \mathcal{C})^{\mathrm{op}} \to \mathrm{B}(\alpha \downarrow \mathcal{C})^{\mathrm{op}}$ with the natural injection into the coproduct. We define a map from $\mathrm{B}(\alpha \downarrow \mathcal{C})^{\mathrm{op}}$ to $\mathrm{B}\mathcal{C}^{\mathrm{op}}$ by sending the simplex

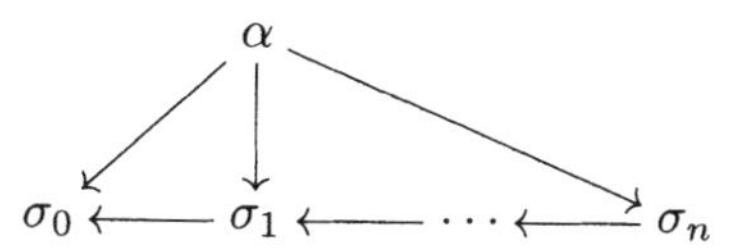

of $\mathrm{B}(\alpha \downarrow \mathcal{C})^{\mathrm{op}}$ to the simplex $\sigma_0 \leftarrow \sigma_1 \leftarrow \cdots \leftarrow \sigma_n$ of $\mathrm{B}\mathcal{C}^{\mathrm{op}}$. This defines a surjective map $\operatorname{hocolim} \boldsymbol{P} \to \mathrm{B}\mathcal{C}^{\mathrm{op}}$ which is also injective because every simplex of $\coprod_{\sigma \in \mathrm{Ob}(\mathcal{C})} \mathrm{B}(\alpha \downarrow \mathcal{C})^{\mathrm{op}}$ that is mapped to $\sigma_0 \leftarrow \sigma_1 \leftarrow \cdots \leftarrow \sigma_n$ is equal (in $\operatorname{hocolim} \boldsymbol{P}$)

to the simplex

$$\begin{array}{ccccccc} & \sigma_n & & & & & \\ \swarrow & \downarrow & & \searrow^{1_{\sigma_n}} & & & \\ \sigma_0 \longleftarrow & \sigma_1 & \longleftarrow & \cdots & \longleftarrow & \sigma_n & . \end{array}$$

□

18.1.7. Homotopy limits.

DEFINITION 18.1.8. Let $\mathcal{M}$ be a simplicial model category and let $\mathcal{C}$ be a small category. If $\boldsymbol{X}$ is a $\mathcal{C}$-diagram in $\mathcal{M}$ (see Definition 11.5.2), then the *homotopy limit* $\operatorname{holim}\boldsymbol{X}$ of $\boldsymbol{X}$ is defined to be the equalizer of the maps

$$\prod_{\alpha\in\operatorname{Ob}(\mathcal{C})} (\boldsymbol{X}_\alpha)^{\mathrm{B}(\mathcal{C}\downarrow\alpha)} \overset{\phi}{\underset{\psi}{\rightrightarrows}} \prod_{(\sigma\colon\alpha\to\alpha')\in\mathcal{C}} (\boldsymbol{X}_{\alpha'})^{\mathrm{B}(\mathcal{C}\downarrow\alpha)} .$$

(see Definition 14.1.1 and Definition 11.8.1) where the projection of the map ϕ on the factor $\sigma\colon\alpha\to\alpha'$ is the composition of a natural projection from the product with the map

$$\sigma_*^{1_{\mathrm{B}(\mathcal{C}\downarrow\alpha)}}\colon (\boldsymbol{X}_\alpha)^{\mathrm{B}(\mathcal{C}\downarrow\alpha)} \longrightarrow (\boldsymbol{X}_{\alpha'})^{\mathrm{B}(\mathcal{C}\downarrow\alpha)}$$

and the projection of the map ψ on the factor $\sigma\colon\alpha\to\alpha'$ is the composition of a natural projection from the product with the map

$$(1_{\boldsymbol{X}_{\alpha'}})^{\mathrm{B}(\sigma_*)}\colon (\boldsymbol{X}_{\alpha'})^{\mathrm{B}(\mathcal{C}\downarrow\alpha')} \longrightarrow (\boldsymbol{X}_{\alpha'})^{\mathrm{B}(\mathcal{C}\downarrow\alpha)}$$

(where $\sigma_*\colon(\mathcal{C}\downarrow\alpha)\to(\mathcal{C}\downarrow\alpha')$; see Definition 14.7.8).

For a discussion of the relation of our definition of the homotopy limit to that of [**14**], see Remark 18.1.11.

EXAMPLE 18.1.9. If $\mathrm{Spc}_{(*)}$ is one of our categories of spaces (see Notation 7.10.5) and $g\colon X\to Y$ is a map in $\mathrm{Spc}_{(*)}$, then the homotopy limit of the diagram consisting of just the map g is what is usually called the *mapping path space* of g.

THEOREM 18.1.10. *Let $\mathcal{M}$ be a simplicial model category and let $\mathcal{C}$ be a small category. If $\boldsymbol{X}$ is a $\mathcal{C}$-diagram in $\mathcal{M}$ and Y is an object of $\mathcal{M}$ then there is a natural isomorphism of simplicial sets* $\operatorname{Map}(\operatorname{hocolim}_{\mathcal{C}}\boldsymbol{X},Y)\approx\operatorname{holim}_{\mathcal{C}^{\mathrm{op}}}\operatorname{Map}(\boldsymbol{X},Y)$.

PROOF. Definition 18.1.2 and Proposition 9.2.2 imply that $\operatorname{Map}(\operatorname{hocolim}_{\mathcal{C}}\boldsymbol{X},Y)$ is naturally isomorphic to the limit of the diagram

$$\prod_{\alpha\in\operatorname{Ob}(\mathcal{C})} \operatorname{Map}\big(\boldsymbol{X}_\alpha\otimes\mathrm{B}(\alpha\downarrow\mathcal{C})^{\mathrm{op}},Y\big) \rightrightarrows \prod_{(\sigma\colon\alpha\to\alpha')\in\mathcal{C}} \operatorname{Map}\big(\boldsymbol{X}_\alpha\otimes\mathrm{B}(\alpha'\downarrow\mathcal{C})^{\mathrm{op}},Y\big)$$

Axiom M6 (see Definition 9.1.6) and Corollary 14.7.13 imply that there are natural isomorphisms

$$\begin{aligned}\operatorname{Map}\big(\boldsymbol{X}_\alpha\otimes\mathrm{B}(\alpha'\downarrow\mathcal{C})^{\mathrm{op}},Y\big) &\approx \operatorname{Map}\big(\mathrm{B}(\alpha'\downarrow\mathcal{C})^{\mathrm{op}},\operatorname{Map}(\boldsymbol{X}_\alpha,Y)\big)\\ &\approx \operatorname{Map}\big(\mathrm{B}(\mathcal{C}^{\mathrm{op}}\downarrow\alpha'),\operatorname{Map}(\boldsymbol{X}_\alpha,Y)\big)\\ &\approx \big(\operatorname{Map}(\boldsymbol{X}_\alpha,Y)\big)^{\mathrm{B}(\mathcal{C}^{\mathrm{op}}\downarrow\alpha')} ,\end{aligned}$$

and so $\mathrm{Map}(\mathrm{hocolim}_{\mathcal{C}} \boldsymbol{X}, Y)$ is naturally isomorphic to the limit of the diagram

$$\prod_{\alpha\in\mathrm{Ob}(\mathcal{C}^{\mathrm{op}})} \left(\mathrm{Map}(\boldsymbol{X}_\alpha, Y)\right)^{\mathrm{B}(\mathcal{C}^{\mathrm{op}}\downarrow\alpha)} \rightrightarrows \prod_{(\sigma\colon \alpha'\to\alpha)\in\mathcal{C}^{\mathrm{op}}} \left(\mathrm{Map}(\boldsymbol{X}_\alpha, Y)\right)^{\mathrm{B}(\mathcal{C}^{\mathrm{op}}\downarrow\alpha')}$$

which is the definition of $\mathrm{holim}_{\mathcal{C}^{\mathrm{op}}} \mathrm{Map}(\boldsymbol{X}, Y)$ (see Definition 18.1.8). □

REMARK 18.1.11. There are two respects in which our definitions of the homotopy colimit and homotopy limit differ from those of [**14**] (which uses the term *homotopy direct limit* for the homotopy colimit and *homotopy inverse limit* for the homotopy limit). First, we use the diagrams of simplicial sets $\mathrm{B}(-\downarrow\mathcal{C})^{\mathrm{op}}$ and $\mathrm{B}(\mathcal{C}\downarrow -)$ (see Definition 19.1.2 and Definition 19.1.5) where [**14**] uses the diagrams $\mathrm{B}(-\downarrow\mathcal{C})$ and $\mathrm{B}(\mathcal{C}\downarrow -)$ (see [**14**, Chapter XII, Paragraph 2.1 and Chapter XI, Paragraph 3.2]. Since both $\mathrm{B}(-\downarrow\mathcal{C})^{\mathrm{op}}$ and $\mathrm{B}(-\downarrow\mathcal{C})$ are cofibrant approximations to the constant $\mathcal{C}^{\mathrm{op}}$-diagram at a point (see Corollary 14.8.8), these two choices give definitions that are naturally weakly equivalent for $\mathcal{C}$-diagrams of cofibrant objects (see Theorem 19.4.7), but our definition was chosen to make Theorem 18.1.10 true. It is incorrectly stated in [**14**, Chapter XII, Proposition 4.1] that this is true for the definitions used in [**14**]; this is due to an error in the proof of [**14**, Chapter XII, Proposition 4.1]. This error is a minor one, since the spaces claimed there to be isomorphic are in fact naturally weakly equivalent, which is all that was needed.

The second difference between our definitions and those of [**14**] is that the definition of the classifying space (i.e., the nerve) of a category used in [**14**] is "opposite" to our definition (see Definition 14.1.1 and [**14**, Chapter XI, Paragraph 2.1]), i.e., if $\mathcal{C}$ is a small category, then the definition of $\mathrm{B}\mathcal{C}$ used in [**14**] (which is called there the *underlying space* of the category) is isomorphic to our definition of $\mathrm{B}\mathcal{C}^{\mathrm{op}}$.

The combined effect of the above two differences is that our definition of the homotopy colimit is isomorphic to that of [**14**], but our definition of the homotopy limit is different. Since the $\mathcal{C}$-diagrams of simplicial sets $\mathrm{B}(\mathcal{C}\downarrow -)$ and $\mathrm{B}(\mathcal{C}\downarrow -)^{\mathrm{op}}$ are both free cell complexes (see Definition 11.5.35), these two definitions of the homotopy limit are naturally weakly equivalent for diagrams of fibrant objects (see Theorem 19.4.7).

18.2. The homotopy limit of a diagram of spaces

Each of our categories of spaces (see Notation 7.10.5) has an internal mapping space, and these can be used to describe the homotopy limit of a diagram of spaces as a space of maps.

DEFINITION 18.2.1 (Internal mapping spaces).

- If X and Y are objects of SS, then the internal mapping space Y^X equals the simplicial mapping space $\mathrm{Map}(X, Y)$ (see Example 9.1.13).
- If X and Y are objects of SS_*, then the internal mapping space Y^X is the pointed simplicial set with n-simplices the basepoint preserving simplicial maps $X \wedge \Delta[n]^+ \to Y$, and face and degeneracy maps induced by the standard maps between the $\Delta[n]$. When we need to emphasize the category in which we work, we will use the notation $\mathrm{Map}_*(X, Y)$ for the pointed simplicial set of basepoint preserving maps.
- If X and Y are objects of Top, then the internal mapping space Y^X is the topological space (see Notation 7.10.2) of continuous functions from X to Y. When we need to emphasize the category in which we work, we

will use the notation $\mathrm{map}(X, Y)$ for the unpointed topological space of continuous functions.
- If X and Y are objects of Top_*, then the internal mapping space Y^X is the pointed topological space (see Notation 7.10.2) of basepoint preserving continuous functions from X to Y. When we need to emphasize the category in which we work, we will use the notation $\mathrm{map}_*(X, Y)$ for the pointed topological space of basepoint preserving continuous functions.

PROPOSITION 18.2.2. *The internal mapping spaces Y^X of Definition 18.2.1 are related to the simplicial mapping spaces $\mathrm{Map}(X, Y)$ of Example 9.1.13, Example 9.1.14, Example 9.1.15, and Example 9.1.16 as follows:*
- *If X and Y are objects of SS, then $\mathrm{Map}(X, Y)$ equals Y^X.*
- *If X and Y are objects of SS_*, then $\mathrm{Map}(X, Y)$ is obtained from Y^X by forgetting the basepoint.*
- *If X and Y are objects of Top, then the simplicial set $\mathrm{Map}(X, Y)$ is the total singular complex of Y^X.*
- *If X and Y are objects of Top_*, then the simplicial set $\mathrm{Map}(X, Y)$ is the total singular complex of the unpointed space obtained from Y^X by forgetting the basepoint.*

PROOF. This follows from the natural isomorphisms of sets

$$\mathrm{Top}\big(|\Delta[n]|, Y^X\big) \approx \mathrm{Top}\big(X \times |\Delta[n]|, Y\big)$$
$$\mathrm{Top}_*\big(|\Delta[n]|^+, Y^X\big) \approx \mathrm{Top}_*\big(X \wedge |\Delta[n]|^+, Y\big) \ .$$

□

DEFINITION 18.2.3. Let $\mathcal{C}$ be a small category.

(1) If $\boldsymbol{X}$ and $\boldsymbol{Y}$ are $\mathcal{C}$-diagrams of unpointed simplicial sets, then $\boldsymbol{Y}^{\boldsymbol{X}}$ is the unpointed simplicial set of maps of diagrams (i.e., natural transformations) from $\boldsymbol{X}$ to $\boldsymbol{Y}$ whose set of n-simplices is the set of maps of diagrams from $\boldsymbol{X} \otimes \Delta[n]$ to $\boldsymbol{Y}$ (see Definition 11.7.1). When we need to emphasize the category in which we work, we will use the notation $\mathrm{Map}(\boldsymbol{X}, \boldsymbol{Y})$ for the unpointed simplicial set of maps from $\boldsymbol{X}$ to $\boldsymbol{Y}$.

(2) If $\boldsymbol{X}$ and $\boldsymbol{Y}$ are $\mathcal{C}$-diagrams of pointed simplicial sets, then $\boldsymbol{Y}^{\boldsymbol{X}}$ is the pointed simplicial set of maps of diagrams (i.e., natural transformations) from $\boldsymbol{X}$ to $\boldsymbol{Y}$ whose set of n-simplices is the set of maps of diagrams from $\boldsymbol{X} \otimes \Delta[n]$ to $\boldsymbol{Y}$ (see Definition 11.7.1). When we need to emphasize the category in which we work, we will use the notation $\mathrm{Map}_*(X, Y)$ for the pointed simplicial set of maps from $\boldsymbol{X}$ to $\boldsymbol{Y}$.

(3) If $\boldsymbol{X}$ and $\boldsymbol{Y}$ are $\mathcal{C}$-diagrams of unpointed topological spaces, then $\boldsymbol{Y}^{\boldsymbol{X}}$ is the unpointed topological space of maps of diagrams (i.e., natural transformations) from $\boldsymbol{X}$ to $\boldsymbol{Y}$ topologized as a subset of the product $\prod_{\alpha \in \mathrm{Ob}(\mathcal{C})} \mathrm{map}(\boldsymbol{X}_\alpha, \boldsymbol{Y}_\alpha)$ (see Definition 18.2.1). When we need to emphasize the category in which we work, we will use the notation $\mathrm{map}(\boldsymbol{X}, \boldsymbol{Y})$ for the unpointed topological space of maps from $\boldsymbol{X}$ to $\boldsymbol{Y}$.

(4) If $\boldsymbol{X}$ and $\boldsymbol{Y}$ are $\mathcal{C}$-diagrams of pointed topological spaces, then $\boldsymbol{Y}^{\boldsymbol{X}}$ is the pointed topological space of maps of diagrams (i.e., natural transformations) from $\boldsymbol{X}$ to $\boldsymbol{Y}$ topologized as a subset of $\prod_{\alpha \in \mathrm{Ob}(\mathcal{C})} \mathrm{map}_*(\boldsymbol{X}_\alpha, \boldsymbol{Y}_\alpha)$ (see Definition 18.2.1). When we need to emphasize the category in which

we work, we will use the notation $\operatorname{map}_*(\boldsymbol{X},\boldsymbol{Y})$ for the pointed topological space of maps from $\boldsymbol{X}$ to $\boldsymbol{Y}$.

LEMMA 18.2.4. *If $\mathcal{C}$ is a small category and $\boldsymbol{X}$ and $\boldsymbol{Y}$ are $\mathcal{C}$-diagrams of unpointed topological spaces, then there is a natural isomorphism of simplicial sets*

$$\operatorname{Sing}(\boldsymbol{Y}^{\boldsymbol{X}}) \approx \operatorname{Map}(\boldsymbol{X},\boldsymbol{Y})$$

(see Definition 11.7.2 and Definition 18.2.3).

PROOF. Since the total singular complex functor is a right adjoint it commutes with limits, and so the result follows from Proposition 18.2.2. □

PROPOSITION 18.2.5. *If $\mathcal{C}$ is a small category, $\mathrm{Spc}_{(*)}$ is one of our categories of spaces (see Notation 7.10.5), and $\boldsymbol{X}$ and $\boldsymbol{Y}$ are $\mathcal{C}$-diagrams in $\mathrm{Spc}_{(*)}$, then the internal mapping spaces $\boldsymbol{Y}^{\boldsymbol{X}}$ of Definition 18.2.3 are related to the simplicial mapping spaces $\operatorname{Map}(\boldsymbol{X},\boldsymbol{Y})$ of Definition 11.7.2 as follows:*

- *If $\mathrm{Spc}_{(*)} = \mathrm{Top}$, then the simplicial set $\operatorname{Map}(\boldsymbol{X},\boldsymbol{Y})$ is the total singular complex of $\boldsymbol{Y}^{\boldsymbol{X}}$.*
- *If $\mathrm{Spc}_{(*)} = \mathrm{Top}_*$, the simplicial set $\operatorname{Map}(\boldsymbol{X},\boldsymbol{Y})$ is the total singular complex of the unpointed space obtained from $\boldsymbol{Y}^{\boldsymbol{X}}$ by forgetting the basepoint.*
- *If $\mathrm{Spc}_{(*)} = \mathrm{SS}$, then $\operatorname{Map}(\boldsymbol{X},\boldsymbol{Y})$ equals $\boldsymbol{Y}^{\boldsymbol{X}}$.*
- *If $\mathrm{Spc}_{(*)} = \mathrm{SS}_*$, then $\operatorname{Map}(\boldsymbol{X},\boldsymbol{Y})$ is obtained from $\boldsymbol{Y}^{\boldsymbol{X}}$ by forgetting the basepoint.*

PROOF. This follows from Proposition 18.2.2 and Lemma 18.2.4. □

PROPOSITION 18.2.6. *If $\mathrm{Spc}_{(*)}$ is one of our categories of spaces (see Notation 7.10.5), $\mathcal{C}$ is a small category, and $\boldsymbol{X}$ is a $\mathcal{C}$-diagram in $\mathrm{Spc}_{(*)}$, then $\operatorname{holim}\boldsymbol{X}$ is naturally isomorphic to the space of maps between diagrams*

$$\begin{array}{ll}
\operatorname{Map}\big(\mathrm{B}(\mathcal{C}\downarrow -),\boldsymbol{X}\big), & \text{if } \mathrm{Spc}_{(*)} = \mathrm{SS}\\
\operatorname{Map}_*\big(\mathrm{B}(\mathcal{C}\downarrow -)^+,\boldsymbol{X}\big), & \text{if } \mathrm{Spc}_{(*)} = \mathrm{SS}_*\\
\operatorname{map}\big(|\mathrm{B}(\mathcal{C}\downarrow -)|,\boldsymbol{X}\big), & \text{if } \mathrm{Spc}_{(*)} = \mathrm{Top}\\
\operatorname{map}_*\big(|\mathrm{B}(\mathcal{C}\downarrow -)|^+,\boldsymbol{X}\big), & \text{if } \mathrm{Spc}_{(*)} = \mathrm{Top}_*
\end{array}$$

(see Definition 18.2.3).

PROOF. For each object α of $\mathcal{C}$ the space $(\boldsymbol{X}_\alpha)^{\mathrm{B}(\mathcal{C}\downarrow\alpha)}$ is a $\mathrm{Spc}_{(*)}$-object of maps in $\mathrm{Spc}_{(*)}$ (see Example 9.1.13, Example 9.1.14, Example 9.1.15, and Example 9.1.16)

$$\boldsymbol{X}^{\mathrm{B}(\mathcal{C}\downarrow\alpha)} \approx \begin{cases}
\operatorname{Map}(\mathrm{B}(\mathcal{C}\downarrow\alpha),\boldsymbol{X}) & \text{if } \mathrm{Spc}_{(*)} = \mathrm{SS}\\
\operatorname{Map}_*(\mathrm{B}(\mathcal{C}\downarrow\alpha)^+,\boldsymbol{X}) & \text{if } \mathrm{Spc}_{(*)} = \mathrm{SS}_*\\
\operatorname{map}\big(|\mathrm{B}(\mathcal{C}\downarrow\alpha)|,\boldsymbol{X}\big) & \text{if } \mathrm{Spc}_{(*)} = \mathrm{Top}\\
\operatorname{map}_*\big(|\mathrm{B}(\mathcal{C}\downarrow\alpha)|^+,\boldsymbol{X}\big) & \text{if } \mathrm{Spc}_{(*)} = \mathrm{Top}_*
\end{cases}$$

and so the result follows from Definition 18.1.8. □

18.3. Coends and ends

In this section we define general constructions (see Definition 18.3.2) that allow us to analyze the colimit and homotopy colimit as two examples of the same construction (and, similarly, the limit and homotopy limit as two examples of the same construction) (see Example 18.3.8). These definitions also enable us to obtain adjointness relations (see Section 18.3.9) that will be used to obtain the homotopy invariance results of Section 18.5.

18.3.1. Definitions.

DEFINITION 18.3.2. Let $\mathcal{M}$ be a simplicial model category and let $\mathcal{C}$ be a small category.

(1) If $\boldsymbol{X}$ is a $\mathcal{C}$-diagram in $\mathcal{M}$ and $\boldsymbol{K}$ is a $\mathcal{C}^{\mathrm{op}}$-diagram of simplicial sets, then $\boldsymbol{X} \otimes_{\mathcal{C}} \boldsymbol{K}$ is defined to be the object of $\mathcal{M}$ that is the coequalizer of the maps

$$\coprod_{(\sigma\colon \alpha \to \alpha') \in \mathcal{C}} \boldsymbol{X}_\alpha \otimes \boldsymbol{K}_{\alpha'} \overset{\phi}{\underset{\psi}{\rightrightarrows}} \coprod_{\alpha \in \mathrm{Ob}(\mathcal{C})} \boldsymbol{X}_\alpha \otimes \boldsymbol{K}_\alpha \tag{18.3.3}$$

where the map ϕ on the summand $\sigma\colon \alpha \to \alpha'$ is the composition of the map

$$\sigma_* \otimes 1_{\boldsymbol{K}_{\alpha'}} \colon \boldsymbol{X}_\alpha \otimes \boldsymbol{K}_{\alpha'} \longrightarrow \boldsymbol{X}_{\alpha'} \otimes \boldsymbol{K}_{\alpha'}$$

(where $\sigma_*\colon \boldsymbol{X}_\alpha \to \boldsymbol{X}_{\alpha'}$) with the natural injection into the coproduct, and the map ψ on the summand $\sigma\colon \alpha \to \alpha'$ is the composition of the map

$$1_{\boldsymbol{X}_\alpha} \otimes \sigma^* \colon \boldsymbol{X}_\alpha \otimes \boldsymbol{K}_{\alpha'} \longrightarrow \boldsymbol{X}_\alpha \otimes \boldsymbol{K}_\alpha$$

(where $\sigma^*\colon \boldsymbol{K}_{\alpha'} \to \boldsymbol{K}_\alpha$) with the natural injection into the coproduct.

The construction of the object $\boldsymbol{X} \otimes_{\mathcal{C}} \boldsymbol{K}$ in $\mathcal{M}$ from the functor $\boldsymbol{X} \otimes \boldsymbol{K}\colon \mathcal{C} \times \mathcal{C}^{\mathrm{op}} \to \mathcal{M}$ is an example of the general construction known as a *coend* (see [**47**, pages 222–223]). In the notation of [**47**], $\boldsymbol{X} \otimes_{\mathcal{C}} \boldsymbol{K} = \int^\alpha \boldsymbol{X}_\alpha \otimes \boldsymbol{K}_\alpha$.

(2) If $\boldsymbol{X}$ is a $\mathcal{C}$-diagram in $\mathcal{M}$ and $\boldsymbol{K}$ is a $\mathcal{C}$-diagram of simplicial sets, then $\hom^{\mathcal{C}}(\boldsymbol{K}, \boldsymbol{X})$ is defined to be the object of $\mathcal{M}$ that is the equalizer of the maps

$$\prod_{\alpha \in \mathrm{Ob}(\mathcal{C})} (\boldsymbol{X}_\alpha)^{\boldsymbol{K}_\alpha} \overset{\phi}{\underset{\psi}{\rightrightarrows}} \prod_{(\sigma\colon \alpha \to \alpha') \in \mathcal{C}} (\boldsymbol{X}_{\alpha'})^{\boldsymbol{K}_\alpha} \tag{18.3.4}$$

where the projection of the map ϕ on the factor $\sigma\colon \alpha \to \alpha'$ is the composition of a natural projection from the product with the map

$$\sigma_*^{1_{\boldsymbol{K}_\alpha}} \colon (\boldsymbol{X}_\alpha)^{\boldsymbol{K}_\alpha} \longrightarrow (\boldsymbol{X}_{\alpha'})^{\boldsymbol{K}_\alpha}$$

(where $\sigma_*\colon \boldsymbol{X}_\alpha \to \boldsymbol{X}_{\alpha'}$) and the projection of the map ψ on the factor $\sigma\colon \alpha \to \alpha'$ is the composition of a natural projection from the product with the map

$$(1_{\boldsymbol{X}_{\alpha'}})^{\boldsymbol{K}_{\sigma_*}} \colon (\boldsymbol{X}_{\alpha'})^{\boldsymbol{K}_{\alpha'}} \longrightarrow (\boldsymbol{X}_{\alpha'})^{\boldsymbol{K}_\alpha}$$

(where $\sigma_*\colon (\mathcal{C} \downarrow \alpha) \to (\mathcal{C} \downarrow \alpha')$; see Definition 14.7.8).

The construction of the object $\hom^{\mathcal{C}}(\boldsymbol{K}, \boldsymbol{X})$ of $\mathcal{M}$ from the functor $\boldsymbol{X}^{\boldsymbol{K}}\colon \mathcal{C} \times \mathcal{C}^{\mathrm{op}} \to \mathcal{M}$ is an example of the general construction known as an *end* (see [**47**, pages 218–223] or [**7**, page 329]). In the notation of [**47**], $\hom^{\mathcal{C}}(\boldsymbol{K}, \boldsymbol{X}) = \int_\alpha (\boldsymbol{X}_\alpha)^{\boldsymbol{K}_\alpha}$.

REMARK 18.3.5. The tensor product of functors (see Definition 18.3.2) is a special case of a coend of a functor $\mathrm{H}\colon \mathcal{C} \times \mathcal{C}^{\mathrm{op}} \to \mathrm{SS}$, where $\mathrm{H}(K, L) = K \times L$ (see Definition 18.3.2). We use the name "tensor product" because of the similarity to the case in which a ring R is viewed as an additive category (with one object, and with morphisms equal to the elements of R). In that case, a left R-module is just an additive functor $\mathrm{G}\colon R \to \mathcal{A}$ from R to the category of abelian groups, and a right R-module is an additive functor $\mathrm{F}\colon R^{\mathrm{op}} \to \mathcal{A}$. If $\mathrm{H}\colon R^{\mathrm{op}} \times R \to \mathcal{A}$ is defined by $\mathrm{H}(\alpha, \alpha) = \mathrm{F}(\alpha) \otimes \mathrm{G}(\alpha)$, then $\mathrm{F} \otimes_{R^{\mathrm{op}}} \mathrm{G}$ is the usual tensor product of a right R-module F with a left R-module G.

EXAMPLE 18.3.6. Let $\mathcal{M}$ be a simplicial model category and let $\mathcal{C}$ be a small category.

(1) If $\boldsymbol{X}$ is a $\mathcal{C}$-diagram in $\mathcal{M}$, then $\boldsymbol{X} \otimes_{\mathcal{C}} \mathrm{B}(- \downarrow \mathcal{C})^{\mathrm{op}}$ (see Definition 14.7.2) is the homotopy colimit of $\boldsymbol{X}$ (see Definition 18.1.2).
(2) If $\boldsymbol{X}$ is a $\mathcal{C}$-diagram in $\mathcal{M}$, then $\hom^{\mathcal{C}}(\mathrm{B}(\mathcal{C} \downarrow -), \boldsymbol{X})$ (see Definition 14.7.8) is the homotopy limit of $\boldsymbol{X}$ (see Definition 18.1.8).

PROPOSITION 18.3.7. *Let $\mathcal{M}$ be a simplicial model category and let $\mathcal{C}$ be a small category.*

(1) *If $\boldsymbol{X}$ is a $\mathcal{C}$-diagram in $\mathcal{M}$ and $\boldsymbol{P}\colon \mathcal{C}^{\mathrm{op}} \to \mathrm{SS}$ is a single point for every object α of $\mathcal{C}$, then $\boldsymbol{X} \otimes_{\mathcal{C}} \boldsymbol{P}$ is naturally isomorphic to $\operatorname{colim} \boldsymbol{X}$.*
(2) *If $\boldsymbol{X}$ is a $\mathcal{C}$-diagram in $\mathcal{M}$ and $\boldsymbol{P}\colon \mathcal{C} \to \mathrm{SS}$ is a single point for every object α of $\mathcal{C}$, then $\hom^{\mathcal{C}}(\boldsymbol{P}, \boldsymbol{X})$ is naturally isomorphic to $\lim \boldsymbol{X}$.*

PROOF. For part 1, $\boldsymbol{P}_\alpha$ is naturally isomorphic to $\Delta[0]$ for every object α of $\mathcal{C}^{\mathrm{op}}$, and so we have natural isomorphisms

$$\boldsymbol{X}_\alpha \otimes \boldsymbol{P}_\alpha \approx \boldsymbol{X}_\alpha \otimes \Delta[0] \approx \boldsymbol{X}_\alpha \ .$$

Under these isomorphisms, the map ϕ of Definition 18.3.2 is defined by $\sigma_*\colon \boldsymbol{X}_\alpha \to \boldsymbol{X}_{\alpha'}$ and the map ψ is the identity.

For part 2, $\boldsymbol{P}_\alpha$ is naturally isomorphic to $\Delta[0]$ for every object α of $\mathcal{C}$, and so we have natural isomorphisms

$$\boldsymbol{X}_\alpha^{\boldsymbol{P}_\alpha} \approx \boldsymbol{X}_\alpha^{\Delta[0]} \approx \boldsymbol{X}_\alpha \ .$$

Under these isomorphisms, the map ϕ of Definition 18.3.2 is defined by $\sigma_*\colon \boldsymbol{X}_\alpha \to \boldsymbol{X}_{\alpha'}$ and the map ψ is the identity. □

EXAMPLE 18.3.8. Let $\mathcal{M}$ be a simplicial model category and let $\mathcal{C}$ be a small category.

(1) If $\boldsymbol{P}$ is the $\mathcal{C}^{\mathrm{op}}$-diagram of simplicial sets that is a single point for every object α of $\mathcal{C}^{\mathrm{op}}$, then the unique map of $\mathcal{C}^{\mathrm{op}}$-diagrams $\mathrm{B}(- \downarrow \mathcal{C})^{\mathrm{op}} \to \boldsymbol{P}$ induces a natural map
$$\operatorname{hocolim} \boldsymbol{X} = \boldsymbol{X} \otimes_{\mathcal{C}} \mathrm{B}(- \downarrow \mathcal{C})^{\mathrm{op}} \longrightarrow \boldsymbol{X} \otimes_{\mathcal{C}} \boldsymbol{P} = \operatorname{colim} \boldsymbol{X}$$
for all $\mathcal{C}$-diagrams $\boldsymbol{X}$ in $\mathcal{M}$ (see Example 18.3.6 and Proposition 18.3.7).
(2) If $\boldsymbol{P}$ is the $\mathcal{C}$-diagram of simplicial sets that is a single point for every object α of $\mathcal{C}$, then the unique map of $\mathcal{C}$-diagrams $\mathrm{B}(\mathcal{C} \downarrow -) \to \boldsymbol{P}$ induces a natural map
$$\lim \boldsymbol{X} = \hom^{\mathcal{C}}(\boldsymbol{P}, \boldsymbol{X}) \longrightarrow \hom^{\mathcal{C}}\big(\mathrm{B}(\mathcal{C} \downarrow -), \boldsymbol{X}\big) = \operatorname{holim} \boldsymbol{X}$$
for all $\mathcal{C}$-diagrams $\boldsymbol{X}$ in $\mathcal{M}$.

18.3.9. Adjointness.

PROPOSITION 18.3.10. *Let $\mathcal{M}$ be a simplicial model category and let $\mathcal{C}$ be a small category.*

1. *If $\boldsymbol{X}$ is a $\mathcal{C}$-diagram in $\mathcal{M}$, $\boldsymbol{K}$ is a $\mathcal{C}^{\mathrm{op}}$-diagram of simplicial sets, and Z is an object of $\mathcal{M}$, then there is a natural isomorphism of sets*
$$\mathcal{M}(\boldsymbol{X} \otimes_{\mathcal{C}} \boldsymbol{K}, Z) \approx \mathrm{SS}^{\mathcal{C}^{\mathrm{op}}}\big(\boldsymbol{K}, \mathcal{M}(\boldsymbol{X}, Z)\big)$$
(where $\boldsymbol{X} \otimes_{\mathcal{C}} \boldsymbol{K}$ is as in Definition 18.3.2).
2. *If $\boldsymbol{X}$ is a $\mathcal{C}$-diagram in $\mathcal{M}$, $\boldsymbol{K}$ is a $\mathcal{C}$-diagram of simplicial sets, and W is an object of $\mathcal{M}$, then there is a natural isomorphism of sets*
$$\mathcal{M}\big(W, \hom^{\mathcal{C}}(\boldsymbol{K}, \boldsymbol{X})\big) \approx \mathrm{SS}^{\mathcal{C}}\big(\boldsymbol{K}, \mathcal{M}(W, \boldsymbol{X})\big)$$
(where $\hom^{\mathcal{C}}(\boldsymbol{K}, \boldsymbol{X})$ is as in Definition 18.3.2).

PROOF. We will prove part 1; the proof of part 2 is similar.

The object $\boldsymbol{X} \otimes_{\mathcal{C}} \boldsymbol{K}$ is defined as the colimit of Diagram 18.3.3, and so $\mathcal{M}(\boldsymbol{X} \otimes_{\mathcal{C}} \boldsymbol{K}, Z)$ is naturally isomorphic to the limit of the diagram

$$\prod_{\alpha \in \mathrm{Ob}(\mathcal{C})} \mathcal{M}(\boldsymbol{X}_\alpha \otimes \boldsymbol{K}_\alpha, Z) \overset{\phi^*}{\underset{\psi^*}{\rightrightarrows}} \prod_{(\sigma\colon \alpha \to \alpha') \in \mathcal{C}} \mathcal{M}(\boldsymbol{X}_\alpha \otimes \boldsymbol{K}_{\alpha'}, Z) .$$

Axiom M6 (see Definition 9.1.6) implies that this limit is naturally isomorphic to the limit of the diagram

$$\prod_{\alpha \in \mathrm{Ob}(\mathcal{C})} \mathrm{SS}\big(\boldsymbol{K}_\alpha, \mathcal{M}(\boldsymbol{X}_\alpha, Z)\big) \overset{\phi^*}{\underset{\psi^*}{\rightrightarrows}} \prod_{(\sigma\colon \alpha \to \alpha') \in \mathcal{C}} \mathrm{SS}\big(\boldsymbol{K}_{\alpha'}, \mathcal{M}(\boldsymbol{X}_\alpha, Z)\big) ,$$

which is the definition of $\mathrm{SS}^{\mathcal{C}^{\mathrm{op}}}\big(\boldsymbol{K}, \mathcal{M}(\boldsymbol{X}, Z)\big)$. □

LEMMA 18.3.11. *Let $\mathcal{M}$ be a simplicial model category and let $\mathcal{C}$ be a small category.*

1. *If $\boldsymbol{A} \to \boldsymbol{B}$ is a map of $\mathcal{C}$-diagrams in $\mathcal{M}$, $\boldsymbol{K} \to \boldsymbol{L}$ is a map of $\mathcal{C}^{\mathrm{op}}$-diagrams of simplicial sets, and $X \to Y$ is a map of objects in $\mathcal{M}$, then the dotted arrow exists in every solid arrow diagram of the form*
$$\begin{array}{ccc} \boldsymbol{A} \otimes_{\mathcal{C}} \boldsymbol{L} \amalg_{\boldsymbol{A} \otimes_{\mathcal{C}} \boldsymbol{K}} \boldsymbol{B} \otimes_{\mathcal{C}} \boldsymbol{K} & \longrightarrow & X \\ \big\downarrow & \nearrow & \big\downarrow \\ \boldsymbol{B} \otimes_{\mathcal{C}} \boldsymbol{L} & \longrightarrow & Y \end{array}$$
if and only if the dotted arrow exists in every solid arrow diagram of the form
$$\begin{array}{ccc} \boldsymbol{K} & \longrightarrow & \mathrm{Map}(\boldsymbol{B}, X) \\ \big\downarrow & \nearrow & \big\downarrow \\ \boldsymbol{L} & \longrightarrow & \mathrm{Map}(\boldsymbol{A}, X) \times_{\mathrm{Map}(\boldsymbol{A}, Y)} \mathrm{Map}(\boldsymbol{B}, Y) . \end{array}$$

(2) *If $\boldsymbol{X} \to \boldsymbol{Y}$ is a map of $\mathcal{C}$-diagrams in $\mathcal{M}$, $\boldsymbol{K} \to \boldsymbol{L}$ is a map of $\mathcal{C}$-diagrams of simplicial sets, and $A \to B$ is a map of objects in $\mathcal{M}$, then the dotted arrow exists in every solid arrow diagram of the form*

$$\begin{array}{ccc} A & \longrightarrow & \hom^{\mathcal{C}}(\boldsymbol{L}, \boldsymbol{X}) \\ \downarrow & \nearrow & \downarrow \\ B & \longrightarrow & \hom^{\mathcal{C}}(\boldsymbol{K}, \boldsymbol{X}) \times_{\hom^{\mathcal{C}}(\boldsymbol{K}, \boldsymbol{Y})} \hom^{\mathcal{C}}(\boldsymbol{L}, \boldsymbol{Y}) \end{array}$$

if and only if the dotted arrow exists in every solid arrow diagram of the form

$$\begin{array}{ccc} \boldsymbol{K} & \longrightarrow & \operatorname{Map}(B, \boldsymbol{X}) \\ \downarrow & \nearrow & \downarrow \\ \boldsymbol{L} & \longrightarrow & \operatorname{Map}(A, \boldsymbol{X}) \times_{\operatorname{Map}(A, \boldsymbol{Y})} \operatorname{Map}(B, \boldsymbol{Y}) \end{array} .$$

PROOF. This follows from Proposition 18.3.10. □

18.4. Consequences of adjointness

In this section we combine the adjointness relations of Section 18.3.9 with the homotopy lifting extension theorem (see Remark 9.1.7) to obtain the technical results that imply the homotopy invariance results of Section 18.5.

THEOREM 18.4.1. *Let $\mathcal{M}$ be a simplicial model category and let $\mathcal{C}$ be a small category.*

(1) *If $j\colon \boldsymbol{A} \to \boldsymbol{B}$ is an objectwise cofibration of $\mathcal{C}$-diagrams in $\mathcal{M}$ and $i\colon \boldsymbol{K} \to \boldsymbol{L}$ is a cofibration of $\mathcal{C}^{\mathrm{op}}$-diagrams of simplicial sets (see Theorem 11.6.1), then the pushout corner map*

$$\boldsymbol{A} \otimes_{\mathcal{C}} \boldsymbol{L} \amalg_{\boldsymbol{A} \otimes_{\mathcal{C}} \boldsymbol{K}} \boldsymbol{B} \otimes_{\mathcal{C}} \boldsymbol{K} \longrightarrow \boldsymbol{B} \otimes_{\mathcal{C}} \boldsymbol{L}$$

is a cofibration in $\mathcal{M}$ that is a weak equivalence if either j is an objectwise weak equivalence or i is a weak equivalence.

(2) *If $p\colon \boldsymbol{X} \to \boldsymbol{Y}$ is an objectwise fibration of $\mathcal{C}$-diagrams in $\mathcal{M}$ and $i\colon \boldsymbol{K} \to \boldsymbol{L}$ is a cofibration of $\mathcal{C}$-diagrams of simplicial sets (see Theorem 11.6.1), then the pullback corner map*

$$\hom^{\mathcal{C}}(\boldsymbol{L}, \boldsymbol{X}) \longrightarrow \hom^{\mathcal{C}}(\boldsymbol{K}, \boldsymbol{X}) \times_{\hom^{\mathcal{C}}(\boldsymbol{K}, \boldsymbol{Y})} \hom^{\mathcal{C}}(\boldsymbol{L}, \boldsymbol{Y})$$

is a fibration in $\mathcal{M}$ that is a weak equivalence if either p is an objectwise weak equivalence or i is a weak equivalence.

PROOF. We will prove part 1; the proof of part 2 is similar.

If $p\colon X \to Y$ is a fibration in $\mathcal{M}$, then axiom M7 (see Definition 9.1.6) implies that the map of $\mathcal{C}^{\mathrm{op}}$-diagrams of simplicial sets $\operatorname{Map}(\boldsymbol{B}, X) \to \operatorname{Map}(\boldsymbol{A}, X) \times_{\operatorname{Map}(\boldsymbol{A}, Y)} \operatorname{Map}(\boldsymbol{B}, Y)$ is an objectwise fibration that is an objectwise weak equivalence if either j is an objectwise weak equivalence or p is a weak equivalence. The result now follows from Lemma 18.3.11, Proposition 7.2.3, and Theorem 11.6.1. □

COROLLARY 18.4.2. *Let $\mathcal{M}$ be a simplicial model category and let $\mathcal{C}$ be a small category.*

(1) *If $\boldsymbol{K}$ is a cofibrant $\mathcal{C}^{\mathrm{op}}$-diagram of simplicial sets (see Theorem 11.6.1) and $j\colon \boldsymbol{A} \to \boldsymbol{B}$ is an objectwise cofibration of $\mathcal{C}$-diagrams in $\mathcal{M}$, then the map $\boldsymbol{A} \otimes_{\mathcal{C}} \boldsymbol{K} \to \boldsymbol{B} \otimes_{\mathcal{C}} \boldsymbol{K}$ is a cofibration in $\mathcal{M}$ that is a weak equivalence if j is an objectwise weak equivalence.*
(2) *If $\boldsymbol{K}$ is a cofibrant $\mathcal{C}$-diagram of simplicial sets (see Theorem 11.6.1) and $p\colon \boldsymbol{X} \to \boldsymbol{Y}$ is an objectwise fibration of $\mathcal{C}$-diagrams in $\mathcal{M}$, then the map $\hom^{\mathcal{C}}(\boldsymbol{K}, \boldsymbol{X}) \to \hom^{\mathcal{C}}(\boldsymbol{K}, \boldsymbol{Y})$ is a fibration in $\mathcal{M}$ that is a weak equivalence if p is an objectwise weak equivalence.*

PROOF. This follows from Theorem 18.4.1. □

COROLLARY 18.4.3. *Let $\mathcal{M}$ be a simplicial model category and let $\mathcal{C}$ be a small category.*

(1) *If $\boldsymbol{K}$ is a cofibrant $\mathcal{C}^{\mathrm{op}}$-diagram of simplicial sets (see Theorem 11.6.1) and $\boldsymbol{X}$ is an objectwise cofibrant $\mathcal{C}$-diagram in $\mathcal{M}$, then $\boldsymbol{X} \otimes_{\mathcal{C}} \boldsymbol{K}$ is a cofibrant object of $\mathcal{M}$.*
(2) *If $\boldsymbol{K}$ is a cofibrant $\mathcal{C}$-diagram of simplicial sets (see Theorem 11.6.1) and $\boldsymbol{X}$ is an objectwise fibrant $\mathcal{C}$-diagram in $\mathcal{M}$, then $\hom^{\mathcal{C}}(\boldsymbol{K}, \boldsymbol{X})$ is a fibrant object of $\mathcal{M}$.*

PROOF. This follows from Theorem 18.4.1. □

COROLLARY 18.4.4. *Let $\mathcal{M}$ be a simplicial model category and let $\mathcal{C}$ be a small category.*

(1) *If $\boldsymbol{K}$ is a cofibrant $\mathcal{C}^{\mathrm{op}}$-diagram of simplicial sets (see Theorem 11.6.1) and $f\colon \boldsymbol{X} \to \boldsymbol{Y}$ is an objectwise weak equivalence of objectwise cofibrant $\mathcal{C}$-diagrams in $\mathcal{M}$, then the induced map $f_*\colon \boldsymbol{X} \otimes_{\mathcal{C}} \boldsymbol{K} \to \boldsymbol{Y} \otimes_{\mathcal{C}} \boldsymbol{K}$ is a weak equivalence.*
(2) *If $\boldsymbol{K}$ is a cofibrant $\mathcal{C}$-diagram of simplicial sets (see Theorem 11.6.1) and $f\colon \boldsymbol{X} \to \boldsymbol{Y}$ is an objectwise weak equivalence of objectwise fibrant $\mathcal{C}$-diagrams in $\mathcal{M}$, then the induced map $f_*\colon \hom^{\mathcal{C}}(\boldsymbol{K}, \boldsymbol{X}) \to \hom^{\mathcal{C}}(\boldsymbol{K}, \boldsymbol{Y})$ is a weak equivalence.*

PROOF. This follows from Corollary 18.4.2 and Lemma 7.7.1. □

COROLLARY 18.4.5. *Let $\mathcal{M}$ be a simplicial model category and let $\mathcal{C}$ be a small category.*

(1) *If $\boldsymbol{X}$ is an objectwise cofibrant $\mathcal{C}$-diagram in $\mathcal{M}$ and $f\colon \boldsymbol{K} \to \boldsymbol{K}'$ is a weak equivalence of cofibrant $\mathcal{C}^{\mathrm{op}}$-diagrams of simplicial sets (see Theorem 11.6.1), then the induced map $f_*\colon \boldsymbol{X} \otimes_{\mathcal{C}} \boldsymbol{K} \to \boldsymbol{X} \otimes_{\mathcal{C}} \boldsymbol{K}'$ is a weak equivalence of cofibrant objects in $\mathcal{M}$.*
(2) *If $\boldsymbol{X}$ is an objectwise fibrant $\mathcal{C}$-diagram in $\mathcal{M}$ and $f\colon \boldsymbol{K} \to \boldsymbol{K}'$ is a weak equivalence of cofibrant $\mathcal{C}$-diagrams of simplicial sets (see Theorem 11.6.1), then the induced map $f^*\colon \hom^{\mathcal{C}}(\boldsymbol{K}', \boldsymbol{X}) \to \hom^{\mathcal{C}}(\boldsymbol{K}, \boldsymbol{X})$ is a weak equivalence of fibrant objects in $\mathcal{M}$.*

PROOF. This follows from Theorem 18.4.1 and Corollary 7.7.2. □

COROLLARY 18.4.6. *Let $\mathcal{M}$ be a simplicial model category and let $\mathcal{C}$ be a small category.*

(1) *If $\boldsymbol{K}$ is a $\mathcal{C}^{\mathrm{op}}$-diagram of simplicial sets and both $\widetilde{\boldsymbol{K}} \to \boldsymbol{K}$ and $\widetilde{\boldsymbol{K}}' \to \boldsymbol{K}$ are cofibrant approximations to $\boldsymbol{K}$ (see Theorem 11.6.1), then for every objectwise cofibrant $\mathcal{C}$-diagram $\boldsymbol{X}$ in $\mathcal{M}$ there is an essentially unique (see Definition 14.4.2) natural zig-zag of weak equivalences (see Definition 14.4.1) in $\mathcal{M}$ from $\boldsymbol{X} \otimes_{\mathcal{C}} \widetilde{\boldsymbol{K}}$ to $\boldsymbol{X} \otimes_{\mathcal{C}} \widetilde{\boldsymbol{K}}'$.*

(2) *If $\boldsymbol{K}$ is a $\mathcal{C}$-diagram of simplicial sets and both $\widetilde{\boldsymbol{K}} \to \boldsymbol{K}$ and $\widetilde{\boldsymbol{K}}' \to \boldsymbol{K}$ are cofibrant approximations to $\boldsymbol{K}$ (see Theorem 11.6.1), then for every objectwise fibrant $\mathcal{C}$-diagram $\boldsymbol{X}$ in $\mathcal{M}$ there is an essentially unique (see Definition 14.4.2) natural zig-zag of weak equivalences (see Definition 14.4.1) in $\mathcal{M}$ from $\hom^{\mathcal{C}}(\widetilde{\boldsymbol{K}}, \boldsymbol{X})$ to $\hom^{\mathcal{C}}(\widetilde{\boldsymbol{K}}', \boldsymbol{X})$.*

PROOF. This follows from Corollary 18.4.5 and Proposition 14.6.3. $\square$

18.4.7. Reedy diagrams.

LEMMA 18.4.8. *Let $\mathcal{C}$ be a Reedy category and let $\mathcal{M}$ be a simplicial model category.*

(1) *If $\boldsymbol{B}$ is a $\mathcal{C}$-diagram in $\mathcal{M}$ and X is an object of $\mathcal{M}$, then $\operatorname{Map}(\boldsymbol{B}, X)$ is a $\mathcal{C}^{\mathrm{op}}$-diagram of simplicial sets and for every object α of $\mathcal{C}$ there is a natural isomorphism $\mathrm{M}_\alpha \operatorname{Map}(\boldsymbol{B}, X) \approx \operatorname{Map}(\mathrm{L}_\alpha \boldsymbol{B}, X)$.*

(2) *If B is an object of $\mathcal{M}$ and $\boldsymbol{X}$ is a $\mathcal{C}$-diagram in $\mathcal{M}$, then $\operatorname{Map}(B, \boldsymbol{X})$ is a $\mathcal{C}$-diagram of simplicial sets and for every object α of $\mathcal{C}$ there is a natural isomorphism $\mathrm{M}_\alpha \operatorname{Map}(B, \boldsymbol{X}) \approx \operatorname{Map}(B, \mathrm{M}_\alpha \boldsymbol{X})$.*

PROOF. We will prove part 1; the proof of part 2 is similar.

We have natural isomorphisms

$$
\begin{aligned}
\mathrm{M}_\alpha \operatorname{Map}(\boldsymbol{B}, X) &= \lim_{\partial(\alpha\downarrow\overleftarrow{\mathcal{C}})} \operatorname{Map}(\boldsymbol{B}, X) && \text{(see Definition 15.2.5)}\\
&\approx \operatorname{Map}\Big(\operatorname*{colim}_{(\partial(\alpha\downarrow\overleftarrow{\mathcal{C}}))^{\mathrm{op}}} \boldsymbol{B}, X\Big) && \text{(see Proposition 9.2.2)}\\
&\approx \operatorname{Map}\Big(\operatorname*{colim}_{\partial(\overrightarrow{\mathcal{C}}\downarrow\alpha)} \boldsymbol{B}, X\Big) && \text{(see Proposition 15.2.4)}\\
&= \operatorname{Map}(\mathrm{L}_\alpha \boldsymbol{B}, X)\ .
\end{aligned}
$$

$\square$

THEOREM 18.4.9 (The Reedy diagram homotopy lifting extension theorem). *Let $\mathcal{C}$ be a Reedy category and let $\mathcal{M}$ be a simplicial model category.*

(1) *If $i\colon \boldsymbol{A} \to \boldsymbol{B}$ is a Reedy cofibration of $\mathcal{C}$-diagrams in $\mathcal{M}$ and $p\colon X \to Y$ is a fibration in $\mathcal{M}$, then the map of $\mathcal{C}^{\mathrm{op}}$-diagrams of simplicial sets*
$$\operatorname{Map}(\boldsymbol{B}, X) \longrightarrow \operatorname{Map}(\boldsymbol{A}, X) \times_{\operatorname{Map}(\boldsymbol{A}, Y)} \operatorname{Map}(\boldsymbol{B}, Y)$$
is a Reedy fibration that is a Reedy weak equivalence if either of i or p is a weak equivalence.

(2) *If $i\colon A \to B$ is a cofibration in $\mathcal{M}$ and $p\colon \boldsymbol{X} \to \boldsymbol{Y}$ is a Reedy fibration of $\mathcal{C}$-diagrams in $\mathcal{M}$, then the map of $\mathcal{C}$-diagrams of simplicial sets*
$$\operatorname{Map}(B, \boldsymbol{X}) \longrightarrow \operatorname{Map}(A, \boldsymbol{X}) \times_{\operatorname{Map}(A, \boldsymbol{Y})} \operatorname{Map}(B, \boldsymbol{Y})$$
is a Reedy fibration that is a weak equivalence if either i or p is a weak equivalence.

PROOF. We will prove part 1; the proof of part 2 is similar.

We must show that for every object α of $\mathcal{C}$ the map of simplicial sets

(18.4.10)
$$\operatorname{Map}(\boldsymbol{B}_\alpha, X) \longrightarrow \text{Pullback of}\left(\begin{array}{c} \operatorname{Map}(\boldsymbol{A}_\alpha, X) \times_{\operatorname{Map}(\boldsymbol{A}_\alpha, Y)} \operatorname{Map}(\boldsymbol{B}_\alpha, Y) \\ \downarrow \\ \mathrm{M}_\alpha\big(\operatorname{Map}(\boldsymbol{A}, X) \times_{\operatorname{Map}(\boldsymbol{A}, Y)} \operatorname{Map}(\boldsymbol{B}, Y)\big) \\ \uparrow \\ \mathrm{M}_\alpha \operatorname{Map}(\boldsymbol{B}, X) \end{array}\right)$$

is a fibration that is a weak equivalence if either i or p is a weak equivalence. Lemma 18.4.8 implies that this map is isomorphic to the map

$$\operatorname{Map}(\boldsymbol{B}_\alpha, X) \longrightarrow \text{Pullback of}\left(\begin{array}{c} \operatorname{Map}(\boldsymbol{A}_\alpha, X) \times_{\operatorname{Map}(\boldsymbol{A}_\alpha, Y)} \operatorname{Map}(\boldsymbol{B}_\alpha, Y) \\ \downarrow \\ \operatorname{Map}(\mathrm{L}_\alpha \boldsymbol{A}, X) \times_{\operatorname{Map}(\mathrm{L}_\alpha \boldsymbol{A}, Y)} \operatorname{Map}(\mathrm{L}_\alpha \boldsymbol{B}, Y) \\ \uparrow \\ \operatorname{Map}(\mathrm{L}_\alpha \boldsymbol{B}, X) \end{array}\right),$$

and the codomain of this map is the limit of the diagram

$$\begin{array}{ccccc} \operatorname{Map}(\boldsymbol{A}_\alpha, X) & \longrightarrow & \operatorname{Map}(\boldsymbol{A}_\alpha, Y) & \longleftarrow & \operatorname{Map}(\boldsymbol{B}_\alpha, Y) \\ \downarrow & & \downarrow & & \downarrow \\ \operatorname{Map}(\mathrm{L}_\alpha \boldsymbol{A}, X) & \longrightarrow & \operatorname{Map}(\mathrm{L}_\alpha \boldsymbol{A}, Y) & \longleftarrow & \operatorname{Map}(\mathrm{L}_\alpha \boldsymbol{B}, Y) \\ & \nwarrow & \uparrow & \nearrow & \\ & & \operatorname{Map}(\mathrm{L}_\alpha \boldsymbol{B}, X) & & \end{array} .$$

Thus, the map (18.4.10) is isomorphic to the map

$$\operatorname{Map}(\boldsymbol{B}_\alpha, X) \longrightarrow \text{Pullback of}\left(\begin{array}{c} \operatorname{Map}(\boldsymbol{A}_\alpha, X) \times_{\operatorname{Map}(\mathrm{L}_\alpha \boldsymbol{A}, X)} \operatorname{Map}(\mathrm{L}_\alpha \boldsymbol{B}, X) \\ \downarrow \\ \operatorname{Map}(\boldsymbol{A}_\alpha, Y) \times_{\operatorname{Map}(\mathrm{L}_\alpha \boldsymbol{A}, Y)} \operatorname{Map}(\mathrm{L}_\alpha \boldsymbol{B}, Y) \\ \uparrow \\ \operatorname{Map}(\boldsymbol{B}_\alpha, Y) \end{array}\right).$$

Since $\boldsymbol{A}_\alpha \amalg_{\mathrm{L}_\alpha \boldsymbol{A}} \mathrm{L}_\alpha \boldsymbol{B} \to \boldsymbol{B}_\alpha$ is a cofibration that is a weak equivalence if i is a weak equivalence (see Theorem 15.3.15), the result now follows from axiom M7 (see Definition 9.1.6). □

THEOREM 18.4.11. *Let $\mathcal{C}$ be a Reedy category and let $\mathcal{M}$ be a simplicial model category.*

(1) *If* $j\colon \boldsymbol{A} \to \boldsymbol{B}$ *is a Reedy cofibration of* $\mathcal{C}$*-diagrams in* $\mathcal{M}$ *and* $i\colon \boldsymbol{K} \to \boldsymbol{L}$ *is a Reedy cofibration of* $\mathcal{C}^{\mathrm{op}}$*-diagrams of simplicial sets, then the pushout corner map* $\boldsymbol{A} \otimes_{\mathcal{C}} \boldsymbol{L} \amalg_{\boldsymbol{A} \otimes_{\mathcal{C}} \boldsymbol{K}} \boldsymbol{B} \otimes_{\mathcal{C}} \boldsymbol{K} \to \boldsymbol{B} \otimes_{\mathcal{C}} \boldsymbol{L}$ *is a cofibration in* $\mathcal{M}$ *that is a weak equivalence if either* i *or* j *is a Reedy weak equivalence.*

(2) *If* $p\colon \boldsymbol{X} \to \boldsymbol{Y}$ *is a Reedy fibration of* $\mathcal{C}$*-diagrams in* $\mathcal{M}$ *and* $i\colon \boldsymbol{K} \to \boldsymbol{L}$ *is a Reedy cofibration of* $\mathcal{C}$*-diagrams of simplicial sets, then the pullback corner map* $\hom^{\mathcal{C}}(\boldsymbol{L}, \boldsymbol{X}) \to \hom^{\mathcal{C}}(\boldsymbol{K}, \boldsymbol{X}) \times_{\hom^{\mathcal{C}}(\boldsymbol{K}, \boldsymbol{Y})} \hom^{\mathcal{C}}(\boldsymbol{L}, \boldsymbol{Y})$ *is a fibration in* $\mathcal{M}$ *that is a weak equivalence if either* i *or* p *is a Reedy weak equivalence.*

PROOF. This is similar to the proof of Theorem 18.4.1, using Theorem 18.4.9 in place of axiom M7. □

COROLLARY 18.4.12. *Let* $\mathcal{C}$ *be a Reedy category and let* $\mathcal{M}$ *be a simplicial model category.*

(1) *If* $\boldsymbol{K}$ *is a Reedy cofibrant* $\mathcal{C}^{\mathrm{op}}$*-diagram of simplicial sets and* $\boldsymbol{X}$ *is a Reedy cofibrant diagram in* $\mathcal{M}$*, then* $\boldsymbol{X} \otimes_{\mathcal{C}} \boldsymbol{K}$ *is a cofibrant object in* $\mathcal{M}$*.*

(2) *If* $\boldsymbol{K}$ *is a Reedy cofibrant* $\mathcal{C}$*-diagram of simplicial sets and* $\boldsymbol{X}$ *is a Reedy fibrant* $\mathcal{C}$*-diagram in* $\mathcal{M}$*, then* $\hom^{\mathcal{C}}(\boldsymbol{K}, \boldsymbol{X})$ *is a fibrant object in* $\mathcal{M}$*.*

PROOF. This follows from Theorem 18.4.11. □

COROLLARY 18.4.13. *Let* $\mathcal{C}$ *be a Reedy category and let* $\mathcal{M}$ *be a simplicial model category.*

(1) *If* $\boldsymbol{K}$ *is a Reedy cofibrant* $\mathcal{C}^{\mathrm{op}}$*-diagram of simplicial sets and* $f\colon \boldsymbol{X} \to \boldsymbol{Y}$ *is a weak equivalence of Reedy cofibrant* $\mathcal{C}$*-diagrams in* $\mathcal{M}$*, then the induced map* $f_*\colon \boldsymbol{X} \otimes_{\mathcal{C}} \boldsymbol{K} \to \boldsymbol{Y} \otimes_{\mathcal{C}} \boldsymbol{K}$ *is a weak equivalence of cofibrant objects in* $\mathcal{M}$*.*

(2) *If* $\boldsymbol{K}$ *is a Reedy cofibrant* $\mathcal{C}$*-diagram of simplicial sets and* $f\colon \boldsymbol{X} \to \boldsymbol{Y}$ *is a weak equivalence of Reedy fibrant* $\mathcal{C}$*-diagrams in* $\mathcal{M}$*, then the induced map* $f_*\colon \hom^{\mathcal{C}}(\boldsymbol{K}, \boldsymbol{X}) \to \hom^{\mathcal{C}}(\boldsymbol{K}, \boldsymbol{Y})$ *is a weak equivalence of fibrant objects in* $\mathcal{M}$*.*

PROOF. This follows from Corollary 18.4.12, Theorem 18.4.11, and Corollary 7.7.2. □

COROLLARY 18.4.14. *Let* $\mathcal{C}$ *be a Reedy category and let* $\mathcal{M}$ *be a simplicial model category.*

(1) *If* $\boldsymbol{X}$ *is a Reedy cofibrant* $\mathcal{C}$*-diagram in* $\mathcal{M}$ *and* $f\colon \boldsymbol{K} \to \boldsymbol{K}'$ *is a weak equivalence of Reedy cofibrant* $\mathcal{C}^{\mathrm{op}}$*-diagrams of simplicial sets, then the induced map* $f_*\colon \boldsymbol{X} \otimes_{\mathcal{C}} \boldsymbol{K} \to \boldsymbol{X} \otimes_{\mathcal{C}} \boldsymbol{K}'$ *is a weak equivalence of cofibrant objects in* $\mathcal{M}$*.*

(2) *If* $\boldsymbol{X}$ *is a Reedy fibrant* $\mathcal{C}$*-diagram in* $\mathcal{M}$ *and* $f\colon \boldsymbol{K} \to \boldsymbol{K}'$ *is a weak equivalence of Reedy cofibrant* $\mathcal{C}$*-diagrams of simplicial sets, then the induced map* $f^*\colon \hom^{\mathcal{C}}(\boldsymbol{K}', \boldsymbol{X}) \to \hom^{\mathcal{C}}(\boldsymbol{K}, \boldsymbol{X})$ *is a weak equivalence of fibrant objects in* $\mathcal{M}$*.*

PROOF. This follows from Corollary 18.4.12, Theorem 18.4.11, and Corollary 7.7.2. □

COROLLARY 18.4.15. *Let* $\mathcal{C}$ *be a Reedy category and let* $\mathcal{M}$ *be a simplicial model category.*

(1) *If $\boldsymbol{K}$ is a $\mathcal{C}^{\mathrm{op}}$-diagram of simplicial sets and both $\widetilde{\boldsymbol{K}} \to \boldsymbol{K}$ and $\widetilde{\boldsymbol{K}}' \to \boldsymbol{K}$ are Reedy cofibrant approximations to $\boldsymbol{K}$, then for every Reedy cofibrant $\mathcal{C}$-diagram $\boldsymbol{X}$ in $\mathcal{M}$ there is an essentially unique (see Definition 14.4.2) natural zig-zag of weak equivalences in $\mathcal{M}$ from $\boldsymbol{X} \otimes_{\mathcal{C}} \widetilde{\boldsymbol{K}}$ to $\boldsymbol{X} \otimes_{\mathcal{C}} \widetilde{\boldsymbol{K}}'$.*

(2) *If $\boldsymbol{K}$ is a $\mathcal{C}$-diagram of simplicial sets and both $\widetilde{\boldsymbol{K}} \to \boldsymbol{K}$ and $\widetilde{\boldsymbol{K}}' \to \boldsymbol{K}$ are cofibrant approximations to $\boldsymbol{K}$, then for every Reedy fibrant $\mathcal{C}$-diagram $\boldsymbol{X}$ in $\mathcal{M}$ there is an essentially unique (see Definition 14.4.2) natural zig-zag of weak equivalences from $\hom^{\mathcal{C}}(\widetilde{\boldsymbol{K}}, \boldsymbol{X})$ to $\hom^{\mathcal{C}}(\widetilde{\boldsymbol{K}}', \boldsymbol{X})$.*

PROOF. This follows from Corollary 18.4.14 and Proposition 14.6.3. □

THEOREM 18.4.16. *Let $\mathcal{C}$ be a Reedy category and let $\mathcal{M}$ be a simplicial model category.*

(1) *If $\boldsymbol{P}$ is a Reedy cofibrant $\mathcal{C}^{\mathrm{op}}$-diagram of simplicial sets such that $\boldsymbol{P}_\alpha$ is contractible for every object α of $\mathcal{C}$, then for every Reedy cofibrant $\mathcal{C}$-diagram $\boldsymbol{X}$ in $\mathcal{M}$ the object $\boldsymbol{X} \otimes_{\mathcal{C}} \boldsymbol{P}$ is naturally weakly equivalent to $\operatorname{hocolim} \boldsymbol{X}$.*

(2) *If $\boldsymbol{P}$ is a Reedy cofibrant $\mathcal{C}$-diagram of simplicial sets such that $\boldsymbol{P}_\alpha$ is contractible for every object α of $\mathcal{C}$, then for every Reedy fibrant $\mathcal{C}$-diagram $\boldsymbol{X}$ in $\mathcal{M}$ the object $\hom^{\mathcal{C}}(\boldsymbol{P}, \boldsymbol{X})$ is naturally weakly equivalent to $\operatorname{holim} \boldsymbol{X}$.*

PROOF. This follows from Corollary 18.4.15 and Corollary 15.6.7. □

18.5. Homotopy invariance

THEOREM 18.5.1. *Let $\mathcal{M}$ be a simplicial model category and let $\mathcal{C}$ be a small category.*

(1) *If $f\colon \boldsymbol{X} \to \boldsymbol{Y}$ is a map of $\mathcal{C}$-diagrams in $\mathcal{M}$ that is an objectwise cofibration, then the induced map of homotopy colimits $f_*\colon \operatorname{hocolim} \boldsymbol{X} \to \operatorname{hocolim} \boldsymbol{Y}$ is a cofibration that is a weak equivalence if f is an objectwise weak equivalence.*

(2) *If $f\colon \boldsymbol{X} \to \boldsymbol{Y}$ is a map of $\mathcal{C}$-diagrams in $\mathcal{M}$ that is an objectwise fibration, then the induced map of homotopy limits $f_*\colon \operatorname{holim} \boldsymbol{X} \to \operatorname{holim} \boldsymbol{Y}$ is a fibration that is a weak equivalence if f is an objectwise weak equivalence.*

PROOF. This follows from Corollary 18.4.2 and Corollary 14.8.8. □

THEOREM 18.5.2. *Let $\mathcal{M}$ be a simplicial model category and let $\mathcal{C}$ be a small category.*

(1) *If $\boldsymbol{X}$ is an objectwise cofibrant $\mathcal{C}$-diagram in $\mathcal{M}$, then $\operatorname{hocolim} \boldsymbol{X}$ is cofibrant.*

(2) *If $\boldsymbol{X}$ is an objectwise fibrant $\mathcal{C}$-diagram in $\mathcal{M}$, then $\operatorname{holim} \boldsymbol{X}$ is fibrant.*

PROOF. This follows from Corollary 18.4.3 and Corollary 14.8.8. □

THEOREM 18.5.3. *Let $\mathcal{M}$ be a simplicial model category and let $\mathcal{C}$ be a small category.*

(1) *If $f\colon \boldsymbol{X} \to \boldsymbol{Y}$ is a map of $\mathcal{C}$-diagrams in $\mathcal{M}$ that is an objectwise weak equivalence of cofibrant objects, then the induced map of homotopy colimits $f_*\colon \operatorname{hocolim} \boldsymbol{X} \to \operatorname{hocolim} \boldsymbol{Y}$ is a weak equivalence of cofibrant objects of $\mathcal{M}$.*

(2) *If* $f\colon \boldsymbol{X} \to \boldsymbol{Y}$ *is a map of* $\mathcal{C}$*-diagrams in* $\mathcal{M}$ *that is an objectwise weak equivalence of fibrant objects, then the induced map of homotopy limits* $f_*\colon \operatorname{holim} \boldsymbol{X} \to \operatorname{holim} \boldsymbol{Y}$ *is a weak equivalence of fibrant objects of* $\mathcal{M}$.

PROOF. This follows from Corollary 18.4.4 and Theorem 18.5.2. □

REMARK 18.5.4. D. Dugger and D. Isaksen [**28**] have proved that if $\mathcal{C}$ is a small category and $\boldsymbol{X}$ is a $\mathcal{C}$-diagram of topological spaces, then $\operatorname{hocolim} \boldsymbol{X}$ has the "correct" weak equivalence type even if $\boldsymbol{X}$ is not an objectwise cofibrant diagram. That is, they prove that if $\widetilde{\boldsymbol{X}}$ is an objectwise cofibrant diagram of topological spaces and $j\colon \widetilde{\boldsymbol{X}} \to \boldsymbol{X}$ is an objectwise weak equivalence, then the map $\operatorname{hocolim} j\colon \operatorname{hocolim} \widetilde{\boldsymbol{X}} \to \operatorname{hocolim} \boldsymbol{X}$ is a weak equivalence. Thus, their results imply that if $f\colon \boldsymbol{X} \to \boldsymbol{Y}$ is an objectwise weak equivalence of $\mathcal{C}$-diagrams of topological spaces, then $\operatorname{hocolim} f\colon \operatorname{hocolim} \boldsymbol{X} \to \operatorname{hocolim} \boldsymbol{Y}$ is a weak equivalence even if none of the spaces involved are cofibrant.

18.6. Simplicial objects and cosimplicial objects

18.6.1. Definitions.

DEFINITION 18.6.2. If $\mathcal{M}$ is a simplicial model category and $\boldsymbol{X}$ is a simplicial object in $\mathcal{M}$ (see Definition 15.1.10), then the *realization* $|\boldsymbol{X}|$ of $\boldsymbol{X}$ is the coequalizer of the maps

$$\coprod_{(\sigma\colon [n]\to[k])\in\boldsymbol{\Delta}} \boldsymbol{X}_n \otimes \Delta[k] \overset{\phi}{\underset{\psi}{\rightrightarrows}} \coprod_{[n]\in\mathrm{Ob}(\boldsymbol{\Delta})} \boldsymbol{X}_n \otimes \Delta[n]$$

where the map ϕ on the summand $\sigma\colon [n] \to [k]$ is the composition of the map

$$\sigma_* \otimes 1_{\Delta[k]}\colon \boldsymbol{X}_n \otimes \Delta[k] \longrightarrow \boldsymbol{X}_k \otimes \Delta[k]$$

(where $\sigma_*\colon \boldsymbol{X}_n \to \boldsymbol{X}_k$) with the natural injection into the coproduct and the map ψ on the summand $\sigma\colon [n] \to [k]$ is the composition of the map

$$1_{\boldsymbol{X}_n} \otimes \sigma^*\colon \boldsymbol{X}_n \otimes \Delta[k] \longrightarrow \boldsymbol{X}_n \otimes \Delta[n]$$

(where $\sigma^*\colon \Delta[k] \to \Delta[n]$) with the natural injection into the coproduct. In the notation of Definition 18.3.2, $|\boldsymbol{X}| = \boldsymbol{X} \otimes_{\boldsymbol{\Delta}^{\mathrm{op}}} \Delta$ (see Definition 15.1.15).

DEFINITION 18.6.3. If $\mathcal{M}$ is a simplicial model category and $\boldsymbol{X}$ is a cosimplicial object in $\mathcal{M}$ (see Definition 15.1.10), then the *total object* $\operatorname{Tot} \boldsymbol{X}$ of $\boldsymbol{X}$ is the equalizer of the maps

$$\prod_{[n]\in\mathrm{Ob}(\boldsymbol{\Delta})} (\boldsymbol{X}^n)^{\Delta[n]} \overset{\phi}{\underset{\psi}{\rightrightarrows}} \prod_{(\sigma\colon [n]\to[k])\in\boldsymbol{\Delta}} (\boldsymbol{X}^k)^{\Delta[n]}$$

where the projection of the map ϕ on the factor $\sigma\colon [n] \to [k]$ is the composition of the natural projection from the product with the map

$$\sigma_*^{(1_{\Delta[n]})}\colon (\boldsymbol{X}^n)^{\Delta[n]} \longrightarrow (\boldsymbol{X}^k)^{\Delta[n]}$$

and the projection of the map ψ on the factor $\sigma\colon [n] \to [k]$ is the composition of the natural projection from the product with the map

$$(1_{\boldsymbol{X}^k})^{\sigma_*}\colon (\boldsymbol{X}^k)^{\Delta[k]} \longrightarrow (\boldsymbol{X}^k)^{\Delta[n]}$$

(where $\sigma_*\colon \Delta[n] \to \Delta[k]$). In the notation of Definition 18.3.2, $\operatorname{Tot} \boldsymbol{X} = \hom^{\boldsymbol{\Delta}}(\Delta, \boldsymbol{X})$ (see Definition 15.1.15).

REMARK 18.6.4. If $\mathrm{Spc}_{(*)}$ is one of our categories of spaces (see Notation 7.10.5), then the space $(\boldsymbol{X}^n)^{\Delta[n]}$ is a space of maps:

$$(\boldsymbol{X}^n)^{\Delta[n]} = \begin{cases} \operatorname{map}(|\Delta[n]|, \boldsymbol{X}^n) & \text{if } \mathrm{Spc}_{(*)} = \mathrm{Top} \\ \operatorname{map}_*(|\Delta[n]|^+, \boldsymbol{X}^n) & \text{if } \mathrm{Spc}_{(*)} = \mathrm{Top}_* \\ \operatorname{Map}(\Delta[n], \boldsymbol{X}^n) & \text{if } \mathrm{Spc}_{(*)} = \mathrm{SS} \\ \operatorname{Map}_*(\Delta[n]^+, \boldsymbol{X}^n) & \text{if } \mathrm{Spc}_{(*)} = \mathrm{SS}_* \end{cases}$$

(see Example 9.1.13, Example 9.1.14, Example 9.1.15, and Example 9.1.16). Thus, in each case the total space is constructed by first taking the codegreewise mapping space from the cosimplicial space $|\Delta|$ (or $|\Delta|^+$, or Δ, or Δ^+) to the cosimplicial space $\boldsymbol{X}$, and then taking a subspace of the product of these mapping spaces. In the notation of Definition 18.2.3, the total space of a cosimplicial space is $\operatorname{Tot} \boldsymbol{X} = \boldsymbol{X}^{\boldsymbol{\Delta}}$.

18.6.5. Homotopy invariance.

THEOREM 18.6.6. *Let $\mathcal{M}$ be a simplicial model category.*

1. *If $g\colon \boldsymbol{X} \to \boldsymbol{Y}$ is a level-wise weak equivalence of Reedy cofibrant simplicial objects in $\mathcal{M}$, then the induced map of realizations $g_*\colon |\boldsymbol{X}| \to |\boldsymbol{Y}|$ is a weak equivalence of cofibrant objects in $\mathcal{M}$.*
2. *If $g\colon \boldsymbol{X} \to \boldsymbol{Y}$ is a level-wise weak equivalence of Reedy fibrant cosimplicial objects in $\mathcal{M}$, then the induced map of total objects $g_*\colon \operatorname{Tot} \boldsymbol{X} \to \operatorname{Tot} \boldsymbol{Y}$ is a weak equivalence of fibrant objects in $\mathcal{M}$.*

PROOF. This follows from Corollary 18.4.13 and Corollary 15.9.11. □

THEOREM 18.6.7. *Let $\mathcal{M}$ be a simplicial model category.*

1. *if $i\colon \boldsymbol{A} \to \boldsymbol{B}$ is a Reedy cofibration of simplicial objects in $\mathcal{M}$, then the induced map of realizations $|i|\colon |\boldsymbol{A}| \to |\boldsymbol{B}|$ is a cofibration in $\mathcal{M}$ that is a trivial cofibration if i is a trivial cofibration.*
2. *If $p\colon \boldsymbol{X} \to \boldsymbol{Y}$ is a Reedy fibration of cosimplicial objects in $\mathcal{M}$, then the induced map of total objects $\operatorname{Tot} p\colon \operatorname{Tot} \boldsymbol{X} \to \operatorname{Tot} \boldsymbol{Y}$ is a fibration in $\mathcal{M}$ that is a trivial fibration if p is a trivial fibration.*

PROOF. This follows from Theorem 18.4.11 and Corollary 15.9.11. □

18.7. The Bousfield-Kan map

If $\mathcal{M}$ is a simplicial model category and $\boldsymbol{X}$ is a cosimplicial object in $\mathcal{M}$ (see Definition 15.1.10), then

- the total object $\operatorname{Tot} \boldsymbol{X}$ of $\boldsymbol{X}$ is defined (see Definition 18.6.3) as the end $\hom^{\boldsymbol{\Delta}}(\Delta, \boldsymbol{X})$, using the cosimplicial standard simplex Δ (see Definition 15.1.15), and
- the homotopy limit $\operatorname{holim} \boldsymbol{X}$ is defined (see Example 18.3.6) as the end $\hom^{\boldsymbol{\Delta}}(\mathrm{B}(\boldsymbol{\Delta}\downarrow -), \boldsymbol{X})$, using the $\boldsymbol{\Delta}$-diagram of classifying spaces of overcategories $\mathrm{B}(\boldsymbol{\Delta}\downarrow -)$.

Both the cosimplicial standard simplex Δ and the $\boldsymbol{\Delta}$-diagram of classifying spaces of overcategories $\mathrm{B}(\boldsymbol{\Delta}\downarrow -)$ are Reedy cofibrant approximations to the constant $\boldsymbol{\Delta}$-diagram at a point (see Corollary 15.9.12 and Corollary 15.6.8), but since neither of these diagrams is a *fibrant* cofibrant approximation we cannot apply Proposition 8.1.7 to obtain a map between them. Bousfield and Kan, however, directly defined such a map in [**14**, Chapter XI, Example 2.6]. In this section we define the Bousfield-Kan map, modified from [**14**] to accommodate our different definition of the classifying space of a category (see Remark 18.1.11).

DEFINITION 18.7.1. The *Bousfield-Kan map of cosimplicial simplicial sets* is the map $\phi\colon \mathrm{B}(\mathbf{\Delta}\downarrow -) \to \Delta$ (see Definition 14.7.8 and Definition 15.1.15) that for $k \geq 0$ and $n \geq 0$ takes the n-simplex

$$\Bigl(\bigl([i_0] \xrightarrow{\sigma_0} [i_1] \xrightarrow{\sigma_1} \cdots \xrightarrow{\sigma_{n-1}} [i_n]\bigr), \tau\colon [i_n] \to [k]\Bigr)$$

of $\mathrm{B}(\mathbf{\Delta}\downarrow k)$ to the n-simplex

$$[\tau\sigma_{n-1}\sigma_{n-2}\cdots\sigma_0(i_0), \tau\sigma_{n-1}\sigma_{n-2}\cdots\sigma_1(i_1), \ldots, \tau\sigma_{n-1}(i_{n-1}), \tau(i_n)]$$

of $\Delta[k]$. We will also abuse notation and let $\phi\colon \mathrm{B}(-\downarrow \mathbf{\Delta}^{\mathrm{op}})^{\mathrm{op}} \to \Delta$ denote the composition of the isomorphism $\mathrm{B}(-\downarrow \mathbf{\Delta}^{\mathrm{op}})^{\mathrm{op}} \approx \mathrm{B}(\mathbf{\Delta}\downarrow -)$ (see Corollary 14.7.13) with the map $\phi\colon \mathrm{B}(\mathbf{\Delta}\downarrow -) \to \Delta$.

PROPOSITION 18.7.2. *The Bousfield-Kan map of cosimplicial simplicial sets is a weak equivalence of Reedy cofibrant cosimplicial simplicial sets.*

PROOF. For every $k \geq 0$ both $\mathrm{B}(\mathbf{\Delta}\downarrow [k])$ and $\Delta[k]$ are contractible simplicial sets (see Lemma 14.7.10), and so every map $\mathrm{B}(\mathbf{\Delta}\downarrow [k]) \to \Delta[k]$ is a weak equivalence. The result now follows from Corollary 15.6.7 and Corollary 15.9.11. □

DEFINITION 18.7.3. Let $\mathcal{M}$ be a simplicial model category.

(1) If $\boldsymbol{X}$ is a simplicial object in $\mathcal{M}$, then the *Bousfield-Kan map* is the map

$$\phi_*\colon \operatorname{hocolim} \boldsymbol{X} \longrightarrow \bigl|\boldsymbol{X}\bigr|\ ,$$

natural in $\boldsymbol{X}$, that is the composition

$$\operatorname{hocolim} \boldsymbol{X} \approx \boldsymbol{X} \otimes_{\mathbf{\Delta}^{\mathrm{op}}} \mathrm{B}(-\downarrow \mathbf{\Delta}^{\mathrm{op}})^{\mathrm{op}} \xrightarrow{1_{\boldsymbol{X}} \otimes_{\mathbf{\Delta}^{\mathrm{op}}} \phi} \boldsymbol{X} \otimes_{\mathbf{\Delta}^{\mathrm{op}}} \Delta \approx \bigl|\boldsymbol{X}\bigr|$$

where ϕ is the Bousfield-Kan map of cosimplicial simplicial sets (see Definition 18.7.1).

(2) If $\boldsymbol{X}$ is a cosimplicial object in $\mathcal{M}$, then the *Bousfield-Kan map* is the map

$$\phi^*\colon \operatorname{Tot} \boldsymbol{X} \longrightarrow \operatorname{holim} \boldsymbol{X}\ ,$$

natural in $\boldsymbol{X}$, that is the composition

$$\operatorname{Tot} \boldsymbol{X} \approx \hom^{\mathbf{\Delta}}(\Delta, \boldsymbol{X}) \xrightarrow{\hom^{\mathbf{\Delta}}(\phi, 1_{\boldsymbol{X}})} \hom^{\mathbf{\Delta}}\bigl(\mathrm{B}(\mathbf{\Delta}\downarrow -), \boldsymbol{X}\bigr) \approx \operatorname{holim} \boldsymbol{X}$$

where ϕ is the Bousfield-Kan map of cosimplicial simplicial sets.

THEOREM 18.7.4. *Let $\mathcal{M}$ be a simplicial model category.*

(1) *If $\boldsymbol{X}$ is a Reedy cofibrant simplicial object in $\mathcal{M}$, then the Bousfield-Kan map $\phi_*\colon \operatorname{hocolim} \boldsymbol{X} \to \bigl|\boldsymbol{X}\bigr|$ is a natural weak equivalence.*
(2) *If $\boldsymbol{X}$ is a Reedy fibrant cosimplicial object in $\mathcal{M}$, then the Bousfield-Kan map $\phi^*\colon \operatorname{Tot} \boldsymbol{X} \to \operatorname{holim} \boldsymbol{X}$ is a natural weak equivalence.*

PROOF. This follows from Corollary 18.4.14 and Proposition 18.7.2. □

COROLLARY 18.7.5. *If $\boldsymbol{X}$ is a simplicial simplicial set (either pointed or unpointed), then the Bousfield-Kan map $\phi_*\colon \operatorname{hocolim} \boldsymbol{X} \to \bigl|\boldsymbol{X}\bigr|$ is a weak equivalence.*

PROOF. This follows from Theorem 18.7.4 and Corollary 15.8.8. □

THEOREM 18.7.6. *Let $\mathcal{M}$ be a simplicial model category.*

(1) *If $\boldsymbol{X}$ is a Reedy cofibrant simplicial object in $\mathcal{M}$, then there is a natural essentially unique zig-zag of weak equivalences from the realization of $\boldsymbol{X}$ to the homotopy colimit of $\boldsymbol{X}$, containing the Bousfield-Kan map.*
(2) *If $\boldsymbol{X}$ is a Reedy fibrant cosimplicial object in $\mathcal{M}$, then there is a natural essentially unique zig-zag of weak equivalences from the total object of $\boldsymbol{X}$ to the homotopy limit of $\boldsymbol{X}$, containing the Bousfield-Kan map.*

PROOF. This follows from Corollary 18.4.15, Corollary 15.9.11, and Corollary 15.6.7. □

COROLLARY 18.7.7. *If $\boldsymbol{X}$ is a simplicial simplicial set (either pointed or unpointed), then there is a natural weak equivalence* $\operatorname{hocolim} \boldsymbol{X} \to \operatorname{diag} \boldsymbol{X}$ *(see Definition* 15.11.3*).*

PROOF. This follows from Corollary 18.7.5 and Theorem 15.11.6. □

18.8. Diagrams of pointed or unpointed spaces

Given a small category $\mathcal{C}$ and a $\mathcal{C}$-diagram of pointed spaces $\boldsymbol{X}$ (where by a "space" we mean either a topological space or a simplicial set), there are two ways to construct an unpointed version of its homotopy limit:

(1) Take the homotopy limit of the diagram in the category of pointed spaces and then forget the basepoint, or
(2) forget the basepoints of the spaces in the diagram and take the homotopy limit in the category of unpointed spaces.

These two homotopy limits will be homeomorphic (respectively, isomorphic) because if X is an unpointed space and Y is a pointed space, then the space of pointed maps $\operatorname{map}_*(X^+, Y)$ (respectively, $\operatorname{Map}_*(X^+, Y)$) (see Definition 18.2.1) is homeomorphic (respectively, isomorphic) to the space of unpointed maps $\operatorname{map}(X, Y)$ (respectively, $\operatorname{Map}(X, Y)$) once we've forgotten the basepoint of the pointed mapping space. On the other hand, the homotopy colimit of $\boldsymbol{X}$ will generally have different homotopy types when taken in the categories of pointed and unpointed spaces (see Proposition 18.8.4). In this section, we describe the difference between the pointed and unpointed homotopy colimit.

NOTATION 18.8.1. In this section only, if $\boldsymbol{X}$ is a diagram of pointed spaces, then

- $\operatorname{hocolim}_* \boldsymbol{X}$ will denote the homotopy colimit formed in the category of pointed spaces and
- $\operatorname{hocolim} \boldsymbol{X}$ will denote the homotopy colimit formed in the category of unpointed spaces after forgetting the basepoints of the spaces in the diagram.

DEFINITION 18.8.2. A pointed space X will be called *well pointed* if the inclusion of the basepoint into the space is a cofibration in the model category Spc_* (see Notation 7.10.5). Since the one point space is the initial object of Spc_*, a pointed space X is well pointed if and only if it is a cofibrant space.

PROPOSITION 18.8.3. *Every pointed simplicial set is well pointed.*

PROOF. Every inclusion of simplicial sets is a cofibration. □

The following proposition is due to E. Dror Farjoun ([**23**]).

PROPOSITION 18.8.4. *Let* $\mathcal{C}$ *be a small category, let* Spc_* *denote either* SS_* *or* Top_* *(see Notation 7.10.5), let* $\boldsymbol{X}$ *be a* $\mathcal{C}$*-diagram in* Spc_**, and let* $\mathrm{B}\mathcal{C}^{\mathrm{op}}$ *be the classifying space of the category* $\mathcal{C}^{\mathrm{op}}$ *(see Definition 14.1.1).*

- *If* $\mathrm{Spc}_* = \mathrm{SS}_*$*, then there is a natural cofibration* $\mathrm{B}\mathcal{C}^{\mathrm{op}} \to \mathrm{hocolim}\, \boldsymbol{X}$ *and a natural isomorphism* $(\mathrm{hocolim}\, \boldsymbol{X})/(\mathrm{B}\mathcal{C}^{\mathrm{op}}) \approx \mathrm{hocolim}_* \boldsymbol{X}$.
- *If* $\mathrm{Spc}_* = \mathrm{Top}_*$*, then there is a natural inclusion* $|\mathrm{B}\mathcal{C}^{\mathrm{op}}| \to \mathrm{hocolim}\, \boldsymbol{X}$ *that is a cofibration if* $\boldsymbol{X}$ *is a diagram of well pointed spaces and a natural homeomorphism* $(\mathrm{hocolim}\, \boldsymbol{X})/(|\mathrm{B}\mathcal{C}^{\mathrm{op}}|) \approx \mathrm{hocolim}_* \boldsymbol{X}$.

PROOF. This follows from the definition of the homotopy colimit (see Definition 19.1.2), Proposition 18.1.6, and Theorem 18.5.1. □

COROLLARY 18.8.5. *Let* Spc_* *denote either* SS_* *or* Top_* *(see Notation 7.10.5). If* $\mathcal{C}$ *is a small category and* $\boldsymbol{X}$ *is a* $\mathcal{C}$*-diagram of well pointed spaces in* Spc_* *such that the space* $\boldsymbol{X}_\alpha$ *is contractible for every object* α *in* $\mathcal{C}$*, then* $\mathrm{hocolim}_* \boldsymbol{X}$ *is contractible (see Notation 18.8.1).*

PROOF. We will prove this in the case $\mathrm{Spc}_* = \mathrm{Top}_*$; the case $\mathrm{Spc}_* = \mathrm{SS}_*$ is similar.

Proposition 18.8.4, Proposition 18.1.6 and Theorem 18.5.3 imply that the map $|\mathrm{B}\mathcal{C}^{\mathrm{op}}| \to \mathrm{hocolim}\, \boldsymbol{X}$ is a trivial cofibration. Since the quotient space $(\mathrm{hocolim}\, \boldsymbol{X})/(|\mathrm{B}\mathcal{C}^{\mathrm{op}}|)$ is naturally homeomorphic to the pushout of the diagram $* \leftarrow |\mathrm{B}\mathcal{C}^{\mathrm{op}}| \to \mathrm{hocolim}\, \boldsymbol{X}$, this implies that the map $* \to \mathrm{hocolim}_* \boldsymbol{X}$ (see Notation 18.8.1) is a trivial cofibration.

□

PROPOSITION 18.8.6. *Let* Spc_* *denote either* SS_* *or* Top_* *(see Notation 7.10.5). If* $\mathcal{C}$ *is a small category such that the classifying space of* $\mathcal{C}$ *is contractible, then for any* $\mathcal{C}$*-diagram of well pointed spaces* $\boldsymbol{X}$ *in* Spc_* *the natural map (see Proposition 18.8.4)* $\mathrm{hocolim}\, \boldsymbol{X} \to \mathrm{hocolim}_* \boldsymbol{X}$ *is a weak equivalence.*

PROOF. We will prove this in the case $\mathrm{Spc}_* = \mathrm{Top}_*$; the case $\mathrm{Spc}_* = \mathrm{SS}_*$ is similar.

The quotient space $(\mathrm{hocolim}\, \boldsymbol{X})/(|\mathrm{B}\mathcal{C}^{\mathrm{op}}|)$ is naturally homeomorphic to the pushout of the diagram $* \leftarrow |\mathrm{B}\mathcal{C}^{\mathrm{op}}| \to \mathrm{hocolim}\, \boldsymbol{X}$. Since Spc is a proper model category (see Theorem 13.1.11 and Theorem 13.1.13), the result now follows from Proposition 18.8.4. □

EXAMPLE 18.8.7. If $\mathcal{C}$ is the category $\cdot \leftarrow \cdot \to \cdot$ then the homotopy colimit of a $\mathcal{C}$-diagram of well pointed spaces has the same weak homotopy type whether formed in the category of pointed spaces or in the category of unpointed spaces.

EXAMPLE 18.8.8. If $\mathcal{C}$ is the category $\cdot \to \cdot \to \cdot \to \cdots$, then the homotopy colimit of a $\mathcal{C}$-diagram of well pointed spaces has the same weak homotopy type whether formed in the category of pointed spaces or in the category of unpointed spaces.

EXAMPLE 18.8.9. The homotopy colimit of a diagram indexed by a discrete group does not, in general, have the same weak homotopy type when formed in the category of pointed spaces as it does when formed in the category of unpointed spaces since the classifying space of a nontrivial discrete group is not contractible (see Example 14.1.4, Proposition 14.1.6, and Proposition 18.8.4.)

18.9. Diagrams of simplicial sets

LEMMA 18.9.1. *Let $\mathcal{M}$ be a simplicial model category, let X be a simplicial set, and let ΔX be the category of simplices of X (see Definition 15.1.16).*

(1) *If $\boldsymbol{Y}$ is a $(\Delta^{\mathrm{op}}X)$-diagram in $\mathcal{M}$, then the homotopy colimit of $\boldsymbol{Y}$ is naturally isomorphic to the homotopy colimit of the simplicial object $\boldsymbol{Z}$ in $\mathcal{M}$ for which $\boldsymbol{Z}_n = \coprod_{\sigma \in X_n} \boldsymbol{Y}_\sigma$.*
(2) *If $\boldsymbol{Y}$ is a (ΔX)-diagram in $\mathcal{M}$, then the homotopy limit of $\boldsymbol{Y}$ is naturally isomorphic to the homotopy limit of the cosimplicial object $\boldsymbol{Z}$ in $\mathcal{M}$ for which $\boldsymbol{Z}^n = \prod_{\sigma \in X_n} \boldsymbol{Y}_\sigma$.*

PROOF. We will prove part 1; the proof of part 2 is similar.

Definition 18.1.2 describes $\operatorname{hocolim} \boldsymbol{Y}$ as the colimit of the diagram

$$\coprod_{(\sigma\to\sigma')\in\Delta^{\mathrm{op}}X} \boldsymbol{Y}_\sigma \otimes \mathrm{B}\big(\sigma' \downarrow (\Delta^{\mathrm{op}}X)\big)^{\mathrm{op}} \rightrightarrows \coprod_{\sigma\in\mathrm{Ob}(\Delta^{\mathrm{op}}X)} \boldsymbol{Y}_\sigma \otimes \mathrm{B}\big(\sigma \downarrow (\Delta^{\mathrm{op}}X)\big)^{\mathrm{op}} .$$

Lemma 9.2.1 implies that there are natural isomorphisms

$$\begin{aligned}
\coprod_{(\sigma\to\sigma')\in\Delta^{\mathrm{op}}X} \boldsymbol{Y}_\sigma \otimes \mathrm{B}\big(\sigma' \downarrow (\Delta^{\mathrm{op}}X)\big)^{\mathrm{op}} &\approx \coprod_{\substack{n\ge 0\\ k\ge 0}} \coprod_{\substack{\sigma\in X_n\\ \delta\colon X_n\to X_k}} \boldsymbol{Y}_\sigma \otimes \mathrm{B}\big([k]\downarrow \boldsymbol{\Delta}^{\mathrm{op}}\big)^{\mathrm{op}}\\
&\approx \coprod_{\substack{n\ge 0\\ k\ge 0}} \coprod_{\delta\colon X_n\to X_k} \Big(\coprod_{\sigma\in X_n} \boldsymbol{Y}_\sigma\Big) \otimes \mathrm{B}\big([k]\downarrow \boldsymbol{\Delta}^{\mathrm{op}}\big)^{\mathrm{op}}\\
&\approx \coprod_{\substack{n\ge 0\\ k\ge 0}} \coprod_{\boldsymbol{\Delta}^{\mathrm{op}}([n],[k])} \boldsymbol{Z}_n \otimes \mathrm{B}\big([k]\downarrow \boldsymbol{\Delta}^{\mathrm{op}}\big)^{\mathrm{op}}
\end{aligned}$$

and natural isomorphisms

$$\begin{aligned}
\coprod_{\sigma\in\mathrm{Ob}(\Delta^{\mathrm{op}}X)} \boldsymbol{Y}_\sigma \otimes \mathrm{B}\big(\sigma \downarrow (\Delta^{\mathrm{op}}X)\big)^{\mathrm{op}} &\approx \coprod_{n\ge 0}\Big(\coprod_{\sigma\in X_n} \boldsymbol{Y}_\sigma\Big) \otimes \mathrm{B}\big(\sigma \downarrow (\Delta^{\mathrm{op}}X)\big)^{\mathrm{op}}\\
&\approx \coprod_{n\ge 0} \boldsymbol{Z}_n \otimes \mathrm{B}\big([n]\downarrow \boldsymbol{\Delta}^{\mathrm{op}}\big)^{\mathrm{op}} ,
\end{aligned}$$

and so $\operatorname{hocolim} \boldsymbol{Y}$ is naturally isomorphic to the colimit of the diagram

$$\coprod_{\substack{n\ge 0\\ k\ge 0}} \coprod_{\boldsymbol{\Delta}^{\mathrm{op}}([n],[k])} \boldsymbol{Z}_n \otimes \mathrm{B}\big([k]\downarrow \boldsymbol{\Delta}^{\mathrm{op}}\big)^{\mathrm{op}} \rightrightarrows \coprod_{[n]\in\mathrm{Ob}(\boldsymbol{\Delta}^{\mathrm{op}})} \boldsymbol{Z}_n \otimes \mathrm{B}\big([n]\downarrow \boldsymbol{\Delta}^{\mathrm{op}}\big)^{\mathrm{op}} ,$$

which is the definition of $\operatorname{hocolim} \boldsymbol{Z}$. □

PROPOSITION 18.9.2. *If X is a simplicial set, ΔX is the category of simplices of X (see Definition 15.1.16), and $\boldsymbol{P}$ is the $(\Delta^{\mathrm{op}}X)$-diagram of simplicial sets such that $\boldsymbol{P}_\sigma$ is a single point for every object σ of $\Delta^{\mathrm{op}}X$, then there is a natural weak equivalence from the homotopy colimit of $\boldsymbol{P}$ to X.*

PROOF. Lemma 18.9.1 implies that the homotopy colimit of $\boldsymbol{P}$ is naturally isomorphic to hocolim $\boldsymbol{Z}$ where $\boldsymbol{Z}$ is the simplicial simplicial set such that

$$\boldsymbol{Z}_n = \coprod_{\sigma \in X_n} \boldsymbol{P}_\sigma = \coprod_{\sigma \in X_n} * = X_n$$

(where we view the set X_n as a constant (i.e., discrete) simplicial set). Since the diagonal of $\boldsymbol{Z}$ is naturally isomorphic to the original simplicial set X, the result follows from Corollary 18.7.5. □

THEOREM 18.9.3. *If X is a simplicial set and ΔX is the category of simplices of X (see Definition* 15.1.16*) then there is a natural weak equivalence from* $\mathrm{B}(\Delta X)$ *to X.*

PROOF. Proposition 18.1.6 implies that if $\boldsymbol{P}$ is the $(\Delta^{\mathrm{op}} X)$-diagram of simplicial sets such that $\boldsymbol{P}_\sigma$ is a single point for every object σ of $\Delta^{\mathrm{op}} X$, then $\mathrm{B}(\Delta X)$ is naturally isomorphic to hocolim $\boldsymbol{P}$. The result now follows from Proposition 18.9.2. □

PROPOSITION 18.9.4. *If $\mathcal{C}$ is a small category and $\boldsymbol{X}$ is a $\mathcal{C}$-diagram of simplicial sets that is cofibrant in the model category structure of Theorem* 11.6.1, *then the natural map* hocolim $\boldsymbol{X} \to$ colim $\boldsymbol{X}$ *is a weak equivalence.*

PROOF. This follows from Corollary 18.4.4, Example 18.3.8, Lemma 14.7.4, and the isomorphism between the map $\boldsymbol{X} \otimes_{\mathcal{C}} \mathrm{B}(- \downarrow \mathcal{C})^{\mathrm{op}} \to \boldsymbol{X} \otimes_{\mathcal{C}} \boldsymbol{P}$ and the map $\mathrm{B}(- \downarrow \mathcal{C})^{\mathrm{op}} \otimes_{\mathcal{C}^{\mathrm{op}}} \boldsymbol{X} \to \boldsymbol{P} \otimes_{\mathcal{C}^{\mathrm{op}}} \boldsymbol{X}$. □

PROPOSITION 18.9.5. *If $\mathcal{C}$ and $\mathcal{D}$ are small categories and $\mathrm{F}\colon \mathcal{C} \to \mathcal{D}$ is a functor, then the homotopy colimit of the $\mathcal{D}^{\mathrm{op}}$-diagram of classifying spaces of undercategories* $\mathrm{colim}_{\mathcal{D}^{\mathrm{op}}} \mathrm{B}(- \downarrow \mathrm{F})$ *is naturally weakly equivalent to* $\mathrm{B}\mathcal{C}$.

PROOF. This follows from Proposition 14.7.5, Proposition 18.9.4, and Proposition 14.8.5. □

EXAMPLE 18.9.6. If $p\colon E \to B$ is a map of simplicial sets, we will decompose E into a (ΔB)-diagram (see Definition 15.1.16) of simplicial sets $\tilde{p}$. If σ is an n-simplex of B, then the characteristic map of σ is the unique map $\chi_\sigma\colon \Delta[n] \to B$ that takes the non-degenerate n-simplex of $\Delta[n]$ to σ, and we let $\tilde{p}(\sigma)$ be the pullback of the diagram

$$\begin{array}{ccc} & & E \\ & & \big\downarrow{\scriptstyle p} \\ \Delta[n] & \xrightarrow[\chi_\sigma]{} & B \end{array}$$

If $\delta\colon B_n \to B_k$ is a simplicial operator, then δ corresponds to a map $\Delta[k] \to \Delta[n]$, and so we get a map $\tilde{p}(\sigma, \delta)\colon \tilde{p}(\delta(\sigma)) \to \tilde{p}(\sigma)$. For each simplex σ in B there is an obvious map $\tilde{p}(\sigma) \to E$, and these induce an isomorphism of simplicial sets $\mathrm{colim}_{\sigma \in \mathrm{Ob}(\Delta B)} \tilde{p}(\sigma) \approx E$. We will show in Corollary 19.9.2 that the natural map $\mathrm{hocolim}_{\sigma \in \mathrm{Ob}(\Delta B)} \tilde{p}(\sigma) \to \mathrm{colim}_{\sigma \in \mathrm{Ob}(\Delta B)} \tilde{p}(\sigma) \approx E$ is a weak equivalence.

PROPOSITION 18.9.7. *If $p\colon E \to B$ is a map of simplicial sets, then the (ΔB)-diagram of simplicial sets constructed in Example* 18.9.6 *is Reedy cofibrant.*

PROOF. The latching map at the n-simplex σ is the inclusion of the part of $\tilde{p}(\sigma)$ above $\partial\Delta[n]$ into $\tilde{p}(\sigma)$. □

Proposition 18.9.7 allows us to prove in Corollary 19.9.2 that if $p\colon E \to B$ is a map of simplicial sets and $\tilde{p}\colon \Delta B \to \mathrm{SS}$ is the diagram constructed in Example 18.9.6, then the natural map $\operatorname{hocolim} \tilde{p} \to E$ is a weak equivalence.

18.9.8. Geometric realization and total singular complex.

PROPOSITION 18.9.9. *Let $\mathcal{C}$ be a small category.*

(1) *If $\boldsymbol{X}$ is a $\mathcal{C}$-diagram of unpointed (respectively, pointed) simplicial sets and $\boldsymbol{K}$ is a $\mathcal{C}^{\mathrm{op}}$-diagram of unpointed simplicial sets, then there is a natural homeomorphism*

$$\left|\boldsymbol{X} \otimes_{\mathcal{C}} \boldsymbol{K}\right| \approx \left|\boldsymbol{X}\right| \otimes_{\mathcal{C}} \boldsymbol{K}$$

of unpointed (respectively, pointed) topological spaces.

(2) *If $\boldsymbol{X}$ is a $\mathcal{C}$-diagram of unpointed (respectively, pointed) topological spaces and $\boldsymbol{K}$ is a $\mathcal{C}$-diagram of unpointed simplicial sets, then there is a natural isomorphism*

$$\operatorname{Sing}\bigl(\hom^{\mathcal{C}}(\boldsymbol{K}, \boldsymbol{X})\bigr) \approx \hom^{\mathcal{C}}(\boldsymbol{K}, \operatorname{Sing} \boldsymbol{X})$$

of unpointed (respectively, pointed) simplicial sets.

PROOF. We will prove part 1; the proof of part 2 is similar.

Definition 18.3.2 defines $\boldsymbol{X} \otimes_{\mathcal{C}} \boldsymbol{K}$ as a colimit. Since the geometric realization functor is a left adjoint, it commutes with colimits, and so the result follows from Lemma 1.1.9. □

THEOREM 18.9.10. *Let $\mathcal{C}$ be a small category.*

(1) *If $\boldsymbol{X}$ is a $\mathcal{C}$-diagram of unpointed (respectively, pointed) simplicial sets, then there is a natural homeomorphism*

$$\left|\operatorname{hocolim} \boldsymbol{X}\right| \approx \operatorname{hocolim}\left|\boldsymbol{X}\right|$$

of unpointed (respectively, pointed) topological spaces.

(2) *If $\boldsymbol{X}$ is a $\mathcal{C}$-diagram of unpointed (respectively, pointed) topological spaces then there is a natural isomorphism*

$$\operatorname{Sing}(\operatorname{holim} \boldsymbol{X}) \approx \operatorname{holim}(\operatorname{Sing} \boldsymbol{X})$$

of unpointed (respectively, pointed) simplicial sets.

PROOF. This follows from Proposition 18.9.9 and Example 18.3.6. □

THEOREM 18.9.11.

(1) *If $\boldsymbol{X}$ is a simplicial unpointed (respectively, pointed) simplicial set, then there is a natural homeomorphism from the geometric realization of the simplicial set $\left|\boldsymbol{X}\right|$ (see Definition 18.6.2) to the realization (see Definition 18.6.2) of the simplicial unpointed (respectively, pointed) topological space obtained by taking the geometric realization in each simplicial degree.*

(2) *If $\boldsymbol{X}$ is a cosimplicial unpointed (respectively, pointed) topological space, then there is a natural isomorphism $\operatorname{Sing}(\operatorname{Tot} \boldsymbol{X}) \approx \operatorname{Tot}(\operatorname{Sing} \boldsymbol{X})$.*

PROOF. This follows from Proposition 18.9.9. □

PROPOSITION 18.9.12. *Let $\mathcal{C}$ be a small category.*

(1) *If $\boldsymbol{X}$ is an objectwise cofibrant $\mathcal{C}$-diagram of unpointed (respectively, pointed) topological spaces, then there is a natural weak equivalence* $\operatorname{hocolim}(\operatorname{Sing}\boldsymbol{X}) \to \operatorname{Sing}(\operatorname{hocolim}\boldsymbol{X})$.
(2) *If $\boldsymbol{X}$ is an objectwise fibrant $\mathcal{C}$-diagram of unpointed (respectively, pointed) simplicial sets, then there is a natural weak equivalence* $|\operatorname{holim}\boldsymbol{X}| \to \operatorname{holim}|\boldsymbol{X}|$.

PROOF. We will prove part 1; the proof of part 2 is similar.

The natural map of diagrams $|\operatorname{Sing}\boldsymbol{X}| \to \boldsymbol{X}$ induces a natural weak equivalence $\operatorname{hocolim}|\operatorname{Sing}\boldsymbol{X}| \to \operatorname{hocolim}\boldsymbol{X}$ (see Theorem 19.4.2). Theorem 18.9.10 implies that this is isomorphic to a natural weak equivalence $|\operatorname{hocolim}\operatorname{Sing}\boldsymbol{X}| \to \operatorname{hocolim}\boldsymbol{X}$, which corresponds (under the standard adjunction) to a natural weak equivalence $\operatorname{hocolim}(\operatorname{Sing}\boldsymbol{X}) \to \operatorname{Sing}(\operatorname{hocolim}\boldsymbol{X})$. □

CHAPTER 19

Homotopy Limits in General Model Categories

In this chapter we generalize the definitions of homotopy colimit and homotopy limit from diagrams in a simplicial model category to diagrams in an arbitrary model category, using frames (see Definition 16.6.21) in place of the simplicial model category structure. Our definitions are due to D. M. Kan, who also established their properties (using methods different from the ones used here).

We define homotopy colimits and homotopy limits in Section 19.1. In Section 19.2 we define coends and ends, which are constructions that generalize the definitions of the homotopy colimit and the homotopy limit, and we establish some adjointness results. In Section 19.3 we use the adjointness results to obtain homotopy invariance results for the pushout corner map of coends and the pullback corner map of ends.

In Section 19.4 we establish our homotopy invariance results for homotopy colimits and homotopy limits and show that changing the choice of frame results in naturally weakly equivalent homotopy colimit and homotopy limit functors. We also prove an adjointness result connecting the homotopy colimit and the homotopy limit functors. In Section 19.5 we show that a homotopy pullback in a right proper model category is weakly equivalent to the homotopy limit of the diagram if each of the objects in the diagram is fibrant.

In Section 19.6 we define what it means for a functor between small categories to be *homotopy left cofinal* or *homotopy right cofinal*. We show that such a functor is homotopy left cofinal if and only if it induces a weak equivalence of homotopy limits for all diagrams of fibrant objects, and that it is homotopy right cofinal if and only if it induces a weak equivalence of homotopy colimits for all diagrams of cofibrant objects. In Section 19.7 we prove a homotopy lifting extension theorem for diagrams indexed by a Reedy category. In Section 19.8 we discuss the Bousfield-Kan maps from the homotopy colimit of a simplicial object to its realization and from the total object of a cosimplicial object to its homotopy limit.

In Section 19.9 we discuss diagrams indexed by a Reedy category. We prove that the homotopy colimit of a Reedy cofibrant diagram indexed by a Reedy category with fibrant constants is naturally weakly equivalent to its colimit, and that the homotopy limit of a Reedy fibrant diagram indexed by a Reedy category with cofibrant constants is naturally weakly equivalent to its limit.

19.1. Homotopy colimits and homotopy limits

In this section we define the homotopy colimit and homotopy limit of a diagram in a model category; this generalizes the definitions of Chapter 18 for diagrams in a simplicial model category (see Remark 19.1.3 and Remark 19.1.6). Our definitions depend on the choice of a framing for the model category, but we show in Theorem 19.4.3 that two different framings yield naturally weakly equivalent homotopy colimit and homotopy limit functors.

19.1.1. Homotopy colimits.

DEFINITION 19.1.2. Let $\mathcal{M}$ be a framed model category (see Definition 16.6.21) and let $\mathcal{C}$ be a small category. If $\boldsymbol{X}$ is a $\mathcal{C}$-diagram in $\mathcal{M}$ (see Definition 11.5.2), then the *homotopy colimit* $\operatorname{hocolim} \boldsymbol{X}$ of $\boldsymbol{X}$ is defined to be the coequalizer of the maps

$$\coprod_{(\sigma\colon \alpha\to\alpha')\in\mathcal{C}} \widetilde{\boldsymbol{X}}_\alpha \otimes \mathrm{B}(\alpha'\downarrow\mathcal{C})^{\mathrm{op}} \overset{\phi}{\underset{\psi}{\rightrightarrows}} \coprod_{\alpha\in\mathrm{Ob}(\mathcal{C})} \widetilde{\boldsymbol{X}}_\alpha \otimes \mathrm{B}(\alpha\downarrow\mathcal{C})^{\mathrm{op}}$$

(see Definition 16.3.1, Definition 14.1.1, and Definition 11.8.3) where $\widetilde{\boldsymbol{X}}_\alpha$ is the natural cosimplicial frame on $\boldsymbol{X}_\alpha$, the map ϕ on the summand $\sigma\colon \alpha\to\alpha'$ is the composition of the map

$$\sigma_*\otimes 1_{\mathrm{B}(\alpha'\downarrow\mathcal{C})}\colon \widetilde{\boldsymbol{X}}_\alpha \otimes \mathrm{B}(\alpha'\downarrow\mathcal{C})^{\mathrm{op}} \longrightarrow \widetilde{\boldsymbol{X}}_{\alpha'} \otimes \mathrm{B}(\alpha'\downarrow\mathcal{C})^{\mathrm{op}}$$

with the natural injection into the coproduct, and the map ψ on the summand $\sigma\colon \alpha\to\alpha'$ is the composition of the map

$$1_{\widetilde{\boldsymbol{X}}_\alpha}\otimes \mathrm{B}(\sigma^*)\colon \widetilde{\boldsymbol{X}}_\alpha \otimes \mathrm{B}(\alpha'\downarrow\mathcal{C})^{\mathrm{op}} \longrightarrow \widetilde{\boldsymbol{X}}_\alpha \otimes \mathrm{B}(\alpha\downarrow\mathcal{C})^{\mathrm{op}}$$

(where $\sigma^*\colon (\alpha'\downarrow\mathcal{C})^{\mathrm{op}} \to (\alpha\downarrow\mathcal{C})^{\mathrm{op}}$; see Definition 14.7.2) with the natural injection into the coproduct.

For a discussion of the relation of our definition of the homotopy colimit to that of [**14**], see Remark 18.1.11.

REMARK 19.1.3. If $\mathcal{M}$ is a simplicial model category, then for every object X of $\mathcal{M}$ and every simplicial set K the object $\widetilde{\boldsymbol{X}}\otimes K$ (where $\widetilde{\boldsymbol{X}}$ is the natural cosimplicial frame on X; see Proposition 16.6.23) is naturally isomorphic to $X\otimes K$ (see Proposition 16.6.6). Thus, if $\mathcal{M}$ is a simplicial model category with the natural framing induced by the simplicial structure, Definition 19.1.2 agrees with Definition 18.1.2.

19.1.4. Homotopy limits.

DEFINITION 19.1.5. Let $\mathcal{M}$ be a framed model category (see Definition 16.6.21) and let $\mathcal{C}$ be a small category. If $\boldsymbol{X}$ is a $\mathcal{C}$-diagram in $\mathcal{M}$ (see Definition 11.5.2), then the *homotopy limit* $\operatorname{holim} \boldsymbol{X}$ of $\boldsymbol{X}$ is defined to be the equalizer of the maps

$$\prod_{\alpha\in\mathrm{Ob}(\mathcal{C})} (\widehat{\boldsymbol{X}}_\alpha)^{\mathrm{B}(\mathcal{C}\downarrow\alpha)} \overset{\phi}{\underset{\psi}{\rightrightarrows}} \prod_{(\sigma\colon \alpha\to\alpha')\in\mathcal{C}} (\widehat{\boldsymbol{X}}_{\alpha'})^{\mathrm{B}(\mathcal{C}\downarrow\alpha)}$$

(see Definition 16.3.1, Definition 14.1.1, and Definition 11.8.1) where $\widehat{\boldsymbol{X}}_\alpha$ is the natural simplicial frame on $\boldsymbol{X}_\alpha$, the projection of the map ϕ on the factor $\sigma\colon \alpha\to\alpha'$ is the composition of a natural projection from the product with the map

$$\sigma_*^{1_{\mathrm{B}(\mathcal{C}\downarrow\alpha)}}\colon (\widehat{\boldsymbol{X}}_\alpha)^{\mathrm{B}(\mathcal{C}\downarrow\alpha)} \longrightarrow (\widehat{\boldsymbol{X}}_{\alpha'})^{\mathrm{B}(\mathcal{C}\downarrow\alpha)}$$

and the projection of the map ψ on the factor $\sigma\colon \alpha \to \alpha'$ is the composition of a natural projection from the product with the map

$$(1_{\widehat{\boldsymbol{X}}_{\alpha'}})^{\mathrm{B}(\sigma_*)}\colon (\widehat{\boldsymbol{X}}_{\alpha'})^{\mathrm{B}(\mathcal{C}\downarrow\alpha')} \longrightarrow (\widehat{\boldsymbol{X}}_{\alpha'})^{\mathrm{B}(\mathcal{C}\downarrow\alpha)}$$

(where $\sigma_*\colon (\mathcal{C}\downarrow\alpha) \to (\mathcal{C}\downarrow\alpha')$; see Definition 14.7.8).

REMARK 19.1.6. If $\mathcal{M}$ is a simplicial model category then for every object X of $\mathcal{M}$ and every simplicial set K the object $\widehat{\boldsymbol{X}}^K$ (where $\widehat{\boldsymbol{X}}$ is the natural simplicial frame on X; see Proposition 16.6.23) is naturally isomorphic to $\boldsymbol{X}^K$ (see Proposition 16.6.6). Thus, if $\mathcal{M}$ is a simplicial model category with the natural framing induced by the simplicial structure, Definition 19.1.5 agrees with Definition 18.1.8.

19.1.7. Induced diagrams.

PROPOSITION 19.1.8. *Let $\mathcal{M}$ be a framed model category. If $\mathcal{C}$ and $\mathcal{D}$ are small categories, $\mathrm{F}\colon \mathcal{C} \to \mathcal{D}$ is a functor, $\boldsymbol{X}$ is a $\mathcal{D}$-diagram in $\mathcal{M}$, and $\mathrm{F}^*\boldsymbol{X}$ is the $\mathcal{C}$-diagram in $\mathcal{M}$ induced by F (see Definition 11.5.5), then*

(1) *there is a natural map*

$$\operatorname*{hocolim}_{\mathcal{C}} \mathrm{F}^*\boldsymbol{X} \longrightarrow \operatorname*{hocolim}_{\mathcal{D}} \boldsymbol{X}$$

defined by sending $\widetilde{\mathrm{F}^*\boldsymbol{X}}_\alpha \otimes \mathrm{B}(\alpha\downarrow\mathcal{C})^{\mathrm{op}} = \widetilde{\boldsymbol{X}}_{\mathrm{F}\alpha} \otimes \mathrm{B}(\alpha\downarrow\mathcal{C})^{\mathrm{op}}$ to $\widetilde{\boldsymbol{X}}_{\mathrm{F}\alpha} \otimes \mathrm{B}(\mathrm{F}\alpha\downarrow\mathcal{D})^{\mathrm{op}}$ by the map $1_{\widetilde{\boldsymbol{X}}_{\mathrm{F}\alpha}} \otimes \mathrm{F}_*$ (see Lemma 14.7.3), and

(2) *there is a natural map*

$$\operatorname*{holim}_{\mathcal{D}} \boldsymbol{X} \longrightarrow \operatorname*{holim}_{\mathcal{C}} \mathrm{F}^*\boldsymbol{X}$$

induced by the natural map $\mathrm{F}_*\colon \mathrm{B}(\mathcal{C}\downarrow\alpha) \to \mathrm{B}(\mathcal{D}\downarrow\mathrm{F}\alpha)$ (see Lemma 14.7.9).

PROOF. This follows directly from the definitions. □

It is often of interest to know conditions on a functor F that ensure that the natural map of homotopy colimits of Proposition 19.1.8 part 1 is a weak equivalence for all $\mathcal{D}$-diagrams of cofibrant objects, or that the natural map of homotopy limits of Proposition 19.1.8 part 2 is a weak equivalence for all $\mathcal{D}$-diagrams of fibrant objects. For this, see Theorem 19.6.13.

19.2. Coends and ends

In this section we define general constructions (see Definition 19.2.2) that allow us to analyze the colimit and homotopy colimit as two examples of the same construction and, dually, the limit and homotopy limit as two examples of the same construction (see Example 19.2.10). These definitions also enable us to obtain adjointness relations (see Section 19.2.12) that will be used to obtain the homotopy invariance results of Section 19.4.

19.2.1. Definitions.

DEFINITION 19.2.2. Let $\mathcal{M}$ be a model category and let $\mathcal{C}$ be a small category.

(1) If $\boldsymbol{X}$ is a $\mathcal{C}$-diagram in $\mathcal{M}$, $\widetilde{\boldsymbol{X}}$ is a cosimplicial frame on $\boldsymbol{X}$ (see Definition 16.7.2), and $\boldsymbol{K}$ is a $\mathcal{C}^{\mathrm{op}}$-diagram of simplicial sets, then $\boldsymbol{X} \otimes_{\mathcal{C}}^{\widetilde{\boldsymbol{X}}} \boldsymbol{K}$ is

defined to be the object of $\mathcal{M}$ that is the coequalizer of the maps

$$\coprod_{(\sigma\colon \alpha\to\alpha')\in\mathcal{C}} \widetilde{\boldsymbol{X}}_\alpha \otimes \boldsymbol{K}_{\alpha'} \overset{\phi}{\underset{\psi}{\rightrightarrows}} \coprod_{\alpha\in\mathrm{Ob}(\mathcal{C})} \widetilde{\boldsymbol{X}}_\alpha \otimes \boldsymbol{K}_\alpha \tag{19.2.3}$$

(see Definition 16.3.1) where the map ϕ on the summand $\sigma\colon \alpha \to \alpha'$ is the composition of the map

$$\sigma_* \otimes 1_{\boldsymbol{K}_{\alpha'}} \colon \widetilde{\boldsymbol{X}}_\alpha \otimes \boldsymbol{K}_{\alpha'} \longrightarrow \widetilde{\boldsymbol{X}}_{\alpha'} \otimes \boldsymbol{K}_{\alpha'}$$

(where $\sigma_*\colon \widetilde{\boldsymbol{X}}_\alpha \to \widetilde{\boldsymbol{X}}_{\alpha'}$) with the natural injection into the coproduct, and the map ψ on the summand $\sigma\colon \alpha \to \alpha'$ is the composition of the map

$$1_{\widetilde{\boldsymbol{X}}_\alpha} \otimes \sigma^* \colon \widetilde{\boldsymbol{X}}_\alpha \otimes \boldsymbol{K}_{\alpha'} \longrightarrow \widetilde{\boldsymbol{X}}_\alpha \otimes \boldsymbol{K}_\alpha$$

(where $\sigma^*\colon \boldsymbol{K}_{\alpha'} \to \boldsymbol{K}_\alpha$) with the natural injection into the coproduct.

(2) If $\boldsymbol{X}$ is a $\mathcal{C}$-diagram in $\mathcal{M}$, $\widehat{\boldsymbol{X}}$ is a simplicial frame on $\boldsymbol{X}$, and $\boldsymbol{K}$ is a $\mathcal{C}$-diagram of simplicial sets, then $\hom^{\mathcal{C}}_{\widehat{\boldsymbol{X}}}(\boldsymbol{K}, \boldsymbol{X})$ is defined to be the object of $\mathcal{M}$ that is the equalizer of the maps

$$\prod_{\alpha\in\mathrm{Ob}(\mathcal{C})} (\widehat{\boldsymbol{X}}_\alpha)^{\boldsymbol{K}_\alpha} \overset{\phi}{\underset{\psi}{\rightrightarrows}} \prod_{(\sigma\colon \alpha\to\alpha')\in\mathcal{C}} (\widehat{\boldsymbol{X}}_{\alpha'})^{\boldsymbol{K}_\alpha} \tag{19.2.4}$$

(see Definition 16.3.1) where the projection of the map ϕ on the factor $\sigma\colon \alpha \to \alpha'$ is the composition of a natural projection from the product with the map

$$\sigma_*^{1_{\boldsymbol{K}_\alpha}} \colon (\widehat{\boldsymbol{X}}_\alpha)^{\boldsymbol{K}_\alpha} \longrightarrow (\widehat{\boldsymbol{X}}_{\alpha'})^{\boldsymbol{K}_\alpha}$$

(where $\sigma_*\colon \widehat{\boldsymbol{X}}_\alpha \to \widehat{\boldsymbol{X}}_{\alpha'}$) and the projection of the map ψ on the factor $\sigma\colon \alpha \to \alpha'$ is the composition of a natural projection from the product with the map

$$(1_{\widehat{\boldsymbol{X}}_{\alpha'}})^{\boldsymbol{K}_{\sigma_*}} \colon (\widehat{\boldsymbol{X}}_{\alpha'})^{\boldsymbol{K}_{\alpha'}} \longrightarrow (\widehat{\boldsymbol{X}}_{\alpha'})^{\boldsymbol{K}_\alpha}$$

(where $\sigma_*\colon (\mathcal{C}\downarrow\alpha) \to (\mathcal{C}\downarrow\alpha')$; see Definition 14.7.8).

PROPOSITION 19.2.5. *Let $\mathcal{M}$ be a simplicial model category, let $\mathcal{C}$ be a small category, and let $\boldsymbol{X}$ be a $\mathcal{C}$-diagram in $\mathcal{M}$.*

1. *If $\widetilde{\boldsymbol{X}}$ is the natural cosimplicial frame on $\boldsymbol{X}$ defined by the simplicial structure on $\mathcal{M}$ and $\boldsymbol{K}$ is a $\mathcal{C}^{\mathrm{op}}$-diagram of simplicial sets, then $\boldsymbol{X} \otimes^{\widetilde{\boldsymbol{X}}}_{\mathcal{C}} \boldsymbol{K}$ (see Definition 19.2.2) is naturally isomorphic to $\boldsymbol{X} \otimes_{\mathcal{C}} \boldsymbol{K}$ (see Definition 18.3.2).*
2. *If $\widehat{\boldsymbol{X}}$ is the natural simplicial frame on $\boldsymbol{X}$ defined by the simplicial structure on $\mathcal{M}$ and $\boldsymbol{K}$ is a $\mathcal{C}$-diagram of simplicial sets, then $\hom^{\mathcal{C}}_{\widehat{\boldsymbol{X}}}(\boldsymbol{K}, \boldsymbol{K})$ (see Definition 19.2.2) is naturally isomorphic to $\hom^{\mathcal{C}}(\boldsymbol{K}, \boldsymbol{X})$ (see Definition 18.3.2).*

PROOF. This follows directly from the definitions. □

19.2.6. Framed model categories.

NOTATION 19.2.7. Let $\mathcal{M}$ be a framed model category, let $\mathcal{C}$ be a small category, and let $\boldsymbol{X}$ be a $\mathcal{C}$-diagram in $\mathcal{M}$.

(1) If $\widetilde{\boldsymbol{X}}$ is the natural cosimplicial frame on $\boldsymbol{X}$ induced by the framing on $\mathcal{M}$ (see Example 16.7.3), then $\boldsymbol{X} \otimes^{\widetilde{\boldsymbol{X}}}_{\mathcal{C}} \boldsymbol{K}$ will be denoted $\boldsymbol{X} \otimes_{\mathcal{C}} \boldsymbol{K}$ (see Proposition 19.2.5).

(2) If $\widehat{\boldsymbol{X}}$ is the natural simplicial frame induced by the framing on $\mathcal{M}$, then $\hom^{\mathcal{C}}_{\widehat{\boldsymbol{X}}}(\boldsymbol{K}, \boldsymbol{X})$ will be denoted $\hom^{\mathcal{C}}(\boldsymbol{K}, \boldsymbol{X})$ (see Proposition 19.2.5).

EXAMPLE 19.2.8. Let $\mathcal{M}$ be a framed model category and let $\mathcal{C}$ be a small category.

(1) If $\boldsymbol{X}$ is a $\mathcal{C}$-diagram in $\mathcal{M}$, then $\boldsymbol{X} \otimes_{\mathcal{C}} \mathrm{B}(-\downarrow \mathcal{C})^{\mathrm{op}}$ (see Notation 19.2.7) is the homotopy colimit of $\boldsymbol{X}$ (see Definition 19.1.2).

(2) If $\boldsymbol{X}$ is a $\mathcal{C}$-diagram in $\mathcal{M}$, then $\hom^{\mathcal{C}}(\mathrm{B}(\mathcal{C}\downarrow -), \boldsymbol{X})$ is the homotopy limit of $\boldsymbol{X}$ (see Definition 19.1.5).

PROPOSITION 19.2.9. *Let $\mathcal{M}$ be a framed model category and let $\mathcal{C}$ be a small category.*

(1) *If $\boldsymbol{X}$ is a $\mathcal{C}$-diagram in $\mathcal{M}$ and $\boldsymbol{P}\colon \mathcal{C}^{\mathrm{op}} \to \mathrm{SS}$ is a single point for every object α of $\mathcal{C}$, then $\boldsymbol{X} \otimes_{\mathcal{C}} \boldsymbol{P}$ (see Notation 19.2.7) is naturally isomorphic to $\operatorname{colim} \boldsymbol{X}$.*

(2) *If $\boldsymbol{X}$ is a $\mathcal{C}$-diagram in $\mathcal{M}$ and $\boldsymbol{P}\colon \mathcal{C} \to \mathrm{SS}$ is a single point for every object α of $\mathcal{C}$, then $\hom^{\mathcal{C}}(\boldsymbol{P}, \boldsymbol{X})$ is naturally isomorphic to $\lim \boldsymbol{X}$.*

PROOF. For part 1, $\boldsymbol{P}_\alpha$ is naturally isomorphic to $\Delta[0]$ for every object α of $\mathcal{C}^{\mathrm{op}}$, and so we have natural isomorphisms

$$\widetilde{\boldsymbol{X}}_\alpha \otimes \boldsymbol{P}_\alpha \approx \widetilde{\boldsymbol{X}}_\alpha \otimes \Delta[0] \approx (\widetilde{\boldsymbol{X}}_\alpha)_0 \approx \boldsymbol{X}_\alpha$$

(where $\widetilde{\boldsymbol{X}}$ is the natural cosimplicial frame on $\boldsymbol{X}$ induced by the framing on $\mathcal{M}$) (see Lemma 16.3.6). Under these isomorphisms, the map ϕ of Definition 19.2.2 is defined by $\sigma_*\colon \boldsymbol{X}_\alpha \to \boldsymbol{X}_{\alpha'}$ and the map ψ is the identity.

For part 2, $\boldsymbol{P}_\alpha$ is naturally isomorphic to $\Delta[0]$ for every object α of $\mathcal{C}$, and so we have natural isomorphisms

$$\widehat{\boldsymbol{X}}_\alpha^{\boldsymbol{P}_\alpha} \approx \widehat{\boldsymbol{X}}_\alpha^{\Delta[0]} \approx (\widehat{\boldsymbol{X}}_\alpha)^0 \approx \boldsymbol{X}_\alpha$$

(where $\widehat{\boldsymbol{X}}$ is the natural simplicial frame on $\boldsymbol{X}$ induced by the framing on $\mathcal{M}$) (see Lemma 16.3.6). Under these isomorphisms, the map ϕ of Definition 19.2.2 is defined by $\sigma_*\colon \boldsymbol{X}_\alpha \to \boldsymbol{X}_{\alpha'}$ and the map ψ is the identity. □

EXAMPLE 19.2.10. Let $\mathcal{M}$ be a framed model category and let $\mathcal{C}$ be a small category.

(1) If $\boldsymbol{P}\colon \mathcal{C}^{\mathrm{op}} \to \mathrm{SS}$ is a single point for every object α of $\mathcal{C}^{\mathrm{op}}$, then the unique map of $\mathcal{C}^{\mathrm{op}}$-diagrams $\mathrm{B}(-\downarrow \mathcal{C})^{\mathrm{op}} \to \boldsymbol{P}$ induces a natural map

$$\operatorname{hocolim} \boldsymbol{X} = \boldsymbol{X} \otimes^{\widetilde{\boldsymbol{X}}}_{\mathcal{C}} \mathrm{B}(-\downarrow \mathcal{C})^{\mathrm{op}} \to \boldsymbol{X} \otimes^{\widetilde{\boldsymbol{X}}}_{\mathcal{C}} \boldsymbol{P} = \operatorname{colim} \boldsymbol{X}$$

for all $\mathcal{C}$-diagrams $\boldsymbol{X}$ in $\mathcal{M}$ (see Example 19.2.8 and Proposition 19.2.9).

(2) If $\boldsymbol{P}\colon \mathcal{C} \to \mathrm{SS}$ is a single point for every object α of $\mathcal{C}$, then the unique map of $\mathcal{C}$-diagrams $\mathrm{B}(\mathcal{C}\downarrow -) \to \boldsymbol{P}$ induces a natural map

$$\lim \boldsymbol{X} = \hom^{\mathcal{C}}_{\widehat{\boldsymbol{X}}}(\boldsymbol{P}, \boldsymbol{X}) \to \hom^{\mathcal{C}}_{\widehat{\boldsymbol{X}}}(\mathrm{B}(\mathcal{C}\downarrow -), \boldsymbol{X}) = \operatorname{holim} \boldsymbol{X}$$

for all $\mathcal{C}$-diagrams $\boldsymbol{X}$ in $\mathcal{M}$.

PROPOSITION 19.2.11. *Let $\mathcal{M}$ and $\mathcal{N}$ be model categories and let $\mathrm{F}\colon \mathcal{M} \rightleftarrows \mathcal{N} : \mathrm{U}$ be a Quillen pair (see Definition 8.5.2).*

(1) *If $\mathcal{C}$ is a small category, $\boldsymbol{K}$ is a $\mathcal{C}^{\mathrm{op}}$-diagram of simplicial sets, $\boldsymbol{X}$ is a $\mathcal{C}$-diagram in $\mathcal{M}$, and $\widetilde{\boldsymbol{X}}$ is a cosimplicial frame on $\boldsymbol{X}$ (see Definition 16.7.2), then $\mathrm{F}\widetilde{\boldsymbol{X}}\colon \mathcal{C} \to \mathcal{N}^{\boldsymbol{\Delta}}$ is a cosimplicial frame on $\mathrm{F}\boldsymbol{X}$ and there is a natural isomorphism*

$$\mathrm{F}(\boldsymbol{X} \otimes_{\mathcal{C}}^{\widetilde{\boldsymbol{X}}} \boldsymbol{K}) \approx (\mathrm{F}\boldsymbol{X}) \otimes_{\mathcal{C}}^{\mathrm{F}\widetilde{\boldsymbol{X}}} \boldsymbol{K} \ .$$

(2) *If $\mathcal{C}$ is a small category, $\boldsymbol{K}$ is a $\mathcal{C}$-diagram of simplicial sets, $\boldsymbol{X}$ is a $\mathcal{C}$-diagram in $\mathcal{N}$, and $\widehat{\boldsymbol{X}}$ is a simplicial frame on $\boldsymbol{X}$ (see Definition 16.7.2), then $\mathrm{U}\widehat{\boldsymbol{X}}$ is a simplicial frame on $\mathrm{U}\boldsymbol{X}$ and there is a natural isomorphism*

$$\mathrm{U}\big(\hom_{\widehat{\boldsymbol{X}}}^{\mathcal{C}}(\boldsymbol{K}, \boldsymbol{X})\big) \approx \hom_{\mathrm{U}\widehat{\boldsymbol{X}}}^{\mathcal{C}}(\boldsymbol{K}, \mathrm{U}\boldsymbol{X}) \ .$$

PROOF. We will prove part 1; the proof of part 2 is dual.

The coend $\boldsymbol{X} \otimes_{\mathcal{C}}^{\widetilde{\boldsymbol{X}}} \boldsymbol{K}$ is the coequalizer of Diagram 19.2.3. As a functor of $\boldsymbol{X}$, this is a composition of functors that commute with left adjoints, and so it commutes with left adjoints. The result now follows from Proposition 16.6.19. □

19.2.12. Adjointness.

PROPOSITION 19.2.13. *Let $\mathcal{M}$ be a model category and let $\mathcal{C}$ be a small category.*

(1) *If $\boldsymbol{X}$ is a $\mathcal{C}$-diagram in $\mathcal{M}$, $\widetilde{\boldsymbol{X}}$ is a cosimplicial frame on $\boldsymbol{X}$ (see Definition 16.7.2), $\boldsymbol{K}$ is a $\mathcal{C}^{\mathrm{op}}$-diagram of simplicial sets, and Z is an object of $\mathcal{M}$, then there is a natural isomorphism of sets*

$$\mathcal{M}(\boldsymbol{X} \otimes_{\mathcal{C}}^{\widetilde{\boldsymbol{X}}} \boldsymbol{K}, Z) \approx \mathrm{SS}^{\mathcal{C}^{\mathrm{op}}}\big(\boldsymbol{K}, \mathcal{M}(\widetilde{\boldsymbol{X}}, Z)\big)$$

(where $\boldsymbol{X} \otimes_{\mathcal{C}}^{\widetilde{\boldsymbol{X}}} \boldsymbol{K}$ is as in Definition 19.2.2).

(2) *If $\boldsymbol{X}$ is a $\mathcal{C}$-diagram in $\mathcal{M}$, $\widehat{\boldsymbol{X}}$ is a simplicial frame on $\boldsymbol{X}$ (see Definition 16.7.2), $\boldsymbol{K}$ is a $\mathcal{C}$-diagram of simplicial sets, and W is an object of $\mathcal{M}$, then there is a natural isomorphism of sets*

$$\mathcal{M}\big(W, \hom_{\widehat{\boldsymbol{X}}}^{\mathcal{C}}(\boldsymbol{K}, \boldsymbol{X})\big) \approx \mathrm{SS}^{\mathcal{C}}\big(\boldsymbol{K}, \mathcal{M}(W, \widehat{\boldsymbol{X}})\big)$$

(where $\hom_{\widehat{\boldsymbol{X}}}^{\mathcal{C}}(\boldsymbol{K}, \boldsymbol{X})$ is as in Definition 19.2.2).

PROOF. We will prove part 1; the proof of part 2 is similar.

The object $\boldsymbol{X} \otimes_{\mathcal{C}}^{\widetilde{\boldsymbol{X}}} \boldsymbol{K}$ is defined as the colimit of Diagram 19.2.3, and so $\mathcal{M}(\boldsymbol{X} \otimes_{\mathcal{C}}^{\widetilde{\boldsymbol{X}}} \boldsymbol{K}, Z)$ is naturally isomorphic to the limit of the diagram

$$\prod_{\alpha \in \mathrm{Ob}(\mathcal{C})} \mathcal{M}(\widetilde{\boldsymbol{X}}_\alpha \otimes \boldsymbol{K}_\alpha, Z) \overset{\phi^*}{\underset{\psi^*}{\rightrightarrows}} \prod_{(\sigma\colon \alpha \to \alpha') \in \mathcal{C}} \mathcal{M}(\widetilde{\boldsymbol{X}}_\alpha \otimes \boldsymbol{K}_{\alpha'}, Z) \ .$$

Theorem 16.4.2 implies that this limit is naturally isomorphic to the limit of the diagram

$$\prod_{\alpha \in \mathrm{Ob}(\mathcal{C})} \mathrm{SS}\big(\boldsymbol{K}_\alpha, \mathcal{M}(\widetilde{\boldsymbol{X}}_\alpha, Z)\big) \overset{\phi^*}{\underset{\psi^*}{\rightrightarrows}} \prod_{(\sigma\colon \alpha \to \alpha') \in \mathcal{C}} \mathrm{SS}\big(\boldsymbol{K}_{\alpha'}, \mathcal{M}(\widetilde{\boldsymbol{X}}_\alpha, Z)\big) \ ,$$

which is the definition of $\mathrm{SS}^{\mathcal{C}^{\mathrm{op}}}\big(\boldsymbol{K}, \mathcal{M}(\widetilde{\boldsymbol{X}}, Z)\big)$. □

LEMMA 19.2.14. *Let $\mathcal{M}$ be a model category and let $\mathcal{C}$ be a small category.*

(1) *If $j\colon \boldsymbol{A} \to \boldsymbol{B}$ is a map of $\mathcal{C}$-diagrams in $\mathcal{M}$, $\tilde{j}\colon \widetilde{\boldsymbol{A}} \to \widetilde{\boldsymbol{B}}$ is a cosimplicial frame on j (see Definition 16.7.2), $\boldsymbol{K} \to \boldsymbol{L}$ is a map of $\mathcal{C}^{\mathrm{op}}$-diagrams of simplicial sets, and $X \to Y$ is a map of objects in $\mathcal{M}$, then the dotted arrow exists in every solid arrow diagram of the form*

$$\begin{array}{ccc} \boldsymbol{A} \otimes_{\mathcal{C}}^{\widetilde{\boldsymbol{A}}} \boldsymbol{L} \amalg_{\boldsymbol{A} \otimes_{\mathcal{C}}^{\widetilde{\boldsymbol{A}}} \boldsymbol{K}} \boldsymbol{B} \otimes_{\mathcal{C}}^{\widetilde{\boldsymbol{B}}} \boldsymbol{K} & \longrightarrow & X \\ \downarrow & \nearrow & \downarrow \\ \boldsymbol{B} \otimes_{\mathcal{C}}^{\widetilde{\boldsymbol{B}}} \boldsymbol{L} & \longrightarrow & Y \end{array}$$

if and only if the dotted arrow exists in every solid arrow diagram of the form

$$\begin{array}{ccc} \boldsymbol{K} & \longrightarrow & \mathcal{M}(\widetilde{\boldsymbol{B}}, X) \\ \downarrow & \nearrow & \downarrow \\ \boldsymbol{L} & \longrightarrow & \mathcal{M}(\widetilde{\boldsymbol{A}}, X) \times_{\mathcal{M}(\widetilde{\boldsymbol{A}}, Y)} \mathcal{M}(\widetilde{\boldsymbol{B}}, Y) \end{array} .$$

(2) *If $p\colon \boldsymbol{X} \to \boldsymbol{Y}$ is a map of $\mathcal{C}$-diagrams in $\mathcal{M}$, $\hat{p}\colon \widehat{\boldsymbol{X}} \to \widehat{\boldsymbol{Y}}$ is a simplicial frame on p (see Definition 16.7.2), $\boldsymbol{K} \to \boldsymbol{L}$ is a map of $\mathcal{C}$-diagrams of simplicial sets, and $A \to B$ is a map of objects in $\mathcal{M}$, then the dotted arrow exists in every solid arrow diagram of the form*

$$\begin{array}{ccc} A & \longrightarrow & \hom^{\mathcal{C}}(\boldsymbol{L}, \boldsymbol{X}) \\ \downarrow & \nearrow & \downarrow \\ B & \longrightarrow & \hom^{\mathcal{C}}_{\widehat{\boldsymbol{X}}}(\boldsymbol{K}, \boldsymbol{X}) \times_{\hom^{\mathcal{C}}_{\widehat{\boldsymbol{Y}}}(\boldsymbol{K}, \boldsymbol{Y})} \hom^{\mathcal{C}}_{\widehat{\boldsymbol{Y}}}(\boldsymbol{L}, \boldsymbol{Y}) \end{array}$$

if and only if the dotted arrow exists in every solid arrow diagram of the form

$$\begin{array}{ccc} \boldsymbol{K} & \longrightarrow & \mathcal{M}(\mathrm{B}, \widehat{\boldsymbol{X}}) \\ \downarrow & \nearrow & \downarrow \\ \boldsymbol{L} & \longrightarrow & \mathcal{M}(A, \widehat{\boldsymbol{X}}) \times_{\mathcal{M}(A, \widehat{\boldsymbol{Y}})} \mathcal{M}(B, \widehat{\boldsymbol{Y}}) \end{array} .$$

PROOF. This follows from Proposition 19.2.13. □

19.3. Consequences of adjointness

In this section we combine the adjointness results of Section 19.2 with the homotopy lifting extension theorem to obtain results for the pushout corner map of coends and the pullback corner map of ends, which we will use in Section 19.4 to obtain our homotopy invariance results.

THEOREM 19.3.1. *Let $\mathcal{M}$ be a model category and let $\mathcal{C}$ be a small category.*

(1) *If $j\colon \boldsymbol{A} \to \boldsymbol{B}$ is an objectwise cofibration of $\mathcal{C}$-diagrams in $\mathcal{M}$, $\tilde{j}\colon \widetilde{\boldsymbol{A}} \to \widetilde{\boldsymbol{B}}$ is a cosimplicial frame on j (see Definition 16.7.2) that is an objectwise Reedy cofibration, and $i\colon \boldsymbol{K} \to \boldsymbol{L}$ is a cofibration of $\mathcal{C}^{\mathrm{op}}$-diagrams of simplicial sets (see Theorem 11.6.1), then the pushout corner map*

$$\boldsymbol{A} \otimes_{\mathcal{C}}^{\widetilde{\boldsymbol{A}}} \boldsymbol{L} \amalg_{\boldsymbol{A} \otimes_{\mathcal{C}}^{\widetilde{\boldsymbol{A}}} \boldsymbol{K}} \boldsymbol{B} \otimes_{\mathcal{C}}^{\widetilde{\boldsymbol{B}}} \boldsymbol{K} \longrightarrow \boldsymbol{B} \otimes_{\mathcal{C}}^{\widetilde{\boldsymbol{B}}} \boldsymbol{L}$$

is a cofibration in $\mathcal{M}$ that is a weak equivalence if either j is an objectwise weak equivalence or i is a weak equivalence.

(2) *If $p\colon \boldsymbol{X} \to \boldsymbol{Y}$ is an objectwise fibration of $\mathcal{C}$-diagrams in $\mathcal{M}$, $\hat{p}\colon \widehat{\boldsymbol{X}} \to \widehat{\boldsymbol{Y}}$ is a simplicial frame on p (see Definition 16.7.2) that is an objectwise Reedy fibration, and $i\colon \boldsymbol{K} \to \boldsymbol{L}$ is a cofibration of $\mathcal{C}^{\mathrm{op}}$-diagrams of simplicial sets (see Theorem 11.6.1), then the pullback corner map*

$$\hom_{\widehat{\boldsymbol{X}}}^{\mathcal{C}}(\boldsymbol{L}, \boldsymbol{X}) \longrightarrow \hom_{\widehat{\boldsymbol{X}}}^{\mathcal{C}}(\boldsymbol{K}, \boldsymbol{X}) \times_{\hom_{\widehat{\boldsymbol{Y}}}^{\mathcal{C}}(\boldsymbol{K}, \boldsymbol{Y})} \hom_{\widehat{\boldsymbol{Y}}}^{\mathcal{C}}(\boldsymbol{L}, \boldsymbol{Y})$$

is a fibration in $\mathcal{M}$ that is a weak equivalence if either p is an objectwise weak equivalence or i is a weak equivalence.

PROOF. We will prove part 1; the proof of part 2 is similar.

If $p\colon X \to Y$ is a fibration in $\mathcal{M}$, then Theorem 16.5.2 implies that the map of $\mathcal{C}^{\mathrm{op}}$-diagrams of simplicial sets $\mathcal{M}(\widetilde{\boldsymbol{B}}, X) \to \mathcal{M}(\widetilde{\boldsymbol{A}}, X) \times_{\mathcal{M}(\widetilde{\boldsymbol{A}}, Y)} \mathcal{M}(\widetilde{\boldsymbol{B}}, Y)$ is an objectwise fibration that is an objectwise weak equivalence if either j is an objectwise weak equivalence or p is a weak equivalence. The result now follows from Lemma 19.2.14, Proposition 7.2.3, and Theorem 11.6.1. □

PROPOSITION 19.3.2. *Let $\mathcal{M}$ be a model category and let $\mathcal{C}$ be a small category.*

(1) *If $\boldsymbol{X}$ is an objectwise cofibrant $\mathcal{C}$-diagram in $\mathcal{M}$, $\widetilde{\boldsymbol{X}}$ is a cosimplicial frame on $\boldsymbol{X}$, and $\boldsymbol{K}$ is a $\mathcal{C}^{\mathrm{op}}$-diagram of simplicial sets that is a cofibrant object of $\mathrm{SS}^{\mathcal{C}^{\mathrm{op}}}$ (see Theorem 11.6.1), then $\boldsymbol{X} \otimes_{\mathcal{C}}^{\widetilde{\boldsymbol{X}}} \boldsymbol{K}$ is a cofibrant object of $\mathcal{M}$.*

(2) *If $\boldsymbol{X}$ is an objectwise fibrant $\mathcal{C}$-diagram in $\mathcal{M}$, $\widehat{\boldsymbol{X}}$ is a simplicial frame on $\boldsymbol{X}$, and $\boldsymbol{K}$ is a $\mathcal{C}$-diagram of simplicial sets that is a cofibrant object of $\mathrm{SS}^{\mathcal{C}}$ (see Theorem 11.6.1), then $\hom_{\widehat{\boldsymbol{X}}}^{\mathcal{C}}(\boldsymbol{K}, \boldsymbol{X})$ is a fibrant object of $\mathcal{M}$.*

PROOF. This follows form Theorem 19.3.1. □

PROPOSITION 19.3.3. *Let $\mathcal{M}$ be a model category and let $\mathcal{C}$ be a small category.*

(1) *If $\boldsymbol{X}$ is an objectwise cofibrant $\mathcal{C}$-diagram in $\mathcal{M}$, $\widetilde{\boldsymbol{X}}$ is a cosimplicial frame on $\boldsymbol{X}$, and $f\colon \boldsymbol{K} \to \boldsymbol{K}'$ is a weak equivalence of cofibrant $\mathcal{C}^{\mathrm{op}}$-diagrams of simplicial sets (see Theorem 11.6.1), then the induced map $f_*\colon \boldsymbol{X} \otimes_{\mathcal{C}}^{\widetilde{\boldsymbol{X}}} \boldsymbol{K} \to \boldsymbol{X} \otimes_{\mathcal{C}}^{\widetilde{\boldsymbol{X}}} \boldsymbol{K}'$ is a weak equivalence of cofibrant objects in $\mathcal{M}$.*

(2) *If $\boldsymbol{X}$ is an objectwise fibrant $\mathcal{C}$-diagram $\mathcal{M}$, $\widehat{\boldsymbol{X}}$ is a simplicial frame on $\boldsymbol{X}$, and $f\colon \boldsymbol{K} \to \boldsymbol{K}'$ is a weak equivalence of cofibrant $\mathcal{C}$-diagrams of simplicial sets (see Theorem 11.6.1), then the induced map $f^*\colon \hom_{\widehat{\boldsymbol{X}}}^{\mathcal{C}}(\boldsymbol{K}', \boldsymbol{X}) \to \hom_{\widehat{\boldsymbol{X}}}^{\mathcal{C}}(\boldsymbol{K}, \boldsymbol{X})$ is a weak equivalence of fibrant objects in $\mathcal{M}$.*

PROOF. This follows from Theorem 19.3.1 and Corollary 7.7.2. □

PROPOSITION 19.3.4. *Let $\mathcal{M}$ be a model category and let $\mathcal{C}$ be a small category.*

(1) *If $\boldsymbol{K}$ is a $\mathcal{C}^{\mathrm{op}}$-diagram of simplicial sets and both $\widetilde{\boldsymbol{K}} \to \boldsymbol{K}$ and $\widetilde{\boldsymbol{K}'} \to \boldsymbol{K}$ are cofibrant approximations to $\boldsymbol{K}$ (see Theorem 11.6.1), then for every*

objectwise cofibrant $\mathcal{C}$-diagram $\boldsymbol{X}$ in $\mathcal{M}$ and every cosimplicial frame $\widetilde{\boldsymbol{X}}$ on $\boldsymbol{X}$ there is an essentially unique natural zig-zag of weak equivalences (see Definition 14.4.1) in $\mathcal{M}$ from $\boldsymbol{X} \otimes_{\mathcal{C}}^{\widetilde{\boldsymbol{X}}} \boldsymbol{K}$ to $\boldsymbol{X} \otimes_{\mathcal{C}}^{\widetilde{\boldsymbol{X}}} \boldsymbol{K}'$ that is natural up to weak equivalence.

(2) *If $\boldsymbol{K}$ is a $\mathcal{C}$-diagram of simplicial sets and both $\widetilde{\boldsymbol{K}} \to \boldsymbol{K}$ and $\widetilde{\boldsymbol{K}}' \to \boldsymbol{K}$ are cofibrant approximations to $\boldsymbol{K}$ (see Theorem 11.6.1), then for every objectwise fibrant $\mathcal{C}$-diagram $\boldsymbol{X}$ in $\mathcal{M}$ and every simplicial frame $\widehat{\boldsymbol{X}}$ on $\boldsymbol{X}$ there is an essentially unique natural zig-zag of weak equivalences (see Definition 14.4.1) in $\mathcal{M}$ from $\hom_{\widehat{\boldsymbol{X}}}^{\mathcal{C}}(\widetilde{\boldsymbol{K}}, \boldsymbol{X})$ to $\hom_{\widehat{\boldsymbol{X}}}^{\mathcal{C}}(\widetilde{\boldsymbol{K}}', \boldsymbol{X})$ that is natural up to weak equivalence.*

PROOF. This follows from Proposition 19.3.3 and Proposition 14.6.3. □

PROPOSITION 19.3.5. *Let $\mathcal{M}$ be a model category and let $\mathcal{C}$ be a small category.*

(1) *If $\boldsymbol{K}$ is a cofibrant $\mathcal{C}^{\mathrm{op}}$-diagram of simplicial sets (see Theorem 11.6.1), $f\colon \boldsymbol{X} \to \boldsymbol{Y}$ is a map of $\mathcal{C}$-diagrams in $\mathcal{M}$ that is an objectwise weak equivalence of cofibrant objects, and $\tilde{f}\colon \widetilde{\boldsymbol{X}} \to \widetilde{\boldsymbol{Y}}$ is a cosimplicial frame on f (see Definition 16.6.12), then the induced map of coends $f_*\colon \boldsymbol{X} \otimes_{\mathcal{C}}^{\widetilde{\boldsymbol{X}}} \boldsymbol{K} \to \boldsymbol{Y} \otimes_{\mathcal{C}}^{\widetilde{\boldsymbol{Y}}} \boldsymbol{K}$ is a weak equivalence.*

(2) *If $\boldsymbol{K}$ is a cofibrant $\mathcal{C}$-diagram of simplicial sets (see Theorem 11.6.1), $f\colon \boldsymbol{X} \to \boldsymbol{Y}$ is a map of $\mathcal{C}$-diagrams in $\mathcal{M}$ that is an objectwise weak equivalence of fibrant objects, and $\hat{f}\colon \widehat{\boldsymbol{X}} \to \widehat{\boldsymbol{Y}}$ is a simplicial frame on f (see Definition 16.6.12), then the induced map $f_*\colon \hom_{\widehat{\boldsymbol{X}}}^{\mathcal{C}}(\boldsymbol{K}, \boldsymbol{X}) \to \hom_{\widehat{\boldsymbol{Y}}}^{\mathcal{C}}(\boldsymbol{K}, \boldsymbol{Y})$ is a weak equivalence.*

PROOF. This follows from Theorem 19.3.1 and Corollary 7.7.2. □

COROLLARY 19.3.6. *Let $\mathcal{M}$ be a model category and let $\mathcal{C}$ be a small category.*

(1) *If $\boldsymbol{K}$ is a cofibrant $\mathcal{C}^{\mathrm{op}}$-diagram of simplicial sets (see Theorem 11.6.1), $\boldsymbol{X}$ is an objectwise cofibrant $\mathcal{C}$-diagram in $\mathcal{M}$, and $f\colon \widetilde{\boldsymbol{X}} \to \widetilde{\boldsymbol{X}}'$ is a map of cosimplicial frames on $\boldsymbol{X}$, then the induced map of coends $f_*\colon \boldsymbol{X} \otimes_{\mathcal{C}}^{\widetilde{\boldsymbol{X}}} \boldsymbol{K} \to \boldsymbol{X} \otimes_{\mathcal{C}}^{\widetilde{\boldsymbol{X}}'} \boldsymbol{K}$ is a weak equivalence.*

(2) *If $\boldsymbol{K}$ is a cofibrant $\mathcal{C}$-diagram of simplicial sets (see Theorem 11.6.1), $\boldsymbol{X}$ is an objectwise fibrant $\mathcal{C}$-diagram in $\mathcal{M}$, and $f\colon \widehat{\boldsymbol{X}} \to \widehat{\boldsymbol{X}}'$ is a map of simplicial frames on $\boldsymbol{X}$, then the induced map of ends $f_*\colon \hom_{\widehat{\boldsymbol{X}}}^{\mathcal{C}}(\boldsymbol{K}, \boldsymbol{X}) \to \hom_{\widehat{\boldsymbol{X}}'}^{\mathcal{C}}(\boldsymbol{K}, \boldsymbol{X})$ is a weak equivalence.*

PROOF. This follows from Proposition 19.3.5. □

COROLLARY 19.3.7. *Let $\mathcal{M}$ be a model category and let $\mathcal{C}$ be a small category.*

(1) *If $\boldsymbol{K}$ is a cofibrant $\mathcal{C}^{\mathrm{op}}$-diagram of simplicial sets (see Theorem 11.6.1), $\boldsymbol{X}$ is an objectwise cofibrant $\mathcal{C}$-diagram in $\mathcal{M}$, and both $\widetilde{\boldsymbol{X}}$ and $\widetilde{\boldsymbol{X}}'$ are cosimplicial frames on $\boldsymbol{X}$, then there is an essentially unique zig-zag (see Definition 14.4.2) of weak equivalences induced by maps of frames from the coend $\boldsymbol{X} \otimes_{\mathcal{C}}^{\widetilde{\boldsymbol{X}}} \boldsymbol{K}$ to the coend $\boldsymbol{X} \otimes_{\mathcal{C}}^{\widetilde{\boldsymbol{X}}'} \boldsymbol{K}$.*

(2) *If $\boldsymbol{K}$ is a cofibrant $\mathcal{C}$-diagram of simplicial sets (see Theorem 11.6.1), $\boldsymbol{X}$ is an objectwise fibrant $\mathcal{C}$-diagram in $\mathcal{M}$, and both $\widehat{\boldsymbol{X}}$ and $\widehat{\boldsymbol{X}}'$ are simplicial frames on $\boldsymbol{X}$, then there is an essentially unique zig-zag of*

weak equivalences induced by maps of frames from the end $\hom^{\mathcal{C}}_{\widetilde{\boldsymbol{X}}}(\boldsymbol{K}, \boldsymbol{X})$ *to the end* $\hom^{\mathcal{C}}_{\widetilde{\boldsymbol{X}}'}(\boldsymbol{K}, \boldsymbol{X})$

PROOF. This follows from Corollary 19.3.6 and Theorem 16.7.6. □

PROPOSITION 19.3.8. *Let* $\mathcal{M}$ *and* $\mathcal{N}$ *be framed model categories and let* $\mathrm{F} : \mathcal{M} \rightleftarrows \mathcal{N} : \mathrm{U}$ *be a Quillen pair (see Definition 8.5.2).*

1. *If* $\mathcal{C}$ *is a small category and* $\boldsymbol{K}$ *is a cofibrant* $\mathcal{C}^{\mathrm{op}}$*-diagram of simplicial sets (see Theorem 11.6.1), then there is an essentially unique zig-zag (see Definition 11.7.2) of natural transformations of functors* $\mathcal{M}^{\mathcal{C}} \to \mathcal{N}$ *induced by maps of frames from* $\mathrm{F}(- \otimes_{\mathcal{C}} \boldsymbol{K})$ *to* $(\mathrm{F}-) \otimes_{\mathcal{C}} \boldsymbol{K}$ *that is a zig-zag of natural weak equivalences on objectwise cofibrant diagrams.*
2. *If* $\mathcal{C}$ *is a small category and* $\boldsymbol{K}$ *is a cofibrant* $\mathcal{C}$*-diagram of simplicial sets (see Theorem 11.6.1), then there is an essentially unique zig-zag (see Definition 11.7.2) of natural transformations of functors* $\mathcal{N}^{\mathcal{C}} \to \mathcal{M}$ *induced by maps of frames from* $\mathrm{U}\hom^{\mathcal{C}}(\boldsymbol{K}, -)$ *to* $\hom^{\mathcal{C}}(\boldsymbol{K}, \mathrm{U}-)$ *that is a zig-zag of natural weak equivalences on objectwise fibrant diagrams.*

PROOF. This follows from Corollary 19.3.6, Proposition 19.2.11, and Theorem 16.6.10. □

19.4. Homotopy invariance

This section contains our homotopy invariance results for the homotopy colimit and homotopy limit functors (see Theorem 19.4.2). We also show that changing the framing of the model category results in naturally weakly equivalent homotopy colimit and homotopy limit functors (see Theorem 19.4.3), and we establish an adjointness result connecting the homotopy colimit and homotopy limit functors (see Theorem 19.4.4).

THEOREM 19.4.1. *Let* $\mathcal{M}$ *be a framed model category and let* $\mathcal{C}$ *be a small category.*

1. *If* $\boldsymbol{X}$ *is an objectwise cofibrant* $\mathcal{C}$*-diagram in* $\mathcal{M}$*, then* $\operatorname{hocolim} \boldsymbol{X}$ *is cofibrant.*
2. *If* $\boldsymbol{X}$ *is an objectwise fibrant* $\mathcal{C}$*-diagram in* $\mathcal{M}$*, then* $\operatorname{holim} \boldsymbol{X}$ *is fibrant.*

PROOF. This follows from Proposition 19.3.2 and Corollary 14.8.8. □

THEOREM 19.4.2. *Let* $\mathcal{M}$ *be a framed model category, and let* $\mathcal{C}$ *be a small category.*

1. *If* $f\colon \boldsymbol{X} \to \boldsymbol{Y}$ *is a map of* $\mathcal{C}$*-diagrams in* $\mathcal{M}$ *such that* $f_\alpha\colon \boldsymbol{X}_\alpha \to \boldsymbol{Y}_\alpha$ *is a weak equivalence of cofibrant objects for every object* α *of* $\mathcal{C}$*, then the induced map of homotopy colimits* $f_*\colon \operatorname{hocolim} \boldsymbol{X} \to \operatorname{hocolim} \boldsymbol{Y}$ *is a weak equivalence of cofibrant objects of* $\mathcal{M}$*.*
2. *If* $f\colon \boldsymbol{X} \to \boldsymbol{Y}$ *is a map of* $\mathcal{C}$*-diagrams in* $\mathcal{M}$ *such that* $f_\alpha\colon \boldsymbol{X}_\alpha \to \boldsymbol{Y}_\alpha$ *is a weak equivalence of fibrant objects for every object* α *of* $\mathcal{C}$*, then the induced map of homotopy limits* $f_*\colon \operatorname{holim} \boldsymbol{X} \to \operatorname{holim} \boldsymbol{Y}$ *is a weak equivalence of fibrant objects of* $\mathcal{M}$*.*

PROOF. This follows from Proposition 19.3.5 and Theorem 19.4.1. □

THEOREM 19.4.3. *Let $\mathcal{M}$ be a model category, let $\mathcal{M}'$ and $\mathcal{M}''$ be framed model categories for which the underlying model category is $\mathcal{M}$, and let $\mathcal{C}$ be a small category.*

(1) *There is an essentially unique zig-zag (see Definition 14.4.2) of natural transformations induced by maps of framings connecting the homotopy colimit constructed using the framing of $\mathcal{M}'$ to the homotopy colimit constructed using the framing of $\mathcal{M}''$ that is a zig-zag of weak equivalences on objectwise cofibrant diagrams.*
(2) *There is an essentially unique zig-zag (see Definition 14.4.2) of natural transformations induced by maps of framings connecting the homotopy limit constructed using the framing of $\mathcal{M}'$ to the homotopy limit constructed using the framing of $\mathcal{M}''$ that is a zig-zag of weak equivalences on objectwise fibrant diagrams.*

PROOF. This follows from Example 19.2.8, Corollary 19.3.6, and Theorem 16.6.10. □

THEOREM 19.4.4. *Let $\mathcal{M}$ be a framed model category and let $\mathcal{C}$ be a small category.*

(1) *If $\boldsymbol{X}$ is an objectwise cofibrant $\mathcal{C}$-diagram in $\mathcal{M}$ and Y is a fibrant object of $\mathcal{M}$, then* $\operatorname{map}(\operatorname{hocolim}\boldsymbol{X}, Y)$ *(see Notation 17.4.2) is naturally weakly equivalent to* $\operatorname{holim}\operatorname{map}(\boldsymbol{X}, Y)$.
(2) *If X is a cofibrant object of $\mathcal{M}$ and $\boldsymbol{Y}$ is an objectwise fibrant $\mathcal{C}$-diagram in $\mathcal{M}$, then* $\operatorname{map}(X, \operatorname{holim}\boldsymbol{Y})$ *(see Notation 17.4.2) is naturally weakly equivalent to* $\operatorname{holim}\operatorname{map}(X, \boldsymbol{Y})$.

PROOF. We will prove part 1; the proof of part 2 is similar.

Since Y is fibrant, the simplicial frame $\widehat{\boldsymbol{Y}}$ on Y defined by the framing of $\mathcal{M}$ is a simplicial resolution of Y, and so $\mathcal{M}(\operatorname{hocolim}\boldsymbol{X}, \widehat{\boldsymbol{Y}})$ is a homotopy function complex $\operatorname{map}(\operatorname{hocolim}\boldsymbol{X}, Y)$. Example 19.2.8, Proposition 19.2.13, and Corollary 11.8.7 imply that $\mathcal{M}(\operatorname{hocolim}\boldsymbol{X}, \widehat{\boldsymbol{Y}})$ is naturally isomorphic to $\mathrm{SS}^{\mathcal{C}^{\mathrm{op}}}\big(\mathrm{B}(\mathcal{C}^{\mathrm{op}}\downarrow -), \mathcal{M}(\widetilde{\boldsymbol{X}}, \widehat{\boldsymbol{Y}})\big)$ (where $\widetilde{\boldsymbol{X}}\colon \mathcal{C} \to \mathcal{M}^{\boldsymbol{\Delta}}$ is the frame on $\boldsymbol{X}$ induced by the framing of $\mathcal{M}$). Since each $\boldsymbol{X}_\alpha$ is cofibrant, we can take the $\mathcal{M}(\widetilde{\boldsymbol{X}}_\alpha, Y)$ as our homotopy function complexes $\operatorname{map}(\boldsymbol{X}_\alpha, Y)$, and Corollary 16.5.16 implies that the bisimplicial set $\mathcal{M}(\widetilde{\boldsymbol{X}}_\alpha, \widehat{\boldsymbol{Y}})$ is a resolution of the simplicial set $\mathcal{M}(\widetilde{\boldsymbol{X}}_\alpha, Y)$. Example 17.2.4 and Theorem 17.5.14 imply that each simplicial set $\mathrm{SS}\big(\mathrm{B}(\mathcal{C}^{\mathrm{op}}\downarrow\alpha), \mathcal{M}(\widetilde{\boldsymbol{X}}_\alpha, \widehat{\boldsymbol{Y}})\big)$ is naturally weakly equivalent to $\operatorname{Map}\big(\mathrm{B}(\mathcal{C}^{\mathrm{op}}\downarrow\alpha), \operatorname{map}(\boldsymbol{X}_\alpha, Y)\big)$, and so the result follows from Proposition 18.2.6. □

THEOREM 19.4.5. *Let $\mathcal{M}$ and $\mathcal{N}$ be framed model categories and let* $\mathrm{F}\colon \mathcal{M} \rightleftarrows \mathcal{N} :\mathrm{U}$ *be a Quillen pair (see Definition 8.5.2).*

(1) *If $\mathcal{C}$ is a small category, then there is an essentially unique zig-zag of natural transformations of functors $\mathcal{M}^{\mathcal{C}} \to \mathcal{N}$ induced by maps of frames from* $\mathrm{F}\operatorname{hocolim}$ *to* $\operatorname{hocolim}\mathrm{F}$ *that is a natural zig-zag of weak equivalences on objectwise cofibrant diagrams.*
(2) *If $\mathcal{C}$ is a small category, then there is an essentially unique zig-zag of natural transformations of functors $\mathcal{N}^{\mathcal{C}} \to \mathcal{M}$ induced by maps of frames from* $\mathrm{U}\operatorname{holim}$ *to* $\operatorname{holim}\mathrm{U}$ *that is a natural zig-zag of weak equivalences on objectwise fibrant diagrams.*

PROOF. This follows from Proposition 19.3.8 and Corollary 14.8.8. □

19.4.6. The significance of overcategories and undercategories.

THEOREM 19.4.7. *Let $\mathcal{C}$ be a small category and let $\mathcal{M}$ be a model category.*

(1) *If $\boldsymbol{P}$ is a $\mathcal{C}^{\mathrm{op}}$-diagram of simplicial sets that is a cofibrant approximation (see Definition 8.1.2) to the constant $\mathcal{C}^{\mathrm{op}}$-diagram at a point, then for every objectwise cofibrant $\mathcal{C}$-diagram $\boldsymbol{X}$ in $\mathcal{M}$ there is an essentially unique natural zig-zag of weak equivalences induced by maps of cofibrant approximations from $\boldsymbol{X} \otimes_{\mathcal{C}} \boldsymbol{P}$ to* hocolim $\boldsymbol{X}$.
(2) *If $\boldsymbol{P}$ is a $\mathcal{C}$-diagram of simplicial sets that is a cofibrant approximation (see Definition 8.1.2) to the constant $\mathcal{C}$-diagram at a point, then for every objectwise fibrant $\mathcal{C}$-diagram $\boldsymbol{X}$ in $\mathcal{M}$ there is an essentially unique natural zig-zag of weak equivalences induced by maps of cofibrant approximations from* $\hom^{\mathcal{C}}(\boldsymbol{P}, \boldsymbol{X})$ *to* holim $\boldsymbol{X}$.

PROOF. This follows from Proposition 19.3.4, Example 19.2.8, and Proposition 14.8.9. □

19.5. Homotopy pullbacks and homotopy pushouts

If $\mathcal{M}$ is a *right proper* framed model category, then the diagram $X \to Z \leftarrow Y$ has both a homotopy pullback (see Definition 13.3.2) and a homotopy limit (see Definition 19.1.5). We will show that for *fibrant* X, Y, and Z, the homotopy pullback of the diagram $X \to Z \leftarrow Y$ is naturally weakly equivalent to the homotopy limit of that diagram (see Proposition 19.5.3). We begin by showing that, for a map of fibrant objects, the "classical" method of converting a map into a fibration does provide a factorization into a weak equivalence followed by a fibration.

LEMMA 19.5.1. *Let $\mathcal{M}$ be a framed model category and let $g\colon X \to Z$ be a map of fibrant objects. If $\mathrm{ev}_0\colon \widehat{\boldsymbol{Z}}^{\Delta[1]} \to Z$ is the composition $\widehat{\boldsymbol{Z}}^{\Delta[1]} \xrightarrow{(d^1)^*} \widehat{\boldsymbol{Z}}^{\Delta[0]} \approx \widehat{\boldsymbol{Z}}_0 \approx Z$ (see Lemma 16.3.6) and the square*

$$\begin{array}{ccc} W & \xrightarrow{\tilde{g}} & \widehat{\boldsymbol{Z}}^{\Delta[1]} \\ {\scriptstyle k}\downarrow & & \downarrow{\scriptstyle \mathrm{ev}_0} \\ X & \xrightarrow{g} & Z \end{array}$$

is a pullback, then

(1) *the map $\mathrm{ev}_1 \tilde{g}\colon W \to Z$ is a fibration (where $\mathrm{ev}_1\colon \widehat{\boldsymbol{Z}}^{\Delta[1]} \to Z$ is the composition $\widehat{\boldsymbol{Z}}^{\Delta[1]} \xrightarrow{(d^0)^*} \widehat{\boldsymbol{Z}}^{\Delta[0]} \approx \widehat{\boldsymbol{Z}}_0 \approx Z$),*
(2) *if $j\colon X \to W$ is defined by the requirements that $kj = 1_X$ and $\tilde{g}j\colon X \to \widehat{\boldsymbol{Z}}^{\Delta[1]}$ equals the composition $X \xrightarrow{g} Z \approx \widehat{\boldsymbol{Z}}_0 \approx \widehat{\boldsymbol{Z}}^{\Delta[0]} \xrightarrow{(s^0)^*} \widehat{\boldsymbol{Z}}^{\Delta[1]}$, then j is a weak equivalence, and*
(3) $(\mathrm{ev}_1 \tilde{g}) \circ j = g$.

PROOF. Since Z is fibrant, $\widehat{\boldsymbol{Z}}$ is Reedy fibrant, and so ev_0 is a trivial fibration. Thus, k is a trivial fibration. Since $kj = 1_X$, this implies that j is a weak equivalence. Since the composition $\widehat{\boldsymbol{Z}}^{\Delta[0]} \xrightarrow{(s^0)^*} \widehat{\boldsymbol{Z}}^{\Delta[1]} \xrightarrow{(d^0)^*} \widehat{\boldsymbol{Z}}^{\Delta[0]}$ is the identity map, it follows that $(\mathrm{ev}_1 \tilde{g})j = g$, and so it remains only to show that $\mathrm{ev}_1 \tilde{g}$ is a fibration.

Let $i\colon A \to B$ be a trivial cofibration, and suppose that we have the solid arrow diagram

(19.5.2)
$$\begin{array}{ccc} A & \xrightarrow{r} & W \\ {\scriptstyle i}\downarrow & \nearrow & \downarrow{\scriptstyle \mathrm{ev}_1 \tilde{g}} \\ B & \xrightarrow{s} & Z \end{array} ;$$

Proposition 7.2.3 implies that it is sufficient to show that there exists a dotted arrow making both triangles commute. Since X is fibrant, the map $kr\colon A \to X$ can be extended over B to a map $t\colon B \to X$ such that $ti = kr$. We thus have a map $(gt \times s)\colon B \to Z \times Z \approx \widehat{\boldsymbol{Z}}^{\Delta[0]} \times \widehat{\boldsymbol{Z}}^{\Delta[0]} \approx \widehat{\boldsymbol{Z}}^{\partial\Delta[1]}$ that fits into the commutative solid arrow diagram

$$\begin{array}{ccc} A & \xrightarrow{\tilde{g}r} & \widehat{\boldsymbol{Z}}^{\Delta[1]} \\ \downarrow & \nearrow & \downarrow \\ B & \longrightarrow & \widehat{\boldsymbol{Z}}^{\partial\Delta[1]} \end{array}$$

Since $\widehat{\boldsymbol{Z}}$ is Reedy fibrant, Proposition 16.3.8 implies that $\widehat{\boldsymbol{Z}}^{\Delta[1]} \to \widehat{\boldsymbol{Z}}^{\partial\Delta[1]}$ is a fibration, and so the dotted arrow exists in this diagram. This dotted arrow combines with the map $t\colon B \to X$ to define the dotted arrow in Diagram 19.5.2. □

PROPOSITION 19.5.3. *Let $\mathcal{M}$ be a right proper framed model category. If X, Y, and Z are fibrant objects, then the homotopy pullback (see Definition 13.3.2) of the diagram $X \xrightarrow{g} Z \xleftarrow{h} Y$ is naturally weakly equivalent to the homotopy limit (see Definition 19.1.5) of that diagram.*

PROOF. If K is the simplicial set that is the union of two copies of $\Delta[1]$ with vertex 1 of both copies identified to a single point, then the homotopy limit of the diagram $X \xrightarrow{g} Z \xleftarrow{h} Y$ is naturally isomorphic to the limit of the diagram

$$\begin{array}{ccccccc} X & & & \widehat{\boldsymbol{Z}}^{K} & & & Y \\ & {\scriptstyle g}\searrow & \swarrow & & \searrow & \swarrow{\scriptstyle h} & \\ & & Z & & Z & & \end{array}$$

where the two maps with domain $\widehat{\boldsymbol{Z}}^{K}$ are defined by evaluation on vertex 0 of the two copies of $\Delta[1]$ (see Definition 19.1.5). The limit of this last diagram is naturally isomorphic to the limit of the diagram

(19.5.4)
$$\begin{array}{ccccccccc} X & & \widehat{\boldsymbol{Z}}^{\Delta[1]} & & & & \widehat{\boldsymbol{Z}}^{\Delta[1]} & & Y \\ & {\scriptstyle g}\searrow \ \swarrow{\scriptstyle \mathrm{ev}_0} & & {\scriptstyle \mathrm{ev}_1}\searrow & & \swarrow{\scriptstyle \mathrm{ev}_1} & & {\scriptstyle \mathrm{ev}_0}\searrow \ \swarrow{\scriptstyle h} & \\ & Z & & & Z & & & Z & \end{array}$$

(see Proposition 16.4.3). If W_g is the pullback of the diagram $X \xrightarrow{g} Z \xleftarrow{\mathrm{ev}_0} \widehat{\boldsymbol{Z}}^{\Delta[1]}$ and W_h is the pullback of the diagram $Y \xrightarrow{h} Z \xleftarrow{\mathrm{ev}_0} \widehat{\boldsymbol{Z}}^{\Delta[1]}$, then the limit of Diagram 19.5.4 is naturally isomorphic to the pullback of the diagram $W_g \to Z \leftarrow W_h$. Lemma 19.5.1 implies that the maps $W_g \to Z$ and $W_h \to Z$ arise as factorizations of, respectively, g and h into a weak equivalence followed by a fibration, and so the result follows from Proposition 13.3.7. □

19.6. Homotopy cofinal functors

In this section we describe those functors of small categories that induce weak equivalences of homotopy colimits for all objectwise cofibrant diagrams and, dually, those that induce weak equivalences of homotopy limits for all objectwise fibrant diagrams.

DEFINITION 19.6.1. Let $\mathcal{C}$ and $\mathcal{D}$ be small categories and let $F\colon \mathcal{C} \to \mathcal{D}$ be a functor.

- The functor F is *homotopy left cofinal* (or *homotopy initial*) if for every object α of $\mathcal{D}$ the space $B(F \downarrow \alpha)$ (see Definition 14.1.1 and Definition 11.8.1) is contractible.
- The functor F is *homotopy right cofinal* (or *homotopy terminal*) if for every object α of $\mathcal{D}$ the space $B(\alpha \downarrow F)$ (see Definition 11.8.3) is contractible.

If $\mathcal{C}$ is a subcategory of $\mathcal{D}$ and F is the inclusion, then if F is homotopy left cofinal or homotopy right cofinal we will say that $\mathcal{C}$ is, respectively, *a homotopy left cofinal subcategory* or a *homotopy right cofinal subcategory* of $\mathcal{D}$.

We will show in Theorem 19.6.13 that these are the correct notions when considering homotopy limits and homotopy colimits.

REMARK 19.6.2. The reader should be aware that there are conflicting uses of the above terms in the literature. The definitions of Heller ([**41**, page 54]) agree with ours, but Bousfield and Kan ([**14**, page 316]) use the terms *initial* and *left cofinal* for what we call *homotopy initial* and *homotopy left cofinal* and the terms *final* and *right cofinal* for what we call *homotopy final* and *homotopy right cofinal.*

PROPOSITION 19.6.3. *Let $\mathcal{C}$ and $\mathcal{D}$ be small categories, and let $F\colon \mathcal{C} \to \mathcal{D}$ be a functor.*

(1) *If F is homotopy left cofinal, then it is left cofinal (see Definition 14.2.1).*
(2) *If F is homotopy right cofinal, then it is right cofinal (see Definition 14.2.1).*

PROOF. This follows directly from the definitions. □

COROLLARY 19.6.4. *Let $\mathcal{M}$ be a category that is complete and cocomplete.*

(1) *If $\mathcal{C}$ and $\mathcal{D}$ are small categories and $F\colon \mathcal{C} \to \mathcal{D}$ is a homotopy left cofinal functor, then for every $\mathcal{D}$-diagram $\boldsymbol{X}$ in $\mathcal{M}$ the natural map $\lim_{\mathcal{D}} \boldsymbol{X} \to \lim_{\mathcal{C}} F^{*}\boldsymbol{X}$ is an isomorphism.*
(2) *If $\mathcal{C}$ and $\mathcal{D}$ are small categories and $F\colon \mathcal{C} \to \mathcal{D}$ is a homotopy right cofinal functor, then for every $\mathcal{D}$-diagram $\boldsymbol{X}$ in $\mathcal{M}$ the natural map $\operatorname{colim}_{\mathcal{C}} F^{*}\boldsymbol{X} \to \operatorname{colim}_{\mathcal{D}} \boldsymbol{X}$ is an isomorphism.*

PROOF. This follows from Proposition 19.6.3 and Theorem 14.2.5. □

19.6.5. Homotopy colimits and homotopy limits. In this section, we prove that a homotopy right cofinal functor induces a weak equivalence of homotopy colimits for every objectwise cofibrant diagram (see Theorem 19.6.7), and that any functor with this property must be homotopy right cofinal (see Theorem 19.6.13). Dually, we prove that a homotopy left cofinal functor induces a weak equivalence of homotopy limits for every objectwise fibrant diagram, and that any functor with this property must be homotopy left cofinal.

PROPOSITION 19.6.6. *Let $\mathcal{M}$ be a framed model category, let $\mathcal{C}$ and $\mathcal{D}$ be small categories, let $F\colon \mathcal{C} \to \mathcal{D}$ be a functor, let $\boldsymbol{X}$ be a $\mathcal{D}$-diagram in $\mathcal{M}$, and let $F^*\boldsymbol{X}$ be the $\mathcal{C}$-diagram in $\mathcal{M}$ induced by* F *(see Definition* 11.5.5*).*

(1) *There is a natural isomorphism of objects of $\mathcal{M}$*

$$\operatorname*{hocolim}_{\mathcal{C}} F^*\boldsymbol{X} \approx \boldsymbol{X} \otimes_{\mathcal{D}} B(-\downarrow F)^{\mathrm{op}}$$

(Definition 14.7.2*).*

(2) *There is a natural isomorphism of objects of $\mathcal{M}$*

$$\operatorname*{holim}_{\mathcal{C}} F^*\boldsymbol{X} \approx \hom^{\mathcal{D}}\bigl(B(F\downarrow -), \boldsymbol{X}\bigr)$$

(see Definition 19.2.2 *and Definition* 14.7.8*).*

PROOF. Example 19.2.8 implies that $\operatorname{hocolim}_{\mathcal{C}} F^*\boldsymbol{X}$ is naturally isomorphic to $(F^*\boldsymbol{X}) \otimes_{\mathcal{C}} B(-\downarrow \mathcal{C})^{\mathrm{op}}$. For every object α of $\mathcal{C}$ we have a map $F_*\colon B(\alpha\downarrow\mathcal{C})^{\mathrm{op}} \to B(F\alpha\downarrow F)^{\mathrm{op}}$ (see Example 14.1.8), and we use this to define a map $\phi\colon (F^*\boldsymbol{X}) \otimes_{\mathcal{C}} B(-\downarrow\mathcal{C})^{\mathrm{op}} \to \boldsymbol{X} \otimes_{\mathcal{D}} B(-\downarrow F)^{\mathrm{op}}$ as the map induced by the composition

$$\coprod_{\alpha\in\mathrm{Ob}(\mathcal{C})} (\widetilde{F^*\boldsymbol{X}}) \otimes B(\alpha\downarrow\mathcal{C})^{\mathrm{op}} = \coprod_{\alpha\in(\mathrm{Ob}\,\mathcal{C})} \widetilde{\boldsymbol{X}}_{F\alpha} \otimes B(\alpha\downarrow\mathcal{C})^{\mathrm{op}}$$
$$\xrightarrow{1_{\widetilde{\boldsymbol{X}}_{F\alpha}}\otimes F_*} \coprod_{\alpha\in(\mathrm{Ob}\,\mathcal{C})} \widetilde{\boldsymbol{X}}_{F\alpha} \otimes B(F\alpha\downarrow F)^{\mathrm{op}} \to \coprod_{\gamma\in(\mathrm{Ob}\,\mathcal{D})} \widetilde{\boldsymbol{X}}_{\gamma} \otimes B(\gamma\downarrow F)^{\mathrm{op}}$$

(see Definition 19.2.2). We will show that ϕ is an isomorphism by showing that for every object Z of $\mathcal{M}$ the map

$$\phi^*\colon \mathcal{M}\bigl(\boldsymbol{X}\otimes_{\mathcal{D}} B(-\downarrow F)^{\mathrm{op}}, Z\bigr) \longrightarrow \mathcal{M}\bigl((F^*\boldsymbol{X})\otimes_{\mathcal{C}} B(-\downarrow\mathcal{C})^{\mathrm{op}}, Z\bigr)$$

is an isomorphism of sets. Proposition 19.2.13 implies that the map ϕ^* is isomorphic to the map

$$\tilde{\phi}\colon \mathrm{SS}^{\mathcal{D}^{\mathrm{op}}}\bigl(B(-\downarrow F)^{\mathrm{op}}, \mathcal{M}(\widetilde{\boldsymbol{X}}, Z)\bigr) \longrightarrow \mathrm{SS}^{\mathcal{C}^{\mathrm{op}}}\bigl(B(-\downarrow\mathcal{C})^{\mathrm{op}}, \mathcal{M}(\widetilde{F^*\boldsymbol{X}}, Z)\bigr)$$

(where $\widetilde{\boldsymbol{X}}$ and $\widetilde{F^*\boldsymbol{X}}$ are defined using the natural cosimplicial frame on $\mathcal{M}$). We will show that $\tilde{\phi}$ is an isomorphism by using Proposition 14.8.15 to define an inverse ψ to ϕ.

If $g \in \mathrm{SS}^{\mathcal{C}^{\mathrm{op}}}\bigl(B(-\downarrow\mathcal{C})^{\mathrm{op}}, \mathcal{M}(\widetilde{F^*\boldsymbol{X}}, Z)\bigr)$, then we will define $\psi(g)$ inductively over the skeleta of $B(-\downarrow F)^{\mathrm{op}}$. Corollary 14.8.10 defines a one to one correspondence between the elements of a basis of $B(-\downarrow F)^{\mathrm{op}}$ (a typical element of which is the simplex $\bigl((\beta_0 \xleftarrow{\sigma_0} \beta_1 \xleftarrow{\sigma_1} \cdots \xleftarrow{\sigma_{n-1}} \beta_n), 1_{F\beta_n}\colon F\beta_n \to F\beta_n)\bigr)$ and the elements of a basis of $B(-\downarrow\mathcal{C})^{\mathrm{op}}$ (in which the corresponding simplex is $\bigl((\beta_0 \xleftarrow{\sigma_0} \beta_1 \xleftarrow{\sigma_1} \cdots \xleftarrow{\sigma_{n-1}} \beta_n), 1_{\beta_n}\colon \beta_n \to \beta_n)\bigr)$. The map g takes this simplex to an n-simplex of $\mathcal{M}(\widetilde{(F^*\boldsymbol{X})}_{\beta_n}, Z) = \mathcal{M}(\widetilde{\boldsymbol{X}}_{F\beta_n}, Z) = \mathcal{M}(\widetilde{\boldsymbol{X}}_{\alpha}, Z)$; since the correspondence of Corollary 14.8.10 commutes with face operators, Proposition 14.8.15 implies that we can define $\psi(g)$ by letting $\psi(g)$ take the first simplex above to this simplex of $\mathcal{M}(\widetilde{\boldsymbol{X}}_{\alpha}, Z)$. If $g \in \mathrm{SS}^{\mathcal{C}^{\mathrm{op}}}\bigl(B(-\downarrow\mathcal{C})^{\mathrm{op}}, \mathcal{M}(\widetilde{F^*\boldsymbol{X}}, Z)\bigr)$, then $\tilde{\phi}\psi(g)$ agrees with g on our basis of $B(-\downarrow\mathcal{C})^{\mathrm{op}}$, and if $h \in \mathrm{SS}^{\mathcal{D}^{\mathrm{op}}}\bigl(B(-\downarrow F)^{\mathrm{op}}, \mathcal{M}(\widetilde{\boldsymbol{X}}, Z)\bigr)$, then $\psi\tilde{\phi}(h)$ agrees with h on our basis of $B(-\downarrow F)^{\mathrm{op}}$. Thus, Proposition 14.8.15 implies that $\tilde{\phi}\psi = 1$ and $\psi\tilde{\phi} = 1$. □

THEOREM 19.6.7. *Let $\mathcal{M}$ be a framed model category, let $\mathcal{C}$ and $\mathcal{D}$ be small categories, and let $\mathrm{F}\colon \mathcal{C} \to \mathcal{D}$ be a functor.*

(1) *If F is homotopy right cofinal (see Definition 19.6.1), then for every objectwise cofibrant $\mathcal{D}$-diagram $\boldsymbol{X}$ in $\mathcal{M}$, the natural map of homotopy colimits (see Proposition 19.1.8)*

$$\operatorname*{hocolim}_{\mathcal{C}} \mathrm{F}^*\boldsymbol{X} \longrightarrow \operatorname*{hocolim}_{\mathcal{D}} \boldsymbol{X}$$

is a weak equivalence.

(2) *If F is homotopy left cofinal (see Definition 19.6.1), then for every objectwise fibrant $\mathcal{D}$-diagram $\boldsymbol{X}$ in $\mathcal{M}$, the natural map of homotopy limits (see Proposition 19.1.8)*

$$\operatorname*{holim}_{\mathcal{D}} \boldsymbol{X} \longrightarrow \operatorname*{holim}_{\mathcal{C}} \mathrm{F}^*\boldsymbol{X}$$

is a weak equivalence.

PROOF. We will prove part 1; the proof of part 2 is similar.

Proposition 19.6.6 and Example 19.2.8 imply that our map of homotopy colimits is isomorphic to the map

$$\boldsymbol{X} \otimes_{\mathcal{D}} \mathrm{B}(- \downarrow \mathrm{F})^{\mathrm{op}} \longrightarrow \boldsymbol{X} \otimes_{\mathcal{D}} \mathrm{B}(- \downarrow \mathcal{D})^{\mathrm{op}} \ .$$

Proposition 14.8.5 and Corollary 14.8.8 imply that both of the $\mathcal{D}^{\mathrm{op}}$-diagrams of simplicial sets $\mathrm{B}(- \downarrow \mathrm{F})^{\mathrm{op}}$ and $\mathrm{B}(- \downarrow \mathcal{D})^{\mathrm{op}}$ are free cell complexes, and are thus cofibrant objects in $\mathrm{SS}^{\mathcal{D}^{\mathrm{op}}}$. Lemma 14.7.4 implies that $\mathrm{B}(\alpha \downarrow \mathcal{D})^{\mathrm{op}}$ is contractible for every object α of $\mathcal{D}$, and so F is homotopy right cofinal if and only if the map $\mathrm{B}(- \downarrow \mathrm{F})^{\mathrm{op}} \to \mathrm{B}(- \downarrow \mathcal{D})^{\mathrm{op}}$ is a weak equivalence of cofibrant objects in $\mathrm{SS}^{\mathcal{D}^{\mathrm{op}}}$. The result now follows from Proposition 19.3.3. □

COROLLARY 19.6.8. *Let $\mathcal{M}$ be a framed model category and let $\mathcal{C}$ be a small category.*

(1) *If α is a terminal object of $\mathcal{C}$ and $\boldsymbol{X}$ is an objectwise cofibrant $\mathcal{C}$-diagram in $\mathcal{M}$, then the natural map $\boldsymbol{X}_\alpha \to \operatorname{hocolim} \boldsymbol{X}$ is a weak equivalence.*
(2) *If α is an initial object of $\mathcal{C}$ and $\boldsymbol{X}$ is an objectwise fibrant $\mathcal{C}$-diagram in $\mathcal{M}$, then the natural map $\operatorname{holim} \boldsymbol{X} \to \boldsymbol{X}_\alpha$ is a weak equivalence.*

PROOF. We will prove part 1; the proof of part 2 is similar.

Let α be a terminal object of $\mathcal{C}$ and let $\mathbf{1}$ be the category with one object and no non-identity maps. If $\mathrm{F}_\alpha\colon \mathbf{1} \to \mathcal{C}$ is the functor that takes the object of $\mathbf{1}$ to α, then for every object β of $\mathcal{C}$ the undercategory $(\beta \downarrow \mathrm{F}_\alpha)$ has one object and no non-identity maps, and so the result follows from Theorem 19.6.7. □

We are indebted to W. G. Dwyer for the following proposition.

PROPOSITION 19.6.9. *Let $\mathcal{C}$ be a small category, let α be an object of $\mathcal{C}$, and let $\mathbf{F}^\alpha_*$ be the free $\mathcal{C}$-diagram of sets generated at α (see Definition 11.5.7) regarded as a $\mathcal{C}$-diagram of discrete simplicial sets.*

(1) *If $\boldsymbol{K}\colon \mathcal{C}^{\mathrm{op}} \to \mathrm{SS}$ is a $\mathcal{C}^{\mathrm{op}}$-diagram of simplicial sets, then the natural map*

$$\boldsymbol{K}_\alpha \approx \{1_\alpha\} \times \boldsymbol{K}_\alpha \subset \mathbf{F}^\alpha_*(\alpha) \times \boldsymbol{K}_\alpha \longrightarrow \mathbf{F}^\alpha_* \otimes_{\mathcal{C}} \boldsymbol{K}$$

(see Definition 19.2.2) is an isomorphism.

(2) *If $\boldsymbol{K}\colon \mathcal{C} \to \mathrm{SS}$ is a $\mathcal{C}$-diagram of simplicial sets, then the natural map*

$$\hom^{\mathcal{C}}(\mathbf{F}^\alpha_*, \boldsymbol{K}) \longrightarrow \boldsymbol{K}_\alpha^{\mathbf{F}^\alpha_*(\alpha)} \xrightarrow{1_{\boldsymbol{K}_\alpha}^{(1_\alpha \subset \mathbf{F}^\alpha_*(\alpha))}} \boldsymbol{K}_\alpha^{\{1_\alpha\}} \approx \boldsymbol{K}_\alpha$$

is an isomorphism.

PROOF. We will prove part 1; the proof of part 2 is similar.

We let $f\colon \boldsymbol{K}_\alpha \to \mathbf{F}^\alpha_* \otimes_{\mathcal{C}} \boldsymbol{K}$ denote our natural map, and we will construct on inverse g to f. If β is an object of $\mathcal{C}$, then (since $\mathbf{F}^\alpha_*(\beta)$ is discrete) $\mathbf{F}^\alpha_*(\beta) \otimes \boldsymbol{K}_\beta \approx \coprod_{\mathcal{C}(\alpha,\beta)} \boldsymbol{K}_\beta$ (see Lemma 9.2.1) and so

$$\coprod_{\beta \in \mathrm{Ob}(\mathcal{C})} \mathbf{F}^\alpha_*(\beta) \otimes \boldsymbol{K}_\beta \approx \coprod_{\beta \in \mathrm{Ob}(\mathcal{C})} \coprod_{\mathcal{C}(\alpha,\beta)} \boldsymbol{K}_\beta \ .$$

We define $h\colon \coprod_{\substack{\beta\in\mathrm{Ob}(\mathcal{C})\\ \mathcal{C}(\alpha,\beta)}} \boldsymbol{K}_\beta \to \boldsymbol{K}_\alpha$ by defining h on the summand $\sigma\colon \alpha \to \beta$ to be $\sigma^*\colon \boldsymbol{K}_\beta \to \boldsymbol{K}_\alpha$. To show that h defines a map $g\colon \mathbf{F}^\alpha_* \otimes_{\mathcal{C}} \boldsymbol{K} \to \boldsymbol{K}_\alpha$ (see Definition 19.2.2 and Proposition 19.2.5) we must show that $h\phi = h\psi$ (see (19.2.3)). We have

$$\coprod_{(\sigma\colon \beta \to \beta') \in \mathcal{C}} \mathbf{F}^\alpha_*(\beta) \otimes \boldsymbol{K}_{\beta'} \approx \coprod_{(\sigma\colon \beta\to\beta')\in\mathcal{C}} \coprod_{\mathcal{C}(\alpha,\beta)} \boldsymbol{K}_{\beta'} \ ,$$

and on the summand indexed by $(\sigma\colon \beta \to \beta', \tau\colon \alpha \to \beta)$ the map ϕ takes $\boldsymbol{K}_{\beta'}$ by the identity map onto the summand indexed by $\sigma\tau\colon \alpha \to \beta'$, and so $h\phi$ on that summand is $(\sigma\tau)^*\colon \boldsymbol{K}_{\beta'} \to \boldsymbol{K}_\alpha$. On that same summand, the map ψ is $\sigma^*\colon \boldsymbol{K}_{\beta'} \to \boldsymbol{K}_\beta$ where the $\boldsymbol{K}_\beta$ is the summand indexed by $\tau\colon \alpha \to \beta$, and so $h\psi$ on that summand is $\tau^*\sigma^*\colon \boldsymbol{K}_{\beta'} \to \boldsymbol{K}_\alpha$. Since $(\sigma\tau)^* = \tau^*\sigma^*$, we have a well defined map $g\colon \mathbf{F}^\alpha_* \otimes_{\mathcal{C}} \boldsymbol{K} \to \boldsymbol{K}_\alpha$.

The composition $gf\colon \boldsymbol{K}_\alpha \to \boldsymbol{K}_\alpha$ is $(1_\alpha)^* = 1_{\boldsymbol{K}_\alpha}$. The composition fg on the summand $\boldsymbol{K}_\beta$ indexed by $\sigma\colon \alpha \to \beta$ is induced by

$$\boldsymbol{K}_\beta \xrightarrow{\sigma^*} \boldsymbol{K}_\alpha \approx \{1_\alpha\} \times \boldsymbol{K}_\alpha \subset \big(\mathbf{F}^\alpha_*(\alpha)\big) \times \boldsymbol{K}_\alpha \ ,$$

which the relations defined by (19.2.3) imply is the identity map. □

COROLLARY 19.6.10. *If $\mathcal{C}$ is a small category, α is an object of $\mathcal{C}$, and we regard $\mathbf{F}^\alpha_*$ (see Definition 11.5.7) as a diagram of discrete simplicial sets, then there are natural isomorphisms*

$$\operatorname{hocolim} \mathbf{F}^\alpha_* \approx \mathrm{B}(\alpha\downarrow\mathcal{C})^{\mathrm{op}}$$
$$\operatorname{holim} \mathbf{F}^\alpha_* \approx \mathrm{B}(\mathcal{C}\downarrow\alpha) \ .$$

PROOF. This follows from Proposition 19.6.9 and Example 19.2.8. □

COROLLARY 19.6.11. *If $\mathcal{C}$ and $\mathcal{D}$ are small categories, $\mathrm{F}\colon \mathcal{C} \to \mathcal{D}$ is a functor, α is an object of $\mathcal{D}$, and we regard $\mathbf{F}^\alpha_*$ (see Definition 11.5.7) as a diagram of discrete simplicial sets, then there are natural isomorphisms*

$$\mathbf{F}^\alpha_* \otimes_{\mathcal{D}} \mathrm{B}(-\downarrow F)^{\mathrm{op}} \approx \mathrm{B}(\alpha\downarrow F)^{\mathrm{op}}$$
$$\hom^{\mathcal{D}}\big(\mathbf{F}^\alpha_*, \mathrm{B}(F\downarrow -)\big) \approx \mathrm{B}(F\downarrow\alpha) \ .$$

PROOF. This follows from Proposition 19.6.9. □

PROPOSITION 19.6.12. *Let $\mathcal{C}$ and $\mathcal{D}$ be small categories and let $\mathrm{F}\colon \mathcal{C} \to \mathcal{D}$ be a functor.*

(1) *If for every $\mathcal{D}$-diagram $\boldsymbol{X}$ of simplicial sets the induced map of homotopy colimits*

$$\operatorname*{hocolim}_{\mathcal{C}} \mathrm{F}^*\boldsymbol{X} \longrightarrow \operatorname*{hocolim}_{\mathcal{D}} \boldsymbol{X}$$

is a weak equivalence, then F *is a homotopy right cofinal functor.*

(2) *If for every $\mathcal{D}$-diagram $\boldsymbol{X}$ of fibrant simplicial sets the induced map of homotopy limits*

$$\operatorname*{holim}_{\mathcal{D}} \boldsymbol{X} \longrightarrow \operatorname*{holim}_{\mathcal{C}} \mathrm{F}^*\boldsymbol{X}$$

is a weak equivalence, then F *is a homotopy left cofinal functor.*

PROOF. For part 1, if α is an object of $\mathcal{D}$, we can let $\boldsymbol{X} = \mathbf{F}^{\alpha}_{*}$ (see Definition 11.5.7), regarded as a diagram of discrete simplicial sets. Proposition 19.6.6, Corollary 19.6.10, and Corollary 19.6.11 imply that $\mathrm{B}(\alpha\downarrow \mathrm{F})$ and $\mathrm{B}(\alpha\downarrow \mathcal{D})$ are weakly equivalent. Since $\mathrm{B}(\alpha\downarrow \mathcal{D})^{\mathrm{op}}$ is always contractible (see Lemma 14.7.4), Proposition 14.1.6 implies that F is homotopy right cofinal.

For part 2, Example 19.2.8 and Proposition 19.6.6 imply that our natural map of homotopy limits is isomorphic to the map

$$\hom^{\mathcal{D}}\bigl(\mathrm{B}(\mathcal{D}\downarrow -), \boldsymbol{X}\bigr) \longrightarrow \hom^{\mathcal{D}}\bigl(\mathrm{B}(\mathrm{F}\downarrow -), \boldsymbol{X}\bigr) \ ;$$

Proposition 19.2.5 and Example 9.1.13 imply that this is isomorphic to the map

$$\mathrm{Map}\bigl(\mathrm{B}(\mathcal{D}\downarrow -), \boldsymbol{X}\bigr) \longrightarrow \mathrm{Map}\bigl(\mathrm{B}(\mathrm{F}\downarrow -), \boldsymbol{X}\bigr) \ .$$

The $\mathcal{D}$-diagrams of simplicial sets $\mathrm{B}(\mathrm{F}\downarrow -)$ and $\mathrm{B}(\mathcal{D}\downarrow -)$ are always free cell complexes (see Proposition 14.8.5 and Corollary 14.8.8), and are thus cofibrant $\mathcal{D}$-diagrams (see Proposition 11.6.2). Since $\mathrm{B}(\mathcal{D}\downarrow -)$ is a diagram of contractible simplicial sets (see Lemma 14.7.10), the map $\mathrm{B}(\mathrm{F}\downarrow -) \to \mathrm{B}(\mathcal{D}\downarrow -)$ is a weak equivalence of $\mathcal{D}$-diagrams if and only if the functor F is homotopy left cofinal. Since a $\mathcal{D}$-diagram of simplicial sets is fibrant exactly when it is a diagram of fibrant simplicial sets (see Theorem 11.6.1), we are trying to prove that a map of cofibrant diagrams is a weak equivalence if it induces a weak equivalence of simplicial mapping spaces to an arbitrary fibrant object. This follows from Corollary 9.7.5. □

THEOREM 19.6.13. *Let $\mathcal{C}$ and $\mathcal{D}$ be small categories.*

(1) *A functor $\mathrm{F}\colon \mathcal{C} \to \mathcal{D}$ is homotopy right cofinal (see Definition 19.6.1) if and only if for every framed model category $\mathcal{M}$ and every objectwise cofibrant $\mathcal{D}$-diagram $\boldsymbol{X}$ in $\mathcal{M}$, the natural map*

$$\operatorname*{hocolim}_{\mathcal{C}} \mathrm{F}^*\boldsymbol{X} \longrightarrow \operatorname*{hocolim}_{\mathcal{D}} \boldsymbol{X}$$

(see Proposition 19.1.8) is a weak equivalence.

(2) *A functor $\mathrm{F}\colon \mathcal{C} \to \mathcal{D}$ is homotopy left cofinal if and only if for every framed model category $\mathcal{M}$ and every objectwise fibrant $\mathcal{D}$-diagram $\boldsymbol{X}$ in $\mathcal{M}$, the natural map*

$$\operatorname*{holim}_{\mathcal{D}} \boldsymbol{X} \longrightarrow \operatorname*{holim}_{\mathcal{C}} \mathrm{F}^*\boldsymbol{X}$$

(see Proposition 19.1.8) is a weak equivalence.

PROOF. This follows from Theorem 19.6.7 and Proposition 19.6.12. □

As a corollary, we obtain Quillen's "Theorem A" (see [**56**, Page 93]).

THEOREM 19.6.14 (Quillen). *If $\mathcal{C}$ and $\mathcal{D}$ are small categories and $\mathrm{F}\colon \mathcal{C} \to \mathcal{D}$ is a homotopy right cofinal functor, then* F *induces a weak equivalence of classifying spaces* $\mathrm{B}\mathcal{C} \cong \mathrm{B}\mathcal{D}$.

PROOF. This follows from Theorem 19.6.13, Proposition 18.1.6, and Proposition 14.1.6. □

19.7. The Reedy diagram homotopy lifting extension theorem

THEOREM 19.7.1 (The Reedy diagram homotopy lifting extension theorem). *Let $\mathcal{C}$ be a Reedy category and let $\mathcal{M}$ be a model category.*

(1) *If $i\colon \boldsymbol{A} \to \boldsymbol{B}$ is a Reedy cofibration of Reedy cofibrant $\mathcal{C}$-diagrams in $\mathcal{M}$, $\widetilde{\imath}\colon \widetilde{\boldsymbol{A}} \to \widetilde{\boldsymbol{B}}$ is a Reedy cosimplicial frame on i that is a Reedy cofibration, and $p\colon X \to Y$ is a fibration in $\mathcal{M}$, then the map of $\mathcal{C}^{\mathrm{op}}$-diagrams of simplicial sets*
$$\mathcal{M}(\widetilde{\boldsymbol{B}}, X) \longrightarrow \mathcal{M}(\widetilde{\boldsymbol{A}}, X) \times_{\mathcal{M}(\widetilde{\boldsymbol{A}},Y)} \mathcal{M}(\widetilde{\boldsymbol{B}}, Y)$$
is a Reedy fibration (see Proposition 15.1.5) that is a Reedy trivial fibration if at least one of i and p is a weak equivalence.

(2) *If $i\colon A \to B$ is a cofibration in $\mathcal{M}$, $p\colon \boldsymbol{X} \to \boldsymbol{Y}$ is a Reedy fibration of Reedy fibrant $\mathcal{C}$-diagrams in $\mathcal{M}$, and $\widehat{p}\colon \widehat{\boldsymbol{X}} \to \widehat{\boldsymbol{Y}}$ is a Reedy simplicial frame on p that is a Reedy fibration, then the map of $\mathcal{C}$-diagrams of simplicial sets*
$$\mathcal{M}(B, \widehat{\boldsymbol{X}}) \longrightarrow \mathcal{M}(A, \widehat{\boldsymbol{X}}) \times_{\mathcal{M}(A,\widehat{\boldsymbol{Y}})} \mathcal{M}(B, \widehat{\boldsymbol{Y}})$$
is a Reedy fibration that is a Reedy trivial fibration if at least one of i and p is a weak equivalence.

PROOF. We will prove part 1; the proof of part 2 is similar.

Theorem 15.3.15 implies that it is sufficient to show that for every object α of $\mathcal{C}^{\mathrm{op}}$ the map
$$\mathcal{M}(\widetilde{\boldsymbol{B}}_\alpha, X) \longrightarrow \\ \bigl(\mathcal{M}(\widetilde{\boldsymbol{A}}_\alpha, X) \times_{\mathcal{M}(\widetilde{\boldsymbol{A}}_\alpha,Y)} \mathcal{M}(\widetilde{\boldsymbol{B}}_\alpha, Y)\bigr) \times_{\mathrm{M}_\alpha(\mathcal{M}(\widetilde{\boldsymbol{A}},X)\times_{\mathcal{M}(\widetilde{\boldsymbol{A}},Y)}\mathcal{M}(\widetilde{\boldsymbol{B}},Y))} \mathrm{M}_\alpha\mathcal{M}(\widetilde{\boldsymbol{B}}, X)$$
is a fibration of simplicial sets that is a trivial fibration if at least one of i and p is a weak equivalence. Lemma 16.5.9 implies that this map is isomorphic to the map
$$\mathcal{M}(\widetilde{\boldsymbol{B}}_\alpha, X) \longrightarrow \\ \bigl(\mathcal{M}(\widetilde{\boldsymbol{A}}_\alpha, X) \times_{\mathcal{M}(\widetilde{\boldsymbol{A}}_\alpha,Y)} \mathcal{M}(\widetilde{\boldsymbol{B}}_\alpha, Y)\bigr) \times_{(\mathcal{M}(\mathrm{L}_\alpha\widetilde{\boldsymbol{A}},X)\times_{\mathcal{M}(\mathrm{L}_\alpha\widetilde{\boldsymbol{A}},Y)}\mathcal{M}(\mathrm{L}_\alpha\widetilde{\boldsymbol{B}},Y))} \mathcal{M}(\mathrm{L}_\alpha\widetilde{\boldsymbol{B}}, X)\ .$$
The codomain of this map is the limit of the diagram
$$\begin{array}{ccccc}
\mathcal{M}(\widetilde{\boldsymbol{A}}_\alpha, X) & \longrightarrow & \mathcal{M}(\widetilde{\boldsymbol{A}}_\alpha, Y) & \longleftarrow & \mathcal{M}(\widetilde{\boldsymbol{B}}_\alpha, Y) \\
\downarrow & & \downarrow & & \downarrow \\
\mathcal{M}(\mathrm{L}_\alpha\widetilde{\boldsymbol{A}}, X) & \longrightarrow & \mathcal{M}(\mathrm{L}_\alpha\widetilde{\boldsymbol{A}}, Y) & \longleftarrow & \mathcal{M}(\mathrm{L}_\alpha\widetilde{\boldsymbol{B}}, Y) \\
& \nwarrow & \uparrow & \nearrow & \\
& & \mathcal{M}(\mathrm{L}_\alpha\widetilde{\boldsymbol{B}}, X) & &
\end{array}$$

and so our map is isomorphic to the map

$$\mathcal{M}(\widetilde{\boldsymbol{B}}_\alpha, X) \longrightarrow$$
$$\mathcal{M}(\widetilde{\boldsymbol{B}}_\alpha, Y) \times_{(\mathcal{M}(\widetilde{\boldsymbol{A}}_\alpha, Y) \times_{\mathcal{M}(\mathrm{L}_\alpha \widetilde{\boldsymbol{A}}, Y)} \mathcal{M}(\mathrm{L}_\alpha \widetilde{\boldsymbol{B}}, Y))} \left(\mathcal{M}(\widetilde{\boldsymbol{A}}_\alpha, X) \times_{\mathcal{M}(\mathrm{L}_\alpha \widetilde{\boldsymbol{A}}, X)} \mathcal{M}(\mathrm{L}_\alpha \widetilde{\boldsymbol{B}}, X)\right) .$$

Proposition 16.7.16 implies that $\mathrm{L}_\alpha \widetilde{\boldsymbol{A}}$ is a cosimplicial frame on $\mathrm{L}_\alpha \boldsymbol{A}$ and that $\mathrm{L}_\alpha \widetilde{\boldsymbol{B}}$ is a cosimplicial frame on $\mathrm{L}_\alpha \boldsymbol{B}$. Since $\tilde{\imath}$ is a Reedy cofibration, this implies that the relative latching map $\widetilde{\boldsymbol{A}}_\alpha \amalg_{\mathrm{L}_\alpha \widetilde{\boldsymbol{A}}} \mathrm{L}_\alpha \widetilde{\boldsymbol{B}} \to \widetilde{\boldsymbol{B}}_\alpha$ is a cofibration that is a weak equivalence if i is a weak equivalence. Theorem 16.5.2 now implies that our map is a fibration that is a weak equivalence if at least one of i and p is a weak equivalence. □

THEOREM 19.7.2. *Let $\mathcal{C}$ be a Reedy category and let $\mathcal{M}$ be a model category.*

(1) *If $i\colon \boldsymbol{A} \to \boldsymbol{B}$ is a Reedy cofibration of Reedy cofibrant $\mathcal{C}$-diagrams in $\mathcal{M}$, $\tilde{\imath}\colon \widetilde{\boldsymbol{A}} \to \widetilde{\boldsymbol{B}}$ is a Reedy cosimplicial frame on i that is a Reedy cofibration, and $j\colon \boldsymbol{K} \to \boldsymbol{L}$ is a Reedy cofibration of $\mathcal{C}^{\mathrm{op}}$-diagrams of simplicial sets, then the pushout corner map*
$$\boldsymbol{A} \otimes_{\mathcal{C}}^{\widetilde{\boldsymbol{A}}} \boldsymbol{L} \amalg_{\boldsymbol{A} \otimes_{\mathcal{C}}^{\widetilde{\boldsymbol{B}}} \boldsymbol{K}} \boldsymbol{B} \otimes_{\mathcal{C}}^{\widetilde{\boldsymbol{B}}} \boldsymbol{K} \longrightarrow \boldsymbol{B} \otimes_{\mathcal{C}}^{\widetilde{\boldsymbol{B}}} \boldsymbol{L}$$
is a cofibration in $\mathcal{M}$ that is a weak equivalence if at least one of i and j is a weak equivalence.

(2) *If $p\colon \boldsymbol{X} \to \boldsymbol{Y}$ is a Reedy fibration of Reedy fibrant $\mathcal{C}$-diagrams in $\mathcal{M}$, $\hat{p}\colon \widehat{\boldsymbol{X}} \to \widehat{\boldsymbol{Y}}$ is a Reedy simplicial frame on p that is a Reedy fibration, and $j\colon \boldsymbol{K} \to \boldsymbol{L}$ is a Reedy cofibration of $\mathcal{C}$-diagrams of simplicial sets, then the pullback corner map*
$$\hom_{\widehat{\boldsymbol{X}}}^{\mathcal{C}}(\boldsymbol{L}, \boldsymbol{X}) \longrightarrow \hom_{\widehat{\boldsymbol{X}}}^{\mathcal{C}}(\boldsymbol{K}, \boldsymbol{X}) \times_{\hom_{\widehat{\boldsymbol{Y}}}^{\mathcal{C}}(\boldsymbol{K}, \boldsymbol{Y})} \hom_{\widehat{\boldsymbol{Y}}}^{\mathcal{C}}(\boldsymbol{L}, \boldsymbol{Y})$$
is a fibration that is a weak equivalence if at least one of p and j is a weak equivalence.

PROOF. The result follows from Proposition 7.2.3, Lemma 19.2.14, and Theorem 19.7.1. □

COROLLARY 19.7.3. *Let $\mathcal{C}$ be a Reedy category and let $\mathcal{M}$ be a model category.*

(1) *If $\boldsymbol{X}$ is a Reedy cofibrant $\mathcal{C}$-diagram in $\mathcal{M}$, $\widetilde{\boldsymbol{X}}$ is a Reedy cosimplicial frame on $\boldsymbol{X}$, and $\boldsymbol{K}$ is a Reedy cofibrant $\mathcal{C}^{\mathrm{op}}$-diagram of simplicial sets, then $\boldsymbol{X} \otimes_{\mathcal{C}}^{\widetilde{\boldsymbol{X}}} \boldsymbol{K}$ is a cofibrant object of $\mathcal{M}$.*

(2) *If $\boldsymbol{X}$ is a Reedy fibrant $\mathcal{C}$-diagram in $\mathcal{M}$, $\widehat{\boldsymbol{X}}$ is a Reedy simplicial frame on $\boldsymbol{X}$, and $\boldsymbol{K}$ is a Reedy cofibrant $\mathcal{C}$-diagram of simplicial sets, then $\hom_{\widehat{\boldsymbol{X}}}^{\mathcal{C}}(\boldsymbol{K}, \boldsymbol{X})$ is a fibrant object of $\mathcal{M}$.*

PROOF. This follows from Theorem 19.7.2. □

COROLLARY 19.7.4. *Let $\mathcal{C}$ be a Reedy category and let $\mathcal{M}$ be a model category.*

(1) *If $\boldsymbol{K}$ is a Reedy cofibrant $\mathcal{C}^{\mathrm{op}}$-diagram of simplicial sets, $f\colon \boldsymbol{X} \to \boldsymbol{Y}$ is a weak equivalence of Reedy cofibrant $\mathcal{C}$-diagrams in $\mathcal{M}$, and $\tilde{f}\colon \widetilde{\boldsymbol{X}} \to \widetilde{\boldsymbol{Y}}$ is a Reedy cosimplicial frame on f, then the induced map $f_*\colon \boldsymbol{X} \otimes_{\mathcal{C}}^{\widetilde{\boldsymbol{X}}} \boldsymbol{K} \to \boldsymbol{Y} \otimes_{\mathcal{C}}^{\widetilde{\boldsymbol{Y}}} \boldsymbol{K}$ is a weak equivalence of cofibrant objects in $\mathcal{M}$.*

(2) *If $\boldsymbol{K}$ is a Reedy cofibrant $\mathcal{C}$-diagram of simplicial sets, $f\colon \boldsymbol{X} \to \boldsymbol{Y}$ is a weak equivalence of Reedy fibrant $\mathcal{C}$-diagrams in $\mathcal{M}$, and $\hat{f}\colon \widehat{\boldsymbol{X}} \to \widehat{\boldsymbol{Y}}$ is a Reedy simplicial frame on f, then the induced map $f_*\colon \hom^{\mathcal{C}}_{\widehat{\boldsymbol{X}}}(\boldsymbol{K}, \boldsymbol{X}) \to \hom^{\mathcal{C}}_{\widehat{\boldsymbol{Y}}}(\boldsymbol{K}, \boldsymbol{Y})$ is a weak equivalence of fibrant objects in $\mathcal{M}$.*

PROOF. This follows from Corollary 19.7.3, Theorem 19.7.2, and Corollary 7.7.2. $\square$

COROLLARY 19.7.5. *Let $\mathcal{C}$ be a Reedy category and let $\mathcal{M}$ be a model category.*

(1) *If $\boldsymbol{X}$ is a Reedy cofibrant $\mathcal{C}$-diagram in $\mathcal{M}$, $\widetilde{\boldsymbol{X}}$ is a Reedy cosimplicial frame on $\boldsymbol{X}$, and $f\colon \boldsymbol{K} \to \boldsymbol{K}'$ is a weak equivalence of Reedy cofibrant $\mathcal{C}^{\mathrm{op}}$-diagrams of simplicial sets, then the induced map $f_*\colon \boldsymbol{X} \otimes^{\widetilde{\boldsymbol{X}}}_{\mathcal{C}} \boldsymbol{K} \to \boldsymbol{X} \otimes^{\widetilde{\boldsymbol{X}}}_{\mathcal{C}} \boldsymbol{K}'$ is a weak equivalence of cofibrant objects in $\mathcal{M}$.*

(2) *If $\boldsymbol{X}$ is a Reedy fibrant $\mathcal{C}$-diagram in $\mathcal{M}$, $\widehat{\boldsymbol{X}}$ is a Reedy simplicial frame on $\boldsymbol{X}$, and $f\colon \boldsymbol{K} \to \boldsymbol{K}'$ is a weak equivalence of Reedy cofibrant $\mathcal{C}$-diagrams of simplicial sets, then the induced map $f^*\colon \hom^{\mathcal{C}}_{\widehat{\boldsymbol{X}}}(\boldsymbol{K}', \boldsymbol{X}) \to \hom^{\mathcal{C}}_{\widehat{\boldsymbol{X}}}(\boldsymbol{K}, \boldsymbol{X})$ is a weak equivalence of fibrant objects in $\mathcal{M}$.*

PROOF. This follows from Theorem 19.7.2 and Corollary 7.7.2. $\square$

COROLLARY 19.7.6. *Let $\mathcal{C}$ be a Reedy category and let $\mathcal{M}$ be a model category.*

(1) *If $\boldsymbol{K}$ is a $\mathcal{C}^{\mathrm{op}}$-diagram of simplicial sets and both $\widetilde{\boldsymbol{K}} \to \boldsymbol{K}$ and $\widetilde{\boldsymbol{K}'} \to \boldsymbol{K}$ are Reedy cofibrant approximations to $\boldsymbol{K}$, then for every Reedy cofibrant $\mathcal{C}$-diagram $\boldsymbol{X}$ in $\mathcal{M}$ and Reedy cosimplicial frame $\widetilde{\boldsymbol{X}}$ on $\boldsymbol{X}$ there is an essentially unique natural zig-zag of weak equivalences in $\mathcal{M}$ from $\boldsymbol{X} \otimes^{\widetilde{\boldsymbol{X}}}_{\mathcal{C}} \widetilde{\boldsymbol{K}}$ to $\boldsymbol{X} \otimes^{\widetilde{\boldsymbol{X}}}_{\mathcal{C}} \widetilde{\boldsymbol{K}'}$.*

(2) *If $\boldsymbol{K}$ is a $\mathcal{C}$-diagram of simplicial sets and both $\widetilde{\boldsymbol{K}} \to \boldsymbol{K}$ and $\widetilde{\boldsymbol{K}'} \to \boldsymbol{K}$ are Reedy cofibrant approximations to $\boldsymbol{K}$, then for every Reedy fibrant $\mathcal{C}$-diagram $\boldsymbol{X}$ in $\mathcal{M}$ and Reedy simplicial frame $\widehat{\boldsymbol{X}}$ on $\boldsymbol{X}$ there is an essentially unique natural zig-zag of weak equivalences in $\mathcal{M}$ from $\hom^{\mathcal{C}}_{\widehat{\boldsymbol{X}}}(\widetilde{\boldsymbol{K}}, \boldsymbol{X})$ to $\hom^{\mathcal{C}}_{\widehat{\boldsymbol{X}}}(\widetilde{\boldsymbol{K}'}, \boldsymbol{X})$.*

PROOF. This follows from Corollary 19.7.5 and Proposition 14.6.3. $\square$

PROPOSITION 19.7.7. *Let $\mathcal{M}$ be a model category and let $\mathcal{C}$ be a Reedy category. If $\boldsymbol{B}$ is a $\mathcal{C}$-diagram in $\mathcal{M}$ that is Reedy cofibrant and $\boldsymbol{X}$ is a simplicial resolution in $\mathcal{M}$, then the $\mathcal{C}^{\mathrm{op}}$-diagram of simplicial sets $\mathcal{M}(\boldsymbol{B}, \boldsymbol{X})$ (which on an object α of $\mathcal{C}$ is $\mathcal{M}(\boldsymbol{B}_\alpha, \boldsymbol{X})$) is Reedy fibrant.*

PROOF. If α is an object of $\mathcal{C}$ and $\mathrm{L}_\alpha \boldsymbol{B}$ is the latching object of $\boldsymbol{B}$ at α (see Definition 15.2.5), then Proposition 15.2.4 implies that

$$\begin{aligned}
\mathcal{M}(\mathrm{L}_\alpha \boldsymbol{B}, \boldsymbol{X}) &= \mathcal{M}\bigl(\operatorname*{colim}_{\partial(\overrightarrow{\mathcal{C}}\downarrow\alpha)} \boldsymbol{B}, \boldsymbol{X}\bigr) \\
&\approx \lim_{\partial(\overrightarrow{\mathcal{C}}\downarrow\alpha)^{\mathrm{op}}} \mathcal{M}(\boldsymbol{B}, \boldsymbol{X}) \\
&\approx \lim_{\partial(\alpha\downarrow\overleftarrow{\mathcal{C}^{\mathrm{op}}})} \mathcal{M}(\boldsymbol{B}, \boldsymbol{X}) \\
&\approx \mathrm{M}_\alpha \mathcal{M}(\boldsymbol{B}, \boldsymbol{X})
\end{aligned}$$

and so $\mathcal{M}(\mathrm{L}_\alpha \boldsymbol{B}, \boldsymbol{X})$ is naturally isomorphic to the matching object at α of the $\mathcal{C}^{\mathrm{op}}$-diagram of simplicial sets $\mathcal{M}(\boldsymbol{B}, \boldsymbol{X})$. Since the latching map $\mathrm{L}_\alpha \boldsymbol{B} \to \boldsymbol{B}_\alpha$ is a cofibration, Corollary 16.5.4 implies that the matching map $\mathcal{M}(\boldsymbol{B}_\alpha, \boldsymbol{X}) \to \mathrm{M}_\alpha \mathcal{M}(\boldsymbol{B}, \boldsymbol{X})$ is a fibration, and so $\mathcal{M}(\boldsymbol{B}, \boldsymbol{X})$ is a Reedy fibrant diagram. □

19.8. Realizations and total objects

DEFINITION 19.8.1. Let $\mathcal{M}$ be a model category.

(1) If $(\boldsymbol{\Delta}^{\mathrm{op}}, \mathcal{M})$ is a Reedy framed diagram category structure (see Definition 16.7.15) on the category of simplicial objects in $\mathcal{M}$, then the *realization functor* $\mathcal{M}^{\boldsymbol{\Delta}^{\mathrm{op}}} \to \mathcal{M}$ is defined to be the functor that takes the simplicial object $\boldsymbol{X}$ in $\mathcal{M}$ to $|\boldsymbol{X}| = \boldsymbol{X} \otimes_{\boldsymbol{\Delta}^{\mathrm{op}}} \Delta$ where Δ is the cosimplicial standard simplex (see Definition 15.1.15) and the tensor product is defined relative to the functorial Reedy cosimplicial frame on $\mathcal{M}^{\boldsymbol{\Delta}^{\mathrm{op}}}$.
(2) If $(\boldsymbol{\Delta}, \mathcal{M})$ is a Reedy framed diagram category structure on the category of cosimplicial objects in $\mathcal{M}$, then the *total object functor* $\mathcal{M}^{\boldsymbol{\Delta}} \to \mathcal{M}$ is defined to be the functor that takes the cosimplicial object $\boldsymbol{X}$ in $\mathcal{M}$ to $\operatorname{Tot} \boldsymbol{X} = \hom^{\boldsymbol{\Delta}}(\Delta, \boldsymbol{X})$ where Δ is the cosimplicial standard simplex and the hom is defined relative to the functorial Reedy simplicial frame on $\mathcal{M}^{\boldsymbol{\Delta}}$.

THEOREM 19.8.2. *Let $\mathcal{M}$ be a model category.*

(1) *If $(\boldsymbol{\Delta}^{\mathrm{op}}, \mathcal{M})$ is a Reedy framed diagram category structure on the category of simplicial objects in $\mathcal{M}$ and $\boldsymbol{X}$ is a Reedy cofibrant simplicial object in $\mathcal{M}$, then the realization $|\boldsymbol{X}|$ of $\boldsymbol{X}$ is a cofibrant object of $\mathcal{M}$.*
(2) *If $(\boldsymbol{\Delta}, \mathcal{M})$ is a Reedy framed diagram category structure on the category of cosimplicial objects in $\mathcal{M}$ and $\boldsymbol{X}$ is a cosimplicial object in $\mathcal{M}$, then the total object* $\operatorname{Tot} \boldsymbol{X}$ *of $\boldsymbol{X}$ is a fibrant object of $\mathcal{M}$.*

PROOF. This follows from Corollary 19.7.3 and Corollary 15.9.11. □

THEOREM 19.8.3. *Let $\mathcal{M}$ be a model category.*

(1) *If $(\boldsymbol{\Delta}^{\mathrm{op}}, \mathcal{M})$ is a Reedy framed diagram category structure on the category of simplicial objects in $\mathcal{M}$ and $f\colon \boldsymbol{X} \to \boldsymbol{Y}$ is an objectwise weak equivalence of Reedy cofibrant simplicial objects in $\mathcal{M}$, then the induced map of realizations $|f|\colon |\boldsymbol{X}| \to |\boldsymbol{Y}|$ is a weak equivalence of cofibrant objects of $\mathcal{M}$.*
(2) *If $(\boldsymbol{\Delta}, \mathcal{M})$ is a Reedy framed diagram category structure on the category of cosimplicial objects in $\mathcal{M}$ and $f\colon \boldsymbol{X} \to \boldsymbol{Y}$ is an objectwise weak equivalence of Reedy fibrant cosimplicial objects in $\mathcal{M}$, then the induced map of total objects* $(\operatorname{Tot} f)\colon \operatorname{Tot} \boldsymbol{X} \to \operatorname{Tot} \boldsymbol{Y}$ *is a weak equivalence of fibrant objects of $\mathcal{M}$.*

PROOF. This follows from Corollary 19.7.4 and Corollary 15.9.11. □

THEOREM 19.8.4. *Let $\mathcal{M}$ be a model category.*

(1) *If $(\boldsymbol{\Delta}^{\mathrm{op}}, \mathcal{M})$ is a Reedy framed diagram category structure on the category of simplicial objects in $\mathcal{M}$ and $\boldsymbol{X}$ is a Reedy cofibrant simplicial object in $\mathcal{M}$, then the realization $|\boldsymbol{X}|$ of $\boldsymbol{X}$ is naturally weakly equivalent to the homotopy colimit* $\operatorname{hocolim} \boldsymbol{X}$ *of $\boldsymbol{X}$.*

(2) *If* $(\mathbf{\Delta}, \mathcal{M})$ *is a Reedy framed diagram category structure on the category of cosimplicial objects in* $\mathcal{M}$ *and* $\boldsymbol{X}$ *is a cosimplicial object in* $\mathcal{M}$*, then the total object* $\operatorname{Tot} \boldsymbol{X}$ *of* $\boldsymbol{X}$ *is naturally weakly equivalent to the homotopy limit* $\operatorname{holim} \boldsymbol{X}$ *of* $\boldsymbol{X}$*.*

19.8.5. The Bousfield-Kan map.

DEFINITION 19.8.6. Let $\mathcal{M}$ be a model category.

(1) If $(\mathbf{\Delta}^{\mathrm{op}}, \mathcal{M})$ is a Reedy framed diagram category structure on the category of simplicial objects in $\mathcal{M}$ and $\boldsymbol{X}$ is a simplicial object in $\mathcal{M}$, then the *Bousfield-Kan map* is the map
$$\phi_*\colon \operatorname{hocolim} \boldsymbol{X} \longrightarrow \left|\boldsymbol{X}\right| \ ,$$
natural in $\boldsymbol{X}$, that is the composition
$$\operatorname{hocolim} \boldsymbol{X} \approx \boldsymbol{X} \otimes_{\mathbf{\Delta}^{\mathrm{op}}} \mathrm{B}(- \downarrow \mathbf{\Delta}^{\mathrm{op}})^{\mathrm{op}} \xrightarrow{1_{\boldsymbol{X}} \otimes_{\mathbf{\Delta}^{\mathrm{op}}} \phi} \boldsymbol{X} \otimes_{\mathbf{\Delta}^{\mathrm{op}}} \Delta \approx \left|\boldsymbol{X}\right|$$
where ϕ is the Bousfield-Kan map of cosimplicial simplicial sets (see Definition 18.7.1).

(2) If $(\mathbf{\Delta}, \mathcal{M})$ is a Reedy framed diagram category structure on the category of cosimplicial objects in $\mathcal{M}$ and $\boldsymbol{X}$ is a cosimplicial object in $\mathcal{M}$, then the *Bousfield-Kan map* is the map
$$\phi^*\colon \operatorname{Tot} \boldsymbol{X} \longrightarrow \operatorname{holim} \boldsymbol{X} \ ,$$
natural in $\boldsymbol{X}$, that is the composition
$$\operatorname{Tot} \boldsymbol{X} \approx \hom^{\mathbf{\Delta}}(\Delta, \boldsymbol{X}) \xrightarrow{\hom^{\mathbf{\Delta}}(\phi, 1_{\boldsymbol{X}})} \hom^{\mathbf{\Delta}}\big(\mathrm{B}(\mathbf{\Delta} \downarrow -), \boldsymbol{X}\big) \approx \operatorname{holim} \boldsymbol{X}$$
where ϕ is the Bousfield-Kan map of cosimplicial simplicial sets.

THEOREM 19.8.7. *Let* $\mathcal{M}$ *be a simplicial model category.*

(1) *If* $\boldsymbol{X}$ *is a Reedy cofibrant simplicial object in* $\mathcal{M}$*, then the Bousfield-Kan map* $\phi_*\colon \operatorname{hocolim} \boldsymbol{X} \to \left|\boldsymbol{X}\right|$ *is a weak equivalence.*

(2) *If* $\boldsymbol{X}$ *is a Reedy fibrant cosimplicial object in* $\mathcal{M}$*, then the Bousfield-Kan map* $\phi^*\colon \operatorname{Tot} \boldsymbol{X} \to \operatorname{holim} \boldsymbol{X}$ *is a weak equivalence.*

PROOF. This follows from Corollary 19.7.5. □

19.9. Reedy cofibrant diagrams and Reedy fibrant diagrams

THEOREM 19.9.1. *Let* $\mathcal{M}$ *be a framed model category.*

(1) *If* $\mathcal{C}$ *is a Reedy category with fibrant constants (see Definition* 15.10.1*) and* $\boldsymbol{X}$ *is a Reedy cofibrant* $\mathcal{C}$*-diagram in* $\mathcal{M}$*, then the natural map* $\operatorname{hocolim} \boldsymbol{X} \to \operatorname{colim} \boldsymbol{X}$ *(see Example* 19.2.10*) is a weak equivalence.*

(2) *If* $\mathcal{C}$ *is a Reedy category with cofibrant constants and* $\boldsymbol{X}$ *is a Reedy fibrant* $\mathcal{C}$*-diagram in* $\mathcal{M}$*, then the natural map* $\lim \boldsymbol{X} \to \operatorname{holim} \boldsymbol{X}$ *(see Example* 19.2.10*) is a weak equivalence.*

PROOF. We will prove part 1; the proof of part 2 is similar.

The map $\operatorname{hocolim} \boldsymbol{X} \to \operatorname{colim} \boldsymbol{X}$ is naturally isomorphic to the map $\boldsymbol{X} \otimes_{\mathcal{C}}^{\widetilde{\boldsymbol{X}}} (- \downarrow \mathcal{C})^{\mathrm{op}} \to \boldsymbol{X} \otimes_{\mathcal{C}}^{\widetilde{\boldsymbol{X}}} \boldsymbol{P}$ where $\widetilde{\boldsymbol{X}}$ is the cosimplicial frame on $\boldsymbol{X}$ induced by the framing on $\mathcal{M}$ and $\boldsymbol{P}\colon \mathcal{C}^{\mathrm{op}} \to \mathrm{SS}$ is the constant diagram at a point. If $\widetilde{\boldsymbol{X}}'$ is some other cosimplicial frame on $\boldsymbol{X}$, then there is a zig-zag of weak equivalences from this map to the map $\boldsymbol{X} \otimes_{\mathcal{C}}^{\widetilde{\boldsymbol{X}}'} (-\downarrow \mathcal{C})^{\mathrm{op}} \to \boldsymbol{X} \otimes_{\mathcal{C}}^{\widetilde{\boldsymbol{X}}'} \boldsymbol{P}$ (see Corollary 19.3.7), and so it is sufficient to show that this map is a weak equivalence for some cosimplicial frame $\widetilde{\boldsymbol{X}}'$ on $\boldsymbol{X}$. If we let $\widetilde{\boldsymbol{X}}'$ be a Reedy cosimplicial frame on $\boldsymbol{X}$ (see Proposition 16.7.11), then (since $\mathcal{C}$ has fibrant constants) the result follows from Corollary 19.7.5 and Proposition 15.10.3. □

I am indebted to E. Dror Farjoun for the following result. (For an alternative approach to the same result, see [**44**, Lemma 2.7]):

COROLLARY 19.9.2. *If $p\colon E \to B$ is a map of simplicial sets and $\tilde{p}\colon \Delta B \to \mathrm{SS}$ is the diagram constructed in Example* 18.9.6, *then the natural map* $\operatorname{hocolim} \tilde{p} \to E$ *is a weak equivalence.*

PROOF. This follows from Theorem 19.9.1 and Proposition 18.9.7. □

PROPOSITION 19.9.3. *Let $\mathcal{M}$ be a framed model category. If the object X is a retract of the cofibrant object Y (with inclusion $i\colon X \to Y$ and retraction $r\colon Y \to X$), then X is weakly equivalent to the homotopy colimit of the diagram*

$$Y \xrightarrow{ir} Y \xrightarrow{ir} Y \xrightarrow{ir} \cdots$$

PROOF. We have the ω-sequence (where ω is the first infinite ordinal)

$$X \xrightarrow{i} Y \xrightarrow{r} X \xrightarrow{i} Y \xrightarrow{r} X \xrightarrow{i} Y \cdots$$

which has the two subdiagrams

$$X \xrightarrow{1_X} X \xrightarrow{1_X} X \xrightarrow{1_X} \cdots$$

and

$$Y \xrightarrow{ir} Y \xrightarrow{ir} Y \xrightarrow{ir} \cdots .$$

Both of the subdiagrams are homotopy right cofinal because all of the undercategories have an initial object (see Proposition 14.3.14). Thus, Theorem 19.6.13 implies that the homotopy colimits of the three diagrams are all weakly equivalent. Since X is a retract of a cofibrant object, it is cofibrant, and so the second diagram is a Reedy cofibrant diagram. Since this diagram shape has fibrant constants (see Definition 15.10.1), the homotopy colimit of this diagram is weakly equivalent to its colimit (see Theorem 19.9.1), which is isomorphic to X. □

PROPOSITION 19.9.4. *Let $\mathcal{M}$ be a framed model category.*

(1) *If $\boldsymbol{X}$ is the diagram $C \leftarrow A \to B$ in $\mathcal{M}$, the objects A, B, and C are cofibrant, and at least one of the maps $A \to B$ and $A \to C$ is a cofibration, then the natural map* $\operatorname{hocolim} \boldsymbol{X} \to \operatorname{colim} \boldsymbol{X}$ *(see Example* 19.2.10*) is a weak equivalence.*
(2) *If $\boldsymbol{X}$ is the diagram $C \to A \leftarrow B$ in $\mathcal{M}$, the objects A, B, and C are fibrant, and at least one of the maps $B \to A$ and $C \to A$ is a fibration, then the natural map* $\lim \boldsymbol{X} \to \operatorname{holim} \boldsymbol{X}$ *is a weak equivalence.*

PROOF. We will prove part 1; the proof of part 2 is dual.

We will assume that the map $A \to B$ is a cofibration (the other case is similar). If the indexing category for the diagram $\boldsymbol{X}$ is $\gamma \leftarrow \alpha \to \beta$ and we let $\deg(\alpha) = 2$, $\deg(\beta) = 3$, and $\deg(\gamma) = 1$, then the indexing category is a Reedy category with fibrant constants (see Definition 15.10.1) and $\boldsymbol{X}$ is a cofibrant diagram. The result now follows from Theorem 19.9.1. □

Index

Page numbers in bold point to the definitions of the indexed terms.

Bibliography

1. J. Adámek and J. Rosický, *Locally presentable and accessible categories*, London Math. Soc. Lecture Note Series, vol. 189, Cambridge University Press, Cambridge, 1994.
2. J. F. Adams, *Idempotent functors in homotopy theory*, Manifolds—Tokyo 1973 (Proc. Internat. Conf., Tokyo, 1973), Univ. Tokyo Press, Tokyo, 1975, pp. 247–253.
3. M. Barr and J. Beck, *Acyclic models and triples*, Proceedings of the Conference on Categorical Algebra, La Jolla, 1965 (S. Eilenberg, D. K. Harrison, S. MacLane, and H. Röhrl, eds.), Springer-Verlag, 1966, pp. 336–343.
4. M. G. Barratt, V. K. A. M. Guggenheim, and J. C. Moore, *On semisimplicial fibre bundles*, Amer. J. Math **81** (1959), 639–657.
5. F.-W. Bauer and J. Dugundji, *Categorical homotopy and fibrations*, Trans. Amer. Math. Soc. **140** (1969), 239–256.
6. F. Borceux, *Handbook of categorical algebra 1, basic category theory*, Encyclopedia of Mathematics and its Applications, vol. 50, Cambridge University Press, 1994.
7. ———, *Handbook of categorical algebra 2, categories and structures*, Encyclopedia of Mathematics and its Applications, vol. 51, Cambridge University Press, 1994.
8. A. K. Bousfield, *The localization of spaces with respect to homology*, Topology **14** (1975), 133–150.
9. ———, *Constructions of factorization systems in categories*, J. Pure Appl. Algebra **9** (1977), 207–220.
10. ———, *K-localizations and K-equivalences of infinite loop spaces*, Proc. London Math. Soc. **44** (1982), 291–311.
11. ———, *Localization and periodicity in unstable homotopy theory*, J. Amer. Math. Soc. **7** (1994), 831–873.
12. ———, *Homotopical localizations of spaces*, Amer. J. Math. **119** (1997), no. 6, 1321–1354.
13. A. K. Bousfield and E. M. Friedlander, *Homotopy theory of Γ-spaces, spectra, and bisimplicial sets*, Geometric Applications of Homotopy Theory II, Lect. Notes in Math., vol. 658, Springer-Verlag, New York, 1978, pp. 80–130.
14. A. K. Bousfield and D. M. Kan, *Homotopy limits, completions and localizations*, Lect. Notes in Math., vol. 304, Springer-Verlag, New York, 1972.
15. K. S. Brown, *Abstract homotopy theory and generalized sheaf cohomology*, Trans. Amer. Math. Soc. **186** (1973), 419–458.
16. C. Casacuberta, G. Peschke, and M. Pfenninger, *On orthogonal pairs in categories and localisation*, Adams Memorial Symposium on Algebraic Topology, 1 (Manchester, 1990), Cambridge Univ. Press, Cambridge, 1992, pp. 211–223.
17. K. Ciesielski, *Set theory for the working mathematician*, London Math. Soc. Student Texts, no. 39, Cambridge University Press, Cambridge, UK, 1997.
18. E. B. Curtis, *Simplicial homotopy theory*, Advances in Math. **6** (1971), 107–209 (1971).
19. E. Dror Farjoun, *Homotopy and homology of diagrams of spaces*, Algebraic Topology, Proceedings, Seattle 1985 (H. R. Miller and D. C. Ravenel, eds.), Lect. Notes in Math., vol. 1286, Springer-Verlag, 1985, pp. 93–134.
20. ———, *Localizations, fibrations, and conic structures*, Preprint, 1991.
21. ———, *Cellular spaces*, Preprint, 1992.
22. ———, *Homotopy localization and v_1-periodic spaces*, Algebraic Topology: Homotopy and Group Cohomology, Proceedings of the 1990 Barcelona Conference on Algebraic Topology (J. Aguade, M. Castellet, and F. R. Cohen, eds.), Lect. Notes in Math., vol. 1509, Springer-Verlag, 1992, pp. 104–113.

23. ______, *Cellular inequalities*, The Čech Centennial (M. Cenkl and H. Miller, eds.), Contemporary Mathematics, vol. 181, American Mathematical Society, 1995, pp. 159–181.
24. ______, *Cellular spaces, null spaces and homotopy localization*, Lect. Notes in Math., vol. 1622, Springer-Verlag, Berlin, 1996.
25. E. Dror Farjoun and J. H. Smith, *Homotopy localization nearly preserves fibrations*, Topology **34** (1995), 359–375.
26. D. Dugger, *Replacing model categories with simplicial ones*, Trans. Amer. Math. Soc. **353** (2001), no. 12, 5003–5027.
27. ______, *Universal homotopy theories*, Adv. Math. **164** (2001), no. 1, 144–176.
28. D. Dugger and D. C. Isaksen, *Hypercovers in topology*, Preprint, 2001.
29. J. Dugundji, *Topology*, Allyn and Bacon, Boston, 1966.
30. W. G. Dwyer, P. S. Hirschhorn, and D. M. Kan, *Model categories and more general abstract homotopy theory*, preprint, available at `http://www-math.mit.edu/~psh`.
31. W. G. Dwyer and D. M. Kan, *Calculating simplicial localizations*, J. Pure Appl. Algebra **18** (1980), 17–35.
32. ______, *Function complexes in homotopical algebra*, Topology **19** (1980), 427–440.
33. ______, *Simplicial localizations of categories*, J. Pure Appl. Algebra **17** (1980), 267–284.
34. W. G. Dwyer, D. M. Kan, and C. R. Stover, *An E^2 model category structure for pointed simplicial spaces*, J. Pure Appl. Algebra **90** (1993), 137–152.
35. W. G. Dwyer and J. Spalinski, *Homotopy theories and model categories*, Handbook of Algebraic Topology (I. M. James, ed.), Elsevier Science B.V., 1995.
36. S. Eilenberg and J. C. Moore, *Foundations of relative homological algebra*, Mem. Amer. Math. Soc. No. **55** (1965), 39.
37. R. Fritsch and R. A. Piccinini, *Cellular structures in topology*, Cambridge Studies in Advanced Mathematics, vol. 19, Cambridge University Press, Cambridge, 1990.
38. P. Gabriel and M. Zisman, *Calculus of fractions and homotopy theory*, Ergeb. der Math. und ihrer Grenzgebiete, vol. 35, Springer-Verlag, New York, 1967.
39. P. G. Goerss and J. F. Jardine, *Simplicial homotopy theory*, Progress in Math., vol. 174, Birkhäuser Verlag, Basel, 1999.
40. A. G. Hamilton, *Numbers, sets and axioms, the apparatus of mathematics*, Cambridge University Press, 1982.
41. A. Heller, *Homotopy theories*, vol. 71, Memoirs Amer. Math. Soc., no. 383, Amer. Math. Soc., 1988.
42. M. Hovey, *Model categories*, Mathematical Surveys and Monographs, vol. 63, Amer. Math. Soc., Providence, RI, 1999.
43. M. Hovey, B. Shipley, and J. Smith, *Symmetric spectra*, J. Amer. Math. Soc. **13** (2000), no. 1, 149–208.
44. J. F. Jardine, *The homotopical foundations of algebraic K-theory*, Algebraic K-Theory and Algebraic Number Theory, Contemporary Mathematics, vol. 83, Amer. Math. Soc., 1989, pp. 57–82.
45. D. M. Kan, *On c.s.s. categories*, Bol. Soc. Mat. Mexicana **2** (1957), 82–94.
46. ______, *On homotopy theory and c.s.s. groups*, Ann. Math. **68** (1958), 38–53.
47. S. Mac Lane, *Categories for the working mathematician*, Grad. Texts in Math., vol. 5, Springer-Verlag, New York, 1971.
48. S. Mac Lane and I. Moerdijk, *Sheaves in geometry and logic, a first introduction to category theory*, Universitext, Springer-Verlag, New York, 1992.
49. J. P. May, *Simplicial objects in algebraic topology*, Van Nostrand Math. Studies, vol. 11, D. Van Nostrand, 1967.
50. J. R. Munkres, *Topology: a first course*, second ed., Prentice-Hall, 2000.
51. A. Nofech, *On localization of inverse limits*, Ph.D. thesis, The Hebrew University of Jerusalem, 1993.
52. D. G. Quillen, *Homotopical algebra*, Lect. Notes in Math., vol. 43, Springer-Verlag, Berlin, 1967.
53. ______, *The geometric realization of a Kan fibration is a Serre fibration*, Proc. Amer. Math. Soc. **19** (1968), 1499–1500.
54. ______, *Rational homotopy theory*, Ann. Math. **90** (1969), 205–295.

55. ———, *On the (co-) homology of commutative rings*, Applications of Categorical Algebra (Proc. Sympos. Pure Math., Vol. XVII, New York, 1968), Amer. Math. Soc., Providence, R.I., 1970, pp. 65–87.
56. ———, *Higher algebraic K-theory I*, Higher K-Theories (Berlin), Lect. Notes in Math., vol. 341, Springer-Verlag, 1973, pp. 85–147.
57. C. L. Reedy, *Homotopy theory of model categories*, Available at the Hopf Topology Archive as `ftp://hopf.math.purdue.edu/pub/Reedy/reedy.dvi`, 1974, Unpublished manuscript.
58. C. Rezk, *A model for the homotopy theory of homotopy theory*, Trans. Amer. Math. Soc. **353** (2001), no. 3, 973–1007 (electronic).
59. ———, *Every homotopy theory of simplicial algebras admits a proper model*, Topology and its Appl. **119** (2002), no. 1, 65–94.
60. H. Schubert, *Categories*, Springer-Verlag, New York, 1972.
61. J. H. Smith, *Specters of symmetry*, Unpublished manuscript, 1994.
62. N. E. Steenrod, *A convenient category of topological spaces*, Michigan Math. J. **14** (1967), 133–152.
63. R. M. Vogt, *Convenient categories of topological spaces for homotopy theory*, Arch. Math. (Basel) **22** (1971), 545–555.

Titles in This Series

66 **Yu. Ilyashenko and Weigu Li,** Nonlocal bifurcations, 1999

65 **Carl Faith,** Rings and things and a fine array of twentieth century associative algebra, 1999

64 **Rene A. Carmona and Boris Rozovskii, Editors,** Stochastic partial differential equations: Six perspectives, 1999

63 **Mark Hovey,** Model categories, 1999

62 **Vladimir I. Bogachev,** Gaussian measures, 1998

61 **W. Norrie Everitt and Lawrence Markus,** Boundary value problems and symplectic algebra for ordinary differential and quasi-differential operators, 1999

60 **Iain Raeburn and Dana P. Williams,** Morita equivalence and continuous-trace C^*-algebras, 1998

59 **Paul Howard and Jean E. Rubin,** Consequences of the axiom of choice, 1998

58 **Pavel I. Etingof, Igor B. Frenkel, and Alexander A. Kirillov, Jr.,** Lectures on representation theory and Knizhnik-Zamolodchikov equations, 1998

57 **Marc Levine,** Mixed motives, 1998

56 **Leonid I. Korogodski and Yan S. Soibelman,** Algebras of functions on quantum groups: Part I, 1998

55 **J. Scott Carter and Masahico Saito,** Knotted surfaces and their diagrams, 1998

54 **Casper Goffman, Togo Nishiura, and Daniel Waterman,** Homeomorphisms in analysis, 1997

53 **Andreas Kriegl and Peter W. Michor,** The convenient setting of global analysis, 1997

52 **V. A. Kozlov, V. G. Maz′ya, and J. Rossmann,** Elliptic boundary value problems in domains with point singularities, 1997

51 **Jan Malý and William P. Ziemer,** Fine regularity of solutions of elliptic partial differential equations, 1997

50 **Jon Aaronson,** An introduction to infinite ergodic theory, 1997

49 **R. E. Showalter,** Monotone operators in Banach space and nonlinear partial differential equations, 1997

48 **Paul-Jean Cahen and Jean-Luc Chabert,** Integer-valued polynomials, 1997

47 **A. D. Elmendorf, I. Kriz, M. A. Mandell, and J. P. May (with an appendix by M. Cole),** Rings, modules, and algebras in stable homotopy theory, 1997

46 **Stephen Lipscomb,** Symmetric inverse semigroups, 1996

45 **George M. Bergman and Adam O. Hausknecht,** Cogroups and co-rings in categories of associative rings, 1996

44 **J. Amorós, M. Burger, K. Corlette, D. Kotschick, and D. Toledo,** Fundamental groups of compact Kähler manifolds, 1996

43 **James E. Humphreys,** Conjugacy classes in semisimple algebraic groups, 1995

42 **Ralph Freese, Jaroslav Ježek, and J. B. Nation,** Free lattices, 1995

41 **Hal L. Smith,** Monotone dynamical systems: an introduction to the theory of competitive and cooperative systems, 1995

40.5 **Daniel Gorenstein, Richard Lyons, and Ronald Solomon,** The classification of the finite simple groups, number 5, 2002

40.4 **Daniel Gorenstein, Richard Lyons, and Ronald Solomon,** The classification of the finite simple groups, number 4, 1999

40.3 **Daniel Gorenstein, Richard Lyons, and Ronald Solomon,** The classification of the finite simple groups, number 3, 1998